The Mycobacterial Cell Envelope

The Mycobacterial Cell Envelope

Edited by

Mamadou Daffé
Department of Molecular Mechanisms of Mycobacterial Infections
Institut de Pharmacologie et de Biologie Structurale du
Centre National de la Recherche Scientifique (CNRS)
and Université Paul Sabatier
Toulouse, France

Jean-Marc Reyrat
Faculté de Médecine René Descartes
Université Paris Descartes
and
Groupe Avenir
Unité de Pathogénie des Infections Systémiques—U570
INSERM
Paris, France

ASM PRESS

Washington, DC

Address editorial correspondence to ASM Press, 1752 N St. NW, Washington, DC 20036-2904, USA

Send orders to ASM Press, P.O. Box 605, Herndon, VA 20172, USA
Phone: (800) 546-2416 or (703) 661-1593
Fax: (703) 661-1501
E-mail: books@asmusa.org
Online: estore.asm.org

Library of Congress Cataloging-in-Publication Data

The mycobacterial cell envelope / edited by Mamadou Daffé, Jean-Marc Reyrat.
     p. ; cm.
  Includes bibliographical references and index.
  ISBN 978-1-55581-468-7 (hardcover)
  1. Mycobacteria. 2. Bacterial cell walls. I. Daffé, Mamadou. II. Reyrat, Jean-Marc.
  [DNLM: 1. Mycobacterium—cytology. 2. Cell Membrane—metabolism. 3. Cell Wall—
  metabolism. 4. Mycobacterium—pathogenicity. QW 125.5.M9 M9953 2008]

QR82.M8M938 2008
579.3'74—dc22   2007030720

10 9 8 7 6 5 4 3 2 1

ISBN 978-1-55581-468-7

Cover design by Meral Dabcovich, VisPer

Cover images: (Upper photo) Photomicrograph of colonies of *erp::aph Mycobacterium smegmatis*. See chapter 8. (Lower photo) High magnification of the cell envelope of a freeze-substituted *M. kansasii* cell. See chapter 2.

CONTENTS

CONTRIBUTORS

Luke Alderwick
School of Biosciences, University of Birmingham, Edgbaston, Birmingham B15 2TT, United Kingdom

Peter Andersen
Department of Infectious Disease Immunology, Statens Serum Institut, Artillerivej 5, DK-2300 Copenhagen S, Denmark

Fabienne Bardou
Department of Molecular Mechanisms of Mycobacterial Infections, Institut de Pharmacologie et de Biologie Structurale du Centre National de la Recherche Scientifique (CNRS) and Université Paul Sabatier, 205 route de Narbonne, 31077 Toulouse Cedex 4, France

Carolyn R. Bertozzi
Departments of Chemistry and Molecular and Cell Biology, Howard Hughes Medical Institute, University of California, Berkeley, Berkeley, CA 94720-1460

Gurdyal S. Besra
School of Biosciences, University of Birmingham, Edgbaston, Birmingham B15 2TT, United Kingdom

Terry J. Beveridge (*deceased*)
Department of Molecular and Cellular Biology and Advanced Food and Materials Network—Networks Centers of Excellence, College of Biological Science, University of Guelph, Guelph, Ontario, Canada N1G 2W1

Veemal Bhowruth
School of Biosciences, University of Birmingham, Edgbaston, Birmingham B15 2TT, United Kingdom

Fabiana Bigi
Institute of Biotechnology, National Institute of Agricultural Research, Los Reseros y Las Cabañas, 1712, Castelar, Argentina

Daria Bottai
Unité de Génétique Moléculaire Bactérienne, Institut Pasteur, 25-28 rue du Dr. Roux, 75724 Paris Cedex 15, France, and Dipartimento di Patologia Sperimentale, Biotecnologie Mediche, Infettivologia ed Epidemiologia, University of Pisa, Via San Zeno 35/39, 56127 Pisa, Italy

Patrick J. Brennan
Department of Microbiology, Immunology and Pathology, Colorado State University, Fort Collins, CO 80523

Warwick J. Britton
Centenary Institute of Cancer Medicine and Cell Biology, Locked Bag No. 6, Newtown, NSW 2042, and Department of Medicine, University of Sydney (D06), Sydney, NSW 2006, Australia

Roland Brosch
Unité de Génétique Moléculaire Bactérienne and Unité Postulante de Pathogénomique Mycobactérienne Intégrée, Institut Pasteur, 25-28 rue du Dr. Roux, 75724 Paris Cedex 15, France

Alistair K. Brown
School of Biosciences, University of Birmingham, Edgbaston, Birmingham B15 2TT, United Kingdom

Alessandro Cascioferro
Department of Histology, Microbiology and Medical Biotechnologies, University of Padova Medical School, Via Gabelli 63, 35121 Padua, Italy

Angel Cataldi
Institute of Biotechnology, National Institute of Agricultural Research, Los Reseros y Las Cabañas, 1712, Castelar, Argentina

Christian Chalut
Department of Molecular Mechanisms of Mycobacterial Infections, Institut de Pharmacologie et de Biologie Structurale du Centre National de la Recherche Scientifique (CNRS), and Université Paul Sabatier, 205 route de Narbonne, 31077 Toulouse Cedex 4, France

Eric D. Chow
Department of Microbiology and Immunology, University of California, San Francisco, San Francisco, CA 94143-2200

Jean Content
Molecular Microbiology Unit, Pasteur Institute of Brussels, 642, rue Engeland, B-1180 Brussels, Belgium

Jeffery S. Cox
Department of Microbiology and Immunology, University of California, San Francisco, San Francisco, CA 94143-2200

Dean C. Crick
Department of Microbiology, Immunology and Pathology, Colorado State University, Fort Collins, CO 80523

Mamadou Daffé
Department of Molecular Mechanisms of Mycobacterial Infections, Institut de Pharmacologie et de Biologie Structurale du Centre National de la Recherche Scientifique (CNRS), and Université Paul Sabatier, 205 route de Narbonne, 31077 Toulouse Cedex 4, France

Giovanni Delogu
Institute of Microbiology, Catholic University of the Sacred Heart, Largo A. Gemelli, 8, 00168 Rome, Italy

Caroline Deshayes
Faculté de Médecine René Descartes, Université Paris Descartes, and Groupe Avenir, Unité de Pathogénie des Infections Systémiques—U570, INSERM, F-75730 Paris Cedex 15, France

Lynn G. Dover
Biomedical Sciences Division, School of Applied Sciences, Ellison Building, Northumbria University, Newcastle upon Tyne NE1 8ST, United Kingdom

Gilles Etienne
Department of Molecular Mechanisms of Mycobacterial Infections, Institut de Pharmacologie et de Biologie Structurale du Centre National de la Recherche Scientifique (CNRS), and Université Paul Sabatier, 205 route de Narbonne, 31077 Toulouse Cedex 4, France

Christophe Genisset
INSERM U629, Lille, Institut Pasteur de Lille, 1, rue du Prof. Calmette, F-59019 Lille Cedex, France

Martine Gilleron
Department of Molecular Mechanisms of Mycobacterial Infections, Institut de Pharmacologie et de Biologie Structurale, du Centre National de la Recherche Scientifique (CNRS) and Université Paul Sabatier, 205 route de Narbonne, 31077 Toulouse Cedex 4, France

Michael S. Glickman
Infectious Diseases Service, Immunology Program, Sloan Kettering Institute, 415 E. 68th Street, Z1605, New York, NY 10065

Rajesh S. Gokhale
National Institute of Immunology, Aruna Asaf Ali Marg, New Delhi 110 067, India

Christophe Guilhot
Department of Molecular Mechanisms of Mycobacterial Infections, Institut de Pharmacologie et de Biologie Structurale du Centre National de la Recherche Scientifique (CNRS), and Université Paul Sabatier, 205 route de Narbonne, 31077 Toulouse Cedex 4, France

Vibha Gupta
Department of Biochemistry, University of Delhi South Campus, Benito Juarcz Road, New Delhi 110 021, India

Seyed E. Hasnain
Vice-Chancellor's Office, University of Hyderabad, Central University P.O., Hyderabad 500 046, India

Robert N. Husson
Children's Hospital Boston and Division of Infectious Diseases, Harvard Medical School, 300 Longwood Avenue, Boston, MA 02115

Mary Jackson
Department of Microbiology, Immunology and Pathology, Colorado State University, Fort Collins, CO 80523

Madhulika Jain
Department of Microbiology and Immunology, University of California, San Francisco, San Francisco, CA 94143-2200

Dana Kocíncová
Faculté de Médecine René Descartes, Université Paris Descartes, and Groupe Avenir, Unité de Pathogénie des Infections Systémiques—U570, INSERM, F-75730 Paris Cedex 15, France

Laurent Kremer
CNRS UMR 5235, Dynamique des Interactions,
Membranaires Normales et Pathologiques,
Université de Montpellier II, Case 107, Place Eugène
Bataillon, 34095 Montpellier Cedex 5, France

Marie-Antoinette Lanéelle
Department of Molecular Mechanisms of Mycobac-
terial Infections, Institut de Pharmacologie et de
Biologie Structurale du Centre National de la
Recherche Scientifique (CNRS) and Université Paul
Sabatier, 205 route de Narbonne, 31077 Toulouse
Cedex 4, France

Camille Locht
INSERM U629, Lille, Institut Pasteur de Lille, 1, rue
du Prof. Calmette, F-59019 Lille Cedex, France

Riccardo Manganelli
Department of Histology, Microbiology and
Medical Biotechnologies, University of Padova
Medical School, Via Gabelli 63, 35121 Padua, Italy

Hedia Marrakchi
Department of Molecular Mechanisms of Mycobac-
terial Infections, Institut de Pharmacologie et de
Biologie Structurale du Centre National de la
Recherche Scientifique (CNRS) and Université Paul
Sabatier, 205 route de Narbonne, 31077 Toulouse
Cedex 4, France

Françoise Mascart
Laboratory of Vaccinology and Mucosal Immunity,
Erasme Hospital, Université Libre de Bruxelles,
808, route de Lennik, B-1070 Brussels, Belgium

Debasisa Mohanty
National Institute of Immunology, Aruna Asaf Ali
Marg, New Delhi 110 067, India

Vivek T. Natarajan
National Institute of Immunology, Aruna Asaf Ali
Marg, New Delhi 110 067, India

Michael Niederweis
Department of Microbiology, University of Alabama
at Birmingham, Birmingham, AL 35294

Jérôme Nigou
Department of Molecular Mechanisms of Mycobac-
terial Infections, Institut de Pharmacologie et de
Biologie Structurale, du Centre National de la
Recherche Scientifique (CNRS) and Université Paul
Sabatier, 205 route de Narbonne, 31077 Toulouse
Cedex 4, France

Priska Peirs
Molecular Microbiology Unit, Pasteur Institute of
Brussels, 642, rue Engeland, B-1180 Brussels, Belgium

Germain Puzo
Department of Molecular Mechanisms of Mycobac-
terial Infections, Institut de Pharmacologie et de
Biologie Structurale, du Centre National de la
Recherche Scientifique (CNRS) and Université Paul
Sabatier, 205 route de Narbonne, 31077 Toulouse
Cedex 4, France

Luis E. N. Quadri
Department of Microbiology and Immunology,
Weill Medical College of Cornell University, New
York, NY 10021

Sahadevan Raman
Children's Hospital Boston and Division of
Infectious Diseases, Harvard Medical School,
300 Longwood Avenue, Boston, MA 02115

Dominique Raze
INSERM U629, Lille, Institut Pasteur de Lille, 1,
rue du Prof. Calmette, F-59019 Lille Cedex, France

Jean-Marc Reyrat
Faculté de Médecine René Descartes, Université
Paris Descartes, and Groupe Avenir, Unité de
Pathogénie des Infections Systémiques—U570,
INSERM, F-75730 Paris Cedex 15, France

Ida Rosenkrands
Department of Infectious Disease Immunology,
Statens Serum Institut, Artillerivej 5, DK-2300
Copenhagen S, Denmark

Carine Rouanet
INSERM U629, Lille, Institut Pasteur de Lille, 1,
rue du Prof. Calmette, F-59019 Lille Cedex,
France

Michael W. Schelle
Department of Chemistry, University of California,
Berkeley, Berkeley, CA 94720-1460

Jérôme Segers
INSERM U629, Lille, Institut Pasteur de Lille, 1,
rue du Prof. Calmette, F-59019 Lille Cedex,
France

Ramandeep Singh
Laboratory of Immunogenetics, Tuberculosis
Research Section, NIAID, National Institutes of
Health, Rockville, MD 20852

Pamela L. C. Small
Department of Microbiology, Walters Life Sciences,
University of Tennessee, Knoxville, TN 37996-0845

Timothy P. Stinear
Department of Microbiology, Monash University,
Wellington Road, Clayton 3800, Australia

James A. Triccas
Department of Infectious Diseases and Immunology,
University of Sydney (D06), Sydney, NSW 2006,
Australia

Anil K. Tyagi
Department of Biochemistry, University of Delhi
South Campus, Benito Juarez Road, New Delhi 110
021, India

INTRODUCTION

Right from the time that the acid-fast reaction was discovered by Paul Ehrlich, it has been clear that there is something special about the mycobacterial envelope: some feature that causes this large group of bacteria to behave differently from all other types. The potential importance of this biologically interesting property is increased by the fact that one of the mycobacteria, *Mycobacterium tuberculosis*, causes a fatal human disease.

Formerly common worldwide, tuberculosis kills about 40% of people in whom it occurs if it is not treated, and until the discovery of streptomycin about 60 years ago, there was no effective treatment. Now it occurs mostly in the developing world, where it kills nearly two million people each year, but the currently low incidence in developed countries is rising. Although there are several excellent drugs that are used in combination to cure the disease, these are slow to act and often too expensive for patients in the Third World. In any case, the appearance in several parts of the world of multidrug-resistant strains of *M. tuberculosis* threatens to compromise existing treatments. The attenuated live vaccine strain, *Mycobacterium bovis* bacillus Calmette-Guérin (BCG), is effective only in a subset of countries in which the disease occurs. So it is unsurprising that research on *M. tuberculosis* is active.

Although research on *M. tuberculosis* dominates the mycobacterial field today, it is important to remember that there are several other significant mycobacterial pathogens as well as a large number of known environmental species (and probably many as yet undiscovered ones). Information about these other species is likely to contribute to understanding of the physiology of the whole genus, and of what properties have enabled some species to become major pathogens. Many of the chapters in this book deal with mycobacteria in general, and some with interesting aspects of nontuberculosis species.

A good deal of such research is, and has long been, concerned with the mycobacterial envelope. There are good reasons for this: the envelope is the interface between the pathogen and its human host, where it occurs intracellularly, at least until the last stages of the disease. The property of acid fastness seems to reflect the very low permeability of the enve-lope. That low permeability, along with a chemical structure capable of resisting lysis by host-cell enzymes, seems to be responsible for the characteristic resistance of the mycobacterial cell to destruction by the host. In addition, envelope components are apparently involved in bringing about the uptake of the mycobacterium by the cell, and also in modifying the responses of the cell and of the host's immune system—particularly the cell-mediated immune system—to improve conditions for bacterial survival and multiplication.

The mycobacterial envelope is a complex arrangement of lipids and carbohydrates. Some of these are linked covalently into a bacterium-shaped container—the mycobacterial version of the envelope structure found in most bacteria. Others are physically associated with the covalent structure and more or less firmly entangled with it. The chemical nature of the components is rather well worked out, despite their complexity, and there is a plausible model of their arrangement. These matters are described in chapter 1.

Many mycobacterial envelope components are highly unusual, and their biosynthesis occurs by way of novel pathways. The sequencing of the genome of *M. tuberculosis* has hugely assisted discovery of such pathways. Chapter 3 describes the biosynthesis of the arabinogalactan, the high-molecular-weight heteropolysaccharide that links the peptidoglycan (the "common" bacterial envelope polymer) to the major distinctive lipids of mycobacteria, mycolic acids, whereas chapter 4 explains the biosynthetic route for these lipids. These three components comprise the majority of the envelope and the whole of its covalently combined part.

The genome of *M. tuberculosis* possesses many lipid-biosynthetic enzymes in addition to the ones used to construct mycolic acids, including a number of polyketide synthases (PKSs), enzymes able to assemble complex lipids from acetate and propionate units. Several important mycobacterial lipids, notably phthiocerol dimycocerosate of *M. tuberculosis* and the various phenolic glycolipids (PGLs), are synthesized using PKSs, and some of these enzymes use novel reaction mechanisms. Some characteristics of mycobacterial PKSs are discussed in chapters 15 and 17.

There is good reason to believe that the mycolic acids of the envelope form a close-packed lipid monolayer that is responsible for the very low permeability of these bacteria. This layer is an obvious functional analogue of the outer membrane of gram-negative bacteria, although it is completely different chemically and stabilized by being covalently bonded to the rest of the envelope. The existence of this outer layer has two consequences: first, it produces a special compartment of the envelope, analogous to the periplasm of gram-negative bacteria, between the mycolic acid layer and the plasma membrane. Access to this compartment from the inside of the bacterium and from the environment is restricted; it is discussed in chapter 2. Second, the presence of the layer makes it very difficult for small hydrophilic molecules to pass between the environment and the interior of the mycobacterial cell, with potential problems for nutrition and waste disposal. Transport across the plasma membrane can be facilitated and controlled by mechanisms driven by energy from biochemical processes within the bacterium, but the outer lipid layer is remote from such a source of energy. In gram-negative bacteria, passage across the outer membrane by small hydrophilic molecules is mediated by specialized pore-forming proteins called porins; chapter 9 discusses current knowledge of their mycobacterial equivalents.

In addition to the general transport function of porins, specialized transport mechanisms exist for several components. Iron transport, which is complicated by very restricted supplies of this essential mineral in the environment of the mycobacterium, is discussed in chapter 10. Chapter 12 explains the MmpL family of proteins, which appear to be involved in lipid transport, and chapter 11 describes ABC transporters, which are involved in the export of proteins.

The glycolipids, called PIMs, and the related lipoarabinomannan may be among the components whose function is to cause the mycobacterial cell to be recognized and taken up by host macrophages; these substances are discussed in chapter 6. Chapter 5 concerns cord factor, another abundant and biologically active component of the *M. tuberculosis* envelope.

Numerous mycobacterial components have been found experimentally to affect the host, and particularly the host immune system, although the physiological or pathological significance of the effects has not always been clear. However, this unsatisfactory situation is now changing. Chapter 13 considers the ESAT secretion system, which is one of the mechanisms by which host-modifying mycobacterial proteins are exported, while chapter 8 reviews several surface proteins that appear to be virulence factors but whose precise function and mode of action are not yet known. A general review of the effects of mycobacterial envelope constituents on the host immune system, and how the mycobacterium uses the effects to modify the behavior of the host to the advantage of the pathogen, is provided by chapter 16.

The multifunctional behavior of the envelope in mediating host-pathogen interactions suggests that the envelope is a good target for drugs, and the efficacy of isoniazid, which inhibits mycolic acid biosynthesis, shows how successful they can be in the control of mycobacterial diseases. Chapter 7 considers substances that might interfere with the construction of the mycobacterial envelope and so be useful as novel antimycobacterial drugs. Development of novel drugs is essential if the spread of essentially untreatable tuberculosis caused by drug-resistant organisms is to be prevented.

Although the first part of the book concerns features of the mycobacterial envelope that are common to many sorts of mycobacteria, the second part deals with components that are possessed by only one or a few species. Chapters 17 and 18 deal, respectively, with PGL biosynthesis and sulfolipids of *M. tuberculosis*. PGLs, varying in molecular detail but conforming to the same general type, are found in several types of mycobacteria, and are produced in massive amounts by *Mycobacterium leprae*, but not all members of the *M. tuberculosis* complex form PGLs. However, enzymes for the biosynthetic pathway are present and various probably related substances are synthesized. Sulfolipids have been included among mycobacterial products that may affect the host cell, but the evidence is equivocal and there is a need to clarify their role.

Chapter 19 describes a protein adhesin of *M. tuberculosis*, which seems to be an important virulence factor, and chapter 20 discusses mycobacterial kinases and phosphatases and their importance in pathogenesis of mycobacterial diseases.

Chapter 21 describes the biosynthesis of a family of unusual glycopeptidolipids (GPLs) synthesized by a subset of mycobacteria, including pathogens and environmental species, but not by members of the *M. tuberculosis* group. Species related to *Mycobacterium avium*, which produce GPLs, cause human disease relatively rarely, but such infections that do occur are extremely difficult to treat with existing drugs. The infections are particularly threatening to individuals whose immune system is compromised.

The final chapter, chapter 22, concerns a unique chemical, a mycobacterial toxin. In general, pathogenic mycobacteria seem not to produce toxins, depending more for their success in manipulating the host. However, *Mycobacterium ulcerans*, which causes a progressive and difficult-to-treat skin ulcer, does make a toxin. The disease remains rather mysterious but appears to be becoming more common, so it is important to understand it properly. Many strains of *Mycobacterium marinum*, which can cause skin infections, produce related molecules; it is believed that the ability to synthesize the toxin depends on the presence of a plasmid.

Altogether this book provides an exciting menu of modern information about the mycobacterial envelope, and demonstrates that the field is of continuing importance. The editors and ASM Press are to be congratulated on assembling such a collection; it deserves to be a great success.

Philip Draper

I. COMMON FEATURES

Chapter 1

The Global Architecture of the Mycobacterial Cell Envelope

MAMADOU DAFFÉ

The cell envelope of mycobacteria, i.e., the compounds that surround the cytoplasm and protect the micro-organisms from their environment, is important for the bacterial physiology because inhibition of the production of some of its constituents, e.g., mycolic acids and arabinogalactan, kills the cells. It is also this structure that controls the transfer of materials into and out of the mycobacterium. A major discovery of the last decade has been the demonstration of the notably low permeability of mycobacteria to nutrients and antibacterial drugs, which slows down the growth of mycobacteria and makes disease caused by pathogenic species difficult to treat. The permeability of cell walls of mycobacteria was found to be 10- to 100-fold lower than that of the notably impermeable bacillus *P. aeruginosa* (Jarlier and Nikaido, 1990). This property is explained by the singularity of the cell envelope of mycobacteria and related micro-organisms, namely, corynebacteria, nocardiae, rhodococci, and other bacteria grouped in the suborder *Corynebacterineae*. These genera belong to one of the two groups of gram-positive bacteria, i.e., the division *Actinobacteria* (formerly the group of high G + C content), the other being the division of *Firmicutes*. Importantly, however, the chemical nature of the envelope is different from those of both groups of gram-positive and from gram-negative bacteria; for instance, the lipid content of the cell envelope of mycobacteria may represent up to 40% of the cell dry mass, compared to less than 5% in other gram-positive bacteria and only 10% in gram-negative bacteria (Goren and Brennan, 1979). Such a lipid-rich coat could explain both the tendency of mycobacteria to grow in clumps and their distinctive property of acid fastness. The low permeability of the mycobacterial envelope also justifies the occurrence of pore-forming proteins (porins) in the cell walls of members of the *Corynebacterineae* group of bacteria (see chapter 9).

This unusual structure certainly makes it difficult for the host to damage the mycobacterial envelope, and while it is intact, the impermeable cell envelope protects the mycobacterium from damage. In addition, being at the interface between the mycobacterium and its host, and while the mycobacterium remains viable, only components of the envelope, and perhaps some secreted molecules, are readily accessible to the host. Accordingly, several studies have been and are still devoted to the biological activities of cell surface components involved in the cross-talk with host cells (see chapter 16). In addition, many of the structures of the envelope major components have been shown to be chemically very different from anything found in animal cells, and so proteins involved in important and/or essential metabolisms represent good targets for selective inhibition.

An overall view of the mycobacterial envelope is illustrated in Fig. 1, which shows a thin section of *Mycobacterium smegmatis* by transmission electron microscopy and provides an interpretation of the electron microscopic image. This thin section is representative of what is obtained with other mycobacterial strains, including pathogenic strains, by this technique in which the mycobacterial cell envelope appears as a multilayered structure; however, there is no indication that the 'native' structure is really layered. This is due to the fact that the different microscopic techniques used various solvents that may extract more of less material; also, the shrinkage and distortion caused by dehydration of formerly hydrated structures during processing certainly affect the final picture of the structure (see chapter 2). For instance, in ultrathin sectioning, the water in the fixed specimen is removed by dehydration in solvents (usually ethanol), causing collapse of structures that cannot be cross-linked by fixatives, notably carbohydrates, which are important components of the mycobacterial envelope (Daffé and

Mamadou Daffé • Department of Molecular Mechanisms of Mycobacterial Infections, Institut de Pharmacologie et de Biologie Structurale du Centre National de la Recherche Scientifique (CNRS), and Université Paul Sabatier, 205 route de Narbonne, 31077 Toulouse Cedex 4, France.

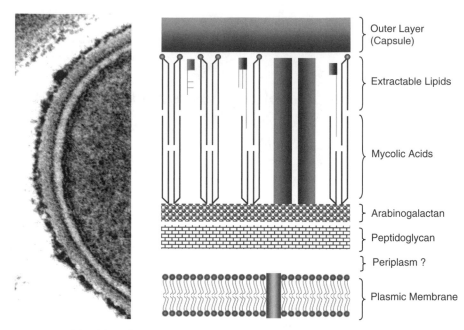

Figure 1. Electron micrograph of the cell envelope of mycobacteria. Left: ultrathin section of *Mycobacterium smegmatis* fixed with glutaraldehyde and lysine in buffer containing ruthenium red and postfixed with osmium tetroxide; bacteria were suspended in uranyl acetate, and cells were embedded in agar before dehydration with ethanol. Right: interpretation of the image showing the proposed arrangement of plasma membrane, hypothetical periplasm, cell wall core components (peptidoglycan, arabinogalactan, mycolic acids) and associated (extractable) lipids, and outer layer (called the "capsule" in the case of pathogenic species). Note that proteins are not represented in this figure, with notable exception of the cell wall pore-forming proteins (porins). This model of the arrangement of the mycobacterial cell envelope is based on that of Minnikin (1982), modified by Daffé and Draper (Daffé and Draper, 1998).

Draper, 1998). Comparative microscopic analysis of the bacterial cell envelopes (see chapter 2) indicates that the layered architecture of the mycobacterial cell envelope is more related to that of gram-negative bacteria, such as species of the division *Proteobacteria* typified by *Pseudomonas,* than that of classical gram-positive bacteria, such as *Micrococcus,* which possess a simple basal structure in which the cytoplasm is surrounded by the plasma membrane and a thick peptidoglycan.

Although the interpretation drawn on chemical and microscopic data is, in part, still speculative, it is well established that the envelope consists of a plasma membrane, which is apparently homologous to plasma membranes of other bacteria, surrounded by a complex wall of carbohydrate and lipid, which is, in turn, surrounded by an outermost layer, called the "capsule" in the case of pathogenic species, of polysaccharide and protein with relatively small quantities of lipid. The wall possesses a fundamental, covalently linked "cell-wall skeleton" (CWS) associated with a great variety of noncovalently linked substances, of which a majority are lipids and glycolipids (Daffé and Draper, 1998). Although not directly demonstrated, a compartment analogous to the periplasmic space in gram-negative bacteria would exist in mycobacteria, between the membrane and the peptidoglycan, as previously proposed (Daffé and Draper, 1998).

THE PLASMA MEMBRANE

The basic structure of the plasma membrane of the mycobacterial cell envelope does not seems to differ from that of the other plasma biological membranes, as judged from their appearance in ultrathin sections (Silva and Macedo, 1983a, Daffé et al., 1989), the characterization of classic metabolic functions (Brodie et al., 1979) and their known chemical composition (Akamatsu et al., 1966; Goldman, 1970; Kumar et al., 1979; Oka et al., 1968). Isolated plasma membranes are typically obtained by breaking the cells by mechanical stress, e.g., sonication or shearing in the French pressure cell, followed by fractionation using differential centrifugation or density gradients (see Rezwan et al., 2007). Chemical and biochemical analyses of the fractions are essential to ensure their purity. As far as lipid composition of the plasma membranes is concerned, no obvious difference was found between those of rapid- and slow-growing species examined (see Minnikin et al., 1982). Polar lipids, mainly phospholipids, assemble themselves, in association with proteins, into a lipid bilayer.

The polar lipids of the mycobacterial plasma membrane are composed of hydrophilic head groups and fatty acid chains that usually consist of mixtures of straight-chain, unsaturated and mono-methyl branched fatty acid residues having less than 20 carbons. Palmitic ($C_{16:0}$), octadecenoic ($C_{18:1}$), and 10-methyloctadecanoic (or tuberculostearic, C_{19r}) are the major fatty acid constituents of the isolated plasma membranes. The main phospholipids of the plasma membrane are phosphatidylinositol mannosides (PIM), phosphatidylglycerol, cardiolipin and phosphatidylethanolamine; phosphatidylinositol occurs in small amounts. Incidentally, PIM and phosphatidylethanolamine have also been identified among the lipids extracted from the mycobacterial cell surfaces by means of gentle mechanical treatment of cells by glass beads (Ortalo-Magné et al., 1996b). It should be noted, however, that thin sections of mycobacteria show no indication of an additional outer membrane such as the one found in gram-negative bacteria. Thus, the function of the phospholipids in the other envelope layers is unknown.

Besides the phospholipids, other mycobacterial lipids may be present in the plasma membrane. Menaquinones are known to be an essential functional part of the respiratory system in bacterial membranes (see Minnikin, 1982). Ornithine lipids have also been characterized in mycobacteria (Lanéelle et al., 1990), and a structural role in natural membranes for carotenoids has also been considered (see Minnikin, 1982). However, no substantive evidence exists to support the location of the two latter classes of lipids in the plasma membrane. The mycobacterial lipoarabinomannan (LAM) may be a membrane component because it was not detected in the cell outermost material of mycobacteria (Ortalo-Magné et al., 1996a), although in fact its localization in the mycobacterial envelope is unknown. LAM is composed of a phosphatidylinositol group (Hunter and Brennan, 1990) covalently linked to arabinomannan. The phosphatidylinositol is similar to the "anchor" possessed by lipoteichoic acid, which is a known membrane component of some gram-positive bacteria. In this case, the anchor is embedded in the plasma membrane, and it seems likely that LAM is similarly situated (Daffé and Draper, 1998). As an alternative, LAM may serve to link the plasma membrane and the CWS or be located in the outer leaflet of cell wall pseudomembrane of mycobacteria, as discussed elsewhere in this book (see chapter 7). In some models of the envelope (McNeil and Brennan, 1991), LAM is shown spanning the CWS, including the mycolate monolayer, but there is no direct evidence for this. The reported serological detection of the substance on the cell surface is compromised by cross-reactivities between LAM and capsular arabinomannan.

The mycobacterial plasma membrane appears asymmetrical in thin sections of viable bacterial cells, with the accumulation of material in its outer leaflet (Daffé et al., 1989; Silva and Macedo, 1983a; Silva and Macedo, 1984). However, this appearance depends on conditions of fixation and is not seen in sections of bacteria that were dead before fixation or have been subjected to membrane-damaging treatments (Silva and Macedo, 1983b); thus, the symmetrical membrane reported for *M. leprae* is probably an artefact of fixation (Silva et al., 1989). The asymmetrical appearance is attributed to the presence of excess glycoconjugates in the thicker electron-dense outer leaflet, compared with the inner one (Daffé et al., 1989, Silva and Macedo, 1983a; Silva and Macedo, 1984). Indeed, possible candidate molecules exist: both PIMs and the structurally related lipomannan and LAM have been found associated with the plasma membrane fraction (McNeil and Brennan, 1991). Owing to technical problems involved in obtaining purified plasma membrane, however, it is difficult to demonstrate the occurrence of LAM, because high-molecular-weight glycans may sediment with $100,000 \times g$ membrane fractions. Both the mycobacterial glucan and arabinomannan are recoverable from such fractions (unpublished data of Daffé and colleagues), so it is possible that capsular polysaccharides, shed from the cells due to the mechanical stress, and also intracellular glycogen may contaminate the membranes.

THE PERIPLASM

The periplasmic space in gram-negative bacteria is the region located between the plasma membrane and the outer membrane. In addition to the peptidoglycan, it contains numerous enzymes and other proteins. If the model of the mycobacterial envelope described above is correct, then mycobacteria possess an analogous compartment between the plasma membrane and the mycolate layer of the cell wall skeleton. Whether this compartment is a reality remains to be determined (Daffé and Draper, 1998). The application of new electron cryomicroscopy techniques should help in visualizing such space (see chapter 2).

CELL WALL

The wall of mycobacteria consists of a covalently linked CWS, and an abundant variety of wall-associated lipids and a few polypeptides. The CWS is defined as the material remaining after removal of all noncovalently bound wall-associated substances: soluble proteins, lipids, and glycans (Kotani et al.,

1959). It is a giant macromolecule chemically composed of three covalently linked constituents: peptidoglycan, arabinogalactan, and mycolic acids, and defines the shape of the mycobacterial cell. Mycobacterial walls are readily isolated and purified from fragments of plasma membrane and cytoplasmic material by breaking the cells with mechanical stress, followed by purification using differential centrifugation or density gradients to remove unbroken cells (see Rezwan et al., 2007). It can be dissected into its constituent parts by relatively gentle methods, so that each part may be studied separately (see Daffé et al., 1990; McNeil et al., 1990, 1991). The peptidoglycan is composed of repeating units of N-acetylglucosamine and N-acetyl/glycollylmuramic acid cross-linked by short peptides. The arabinogalactan is a complex branched heteropolysaccharide that contains a galactan chain composed of alternating 5- and 6-linked D-galactofuranosyl residues; three D-arabinan chains substituted the D-galactan chain in *M. tuberculosis*. In this species, two thirds of the nonreducing termini pentaarabinosyl motifs are esterified by mycolic acids, whereas half of these are occupied by mycoloyl residues in other mycobacterial species, including *M. leprae* and *M. bovis* bacille Calmette-Guérin (BCG) (McNeil et al., 1991). Mycolic acids, very-long-chain (up to C90) α-branched and β-hydroxylated fatty acids, esterified the four hydroxyl groups at position 5 of both terminal and 2-linked D-arabinofuranosyl of the pentaarabinosyl motifs. The structures and biosynthesis of the three constituents of the mycobacterial CWS are described in detail in other chapters of this book (chapters 3 and 4).

The wall has the same ultrastructural appearance in ultrathin sections of both the intact bacterium and purified walls, i.e., an inner electron-dense layer surrounded by a rather thicker electron-transparent layer (Fig. 1). Removal of lipids, including covalently linked ester-bound mycolic acids, causes the disappearance of the electron-transparent layer (Draper, 1971). This has been interpreted as an evidence for the electron transparent layer being composed of mycolates. However, the occurrence of this layer in thin sections *Corynebacterineae* devoid of mycolic acids, e.g., *Corynebacterium amycolatum* (Puech et al., 2001) clearly indicates that the electron transparency of the layer can not be taken as a proof of the presence of mycolic acids. The electron density of the inner layer makes it likely that it contains the peptidoglycan; for physical reasons, part of the covalently linked arabinogalactan must be in the stained peptidoglycan layer (Daffé and Draper, 1998).

Minnikin was the first to propose a model for the arrangement of the mycobacterial CWS and some of the wall-associated lipids, in which the mycolic acids formed a monolayer near the outer surface of the wall and the fatty acyl chains of the various wall-associated lipids intercalated with the mycolate chains (Minnikin, 1982). This original model implied the existence of a second hydrophobic layer, impermeable to polar molecules, in the wall of mycobacteria, forming an analogue of the outer membrane of gram-negative bacteria. This would provide a good explanation for the additional and main fracture plane observed in freeze fractures of mycobacteria (Barksdale and Kim, 1977; Benedetti et al., 1984; Rulong et al., 1991) in that the fracture would occur through the separation of the mycolates attached to the rest of the CWS and the intercalating polar lipids. This model was later supported by experimental data (see Draper, 1998), notably the work of Nikaido and his colleagues, who showed that an aqueous suspension of purified walls of *M. chelonae* produced an X-ray diffraction pattern with reflections characteristic of close-packed hydrocarbon chains in a crystalline monolayer (Nikaido et al., 1993). Several variants on the basic Minnikin model have been proposed (McNeil and Brennan, 1991; Rastogi, 1991; Liu et al., 1995) and discussed elsewhere (Draper, 1998).

The existence of an outer membrane, as proposed in the various Minnikin-based models, poses the problem of getting small polar molecules, notably needed for nutrition. Gram-negative bacteria solve this problem by producing specialized proteins called porins, which form hydrophilic pores through the structure. Pore-forming proteins have been found in the walls of both slow- (Senaratne et al., 1998) and rapid-growing mycobacterial species such as *M. chelonae* and *M. smegmatis*, two rapidly growing species (Trias et al., 1992; Trias and Benz, 1993). The mycobacterial porins assemble themselves to form pores spontaneously in artificial bilayer membranes and have other properties characteristic of the family (see chapter 9).

THE OUTER LAYER

A Historical Perspective

Chapman was the first who has mentioned the existence of a "capsular space" to design the space between the phagosomal membrane of the infected cell and the wall of the enclosed *M. lepraemurium* (Chapman et al., 1959). Soon after, Hanks used light microscopic techniques (Hanks, 1961b, 1961c; Hanks et al., 1961) to demonstrate the presence of an unstainable "halo" around some pathogenic mycobacteria, and sought to correlate its presence with their permeability to dyes. He also has pointed out that what appeared to be a capsule could be seen in several published electron micrographs of pathogenic mycobacteria living within host cells and commonly surrounded by an "electron-transparent zone" (ETZ) (Hanks, 1961a). Hanks has discussed the evidence for the nature of the capsule and concluded that its physical and chemical properties were consistent with its being of bacterial, rather than of host origin (Hanks, 1961d). An ETZ was always seen around intracellular mycobacterial pathogens but not surrounding nonpathogenic species (Fréhel et al., 1986; Ryter et al., 1984).

For a long time, the chemical identity of the ETZ was available only for two host-grown mycobacterial species, namely M. leprae and M. lepraemurium. In the case of M. lepraemurium the 'fibrillar' substance that surrounded bacteria inside the phagocytic vacuoles of the host cell was isolated by differential centrifugation and density gradients, and found to be a C-type glycopeptidolipid (GPL) (Draper, 1971). One of the major ETZ materials of M. leprae in phagosomes probably corresponds to the abundant phenolic glycolipid (PGL), the massive presence of which was clearly demonstrated in the ETZ by immunoelectron microscopy (Boddingius and Dijkman, 1990). Although in both mycobacterial species ETZ represented an accumulation of distinctive lipids around the bacteria, the simultaneous presence of other components could not be ruled out. However, it was clear that accumulation of glycolipids was not a general explanation of the ETZ, because most strains of M. tuberculosis do not synthesize significant amounts of PGL (Daffé et al., 1987). Moreover, both GPL-positive and GPL-negative strains of M. avium-M. intracellulare elaborate capsules (Rastogi et al., 1989). The subsequent knowledge that all these species can secrete polysaccharides and proteins (Lemassu and Daffé, 1996; Ortalo-Magné et al., 1995) suggests that this material forms the ETZ. Although there is no direct chemical evidence, it seems likely that M. leprae and M. lepraemurium also surround themselves with polysaccharide inside their host cells; the staining of the outermost layer of M. leprae by concanavalin A (Picard et al., 1984) supports this suggestion.

From the above-mentioned data, it is tempting to suggest that the ETZ represents a capsule of polysaccharide and protein, plus in a few cases, specialized glycolipids, which differs from the capsule around mycobacteria grown in vitro only in being more extensive because the material that would be shed into the culture medium is retained by the phagosomal membrane (Daffé and Draper, 1998). The transparency of ETZ may arise because although the zone in the living host cell was occupied by hydrated polymers, these are shrunk during the preparation for microscopy. If this explanation is correct, the zone should have a different appearance in specimens prepared by freeze substitution. Interestingly, mycobacteria embedded by freeze-substitution show an extended capsule (Paul and Beveridge, 1992, 1994) whose thickness varies considerably between species. The above-mentioned hypothesis is also supported by the appearance of an ETZ around in vitro grown bacteria that were coated with either specific antibodies or embedded in gelatin prior to the different drastic treatments involved in conventional electron microscopy studies (Daffé and Etienne, 1999; Fréhel et al., 1988). This latter treatment presumably minimizes the collapsing of capsular material. In these conditions, a capsule-like structure, resembling the ETZ observed in vivo, was observed around the pathogenic mycobacterial species examined but not around the nonpathogenic M. smegmatis and M. aurum (Fréhel et al., 1988). When the same bacteria are instead processed by the conventional method commonly used to observe mycobacteria by transmission electron microscopy, the outer layer probably collapses by dehydration to give a thin dark layer as seen in Fig. 1.

Isolation and Chemical Analysis of the Outer Layer Components

That the outermost layer of the mycobacterial cell envelope is partly polysaccharide in nature could be deduced from the staining of ETZ with peroxidase-conjugated concanavalin A (Picard et al., 1984) and from old data showing that mycobacterial culture filtrates, notably the preparation called tuberculin, contain several polysaccharides (Seibert, 1949). We have subsequently shown that polysaccharides are present in substantial amounts in the medium of young exponentially growing cultures in which the numbers of dead or dying bacteria would be expected to be small (Lemassu and Daffé, 1994; Lemassu et al., 1996). Furthermore, we also showed that the mycobacterial cell is surrounded by an attached layer of polysaccharide (Ortalo-Magné et al., 1995; Lemassu et al., 1996) which is chemically similar to that present in the medium.

The surface-exposed material can be removed by mild mechanical or chemical treatment (Ortalo-Magné et al., 1995; Ortalo-Magné et al., 1996b; Raynaud, 1998). Growth of the cells as a pellicle apparently minimizes the loss of the capsule into the medium, presumably because it eliminates the mechanical stresses on the capsule caused by agitation of the medium (Daffé and Draper, 1998). Again, more extracellular material was recovered from the culture filtrates of the pathogenic species, e.g., M. tuberculosis and M. kansasii, than those of saprophytic and nonpathogenic strains such as M. smegmatis and

M. aurum (Lemassu and Daffé, 1994; Lemassu et al., 1996). Shaking gently the bacterial pellicles with glass beads (Ortalo-Magné et al., 1995, 1996b; Raynaud et al., 1998) extracts the outermost capsular compounds, whereas the use of Tween 80 for short periods of time leads to the progressive extraction of more material from expectedly deeper compartments. Importantly, these treatments do not affect the viability of cells that are obviously declumped, as judged by electron microscopy (Ortalo-Magné et al., 1995, 1996b). However, the use of Tween may affect both the morphology of colonies and the virulence of *M. tuberculosis*, depending of the concentration of the detergent (Middlebrook et al., 1947; Bloch and Noll, 1953). The main components of the outermost capsular layer of slow-growing mycobacterial species, most of which are pathogens (e.g., *M. tuberculosis* and *M. kansasii*), are polysaccharides, whereas the major components of the outer layer of rapid growers, e.g., *M. phlei* and *M. smegmatis*, are proteins (Lemassu et al., 1996; Ortalo-Magné et al., 1995).

Polysaccharides

The major extracellular and capsular component of slow-growing mycobacterial species is a glucan that is composed of repeating units of five or six ->4-α-D-glucosyl residues substituted at position 6 with a mono- or oligoglucosyl residues (Dinadayala et al., 2003; Lemassu and Daffé, 1994; Ortalo-Magné et al., 1995). This glycogen-like polysaccharide probably corresponds to the "highly branched glycogen-type glucan" found associated with cell wall preparations of *M. tuberculosis* (Amar-Nacasch and Vilkas, 1970) and *M. bovis* BCG (Misaki and Yukawa, 1966). Based on its presence in the culture medium of several mycobacterial species (Daniel, 1984) and its similar apparent molecular mass (Kent, 1951), the polysaccharide-II of Seibert et al. (1949) probably corresponds to the extracellular and capsular glucan of Daffé and colleagues despite the different structure proposed earlier (Kent, 1951).

The surface-exposed and extracellular materials also contain a D-arabino-D-mannan, a heteropolysaccharide that exhibits an apparent molecular mass of 13 kDa, and a mannan chain composed of a ->6-α-D-mannosyl-1-> core substituted at some positions 2 with an α-D-mannosyl unit. The arabinomannan of *M. tuberculosis* has a structure identical to the carbohydrate moiety of LAM from the same species (see chapter 6); in slow-growing species, the arabinan segment is capped on the nonreducing termini by oligomannosides (Lemassu and Daffé, 1994; Lemassu et al., 1996; Ortalo-Magné et al., 1995,

1996a), a feature not found in the wall arabinogalactan (Daffé et al., 1990, 1993).

Lipids

The outermost part of the mycobacterial cell envelope contains only a tiny amount of lipid (2% to 3% of the surface-exposed material), and progressive removal of the capsular material shows that most of the lipid is in the inner rather than the outer part of the capsule (Ortalo-Magné et al., 1996b). Some of the species- and type-specific lipids and glycolipids, e.g., phthiocerol dimycocerosates, PGL and GPL, can be found on the surface of the capsule, in agreement with serological and ultrastructural findings. In contrast, other lipids, notably trehalose dimycolates, traditionally considered surface-exposed lipids, are not detected in the outermost layer of *M. tuberculosis* and most mycobacterial species (Ortalo-Magné et al., 1996b). They evidently occur also in the capsule but in deeper compartments because they are extracted with Tween 80 used for 1 h. Similarly, the mycobacterial lipopolysaccharide, LAM, is not detectable in the surface-exposed of *M. tuberculosis* (Ortalo-Magné et al., 1996a); its location within the envelope is unknown, as discussed above (the plasma membrane section) and in chapter 6.

Proteins

The outermost capsular proteins of in vitro-grown *M. tuberculosis* are a complex mixture of polypeptides (Ortalo-Magné et al., 1995). Some proteins seem to correspond to secreted polypeptides found in short-term culture filtrates, whereas others appear to be cell wall associated (Daffé and Etienne, 1999). Although localization index (Wiker et al., 1991), gene fusion methodology and bioinformatics analysis have been used to identify extracytoplasmic proteins, it is not possible to discriminate between secreted, and capsule- and cell wall-associated components. For instance, immunocytochemistry studies have clearly shown that some of the major secreted proteins, e.g., the antigen 85 complex, also occur in substantial amounts in both the cell wall and cell surface (Rambukkana et al., 1991). In fact, most of the hundreds of proteins identified in the short-term culture filtrate of *M. tuberculosis* by 2-D SDS-PAGE (Sonnenberg and Belisle, 1997) are also found in significant amounts in the surface-exposed material extracted by mild mechanical treatment of cells. This observation supports the concept that what is found in the culture filtrate of in vitro-grown cells is probably shed from the surface of the bacilli and, in an in vivo context, would be confined around the cells in the ETZ (Daffé and Draper, 1998).

CONCLUDING REMARKS

The gross structures of the core components of the mycobacterial cell envelope have been established since the discovery of the singularity of this structure, more than half a century ago; the fine structures of these substances, as well as those of new constituents, have also been established more recently. Similarly, the "multilayered" structure of the cell envelope has been observed some 30 years ago, followed by the building of an elegant "chemical" model by Minnikin. Although data that are consistent with the model have been published, no sign of an outer lipid layer had ever been reported in electron micrographs of sectioned mycobacterial cells. In addition, many artifacts are known to occur during the various treatments commonly used to visualize bacterial cells. Cryomicroscopy seems to be a powerful technique that may give a better picture of bacterial cell envelopes; its application to the study of wild types and different isogenic mutants of mycobacteria and related genera affected in the production of various compounds is urgently needed.

The mycobacterial cell envelope is important for the bacterial physiology, as proved by the essentiality of the synthesis of several of its components, e.g., mycolic acids and arabinogalactan. The main components of the mycobacterial cell envelope are produced inside the infected cells as indicated by the isolation and/or characterization of arabinogalactan, lipoarabinomannan and mycolic acids from the "noncultivable" leprosy bacillus (Asselineau et al., 1981; Daffé et al., 1993; Hunter et al., 1986) and of phthiocerol dimycocerosates and glucan from the tubercle bacillus (Kondo and Kanai, 1976; Schwebach et al., 2002). Because the structures of these compounds are exotic, compared with mammalian products, enzymes involved in the building of the cell envelope remain an attractive source for new targets in the context of the development of inhibitors that may serve as lead compounds for the development of antituberculous drugs. These small products have to pass through a very impermeable structure that is well developed for controlling access to the interior of the bacterium.

As far as pathogenicity is concerned, the only obvious difference observed between pathogenic and nonpathogenic mycobacteria species in terms of architecture of their cell envelope is the occurrence of a capsule around pathogens inside host cells, whereas a thin outer layer surrounds nonpathogens in compa-

rable conditions (Hanks, 1961a, 1961b, 1961c; Fréhel et al., 1988; Ryter et al., 1984). However, only little is known of the molecular organization of the capsular constituents. For instance, how lipids are distributed at different levels within the capsule is unknown. It is not clear how lipids interact with polysaccharides and proteins to expose themselves at the cell surface and, as such, influence the surface properties of mycobacteria (e.g., hydrophobicity, colony morphology, formation of biofilms, and sliding).

The polysaccharide/protein matrix in pathogens would also serve as a defense mechanism by facilitating survival in the host (Draper, 1998). This may be achieved by playing a role as a physical barrier; electron microscopy images suggest that the action of potentially harmful host-derived material such as lytic enzymes is impeded by the capsule (Ryter, 1984). The capsule and its constituents can also act by actively modifying the behavior of host cells in some way beneficial to the bacterium, as recently shown for the glucan from *M. tuberculosis* (Gagliardi et al., 2007). Further studies using capsule-null mutants (if viable) are clearly needed to decipher the functions of the capsule.

Acknowledgments. I am grateful to Pierre Gounon (UFR-Médecine, Nice, France) for the electron micrograph, Pierre Roblin (IPBS, Toulouse, France) for drawing the chemical model, and Gilles Etienne (IPBS, Toulouse) for the design of the figure.

REFERENCES

Akamatsu, Y., Y. Ono, and S. Nojima. 1966. Phospholipid patterns in subcellular fractions of *Mycobacterium phlei. J. Biochem.* (Tokyo) **59:**176–182.

Amar-Nacasch, C., and E.Vilkas. 1970. Étude des parois de *Mycobacterium tuberculosis.* II. Mise en évidence d'un mycolate d'arabinobiose et d'un glucane dans les parois de *M. tuberculosis* H37Ra. *Bull. Soc. Chim. Biol.* **52:**145–151.

Asselineau, C., S. Clavel, F. Clément, M. Daffé, H. L. David, M.-A. Lanéelle, and J.-C. Promé. 1981. Constituants lipidiques de *Mycobacterium leprae* isolé de tatou infecté expérimentalement. *Ann. Microbiol.* (*Inst. Pasteur*) **132A:**19–30.

Barksdale, L. and K.-S. Kim. 1977. *Mycobacterium. Bacteriol. Rev.* **41:**217–372.

Benedetti, E. L., L. Dunia, M. A. Ludosky, N. V. Man, D. D. Trach, N. Rastogi, and H. L. David. 1984. Freeze-etching and freeze-fracture structural features of cell envelopes in mycobacteria and leprosy derived corynebacteria. *Acta Leprol.* **95:**237–248.

Bloch, H. and H. Noll. 1953. Studies on the virulence of tubercle bacilli. Variation in virulence effected by Tween 80 and thiosemicarbazone. *J. Exp. Med.* **97:**1–16.

Boddingius, J. and H. Dijkman. 1990. Subcellular localization of *Mycobacterium leprae*-specific phenolic glycolipid (PGL-I) antigen in human leprosy lesions and in *M. leprae* isolated from armadillo liver. *J. Gen. Microbiol.* **136:**2001–2012.

Brodie, A. F., S.-H. Lee, and V. K. Kalra. 1979. Transport and energy transduction mechanism in *Mycobacterium phlei*, p. 46–53. *In* D. Schlessinger (ed.), *Microbiology*. ASM, Washington, DC.

Chapman, G. B., J. H. Hanks, and J. H. Wallace. 1959. An electron microscope study of the disposition and fine structure of

Mycobacterium lepraemurium in mouse spleen. *J. Bacteriol.* **77**:205–211.

Daffé, M., and P. Draper. 1998. The envelope layers of mycobacteria with reference to their pathogenicity. *Adv. Microb. Physiol.* **39**:131–203.

Daffé, M., and G. Etienne. 1999. The capsule of *Mycobacterium tuberculosis* and its implications for pathogenicity. *Tuberc. Lung Dis.* **79**:153–169.

Daffé, M., C. Lacave, M.-A. Lanéelle, and G. Lanéelle. 1987. Structure of the major triglycosyl phenol-phthiocerol of *Mycobacterium tuberculosis* (strain Canetti). *Eur. J. Biochem.* **167**:155–160.

Daffé, M., M.-A. Dupont, and N. Gas. 1989. The cell envelope of *Mycobacterium smegmatis*: cytochemistry and architectural implications. *FEMS Microbiol. Lett.* **61**:89–94.

Daffé, M., P. J. Brennan, and M. McNeil. 1990. Predominant structural features of the cell wall arabinogalactan of *Mycobacterium tuberculosis* as revealed through characterization of oligoglycosyl alditol fragments by gas chromatography/mass spectrometry and by ^{1}H- and ^{13}C-NMR analyses. *J. Biol. Chem.* **265**:6734–6743.

Daffé, M., P. J. Brennan, and M. McNeil. 1993. Major structural features of the cell wall arabinogalactans of *Mycobacterium*, *Rhodococcus*, and *Nocardia* spp. *Carbohydr. Res.* **249**:383–398.

Daniel, T. M. 1984. Soluble mycobacterial antigens, p. 417–465. *In* G. P. Kubica and L. G. Wayne (ed.), *The Mycobacteria: a Sourcebook*, part A. Marcel Dekker, Inc, New York, NY.

Dinadayala, P., F. Laval, C. Raynaud, A. Lemassu, M. A. Lanéelle, G. Lanéelle, and M. Daffé. 2003. Tracking the putative precursors of oxygenated mycolates of *Mycobacterium tuberculosis*. Structural analysis of fatty acids of a mutant devoid of methoxy- and ketomycolates. *J. Biol Chem.* **278**:7310–7319.

Draper, P. 1971. The walls of *Mycobacterium lepraemurium*: chemistry and ultrastructure. *J. Gen. Microbiol.* **69**:313–332.

Draper, P. 1998. The outer parts of the mycobacterial envelope as permeability barriers. *Front. Biosci.* **3**:D1253–D1261.

Fréhel, C., A. Ryter, N. Rastogi, and H. L. David. 1986. The electron-transparent zone in phagocytised *Mycobacterium avium* and other mycobacteria: formation, persistence and role in bacterial survival. *Ann. Inst. Pasteur/Microbiol.* **137B**:239–257.

Fréhel, C., N. Rastogi, J.-C. Bénichou, and A. Ryter. 1988. Do test tube-grown pathogenic mycobacteria possess a protective capsule? *FEMS Microbiol. Lett.* **56**:225–230.

Gagliardi, M. C., L. Lemassu, R. Teloni, S. Mariotti, V. Sargentini, M. Pardini, M. Daffé, and R. Nisini. Cell-wall associated alpha-glucan is instrumental for *Mycobacterium tuberculosis* to block CD1 molecule expression and disable the function of dendritic cell derived from infected monocyte. *Cell Microbiol.* **3**:2081–2092.

Goldman, D. S. 1970. Subcellular localization of individual mannose-containing phospholipids in *Mycobacterium tuberculosis*. *Am. Rev. Respir. Dis.* **102**:543–555.

Goren, M. B., and P. J. Brennan. 1979. Mycobacterial lipids: chemistry and biologic activities, p. 63–193. *In* G. P. Youmans (ed.), *Tuberculosis*. W. B. Saunders Company, Philadelphia, PA.

Hanks, J. H. 1961a. Capsules in electron micrographs of *Mycobacterium leprae*. *Int. J. Lepr.* **29**:84–87.

Hanks, J. H. 1961b. The problem of preserving internal structures in pathogenic mycobacteria by conventional methods of fixation. *Int. J. Lepr.* **29**:175–178.

Hanks, J. H. 1961c. Demonstration of capsules on *M. leprae* during carbol-fuchsin staining mechanism of the Ziehl-Neelsen stain. *Int. J. Lepr.* **26**:179–182.

Hanks, J. H. 1961d. The origin of the capsules on *Mycobacterium leprae* and other tissue-grown mycobcteria. *Int. J. Lepr.* **26**:172–174.

Hanks, J. H., J. T. Moore, and J. E. Michaels. 1961. Significance of capsular components of *Mycobacterium leprae* and other mycobacteria. *Int. J. Lepr.* **26**:74–83.

Hunter, S. W., and P. J. Brennan. 1990. Evidence for the presence of a phosphatidylinositol anchor on the lipoarabinomannan and lipomannan of *Mycobacterium tuberculosis*. *J. Biol. Chem.* **265**:9272–9279.

Hunter, S. W., H. Gaylord, and P. J. Brennan. 1986. Structure and antigenicity of the phosphorylated lipopolysaccharides from the leprosy and tubercle bacilli. *J. Biol. Chem.* **261**:12345–12351.

Jarlier, V., and H. Nikaido, 1990. Permeability barrier to hydrophilic solutes in *Mycobacterium chelonei* (sic). *J. Bacteriol.* **172**:1418–1423.

Kent, P. W. 1951. Structure of an antigenic polysaccharide isolated from tuberculin. *J. Chem. Soc.* **1**:364–368.

Kondo, E., and K. Kanai. 1976. A suggested role of a host-parasite lipid complex in mycobacterial infection. *Jpn. J. Med. Sci. Biol.* **29**:199–210.

Kotani, S., T. Kitaura, T. Hirano, and A. Tanaka. (1959). Isolation and chemical composition of the cells walls of BCG. *Biken J.* **2**:129–141.

Kumar, G., V. K. Kalra, and A. F. Brodie. 1979. Asymmetric distribution of phospholipids in membranes from *Mycobacterium phlei*. *Arch. Biochem. Biophys.* **198**:22–30.

Lanéelle, M. A., D. Promé, G. Lanéelle, and J. C. Promé. 1990. Ornithine lipid of *Mycobacterium tuberculosis*: its distribution in some slow- and fast-growing mycobacteria. *J. Gen. Microbiol.* **136**:773–778.

Lemassu, A., and M. Daffé. 1994. Structural features of the exocellular polysaccharides of *Mycobacterium tuberculosis*. *Biochem. J.* **297**:351–357.

Lemassu, A., A. Ortalo-Magné, F. Bardou, G. Silve, M. A. Lanéelle, and M. Daffé. 1996. Extracellular and surface-exposed polysaccharides of non-tuberculous mycobacteria. *Microbiology* **142**:1513–1520.

Liu, J., E. Y. Rosenberg, and H. Nikaido, 1995. Fluidity of the lipid domain of cell wall from *Mycobacterium chelonae*. *Proc. Natl. Acad. Sci. USA* **92**:11254–11258.

McNeil, M., M. Daffé, and P. J. Brennan. 1990. Evidence for the nature of the link between the arabinogalactan and peptidoglycan of mycobacterial cell walls. *J. Biol. Chem.* **265**:18200–18206.

McNeil, M., M. Daffé, and P. J. Brennan 1991. Location of the mycoloyl ester substituents in the cell walls of mycobacteria. *J. Biol. Chem.* **266**:13217–13223.

McNeil, M. R., and P. J. Brennan. 1991. Structure, function and biogenesis of the cell envelope of mycobacterial in relation to bacterial physiology, pathogenesis and drug resistance; some thoughts and possibilities arising from recent structural information. *Res. Microbiol.* **142**:451–463.

Middlebrook, G., R. J. Dubos, and C. Pierce. 1947. Virulence and morphological characteristics of mammalian tubercle bacilli. *J. Exp. Med.* **86**:175–184.

Minnikin, D. E. 1982. Lipids: complex lipids, their chemistry, biosynthesis and roles, p. 95–184. *In* C. Ratledge and J. Stanford (ed.), *The Biology of the Mycobacteria*, vol. 1. *Physiology, Identification and Classification*. Academic Press, London, United Kingdom.

Misaki, A., and S. Yukawa. 1966. Studies on cell walls of Mycobacteria. II. Constitution of polysaccharides from BCG cell walls. *J. Biochem.* **59**:511–520.

Nikaido, H., S.-H. Kim, and E. Y. Rosenberg. 1993. Physical organization of lipids in the cell wall of *Mycobacterium chelonae*. *Mol. Microbiol.* **8**:1025–1030.

Oka, S., K. Fukushi, M. Fujimoto, H. Sato, and M. Motomiya. 1968. La distribution subcellulaire des phospholipides de la mycobactérie. *C. R. Soc. Franco-Jpn. Biol.* **162**:1648–1650.

Ortalo-Magné, A., M.-A. Dupont, A. Lemassu, Å. B. Andersen, P. Gounon, and M. Daffé. 1995. Molecular composition of the outermost capsular material of the tubercle bacillus. *Microbiology* 141:1609–1620.

Ortalo-Magné, A., Å. B. Andersen, and M. Daffé, 1996a. The outermost capsular arabinomannans and other mannoconjugates of virulent and avirulent tubercle bacilli. *Microbiology* 142:927–935.

Ortalo-Magné, A., A. Lemassu, M. A. Lanéelle, F. Bardou, G. Silve, P. Gounon, G. Marchal, and M. Daffé. 1996b. Identification of the surface-exposed lipids on the cell envelope of *Mycobacterium tuberculosis* and other mycobacterial species. *J. Bacteriol.* 178:456–461.

Paul, T. R., and T. J. Beveridge. 1992. Reevaluation of envelope profiles and cytoplasmic ultrastructure of mycobacteria processed by conventional embedding and freeze-substitution protocols. *J. Bacteriol.* 174:6508–6517.

Paul, T. R., and T. J. Beveridge. 1994. Preservation of surface lipids and determination of ultrastructure of *Mycobacterium kansasii* by freeze-substitution. *Infect. Immun.* 62:1542–1550.

Picard, B., C. Frehel, and N. Rastogi. 1984. Cytochemical characterization of mycobacterial outer surfaces. *Acta Leprol.* 95:227–235.

Puech, V., M. Chami, A. Lemassu, M. A. Lanéelle, B. Schiffler, P. Gounon, N. Bayan, R. Benz, and M. Daffé. 2001. Structure of the cell envelope of corynebacteria: importance of the non-covalently bound lipids in the formation of the cell wall permeability barrier and fracture plane. *Microbiology* 147:1365–1382.

Rambukkana, A., P. K. A. Das, A. Chand, J. G. Baas, D. G. Groothuis, and A. H. J. Kolk. 1991. Subcellular distribution of monoclonal antibody defined epitopes on immunodominant *Mycobacterium tuberculosis* proteins in the 30-kDa region: identification and localization of 29/33-kDa doublet proteins on mycobacterial cell wall. *Scand. J. Immunol.* 33:763–775.

Rastogi, N. 1991. Recent observations concerning structure and function relationships in the mycobacterial cell envelope: elaboration of a model in terms of mycobacterial pathogenicity, virulence and drug-resistance. *Res. Microbiol.* 142:464–476.

Rastogi, N., V. Lévy-Frebault, M. C. Blom-Potar, and H. L. David. 1989. Ability of smooth and rough variants of *Mycobacterium avium* and *M. intracellulare* to multiply and survive intracellularly: role of C-mycosides. *Zentbl. Bakteriol. Mikrobiol. Hyg.* 270:345-360.

Raynaud, C., G. Etienne, P. Peyron, M.-A. Lanéelle, and M. Daffé. 1998. Extracellular enzyme activities potentially involved in the pathogenicity of *Mycobacterium tuberculosis*. *Microbiology* 144:577–587.

Rezwan, M., M.-A. Lanéelle, P. Sander, and M. Daffé. 2007. Breaking down the wall: fractionation of mycobacteria. *J. Microbiol. Methods* 68:32–39.

Rulong, S., A. P. Aguas, P. P. Da Silva, and T. S. Silva. 1991. Intramacrophagic *Mycobacterium avium* bacilli are coated by a multiple lamellar structure: freeze fracture analysis of infected mouse liver. *Infect. Immun.* 59:3895–3902.

Ryter, A., C. Frehel, N. Rastogi, and H. L. David. 1984. Macrophage interaction with mycobacteria including *M. leprae*. *Acta Leprol.* 95:211–235.

Schwebach, J., A. Glatman-Freedman, L. Gunter-Cummins, Z. Dai, J. Robbins, R. Schneerson, and A. Casadevall, 2002. Glucan is a component of the *Mycobacterium tuberculosis* surface that is expressed *in vitro* and *in vivo*. *Infect. Immun.* 70:2566–2575.

Seibert, F. B. 1949. The isolation of three different proteins and two polysaccharides from tuberculin by alcohol fractionation. Their chemical and biological properties. *Am. Rev. Tuberc.* 59:86–101.

Seibert, F. B., M. Stacey, and P. W. Kent. 1949. An antigenic polysaccharide, "polysaccharide II" isolated from tuberculin. *Biochim. Biophys. Acta* 3:632–640.

Senaratne, R. H., H. Mobasheri, K. G. Papavininasasundaram, P. Jenner, E. J. A. Lea, and P. Draper. 1998. Expression of a gene for a porin-like protein of the OmpA family from *Mycobacterium tuberculosis* H37Rv. *J. Bacteriol.* 180:3541–3547.

Silva, M. T., and P. M. Macedo. 1983a. A comparative ultrastructural study of the membranes of *Mycobacterium leprae* and of cultivable *Mycobacteria* (sic). *Biol. Cell.* 47:383–386.

Silva, M. T., and P. M. Macedo. 1983b. The interpretation of the ultrastructure of mycobacterial cells in transmission electron microscopy in ultrathin sections. *Int. J. Lepr.* 51:225–234.

Silva, M. T., and P. M. Macedo. 1984. Ultrastructural characterization of normal and damaged membranes of *Mycobacterium leprae* and cultivable mycobacteria. *J. Gen. Microbiol.* 130:369–380.

Silva, M.T., F. Portaels, and P. M. Macedo. 1989. New data on the ultrastructure of the membrane of *Mycobacterium leprae*. *Int. J. Lepr.* 57:54–64.

Sonnenberg, M. G., and J. T. Belisle. 1997. Definition of Mycobacterium tuberculosis culture filtrate proteins by two-dimensional polyacrylamide gel electrophoresis, N-terminal amino-acid sequencing and electrospray mass spectrometry. *Infect. Immun.* 65:4515–4524.

Trias, J., and R. Benz. 1993. Characterization of the channel formed by the mycobacterial porin in lipid bilayer membranes. Demonstration of voltage gating and of negative point charges at the channel mouth. *J. Biol. Chem.* 268:6234–6240.

Trias, J., V. Jarlier, and R. Benz. 1992. Porins in the cell wall of mycobacteria. *Science* 258:1479–1481.

Wiker, H. G., M. Harboe, and S. Nagai. 1991. A localization index for distinction between extracellular and intracellular antigens of *Mycobacterium tuberculosis*. *J. Gen. Microbiol.* 137:875–884.

The Mycobacterial Cell Envelope
Edited by M. Daffé and J.-M. Reyrat
© 2008 ASM Press, Washington, DC

Chapter 2

Wall Ultrastructure and Periplasm

Terry J. Beveridge

Gram-positive bacteria have robust thick cell walls consisting of peptidoglycan complexed to a variety of macromolecular and polymeric substances. Mycobacteria possess gram-positive walls, and these are among the most complicated in both chemical and ultrastructural terms. Yet, admittedly, little is really known about their true ultrastructure for the simple reason that they are so difficult to study by imaging, especially by electron microscopy. As is explained in several of this book's chapters, mycobacterial walls consist of several complex constituents, such as peptidoglycan, mycolic acids, arabinogalactans and lipoarabinomannan (see chapters 1, 3, 4 and 6). This amalgam of polymers makes for an extremely strong boundary layer around the protoplast, but it also is a relatively unwettable matrix that is difficult to stain and differentiate by the water-soluble heavy metal ions that are used in staining for transmission electron microscopy (TEM). Both negative stains and positively stained thin sections prove stubborn with this uncooperative mycoylated arabinogalactan material. I must admit that this dilemma frustrated my own laboratory, and for this reason, we abandoned the study of mycobacteria several years ago, hoping that new techniques would eventually make high-fidelity study feasible. For this reason, in this chapter, I will concentrate on other more easily studied gram-positive bacteria, such as *Bacillus subtilis* and *Staphylococcus aureus*, that we have elucidated using new cryoTEM technology, which maintains the hydrated nature of the sample. Then, this information will be correlated with older work on mycobacteria so that conclusions may be drawn.

PROBLEMS WITH OLD TECHNIQUES

The conventional techniques for TEM are reliable and interpretable as long as one knows how each step in the preparatory regimen affects the structure of the bacteria that one wants to examine. For example, a typical regiment to produce thin-sectioned material uses chemical fixatives followed by heavy metal staining, dehydration through organic solvents, and eventual infiltration of an embedding plastic followed by curing, which polymerizes the plastic and hardens the specimen so that it can be thin sectioned (Beveridge et al., 2007; Daffe and Draper, 1998). Chemical fixatives, such as glutaraldehyde and osmium tetroxide, are added so as to increase the covalent bonding in structures such as membranes and ribosomes and, it is hoped, better preserve them for the severe procedures that follow. Yet, these fixatives destroy all functionality of structure and presumably condense it. Heavy metal stains, such as uranyl acetate, are highly electropositive (so as to stick to and stain constituent macromolecules) and operate at a pH below 4; they alter charge capacity of structure that they stain, and the low pH denatures constituent proteins. *Dehydration* is now italicized for emphasis because this is possibly the worst insult during the entire preparatory regimen. Water must be extracted from the cells so as to make them compatible with the plastic resin. We must infiltrate into them before thin sectioning can be done. How traumatized these bacteria must be as we extract almost 90% of their substance, especially because all cellular macromolecules depend on an aqueous milieu! Membranes are depleted of lipid by the dehydrating solvent and hydration shells surrounding molecules disturbed. It is remarkable how well bacteria are actually preserved after such treatment (Fig. 1A).

We are presented with an almost unsolvable puzzle when it comes to the electron microscopy of bacteria. Single cells are so small that any form of light microscopy cannot truly resolve their minute structure with high fidelity. There have been great strides with laser scanning confocal microscopy with fluorescent

Terry J. Beveridge (deceased) • Department of Molecular and Cellular Biology and Advanced Food and Materials Network—Networks Centers of Excellence, College of Biological Science, University of Guelph, Guelph, Ontario, Canada N1G 2W1.

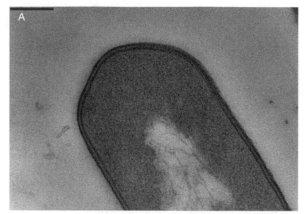

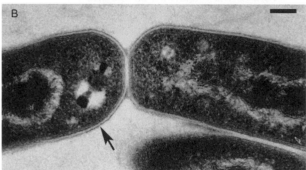

Figure 1. (A) Thin section of a conventionally embedded *Bacillus subtilis* showing a rather featureless gram-positive cell wall. The DNA is condensed in the center of the cell. Scale bar = 200 nm. (B) Thin section of a dividing *Mycobacterium phlei* cell through conventional fixation. Note that this cell possesses a more complex cell wall that possesses an electron dense outer layer and an electron translucent inner layer (arrow). The DNA is (again) condensed in the center of the cell. These cells are more difficult to fix than *B. subtilis* (A) and are not as well preserved. Scale bar = 100 nm. See Paul and Beveridge (1994) for more details.

probes and multiphoton systems or photoactivated light microscopy to pinpoint single molecules, but this does not resolve, more properly, it only detects. Scanning electron microscopy, even with advances in resolution using field-emission guns and low-voltage scanning electron microscopy, still remains a topographical technique and cannot decipher any structure below a surface. TEM of conventional thin sections is really the best technique because it gives us good resolution of internal cell structure as well as giving us a view of the juxtaposition of the bacterium's enveloping structures (Fig. 1B). Yet, as discussed earlier, this thin section technique is fraught with artifact and we must be careful with our interpretations. For mycobacterial cell envelopes, even more care is necessary because it is recognized that several of the wall constituents are soluble or deformable in the organic solvents we use during dehydration (Paul and Beveridge, 1992, 1994). In addition, previous studies have suggested that mycobacterial walls are

organized into distinct structural zones (Barksdale and Kim, 1977; Daffe and Draper, 1998; Imaeda et al., 1968; Rastogi et al., 1993). CryoTEM eliminates the need for chemical fixation and dehydration used in the conventional thin-section technique.

THE NEW TECHNIQUE OF CRYOTRANSMISSION ELECTRON MICROSCOPY

Ultrarapid freezing of cells is the central feature for cryoTEM. This freezing must occur very rapidly (less than milliseconds) so that vitrification of the sample occurs; amorphous ice, having the consistency of a glass, is formed and all macromolecular motion is stopped. Cells are physically preserved in a hydrated condition without any increase in chemical bonding from artificial chemical agents, such as glutaraldehyde or osmium tetroxide. The aqueous cellular milieu is maintained, as are hydration shells surrounding resident molecules. This is as close to the vital condition as we can get because, if flagellated bacteria are thawed, they become motile once again. Accordingly, if we can somehow visualize these frozen cells in thin section we will obtain a much more valid image of their natural structure.

Advantages of Freeze-Substitution

Here is a cryotechnique that offers the advantages of rapid freezing (and its attendant rewards) plus easy manipulation of plastic embeddings (Graham et al., 1991). It provides remarkable preservation of hard-to-maintain bacterial consortia such as biofilms (Fig. 2A) (Hunter and Beveridge, 2005). Using this technique, cells are rapidly frozen as discussed before and, while still vitrified, are subjected to chemical fixation and dehydration; this is the "substitution" process. Care is taken to ensure that chemical substitution occurs below the ice transition temperature (ca. $-80°C$) so that ice crystallization does not occur, which could deform and breakup the cells. Once substitution is complete, the physical fixation (vitrification) has been replaced by a chemical fixation. The specimen is also dehydrated during substitution and can be embedded in plastic so that conventional thin sectioning can proceed. The important point with the freeze-substitution method is that the bacteria have been preserved in a hydrated form by vitrification and, then, gradually chemically fixed while the hydrated form is maintained at low temperature. Water (as amorphous ice) is also removed at this low temperature. Obviously, this is not a perfect technique because, when imaged by TEM, the cells are not actu-

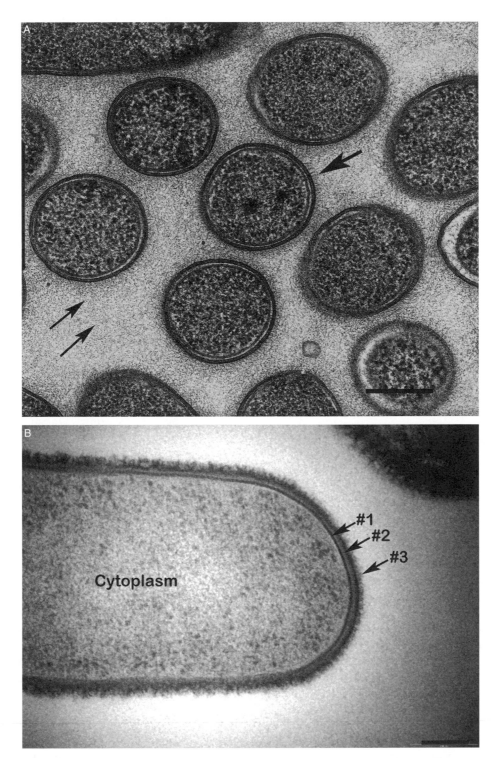

Figure 2. (A) Thin section of a *Pseudomonas aeruginosa* PAO1 biofilm that has been freeze-substituted. This type of bacterial consortium is among the most difficult biomaterials to preserve, and these gram-negative cells complete with their lipopolysaccharide (LPS, large arrow) O-side chains, and exopolymeric substances (EPS, small arrows) are seen at high fidelity. Scale bar = 100 nm. See Hunter and Beveridge (2005) for more details. (B) This *B. subtilis* cell has been freeze-substituted, and the cell has been preserved with high fidelity. Now, the DNA is dispersed throughout the entire cytoplasm, and the cell wall is composed of three separate regions (numbered 1 to 3). As explained in the text, these regions correspond to the way in which this gram-positive wall turns over. Scale bar = 100 nm. See Graham and Beveridge (1994) for more details.

ally hydrated but they are assumed to be close to their hydrated form. Extended polymeric structures, such as lipopolysaccharide O-side chains and exopolymeric substances are maintained (Fig. 2A).

Freeze-substitution also gives improved preservation of cell walls of gram-positive bacteria (Fig. 2B). With *Bacillus subtilis*, instead of seeing a rather featureless wall that is 25-nm thick (Fig. 1A), a tripartite wall segregated into three distinct regions is obtained (Fig. 2B). That closest to the plasma membrane (region 1) is darkly stained by uranyl ions (UO_2^{2+}), indicating that the polymers in this zone have relatively high mass and abundant electronegative sites (e.g., anionic carboxyl or phosphoryl groups) to complex the stain. Above this darkly stained region is a thicker but more translucent region (region 2) that has fewer reactive groups and less mass. The final region (region 3) consists of darkly stained fibrils. These three regions (1 to 3) are all made of the same biomaterials (e.g., peptidoglycan, teichoic and teichuronic acids, and proteins), but they have not been distinguished before the use of freeze-substitution (Graham and Beveridge, 1994). These three regions are consistent with the concept of wall turnover in the *B. subtilis* wall (Beveridge and Matias, 2005; Graham and Beveridge, 1994; Koch, 1988).

Freeze-substitution Substantiates Cell Wall Turnover in *Bacillus subtilis*

In this gram-positive rod, new wall polymers enter the wall immediately above the plasma membrane where they accumulate and are (eventually) covalently bonded into the existing framework via transpeptidation of newly synthesized peptidoglycan strands to older strands (region 1 [Fig. 2B]). This new material contains a rich variety of reactive groups and is under turgor pressure so as to be a rather compressed matrix of peptidoglycan and secondary polymers (in this case teichoic acid with its abundant electronegative phosphate groups). These two traits, generous density and high reactivity, account for its strong staining properties.

The translucent zone (region 2) above region 1 is most critical to the bacterium because its peptidoglycan carries the brunt of turgor pressure (up to 25 atm [Kunin and Rudy, 1991]) and is stretched almost to its breaking point. It does not possess the same profusion of stainable reactive groups or abundance of mass, as in region 1. This region contains older wall material and is strongly covalently bonded so that its fabric does not tear under such tremendous cell-swelling forces. The glycan strands, consisting of repeats of *N*-acetylglucosaminyl–*N*-acetylmuramyl dimers, are covalently bonded along their length and

~50% of their peptide stems link adjacent strands to one another so that a tough three-dimensional bonded fabric is formed. Peptidoglycan is very strong but it is also highly elastic (Stoica et al., 2003); it is a perfect fabric to help conform a bacterium into a rod shape, as well as to contend with the osmotic unequilibrium (between the cell's inside and outside) that generates turgor pressure.

The outside fibrillar zone (region 3) of the wall is unlike the inner two regions previously discussed, even though it consists of the same biomatter. Here, autolysins are active, breaking down the very bonds that hold the peptidoglycan together! One might think that this aspect of cell turnover is entirely counterproductive. It is not. *B. subtilis* has a problem that entangles the two disparate properties of osmosis and cell growth. Its cytoplasm is so concentrated that a semipermeable (plasma) membrane produces high turgor pressure, which must be contained. Yet a cell must also grow and reproduce, and high turgor pressure makes this difficult. For a rod, the cell elongates to a certain length, when it septates and divides. During rod elongation, new material must be inserted into the side (or cylindrical) wall, causing the rod to elongate, and eventually, a septum produced for division. Insertion must be done carefully. New polymers have to be somehow interwoven into the existing wall fabric to expand it and to allow wall extension and cell growth. Old covalent bonds must be broken and new ones built up, all the while maintaining enough strength in the existing framework to resist the ever-present (turgor) pressure. Beta-lactam antibiotics operate by upsetting this critical balance to great advantage. They inhibit the assembly of new peptidoglycan into the old (regions 1 and 2) and allow the cell's own autolysins (region 3) to whittle the existing wall down until it can no longer withstand turgor pressure.

Cell wall turnover is a dynamic process in which new material is inserted at the bottom of the preexisting wall (region 1) and old material is excised at the top (region 3). It is the middle zone (region 2) that holds the cell together. Wall turnover is an ongoing gradual process working from bottom to top, all the while expanding the wall and allowing the cell to grow. As old material is removed from the top, the underlying material replaces it (soon to be excised itself) and the unstressed material at the bottom moves up to be instantly subjected to the lateral force of turgor and stretched. The amount of permissible stretch that is continually added from the lower layer allows the wall to expand and the cell to grow. Presumably this turnover system, in one way or another, is shared by all gram-positive rods. Yet, mycobacteria possess complexly structured cell walls with certain constituents segregated into discrete layers, such as the outer layer, electron transparent layer, and peptidoglycan layer (Paul and Beveridge, 1993; Rastogi et al., 1986). Turnover of cell wall constituents should occur in mycobacteria because the side wall, as with *B. subtilis*, must elongate before division ensues. After all, the cell is shaped as a rod and not a coccus. Unfortunately, the complexity of wall structure and chemistry in mycobacteria should make turnover difficult, and we must await further experimentation to resolve this important aspect. In addition, mycobacteria have to deal with the additional complexity that the arabinogalactan-mycolate part of the envelope (which is a major proportion of the whole) is physically segregated from the peptidoglycan. Also a number of publications have shown images of septation in mycobacteria that indicate that the process differs in detail from that of other gram-positive bacteria.

Frozen hydrated thin sections give a better view of bacteria and their cell envelopes

This technique is different from the previously described techniques and avoids the potential artifacts of conventional processing. Although frozen hydrated sections cannot be stained as conventional sections can, these cryosections rely on the inherent density of the actual structures themselves, which can be very useful. It is apparent when looking at gram-positive bacteria in both conventional (Fig. 1A) and freeze-substitution (Fig. 2B) thin sections that a periplasm is not obvious, even though it is noticeable in gram-negative bacteria (Fig. 2A). Maybe this is not surprising because there is almost a 10-fold difference between the turgor pressures of the two cell types (3 vs. 25 atm). Yet, both conventional and freeze-substitution preparations leave cells anhydrous, and this could be the reason a periplasmic space is not seen in cells of gram-positive organisms. Even though great attempts are made to maintain hydrated structure by rapid freezing in freeze-substitutions, anhydrous plastic sections are obtained. It has been only recently that researchers have discovered how to prepare thin sections of hydrated samples (Al-Amoudi et al., 2004; Matias et al., 2003; Matias and Beveridge, 2005, 2006; Zuber et al., 2006). This is an extremely difficult technique, and again, rapid freezing comes into play. For frozen hydrated thin sections or cryosections, cells are vitrified, as is done in freeze-substitution, but next the frozen block of cells itself is thin sectioned. There is no attempt to alter the cells in any way. They are kept frozen in amorphous ice and held at liquid nitrogen temperature throughout the procedure. A special cryoultramicrotome complete with dedicated diamond knife is used to obtain cryosections, which are then mounted onto a cryospecimen holder to be placed in a cryoTEM. Cryoprotectants such as dextran or glycerol often are used to aid freezing. When the cells are frozen, cryosectioned, and manipulated into the cryoTEM, there can be no water vapor in the surrounding air or it will crystallize onto the specimen and contaminate it. There can be no increase in specimen temperature, otherwise the frozen bacteria will nucleate the transition of amorphous ice to a crystalline form; the growing crystals will deform or even puncture the bacteria, rendering further imaging useless. The vacuum in the cryoTEM must be almost perfect, or any contaminants in it will aggregate onto the frozen section, again rendering imaging useless. It is difficult, demanding, laborious work requiring experienced personnel and expensive unique instrumentation; so far, only a few microbiological laboratories can do it. However, the results justify the cost and effort (Fig. 3A).

These images of cryosectioned material are much different than previous ones shown in this chapter. They are difficult to interpret, and some explanation is required. Unlike other thin sections, they are not stained by heavy metals. Here, instead of the high-energy electrons of the microscope being scattered by heavy metal stains so that structure can be easily seen, it is the natural mass of the cell and its density that we are seeing (Fig. 3A). To be distinguished, the cellular structures must have more scattering power than the frozen water in which they are embedded. Molecules that are made of only C, H, and O, and that are folded loosely (minimal mass per unit volume) have low density and little scattering ability (e.g., polysaccharides and lipids). Molecules with some N and S and that are tightly folded (high mass per unit volume) have good density and stronger scattering power (e.g., proteins). This is why ribosomes are readily distinguished. Molecules with substantial quantities of P (e.g., phospholipids) are even better to see and this is why membranes stand out. DNA can also be distinguished (Eltsov and Zuber, 2006), but it exists as fine fibers in the cytoplasm and can be difficult to see. Indeed, various counter-ions adhering to charged groups, e.g., Mg^{++}, are expected to also contribute to density.

Existence of a gram-positive periplasm

High magnifications of cryosectioned cell walls of Gram-positive bacteria show remarkable detail, especially when one recognizes that it is the actual hydrated molecules that we are seeing (Fig. 3B). The membrane is readily discerned as a bilayer and immediately above is an electron translucent zone, followed by a thicker and darker layer, which obviously

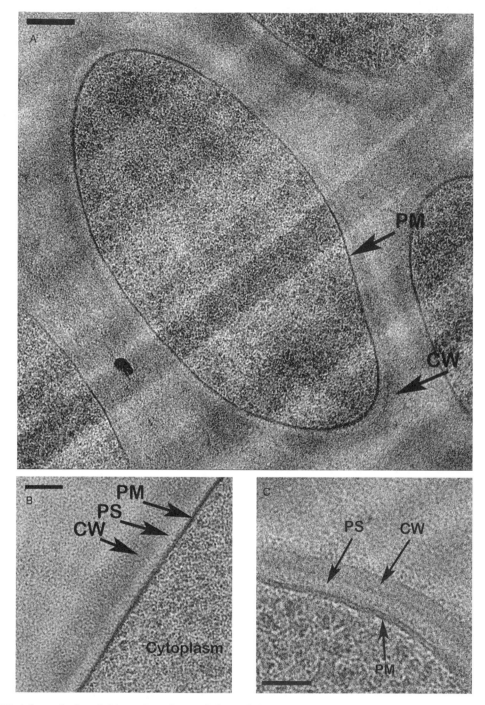

Figure 3. (A) A frozen hydrated thin section of a vitrified *B. subtilis* cell. Here, the cell is unstained, is preserved exactly as it was in life, and is seen simply because of the density of its constituent mass. As in freeze-substitutions, the DNA is dispersed throughout the cytoplasm with the ribosomes appearing as dark particles. The plasma membrane (PM) and cell wall (CW) are pointed out. The large striations running from upper right to lower left are score marks left in the ice by the diamond knife used for sectioning. Scale bar = 100 nm. (B) A higher magnification of a frozen hydrated *B. subtilis* cell showing the plasma membrane and cell wall. Again, it is the actual concentration of mass in the various parts of the cell that provides us with an image. Notice how the cell wall is darker (higher mass) close to the membrane and less dark (lower mass) at the cell walls periphery, confirming the cell wall turnover detected in freeze-substitutions (cf. this Fig with Fig. 4). For the first time, using any thin section method, the periplasmic space (PS) is preserved. Scale bar = 20 nm. See Matias and Beveridge (2005) for more details. (C) High magnification of the plasma membrane (PM) and cell wall (CW) of a frozen hydrated *Staphylococcus aureus* cell. Because this coccoid bacterium produces most of its new wall through septation during division, it does not turnover its cell wall (like *B. subtilis* does), and therefore, the cell wall has a continuous density throughout its thickness. Like *B. subtilis* and other gram-positive bacteria processed by the frozen hydrated thin section method, a periplasmic space (PS) is seen. Scale bar = 20 nm. See Matias and Beveridge (2006b) for more details.

has more mass than the former layer. Experiments whereby the cell is subjected to high external osmotic stress, so as to shrink the protoplast immediately before freezing, have shown that excess membrane, produced by the shrinkage, extrudes into the outer lying translucent zone (Matias and Beveridge, 2005, 2006a). This zone cannot then be a boundary layer composed of substantial mass, such as a cell wall. It resembles a more empty space and likely contains the periplasm. All cells that have so far been studied by cryosectioning in gram-positive bacteria (e.g., *Bacillus*, *Enterococcus*, *Staphylococcus*, and *Streptococcus*) have shown similar cell envelope profiles complete with a periplasmic zone (Matias and Beveridge, 2005, 2006a; Zuber et al., 2006).

Although we have no details of its consistency, we imagine that this gram-positive periplasm should be similar to the periplasmic gel seen in gram-negative bacteria (Graham et al., 1991; Hobot et al., 1984). In this context, the gram-positive periplasm should have similar activity and mass as seen in gram-negative organisms. It should have high enzymatic (protease, lipase, nuclease, etc.), chaperone and transport activity, and it should contain polymers (e.g., membrane-derived oligosaccharides [Kennedy, 1982]) with strong intermolecular interactions so that a gel-like state can be maintained. Interestingly, the two cryoTEM techniques of freeze-substitution and cryosections can be used to great advantage when images of periplasms are correlated. The first technique relies on heavy metal staining to contrast structure and make it visible; these heavy metal stains require reactive groups to be available in the structure with which to bind, thereby revealing reactivity. Cryosections, on the other hand, use no stain and rely on native density and mass. In gram-negative bacteria, the periplasm has low density in cryosections (Matias et al., 2003), but high concentrations of reactive groups in freeze-substitutions (Graham et al., 1991; Hobot et al., 1984). It must be a relatively low-density polymeric matrix with high chemical reactivity. We would expect the same for gram-positive periplasms.

Hydrated gram-positive cell wall matrix

Above the periplasmic zone, there lies a darker zone of higher mass (Fig. 3B). This must be the cell wall constructed of peptidoglycan and secondary polymers. Careful densitometry of *B. subtilis* frozen hydrated walls revealed a graduation of high density to low density from the inside face of the cell wall to the outside (Matias and Beveridge, 2005). We believe that this, as with the images shown in the previous freeze-substitution section, captures the process of cell wall turnover from inside to outside the wall.

The higher density face is the compacted newly exported peptidoglycan that is not under turgor stress or region 1, as discussed previously. The gradual decrease in density from inside to outside in the cell wall denotes regions 2 and 3. Region 2 occurs immediately after region 1 and contains peptidoglycan that is under high turgor pressure and is pulled taut as a consequence. Its polymers must have the same mass as those in region 1 (after all, this zone contains the same material), but they are so stretched that density (mass per unit volume) decreases. The lowest density zone of the wall is on the periphery (region 3), where we can see it blend into the external ice at its periphery (Fig. 3B), mimicking the density of water. In this region, the wall material is being actively solubilized by autolysins. Comparison of Fig. 2B (freeze-substitution) and Fig. 3B (cryosection) reveal gram-positive cell wall turnover, each in a subtly different way.

Interestingly, frozen-hydrated walls of *S. aureus* did not show this gradual decrease in density from inner wall to outer wall (cf. Fig. 3B and 3C), which is expected because the majority of new wall formation in staphylococci occurs at the septum (Matias and Beveridge, 2006a). Because *S. aureus* is a coccus and does not need to elongate like a rod, the old preexisting walls have limited turnover, if at all.

THE GRAM STAIN AND MYCOBACTERIA

There has been some confusion about the gram-reactivity of mycobacteria due to the presence of porins in mycobacteria and the effect on them of *sacB*. Although this does demonstrate that the mycobacterial envelope shares some important properties with the gram-negative cell wall, notably the existence of an outer permeability barrier to small hydrophilic molecules, it was never to be taken as evidence that mycobacteria are biologically part of the gram-negative group. The gram-reactivity of a bacterium (or for that matter an archeon) is based on the ability of the cell to retain the purple primary staining agent crystal violet after undergoing Gram staining (Beveridge, 1990, 2002; Beveridge and Davies, 1983; Beveridge and Schultze-Lam, 1997; Davies et al., 1983). It is not based on the actual structure of the cell wall of the bacterium, although there is often a correlation. All mycobacteria stain gram-positive during the Gram stain because crystal violet enters the cytoplasm and is retained there, even after alcohol decolorization (Beveridge, 1990; Beveridge and Davies, 1983). Mycobacteria are then de facto gram-positive bacteria. By convenience for our classification of prokaryotic cells, most gram-negative bacteria possess cell walls consisting of a periplasm, a thin peptidoglycan layer and an outer membrane (e.g., Fig. 2A) that is readily permeablized by ethanol during the Gram stain, making them gram-negative. Most gram-positive bacteria do not possess an outer membrane but instead have a thicker approximately 25-nm wall consisting of peptidoglycan and secondary polymers (e.g., Fig. 1A) that condenses during ethanol treatment during the Gram stain, retaining the crystal violet in the cell.

Mycobacteria possess an unusual cell wall consisting of an intermediate thickness of peptidoglycan and a set of unique additional components that, together, retain crystal violet within the cell when they undergo Gram staining (Beveridge, 1990). The confusion about their gram-reactivity resides in the way they apparently share some physiological traits that are found in bona fide gram-negative bacteria, such as possessing porins in their cell walls and being killed by the product of *sacB* when this gene is introduced into them (Daffe and Draper, 1998). When this is done with true gram-negative bacteria, they are killed in the presence of sucrose (Pelicic et al., 1996). So-called gram-negative traits, which are shared between mycobacteria and true gram-negative bacteria, are interesting features, but these traits do not alter the undeniable fact that mycobacteria are gram-positive bacteria.

Thin Sections of Mycobacteria Reveal Their Cell Walls To Be Different

Using conventional fixation techniques, mycobacteria are difficult to accurately preserve but even these techniques have shown that these cell walls are more complex than normal gram-positive bacteria (such as *B. subtilis*) (Imaeda et al., 1968; Paul and Beveridge, 1992, 1993, 1994). To this date, there have been no cryosections performed on mycobacteria, and we must rely on conventional preparations and freeze-substitutions for the interpretation of mycobacterial envelope structure (Imaeda et al., 1968; Paul and Beveridge, 1992, 1993, 1994). Cryopreparations of mycobacteria (at least in our hands) are difficult to perform. I suspect this is because of the complex makeup of their surfaces with a high lipid content, which makes them almost unwettable. Their difficulty with freezing is not unreasonable because the very nature of vitrification by rapid freezing depends on the ability of the researcher to solidify all water that is associated with a cell so quickly that it does not have an opportunity to crystallize. Instead, the water must be transformed into a solid glass extending throughout all parts of the bacterium as well as outside the cell. If parts of a mycobacterial cell,

such as specific regions within the cell envelope, contain little water because they are lipoid, these discontinuities could alter homogeneous vitrification. For this reason, the good images of freeze-substituted mycobacteria obtained (Paul and Beveridge, 1992, 1993, 1994) are fortuitous. It is possible, too, that the high lipid content of mycobacterial walls makes instantaneous freezing so difficult that even these freeze-substituted images are also somewhat artifactual.

It is important that high-fidelity images of the mycobacterial cell envelope be obtained because its structural and molecular make up must play a significant role in the cell's infective and pathogenic properties. Furthermore, the complex composition and interdigitation of the molecules within this cell wall are responsible for a number of this pathogen's traits, such as acid fastness (Barksdale and Kim, 1997; Beveridge, 2002), poor penetration of antibiotics, some antibiotic resistance (Brennan and Nikaido, 1995; David, 1981), hydrophobic interactions (Ridgeway et al., 1984), and immunological interactions with the host (Stratford, 1983). It is interesting that both pathogenic and nonpathogenic mycobacteria share the same global cell envelope structure.

Freeze-substituted mycobacteria reveal a typical cytosol in which the chromosomal DNA is spread throughout the cell and intermingles closely with the widely dispersed ribosomes (Paul and Beveridge, 1992, 1993, 1994). (This is in contrast with conventionally fixed cells, in which the DNA is centrally condensed with ribosomes pushed aside [Paul and Beveridge, 1992].) Strikingly, freeze-substituted images of mycobacterial walls are thinner than their conventionally fixed counterparts (Paul and Beveride, 1992, 1994). Yet, freeze-substituted walls seem denser and presumably contained more substance (Fig. 4A). When the extraction of major wall components with organic solvents, e.g., glycolipids, was compared during conventional versus freeze-substitution, it was apparent that the latter protocol was better (Paul and Beveridge, 1992, 1994). Also, when the outer layer and electron transparent layer of walls were compared using both techniques, there were apparent differences, and this applies to both slow- and fast-growing strains (Paul and Beveridge, 1992, 1994).

Do mycobacterial periplasms exist?

Freeze-substituted cell envelopes of mycobacteria do not show periplasmic spaces. This cannot be taken as proof that periplasmic spaces with their constituent periplasms do not exist. So far, freeze-substitution is the only cryotechnique used on mycobacteria and, like other gram-positive bacteria, freeze-substitution

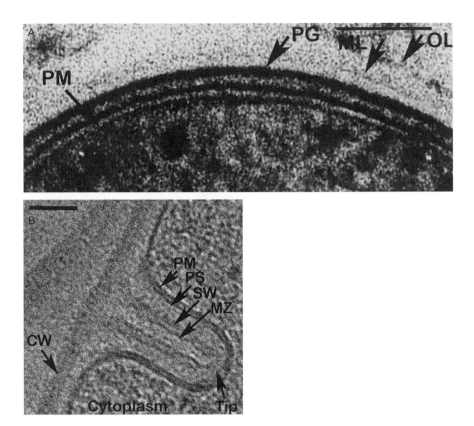

Figure 4. (A) High magnification of the cell envelope of a freeze-substituted *M. kansasii* cell showing a well preserved cell wall structure. The darkly stained peptidoglycan (PG), with a weakly stained midlayer (ML) within the electron transparent layer (the ETL is not labeled), and electron dense outer layer (OL). Scale bar = 100 nm. See Paul and Beveridge (1994) for more details. (B) A frozen hydrated thin section of a *S. aureus* cell revealing that the septum is much more complex than originally thought. PM, plasma membrane; CW, cell wall; PS, periplasmic space; SW, septum wall; MZ, middle zone; Tip, septal tip. Scale bar = 20 nm. See Matias and Beveridge (2007) for more details.

has never revealed periplasmic spaces. With mycobacteria and other gram-positive bacteria, the plasma membrane is often tightly apposed to the inner face of the cell wall (Fig. 4A). When my laboratory first initiated our freeze-substitution studies on gram-positive bacteria, we were puzzled by this observation and had to invoke the concept of there being a periplasm but not a periplasmic space. We knew from the studies of other researchers that periplasmic enzymes existed together with various chaperones and transporters, but we just could not devise (at that time) an electron microscopic technique that could visualize a periplasmic space. Accordingly, we suggested that the periplasm of gram-positives was located within the cell wall matrix and intermingled with the wall polymers (Beveridge, 1995). We were proven wrong as soon as we mastered the art of preparing frozen hydrated thin sections and the use of a cryoTEM (Matias and Beveridge, 2006). Gram-positive bacteria certainly possess a periplasmic space, and it is of lower density than the cell wall that sits above it (Fig. 3B and 3C). This was soon confirmed by other researchers who also used cryoTEM (Zuber et al.,

2006). Because these cells possess high turgor pressure, the periplasmic constituents must be squeezed and compressed between the plasma membrane and the cell wall. Recent information on *S. aureus* using frozen hydrated thin sections has revealed that the periplasmic space extends entirely along the in-growing septum (Matias and Beveridge, 2006) (Fig. 4B). This same study revealed that during septation, the new septum grows from the central inner point of its tip and rapidly separates into two separate septal walls (Fig. 4B). The septum is held together by only a bridge at the cell periphery and the previously mentioned inner growing point. Interestingly, the periplasmic space is also seen in the growing septum (Fig. 4B). Conventional and freeze-substituted preparations of mycobacteria clearly show that the cells divide by septation, and intuitively, the process should actually be similar to that seen in Fig. 4B. This could be verified once frozen hydrated sections are performed on a *Mycobacterium* sp.

Although we do not currently have frozen hydrated thin section data on any mycobacterial species, it would be unlikely that they would differ drastically in their gross cell envelope organization from their more typical gram-positive counterparts (e.g., *Bacillus*, *Enterococcus*, *Staphylococcus*, and *Streptococcus*). It is likely that *Mycobacterium* spp. possess a periplasmic space complete with periplasm, as already suggested (Daffe and Draper, 1998), and that they divide by a similar septation process as outlined above.

Acknowledgments. The work from my laboratory was funded by a Natural Science and Engineering Research Council (NSERC) Discovery Grant and by funds from AFMnet.

The electron microscopy was performed in the NSERC Guelph Regional Integrated Imaging Facility (GRIIF), which is partially funded by an NSERC Major Facilities Access grant to T.J.B. All images are from my laboratory, and some images were supplied by Valerio Matias and Ryan Hunter, for which I am grateful. The older mycobacterial work occurred while Terry Paul was in my laboratory.

REFERENCES

Al-Amoudi, A., J. J. Chang, A. Leforestier, A. McDowall, L. M. Salamin, L. P. Norlen, K. Richter, N. S. Blanc, D. Studer, and J. Dubochet. 2004. Cryo-electron microscopy of vitreous sections. *EMBO J* 23:3583–3588.

Barksdale, L., and K. S. Kim. 1977. *Mycobacterium. Bacteriol. Rev.* 41:217–372.

Beveridge, T. J. 1990. Mechanism of gram variability in select bacteria. *J. Bacteriol.* 172:1609–1620.

Beveridge, T. J. 1995. The periplasmic space and the periplasm in gram-positive and gram-negative bacteria. *ASM News* 61:125–130.

Beveridge, T. J. 2002. Structure of the bacterial surface and its influence on stainability. *J. Histotechnol.* 25:55–60.

Beveridge, T. J., and J. A. Davies. 1983. Cellular response of *Bacillus subtilis* and *Escherichia coli* to the Gram stain. *J. Bacteriol.* 156:846–858.

Beveridge, T. J., and V. R. F. Matias. 2005. Ultrastructure of gram-positive cell walls. p. 3–11. *In* V. A. Fischetti, R. P. Novick, J. J. Ferretti, D. A. Portnoy, and J. I. Rood (ed.), *Gram-Positive Pathogens*, 2nd ed. ASM Press, Washington, DC.

Beveridge, T. J., D. Moyles, and B. Harris. 2007. Electron microscopy, p. 54–81. *In* C. A. Reddy, T. J. Beveridge, J. A. Breznak, G. A. Marzluf, T. M. Schmidt, and L. R. Snyder (ed.), *Methods for General and Molecular Microbiology*, 3rd ed. ASM Press, Washington, DC.

Beveridge, T. J., and S. Schultze-Lam. 1997. The response of selected members of the *Archaea* to the Gram stain. *Microbiology* 142:2887–2895.

Brennan, P. J., and H. Nikaido. 1995. The envelope of mycobacteria. *Annu. Rev. Biochem.* 64:29–63.

Daffe, M., and P. Draper. 1998. The envelope layers of mycobacteria with reference to their pathogenicity. *Adv. Microb. Physiol.* 39:131–203.

David, H. L. 1981. Basis for lack of drug susceptibility of atypical mycobacteria. *Rev. Infect. Dis.* 3:878–884.

Davies, J. A., G. K. Anderson, T. J. Beveridge, and H. C. Clark. 1983. Chemical mechanism of the Gram stain and synthesis of a new electron-opaque marker for electron microscopy which replaces the iodine mordant of the stain. *J. Bacteriol.* 156:837–845.

Eltsov, M., and B. Zuber. 2006. Transmission electron microscopy of the bacterial nucleoid. *J. Struct. Biol.* doi:10.1016/j.jbs.2006.07.007.

Graham, L. L., R. Harris, W. Villiger, and T. J. Beveridge. 1991. Freeze-substitution of gram-negative eubacteria: general cell morphology and envelope profiles. *J. Bacteriol.* 172:1623–1633.

Graham, L. L., and T. J. Beveridge. 1994. Structural differentiation of the *Bacillus subtilis* cell wall. *J. Bacteriol.* 176:1413–1421.

Hobot, J. A., E. Carlemalm, W. Villiger, and E. Kellenberger. 1984. Periplasmic gel: new concept resulting from the reinvestigation of bacterial cell envelope ultrastructure by new methods. *J. Bacteriol.* 160:143–152.

Hunter, R. C., and T. J. Beveridge. 2005. High resolution visualization of *Pseudomonas aeruginosa* PAO1 biofilms by freeze-substitution transmission electron microscopy. *J. Bacteriol.* 187:7619–7630.

Imaeda, T., F. Kanetsuna, and B. Galindo. 1968. Ultrastructure of cell walls of genus Mycobacterium. *J. Ultrastruct. Res.* 25:46–63.

Kennedy, E. P. 1982. Periplasmic membrane-derived oligosaccharides and osmoregulation in *Escherichia coli*, p. 33-43. *In* E. Huber (ed.). *The Cell Membrane.* Plenum Publishing Corporation, New York, NY.

Koch, A. L. 1988. Biophysics of bacterial walls viewed as a stress-bearing fabric. *Microbiol. Rev.* 52:337–353.

Kunin, C. M., and Rudy, J. 1991. Effect of NaCl-induced osmotic stress on intracellular concentrations of glycine betaine and potassium in *Escherichia coli*, *Enterococcus faecalis*, and staphylococci. *J. Lab. Clin. Med.* 118:217–224.

Matias, V. R. F., A. Al-Amoudi, J. Dubochet, and T. J. Beveridge. 2003. Cryo-transmission electron microscopy of frozen-hydrated sections of *Escherichia coli* and *Pseudomonas aeruginosa. J Bacteriol.* 185:6112–6118.

Matias, V. R. F., and T. J. Beveridge. 2005. Cryo-electron microscopy reveals native polymeric cell wall structure in *Bacillus subtilis* 168 and the existence of a periplasmic space. *Mol. Microbiol.* 56:240 251.

Matias, V. R. F., and T. J. Beveridge. 2006. Native cell wall organization shown by cryo-electron microscopy confirms the existence of a periplasmic space in *Staphylococcus aureus. J. Bacteriol.* 188:1011–1021.

Matias, V. R. F., and T. J. Beveridge. 2007. Cryo-electron microscopy of cell division in *Staphylococcus aureus* reveals a soluble mid-zone between nascent cross walls. *Mol. Microbiol.* 64:195–206.

Paul, T. R., and T. J. Beveridge. 1992. Re-evaluation of envelope profiles and cytoplasmic ultrastructure of mycobacteria processed by conventional embedding and freeze-substitution protocols. *J. Bacteriol.* 174:6508–6517.

Paul, T. R., and T. J. Beveridge. 1993. Ultrastructure of mycobacterial surfaces by freeze-substitution. *Zentbl. Bakteriol.* 279:450–457.

Paul, T. R., and T. J. Beveridge. 1994. Preservation of surface lipids and ultrastructure of *Mycobacterium kansasii* using freeze-substitution. *Infect. Immun.* 62:1542–1550.

Pelicic, V., J.-M. Reyrat, and B. Gicquel. 1996. Expression of the *Bacillus subtilis sacB* gene confers sucrose sensitivity on mycobacteria. *J. Bacteriol.* 178:1197–1199.

Rastogi, N., C. Frehel, and H. L. David. 1986. Triple-layered structure of mycobacterial cell wall: evidence for the existence of a polysaccharide-rich outer layer in 18 mycobacterial species. *Curr. Microbiol.* 13:237–242.

Ridgeway, H. F., M. G. Rigby, and D. G. Argo. 1984. Adhesion of a mycobacterial sp. to cellulose diacetate membranes used in reverse osmosis. *Appl. Environ. Microbiol.* 47:61–67.

Stoica, O., A. Tuanyok, X. Yoa, M. H. Jericho, D. Pink, and T. J. Beveridge. 2003. Elasticity of membrane vesicles from *Pseudomonas aeruginosa. Langmuir* 19:10916–10924.

Stratford, J. L. 1983. Immunologically important constituents of mycobacteria: antigens, p. 85-127. *In* C. Ratledge and J. Stratford (ed.), *The Biology of the Mycobacteria*, vol. 2. Academic Press, London, United Kingdom.

Zuber, B., M. Haenni, T. Ribeiro, K. Minnig, F. Lopes, P. Moreillon, and J. Dubochet. 2006. Granular layer in the periplasmic space of Gram-positive bacteria and fine structures of *Enterococcus gallinarum* and *Streptococcus gordonii* septa revealed by cryo-electron microscopy of vitreous sections. *J. Bacteriol.* 188:6652–6660.

The Mycobacterial Cell Envelope
Edited by M. Daffé and J.-M. Reyrat
© 2008 ASM Press, Washington, DC

Chapter 3

Biosynthesis of the Arabinogalactan-Peptidoglycan Complex of *Mycobacterium tuberculosis*

DEAN C. CRICK AND PATRICK J. BRENNAN

The mycobacterial cell envelope is composed of three major constituents; the plasma membrane, the mycolyl-arabinogalactan-peptidoglycan complex (MAPc, Fig. 1), and the extractable, noncovalently linked glycans, lipids, and proteins, some of which make up a capsule-like layer (chapter 1), whereas others may be interspersed within the framework of the MAPc itself. The architecture of the envelope of *Mycobacterium tuberculosis* consists of an inner plasma membrane, peptidoglycan (PG) in phosphodiester linkage to a linear D-galactofuran, to which are attached several strands of a highly branched D-arabinofuran. Linked to the nonreducing termini of the D-arabinan are mycolic acids, which are oriented perpendicular to the plane of the membrane and are thought to provide a permeability barrier that is responsible for many of the physiological and disease-inducing aspects of *M. tuberculosis*. Intercalated with the mycolic acids are the extractable (noncovalently linked) lipids and other components of the capsule-like layer. This extractable material is composed of free lipids (chapters 4 and 13 to 15), such as those containing phthiocerol, the phosphatidylinositol mannosides, lipomannan, lipoarabinomannan, trehalose dimycolate (cord factor), sulfolipids, trehalose monomycolate, and the diacyl- and polyacyl-trehaloses, all presumably intercalating with the α- and meromycolate chains of the mycolic acids and cell wall-associated proteins.

The 1960s and 1970s saw much of the early structural definition of the cell wall of *Mycobacterium* spp. (Lederer et al., 1975; Petit et al., 1969; Petit and Lederer, 1984; Adam et al., 1969; Lederer, 1971; Wietzerbin-Falszpan et al., 1970, 1974; Matsuhashi, 1966; Kotani et al., 1970), followed by a long period of inactivity. Recent developments in genomics, bioinformatics, proteomics, and analytical techniques have resulted in a more detailed and thorough understanding of the structure of the mycobacterial cell wall core and its biosynthesis. Indeed, much of the current interest in its biosynthesis is driven by the need for novel, alternative drugs to counteract drug-resistant tuberculosis.

STRUCTURE OF THE CELL-WALL CORE

Peptidoglycan

PG is a complex polymer forming a rigid layer outside the plasma membrane, providing cellular shape and the strength to withstand osmotic pressure. PG from *M. tuberculosis* has been classified as A1γ, as has that of *Escherichia coli* and *Bacillus* spp., according to the classification system of Schleifer and Kandler (Schleifer and Kandler, 1972), and is composed of linear chains of N-acetyl-α-D-glucosamine (GlcNAc) and modified muramic acid (Mur) substituted with a peptide side-chain that may be cross-linked to that of other glycan strands. The tetrapeptide side chains of PG consist of L-alanyl-D-isoglutaminyl-*meso*-diaminopimelyl-D-alanine (L-Ala-D-Glu-A$_2$pm-D-Ala), with the Glu and A$_2$pm being further amidated (Adam et al., 1969; Lederer et al., 1975; Wietzerbin-Falszpan et al., 1970; Petit et al., 1969; Kotani et al., 1970). In addition, these peptide chains are heavily cross-linked, providing added structural integrity to the bacterium; the overall degree of cross-linking is 70% to 80% in *Mycobacterium* spp. (Matsuhashi, 1966) compared with 50% in *E. coli* (Vollmer, 2004). About two-thirds of the peptide cross-links found in *M. tuberculosis* PG are between the carboxyl group of a terminal D-Ala and the amino group at the D-center of A$_2$pm, resulting in a D,D bond (Wietzerbin et al., 1974), and approximately one-third of the peptide cross-links occur between the carboxyl group of the L-center of one

Dean C. Crick and Patrick J. Brennan • Department of Microbiology, Immunology and Pathology, Colorado State University, Fort Collins, CO 80523.

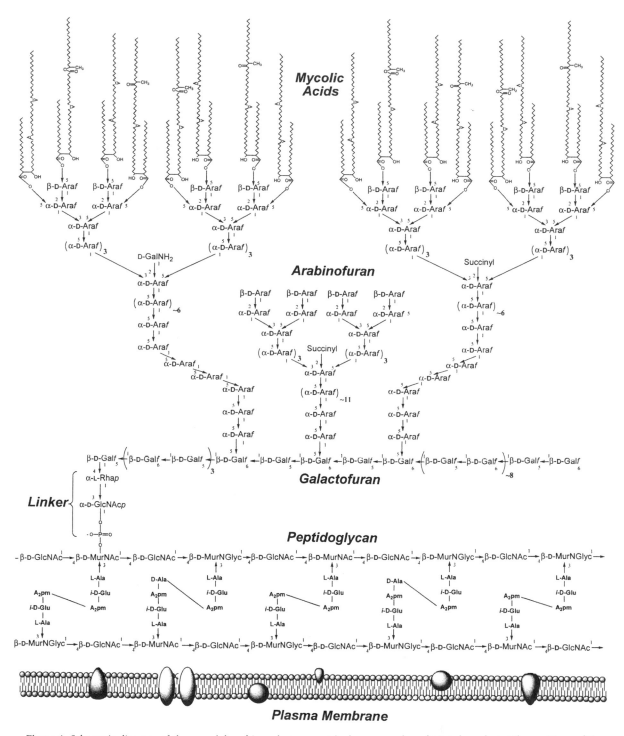

Figure 1. Schematic diagram of the mycolyl-arabinogalactan-peptidoglycan complex of *M. tuberculosis*. The positions of the succinyl and D-galactosamine residues are indicated (Crick et al., 2008).

A₂pm residue and the amino group of the D-center of another A₂pm residue forming a L,D-cross-link (Wietzerbin et al., 1974) (Fig. 1). L,D-Cross-links are relatively rare but recently have also been reported in the PG of *Streptomyces* spp., *Clostridium perfringens* (Leyh-Bouille et al., 1970) and stationary phase *E. coli* (Templin et al., 1999; Goffin and Ghuysen, 2002).

Interestingly, the PG of *Mycobacterium leprae* has a Gly residue substituted for the L-Ala residue found in the stem peptide (Draper et al., 1987).

In addition, in mycobacterial PG, some of the Mur residues are N-acylated with glycolic acid (MurNGlyc) rather than N-acetylated (MurNAc) (Adam et al., 1969; Mahapatra et al., 2005b;

Raymond et al., 2005), and the hydroxyl moiety of C-6 of some of the Mur residues forms phosphodiester bonds to C-1 of GlcNAc, which, in turn, is (1→3) linked to a α-L-rhamnose (Rha) residue providing the "linker unit" between the galactan of arabinogalactan (AG) and PG (McNeil et al., 1991).

Arabinogalactan

This branched-chain AG, with the arabinose residues forming the nonreducing termini, was recognized as the major cell wall polysaccharide in mycobacteria, as early as the 1950s. With the exception of the sugars making up the linker unit and a few galactosamine residues, which are found only in the slow-growing mycobacteria (Draper, 1997), AG is composed entirely of arabinose and galactose residues in the furanose configuration (Ara$f$ and Gal$f$ respectively). Unlike most bacterial polysaccharides, AG lacks repeating units (Anderson and Unger, 1983), being composed of a few distinct structural motifs (Vilkas et al., 1973; Daffe et al., 1990), and it has been established that in *M. tuberculosis* AG:

1. The galactan regions consist of linear alternating 5- and 6-linked β-D-Gal$f$ residues.
2. Arabinan chains are attached to C-5 of the 6-linked Gal$f$ residues of the galactan core.
3. The majority of the arabinan chains consist of approximately 30 5-linked α-D-Ara$f$ residues, with branching introduced by 3,5-α-D-Ara$f$ residues.
4. The nonreducing termini of arabinan consist of the structural motif [β-D-Ara$f$-(1→2)-α-D-Ara$f$]₂-3,5-α-D-Ara$f$-(1→5)-α-D-Ara$f$.
5. Mycolic acids are located in clusters of four on the terminal [β-D-Ara$f$-(1→2)-α-D-Ara$f$]₂-3,5-α-D-Ara$f$-(1→5)-α-D-Ara$f$ hexaarabinofuranoside, but only two-thirds of these acids are mycolated.

There are some variations seen in other mycobacteria. For example in *M. leprae*, *Mycobacterium bovis*, and *Mycobacterium smegmatis*, only 50% of the arabinan termini are mycolylated (Daffe et al., 1993). Overall, this topic has been extensively reviewed (Brennan and Nikaido, 1995; Crick et al., 2001; Lee et al., 1996; Mahapatra et al., 2005a; McNeil et al., 1996; Baulard et al., 1999; Besra and Brennan, 1997; Dover et al., 2004; Tripathi et al., 2005), and the latest results suggest that three arabinan chains are attached to the galactan, specifically on galactosyl residues 8, 10, and 12 (Alderwick et al., 2005). This work was done in *Corynebacterium glutamicum* but is likely directly applicable to *M. tuberculosis* and is consistent with earlier work showing long stretches of linear galactofuran (Besra et al., 1995). It is now believed that the galactofuran is approximately 30 residues long.

Recent mass spectrometric analysis suggests that the arabinan chains released from the cell wall by an endogenous arabinase are also approximately 30 residues long and indicates that galactosamine (GalNH₂) residues in AG isolated from the *M. tuberculosis* CSU20 strain are found on the C2 position of some of the internal 3,5-branched Ara$f$ residues (Khoo, Chatterjee, and McNeil, personal communication). In addition, the C2 positions of other internal 3,5-branched Ara$f$ residues are succinylated (Bhambi and McNeil, personal communication). These structural features are summarized in Fig. 1.

Mycolic acids

The mycolic acids are α-branched β-hydroxylated fatty acids containing 70 to 90 carbon atoms with various modifications of the acyl chains. These large fatty acids constitute a significant proportion of the mass of the cell wall core and form a hydrophobic permeability barrier that surrounds the bacterium through hydrophobic interactions (Brennan and Nikaido, 1995; Minnikin, 1982; Liu et al., 1996; Nikaido and Jarlier, 1991). This feature is certainly responsible for at least part of the endogenous resistance of *M. tuberculosis* to many drugs (Jarlier and Nikaido, 1994; Jarlier et al., 1991; Draper, 1998). The structures, functions, and biosyntheses of these lipids are described in Chapter 2.

Overall structure of the mycolyl-arabinogalactan-peptidoglycan complex

Two opposing models of the secondary structure of the MAPc have been proposed. One predicts that the PG and galactan are parallel to the plasma membrane. This orientation is consistent with traditional models of the PG structure (Ghuysen, 1968, McNeil and Brennan, 1991); however, other modeling studies suggest that the PG and AG strands may be coiled and perpendicular to the plane of the plasma membrane (Dmitriev et al., 1999, 2000). Recently, the relevant studies on the secondary structure of peptidoglycan in gram-negative bacteria were reviewed (Vollmer and Holtje, 2004), and the preponderance of data appears to favor the parallel model in these organisms, but the perpendicular model recently gained support from the three-dimensional solution structure of a synthetic fragment of the cell wall, as determined by nuclear magnetic resonance (NMR). These results indicate that the glycan backbone of the synthetic PG fragment forms a right-handed helix with a periodicity of six sugar residues, leading the authors to conclude that the glycan strand of PG is orthogonal to the plane of the membrane in vivo as opposed to the parallel hypothesis (Meroueh et al., 2006). Thus, the overall structure and topology of the MAPc in *Mycobacterium* spp. remains open to debate, as does the three-dimensional structure of PG in other eubacteria.

Polyprenyl phosphate

Many of the insights to understanding the biosynthesis of the cell wall core derived from earlier structural analysis. For example, a key advance in the context of the biosynthesis of the mycobacterial cell wall core was the characterization of the diglycosyl phosphate bridge linking the PG and the galactan, which resulted in the speculation that the linker unit and AG were synthesized as a unit on a polyprenyl phosphate (Pol-P) carrier lipid in a manner analogous to the synthesis of the teichoic and teichuronic acids and lipid II of PG synthesis. Thus, an important development was the recognition of the central role played by Pol-P in the synthesis of all portions of the cell wall core (Fig. 2). Interestingly, although most bacteria use undecaprenyl phosphate, a di-*E*, poly-*Z*-prenyl phosphate, as a carrier of activated sugars, it has been demonstrated that mycobacteria use at least two and perhaps three forms of Pol-P. *M. smegmatis* contains two major forms of Pol-P, a decaprenyl phosphate (Dec-P) containing one *omega*-, one *E*- and eight *Z*-isoprene units (mono-*E*, poly-*Z*) (Wolucka et al., 1994) and a heptaprenyl phosphate (Takayama et al., 1973) containing four saturated isoprene units on the omega end of the molecule and two *E*- and

one *Z*-isoprene units (Besra et al., 1994) or four saturated and three *Z*-isoprene units (Wolucka and de Hoffmann, 1998). Whereas *M. tuberculosis* uses a single predominant Pol-P (Dec-P) and although the stereochemistry of the individual isoprene units have yet to be determined (Takayama and Goldman, 1970), it is probable that they are similar to those of the *M. smegmatis* Dec-P. These Pol-P molecules are structurally unusual, in that the most common structures of prenol (and therefore Pol-P) tend to be confined to four main groups: (1) all-*E*-prenol, (2) di-*E*, poly-*Z*-prenol, (3) tri-*E*, poly-*Z*-prenol, and (4) all-*Z*-prenol (Anonymous, 1987). Bacteria typically contain a single predominant Pol-P composed of 11 isoprene units, with the di-*E*, poly-*Z* configuration (undecaprenyl phosphate or bactoprenyl phosphate), as indicated above.

These are relatively rare molecules, the paucity of which has been implicated in controlling the rate of cell wall formation in cell free preparations from *Staphylococcus aureus* (Higashi et al., 1970) and *Bacillus* spp. (Anderson et al., 1972); it has also been suggested that the rate of synthesis of lipid I (in PG synthesis) of *E. coli* may be dependent on the pool level of Pol-P (van Heijenoort, 1996) and that the overall rate of bacterial cell wall synthesis in vivo can be regulated by Pol-P levels (Baddiley, 1972). In addition, the antibiotic bacitracin acts by binding the polyprenyl diphosphate (Storm and Strominger, 1973) that is released during transglycosylation and thereby prevents dephosphorylation to form Pol-P,

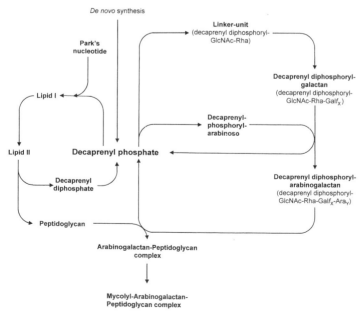

Figure 2. Schematic diagram showing key intermediates in mycolyl-arabinogalactan-peptidoglycan complex synthesis and the central role played by decaprenyl phosphate.

and its reentery into the PG biosynthetic cycle (Higashi et al., 1967; Siewert and Strominger, 1967; van Heijenoort, 2001a). Amphomycin, a lipopeptide antibiotic, acts in a similar fashion. That is, a complex is formed between the lipopeptide and Pol-P (Banerjee, 1989), rendering the lipid unavailable for its role as a carrier of activated sugar residues. Both bacitracin (Rieber et al., 1969) and amphomycin are effective inhibitors of *M. tuberculosis* growth (authors, unpublished data). Thus, with the realization that Pol-P was also a carrier for activated sugar residues in AG synthesis (Wolucka et al., 1994), including the source of Ara*f* residues, and reports that polyprenyl phosphate may be involved in mycolic acid synthesis (Besra et al., 1994; Takayama et al., 2005), the central role played by this molecule in MAPc synthesis became clear (Fig. 2), emphasizing the crucial need to study the synthesis of these molecules in mycobacteria (Crick et al., 2000; Schulbach et al., 2000, 2001; Kaur et al., 2004).

BIOSYNTHESIS OF THE MYCOLYL-ARABINOGALACTAN-PEPTIDOGLYCAN COMPLEX

Polyprenyl Phosphate Biosynthesis

Synthesis of the cell wall core begins with Pol-P because this molecule plays a central role in the overall process. In general, polyprenyl diphosphate (Pol-P-P) molecules are synthesized through sequential condensation of isopentenyl diphosphate (IPP, derived from methylerythritol phosphate in mycobacteria [Bailey et al., 2002; Dhiman et al., 2005; Argyrou and Blanchard, 2004]) with allylic diphosphates, reactions that are catalyzed by prenyl diphosphate synthases (Fig. 3). These enzymes are specific for the type of stereochemistry (*E* or *Z*) of the introduced allylic double bond, and amino acid alignments have shown that there is no amino acid sequence homology between the *E*- and *Z*-polyprenyl families of diphosphate synthases

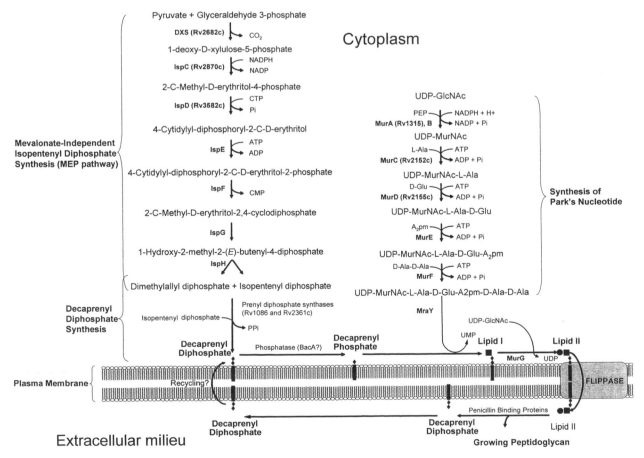

Figure 3. Pathways involved in the biosynthesis of decaprenyl phosphate and peptidoglycan with respect to the topology of the plasma membrane. A simple hypothetical scheme for decaprenyl phosphate recycling has been included; more complex possibilities exist, as is discussed in the text. Gene (Rv) numbers have been provided for enzymes from *M. tuberculosis* which have been biochemically demonstrated to catalyze the indicated reaction. Symbols used: ◆, phosphate; ●, *N*-acylmuramic acid; ■, *N*-acetylglucosamine.

(Fujihashi et al., 1999). In 1998, the first representative of the *Z*-polyprenyl diphosphate synthase family, undecaprenyl diphosphate synthase, was cloned from *Micrococcus luteus* and characterized (Shimizu et al., 1998). Typically, these enzymes use *E,E*-farnesyl diphosphate (*E,E*-FPP) as the allylic acceptor and synthesize long-chain products with the common di*E*, poly*Z* stereoconformation (Apfel et al., 1999; Allen et al., 1976; Baba and Allen, 1978; Muth and Allen, 1984). Thus, the structure of the final Pol-P provides insight into the substrates used by these enzymes during biosynthesis, and the unusual structures of the Pol-P molecules identified in mycobacteria suggested that *E*-geranyl diphosphate (*E*-GPP), rather than the more common *E,E*-FPP, was the likely allylic diphosphate acceptor in mycobacteria.

In fact, it has been shown that *M. tuberculosis* is unusual in that it has two genes (*Rv1086* and *Rv2361c*) that encode proteins with amino acid sequence similarity to undecaprenyl diphosphate synthase. *Rv1086* encodes an *E,Z*-FPP synthase (Schulbach et al., 2000, 2001) and *Rv2361c* encodes a Dec-P synthase (Schulbach et al., 2000; Kaur et al., 2004). Although both Rv1086 and Rv2361c can catalyze the addition of IPP to *E*-GPP, the relative Km values for GPP (38 and 490 μM, respectively [Schulbach et al., 2001; Kaur et al., 2004]) suggest that Rv1086 and Rv2361c act sequentially in the synthesis of *E*,poly*Z* decaprenyl diphosphate (Dec-P-P), the precursor of the *E*,poly*Z* Dec-P (see structure in Fig. 4) found in mycobacteria (Takayama et al., 1973; Takayama and Goldman, 1970; Wolucka and

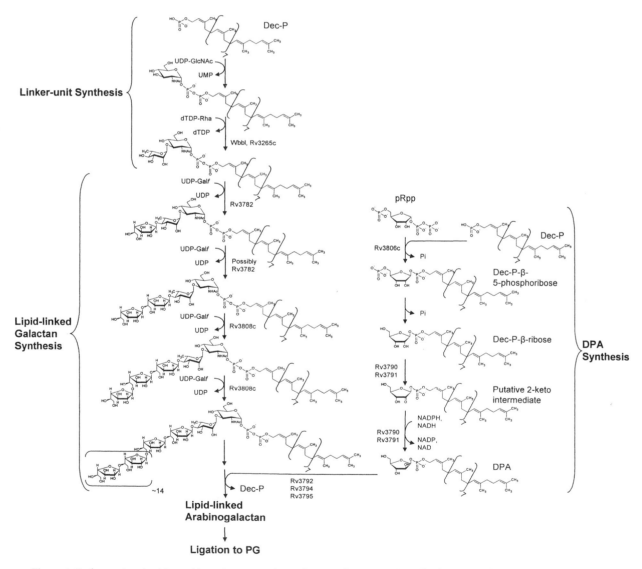

Figure 4. Pathways involved in arabinogalactan synthesis showing the intersection of galactan synthesis and decaprenylphosphorylarabinose (DPA) synthesis. Gene (Rv) numbers have been provided for enzymes from *M. tuberculosis*, which have been biochemically demonstrated to catalyze the indicated reaction (Crick et al., 2008).

de Hoffmann, 1995, 1998; Wolucka et al., 1994), with Rv2361c adding seven isoprene units to the E,Z-FPP synthesized by Rv1086. High-density transposon mutagenesis experiments have shown that Rv2361c is likely essential in M. tuberculosis H37Rv (Sassetti et al., 2003); however, the enzyme encoded by Rv1086 is not essential (Sassetti et al., 2003), even though data indicates that Rv1086 synthesizes the precursor used by Rv2361c for synthesis of Dec-P-P. This is likely because Rv2361c is capable of using E-GPP to synthesize E,Z-FPP, although at a much reduced catalytic efficiency (Kaur et al., 2004), allowing Rv2361c to compensate for the inactivation of Rv1086. Once the Dec-P-P has been synthesized, it must be dephosphorylated to form Dec-P (see Fig. 3). At present, there is little information regarding this biosynthetic transformation; however, an ortholog of BacA, a phosphatase reported to be involved in dephosphorylation of Pol-P-P in E. coli (El Ghachi et al., 2004), may be involved.

Peptidoglycan Biosynthesis

Biosynthesis of activated sugar donors involved in peptidoglycan synthesis

Although there has been intense interest in the PG biosynthetic pathway as a drug target for many years, and underexploited drug targets still exist in this pathway (Wong and Pompliano, 1998; Green, 2002; Chopra et al., 2002; Silver, 2003; El Zoeiby et al., 2003), relatively little is known about the biosynthesis of the PG of M. tuberculosis per se. However, the arrangement of genes responsible for PG synthesis in M. tuberculosis is similar to that in other bacteria (Mahapatra et al., 2000), and in all probability, so is the biochemistry (Mahapatra et al., 2005a). Hence, the excellent reviews on bacterial PG synthesis in general are applicable (van Heijenoort, 1994, 1996, 1998, 2001a, 2001b).

PG synthesis initially involves the synthesis of UDP-MurNAc from UDP-GlcNAc through two separate reactions catalyzed by MurA and MurB. Enoylpyruvate (from phosphoenoylpyruvate) is added to the 3 position of the GlcNAc residue of the UDP-GlcNAc, and subsequently, the enoylpyruvate moiety is reduced to form UDP-MurNAc. The gene encoding M. tuberculosis MurA (Rv1315) has been overexpressed and partially characterized (De Smet et al., 1999), and M. tuberculosis MurA was found to have an aspartate at position 117 instead of the cysteine found in the E. coli enzyme, thus explaining the natural resistance of the tubercle bacillus to the antibiotic fosfomycin (De Smet et al., 1999; Kahan et al., 1974; Marquardt et al., 1994; Skarzynski et al., 1996).

Subsequently, UDP-MurNAc-pentapeptide is formed by the addition of L-Ala to the lactate moiety of the UDP-MurNAc, followed by the sequential addition of D-Glu, A_2pm and a D-alanyl-D-alanine dipeptide forming UDP-MurNAc(pentapeptide) catalyzed by orthologs of MurC, MurD, MurE and MurF of E. coli (Fig. 3). Of these enzymes, MurC (Rv2152c) from M. tuberculosis has been overexpressed and partially characterized (Mahapatra et al., 2000), demonstrating a catalytic activity similar to that of the E. coli enzyme and other known L-Ala ligases but having a lower apparent K_m value for glycine (Mahapatra et al., 2000). The M. tuberculosis ortholog of MurD (Rv2155c) is enzymatically similar to that from E. coli (authors, unpublished data).

In E. coli lipid I is formed by the activity of MraY (sometimes referred to as MurX), which transfers phosphoryl-MurNAc(peptapeptide) to a molecule of undecaprenyl phosphate. A GlcNAc residue is subsequently added to form GlcNAc-MurNAc (pentapeptide)-diphosphoryl-undecaprenol (lipid II) by MurG. The glycopeptide of lipid II is thought to be translocated from the cytoplasmic face of the inner cell membrane to the periplasmic face, where it is used directly in the assembly of PG.

Peptidoglycan assembly

Although the detailed mechanisms of the transbilayer movement of PG intermediates remain unknown, a good deal of information is available regarding the transpeptidation and transglycosylation reactions required to form mature PG. Nascent glycan chains are cross-linked by the formation of peptide cross-bridges. The main cross-linkage in E. coli is between the penultimate D-Ala and the A_2pm residues of an adjacent peptide of a second glycan chain. The reaction is catalyzed by penicillin-binding proteins (PBPs) and involves the cleavage of the D-alanyl-D-alanine bond of the donor peptide, which provides the energy to drive the reaction (Ghuysen, 1991). The situation is similar in mycobacteria, and mycobacterial PBPs are the subject of a recent, extensive review (Goffin and Ghuysen, 2002). Briefly, the PBPs are classified primarily according to their molecular mass and have penicilloyl serine transferase activity that catalyzes the cleavage of the cyclic amide bond of penicillin. The high-molecular-mass PBPs are multimodular enzymes with N-terminal nonpenicillin-binding and C-terminal penicillin binding modules. These are further subdivided into class A and class B based on their primary structures. The high-molecular-mass class A PBPs function as transglycosylases as well as transpeptidases. Chambers and associates (1995) detected four PBPs, including three of high molecular mass in *M. tuberculosis* membranes; however, the transglycosylase or transpeptidase activities of these putative enzymes have not been demonstrated.

Five major PBPs have been detected in *M. smegmatis* (Basu et al., 1992; Billman-Jacobe et al., 1999), and of these two have been disrupted by transposon mutagenesis. One strain had an insertion in the gene encoding a protein that was most similar to Rv0050 (annotated as ponA1 in *M. tuberculosis*); this strain grew slowly and showed increased sensitivity to beta-lactam antibiotics and greater permeability (Billman-Jacobe et al., 1999). The other strain had an insertion in the gene encoding a protein most similar to Rv3682 (annotated as ponA2 in *M. tuberculosis*), which was impaired in stationary-phase survival (Keer et al., 2000). In addition, it has been shown that PBPA from *M. smegmatis* (with greatest similarity to Rv0016c in *M. tuberculosis*) plays an important role in cell division and maintenance of cell shape; signal transduction mediated by PknB and PstP appears to determine the positioning of this PBP at the septum, regulating septal PG biosynthesis (Dasgupta et al., 2006).

Polyprenyl diphosphate recycling

As noted earlier, PG assembly occurs on the outer face of the plasma membrane with the concomitant release of Pol-P-P. This lipid is then reused for subsequent rounds of lipid II formation (Fig. 3). It has been reported that each lipid molecule is reused for lipid II synthesis four to six times and, as mentioned earlier, that the action of the antibiotic bacitracin is due to inhibition of Pol-P-P dephosphorylation (Siewert and Strominger, 1967). Thus, two conclusions can be drawn: (1) Pol-P biosynthesis is insufficient to support bacterial growth in the presence of a compound that inhibits Pol-P-P recycling; (2) because lipid II synthesis takes place on the inside of the plasma membrane whereas PG synthesis occurs on the outside, there must be a mechanism for transbilayer movement of the carrier lipid, analogous to that of the lipid II transporter but operating in the opposite direction. This second conclusion leads to several possible scenarios: (1) released, extracellular Pol-P-P is transported across the membrane, as such, and the phosphatase involved in de novo Pol-P synthesis (presumably a BacA ortholog) converts the Pol-P-P to Pol-P for recycling; (2) extracellular Pol-P-P is dephosphorylated by a second, extracellular Pol-P-P phosphatase activity and the resulting Pol-P is transported across the membrane and then is recycled; (3) extracellular Pol-P-P is dephosphorylated to form Pol-P and then further dephosphorylated to

polyprenol before transport across the membrane. The polyprenol would then be rephosphorylated by a specific polyprenol kinase to form Pol-P once again.

Clearly, the first scenario is the simplest; the second scenario requires multiple specific Pol-P-P phosphatases (with active sites on the inside and the outside of the membrane), and the third scenario requires multiple Pol-P-P phosphatases, as well as a Pol-P phosphatase and a polyprenol kinase. Interestingly, multiple Pol-P-P phosphatase activities have been reported (El Ghachi et al., 2005) in *E. coli*, although the specificity of these activities has not been investigated. On the other hand, specific bacterial Pol-P phosphatase activities have been reported (Willoughby et al., 1972), as well as polyprenol kinase activites (Higashi et al., 1970; Higashi and Strominger, 1970; Kalin and Allen, 1979, 1980). As yet, the function of these activities has not been established; however, it has been hypothesized that a dephosphorylation/rephosphorylation cycle may be used to regulate Pol-P levels in the membrane (Higashi et al., 1970; Willoughby et al., 1972). Alternatively, one could also speculate that these enzymes are involved in the transbilayer movement required by recycling of Pol-P-P.

MODIFICATIONS OF MYCOBACTERIAL PEPTIDOGLYCAN

It has often been reported that mycobacterial glycan chains are composed of alternating units of β 1→4 linked GlcNAc and MurNGlyc, whereas, most other bacteria contain MurNAc (Draper, 1987; Kotani et al., 1970; Lederer, 1971). This conclusion was based on studies of the biosynthetic precursor UDP-N-acylmuramyl-L-Ala-D-Glu-A_2pm from *M. tuberculosis* and *M. phlei* treated with D-cycloserine (Petit et al., 1970; Takayama et al., 1970), which established that all of the muramic acid residues were, apparently, completely NGlyc substituted. Because the UDP-MurNGlyc-L-Ala-D-Glu-A_2pm is the immediate precursor of the UDP-MurNGlyc-L-Ala-D-Glu-A_2pm-D-Ala-D-Ala, all of the UDP-N-acylmuramyl-peptapeptides should be NGlyc substituted, as has been the supposition (Crick et al., 2001; Draper et al., 1987; Kotani et al., 1970). In turn, this scenario suggests that the PG would also contain MurNGlyc exclusively. However, the mass spectra presented in the original identification of NGlyc residues in PG isolated from *M. smegmatis* (Adam et al., 1969) also suggested the presence of MurNAc, an observation that received little attention after it was reported that mycobacterial nucleotide-linked precursors were exclusively NGlyc substituted (Lederer, 1971; Petit et al., 1970; Takayama et al., 1970). More recent analysis of PG from mycobacteria that were not treated with D-cycloserine indicated that the muramic acid residues in both the nucleotide-linked precursors and mature PG consist of a mixture of the NGlyc and NAc derivatives. Furthermore, treatment with D-cycloserine resulted in muropeptides containing exclusively NGlyc substituted muramic acid (Mahapatra et al., 2005b) as previously reported; thus, the new data are consistent with previous but unheralded observations.

The basis of this unexpected effect of D-cycloserine is unknown, and it does not appear to be drug specific in that vancomycin also alters the status of the N-acyl group of the muramic acid (Mahapatra et al., 2005b). Perhaps these drugs directly affect the putative NAc/NGlyc hydroxylase (Essers and Schoop, 1978; Gateau et al., 1976; Raymond et al., 2005). Alternatively, the drugs could affect the equilibrium between peptidoglycan recycling events. The presence of peptidoglycan recycling pathways in *Mycobacterium* spp. have yet to be established, but recycling similar to that first identified in *E. coli* has been demonstrated in other organisms (Jacobs et al., 1995; Mengin-Lecreulx et al., 1996; Park, 2001). Recently, the oxidation of MurNAc to MurNGlyc has been attributed to the action of an enzyme designated NamH (Raymond et al., 2005). Although there is no indication that this modification is essential for *M. tuberculosis* survival, *M. smegmatis* mutants devoid of NamH show increased sensitivity to beta-lactams and lysozyme (Raymond et al., 2005).

The free carboxylic acid groups of the A_2pm or D-isoglutamic acid residues of mycobacterial PG may be amidated, and some of the free carboxylic groups of the D-isoglutamic acid residues may also be modified by the addition of a Gly residue in peptide linkage (Kotani et al., 1970). Mass spectrometric analysis showed that the muropeptides from mycobacterial lipid I are identical to those from *E. coli*, whereas those isolated from mycobacterial lipid II form an unexpectedly complex mixture in which the muramyl residue is present as MurNAc, MurGlyc, and Mur. In addition, the carboxylic functions of the peptide side-chains of lipid II showed three types of modification, with the dominant being amidation (Mahapatra et al., 2005c). The observation that lipid I was unmodified strongly suggests that the lipid II intermediates of *M. smegmatis* are substrates for a variety of enzymes that introduce modifications to the sugar and amino acid residues before synthesis of peptidoglycan. It also seems likely that most, if not all, modifications seen in mycobacterial PG are introduced at the level of lipid II and not at the level of mature PG.

Furthermore, structural analysis of lipid I and II from *M. smegmatis* demonstrated that the lipid moiety is Dec-P; thus, *M. smegmatis* is the first bacterium

reported to use a Pol-P other than undecaprenyl phosphate as the lipid carrier involved in peptidoglycan synthesis (Mahapatra et al., 2005c) and apparently can compartmentalize the various forms of Pol-P found in this species.

A significant number of the peptide cross-links found in *M. tuberculosis* PG are between the carboxyl group at the L-center of a terminal A_2pm of one chain and the amino group at the D-center of the A_2pm of another peptide side chain, resulting in a L,D-cross-link. Similar L,D-peptide cross-links have been reported in *Streptomyces* spp., *Clostridium perfringens* (Leyh-Bouille et al., 1970), and stationary phase *E. coli* (Templin et al., 1999), suggesting the presence of both D,D and L,D-transpeptidase activities in these organisms, although no L,D-transpeptidases have definitively been identified to date. Determination of the importance of these various modifications to mycobacterial PG in the context of the physiology and pathogenesis of the bacillus awaits identification of the enzymes responsible and their biochemical and genetic definition.

Arabinogalactan Biosynthesis

Synthesis of activated sugar donors involved in arabinogalactan biosynthesis

Synthesis of AG begins with the formation of the linker unit, and then there appears to be a concomitant extension of the galactan and arabinan moieties (Fig. 4). In *M. tuberculosis* and *M. smegmatis*, linker unit synthesis is initiated on a Dec-P molecule by synthesis of Dec-P-P-GlcNAc (GL-1). It is predicted that a UDP-GlcNAc GlcNAc 1-phosphate: Dec-P-P GlcNAc transferase is encoded by *Rv1302*, a putative ortholog of WecA (previously known as Rfe) in *M. tuberculosis* (Dal Nogare et al., 1998), and is responsible for this transformation; however, there are as yet no biochemical data to support this supposition. Subsequently, the Rha residue is added to the GlcNAc by Wbbl (Rv3265c), which transfers it to the 3 position of the GlcNAc residue of GL-1, forming GL-2 or the "linker unit" (Mills et al., 2004; McNeil, 1999). The donor of the Rha residue is dTDP-Rha, which, in turn, is synthesized from glucose-1-phosphate through a four-step reaction cascade (Ma et al., 2001; McNeil, 1999). RmlA (Rv0334) initiates the reaction dTTP + α-D-glucose 1-phosphate $\rightarrow$ dTDP-glucose + PP_i (Ma et al., 1997), which is converted to dTDP-Rha in three sequential reactions catalyzed by dTDP-D-glucose 4,6-dehydratase (Rv3464, RmlB), dTDP-4-keto-6-deoxy-D-glucose 3,5 epimerase (Rv3465, RmlC) and dTDP-Rha synthase (Rv3266, RmlD) (Hoang et al., 1999; Ma et al., 2001; Stern et al., 1999).

The Gal$f$ residues are then added to the linker unit from UDP-Gal$f$, the activated donor of the Gal$f$ residues of the galactan (Weston et al., 1997). This molecule is synthesized from UDP-glucose, which is converted to UDP-galactopyranose (UDP-Gal$p$) by a UDP-Gal$p$ epimerase, likely encoded by *Rv3634* in *M. tuberculosis*. Once UDP-Gal$p$ is formed, a UDP-Gal$p$ mutase (Glf) encoded by *Rv3809c* converts UDP-Gal$p$ to UDP-Gal$f$ (Weston et al., 1997) and formation of the galactan can begin.

The Ara$f$ residues of the arabinan are added to the linker unit-galactan polymer from a decaprenyl-phosphoryl-Ara$f$ (DPA) precursor (Mikusova et al., 2000; Wolucka et al., 1994) through a transglycosylation reaction, releasing Pol-P (Fig. 4). The Ara$f$ moiety of DPA originates from the pentose phosphate pathway (Scherman et al., 1995, 1996; Klutts et al., 2002), and the biosynthesis of this compound in mycobacteria is now reasonably well established (Fig. 4). It is proposed that DPA is synthesized by the following sequence of events:

1. 5-Phosphoribose 1-diphosphate (pRpp) is transferred to a Dec-P molecule to form Dec-P-β-D-5-phosphoribose, a reaction which is catalyzed by Rv3806c in *M. tuberculosis* (Huang et al., 2005).

2. Dec-P-β-D-5-phosphoribose is dephosphorylated to form Dec-P-β-D-ribose by an as yet unidentified enzyme.

3. The hydroxyl group at the 2 position of the ribose is oxidized, likely to form Dec-P-2-keto-β-D-*erythro*-pentofuranose.

4. Dec-P-2-keto-β-D-*erythro*-pentofuranose is reduced to form Dec-P-β-D-Ara$f$ (DPA).

Thus, the epimerization of the ribosyl to an arabinosyl residue occurs at the lipid-linked level and constitutes the first example of an epimerase that uses a lipid-linked sugar as a substrate (Mikusova et al., 2005). Based on similarity to proteins implicated in the arabinosylation of the *Azorhizobium caulidans* nodulation factor, two genes (*Rv3790* and *Rv3791*) were cloned from the *M. tuberculosis* genome, expressed, and purified. Together, the proteins are able to catalyze the epimerization, although neither protein individually appears to be sufficient to support the activity (Mikusova et al., 2005).

Synthesis of arabinogalactan proper

Based on the structure of the galactan unit of AG, it was predicted that there would be least two and perhaps several galactosyl transferases involved in its biosynthesis (Crick et al., 2001). One enzyme could transfer the first Gal$f$ residue to the linker unit to form glycolipid 3 (GL-3); formation of the alternating 1→5 and 1→6 links could be catalyzed either by a bifunctional enzyme or two linkage-specific enzymes. Subsequent data have verified this prediction. Recently, it has been demonstrated that *M. tuberculosis Rv3782* encodes a Gal$f$ transferase involved in the first stages of galactan synthesis. Time-course experiments using *M. smegmatis* membranes indicate that the enzyme catalyzes the formation of Dec-P-P-GlcNAc-Rha-Gal$f$ (GL-3), followed rapidly by conversion to Dec-P-P-GlcNAc-Rha-Gal$f$-Gal$f$ (GL-4), suggesting that this is possibly a bifunctional enzyme (Mikusova et al., 2006). Thus, Rv3782 appears to be the initiator of galactan synthesis. A second Gal$f$ transferase gene (Rv3808c, GalT) of *M. tuberculosis* has been cloned, overexpressed, and partially characterized (Kremer et al., 2001; Mikusova et al., 2000) and was shown to catalyze the formation of both 1→5 and 1→6 linkages in cell free assay systems using natural substrates (Mikusova et al., 2000) or synthetic substrate analogs (Kremer et al., 2001). Subsequently, recombinant Rv3808c was purified for kinetic experiments (Rose et al., 2006), and purified Rv3808c is capable of forming both 1→5 and 1→6 linkages. However, a synthetic analog of GL-4 was a poor substrate for the enzyme based on K_m and K_{cat} values; thus, leading to the conclusion that at least three Gal$f$ transferase activities may be required before the action of Rv3808c (Rose et al., 2006). Therefore, it seems possible that there are potentially three to four Gal$f$ transferases involved in mycobacterial galactan synthesis; three if Rv3782 and Rv3808c are both bifunctional, and four if Rv3782 is monofunctional.

Synthesis of the arabinan moiety of the AG appears to be concomitant with synthesis of the galactan (Mikusova et al., 2000; Yagi et al., 2003), which suggests an unusual biosynthetic pathway based on topology, as discussed later. The structural complexity of mycobacterial arabinofuran present in AG is likely to be reflected in a complex biosynthetic pathway, and arabinosyl transferases are now being rapidly identified in *Mycobacterium spp.* The *embA* and *embB* gene products have been reported to be Ara$f$ transferases in *Mycobacterium avium* (Belanger et al., 1996), and the *embA* and *embB* gene products of *M. smegmatis* are involved in the biosynthesis of the terminal β-D-Ara$f$-(1→2)-α-D-Ara$f$ motif characteristic of AG (Escuyer et al., 2001); experiments indicate that it is highly likely that EmbA and EmbB from *M. smegmatis* are, in fact, glycosyltransferases, and that they work in concert to synthesize part or all of the β-D-Ara$f$-(1→2)-α-D-Ara$f$ motif (Khasnobis et al., 2006) characteristic of AG. The concept that Emb proteins are glycosyltransferases is also supported by the observation that point mutations in a putative glycosyltransferase consensus sequence found in EmbC prevents lipoarabinomannan synthesis in *M. smegmatis* (Berg et al., 2005). However, Rv3805c has unambiguously been shown to catalyze the addition of the terminal β-D-Ara*f* residue to the AG (Seidel et al., 2007). In addition, Rv3792 has been reported to catalyze the transfer of the first Ara*f* residue to the galactan in *M. tuberculosis* (Alderwick et al., 2006). Thus, current evidence indicates that AG is sequentially built up on the Pol-P carrier lipid until it reaches mature or near mature size (Mikusova et al., 1996, 2000; Yagi et al., 2003).

Arabinogalactan ligation to peptidoglycan

Once the mature, lipid-linked AG is synthesized, the GlcNAc of the linker unit of AG forms a 1-*O*-phosphoryl linkage with the 6-position a MurNAc residue of peptidoglycan, releasing Pol-P. This reaction is catalyzed by an as yet unidentified ligase. The ligation of AG to peptidoglycan has been demonstrated in cell-free preparations of *M. smegmatis*, and the nature of in vitro synthesized material was confirmed by the observation that the newly ligated AG can be released from peptidoglycan by muramidase treatment (Yagi et al., 2003). In vivo radiolabeling experiments of *M. smegmatis* cell wall demonstrated that the incorporation of AG into the cell wall requires newly synthesized PG and that the PG must be undergoing concomitant cross-linking (Hancock et al., 2002). Thus, ligation likely occurs either at the lipid II level, as seen for other PG modifications, or while the nascent PG is being formed. It seems probable that the ligation reaction occurs at or near the outer surface of the plasma membrane rather than at a more remote physical location that could be envisioned if the ligation occurred on mature PG. Synthesis of the linker-unit between the galactan of AG and PG (McNeil et al., 1991) is clearly essential for mycobacterial survival (Mills et al., 2004; Sassetti et al., 2003). Therefore, identification of the enzyme or enzymes responsible for the ligation of these two macromolecules represents a crucial area of research in the context of drug target identification.

As noted earlier, the arabinosylation of lipid-linked galactan and the ligation reactions both release Pol-P. This is in contrast to PG formation, which releases Pol-P-P. However, like the lipid released in PG synthesis, the Pol-P released by the ligation reaction is oriented toward the outside of the plasma membrane. In addition, it is probable that the Pol-P released by the formation of arabinan is also oriented outward. Therefore, it is anticipated that the Pol-P involved in AG synthesis is also recycled in a manner analogous to the Pol-P-P derived from PG synthesis.

Topology of arabinogalactan synthesis

It is not yet clear whether polymerization to form AG takes place inside or outside the plasma membrane, but it appears that addition of Gal*f* and Ara*f* residues occur simultaneously as described earlier. This poses a significant biosynthetic question. What is the topology of the enzymes involved in AG synthesis, and how is the transbilayer movement accomplished? Because the donor of Gal*f* residues is a nucleotide sugar, this implies that galactan synthesis occurs on the inside of the membrane, unless there is a UDP-Gal*f* transporter to externalize this compound. The donor of the Ara*f* residues is DPA, which suggests that arabinan synthesis takes place on the outside of the membrane. Yet arabinan and galactan synthesis seem to take place simultaneously. Perhaps the fact that the pertinent data were generated using cell-free assays, in which the natural topology of the membrane was disrupted, is the basis of this apparent discrepancy. Nevertheless, it is still unclear how the galactan, presumably synthesized using cytosolic sugar donors, is transported to the extracellular milieu for arabinosylation. Alternatively, if the entire AG moiety is synthesized on the inside of the membrane, how is this large polysaccharide moved to the outside to allow ligation to PG?

CONCLUDING REMARKS

This review of the current status of the structure and synthesis of mycobacterial PG and AG can be regarded as the culmination of more than 40 years of research spanning the initial pioneering structural studies of the complex, through to the present day definition of the diverse biosynthetic pathways and the associated enzymology and genetics. It seems likely that the remaining glycosyltransferases, particularly those responsible for full elaboration of the arabinan component of the AG, will be rapidly identified. Future challenges lie in chemical and enzymatic definition of the mechanism of ligation of the AG to PG and the final acylation of AG described elsewhere in this volume. A major challenge currently being addressed is the topology and organization of the enzymes involved in the synthesis of the MAPc. Another topic now under intense study is the exploitation of this fascinating chapter of novel biochemistry for the purpose of drug development in response to the current epidemic of drug-resistant tuberculosis and the problems associated with current drug regimens. Many of the steps described in this chapter, notably the MEP pathway for IPP synthesis, the pathways for synthesis of the Rha moiety of the linker region, the UDP-Gal mutase (Glf) reaction, responsible for the Gal*f* of AG, and the unique pathways for Ara*f* synthesis are being examined in this context with varying degrees of success. We foresee considerable progress in this area, involving the application of the fruits of 40 years of fundamental research aimed at the alleviation of tuberculosis.

Acknowledgments. Research in our laboratories was supported by grants AI-18357, AI-46393, AI-49151, and AI-06357 from the National Institute of Allergy and Infectious Diseases, National Institutes of Health.

Figure 1 was generously provided by Michael R. McNeil and Michael Scherman of the Department of Microbiology, Immunology and Pathology, Colorado State University.

REFERENCE

Adam, A., J. F. Petit, J. Wietzerbi-Falszpan, P. Sinay, D. W. Thomas, and E. Lederer. 1969. Mass spectrometric identification of *N*-glycolylmuramic acid, a constituent of *Mycobacterium smegmatis* walls. *FEBS Lett.* **4:**87–92.

Alderwick, L. J., E. Radmacher, M. Seidel, R. Gande, P. G. Hitchen, H. R. Morris, A. Dell, H. Sahm, L. Eggeling, and G. S. Besra. 2005. Deletion of Cg-emb in *Corynebacterianeae* leads to a novel truncated cell wall arabinogalactan, whereas inactivation of Cg-ubiA results in an arabinan-deficient mutant with a cell wall galactan core. *J. Biol. Chem.* **280:**32362–32371.

Alderwick, L. J., M. Seidel, H. Sahm, G. S. Besra, and L. Eggeling. 2006. Identification of a novel arabinofuranosyl transferase (AftA) involved in cell wall arabinan biosynthesis in *Mycobacterium tuberculosis.* *J. Biol. Chem.* **281:**15653–15661.

Allen, C. M., M. V. Keenan, and J. Sack. 1976. *Lactobacillus plantarum* undecaprenyl pyrophosphate synthetase: purification and reaction requirements. *Arch. Biochem. Biophys.* **175:**236–248.

Anderson, L., and F. M. Unger. 1983. Bacterial liposaccharides. *In* *ACS Symposium Series 231.* American Chemical Society, Washington, DC.

Anderson, R. G., H. Hussey, and J. Baddiley. 1972. The mechanism of wall synthesis in bacteria. The organization of enzymes and isoprenoid phosphates in the membrane. *Biochem. J.* **127:**11–25.

Anonymous. 1987. Prenol nomenclature. Recommendations 1986. IUPAC-IUB Joint Commission on Biochemical Nomenclature (JCBN). *Eur. J. Biochem.* **167:**181–184.

Apfel, C. M., B. Takacs, M. Fountoulakis, M. Stieger, and W. Keck. 1999. Use of genomics to identify bacterial undecaprenyl pyrophosphate synthetase: cloning, expression, and characterization of the essential uppS gene. *J. Bacteriol.* **181:**483–492.

Argyrou, A., and J. S. Blanchard. 2004. Kinetic and chemical mechanism of *Mycobacterium tuberculosis* 1-deoxy-D-xylulose-5-phosphate isomeroreductase. *Biochemistry* **43:**4375–4384.

Baba, T., and C. M. Allen, Jr. 1978. Substrate specificity of undecaprenyl pyrophosphate synthetase from *Lactobacillus plantarum.* *Biochemistry* **17:**5598–5604.

Baddiley, J. 1972. Teichoic acids in cell walls and membranes of bacteria. *Essays Biochem.* **8:**35–77.

Bailey, A. M., S. Mahapatra, P. J. Brennan, and D. C. Crick. 2002. Identification, cloning, purification, and enzymatic characterization of *Mycobacterium tuberculosis* 1-deoxy-D-xylulose 5-phosphate synthase. *Glycobiology* **12:**813–820.

Banerjee, D. K. 1989. Amphomycin inhibits mannosylphosphoryldolichol synthesis by forming a complex with dolichylmonophosphate. *J. Biol. Chem.* **264:**2024–2028.

Basu, J., R. Chattopadhyay, M. Kundu, and P. Chakrabarti. 1992. Purification and partial characterization of a penicillin-binding protein from *Mycobacterium smegmatis.* *J. Bacteriol.* **174:**4829–4832.

Baulard, A. R., G. S. Besra, and P. J. Brennan. 1999. The cell-wall core of *Mycobacterium*: structure, biogenesis and genetics, p. 240–259. *In* C. Ratledge and J. Dale (ed.), *Mycobacteria: Molecular Biology and Virulence.* Blackwell Science Ltd, London, United Kingdom.

Belanger, A. E., G. S. Besra, M. E. Ford, K. Mikusova, J. T. Belisle, P. J. Brennan, and J. M. Inamine. 1996. The embAB genes of *Mycobacterium avium* encode an arabinosyl transferase involved in cell wall arabinan biosynthesis that is the target for the antimycobacterial drug ethambutol. *Proc. Natl. Acad. Sci. USA* **93:**11919–11924.

Berg, S., J. Starbuck, J. B. Torrelles, V. D. Vissa, D. C. Crick, D. Chatterjee, and P. J. Brennan. 2005. Roles of conserved proline and glycosyltransferase motifs of embC in biosynthesis of lipoarabinomannan. *J. Biol. Chem.* **280:**5651–5663.

Besra, G. S., and P. J. Brennan. 1997. The mycobacterial cell wall: biosynthesis of arabinogalactan and lipoarabinomannan. *Biochem. Soc. Trans.* **25:**845–850.

Besra, G. S., K. H. Khoo, M. R. McNeil, A. Dell, H. R. Morris, and P. J. Brennan. 1995. A new interpretation of the structure of the mycolyl-arabinogalactan complex of *Mycobacterium tuberculosis* as revealed through characterization of oligoglycosylalditol fragments by fast-atom bombardment mass spectrometry and ^{1}H nuclear magnetic resonance spectroscopy. *Biochemistry* **34:**4257–4266.

Besra, G. S., T. Sievert, R. E. Lee, R. A. Slayden, P. J. Brennan, and K. Takayama. 1994. Identification of the apparent carrier in mycolic acid synthesis. *Proc. Natl. Acad. Sci. USA* **91:**12735–12739.

Billman-Jacobe, H., R. E. Haites, and R. L. Coppel. 1999. Characterization of a *Mycobacterium smegmatis* mutant lacking penicillin binding protein 1. *Antimicrob. Agents Chemother.* **43:**3011–3013.

Brennan, P. J., and H. Nikaido. 1995. The envelope of mycobacteria. *Annu. Rev. Biochem.* **64:**29–63.

Chambers, H. F., D. Moreau, D. Yajko, C. Miick, C. Wagner, C. Hackbarth, S. Kocagoz, E. Rosenberg, W. K. Hadley, and H. Nikaido. 1995. Can penicillins and other beta-lactam antibiotics be used to treat tuberculosis? *Antimicrob. Agents Chemother.* **39:**2620–2624.

Chopra, I., L. Hesse, and A. J. O'Neill. 2002. Exploiting current understanding of antibiotic action for discovery of new drugs. *J. Appl. Microbiol.* **92:**4S–15S.

Crick, D. C., S. Mahapatra, and P. J. Brennan. 2001. Biosynthesis of the arabinogalactan-peptidoglycan complex of *Mycobacterium tuberculosis.* *Glycobiology* **11:**107R–118R.

Crick, D. C., L. Quadri, and P. J. Brennan. 2008. p. 1–20. *In* S. H. E. Kaufmann and E. Rubin (ed.), *Handbook of Tuberculosis: Molecular Biology and Biochemistry.* Wiley-VCH, Weinheim, Germany.

Crick, D. C., M. C. Schulbach, E. E. Zink, M. Macchia, S. Barontini, G. S. Besra, and P. J. Brennan. 2000. Polyprenyl phosphate biosynthesis in *Mycobacterium tuberculosis* and *Mycobacterium smegmatis.* *J. Bacteriol.* **182:**5771–5778.

Daffe, M., P. J. Brennan, and M. McNeil. 1990. Predominant structural features of the cell wall arabinogalactan of *Mycobacterium tuberculosis* as revealed through characterization of oligoglycosyl alditol fragments by gas chromatography/mass spectrometry and by ^{1}H and ^{13}C NMR analyses. *J. Biol. Chem.* **265:**6734–6743.

Daffe, M., M. McNeil, and P. J. Brennan. 1993. Major structural features of the cell wall arabinogalactans of *Mycobacterium*, *Rhodococcus*, and *Nocardia* spp. *Carbohydr. Res.* **249:**383–398.

Dal Nogare, A. R., N. Dan, and M. A. Lehrman. 1998. Conserved sequences in enzymes of the UDP-GlcNAc/MurNAc family are essential in hamster UDP-GlcNAc:dolichol-P GlcNAc-1-P transferase. *Glycobiology* 8:625–632.

Dasgupta, A., P. Datta, M. Kundu, and J. Basu. 2006. The serine/threonine kinase PknB of *Mycobacterium tuberculosis* phosphorylates PBPA, a penicillin-binding protein required for cell division. *Microbiology* 152:493–504.

De Smet, K. A., K. E. Kempsell, A. Gallagher, K. Duncan, and D. B. Young. 1999. Alteration of a single amino acid residue reverses fosfomycin resistance of recombinant MurA from *Mycobacterium tuberculosis*. *Microbiology* 145:3177–3184.

Dhiman, R. K., M. L. Schaeffer, A. M. Bailey, C. A. Testa, H. Scherman, and D. C. Crick. 2005. 1-Deoxy-D-xylulose 5-phosphate reductoisomerase (IspC) from *Mycobacterium tuberculosis*: towards understanding mycobacterial resistance to fosmidomycin. *J. Bacteriol.* 187:8395–8402.

Dmitriev, B. A., S. Ehlers, and E. T. Rietschel. 1999. Layered murein revisited: a fundamentally new concept of bacterial cell wall structure, biogenesis and function. *Med. Microbiol. Immunol.* (Berlin) 187:173–181.

Dmitriev, B. A., S. Ehlers, E. T. Rietschel, and P. J. Brennan. 2000. Molecular mechanics of the mycobacterial cell wall: From horizontal layers to vertical scaffolds. *Int. J. Med. Microbiol.* 290:251–258.

Dover, L. G., A. M. Cerdeno-Tarraga, M. J. Pallen, J. Parkhill, and G. S. Besra. 2004. Comparative cell wall core biosynthesis in the mycolated pathogens, *Mycobacterium tuberculosis* and *Corynebacterium diphtheriae*. *FEMS Microbiol. Rev.* 28:225–250.

Draper, P. 1998. The outer parts of the mycobacterial envelope as permeability barriers. *Front. Biosci.* 3:D1253–D1261.

Draper, P., O. Kandler, and A. Darbre. 1987. Peptidoglycan and arabinogalactan of *Mycobacterium leprae*. *J. Gen. Microbiol.* 133:1187–1194.

Draper, P., K. H. Khoo, D. Chatterjee, A. Dell, and H. R. Morris. 1997. Galactosamine in walls of slow-growing mycobacteria. *Biochem. J.* 327:519–525.

El Ghachi, M., A. Bouhss, D. Blanot, and D. Mengin-Lecreulx. 2004. The bacA gene of *Escherichia coli* encodes a undecaprenyl pyrophosphate phosphatase activity. *J. Biol. Chem.* 279:30106–30113

El Ghachi, M., A. Derbise, A. Bouhss, and D. Mengin-Lecreulx. 2005. Identification of multiple genes encoding membrane proteins with undecaprenyl pyrophosphate phosphatase (UppP) activity in *Escherichia coli*. *J. Biol. Chem.* 280:18689–18695.

El Zoeiby, A., F. Sanschagrin, and R. C. Levesque. 2003. Structure and function of the Mur enzymes: development of novel inhibitors. *Mol. Microbiol.* 47:1–12.

Escuyer, V. E., M. A. Lety, J. B. Torrelles, K. H. Khoo, J. B. Tang, C. D. Rithner, C. Frehel, M. R. McNeil, P. J. Brennan, and D. Chatterjee. 2001. The role of the embA and embB gene products in the biosynthesis of the terminal hexaarabinofuranosyl motif of *Mycobacterium smegmatis* arabinogalactan. *J. Biol. Chem.* 276:48854–48862.

Essers, L., and H. J. Schoop. 1978. Evidence for incorporation of molecular oxygen, a pathway in biosynthesis of N-glycolylmuramic acid in *Mycobacterium phlei*. *Biochim. Biophys. Acta* 544:180–184.

Fujihashi, M., N. Shimizu, Y. W. Zhang, T. Koyama, and K. Miki. 1999. Crystallization and preliminary X-ray diffraction studies of undecaprenyl diphosphate synthase from *Micrococcus luteus* B-P 26. *Acta Crystallogr. D.* 55:1606–1607.

Gateau, O., C. Bordet, and G. Michel. 1976. Study of formation of N-glycolylmuramic acid from *Nocardia asteroides*. *Biochim. Biophys. Acta* 421:395–405.

Ghuysen, J. M. 1968. Use of bacteriolytic enzymes in determination of wall structure and their role in cell metabolism. *Bacteriol. Rev.* 32:425–464.

Ghuysen, J. M. 1991. Serine beta-lactamases and penicillin-binding proteins. *Annu. Rev. Microbiol.* 45:37–67.

Goffin, C., and J. M. Ghuysen. 2002. Biochemistry and comparative genomics of SxxK superfamily acyltransferases offer a clue to the mycobacterial paradox: Presence of penicillin-susceptible target proteins versus lack of efficiency of penicillin as therapeutic agent. *Microbiol. Mol. Biol. Rev.* 66:702–738.

Green, D. W. 2002. The bacterial cell wall as a source of antibacterial targets. *Expert. Opin. Ther. Targets* 6:1–19.

Hancock, I. C., S. Carman, G. S. Besra, P. J. Brennan, and E. Waite. 2002. Ligation of arabinogalactan to peptidoglycan in the cell wall of *Mycobacterium smegmatis* requires concomitant synthesis of the two wall polymers. *Microbiology* 148:3059–3067.

Higashi, Y., G. Siewert, and J. L. Strominger. 1970. Biosynthesis of the peptidoglycan of bacterial cell walls. XIX. Isoprenoid alcohol phosphokinase. *J. Biol. Chem.* 245:3683–3690.

Higashi, Y., and J. L. Strominger. 1970. Biosynthesis of the peptidoglycan of bacterial cell walls. XX. Identification of phosphatidylglycerol and cardiolipin as cofactors for isoprenoid alcohol phosphokinase. *J. Biol. Chem.* 245:3691–3696.

Higashi, Y., J. L. Strominger, and C. C. Sweeley. 1967. Structure of a lipid intermediate in cell wall peptidoglycan synthesis: a derivative of a C_{55} isoprenoid alcohol. *Proc. Natl. Acad. Sci. USA* 57:1878–1884.

Hoang, T. T., Y. Ma, R. J. Stern, M. R. McNeil, and H. P. Schweizer. 1999. Construction and use of low-copy number T7 expression vectors for purification of problem proteins: purification of *Mycobacterium tuberculosis* RmlD and *Pseudomonas aeruginosa* LasI and Rh1I proteins, and functional analysis of purified Rh1I. *Gene* 237:361–371.

Huang, H. R., M. S. Scherman, W. D'Haeze, D. Vereecke, M. Holsters, D. C. Crick, and M. R. McNeil. 2005. Identification and active expression of the *Mycobacterium tuberculosis* gene encoding 5-phospho-alpha-D-ribose-1-diphosphate: decaprenyl-phosphate 5-phosphoribosyltransferase, the first enzyme committed to decaprenylphosphoryl-D-arabinose synthesis. *J. Biol. Chem.* 280:24539–24543.

Jacobs, C., B. Joris, M. Jamin, K. Klarsov, J. Vanbeeumen, D. MenginLecreulx, J. vanHeijenoort, J. T. Park, S. Normark, and J. M. Frere. 1995. AmpD, essential for both beta-lactamase regulation and cell-wall recycling, is a novel cytosolic N-acetylmuramyl-L-alanine amidase. *Mol. Microbiol.* 15:553–559.

Jarlier, V., L. Gutmann, and H. Nikaido. 1991. Interplay of cell-wall barrier and beta-lactamase activity determines high-resistance to beta-lactam antibiotics in *Mycobacterium chelonae*. *Antimicrob. Agents Chemother.* 35:1937–1939.

Jarlier, V., and H. Nikaido. 1994. Mycobacterial cell wall: structure and role in natural resistance to antibiotics. *FEMS Microbiol. Lett.* 123:11–18.

Kahan, F. M., J. S. Kahan, P. J. Cassidy, and H. Kropp. 1974. Mechanism of action of fosfomycin (phosphonomycin). *Ann. N. Y. Acad. Sci.* 235:364–386.

Kalin, J. R., and C. M. Allen. 1979. Characterization of undecaprenol kinase from Lactobacillus plantarum. *Biochim. Biophys. Acta* 574:112–122.

Kalin, J. R., and C. M. Allen. 1980. Lipid activation of undecaprenol kinase from Lactobacillus plantarum. *Biochim. Biophys. Acta* 619:76–89.

Kaur, D., P. J. Brennan, and D. C. Crick. 2004. Decaprenyl diphosphate synthesis in *Mycobacterium tuberculosis*. *J. Bacteriol.* 186:7564–7570.

Keer, J., M. J. Smeulders, K. M. Gray, and H. D. Williams. 2000. Mutants of *Mycobacterium smegmatis* impaired in stationary-phase survival. *Microbiology* 146:2209–2217.

Khasnobis, S., J. Zhang, S. K. Angala, A. G. Amin, M. R. McNeil, D. C. Crick, and D. Chatterjee. 2006. Characterization of a specific arabinosyltransferase activity involved in mycobacterial arabinan biosynthesis. *Chem. Biol.* 13:787–795.

Klutts, J. S., K. Hatanaka, Y. T. Pan, and A. D. Elbein. 2002. Biosynthesis of D-arabinose in *Mycobacterium smegmatis*: specific labeling from D-glucose. *Arch. Biochem. Biophys.* 398:229–239.

Kotani, S., I. Yanagida, K. Kato, and T. Matsuda. 1970. Studies on peptides, glycopeptides and antigenic polysaccharide-glycopeptide complexes isolated from an L-11 enzyme lysate of the cell walls of *Mycobacterium tuberculosis* strain H37Rv. *Biken J.* 13:249–275.

Kremer, L., L. G. Dover, C. Morehouse, P. Hitchin, M. Everett, H. R. Morris, A. Dell, P. J. Brennan, M. R. McNeil, C. Flaherty, K. Duncan, and G. S. Besra. 2001. Galactan biosynthesis in *Mycobacterium tuberculosis*: Identification of a bifunctional UDP-galactofuranosyltransferase. *J. Biol. Chem.* 276:26430–26440.

Lederer, E. 1971. The mycobacterial cell wall. *Pure Appl. Chem.* 25:135–165.

Lederer, E., A. Adam, R. Ciorbaru, J. F. Petit, and J. Wietzerbin. 1975. Cell walls of mycobacteria and related organisms; chemistry and immunostimulant properties. *Mol. Cell Biochem.* 7: 87–104.

Lee, R. E., P. J. Brennan, and G. S. Besra. 1996. *Mycobacterium tuberculosis* cell envelope. *Curr. Top. Microbiol. Immunol.* 215: 1–27.

Leyh-Bouille, M., R. Bonaly, J. M. Ghuysen, R. Tinelli, and D. Tipper. 1970. LL-Diaminopimelic acid containing peptidoglycans in walls of *Streptomyces* spp. and of *Clostridium perfringens* (type-A). *Biochemistry* 9:2944–2952.

Liu, J., C. E. Barry, III, G. S. Besra, and H. Nikaido. 1996. Mycolic acid structure determines the fluidity of the mycobacterial cell wall. *J. Biol. Chem.* 271:29545–29551.

Ma, Y., J. A. Mills, J. T. Belisle, V. Vissa, M. Howell, K. Bowlin, M. S. Scherman, and M. McNeil. 1997. Determination of the pathway for rhamnose biosynthesis in mycobacteria: cloning, sequencing and expression of the *Mycobacterium tuberculosis* gene encoding alpha-D-glucose-1-phosphate thymidylyltransferase. *Microbiology* 143:937–945.

Ma, Y., R. J. Stern, M. S. Scherman, V. D. Vissa, W. Yan, V. C. Jones, F. Zhang, S. G. Franzblau, W. H. Lewis, and M. R. McNeil. 2001. Drug targeting *Mycobacterium tuberculosis* cell wall synthesis: Genetics of dTDP-rhamnose synthetic enzymes and development of a microtiter plate-based screen for inhibitors of conversion of dTDP-glucose to dTDP-rhamnose. *Antimicrob. Agents Chemother.* 45:1407–1416.

Mahapatra, S., J. Basu, P. J. Brennan, and D. C. Crick. 2005a. Structure, biosynthesis and genetics of the mycolic acid-arabinogalactan-peptidoglycan complex, p. 275–285. *In* S. T. Cole, K. D. Eisenach, D. N. McMurray, and W. R. Jacobs (ed.), *Tuberculosis and the Tubercle Bacillus*. ASM Press, Washington, DC.

Mahapatra, S., D. C. Crick, and P. J. Brennan. 2000. Comparison of the UDP-N-acetylmuramate: L-alanine ligase enzymes from *Mycobacterium tuberculosis* and *Mycobacterium leprae*. *J. Bacteriol.* 182:6827–6830.

Mahapatra, S., H. Scherman, P. J. Brennan, and D. C. Crick. 2005b. N-glycolylation of the nucleotide precursors of peptidoglycan biosynthesis of *Mycobacterium* spp. is altered by drug treatment. *J. Bacteriol.* 187:2341–2347.

Mahapatra, S., T. Yagi, J. T. Belisle, B. J. Espinosa, P. J. Hill, M. R. McNeil, P. J. Brennan, and D. C. Crick. 2005c. Mycobacterial lipid II is composed of a complex mixture of modified muramyl and peptide moieties linked to decaprenyl phosphate. *J. Bacteriol.* 187:2747–2757.

Marquardt, J. L., E. D. Brown, W. S. Lane, T. M. Haley, Y. Ichikawa, C. H. Wong, and C. T. Walsh. 1994. Kinetics, stoichiometry, and identification of the reactive thiolate in the inactivation of UDP-GlcNAc enolpyruvoyl transferase by the antibiotic fosfomycin. *Biochemistry* 33:10646–10651.

Matsuhashi, M. 1966. Biosynthesis in the bacterial cell wall. *Tanpakushitsu Kakusan Koso* 11:875–886.

McNeil, M. 1999. Arabinogalactan in mycobacteria: structure, biosynthesis, and genetics, p. 207–223. *In* J. B. Goldberg (ed.), *Genetics of Bacterial Polysaccharides*. CRC Press, Washington, DC.

McNeil, M., G. S. Besra, and P. J. Brennan. 1996. Chemistry of the mycobacterial cell wall, p. 171–185. *In* W. N. Rom and S. M. Garay (ed.), *Tuberculosis*. Little, Brown and Company, Boston.

McNeil, M., M. Daffe, and P. J. Brennan. 1991. Location of the mycolyl ester substituents in the cell walls of mycobacteria. *J. Biol. Chem.* 266:13217–13223.

McNeil, M. R., and P. J. Brennan. 1991. Structure, function and biogenesis of the cell envelope of mycobacteria in relation to bacterial physiology, pathogenesis and drug resistance; some thoughts and possibilities arising from recent structural information. *Res. Microbiol.* 142:451–463.

Mengin-Lecreulx, D., J. vanHeijenoort, and J. T. Park. 1996. Identification of the mpl gene encoding UDP-N-acetylmuramate: L-alanyl-gamma-D-glutamyl-meso-diaminopimelate ligase in *Escherichia coli* and its role in recycling of cell wall peptidoglycan. *J. Bacteriol.* 178:5347–5352.

Meroueh, S. O., K. Z. Bencze, D. Hesek, M. Lee, J. F. Fisher, T. L. Stemmler, and S. Mobashery. 2006. Three-dimensional structure of the bacterial cell wall peptidoglycan. *Proc. Natl. Acad. Sci. USA* 103:4404–4409.

Mikusova, K., M. Belanova, J. Kordulakova, K. Honda, M. R. McNeil, S. Mahapatra, D. C. Crick, and P. J. Brennan. 2006. Identification of a novel galactosyl transferase involved in biosynthesis of the mycobacterial cell wall. *J. Bacteriol.* 188: 6592–6598.

Mikusova, K., H. R. Huang, T. Yagi, M. Holsters, D. Vereecke, W. D'Haeze, M. S. Scherman, P. J. Brennan, M. R. McNeil, and D. C. Crick. 2005. Decaprenylphosphoryl arabinofuranose, the donor of the D-arabinofuranosyl residues of mycobacterial arabinan, is formed via a two-step epimerization of decaprenylphosphoryl ribose. *J. Bacteriol.* 187:8020–8025.

Mikusova, K., M. Mikus, G. S. Besra, I. Hancock, and P. J. Brennan. 1996. Biosynthesis of the linkage region of the mycobacterial cell wall. *J. Biol. Chem.* 271:7820–7828.

Mikusova, K., T. Yagi, R. Stern, M. R. McNeil, G. S. Besra, D. C. Crick, and P. J. Brennan. 2000. Biosynthesis of the galactan component of the mycobacterial cell wall. *J. Biol. Chem.* 275:33890–33897.

Mills, J. A., K. Motichka, M. Jucker, H. P. Wu, B. C. Uhlik, R. J. Stern, M. S. Scherman, V. D. Vissa, F. Pan, M. Kundu, Y. F. Ma, and M. McNeil. 2004. Inactivation of the mycobacterial rhamnosyltransferase, which is needed for the formation of the arabinogalactan-peptidoglycan linker, leads to irreversible loss of viability. *J. Biol. Chem.* 279:43540–43546.

Minnikin, D. E. 1982. Lipids: complex lipids, their chemistry biosynthesis and roles, p. 95–184. *In* C. Ratledge and J. Stanford (ed.), *The Biology of Mycobacteria*. Academic Press, London, United Kingdom.

Muth, J. D., and C. M. Allen. 1984. Undecaprenyl pyrophosphate synthetase from *Lactobacillus plantarum*: a dimeric protein. *Arch. Biochem. Biophys.* 230:49–60.

Nikaido, H., and V. Jarlier. 1991. Permeability of the mycobacterial cell-wall. *Res. Microbiol.* 142:437–443.

Park, J. T. 2001. Identification of a dedicated recycling pathway for anhydro-N-acetylmuramic acid and N-acetylglucosamine derived from *Escherichia coli* cell wall murein. *J. Bacteriol.* 183:3842–3847.

Petit, J. F., A. Adam, and J. Wietzerbin-Falszpan. 1970. Constituents of Mycobacteria. 117. Isolation of UDP-N-Glycolylmuramyl-(Ala, Glu, Dap) from Mycobacterium phlei. FEBS Lett. 6: 55–57.

Petit, J. F., A. Adam, J. Wietzerbin-Falszpan, E. Lederer, and J. M. Ghuysen. 1969. Chemical structure of the cell wall of Mycobacterium smegmatis. I. Isolation and partial characterization of the peptidoglycan. Biochem. Biophys. Res. Commun. 35:478–485.

Petit, J. F., and E. Lederer. 1984. The structure of the mycobacterial cell wall, p. 301–322. In G. P. Kubica and L. G. Wayne (ed.), The Mycobacteria, a Source Book. Marcel Dekker, New York, NY.

Raymond, J. B., S. Mahapatra, D. C. Crick, and M. S. Pavelka. 2005. Identification of the namH gene, encoding the hydroxylase responsible for the N-glycolylation of the mycobacterial peptidoglycan. J. Biol. Chem. 280:326–333.

Rieber, M., T. Imaeda, and I. M. Cesari. 1969. Bacitracin action on membranes of mycobacteria. J. Gen. Microbiol. 55:155–159.

Rose, N. L., G. C. Completo, S. J. Lin, M. McNeil, M. M. Palcic, and T. L. Lowary. 2006. Expression, purification, and characterization of a galactofuranosyltransferase involved in Mycobacterium tuberculosis arabinogalactan biosynthesis. J. Am. Chem. Soc. 128:6721–6729.

Sassetti, C. M., D. H. Boyd, and E. J. Rubin. 2003. Genes required for mycobacterial growth defined by high density mutagenesis. Mol. Microbiol. 48:77–84.

Scherman, M., A. Weston, K. Duncan, A. Whittington, R. Upton, L. Deng, R. Comber, J. D. Friedrich, and M. McNeil. 1995. Biosynthetic origin of mycobacterial cell wall arabinosyl residues. J. Bacteriol. 177:7125–7130.

Scherman, M. S., L. Kalbe-Bournonville, D. Bush, Y. Xin, L. Deng, and M. McNeil. 1996. Polyprenylphosphate-pentoses in mycobacteria are synthesized from 5-phosphoribose pyrophosphate. J. Biol. Chem. 271:29652–29658.

Schleifer, K. H., and O. Kandler. 1972. Peptidoglycan types of bacterial cell walls and their taxonomic implications. Bacteriol. Rev. 36:407–477.

Schulbach, M. C., P. J. Brennan, and D. C. Crick. 2000. Identification of a short (C_{15}) chain Z-isoprenyl diphosphate synthase and a homologous long (C_{50}) chain isoprenyl diphosphate synthase in Mycobacterium tuberculosis. J. Biol. Chem. 275:22876–22881.

Schulbach, M. C., S. Mahapatra, M. Macchia, S. Barontini, C. Papi, F. Minutolo, S. Bertini, P. J. Brennan, and D. C. Crick. 2001. Purification, enzymatic characterization, and inhibition of the Z-farnesyl diphosphate synthase from Mycobacterium tuberculosis. J. Biol. Chem. 276:11624–11630.

Seidel, M., L. J. Alderwick, H. L. Birch, H. Sahm, L. Eggeling, and G. S. Besra. 2007. Identification of a novel arabinofuranosyltransferase AftB involved in a terminal step of cell wall arabinan biosynthesis in Corynebacterineae, such as Corynebacterium glutamicum and Mycobacterium tuberculosis. J. Biol. Chem. 282:14729–14740.

Shimizu, N., T. Koyama, and K. Ogura. 1998. Molecular cloning, expression, and characterization of the genes encoding the two essential protein components of Micrococcus luteus B-P 26 hexaprenyl diphosphate synthase. J. Bacteriol. 180:1578–1581.

Siewert, G., and J. L. Strominger. 1967. Bacitracin—an inhibitor of dephosphorylation of lipid pyrophosphate an intermediate in biosynthesis of peptidoglycan of bacterial cell walls. Proc. Natl. Acad. Sci. USA 57:767–773.

Silver, L. L. 2003. Novel inhibitors of bacterial cell wall synthesis. Curr. Opin. Microbiol. 6:431–438.

Skarzynski, T., A. Mistry, A. Wonacott, S. E. Hutchinson, V. A. Kelly, and K. Duncan. 1996. Structure of UDP-N-acetylglucosamine enolpyruvyl transferase, an enzyme essential for the synthesis of bacterial peptidoglycan, complexed with substrate UDP-N-acetylglucosamine and the drug fosfomycin. Structure 4:1465–1474.

Stern, R. J., T. Y. Lee, T. J. Lee, W. Yan, M. S. Scherman, V. D. Vissa, S. K. Kim, B. L. Wanner, and M. R. McNeil. 1999. Conversion of dTDP-4-keto-6-deoxyglucose to free dTDP-4-keto-rhamnose by the rmIC gene products of Escherichia coli and Mycobacterium tuberculosis. Microbiology 145:663–671.

Storm, D. R., and J. L. Strominger. 1973. Complex formation between bacitracin peptides and isoprenyl pyrophosphates. The specificity of lipid-peptide interactions. J. Biol. Chem. 248: 3940–3945.

Takayama, K., H. L. David, L. Wang, and D. S. Goldman. 1970a. Isolation and characterization of uridine diphosphate-N-glycolylmuramyl-L-alanyl-gamma-D-glutamyl-meso-alpha,alpha'-diaminopimelic acid from Mycobacterium tuberculosis. Biochem. Biophys. Res. Commun. 39:7–12.

Takayama, K., and D. S. Goldman. 1970b. Enzymatic synthesis of mannosyl-1-phosphoryl-decaprenol by a cell-free system of Mycobacterium tuberculosis. J. Biol. Chem. 245:6251–6257.

Takayama, K., H. K. Schnoes, and E. J. Semmler. 1973. Characterization of the alkali-stable mannophospholipids of Mycobacterium smegmatis. Biochim. Biophys. Acta 316:212–221.

Takayama, K., C. Wang, and G. S. Besra. 2005. Pathway to synthesis and processing of mycolic acids in Mycobacterium tuberculosis. Clin. Microbiol. Rev. 18:81–101.

Templin, M. F., A. Ursinus, and J. V. Holtje. 1999. A defect in cell wall recycling triggers autolysis during the stationary growth phase of Escherichia coli. EMBO J. 18:4108–4117.

Tripathi, R. P., N. Tewari, N. Dwivedi, and V. K. Tiwari. 2005. Fighting tuberculosis: An old disease with new challenges. Med. Res. Rev. 25:93–131.

van Heijenoort, J. 1994. Biosynthesis of bacterial peptidoglycan unit, p. 39–54. In J. M. Ghuysen and R. Hakenbeck (ed.), Bacterial Cell Wall. Elsevier Medical Press, Amsterdam, The Netherlands.

van Heijenoort, J. 1996. Murein synthesis, p. 1025–1034. In F. C. Neidhardt (ed.), Escherichia coli and Salmonella: Cellular and Molecular Biology, 2nd ed. ASM Press, Washington, DC.

van Heijenoort, J. 1998. Assembly of the monomer unit of bacterial peptidoglycan. Cell. Mol. Life Sci. 54:300–304.

van Heijenoort, J. 2001a. Formation of the glycan chains in the synthesis of bacterial peptidoglycan. Glycobiology 11:25R–36R.

van Heijenoort, J. 2001b. Recent advances in the formation of the bacterial peptidoglycan monomer unit. Nat. Prod. Rep. 18:503–519.

Vilkas, E., C. Amar, J. Markovits, J. Vliegenthart, and J. Kamerling. 1973. Occurrence of a galactofuranose disaccharide in immunoadjuvant fractions of Mycobacterium tuberculosis (cell-walls and wax D). Biochim. Biophys. Acta 297:423–435.

Vollmer, W., and J. V. Holtje. 2004. The architecture of the murein (peptidoglycan) in gram-negative bacteria: Vertical scaffold or horizontal layer(s)? J. Bacteriol. 186:5978–5987.

Weston, A., R. J. Stern, R. E. Lee, P. M. Nassau, D. Monsey, S. L. Martin, M. S. Scherman, G. S. Besra, K. Duncan, and M. R. McNeil. 1997. Biosynthetic origin of mycobacterial cell wall galactofuranosyl residues. Tuber. Lung Dis. 78:123–131.

Wietzerbin, J., B. C. Das, J. F. Petit, E. Lederer, M. Leyh-Bouille, and J. M. Ghuysen. 1974. Occurrence of D-alanyl-(D)-meso-diaminopimelic acid and meso-diaminopimelyl-meso-diaminopimelic acid interpeptide linkages in the peptidoglycan of mycobacteria. Biochemistry 13:3471–3476.

Wietzerbin-Falszpan, J., B. C. Das, I. Azuma, A. Adam, J. F. Petit, and E. Lederer. 1970. Determination of amino acid sequences in peptides by mass spectrometry. 22. Isolation and mass spectrometric identification of peptide subunits of mycobacterial cell walls. *Biochem. Biophys. Res. Commun.* **40**:57–63.

Willoughby, E., Y. Higashi, and J. L. Strominger. 1972. Biosynthesis of peptidoglycan of bacterial cell-wall. 25. Enzymatic dephosphorylation of C55-isoprenylphosphate. *J. Biol. Chem.* **247**:5113–5115.

Wolucka, B. A., and E. de Hoffmann. 1995. The presence of beta-D-ribosyl-1-monophosphodecaprenol in mycobacteria. *J. Biol. Chem.* **270**:20151–20155.

Wolucka, B. A., and E. de Hoffmann. 1998. Isolation and characterization of the major form of polyprenyl-phospho-mannose from *Mycobacterium smegmatis*. *Glycobiology* **8**:955–962.

Wolucka, B. A., M. R. McNeil, E. de Hoffmann, T. Chojnacki, and P. J. Brennan. 1994. Recognition of the lipid intermediate for arabinogalactan/arabinomannan biosynthesis and its relation to the mode of action of ethambutol on mycobacteria. *J. Biol. Chem.* **269**:23328–23335.

Wong, K. K., and D. L. Pompliano. 1998. Peptidoglycan biosynthesis: unexploited targets within a familiar pathway, p. 197–217. *In* B. P. Rosen and S. Mobashery (ed.), *Resolving the Antibiotic Paradox*. Kluwer Academic/Plenum Publishers, New York, NY.

Yagi, T., S. Mahapatra, K. Mikusova, D. C. Crick, and P. J. Brennan. 2003. Polymerization of mycobacterial arabinogalactan and ligation to peptidoglycan. *J. Biol. Chem.* **278**:26497–26504.

Chapter 4

A Comprehensive Overview of Mycolic Acid Structure and Biosynthesis

Hedia Marrakchi, Fabienne Bardou, Marie-Antoinette Lanéelle, and Mamadou Daffé

Mycolic acids are major and very specific components of the cell envelope of *Mycobacterium tuberculosis* and other mycobacterial species, playing a crucial role in its remarkable architecture and impermeability (Daffé and Draper, 1998). They are found either in unbound forms that are extractable with organic solvents, as esters of trehalose or glycerol, or for the most part, esterifying the terminal pentaarabinofuranosyl units of arabinogalactan, the polysaccharide that, together with peptidoglycan, forms the insoluble cell wall skeleton (McNeil et al., 1991; Daffé, 1996; Daffé and Draper, 1998). The importance of the cell wall integrity for the viability of *M. tuberculosis* has raised much interest in the understanding of the pathway and enzymes involved in mycolic acid biosynthesis. Indeed, the inhibition of mycolic acid synthesis is one of the primary effects of the front-line and most efficient antitubercular drug isoniazid (INH) (Takayama et al., 1972). Therefore, this metabolism represents an important reservoir of targets for the development of new antimycobacterial drugs in the context of the reemergence of multidrug-resistant tuberculosis (MDR-TB).

The last 10 years have seen an impressive progress in the development of the genetics, biochemistry, regulation and structure of many of the mycobacterial envelope constituents, including mycolic acids, whose biosynthesis remains an important and active field of research. Several aspects of the development in this area have been reviewed (Daffé and Draper, 1998; Barry et al., 1998; Takayama et al., 2005). This chapter focuses on the major metabolic steps and essential enzymatic players in the mycolic acid biosynthetic pathway, providing a historical perspective and highlighting the key advances of the last few years in this dynamic area.

Two important sources considerably helped researchers to make this progress. First, the advent of molecular technology has propelled *Corynebacterium glutamicum* to the forefront as the model organism for research on mycolic acid metabolism. Members of the *Corynebacterium* species, which share many of the features of the mycobacterial cell envelope (Daffé, 2005), are viable in the absence of these cell wall constituents, unlike mycobacteria. The isolation of genetic mutants of *C. glutamicum* defective in mycolic acid and arabinogalactan synthesis has led to the identification of key genes and enzymes involved in these pathways and thus defined their physiological functions. Second, the availability of genomic sequences for an ever-expanding group of actinomycetales is fuelling the expansion of our knowledge on mycolic acid synthesis beyond the model system of *C. glutamicum* into the realm of pathogenic mycobacteria. In this respect, as more mycobacterial genomic sequences are available, bioinformatic approaches are offering deeper insights into the diversity and function of the mycolic acid biosynthetic enzymes. Specifically, the completion of *M. tuberculosis* (Cole et al., 1998) as well as the closely related *Mycobacterium leprae* (the causative agent of leprosy) genome sequencing projects (Cole et al., 2001) led to the observation that the latter has a large number of pseudogenes that are recognizable remnants of their *M. tuberculosis* orthologues (Cole et al., 2001). It has been suggested that the *M. leprae* genome represents the minimal gene set for mycobacteria because many potentially redundant or overlapping pathways have been inactivated in this species (Cole et al., 2001). Therefore, the conservation of genes in *M. leprae* may be used as a supporting argument in favor of their probable functionality and underlying their important metabolic roles in mycobacteria.

Hedia Marrakchi, Fabienne Bardou, Marie-Antoinette Lanéelle, and Mamadou Daffé • Department of Molecular Mechanisms of Mycobacterial Infections, Institut de Pharmacologie et de Biologie Structurale du Centre National de la Recherche Scientifique (CNRS), and Université Paul Sabatier, 205 route de Narbonne, 31077 Toulouse Cedex 4, France.

STRUCTURE AND DIVERSITY OF MYCOLIC ACIDS

Mycolic acids were first defined in *M. tuberculosis* (Asselineau and Lederer, 1950) as long-chain fatty acids, α-branched, and β-hydroxylated, a structural hallmark feature that confers to the mycolic acid molecule the property to be cleaved at high temperature into a "mero"aldehyde main chain, also called "meromycolic chain," and a "meroacid" or alpha branch, a reaction similar to a reverse Claisen-type condensation (Fig. 1A). Mycolic acids have been characterized in all mycobacterial species and in other members of the actinomycetales regrouped as *Corynebacterineae* or "mycolata" with a few exceptions, e.g., *Corynebacterium amycolatum*. The structure of the simplest mycolic acid, the C_{32} (coryno)mycolic acid of *Corynebacterium diphtheriae*, has been established to be the 2R-tetradecyl, 3R-hydroxy octadecanoic acid (Fig. 1B). The stereochemistry of the centers at positions 2 and 3 has been shown to be conserved in mycolic acids of *Nocardia* (nocardomycolic acids) and *Mycobacterium* (eumycolic acids) genera (Fig. 1C) (Asselineau and Asselineau, 1966; Asselineau et al., 1970b).

Information relative to the mycolic acid structure has been brought through the application of early and modern chemical techniques, in particular thin-layer chromatography (TLC), gas chromatography (GC), high-pressure liquid chromatography, mass spectrometry, and nuclear magnetic resonance spectroscopy. Given the variability in the chain lengths and the degrees of complexity of the structures, mycolic acids have been regarded as genus-

and species-specific compounds and, consequently, were largely used as taxonomic markers; they consist in chains of 22 to 38 carbon atoms in *Corynebacterium* (De Briel et al., 1992), 34 to 38 carbon atoms in *Dietzia* (Rainey et al., 1995), 34 to 52 carbon atoms in *Rhodococcus*, 46 to 60 carbon atoms in *Nocardia* (Nishiuchi et al., 1999), 46 to 66 carbon atoms in *Gordona* (Nishiuchi et al., 2000), 64 to 78 carbon atoms in *Tsukamurella*, and 60 to 90 carbon atoms in *Mycobacterium* (Chun et al., 1997). Additionally, studies on the nature of the esters released by pyrolysis have been of a great interest in clarifying the relationships between these genera. The length of these esters has been characterized as C_8-C_{18} in *Corynebacterium*, C_{12}-C_{18} in *Nocardia*, and C_{22}-C_{24} for most species of the *Mycobacterium* genus, with the notable exception of the *M. tuberculosis* complex and *Mycobacterium xenopi* in which a C_{26} is specifically identified by GC as the alpha chain released by pyrolysis from their mycolates (Daffé et al., 1983).

The structures of mycolic acids of genera other than mycobacteria were found to be relatively simple in terms of chemical functions, being composed only of homologous series with various numbers of double bonds, up to 7 for some *Gordona* species (Tomiyasu and Yano, 1984). In contrast, mycolic acids of mycobacteria display a large diversity of chain length and chemical functions that define the different classes of mycolic acids, leading to more complex TLC patterns (Daffé et al., 1983; Minnikin et al., 1983; Barry et al., 1998) (Fig. 2). The most apolar mycolic acids, referred to as α-mycolic acids, contain 74 to 80 carbon atoms and generally two double bonds (of *cis* or *trans* config-

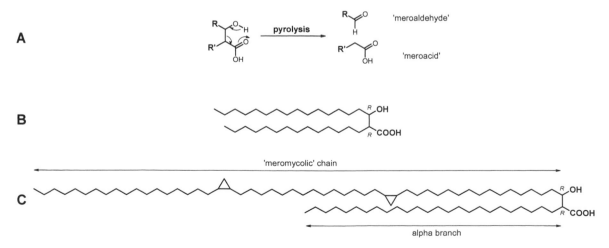

Figure 1. Chemical features of the mycolic acid structure. (A) Scheme of the pyrolytical cleavage of mycolic acid. R and R' represent long hydrocarbon chains. (B) The C_{32} corynomycolic acid: 2R-tetradecyl, 3R-hydroxy octadecanoic acid from *Corynebacterium diphtheriae*, where R indicates the stereochemistry of carbons 2 and 3. The pyrolysis releases C_{16} acid and C_{16} "meroaldehyde." (C) The dicyclopropanated mycolic acid (α-mycolate) from *Mycobacterium tuberculosis*: the pyrolysis of the C_{80} homologue releases C_{26} acid and C_{54} "meroaldehyde."

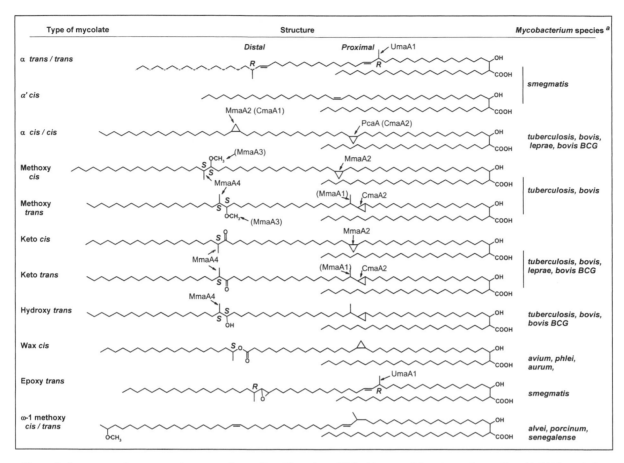

Figure 2. Structures of representative types of mycolic acids and established/proposed functions of mycolic acid SAM-methyltransferases (MA-MTs). [a], representative *Mycobacterium* species where these mycolate types occur are shown. The mycolate types displayed illustrate the functional groups of interest and therefore may not reflect the most abundant mycolate components. The *cis/trans* indicate the configuration of unsaturations (double bonds or cyclopropanes) at the distal/proximal position. R and S, when known, refer to the stereochemistry of the asymmetric centers (i.e., carbons bearing the methyl, methoxyl or hydroxyl groups). Arrows point to the known or predicted action of the mycolic acid SAM-methyltransferases (MA-MTs) whose functions have been assigned by gene knockout or by heterologous expression (into parentheses).

uration) or two *cis* cyclopropyl groups located in the meromycolic chain. A small fraction of α-mycolates may contain more unsaturations, as observed in some strains of the *M. tuberculosis* complex that have been shown to produce minor α-mycolates with three unsaturations and chains longer by 6 to 8 carbons than mycolates of the major series (Watanabe et al., 2001). Polyunsaturated α-mycolates were shown to represent a significant portion of the mycolates in *Mycobacterium fallax* (Rafidinarivo et al., 1985). Mycolic acids with 60 to 62 carbon atoms, known as α'-mycolic acids, and easily separated from the former series by TLC, contain only one or two *cis* double bonds. In addition, mycolic acids of the majority of mycobacteria examined so far contain supplementary oxygen functions located in the distal part of the meromycolic chain, defining the keto, methoxy, wax, epoxy, and hydroxy types of mycolates (Fig. 2). All of these oxygen functions are typified by the occurrence of an adjacent

methyl branch. It is noteworthy that in *M. tuberculosis*, the oxygenated mycolic acids, i.e., keto-, methoxy- and hydroxymycolates contain approximately 84 to 88 carbon atoms, four to six carbons longer than the α-mycolates from the same strains (Laval et al., 2001; Watanabe et al., 2001). The chain length difference may be even greater in minor components with additional cyclopropane rings (Watanabe et al., 2001). In contrast, the chain length of epoxy- and wax ester-mycolates is similar to that of α-mycolates of the same strains that are usually devoid of keto-, methoxy-, and hydroxymycolates (Laval et al., 2001).

MYCOLIC ACID BIOSYNTHESIS

The rich diversity of proteins and enzymes dedicated to mycolic acid biosynthesis reflects the importance of these lipids in *M. tuberculosis* (Table 1).

Table 1. *Mycobacterium tuberculosis* genes involved in fatty acid and mycolic acid synthesis

Enzyme group	ORF[b]	Activity or putative function
De novo fatty acid synthesis		
fas	Rv2524c	Fatty acid synthase
Fatty acid elongation (FAS-II)		
fabD	Rv2243	Malonyl-CoA:ACP transacylase
acpM	Rv2244	Acyl carrier protein
fabH	Rv0533	β-ketoacyl-ACP synthase III
kasA	Rv2245	β-ketoacyl-ACP synthase I
kasB	Rv2246	β-ketoacyl-ACP synthase II
mabA	Rv1483	β-ketoacyl-ACP reductase
inhA	Rv1484	*2-trans* enoyl-ACP reductase
Fatty acid activation		
accD6	Rv2247	Carboxyl transferase subunit of ACCase[a]
accD4 (β4)	Rv3799c	Carboxyl transferase subunit of ACCase
accD5 (β5)	Rv3280	Carboxyl transferase subunit of ACCase
accE (ε)	Rv3281	Epsilon subunit of ACCase
accA3 (α3)	Rv3285	Biotin carboxylase + biotin carboxyl carrier Protein subunits of ACCase
Mycolic acid SAM-dependent methyltransferases		
mmaA1	Rv0645c	Methyl transferase of unknown specificity
mmaA2	Rv0644c	*Cis* cyclopropane synthase of methoxy-, keto- and α-mycolates
mmaA3	Rv0643c	Methoxymycolic acid synthase
hma = *mmaA4*	Rv0642c	Hydroxy mycolic acid synthase
cmaA1	Rv3392c	Methyl transferase of unknown specificity
cmaA2	Rv0503c	*Trans* cyclopropane of methoxy- and ketomycolic acid
umaA1	Rv0469	Methyl transferase of unknown specificity
pcaA = *umaA2*	Rv0470c	Proximal cyclopropane synthase of alpha mycolates
Meromycolic chain activation and mycolic acid condensation		
fadD32	Rv3801c	Fatty acyl-AMP ligase
pks13	Rv3800c	Polyketide synthase
cmrA	Rv2509	β-ketoacyl reductase

[a]ACCase, acyl-CoA carboxylase.
[b]ORF, open reading frame.

Even though the detail of mycolic acid biosynthesis remains to be elucidated, the pathway for synthesis of mycolic acids could be virtually divided into three major steps: (1) the synthesis and elongation of fatty acids to give precursors of both the α-branch and the very long meromycolic chain; (2) the elongation and introduction of functional modifications on the meromycolic chain; and (3) the condensation of two long-chain fatty acids, followed by a reduction to yield the mycolic acid specific motif (Fig. 3).

Synthesis of Mycolic Acid Precursors

Synthesis of malonyl-CoA and carboxyacyl-CoA

The elongation unit used in fatty acid cycles is malonyl-coenzyme A (CoA), which serves as the building block for downstream fatty acid synthesis accomplished by the fatty acid synthase (FAS) systems (Fig. 3). Malonyl-CoA is produced by the carboxylation of acetyl-CoA catalyzed by the acetyl-CoA carboxylase, a key enzyme in most living organisms, being the first committed step in fatty acid biosynthesis (Cronan et al., 2002). The acyl-CoA carboxylation reaction generally proceeds in two steps and is dependent on the prosthetic group biotin. During the first enzymatic step, the biotin carboxylase (BC) catalyzes the adenosine triphosphate (ATP)-dependent carboxylation of biotin using bicarbonate as the carbon dioxide source to form carboxy-biotin. In the second half-reaction, the carboxyltransferase (CT) activity transfers the carboxyl group from biotin to acyl- (or acetyl-)CoA to form carboxyacyl- (or malonyl-)CoA. In mycobacteria, these

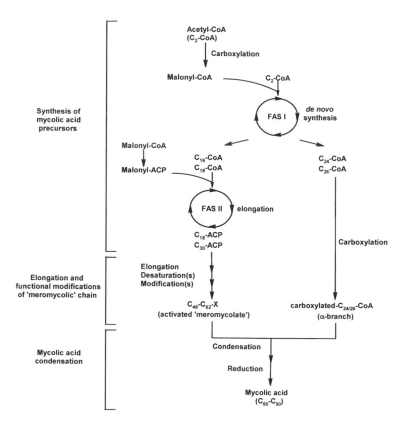

Figure 3. Model of the mycolic acid biosynthetic pathway in mycobacteria. The biosynthesis pathway of mycolic acids starts with the de novo synthesis and elongation of fatty acids operated by the mycobacterial FAS-I and FAS-II synthases, respectively. The FAS-II products have to undergo further elongation and modifications/decorations to produce the very long "meromycolic" chain precursors, whereas the carboxylation of acyl-CoAs (the FAS-I products) provides the activated alpha branch. Condensation of the latter with the activated "meromycolic" chain, followed by reduction, yields the mycolic acid with the characteristic motif.

enzymes display broad substrate specificities (acetyl-CoA, propionyl-CoA, and butyryl-CoA) and therefore are referred to as acyl-CoA carboxylases (ACCases). The BC and CT catalytic activities, as well as the biotin carboxyl carrier protein (BCCP), to which biotin is linked, are necessary for the holoenzyme activity.

1. Carboxy-biotin formation (BC + BCCP)

$$HCO_3^- + Mg\text{-}ATP + BCCP\text{-}biotin \leftrightarrow$$
$$BCCP\text{-}biotin\text{-}CO_2^- + Mg\text{-}ADP + Pi$$

2. Carboxyl transfer to acyl-CoA (CT)

$$BCCP\text{-}biotin\text{-}CO_2^- + acetyl\text{-}CoA \leftrightarrow$$
$$BCCP\text{-}biotin + malonyl\text{-}CoA$$

The ACCase is composed of multiple subunits (forming a complex) in most prokaryotes, whereas it is a large multidomain enzyme in most eukaryotes (Alberts et al., 1971; Cronan et al., 2002; Tong, 2005). In mycobacteria, ACCases adopt an intermediate domain organization: the BC+BCCP domains form the biotinylated α-subunit, whereas the CT do-

main constitutes the β-subunit (Diacovich et al., 2002; Tong, 2005). The latter subunit, being the only activity participating in the recognition of the substrate (second half-reaction), drives the ACCase substrate specificity (Diacovich et al., 2004).

M. tuberculosis displays an unusual number of ACCases in its genome, with three α subunit genes (*accA1-3*) and six ACCase carboxyltransferase domain genes (*accD1-6*). Each ACCase presumably serves a different physiological role and provides various extender units for the biosynthesis of the rich diversity of mycobacterial fatty acids. Recently, a combination of comparative genomics and molecular biology approaches allowed the identification of AccD4 and AccD5 as the CT components of the ACCase involved in mycolic acid biosynthesis in *C. glutamicum* (Gande et al., 2004; Portevin et al., 2005) and *M. tuberculosis* (Portevin et al., 2005). The only *accA* gene found in *M. leprae* is *accA3* encoding the biotinylated AccA3 (α3) subunit, which has been copurified with AccD4 (β4) and AccD5 (β5), and this tridomain complex was proposed to provide substrates for the condensation reaction leading to mycolic acid formation (Portevin et al., 2005). Indeed, the fact that the *accD4* (Rv3799c) gene is clustered with and transcribed in the same orientation as the genes involved in the condensation reaction, *pks13* (Rv3800c) and *fadD32* (Rv3801c), supports this result. The use of two distinct β subunits, AccD4 and AccD5, for mycolic acid synthesis suggested a distinct substrate specificity for each subunit, as reported for *Streptomyces coelicolor* A3(2) ACCases. This organism actually possesses two distinct ACCase complexes that share the same α subunit: an ACC (acetyl-CoA carboxylase) and a propionyl-CoA carboxylase (Diacovich et al., 2002, 2004), each having a discrete β subunit to which a small epsilon (ε) subunit is linked. Interestingly, the search for such subunit in the *M. tuberculosis* genome allowed the characterization of a novel epsilon subunit required for the holo AccA3-AccD5-AccE (α3, β5, ε) complex activity as a unique feature of *Mycobacterium* and *Streptomyces* ACCases (Gago et al., 2006; Lin et al., 2006; Oh et al., 2006). The main locus comprising *accA3*, *birA* (biotin ligase), and *accD5* also includes *accE* (Rv3281) encoding the epsilon subunit.

Concerning the role of AccD6, the organization of the *accD6* gene as part of the FAS-II cluster *fabD-acpM-kasA-kasB-accD6*, whose gene and protein expression profile is affected in response to FAS-II and cell wall inhibitors, strongly suggested the involvement of this CT subunit in the production of malonyl-CoA as the elongation unit for the fatty acid elongation cycles (Boshoff et al., 2004; Hughes et al., 2006). In agreement with a specialized function of AccD6 in FAS-II and within the mycobacteria is the fact that no recognizable orthologue is found in the *Corynebacterium* genomes analyzed, organisms that lack a FAS-II type system and use FAS-I for fatty acid synthesis (Gande et al., 2004). Recently, with the characterization of the AccD6 (β6) subunit as a functional acyl-CoA carboxylase, the various combinations of ACCase activities of *M. tuberculosis* have been successfully reconstituted in vitro from purified subunits of the biotinylated α subunit (AccA3 [α3]), the carboxyltransferase β subunit (AccD4 [β4], AccD5 [β5] or AccD6 [β6]), and the ε subunit AccE of *M. tuberculosis* and their characteristics compared (Gago et al., 2006; Oh et al., 2006; Daniel et al., 2007). The various *M. tuberculosis* ACCases obviously display a high degree of chain-length specificity: the main physiological role of AccA3-AccD5 (epsilon-dependent) would be to generate methylmalonyl-CoA for the biosynthesis of multimethyl-branched fatty acids of the cell envelope and to provide (accessory role) malonyl-CoA as the elongation unit for fatty acid cycles. This latter function would be preferentially achieved by the AccA3-AccD6 activity (epsilon-inhibited), which could thus play an important role in mycolic acid biosynthesis by providing malonyl-CoA to the FAS II complex. The AccA3-AccD4 ACCase would be rather dedicated to synthesize the carboxylated substrate involved in mycolic acid condensation (see Fig. 5). Whether AccD5 (β5) is involved in the AccD4/AccA3-catalyzed reaction in vivo is still unclear. Coimmunoprecipitation experiments have identified an AccA3-AccD4-AccD5 complex (Portevin et al., 2005), whereas another study has reported unsuccessful attempts to reconstitute AccD4-AccD5 hetero-oligomers in vitro (Gago et al., 2006). Recently, a structural study has indicated that AccD4 and AccD5 display similar electronegative surfaces and therefore likely share the same cognate α-subunit (AccA3) in an ACCase holocomplex (Holton et al., 2006). The formation of α/β ACCase complexes and their precise roles still remain to be elucidated.

Owing to their importance in providing building blocks for fatty acid, mycolic, and polyketide biosyntheses, the protein components of ACCases are expected to be essential for viability of mycobacteria. Essentiality studies have been carried out in the *Corynebacterium* model, for which the mycolic acid biosynthesis is not essential, and mutants in this metabolism could be obtained. As expected, AccD4 was found to be essential for mycolic acid synthesis and for the viability of *Mycobacterium* (Portevin et al., 2005), and attempts to delete *accD5* in mycobacteria failed, suggesting its essentiality in survival (Holton et al., 2006). Thus, α/β ACCase complexes should provide good targets for the development of novel antituberculosis drugs.

The fatty acid synthases

Biosynthesis of mycolic acid precursors requires the involvement of at least two types of FAS, namely the multifunctional polypeptide (type I) FAS-I and the FAS-II system, which is composed of a series of discrete soluble enzymes. This nomenclature was first formulated by Bloch (Brindley et al., 1969). Although the two FAS systems differ in their molecular organizations and in their substrate and carrier specificities, they share similar reaction sequences, with iterative series of reactions built on successive additions of a two-carbon (acetate) unit to a nascent acyl group.

De novo fatty acid synthesis (FAS-I). A remarkable characteristic of mycolic acid-producing bacteria (grouped into the *Corynebacterineae* suborder), including mycobacteria, is their use of a multifunctional FAS-I enzyme, as in eukaryotes, for de novo synthesis rather than the usual prokaryotic type II system. The mycobacterial FAS-I protein displays the seven distinct domains corresponding to the catalytic activities required for the synthesis cycle: acyl transferase, enoyl reductase, β-hydroxyacyl dehydratase, malonyl/palmitoyl transferase, acyl carrier protein (ACP), β-ketoacyl reductase, and β-ketoacyl synthase (Fernandes and Kolattukudy, 1996; Bloch and Vance, 1977). Mycobacteria have a single FAS-I-encoding gene (*fas*), which is, in all probability, essential for viability. It is noteworthy that *C. glutamicum*, which lacks a FAS-II elongation system, possesses two functional fatty acid synthases, encoded by *fasA* (FAS-IA) and *fasB* (FAS-IB). FAS-IA appears to be the most relevant FAS of the organism, whereas FAS-IB is supplementary, a hypothesis recently confirmed by mutational analyses showing that *fasA* is essential and *fasB* is dispensable (Radmacher et al., 2005). The *fas* gene from *M. tuberculosis* was found to confer resistance to the antituberculous drug pyrazinamide (PZA) when present on multicopy vectors. PZA markedly inhibited the activity of the *M. tuberculosis* FAS-I and the biosynthesis of C_{16} to $C_{24/26}$ fatty acids from acetyl-CoA, strongly suggesting that FAS-I is a primary target of PZA action in *M. tuberculosis* (Zimhony et al., 2000).

Fatty acid biosynthesis in mycobacteria is initiated by FAS-I, catalyzing the de novo synthesis of long-chain acyl-CoA from acetyl-CoA using malonyl-CoA as the elongation unit. The fatty acids produced are long-chain acyl-CoA with a bimodal distribution, C_{16}-C_{18} and C_{24}-C_{26}, which is a unique feature of the mycobacterial FAS-I among those of *Corynebacterineae* (Bloch and Vance, 1977; Fernandes and Kolattududy, 1996). The FAS-I C_{16}-C_{18} acyl-CoA products serve as substrates for fatty acid elongation by FAS-II to provide meromycolic chain precursors (Qureshi et al., 1984). The long-chain C_{24}-C_{26} acyl-CoAs subsequently participate, after carboxylation by ACCase, in mycolic acid biosynthesis (Fig. 3). Apart from this specific pathway of fatty acid metabolism, the acyl-CoA products of mycobacterial type I FAS are also incorporated, as in all other organisms, into the conventional phospholipids of the plasma membrane. In *M. leprae,* an obligate pathogen that probably uses the conventional fatty acids of its host (Wheeler et al., 1990), the type I FAS system may function primarily as an elongase, converting the host fatty acids into C_{22}-C_{24} products.

Fatty acid elongation (FAS-II). In addition to the type I synthase, mycobacteria contain an ACP-dependent dissociated type II fatty acid synthase consisting of discrete monofunctional proteins, each catalyzing one reaction in the pathway. Unlike the type II synthases of other bacteria, the mycobacterial FAS-II is incapable of de novo fatty acid synthesis from acetyl-CoA; it rather elongates medium-chain-length C_{12}-C_{16} fatty acids to yield C_{18}-C_{30} acyl-ACPs in vitro (Odriozola et al., 1977). The synthesis proceeds through elongation of enzyme-bound intermediates, which are covalently linked to mycobacterial acyl carrier protein (AcpM), by several iterative cycles, each comprising four steps (Bloch, 1977).

Because *Escherichia coli* FAS-II has been extensively studied, the biochemical properties of the individual enzymes are the paradigm of type II fatty acid synthase (Rock and Cronan, 1996). Figure 4 outlines the major steps in the mycobacterial FAS-II. The product of the acetyl-CoA carboxylase (ACC) reaction is malonyl-CoA and the malonate group is transferred to ACP by malonyl-CoA:ACP transacylase (MtFabD) to form malonyl-ACP. Fatty acid synthesis is initiated by the Claisen condensation of malonyl-ACP, with acyl-CoA catalyzed by β-ketoacyl-ACP synthase III (MtFabH) to form β-ketoacyl-ACP. Four enzymes catalyze each cycle of elongation: the *β*-keto group is reduced by the nicotinamide adenine dinucleotide phosphate (NADPH)-dependent β-ketoacyl-ACP reductase (MabA), and the resulting *β*-hydroxyl intermediate is dehydrated by the β-hydroxyacyl-ACP dehydratase to an enoyl-ACP. Next, the reduction of the enoyl chain by the NADPH-dependent enoyl-ACP reductase (InhA) produces an acyl-ACP. Additional cycles of elongation are initiated by the β-ketoacyl-ACP synthase (KasA or KasB), which elongates the acyl-ACP by two carbons to form a β-ketoacyl-ACP. Elongation likely ceases when the acyl-ACP attains the chain length required for meromycolic chain modification or condensation.

The demonstration that the inhibition of FAS-II elongation by the potent antitubercular drug INH is responsible for mycolic acid inhibition proved that FAS-II and the enzymes that it comprises are important and relevant targets for the development of new antitubercular drugs (Marrakchi et al., 2000).

Initiation steps in FAS-II. Like the *E. coli* FAS-II genes (grouped within a cluster at minute 24 of the chromosome), the *M. tuberculosis* main genes involved in the type II system are clustered in two transcriptional units, where genes are transcribed in the same orientation, the *mtfabD-acpM-kasA-kasB-accD6* and the *mabA-inhA* clusters (Cole et al., 1998) (Table 1). Two proteins, AcpM and MtFabD, play a crucial role in initiating the FAS-II biosynthetic cycle. The *acpM* gene encodes the mycobacterial acyl carrier protein (Kremer et al., 2001) that shuttles the acyl groups from one enzyme to another for the completion of the fatty acid synthesis cycle. The FAS-II intermediates are covalently attached to AcpM by a thioester bond

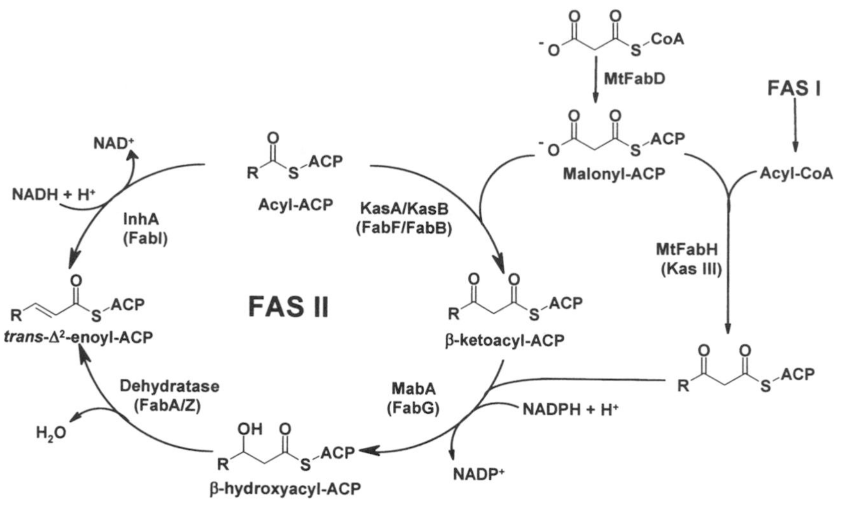

Figure 4. Mycobacterial (FAS-II) fatty acid synthesis. The names of the *Escherichia coli* FAS-II orthologue proteins are indicated into parentheses. The malonyl-CoA:ACP transacylase (MtFabD) converts malonyl-CoA into malonyl-ACP. Cycles of fatty acid elongation are initiated by the condensation of acyl-CoAs (products of FAS-I) with malonyl-ACP catalyzed by β-ketoacyl-ACP synthase III (MtFabH). The second step in the elongation cycle is carried out by the β-ketoacyl-ACP reductase, MabA. The β-hydroxyacyl-ACP intermediate is dehydrated to form *trans*-2-enoyl-ACP. The final step in the elongation is catalyzed by the nicotinamide adenine dinucleotide (NADH)-dependent enoyl-ACP reductase (InhA). Subsequent rounds of elongation are initiated by the elongation condensing enzymes (KasA and KasB) whose substrate specificities govern the structure and distribution of fatty acid products.

to the sulfhydryl of the 4′-phosphopantetheine prosthetic group. Apo-ACP is inactive in fatty acid synthesis and is converted to the active protein by holo-ACP synthase (AcpS) that transfers the 4′-phosphopantetheine from CoA to apo-ACP (Chalut et al., 2006). Compared with the *E. coli* and bacterial ACPs, AcpM is similar at the N-terminus but is unique in possessing an extended C-terminal (Zhang et al., 2003). This added feature, by sequestering the hydrophobic acyl-chain from solvent, likely plays an important role in transferring very long-chain fatty acids in the pathway (Schaeffer et al., 2001a). Although AcpM is hypothesized to be the carrier of the very-long meromycolic chain, no experimental evidence supports this hypothesis and the carrier of the mature meromycolate is still unknown.

Some bacteria, such as *S. coelicolor* A3(2), possess additional specialized ACPs, which contains the regular ACP for FAS-II and two other ACPs involved in polyketide synthesis (Revill et al., 1996). Interestingly, *M. tuberculosis* has, in addition to AcpM, two other putative ACPs (Rv0033 and Rv1344) (Kremer et al., 2001). The Rv1344 has been recently proposed as an acyl carrier protein, but the role of Rv0033 in fatty acid biosynthesis is still to be studied (Huang et al., 2006a). None of these putative ACPs have orthologues in the *M. leprae* genome, thus they are unlikely to participate in the biosynthesis of mycolates.

The *mtfabD* (Rv2243) encodes the malonyl-CoA:AcpM transacylase (MCAT) that provides the

condensing enzymes with the malonyl-AcpM substrate, the carbon donor during the elongation steps (Fig. 4). The MtFabD was shown to catalyze in vitro the transacylation of malonate from malonyl-CoA to activated holo-AcpM (Kremer et al., 2001). Interestingly, another gene (Rv0649, *fabD2*) is present in the *M. tuberculosis* genome that encodes a protein sharing 28% identity with MtFabD and is reported to possess MCAT activity as well (Huang et al., 2006b). However, analysis of the *M. leprae* genome (Cole et al., 2001), the "minimal mycobacterial genome," has revealed that, although *mlfabD* was present and highly homologous to its *M. tuberculosis* counterpart, there was no gene homologous to *fabD2* in *M. leprae*. This suggests that MtFabD represents the essential enzyme possessing MCAT activity involved in fatty acid and mycolic acid biosynthesis in *M. tuberculosis*.

β-Ketoacyl synthases (condensing enzymes). The coupling of the acyl and malonyl groups to form β-ketoacyl derivatives is catalyzed by the β-ketoacyl ACP synthase (KAS) component of the FAS-II system. This reaction is critical in fatty acid synthesis because it accounts for the chain-length of the fatty acids produced. Very early studies on substrate specificity have demonstrated the importance of condensing enzymes in the termination of chain elongation and in the accumulation of specific unsaturated/saturated fatty acids in the cell (Greenspan et al., 1970).

Three distinct KAS, MtFabH (KAS III), KasA, and KasB, operate in the mycobacterial FAS-II pathway. The MtFabH obviously functions as a link between the type I and II FAS systems by initiating the biosynthesis of very long-chain fatty acids used in mycolate biosynthesis (Fig. 4). It catalyzes the condensation of an acyl-CoA substrate (palmitoyl-CoA) with malonyl-ACP to generate a β-ketoacyl-ACP product (Choi et al., 2000). This product is reduced into an acyl-ACP and is extended by condensation catalyzed by one or more different ketoacyl-ACP synthase isozymes. Direct in vivo evidence that initiation KAS (KAS III) catalyzes an essential reaction was lacking until the recent report of the first nonmycobacterial strain lacking KAS III, which provided the genetic evidence that FabH catalyzes the major condensation reaction in the initiation of type II fatty acid biosynthesis in both gram-positive and gram-negative bacteria (Lai and Cronan, 2003). KasA and KasB have been shown to be the β-ketoacyl-AcpM synthases that further elongate the growing acyl chain in the mycobacterial FAS-II system (Schaeffer et al., 2001b; Kremer et al., 2002). Both KasA and KasB exhibit specificities for long-chain substrates (C_{12}-C_{20}). When their activities were assayed using C_{16}-ACP versus C_{16}-CoA, both proteins display a 10-fold higher specific activity with the former substrate, indicating their specificity for ACP-bound acyl substrates and, therefore, functioning as elongation enzymes (Schaeffer et al., 2001b). Even though KasA and KasB share significant sequence similarity, they show different chain-length specificities (Schaeffer et al., 2001b; Kremer et al., 2002), suggesting a distinct role for each enzyme. A recent study showed that cell lysates of M. smegmatis expressing M. tuberculosis kasA could elongate fatty acids up to ~40 carbons, whereas the expression of both kasA and kasB resulted in the production of longer chain fatty acids (~54 carbons), equivalent in length to the meromycolate chains of mycolic acids (Slayden and Barry, 2002). Based on these in vitro biosynthesis studies, it has been proposed that kasA is involved in the initial elongation of mycolates that are extended to their full lengths by kasB (Kremer et al., 2002; Slayden and Barry, 2002). Direct in vivo evidence of this hypothesis was obtained by investigating the role of kasB in mycolate biosynthesis in Mycobacterium marinum using transposon-disrupted kasB mutants; the authors have shown that KasB is required for normal growth of M. marinum in macrophages (Gao et al., 2003) and for full elongation of mycolates to their mature forms because mycolates in kasB mutants were shorter by ~2 to 4 carbon units. Despite this subtle difference in mycolate length, the kasB mutants have markedly increased permeability of

their cell walls and consequently are severely defective in resisting host defense mechanisms and antibiotic action but also showed an impaired growth within infected macrophages, indicating a critical role of KasB in intracellular survival (Gao et al., 2003). Complete recovery of the kasB mutants to wild-type phenotypes is achieved by ectopic expression of the M. tuberculosis kasB gene but not by kasA. These data were recently confirmed and extended to M. tuberculosis, in which the kasB deletion (by allelic exchange) resulted in the shortening (by up to six carbon atoms) of the ketomycolic acids and in the loss/reduction of the trans-cyclopropanated oxygenated mycolates (Bhatt et al., 2007). Moreover, this mutant strain lost its acid-fast staining and was unable to persist in immunocompetent infected mice. These studies identify kasB as an important potential target for novel therapeutic intervention, independent from kasA and support the hypothesis that kasA is involved in initial extension of the FAS-II-mediated growth of the mycolate chains, whereas kasB is involved primarily in extension to full lengths (Kremer et al., 2002; Slayden and Barry, 2002). In addition, the two enzymes are not redundant in vivo because (1) the absence of kasB leads to a specific shortening of the meromycolic chains by two to four carbons and (2) the extrachromosomal addition of kasB, but not kasA, can restore the wild-type phenotype (Gao et al., 2003). This differential role for the KasA and KasB proteins is also supported by a recent in vivo study; null mutants of kasB, but not kasA, could be generated in both Mycobacterium smegmatis and M. tuberculosis, suggesting the sole essentiality of kasA (Bhatt et al., 2005). The conditional depletion of KasA in M. smegmatis leads to the loss of mycolic acid biosynthesis before cell lysis (Bhatt et al., 2005). Finally, it was reported that KasA and KasB undergo a differential post-translational modification (phosphorylation), which was proposed to modulate specifically KasA or KasB enzyme activity in response to environmental changes (Molle et al., 2006).

β-Ketoacyl and enoyl reductases. Two reduction steps take place in the FAS-II cycle. The β-ketoacyl-ACP reductase performs the first, NADPH-dependent, reduction of the β-ketoacyl-ACP substrates to form β-hydroxyacyl-ACP products. The second reductive step in the elongation cycle is catalyzed by the NADH-dependent, *2-trans*-enoyl-ACP reductase (Fig. 4).

M. tuberculosis possesses five annotated fabG (β-ketoacyl reductase) genes in its genome, and MabA (FabG1) was shown to be the β-ketoacyl-ACP reductase of FAS-II (Marrakchi et al., 2002b); the only other ketoacyl reductase studied (FabG3) was determined to be a steroid dehydrogenase (Yang et al., 2002). Interestingly, the genes encoding the FAS-II reductases are found in the *M. tuberculosis* genome in a *mabA-inhA* cluster. The *inhA* gene was identified in 1994 as a putative target for INH and the related drug ethionamide (ETH) in *M. tuberculosis* (Banerjee et al., 1994). Since then, many lines of evidence pointed to InhA as a primary target of INH (Vilcheze et al., 2000; Hazbon et al., 2006), culminating in the demonstration that transfer of an *inhA* mutant allele (Ser 94 to Ala) in *M. tuberculosis* was sufficient for conferring resistance to INH and ETH, thus establishing InhA as the clinically relevant target of INH (Vilcheze et al., 2006). The encoded InhA protein was demonstrated to catalyze the 2-*trans*-enoyl-ACP reduction, with a preference for long-chain substrates (12-24 carbons), exhibiting higher affinity (by two orders of magnitude) for acyl-ACP substrates over their CoA counterparts (Quémard et al., 1995). Consistent with these characteristics, we reported evidence that InhA is part of the elongation system FAS-II, and that this elongation complex is probably involved in the synthesis of mycolic acids (Marrakchi et al., 2000).

The *mabA* and *inhA* genes, which form presumably a single operon in *M. tuberculosis*, encode proteins that are functionally and structurally related. Both proteins belong to the *short-chain dehydrogenase reductase* (SDR) or the *reductase-epimerase-dehydrogenase* (RED) structural superfamily (Labesse et al., 1994; Dessen et al., 1995; Cohen-Gonsaud et al., 2002). MabA was shown to encode a β-ketoacyl-ACP reductase with preference for long-chain substrates and to be involved in the elongation activity of FAS-II (Marrakchi et al., 2002b). Compared with homologous bacterial proteins, MabA has unique functional and structural properties such as a large hydrophobic substrate-binding pocket, which correlates with its preference for long-chain substrates, consistent with its role in mycolic acid biosynthesis (Marrakchi et al., 2002b; Cohen-Gonsaud et al., 2002). Given the common structural features of InhA and MabA, a possible interaction between MabA and a specific inhibitor of InhA, INH, was investigated. Indeed, MabA activity was also shown to be inhibited by INH in vitro, yet the possibility of MabA being a target of INH in vivo has still to be evaluated (Ducasse-Cabanot et al., 2004). These data, together with the coelution of InhA and MabA in a high-molecular-weight protein fraction from *M. smegmatis* displaying the FAS-II activity (Odriozola et al., 1977) and whose activity is inhibited by INH (Marrakchi et al., 2000), suggest that the components of FAS-II might be tightly interconnected. Interestingly, protein-protein interaction analyses have

shown that KasA, KasB, InhA, MabA, and MtFabH (β-ketoacyl-ACP synthase III) interact with each other (Veyron-Churlet et al., 2004). Protein-protein interactions are essential in many cellular processes and, as such, have been used as targets for the search of potent inhibitors.

There have been no successful reports of the generation of any *mabA* or *inhA* gene knockout mutants (despite extensive efforts), strongly suggesting their essentiality for mycobacterial viability. A temperature-sensitive mutation in the *inhA* gene of *M. smegmatis* (that rendered InhA inactive at 42°C) was isolated (Vilcheze et al., 2000). Thermal inactivation of InhA in *M. smegmatis* resulted in the inhibition of mycolic acid biosynthesis and cell lysis in a manner similar to that seen in INH-treated cells (Bardou et al., 1996). The *mabA* and *fabG* genes are predicted to be essential in *M. smegmatis* and *E. coli*, respectively (Zhang and Cronan, 1996; Banerjee et al., 1998). The first documented conditionally lethal temperature sensitive *fabG* mutant was reported recently in *E. coli* and *Salmonella enterica* (Lai and Cronan, 2004). Attempts to construct a *mabA* knockout mutant were unsuccessful, only the presence of a complementing copy of the gene allowed obtaining viable *M. tuberculosis*, indicating that *mabA* is likely to be essential for mycobacterial viability (Parish et al., 2007). Interestingly, both the *mabA* and *inhA* genes were predicted to be non-essential when high-density mutagenesis was applied to identify genes potentially essential for *M. tuberculosis* growth and survival during infection in a mouse model of TB (Sassetti et al., 2003; Sassetti and Rubin, 2003). This example of discrepancy between prediction and experimental evidence provides a word of caution about a strict reliance on transposon mutagenesis experiments and points to the need for further experimental confirmation.

β-Hydroxyacyl dehydratases. Dehydration of the β-hydroxyacyl-ACP intermediate into *trans*-2-enoyl-ACP (in the FAS-II elongation cycle) is catalyzed by β-hydroxyacyl-ACP dehydratase(s) (Fig. 4). In *E. coli*, two such enzymes occur, FabA and FabZ. No recognizable homologues of *fabA* or *fabZ* are found in the *M. tuberculosis* genome (Sacco et al., 2007a). The search for the yeast *S. cerevisiae* dehydratase of mitochondrial FAS-II system by sequence alignments with FabA/FabZ was also unfruitful, and this dehydratase activity was recently demonstrated for a protein belonging rather to the (*R*)-specific hydratase/dehydratase family possessing the hydratase-2 motif (Kastaniotis et al., 2004). Proteins of this family have been recently identified in *M. tuberculosis*. Among these proteins is the product of a putative essential

M. tuberculosis gene (Rv0216) that exhibits a double hotdog (DHD) fold, which is characteristic of a class of enzymes that have thiol esters as substrates (Castell et al., 2005). The Rv0216 structure displays the greatest similarity to bacterial and eukaryotic hydratases that catalyze the *R*-specific hydration of 2-enoyl coenzymeA. Indeed, Rv0216 shows no activity on crotonyl-CoA, and its natural substrate remains to be identified. However, this new structure helped in evaluating other *M. tuberculosis* gene products related to Rv0216 possessing a DHD fold, according to their structure-based sequence alignments. Another candidate protein is the Rv3389c gene product that displays a very well conserved second hotdog domain, clearly indicating *R*-specific enoyl hydratase-like active site. Indeed, the Rv3389c-encoded enzyme was shown to catalyze in vitro the hydration of 2-enoyl-CoA into the β-hydroxyacyl-CoA derivative. Very recently, bioinformatic analyses and an essentiality study allowed the identification of a *M. tuberculosis* candidate protein cluster Rv0635-Rv0636-Rv0637, whose expression in recombinant *E. coli* strains led to the formation of two heterodimers, Rv0635-Rv0636 and Rv0636-Rv0637. Both heterodimers exhibit the enzymatic properties expected for mycobacterial FAS-II dehydratases: a marked specificity for both long chain ($\geq C_{12}$) and ACP-linked substrates. Rv0635-Rv0636 and Rv0636-Rv0637 may therefore be the long sought-after β-hydroxyacyl-ACP dehydratases, and this discovery represents the first description of bacterial monofunctional (3*R*)-hydroxyacyl-ACP dehydratases belonging to the hydratase 2 family (Sacco et al., 2007b).

Fatty acid elongation and formation of double bonds

The presence of *cis* double bonds has been observed in mycolates from all mycolic acid-producing bacteria, from the simplest corynomycolates to the most complex eumycolates. In corynebacteria, the formation of mycolates results from the condensation of two fatty acids, i.e., a palmitic acid (the most abundant in the fatty acid pool) and an oleic acid ($C_{18:1}$*cis*-9, formed by oxidative desaturation of the saturated homologue). The location of the double bond in corynomycolates is in agreement with the involvement of oleic acid in the synthesis of mono- and diethylenic C_{34} and C_{36} corynomycolates, respectively. Little is known about double bond formation of polyunsaturated mycolic acids of *Rhodococcus*, *Nocardia,* and other related mycolata groups.

In mycobacteria, the occurence and positions of *cis* double bonds, which is required for the subsequent introduction of chemical functions by methyltransferases on the meromycolic chain, have generated several hypotheses concerning their formation. The first hypothesis for the formation of the *cis* double bond at the distal position of the meromycolic chain would be explained by the elongation of a *cis-3* enoyl fatty acid (for example a C_{22}/C_{24} fatty acid) resulting from the isomerization of the *trans-2* intermediate in the FAS-II cycle (Fig. 4) (Etemadi and Kolattukudy, 1967; Bloch, 1969). Indeed, some of the dehydratases (FabA but not FabZ) of FAS-II are actually dehydratase/isomerase enzymes and able to catalyze not only the dehydration of β-hydroxyacyl-ACP intermediates but also the isomerization of the *trans-2* enoyl bond into the *cis-3* isomer. The *cis-3*-enoyl intermediate is subsequently not reduced but further elongated by β-ketoacyl-ACP synthase I (FabB) to long-chain monounsaturated fatty acids, thus diverting the nascent acyl-chain into the unsaturated fatty acid synthetic pathway (Rock and Conan, 1996; Marrakchi et al., 2002c). Surprisingly, several groups of bacteria (including mycobacteria) lack both FabA and FabB, although they synthesize unsaturated fatty acids and/or have an anaerobic metabolism. For instance, in *Streptococcus pneumoniae,* the search for another enzyme, in addition to the FabZ dehydratase, that is able to divert fatty acid synthesis to the unsaturated mode led to the discovery of a novel enzyme FabM as a *trans-2-cis-3*-enoyl-ACP isomerase (Marrakchi et al., 2002a). Thus, *S. pneumoniae* uses a combination of FabZ and FabM to introduce the double bond into the growing acyl chain (Marrakchi et al., 2002c). The *fabM* gene was annotated on the *S. pneumoniae* genome as a member of the hydratase/isomerase superfamily (Pfam00378) (Bateman et al., 2000). Candidate proteins belonging to this superfamily have been identified in the *M. tuberculosis* genome, namely EchA10 and EchA11 (Takayama et al., 2005), and may represent potential *trans-2-cis-3*-enoyl-ACP isomerases functioning in a FabM-like manner; yet this hypothesis still needs to be tested.

The presence of Δ^5-unsaturated fatty acids that are structurally similar to the hydrocarbon methyl-terminus of the diethylenic α-mycolates led to the second hypothesis of the existence of a Δ^5-desaturase operating on C_{24} acids to yield $C_{24:1}$, followed by elongation (Asselineau et al., 1970a). Indeed, in various strains of mycobacteria, apart from the abundant oleic and palmitoleic acids, the most frequently encountered fatty acids (such as C_{24} and C_{26}) are accompanied by their analogues with a *cis*-double bond in position 5,6 (Asselineau et al., 1970a). Analysis of the *M. tuberculosis* genome reveals three potential aerobic desaturases encoded by *desA1, desA2* and desA3 (Cole et al., 1998). Overexpression of *desA3* in *M. bovis* BCG and biochemical analysis of the purified enzyme showed that DesA3 (Rv3329c) is responsible for the synthesis of oleic acid (Phetsuksiri et al., 2003). The DesA3 activity thus appears unlikely to be involved in the synthesis of mycolic acids; rather, it most likely corresponds to the desaturase previously described in cell-free extracts of *Mycobacterium phlei* catalyzing the conversion of palmityl-CoA and stearoyl-CoA to their Δ^9-unsaturated derivatives (Fulco and Bloch, 1964). At present, no recognizable mycobacterial Δ^5-desaturase could be identified in the genome; this function may be encoded by another protein family. A third hypothesis, involving the condensation of three common fatty acids (C_{20}, C_{16}, C_{22}) and consistent with the location of the *cis* double bond on the meromycolic chain of α-mycolates, may be considered (Asselineau et al., 2002).

Modification of the Meromycolic Chain

Different types of mycolic acids have been identified in mycobacteria; they differ, in addition to variations in their chain lengths, by the presence of oxygenated functions, cyclopropanes or double bonds at two defined positions, specified as proximal and distal, by reference to the carboxyl group of the meromycolic chain (Fig. 2). Depending on the mycobacterial species, different combinations of mycolic acids are produced (Daffé et al., 1981; Minnikin et al., 1982); for instance, mycolic acids from slow-growing pathogenic mycobacteria members of the *M. tuberculosis* complex possess on their meromycolic chain cyclopropyl rings (in α-mycolates) and cyclopropyl and keto- or methoxy groups. Loss of these specific functions in *M. tuberculosis* deeply affects the virulence and pathogenicity of the mutants, underlying the importance of these modifications (Yuan et al., 1998b; Dubnau et al., 2000; Glickman et al., 2000; Rao et al., 2005, 2006). In contrast, mycolic acids from fast-growing species such as *M. smegmatis* have rather epoxy and/or unsaturations on the meromycolic chain (Fig. 2). Taking advantage of the structural differences between *M. tuberculosis* and *M. smegmatis* mycolates, the heterologous expression of *M. tuberculosis* genes into *M. smegmatis* has led to the discovery of enzymes responsible for methylation, cyclopropanation and introduction of oxygenated groups. These proteins show homology with the cyclopropane fatty acid synthase (CFAS) from *E. coli*, an *S*-adenosylmethionine (SAM)-dependent methyltransferase. More recently, the completion of the *M. tuberculosis* genome sequenc-

ing allowed the identification not only of the CFAS homologue (*ufaA1*, Rv0447c) but also of eight genes encoding putative SAM-dependent methyltransferases dedicated to mycolic acids and named MA-MTs: CmaA1, CmaA2, PcaA (previously UmaA2), MmaA1, MmaA2, MmaA3, MmaA4 (Hma), and UmaA1 (Cole et al., 1998) (Table 1). These methyltransferases transfer the methyl group from SAM to a *cis*-ethylenic precursor to give (1) cyclopropanes and (2) the methyl branch adjacent to both *trans* double bond and *trans* cyclopropane found in mycolates (Fig. 2), probably by a mechanism similar to that proposed by Lederer (Lederer, 1969).

Introduction of cyclopropane functions and methyl groups

Cyclopropane rings are modifications found in membrane lipids of diverse bacterial species. The CFAS of *E. coli* has been extensively studied and shown to introduce a cyclopropane ring into unsaturated fatty acids in the cell membrane (Grogan, 1997). Among the eight MA-MTs identified in *M. tuberculosis*, four display strong sequence similarities with the *E. coli* CFAS: CmaA1, CmaA2, PcaA (UmaA2), and MmaA2, suggesting that these proteins may be involved in cyclopropane formation on the meromycolic chain. Early experiments to investigate the role of CmaA2 used the overexpression strategy. Compared with *M. smegmatis* mycolates (Fig. 2), the expression of *M. tuberculosis* cmaA2 from a multicopy plasmid has led to the production of new types of α-mycolates and epoxymycolates, which possessed a proximal *cis* cyclopropane rather than the regular double bond. Therefore, it was proposed that proximal *cis* cyclopropane of *M. tuberculosis* α-mycolates was introduced by CmaA2 (George et al., 1995). More recently, with the advent of efficient allelic exchange in *M. tuberculosis*, inactivation experiments of the *M. tuberculosis* umaA2 (*un*known *mycolic acid methyltransferase) gene have been performed and showed that the encoded protein is responsible for the formation of the proximal *cis* cyclopropane of *M. tuberculosis* α-mycolates. This gene was thus renamed *pcaA* for *proximal cyclopropanation of *alpha-mycolates (Glickman et al., 2000) (Fig. 2). This unexpected result suggested either an artifact inherent to the overexpression approach employed or, alternatively, an apparent redundancy between *pcaA* and *cmaA2*. This question was elucidated after performing the gene knockout (KO) of *cmaA2* in *M. tuberculosis*. In contrast to the data observed in the heterologous overexpression of the *cmaA2* gene, the KO mutant exhibited a mycolate pattern with modifications exclusively in the oxygenated classes of mycolates (methoxy and keto), where double bonds now occurred instead of the proximal *trans* cyclopropane rings. These results provided evidence that CmaA2 is the *trans*-cyclopropane synthase for the oxygenated mycolates in *M. tuberculosis* (Glickman et al., 2001).

On the other hand, the overexpression of *M. tuberculosis cmaA1* led to the conversion of the distal *cis* double bond into a *cis*-cyclopropane ring in the α-mycolates of *M. smegmatis*. Thus, CmaA1 was proposed to introduce the *cis*-cyclopropane at the distal position (Yuan et al., 1995). Interestingly, more recently, the *cis*-cyclopropanation of α-mycolates has also been attributed to MmaA2, following the report of a *M. tuberculosis* mutant strain inactivated in this gene; mycolate analysis of *mmaA2* null mutants showed the occurrence of double bonds, both in α-mycolates (at the distal position) and in methoxymycolates (at the proximal position), instead of the *cis*-cyclopropanes in these respective positions (Glickman, 2003). Because keto- and methoxymycolates have very likely a filiation (Davidson et al., 1982), it was surprising to observe that despite *mmaA2* inactivation (which reduced the *cis* methoxymycolates content by half), no change in the ratio of *cis/trans* ketomycolates was observed. Therefore, MmaA2 does not seem to be the sole MA-MT responsible for *cis*-cyclopropanation of methoxy- and ketomycolates and for the formation of distal *cis*-cyclopropanes in α-mycolates (Glickman, 2003).

To investigate the role of CmaA1 as an additional MA-MT involved in *cis*-cyclopropanation of α-mycolates, Glickman has constructed a *cmaA1* null mutant and analyzed its mycolate pattern (Glickman, 2003); the α-mycolates of this mutant were found identical to those of the wild-type strain, indicating that CmaA1 is not required for the distal *cis*-cyclopropanation of the α-mycolates, in contrast to a previous report (Yuan et al., 1995). Thus, the CmaA1 function in *M. tuberculosis* still remains to be clarified. Interestingly, comparison of *M. tuberculosis* and *M. leprae* genomes reveals that both *cmaA1* and *mmaA2* are pseudogenes in *M. leprae* (Cole et al., 2001). Because the latter species also possesses *cis/trans* cyclopropanated ketomycolates and *cis*-cyclopropanated α-mycolates (Fig. 2), this observation strongly suggests that the two genes are not necessary for the production of cyclopropanated mycolates in *M. tuberculosis*.

These genetic and biochemical studies have led to important advances in the understanding of how cyclopropanes are introduced on the meromycolic chain and allowed to assign a role for the different CFAS-type MA-MTs, although numerous questions are still pending. For instance, the specific MA-MTs responsible for introducing methyl groups at the proximal position are not yet clearly identified. These methyl groups are adjacent to *trans* cyclopropanes in the oxygenated mycolates of the *M. tuberculosis* complex species and to double bonds in the alpha and epoxymycolates of *M. smegmatis* (Fig. 2). The MmaA1 was proposed to introduce this methyl group for oxygenated mycolates in *M. tuberculosis*: the overexpression of the *mmaA1* gene in *M. tuberculosis* has indeed led to the overproduction of *trans* double bonds or *trans* cyclopropanes in keto- and methoxymycolates, whereas no effect on *M. smegmatis* mycolates was observed when the same gene is overexpressed in this species (Yuan et al., 1997). Recently, the *umaA1* gene KO was performed in both *M. smegmatis* and *M. tuberculosis*, and despite the lack of phenotype in the latter species, the *M. smegmatis* mutant no longer synthesized subtypes of mycolates containing a methyl branch adjacent to either *trans* cyclopropyl group or *trans* double bond at the proximal position of both α- and epoxy-mycolates (Fig. 2). Curiously, however, the *M. tuberculosis umaA1* mutant strain exhibited a hypervirulence phenotype in the severe combined immunodeficient (SCID) mice infection model, pointing to a specific impact of the mutation in an in vivo context (Laval et al., 2007).

Introduction of oxygenated functions

The biosynthesis of both keto- and methoxymycolates is intimately linked to that of the common hydroxymycolate precursor which is catalyzed by the methyltransferase Hma (MmaA4), a SAM-dependent enzyme. The function of Hma has been demonstrated after gene inactivation experiments: the mutant no longer produced keto- or methoxymycolates while still able to synthesize α-mycolates. This result demonstrated that Hma is responsible for both methylation and hydroxylation steps. In fact, like other MA-MTs, Hma is capable of methylation of the *cis* double bond and, specifically in the presence of a water molecule, it catalyzes a concerted hydroxylation. The hydroxylated intermediate is at the branch point to give either the keto- or the methoxymycolate, the latter resulting from the action of the methoxylase MmaA3 (Yuan and Barry, 1996; Dubnau et al., 1997, 1998, 2000; Yuan et al., 1998a; Dinadayala et al., 2003). The α-mycolates of the *hma* mutant strain are of two kinds: having two cyclopropanes (as the parent ones), or bearing a *cis* double bond in the distal position and a *cis* cyclopropane in the proximal part of the molecule. The presence and precise position of this double bond have led to postulate this ethylenic mycolate as the substrate of Hma. However, it has to be noted that the newly synthesized α-mycolate in the *hma* mutant is shorter by four to six carbon atoms compared with the wild-type keto- and methoxymycolates. This discrepancy does not rule out the possibility that the ethylenic molecule be the precursor for Hma because the key transformation may have happened earlier, perhaps at the meromycolate stage. Another hypothesis was postulated for the role of Hma: it would catalyze the introduction of the methyl group on a strongly nucleophilic site such as a β-ketoester generated from a condensation reaction (Lederer, 1969; Asselineau et al., 2002; Dinadayala et al., 2003), the isolation of small amounts of an epoxymycolate devoid of an adjacent methyl branch in the mutant strain would support this interpretation. Recently, the three-dimensional structures of Hma, both in the apoform and in complex with SAM, have been determined (Boissier et al., 2006): this work points out to some differences with the three CFAS-type mycolic acid SAM-methyltransferases previously crystallized (Huang et al., 2002). In particular, it demonstrated some specific features, notably at the α2-α3 structural motif that may play a pivotal role, not only regarding the specific function of this class of enzymes but also in the chemical environment of the active site as no carbonate ion was found for Hma (Boissier et al., 2006).

The genes encoding the proteins involved in the biosynthesis of other oxygenated functions such as wax-, epoxy-, and ω-1 methoxylated mycolic acids remain to be discovered. The *M. tuberculosis* genome contains an abundance of oxidative metabolic pathway genes (Cole et al., 1998). The wax-mycolate has been shown to derive from the keto-mycolic acid by a process of Baeyer-Villiger oxidation consisting in the insertion of molecular oxygen, thus converting the ketone to the corresponding ester (Toriyama et al., 1982), as postulated (Etemadi and Gasche, 1965; Laneelle and Laneelle, 1970). Flavin-containing Baeyer-Villiger monooxygenases use NAD(P)H and molecular oxygen to catalyze the insertion of an oxygen atom into a carbon-carbon bond of a carbonyl substrate. Such enzymes are present in mycobacteria: among these, a flavoprotein monooxygenase, EthA (Rv3854c), which accepts a ketone as substrate, has been recently shown to activate the antituberculous prodrug ETH to its active form (Fraaije et al., 2004). Six additional genes encoding Baeyer-Villiger monooxygenases are present in the *M. tuberculosis* genome. Although ETH has been shown to inhibit the synthesis of oxygenated mycolic acids in *Mycobacterium aurum*, a keto- and wax-mycolate-containing species (Quémard et al., 1992), the annotated genes are not likely to encode the enzymes involved in the oxidation of ketomycolates into

wax-mycolates by a process of Baeyer-Villiger oxidation because *M. tuberculosis* strains are devoid of the latter types of mycolates. It must be noted that a putative Baeyer-Villiger monooxygenase was also detected in clinical isolates of *Mycobacterium avium* (Thierry et al., 1993; Fraaije et al., 2002).

From structural analyses, notably the chain length and stereochemistry of the asymmetric carbons bearing both the methyl branches adjacent to the oxygenated functions as well as those carrying the epoxy-, methoxy- and hydroxyl groups (Daffé et al., 1981; Minnikin et al., 1982; Lacave et al., 1987), it clearly appears that no structural relationship can be established between the epoxy- and the ketomycolates. In contrast, it is reasonable to postulate that the formation of epoxymycolates results from the transformation of an α-mycolate-type intermediate having a *trans* double bond at the distal position with an adjacent methyl branch (Fig. 2).

The ω-1 position of the methoxy group found in a class of mycolates (Fig. 2) raises a new question. Because no adjacent methyl branch to the oxygen function is found in this compound, it is tempting to speculate that the precursor of such product could be an intermediate with a terminal double bond rather than the result of a MmaA3/MmaA4-like process. In this respect, it is interesting to mention that a careful analysis of mycolates from *M. smegmatis* had shown the presence of minor amounts of mycolates with a terminal double bond (Wong and Gray, 1979). Furthermore, *M. smegmatis* was shown to perform ω-1 oxidation of hydrocarbons (Rehm and Reiff, 1981).

Mycolic Acid Condensation

Among the mycolic acid-producing bacteria (*Corynebacterineae*), the *Corynebacterium* species display the shortest and least complex mycolic acids while retaining the characteristic mycolic motif. Since the formation of corynomycolic acid implies only the condensation step, corynebacteria represent an attractive model for studying the mechanism of mycolic condensation. Based on structural considerations, it has been postulated that the C_{32} corynomycolic acid could result from a Claisen condensation between two palmitic acid molecules. This condensation could be explained by two mechanisms: either a condensation of two fatty acyl thioesters or a condensation of palmitoyl-CoA with a carboxylated intermediate, followed by decarboxylation and reduction. In *Corynebacterium diphtheriae*, the use of $[^{14}C]$ palmitic acid allowed to demonstrate the intervention of this precursor in mycolate synthesis in whole cells (Gastambide-Odier and Lederer, 1960). This result was confirmed using a particulate cell-free preparation of *C. diphtheriae* incubated with [1-^{14}C] palmitate, leading to the production of a β-ketoester specifically labeled on carbons 2 and 3 (Walker et al., 1973; Prome et al., 1974). The condensation reaction was inhibited by avidin, indicating the requirement of a carboxylation step (Walker et al., 1973). The condensation products detected in the first seconds in *C. diphtheriae* were always C_{32}-ketoesters linked either to trehalose (Prome et al., 1974; Ahibo-Coffy et al., 1978) or to the phospholipid fraction (Datta and Takayama, 1993). On the other hand, studies using cell-free extracts of *Corynebacterium matruchotii* have led to similar conclusions except that mature corynomycolic acids were directly observed instead of the keto intermediates (Shimakata et al., 1985). In addition, no effect of avidin was observed, whereas cerulenin, a specific inhibitor of β-ketoacyl synthases, inhibited the reaction (Shimakata et al., 1984). What was interpreted as consistent with the above-mentioned conclusion came from labeling experiments with 2,2-^{2}H palmitic acid; the deuterium atom at C-2 position was not conserved in the mature corynomycolic acid, suggesting that palmitate condensation in whole cells of *C. matruchotii* does not implicate a carboxylation step and a mechanism involving a highly activated enolate intermediate has been proposed to explain the reaction (Lee et al., 1997).

The formation of a new carbon-carbon bond by a Claisen condensation is a reaction shared by both fatty acid and polyketide biosyntheses and is catalyzed by condensing enzymes or condensases. Despite extensive efforts in several laboratories, the condensase responsible for synthesis of mycolic acids has remained a mystery for more than 50 years. Given the importance and specificity of this final step, a condensase candidate must be present in all *Corynebacterineae* and conserved in *M. leprae*, the genome of which has a large number of pseudogenes. The polyketide synthase 13 (Pks13), a member of the type I polyketide synthase (Pks) family was identified as the enzyme responsible for the final construction of mycolic acids in both *C. glutamicum* and *M. smegmatis* (Portevin et al., 2004).

The condensing enzyme (Pks13)

The discovery of the condensase for mycolic acid synthesis probably illustrates the remarkable benefit of using the *Corynebacterium* model for exploring the *Mycobacterium* mycolic acid biosynthetic pathway. More than 18 type I *pks* genes are present in the *M. tuberculosis* genome (Cole et al., 1998). However, the search of the *C. diphtheriae* and *C. glutamicum* genomes revealed the occurrence of a single

pks gene conserved in all *Corynebacterineae* species analyzed, leading to the identification of *Mycobacterium* Pks13 as the condensing enzyme (Portevin et al., 2004). In *C. glutamicum*, the inactivation of the *pks13* gene completely abolishes the production of corynomycolates and alters the structure of the envelope. Pks13 is a huge protein displaying five distinct domains corresponding to the catalytic activities required for condensation: peptide carrier protein-like domain (PCP), ketoacyl synthase (KS), acyl transferase (AT), acyl carrier protein domain (ACP), and thioesterase (TE). To be activated, Pks13 needs a post-translational modification, catalyzed by the mycobacterial 4′-phosphopantetheinyl transferase (PptT), which is essential for activation of the ACP and PCP domains (Chalut et al., 2006).

The final "mycolic condensation" involves not only Pks13 but also several enzymes playing a role in the activation of the two condensase substrates. Specifically, activation of the meromycolic chain (through the action of an acyl-AMP ligase) and carboxylation of the alpha branch (by an acyl-CoA carboxylase) are essential for the formation of mycolic acids in *Corynebacterium* and *Mycobacterium* (Portevin et al., 2004, 2005; Gande et al., 2004). These proteins, as well as the condensase, are essential for the viability of mycobacteria and specific for a restricted number of bacterial species. Therefore, they represent new and attractive targets for the development of novel antimycobacterial drugs.

Activation of the meromycolic chain (fatty acyl-AMP ligase)

The *fadD32* gene (Rv3801c) is adjacent to *pks13* (Rv3800c). This genetic organization *fadD32-pks13* is conserved in *M. tuberculosis* and *M. leprae*, and was among the hints suggesting the involvement of the *fadD32*-encoded protein in the condensation reaction. FadD32 catalyzes the production from free fatty acid of acyl-adenylate (acyl-AMP), proposed to be the "activated" substrate for the condensase Pks13 (Trivedi et al., 2004) (Fig. 5). Actually, the *M. tuberculosis* genome displays a significant number (35) of *fadD* genes, which show homology to fatty acyl-CoA synthetases, enzymes converting free fatty acids into acyl-CoA thioesters, the first step in fatty acid degradation. Like *fadD32*, some of these *fadD* are adjacent to *pks* genes on the *M. tuberculosis* genome and, interestingly, were found to encode fatty acyl-AMP ligases rather than having a fatty acyl-CoA ligase activity (Trivedi et al., 2004). These proteins activate long-chain fatty acids as acyl-adenylates, which are then transferred to the multifunctional Pks for further chain extension. The acyl-adenylate synthesized by

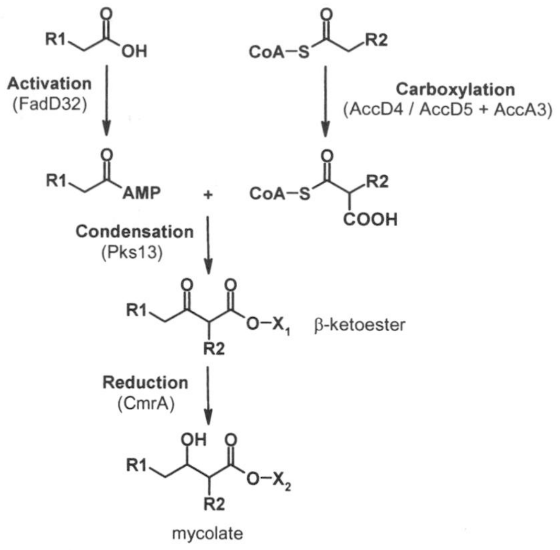

Figure 5. Model of the mycolic acid condensation. R1 and R2 represent long hydrocarbon chains, and X1 and X2 are the unknown intermediate acceptors. The fatty acid molecule bearing R1 is activated as acyl-adenylate (acyl-AMP) by FadD32 to yield the "meromycolate" chain. The other condensation substrate containing the R2 chain is activated by the acyl-CoA carboxylase complex to give the carboxylated intermediate at the origin of the α-branch. Condensation is catalyzed by Pks13, yielding a β-ketoester that gives, after reduction by the ketoacyl reductase CmrA, the mature mycolate.

FadD32 was shown to be transferred onto the condensase Pks13; however, there is no experimental evidence showing that the condensation reaction actually takes place after acyl-adenylate transfer. This mode of activation and transfer of fatty acids is a novel concept and differs from the previously described mechanism involving the formation of acyl-CoA thioesters.

The involvement of FadD32 in mycolic acid synthesis was demonstrated using a *C. glutamicum* knockout mutant in which mycolic acid synthesis was abolished and where severe cell wall impairment were observed (Portevin et al., 2005). Like Pks13 (Portevin et al., 2004), FadD32 was shown to be essential for mycolic acid synthesis and for mycobacterial survival (Portevin et al., 2005).

Activation of the alpha branch (acyl-CoA carboxylase)

The involvement of a carboxylation step in the mycolic acid condensation reaction was reported to vary according to the *Corynebacterium* species studied (as discussed earlier). More recently, the involvement of the AccD4 carboxyltransferase in mycolic acid synthesis (Gande et al., 2004; Portevin et al.,

2005) and its essentiality for mycobacterial survival have been demonstrated (Portevin et al., 2005). The fact that a *fadD32* KO mutation in *C. glutamicum*, which abolishes the synthesis of one of the activated partners for condensation (acyl-AMP), also led to the accumulation of the second condensation substrate (tetradecyl malonic acid) (Portevin et al., 2005) clearly discriminates between the two proposed mechanisms and proves that a carboxylation step, leading to the formation of a malonyl derivative, takes place before the condensation reaction occurs in mycolic acid biosynthesis. An alkyl-malonyl intermediate is also formed in a *C. glutamicum* mutant deficient in the synthesis of Pks13, indicating that carboxylation of the alpha branch chain occurs most likely before the transfer onto the condensase Pks13 (Portevin et al., 2005). The specific involvement of AccD4 in mycolic acid synthesis as well as the role of the other *M. tuberculosis* carboxyltransferases (AccD) is discussed in the section entitled synthesis of malonyl-CoA and carboxyacyl-CoA. The unique specificity of AccD4 for the C_{24}-C_{26} long chains is in agreement with the alpha branch lengths being of 24-26 carbons in mycobacteria and confirms the role of AccD4 in mycolic acid condensation (Oh et al., 2006) (Fig. 5).

Formation of mature mycolic acids and their transfer

Once the condensation is performed, it is not known whether the product of Pks13 is linked to a transporter or released as a thioester before the reduction of the β-keto intermediate takes place. Indeed, the product of Pks13 is the β-ketoester and needs to be reduced to produce the mature mycolate (Fig. 5). It has been shown that a 6-(2-tetradecyl 3-keto octadecanoyl)-α-D-trehalose was detected in the first minute of the synthesis, pointing to the importance of trehalose in the metabolism of corynomycolic acid (Walker et al., 1973; Ahibo-Coffy et al., 1978; Tropis et al., 2005). These data were very recently confirmed in experiments in which *NCgl 2385*, the ortologue of *Rv2509* in *M. tuberculosis*, was inactivated in *C. glutamicum* (Lea-Smith et al., 2007). The resulting mutant synthesizes β-ketoester of trehalose and was deficient in arabinogalactan-linked mycolates. Thus, in addition to identifying the reductase (CmrA), this study indicates that the reduction would take place before the transfer of the mycoloyl residues onto the cell wall arabinogalactan acceptors by the three mycobacterial fibronectin-binding proteins that have been shown to possess the mycoloyl transferase activity in vitro (Belisle et al., 1997) and in vivo (Puech et al., 2000, 2002). Orthologues of these proteins, corynebacterial mycoloyl transferases, have been identified in corynebacteria and functionally characterized (De Sousa-D'Auria et al., 2003; Brand et al., 2003; Kacem et al., 2004).

CONCLUDING REMARKS AND UNRESOLVED QUESTIONS

Even if extensive efforts have been devoted to deciphering the biosynthesis pathway leading to mycolic acids, many relevant questions remain to be solved. For instance, it is still unclear whether FAS-II is the system that carries out the elongation of the acyl chains in vivo until they reach the full-length meromycolates. Actually, FAS-II has been shown to elongate medium-chain-length C_{12} to C_{16} fatty acids to yield C_{18}-C_{30} acyl-ACPs in vitro (Odriozola et al., 1977), which are most likely the precursors of the very long-chain meromycolic acids. The in vitro-observed preference of some of the type II FAS components for longer-chain acyl thioester substrates (KasA, KasB, MabA, and InhA) supports this functional differentiation. Based on the in vitro studies, the FAS-II ketoacyl synthases (KasA and KasB) are involved, with distinct chain length specificities, in the subsequent elongation of AcpM acyl thioesters to C_{40} and C_{54} carbon chains, respectively (Schaeffer et al., 2001b; Slayden and Barry, 2002). Even though fatty acids of up to C_{50} have been shown to be synthesized in cell-free extracts (Qureshi et al., 1984), the extension of C_{18}-C_{30} to 50-60 carbon acyl chains by FAS-II have not been demonstrated formally. Other fatty acid elongation systems (FES) have been reported in mycobacteria (Shimakata et al., 1977; Kikuchi and Kusaka, 1982; Kikuchi et al., 1989) and their possible intervention in providing the mycolic acid precursors cannot be excluded. Finally, no experimental data allow excluding the possibility that the meromycolate chain derives from successive condensations of several fatty acids (Asselineau et al., 2002).

Another unresolved point is related to the elongation/desaturation/modifications that characterize the mature meromycolic acid chain. In *M. tuberculosis*, the characterization of monounsaturated fatty acids ranging from C_{24} to C_{30} with a double bond localized exactly at the position expected for the elongation of a $\Delta 5$ tetracosenoic acid precursor suggested that at least some of the modifications are introduced during the growth of the meromycolic chain (Takayama and Qureshi, 1978); a report supporting this hypothesis has also been published (Yuan et al., 1998a). Nevertheless, conclusive data proving that these modifications are introduced either in the completed mycolic acids or at an earlier stage in their synthesis are still lacking.

Acknowledgment. We are grateful to Sabine Gavalda for her help in preparing the manuscript.

REFERENCES

Ahibo-Coffy, A., H. Aurelle, C. Lacave, J. C. Prome, G. Puzo, and A. Savagnac. 1978. Isolation, structural studies and chemical synthesis of a 'palmitone lipid' from *Corynebacterium diphtheriae. Chem. Phys. Lipids* 22:185–195.

Alberts, A. W., S. G. Gordon, and P. R. Vagelos. 1971. Acetyl CoA carboxylase: the purified transcarboxylase component. *Proc. Natl. Acad. Sci. USA* 68:1259–1263.

Asselineau, C., and J. Asselineau. 1966. Stéréochimie de l'acide corynomycolique. *Bull. Soc. Chim. France* 6:1992–1999.

Asselineau, C., J. Asselineau, G. Laneelle, and M. A. Laneelle. 2002. The biosynthesis of mycolic acids by Mycobacteria: current and alternative hypotheses. *Prog. Lipid Res.* 41:501–523.

Asselineau, C., C. Lacave, H. Montrozier, and J. C. Promé. 1970a. Relations structurales entre les acides mycoliques insaturés et les acides inférieurs insaturés synthétisés par *Mycobacterium phlei*. Implications métaboliques. *Eur. J. Biochem.* 14:406–410.

Asselineau, C., G. Tocanne, and J. F. Tocanne. 1970b. Stéréochimie des acides mycoliques. *Bull. Soc. Chim. France* 4:1455–1459.

Asselineau, J., and E. Lederer. 1950. Structure of the mycolic acids of mycobacteria. *Nature* 166:782–784.

Banerjee, A., E. Dubnau, A. Quemard, V. Balasubramanian, K. S. Um, T. Wilson, D. Collins, G. de Lisle, and W. R. J. Jacobs. 1994. inhA, a gene encoding a target for isoniazid and ethionamide in *Mycobacterium tuberculosis. Science* 263:227–230.

Banerjee, A., M. Sugantino, J. C. Sacchettini, and W. R. J. Jacobs. 1998. The *mabA* gene from the *inhA* operon of *Mycobacterium tuberculosis* encodes a 3-ketoacyl reductase that fails to confer isoniazid resistance. *Microbiology* 144:2697–2704.

Bardou, F., A. Quemard, M. A. Dupont, C. Horn, G. Marchal, and M. Daffe. 1996. Effects of isoniazid on ultrastructure of *Mycobacterium aurum* and *Mycobacterium tuberculosis* and on production of secreted proteins. *Antimicrob. Agents Chemother.* 40:2459–2467.

Barry, C. E., III, R. E. Lee, K. Mdluli, A. E. Sampson, B. G. Schroeder, R. A. Slayden, and Y. Yuan. 1998. Mycolic acids: structure, biosynthesis and physiological functions. *Prog. Lipid Res.* 37:143–179.

Bateman, A., E. Birney, R. Durbin, S. R. Eddy, K. L. Howe, and E. L. Sonnhammer. 2000. The Pfam protein families database. *Nucleic Acids Res.* 28:263–266.

Belisle, J. T., V. D. Vissa, T. Sievert, K. Takayama, P. J. Brennan, and G. S. Besra. 1997. Role of the major antigen of *Mycobacterium tuberculosis* in cell wall biogenesis. *Science* 276:1420–1422.

Bhatt, A., N. Fujiwara, K. Bhatt, S. S. Gurcha, L. Kremer, B. Chen, J. Chan, S. A. Porcelli, K. Kobayashi, G. S. Besra, and W. R. Jacobs, Jr. 2007. Deletion of *kasB* in *Mycobacterium tuberculosis* causes loss of acid-fastness and subclinical latent tuberculosis in immunocompetent mice. *Proc. Natl. Acad. Sci. USA* 104:5157–5162.

Bhatt, A., L. Kremer, A. Z. Dai, J. C. Sacchettini, and W. R. Jacobs, Jr. 2005. Conditional depletion of KasA, a key enzyme of mycolic acid biosynthesis, leads to mycobacterial cell lysis. *J. Bacteriol.* 187:7596–7606.

Bloch, K. 1969. Enzymatic synthesis of monounsaturated fatty acids. *Accounts Chem. Res.* 2:193–202.

Bloch, K. 1977. Control mechanisms for fatty acid synthesis in *Mycobacterium smegmatis. Adv. Enzymol.* 45:1–84.

Bloch, K., and D. Vance. 1977. Control mechanisms in the synthesis of saturated fatty acids. *Annu. Rev. Biochem.* 46:263–298.

Boissier, F., F. Bardou, V. Guillet, S. Uttenweiler-Joseph, M. Daffe, A. Quemard, and L. Mourey. 2006. Further insight into S-adenosylmethionine-dependent methyltransferases: structural characterization of Hma, an enzyme essential for the biosynthesis of oxygenated mycolic acids in *Mycobacterium tuberculosis*. *J. Biol. Chem.* **281**:4434–4445.

Boshoff, H. I., T. G. Myers, B. R. Copp, M. R. McNeil, M. A. Wilson, and C. E. Barry III. 2004. The transcriptional responses of *Mycobacterium tuberculosis* to inhibitors of metabolism: novel insights into drug mechanisms of action. *J. Biol. Chem.* **279**:40174–40184.

Brand, S., K. Niehaus, A. Puhler, and J. Kalinowski. 2003. Identification and functional analysis of six mycolyltransferase genes of *Corynebacterium glutamicum* ATCC 13032: the genes *cop1*, *cmt1*, and *cmt2* can replace each other in the synthesis of trehalose dicorynomycolate, a component of the mycolic acid layer of the cell envelope. *Arch. Microbiol.* **180**:33–44.

Brindley, N., S. Matsumura, and K. Bloch. 1969. *Mycobacterium phlei* fatty acid synthetase-A bacterial multienzyme complex. *Nature* **224**:666–669.

Castell, A., P. Johansson, T. Unge, T. A. Jones, and K. Backbro. 2005. Rv0216, a conserved hypothetical protein from *Mycobacterium tuberculosis* that is essential for bacterial survival during infection, has a double hotdog fold. *Protein Sci.* **14**:1850–1862.

Chalut, C., L. Botella, C. de Sousa-D'Auria, C. Houssin, and C. Guilhot. 2006. The nonredundant roles of two 4'-phosphopantetheinyl transferases in vital processes of Mycobacteria. *Proc. Natl. Acad. Sci. USA* **103**:8511–8516.

Choi, K. H., L. Kremer, G. S. Besra, and C. O. Rock. 2000. Identification and substrate specificity of beta-ketoacyl (acyl carrier protein) synthase III (mtFabH) from *Mycobacterium tuberculosis*. *J. Biol. Chem.* **275**:28201–28207.

Chun, J., L. L. Blackall, S. O. Kang, Y. C. Hah, and M. Goodfellow. 1997. A proposal to reclassify *Nocardia pinensis* Blackall et al. as *Skermania piniformis* gen. nov., comb. nov. *Int. J. Syst. Bacteriol.* **47**:127–131.

Cohen-Gonsaud, M., S. Ducasse, F. Hoh, D. Zerbib, G. Labesse, and A. Quemard. 2002. Crystal structure of MabA from *Mycobacterium tuberculosis*, a reductase involved in long-chain fatty acid biosynthesis. *J. Mol. Biol.* **320**:249–261.

Cole, S., K. Eiglmeier, J. K. Parkhill, N. R. Thomson, P. R. Wheeler, N. Honore, T. Garnier, C. Churcher, D. Harris, K. Mungall, D. Basham, D. Brown, T. Chillingworth, R. Connor, R. M. Davies, K. Devlin, S. Duthoy, T. Feltwell, A. Fraser, N. Hamlin, S. Holroyd, T. Hornsby, K. Jagels, C. Lacroix, J. Maclean, S. Moule, L. Murphy, K. Oliver, M. A. Quail, M. A. Rajandream, K. M. Rutherford, S. Rutter, K. Seeger, S. Simon, M. Simmonds, J. Skelton, R. Squares, S. Squares, K. Stevens, K. Taylor, S. Whitehead, J. R. Woodward, and B. G. Barrell. 2001. Massive gene decay in the leprosy bacillus. *Nature* **409**:1007–1011.

Cole, S. T., R. Brosch, J. Parkhill, T. Garnier, C. Churcher, D. Harris, S. V. Gordon, K. Eiglmeier, S. Gas, C. E. Barry, F. Tekaia, K. Badcock, D. Basham, D. Brown, T. Chillingworth, R. Connor, R. Davies, K. Devlin, T. Feltwell, S. Gentles, N. Hamlin, S. Holroyd, T. Hornsby, K. Jagels, and B. G. Barrell. 1998. Deciphering the biology of *Mycobacterium tuberculosis* from the complete genome sequence. *Nature* **393**:537–544.

Cronan, J. E., Jr., and G. L. Waldrop. 2002. Multi-subunit acetyl-CoA carboxylases. *Prog. Lipid Res.* **41**:407–435.

Daffé, M. 1996. Structure de l'enveloppe de *Mycobacterium tuberculosis*. *Med. Mal. Infect.* **26**:1–7.

Daffé, M. 2005. The cell envelope of corynebacteria, p. 121–148. *In* L. Eggeling and M. Bott (ed.), *Handbook of Corynebacterium glutamicum*. CRC Press, Boca Raton, FL.

Daffé, M., and P. Draper. 1998. The envelope layer of mycobacteria with reference to their pathogenicity. *Adv. Microb. Physiol.* **39**:131–203.

Daffé, M., M. A. Laneelle, C. Asselineau, V. Levy-Frebault, and H. L. David. 1983. Taxonomic value of mycobacterial fatty acids: proposal for a method of analysis. *Ann. Microbiol. (Inst. Pasteur).* **134**:241–256

Daffé, M., M. A. Laneelle, G. Puzo, and C. Asselineau. 1981. Acide mycolique epoxydique : un nouveau type d'acide mycolique. *Tetrahedron Lett.* **22**:4515–4516.

Daniel, J., T. J. Oh, C. M. Lee, and P. E. Kolattukudy. 2007. AccD6, a Member of the Fas II Locus, is a functional carboxyltransferase subunit of the Acyl-CoA carboxylase in *Mycobacterium tuberculosis*. *J. Bacteriol.* **189**:911–917.

Datta, A. K., and K. Takayama. 1993. Biosynthesis of a novel 3-oxo-2-tetradecyloctadecanoate-containing phospholipid by a cell-free extract of *Corynebacterium diphtheriae*. *Biochim. Biophys. Acta* **1169**:135–145.

Davidson, L. A., P. Draper, and D. E. Minnikin. 1982. Studies on the mycolic acids from the walls of *Mycobacterium microti*. *J. Gen. Microbiol.* **128**:823–828.

De Briel, D., F. Couderc, P. Riegel, F. Jehl, and R. Minck. 1992. High-performance liquid chromatography of corynomycolic acids as a tool in identification of *Corynebacterium* species and related organisms. *J. Clin. Microbiol.* **30**:1407–1417.

De Sousa-D'Auria, C., R. Kacem, V. Puech, M. Tropis, G. Leblon, C. Houssin, and M. Daffe. 2003. New insights into the biogenesis of the cell envelope of corynebacteria: identification and functional characterization of five new mycoloyltransferase genes in *Corynebacterium glutamicum*. *FEMS Microbiol. Lett.* **224**:35–44.

Dessen, A., A. Quemard, J. S. Blanchard, W. R. Jacobs, Jr., and J. C. Sacchettini. 1995. Crystal structure and function of the isoniazid target of *Mycobacterium tuberculosis*. *Science* **267**:1638–1641.

Diacovich, L., D. L. Mitchell, H. Pham, G. Gago, M. M. Melgar, C. Khosla, H. Gramajo, and S. C. Tsai. 2004. Crystal structure of the beta-subunit of acyl-CoA carboxylase: structure-based engineering of substrate specificity. *Biochemistry* **43**:14027–14036.

Diacovich, L., S. Peiru, D. Kurth, E. Rodriguez, F. Podesta, C. Khosla, and H. Gramajo. 2002. Kinetic and structural analysis of a new group of Acyl-CoA carboxylases found in *Streptomyces coelicolor* A3(2). *J. Biol. Chem.* **277**:31228–31236.

Dinadayala, P., F. Laval, C. Raynaud, A. Lemassu, M. A. Laneelle, G. Laneelle, and M. Daffe. 2003. Tracking the putative biosynthetic precursors of oxygenated mycolates of *Mycobacterium tuberculosis*: structural analysis of fatty acids of a mutant strain devoid of methoxy- and keto-mycolates. *J. Biol. Chem.* **278**: 7310–7319.

Dubnau, E., J. Chan, C. Raynaud, V. P. Mohan, M. A. Laneelle, K. Yu, A. Quemard, I. Smith, and M. Daffe. 2000. Oxygenated mycolic acids are necessary for virulence of *Mycobacterium tuberculosis* in mice. *Mol. Microbiol.* **36**:630–637.

Dubnau, E., M. A. Laneelle, S. Soares, A. Benichou, T. Vaz, D. Prome, J. C. Prome, M. Daffe, and A. Quemard. 1997. *Mycobacterium bovis* BCG genes involved in the biosynthesis of cyclopropyl keto- and hydroxy-mycolic acids. *Mol. Microbiol.* **23**:313–322.

Dubnau, E., H. Marrakchi, I. Smith, M. Daffe, and A. Quemard. 1998. Mutations in the cmaB gene are responsible for the absence of methoxymycolic acid in *Mycobacterium bovis* BCG Pasteur. *Mol. Microbiol.* **29**:1526–1528.

Ducasse-Cabanot, S., M. Cohen-Gonsaud, H. Marrakchi, M. Nguyen, D. Zerbib, J. Bernadou, M. Daffe, G. Labesse, and A. Quemard. 2004. *In vitro* inhibition of the *Mycobacterium tuberculosis* beta-ketoacyl-acyl carrier protein reductase MabA by isoniazid. *Antimicrob. Agents Chemother.* 48:242–249.

Etemadi, A. H. 1967. Structural and biogenetic correlations of mycolic acids in relation to the phylogenesis of various genera of Actinomycetales. *Bull. Soc. Chim. Biol.* (Paris) 49:695–706.

Etemadi, A. H., and J. Gasche. 1965. On the biogenetic origin of 2-eicosanol and 2-octadecanol of *Mycobacterium avium*. *Bull. Soc. Chim. Biol.* (Paris) 47:2095–2104.

Fernandes, N. D., and P. E. Kolattukudy. 1996. Cloning, sequencing and characterization of a fatty acid synthase-encoding gene from *Mycobacterium tuberculosis* var. bovis BCG. *Gene* 170:95–99.

Fraaije, M. W., N. M. Kamerbeek, A. J. Heidekamp, R. Fortin, and D. B. Janssen. 2004. The prodrug activator EtaA from *Mycobacterium tuberculosis* is a Baeyer-Villiger monooxygenase. *J. Biol. Chem.* 279:3354–3360.

Fraaije, M. W., N. M. Kamerbeek, W. J. van Berkel, and D. B. Janssen. 2002. Identification of a Baeyer-Villiger monooxygenase sequence motif. *FEBS Lett.* 518:43–47.

Fulco, A. J., and K. Bloch. 1964. Cofactor requirements for the formation of Delta-9-unsaturated fatty acids in *Mycobacterium phlei*. *J. Biol. Chem.* 239:993–997.

Gago, G., D. Kurth, L. Diacovich, S. C. Tsai, and H. Gramajo. 2006. Biochemical and structural characterization of an essential acyl coenzyme A carboxylase from *Mycobacterium tuberculosis*. *J. Bacteriol.* 188:477–486.

Gande, R., K. J. Gibson, A. K. Brown, K. Krumbach, L. G. Dover, H. Sahm, S. Shioyama, T. Oikawa, G. S. Besra, and L. Eggeling. 2004. Acyl-CoA carboxylases (*accD2* and *accD3*), together with a unique polyketide synthase (*Cg-pks*), are key to mycolic acid biosynthesis in Corynebacterineae such as *Corynebacterium glutamicum* and *Mycobacterium tuberculosis*. *J. Biol. Chem.* 279:44847–44857.

Gao, L. Y., F. Laval, E. H. Lawson, R. K. Groger, A. Woodruff, J. H. Morisaki, J. S. Cox, M. Daffe, and E. J. Brown. 2003. Requirement for *kasB* in *Mycobacterium* mycolic acid biosynthesis, cell wall impermeability and intracellular survival: implications for therapy. *Mol. Microbiol.* 49:1547–1563.

Gastambide-Odier, M., and E. Lederer. 1960. Biosynthesis of corynomycolic acid from 2 molecules of palmitic acid. *Biochem. Z.* 333:285–295.

George, K. M., Y. Yuan, D. R. Sherman, and C. E. Barry III. 1995. The biosynthesis of cyclopropanated mycolic acids in *Mycobacterium tuberculosis*. Identification and functional analysis of CMAS-2. *J. Biol. Chem.* 270:27292–27298.

Glickman, M. S. 2003. The *mmaA2* gene of *Mycobacterium tuberculosis* encodes the distal cyclopropane synthase of the alpha-mycolic acid. *J. Biol. Chem.* 278:7844–7849.

Glickman, M. S., S. M. Cahill, and W. R. Jacobs, Jr. 2001. The *Mycobacterium tuberculosis cmaA2* gene encodes a mycolic acid *trans*-cyclopropane synthetase. *J. Biol. Chem.* 276:2228–2233.

Glickman, M. S., J. S. Cox, and W. R. Jacobs, Jr. 2000. A novel mycolic acid cyclopropane synthetase is required for cording, persistence, and virulence of *Mycobacterium tuberculosis*. *Mol. Cell* 5:717–727.

Greenspan, M. D., C. H. Birge, G. Powell, W. S. Hancock, and P. R. Vagelos. 1970. Enzyme specificity as a factor in regulation of fatty acid chain length in *Escherichia coli*. *Science* 170:1203–1204.

Grogan, D. W., and J. E. Cronan, Jr. 1997. Cyclopropane ring formation in membrane lipids of bacteria. *Microbiol. Mol. Biol. Rev.* 61:429–441.

Hazbon, M. H., M. Brimacombe, M. Bobadilla del Valle, M. Cavatore, M. I. Guerrero, M. Varma-Basil, H. Billman-Jacobe, C. Lavender, J. Fyfe, L. Garcia-Garcia, C. I. Leon,

M. Bose, F. Chaves, M. Murray, K. D. Eisenach, J. Sifuentes-Osornio, M. D. Cave, A. Ponce de Leon, and D. Alland. 2006. Population genetics study of isoniazid resistance mutations and evolution of multidrug-resistant *Mycobacterium tuberculosis*. *Antimicrob. Agents Chemother.* 50:2640–2649.

Holton, S. J., S. King-Scott, A. N. Eddine, S. H. Kaufmann, and M. Wilmanns. 2006. Structural diversity in the six-fold redundant set of acyl-CoA carboxyltransferases in *Mycobacterium tuberculosis*. *FEBS Lett.* 580:6898–6902.

Huang, C. C., C. V. Smith, M. S. Glickman, W. R. Jacobs, Jr., and J. C. Sacchettini. 2002. Crystal structures of mycolic acid cyclopropane synthases from *Mycobacterium tuberculosis*. *J. Biol. Chem.* 277:11559–11569.

Huang, Y., J. Ge, Y. Yao, Q. Wang, H. Shen, and H. Wang. 2006a. Characterization and site-directed mutagenesis of the putative novel acyl carrier protein Rv0033 and Rv1344 from *Mycobacterium tuberculosis*. *Biochem. Biophys. Res. Commun.* 342:618–624.

Huang, Y. S., J. Ge, H. M. Zhang, J. Q. Lei, X. L. Zhang, and H. H. Wang. 2006b. Purification and characterization of the *Mycobacterium tuberculosis* FabD2, a novel malonyl-CoA:AcpM transacylase of fatty acid synthase. *Protein Expr. Purif.* 45:393–399.

Hughes, M. A., J. C. Silva, S. J. Geromanos, and C. A. Townsend. 2006. Quantitative proteomic analysis of drug-induced changes in mycobacteria. *J. Proteome Res.* 5:54–63.

Kacem, R., C. De Sousa-D'Auria, M. Tropis, M. Chami, P. Gounon, G. Leblon, C. Houssin, and M. Daffe. 2004. Importance of mycoloyltransferases on the physiology of *Corynebacterium glutamicum*. *Microbiology* 150:73–84.

Kastaniotis, A. J., K. J. Autio, R. T. Sormunen, and J. K. Hiltunen. 2004. Htd2p/Yhr067p is a yeast 3-hydroxyacyl-ACP dehydratase essential for mitochondrial function and morphology. *Mol. Microbiol.* 53:1407–1421.

Kikuchi, S., and T. Kusaka. 1982. New malonyl-CoA-dependent fatty acid elongation system in *Mycobacterium smegmatis*. *J. Biochem.* (Tokyo) 92:839–844.

Kikuchi, S., T. Takeuchi, M. Yasui, T. Kusaka, and P. E. Kolattukudy. 1989. A very long-chain fatty acid elongation system in *Mycobacterium avium* and a possible mode of action of isoniazid on the system. *Agric. Biol. Chem.* 53:1689–1698.

Kremer, L., L. G. Dover, S. Carrere, K. M. Nampoothiri, S. Lesjean, A. K. Brown, P. J. Brennan, D. E. Minnikin, C. Locht, and G. S. Besra. 2002. Mycolic acid biosynthesis and enzymic characterization of the beta-ketoacyl-ACP synthase A-condensing enzyme from *Mycobacterium tuberculosis*. *Biochem. J.* 364:423–430.

Kremer, L., K. M. Nampoothiri, S. Lesjean, L. G. Dover, S. Graham, J. Betts, P. J. Brennan, D. E. Minnikin, C. Locht, and G. S. Besra. 2001. Biochemical characterization of acyl carrier protein (AcpM) and malonyl-CoA:AcpM transacylase (mtFabD), two major components of *Mycobacterium tuberculosis* fatty acid synthase II. *J. Biol. Chem.* 276:27967–27974.

Labesse, G., A. Vidal-Cros, J. Chomilier, M. Gaudry, and J.-P. Mornon. 1994. Structural comparisons lead to the definition of a new superfamily of NAD(P)(H)-accepting oxidoreductases: the single-domain reductases/epimerases/dehydrogenases (the RED family). *Biochem. J.* 304:95–99.

Lacave, C., M. A. Laneelle, M. Daffe, H. Montrozier, M. P. Rols, and C. Asselineau. 1987. Structural and metabolic study of the mycolic acids of *Mycobacterium fortuitum*. *Eur. J. Biochem.* 163:369–378.

Lai, C. Y., and J. E. Cronan. 2003. Beta-ketoacyl-acyl carrier protein synthase III (FabH) is essential for bacterial fatty acid synthesis. *J. Biol. Chem.* 278:51494–51503.

Lai, C. Y., and J. E. Cronan. 2004. Isolation and characterization of beta-ketoacyl-acyl carrier protein reductase (*fabG*) mutants of *Escherichia coli* and *Salmonella enterica* serovar Typhimurium. *J. Bacteriol.* **186:**1869–1878.

Laneelle, M. A., and G. Laneelle. 1970. Structure of mycolic acids and an intermediate in the biosynthesis of dicarboxylic mycolic acids. *Eur. J. Biochem.* **12:**296–300.

Laval, F., R. Haites, F. Movahedzadeh, A. Lemassu, C. Y. Wong, N. Stoker, H. Billman-Jacobe, and M. Daffé. 2007. Investigating the function of the putative mycolic-acid methyltransferase UmaA: divergence between the *Mycobacterium smegmatis* and *Mycobacterium tuberculosis* proteins. *J. Biol. Chem.*, in press.

Laval, F., M. A. Laneelle, C. Deon, B. Monsarrat, and M. Daffe. 2001. Accurate molecular mass determination of mycolic acids by MALDI-TOF mass spectrometry. *Anal. Chem.* **73:**4537–4544.

Lea-Smith, D. J., J. S. Pyke, D. Tull, M. J. McConville, R. L. Coppel, and P. K. Crellin. 2007. The reductase that catalyzes mycolic motif synthesis is required for efficient attachment of mycolic acids to arabinogalactan. *J. Biol. Chem.* **282:**11000–11008.

Lederer, E. 1969. Some problems concerning biological C-alkylation reactions and phytosterol biosynthesis. *Q. Rev. Chem. Soc.* **23:**453–481.

Lee, R., J. Armour, K. Takayama, P. Brennan, and G. Besra. 1997. Mycolic acid biosynthesis: Definition and targeting of the Claisen condensation step. *Biochim. Biophys. Acta* **1346:**275–284.

Lin, T. W., M. M. Melgar, D. Kurth, S. J. Swamidass, J. Purdon, T. Tseng, G. Gago, P. Baldi, H. Gramajo, and S. C. Tsai. 2006. Structure-based inhibitor design of AccD5, an essential acyl-CoA carboxylase carboxyltransferase domain of *Mycobacterium tuberculosis*. *Proc. Natl. Acad. Sci. USA* **103:**3072–3077.

Marrakchi, H., K. H. Choi, and C. O. Rock. 2002a. A new mechanism for anaerobic unsaturated fatty acid formation in *Streptococcus pneumoniae*. *J. Biol. Chem.* **277:**44809–44816.

Marrakchi, H., S. Ducasse, G. Labesse, H. Montrozier, E. Margeat, L. Emorine, X. Charpentier, M. Daffe, and A. Quemard. 2002b. MabA (FabG1), a *Mycobacterium tuberculosis* protein involved in the long-chain fatty acid elongation system FAS-II. *Microbiology* **148:**951–960.

Marrakchi, H., G. Laneelle, and A. Quemard. 2000. InhA, a target of the antituberculous drug isoniazid, is involved in a mycobacterial fatty acid elongation system, FAS-II. *Microbiology* **146:**289–296.

Marrakchi, H., Y. M. Zhang, and C. O. Rock. 2002c. Mechanistic diversity and regulation of Type II fatty acid synthesis. *Biochem. Soc. Trans.* **30:**1050–1055.

McNeil, M., M. Daffe, and P. J. Brennan. 1991. Location of the mycolyl ester substituents in the cell walls of mycobacteria. *J. Biol. Chem.* **266:**13217–13223.

Minnikin, D. E., S. M. Minnikin, G. Dobson, M. Goodfellow, F. Portaels, L. van den Breen, and D. Sesardic. 1983. Mycolic acid patterns of four vaccine strains of *Mycobacterium bovis* BCG. *J. Gen. Microbiol.* **129:**889–891.

Minnikin, D. E., S. M. Minnikin, and M. Goodfellow. 1982. The oxygenated mycolic acids of *Mycobacterium fortiutum*, *M. arcinogenes* and *M. senegalense*. *Biochim. Biophys. Acta* **712:**616–620.

Molle, V., A. K. Brown, G. S. Besra, A. J. Cozzone, and L. Kremer. 2006. The condensing activities of the *Mycobacterium tuberculosis* type II fatty acid synthase are differentially regulated by phosphorylation. *J. Biol. Chem.* **281:**30094–30103.

Nishiuchi, Y., T. Baba, H. H. Hotta, and I. Yano. 1999. Mycolic acid analysis in *Nocardia* species. The mycolic acid compositions of *Nocardia asteroides*, *N. farcinica*, and *N. nova*. *J. Microbiol. Methods* **37:**111–122.

Nishiuchi, Y., T. Baba, and I. Yano. 2000. Mycolic acids from *Rhodococcus*, *Gordonia*, and *Dietzia*. *J. Microbiol. Methods* **40:**1–9.

Odriozola, J. M., J. A. Ramos, and K. Bloch. 1977. Fatty acid synthetase activity in *Mycobacterium smegmatis*. Characterization of the acyl carrier protein-dependent elongating system. *Biochim. Biophys. Acta* **488:**207–217.

Oh, T. J., J. Daniel, H. J. Kim, T. D. Sirakova, and P. E. Kolattukudy. 2006. Identification and characterization of Rv3281 as a novel subunit of a biotin-dependent acyl-CoA Carboxylase in *Mycobacterium tuberculosis* H37Rv. *J. Biol. Chem.* **281:**3899–3908.

Parish, T., G. Roberts, F. Laval, M. Schaeffer, M. Daffé, and K. Duncan. 2007. Functional complementation of the essential gene *fabG1* of *Mycobacterium tuberculosis* by *Mycobacterium smegmatis fabG* but not *Escherichia coli fabG*. *J. Bacteriol.* **189:**3721–3728.

Phetsuksiri, B., M. Jackson, H. Scherman, M. McNeil, G. S. Besra, A. R. Baulard, R. A. Slayden, A. E. DeBarber, C. E. Barry III, M. S. Baird, D. C. Crick, and P. J. Brennan. 2003. Unique mechanism of action of the thiourea drug isoxyl on *Mycobacterium tuberculosis*. *J. Biol. Chem.* **278:**53123–53130.

Portevin, D., C. De Sousa-D'Auria, C. Houssin, C. Grimaldi, M. Chami, M. Daffe, and C. Guilhot. 2004. A polyketide synthase catalyzes the last condensation step of mycolic acid biosynthesis in mycobacteria and related organisms. *Proc. Natl. Acad. Sci. USA* **101:**314–319.

Portevin, D., C. de Sousa-D'Auria, H. Montrozier, C. Houssin, A. Stella, M. A. Laneelle, F. Bardou, C. Guilhot, and M. Daffe. 2005. The acyl-AMP ligase FadD32 and AccD4-containing acyl-CoA carboxylase are required for the synthesis of mycolic acids and essential for mycobacterial growth: identification of the carboxylation product and determination of the acyl-CoA carboxylase components. *J. Biol. Chem.* **280:**8862–8874.

Prome, J. C., R. W. Walker, and C. Lacave. 1974. Condensation of two molecules of palmitic acid in *Corynebacterium diphtheriae*: formation of a beta-keto ester of trehalose. *C. R. Acad. Sci. Paris C* **278:**1065–1068.

Puech, V., N. Bayan, K. Salim, G. Leblon, and M. Daffe. 2000. Characterization of the *in vivo* acceptors of the mycoloyl residues transferred by the corynebacterial PS1 and the related mycobacterial antigens 85. *Mol. Microbiol.* **35:**1026–1041.

Puech, V., C. Guilhot, E. Perez, M. Tropis, L. Y. Armitige, B. Gicquel, and M. Daffe. 2002. Evidence for a partial redundancy of the fibronectin-binding proteins for the transfer of mycoloyl residues onto the cell wall arabinogalactan termini of *Mycobacterium tuberculosis*. *Mol. Microbiol.* **44:**1109–1122.

Quémard, A., G. Laneelle, and C. Lacave. 1992. Mycolic acid synthesis: a target for ethionamide in mycobacteria? *Antimicrob. Agents Chemother.* **36:**1316–1321.

Quémard, A., J. C. Sacchettini, A. Dessen, C. Vilcheze, R. Bittman, W. R. Jacobs, Jr., and J. S. Blanchard. 1995. Enzymatic characterization of the target for isoniazid in *Mycobacterium tuberculosis*. *Biochemistry* **34:**8235–8241.

Qureshi, N., N. Sathyamoorthy, and K. Takayama. 1984. Biosynthesis of C_{30} to C_{56} fatty acids by an extract of *Mycobacterium tuberculosis* H37Ra. *J. Bacteriol.* **157:**46–52.

Radmacher, E., L. J. Alderwick, G. S. Besra, A. K. Brown, K. J. Gibson, H. Sahm, and L. Eggeling. 2005. Two functional FAS-I type fatty acid synthases in *Corynebacterium glutamicum*. *Microbiology* **151:**2421–2427.

Rafidinarivo, E., J. C. Prome, and V. Levy-Frebault. 1985. New kinds of unsaturated mycolic acids from *Mycobacterium fallax* sp. nov. *Chem. Phys. Lipids* **36:**215–228.

Rainey, F. A., S. Klatte, R. M. Kroppenstedt, and E. Stackebrandt. 1995. *Dietzia*, a new genus including *Dietzia maris* comb. nov., formerly *Rhodococcus maris*. *Int. J. Syst. Bacteriol.* **45:**32–36.

Rao, V., N. Fujiwara, S. A. Porcelli, and M. S. Glickman. 2005. *Mycobacterium tuberculosis* controls host innate immune activation through cyclopropane modification of a glycolipid effector molecule. *J. Exp. Med.* **201:**535–543.

Rao, V., F. Gao, B. Chen, W. R. Jacobs, Jr., and M. S. Glickman. 2006. *Trans*-cyclopropanation of mycolic acids on trehalose dimycolate suppresses *Mycobacterium tuberculosis*-induced inflammation and virulence. *J. Clin. Investig.* **116:**1660–1667.

Rehm, H. J., and I. Reiff. 1981. Mechanisms and occurence of microbial oxidation of long-chain alkanes. *Adv. Biochem. Eng.* **19:**175–215.

Revill, W. P., M. J. Bibb, and D. A. Hopwood. 1996. Relationships between fatty acid and polyketide synthases from *Streptomyces coelicolor* A3(2): characterization of the fatty acid synthase acyl carrier protein. *J. Bacteriol.* **178:**5660–5667.

Rock, C. O., and J. E. Cronan. 1996. *Escherichia coli* as a model for the regulation of dissociable (type II) fatty acid biosynthesis. *Biochim. Biophys. Acta* **1302:**1–16.

Sacco, E., V. Legendre, F. Laval, D. Zerbib, H. Montrozier, N. Eynard, C. Guilhot, M. Daffe, and A. Quemard. 2007a. Rv3389c from *Mycobacterium tuberculosis*, a member of the (R)-specific hydratase/dehydratase family. *Biochim. Biophys. Acta* **1774:**303–311.

Sacco, E., A. Suarez Covarrubias, H. M. O'Hare, P. Carroll, N. Eynard, T. A. Jones, T. Parish, M. Daffé, K. Bäckbro, and A. Quémard. 2007b. The missing piece of the type II fatty acid synthase system from *Mycobacterium tuberculosis*. *Proc. Natl. Acad. Sci. USA* **104:**14628–14633.

Sassetti, C. M., D. H. Boyd, and E. J. Rubin. 2003. Genes required for mycobacterial growth defined by high density mutagenesis. *Mol. Microbiol.* **48:**77–84.

Sassetti, C. M., and E. J. Rubin. 2003. Genetic requirements for mycobacterial survival during infection. *Proc. Natl. Acad. Sci. USA* **100:**12989–12994.

Schaeffer, M. L., G. Agnihotri, H. Kallender, P. J. Brennan, and J. T. Lonsdale. 2001a. Expression, purification, and characterization of the *Mycobacterium tuberculosis* acyl carrier protein, AcpM. *Biochim. Biophys. Acta* **1532:**67–78.

Schaeffer, M. L., G. Agnihotri, C. Volker, H. Kallender, P. J. Brennan, and J. T. Lonsdale. 2001b. Purification and biochemical characterization of the *Mycobacterium tuberculosis* beta-ketoacyl-acyl carrier protein synthases KasA and KasB. *J. Biol. Chem.* **276:**47029–47037.

Shimakata, T., Y. Fujita, and T. Kusaka. 1977. Acetyl-CoA-dependent elongation of fatty acids in *Mycobacterium smegmatis*. *J. Biochem.* (Tokyo) **82:**725–732.

Shimakata, T., M. Iwaki, and T. Kusaka. 1984. *In vitro* synthesis of mycolic acids by the fluffy layer fraction of *Bacterionema matruchotii*. *Arch. Biochem. Biophys.* **229:**329–339.

Shimakata, T., K. Tsubokura, T. Kusaka, and K. Shizukuishi. 1985. Mass-spectrometric identification of trehalose 6-monomycolate synthesized by the cell-free system of *Bacterionema matruchotii*. *Arch. Biochem. Biophys.* **238:**497–508.

Slayden, R. A., and C. E. Barry III. 2002. The role of KasA and KasB in the biosynthesis of meromycolic acids and isoniazid resistance in *Mycobacterium tuberculosis*. *Tuberculosis* (Edinburgh) **82:**149–160.

Takayama, K., and N. Qureshi. 1978. Isolation and characterization of monounsaturated long chain fatty acids of *Mycobacterium tuberculosis*. *Lipids* **13:**575–579.

Takayama, K., C. Wang, and G. S. Besra. 2005. Pathway to synthesis and processing of mycolic acids in *Mycobacterium tuberculosis*. *Clin. Microbiol. Rev.* **18:**81–101.

Takayama, K., L. Wang, and H. L. David. 1972. Effect of isoniazid on the in vivo mycolic acid synthesis, cell growth, and via-

bility of *Mycobacterium tuberculosis*. *Antimicrob. Agents Chemother.* **2:**29–35.

Thierry, D., V. Vincent, F. Clement, and J. L. Guesdon. 1993. Isolation of specific DNA fragments of *Mycobacterium avium* and their possible use in diagnosis. *J. Clin. Microbiol.* **31:**1048–1054.

Tomiyasu, I., and I. Yano. 1984. Isonicotinic acid hydrazide induced changes and inhibition in mycolic acid synthesis in *Nocardia* and related taxa. *Arch. Microbiol.* **137:**316–323.

Tong, L. 2005. Acetyl-coenzyme A carboxylase: crucial metabolic enzyme and attractive target for drug discovery. *Cell Mol. Life Sci.* **62:**1784-1803.

Toriyama, S., S. Izaizumi, I. Tomiyasu, M. Masui, and I. Yano. 1982. Incorporation of ^{18}O into long-chain secondary alkohols derived from ester mycolic acids in *Mycobacterium phlei*. *Biochim. Biophys. Acta* **712:**427–429.

Trivedi, O. A., P. Arora, V. Sridharan, R. Tickoo, D. Mohanty, and R. S. Gokhale. 2004. Enzymic activation and transfer of fatty acids as acyl-adenylates in mycobacteria. *Nature* **428:**441–445.

Tropis, M., X. Meniche, A. Wolf, H. Gebhardt, S. Strelkov, M. Chami, D. Schomburg, R. Kramer, S. Morbach, and M. Daffe. 2005. The crucial role of trehalose and structurally related oligosaccharides in the biosynthesis and transfer of mycolic acids in Corynebacterineae. *J. Biol. Chem.* **280:**26573–26585.

Veyron-Churlet, R., O. Guerrini, L. Mourey, M. Daffe, and D. Zerbib. 2004. Protein-protein interactions within the Fatty Acid Synthase-II system of *Mycobacterium tuberculosis* are essential for mycobacterial viability. *Mol. Microbiol.* **54:**1161–1172.

Vilcheze, C., H. Morbidoni, T. Weisbrod, H. Iwamoto, M. Kuo, J. Sacchattini, and W. R. Jacobs, Jr. 2000. Inactivation of the *inhA*-encoded fatty acid synthase II (FASII) enoyl-acyl carrier protein reductase induces accumulation of the FASI end products and cell lysis of *Mycobacterium smegmatis*. *J. Bacteriol.* **182:**4059–4067.

Vilcheze, C., F. Wang, M. Arai, M. H. Hazbon, R. Colangeli, L. Kremer, T. R. Weisbrod, D. Alland, J. C. Sacchettini, and W. R. Jacobs, Jr. 2006. Transfer of a point mutation in *Mycobacterium tuberculosis inhA* resolves the target of isoniazid. *Nat. Med.* **12:**1027–1029.

Walker, R. W., J. C. Promé, and C. Lacave. 1973. Biosynthesis of mycolic acids. Formation of a C32-β-ketoester from palmitic acid in a cell-free system of *Corynebacterium diphteriae*. *Biochem. Biophys. Acta* **326:**52–62.

Watanabe, M., Y. Aoyagi, M. Ridell, and D. E. Minnikin. 2001. Separation and characterization of individual mycolic acids in representative mycobacteria. *Microbiology* **147:**1825–1837.

Wheeler, P. R., K. Bulmer, and C. Ratledge. 1990. Enzymes for biosynthesis *de novo* and elongation of fatty acids in mycobacteria grown in host cells: is *Mycobacterium leprae* competent in fatty acid biosynthesis? *J. Gen. Microbiol.* **136:**211–217.

Wong, M. Y., and G. R. Gray. 1979. Structures of the homologous series of monoalkene mycolic acids from *Mycobacterium smegmatis*. *J. Biol. Chem.* **254:**5741–5744.

Yang, J. K., H. J. Yoon, H. J. Ahn, B. I. Lee, S. H. Cho, G. S. Waldo, M. S. Park, and S. W. Suh. 2002. Crystallization and preliminary X-ray crystallographic analysis of the Rv2002 gene product from *Mycobacterium tuberculosis*, a beta-ketoacyl carrier protein reductase homologue. *Acta Crystallogr. D* **58:**303–305.

Yuan, Y., and C. E. Barry. 1996. A common mechanism for the biosynthesis of methoxy and cyclopropyl mycolic acids in *Mycobacterium tuberculosis*. *Proc. Natl. Acad. Sci. USA* **93:**12828–12833.

Yuan, Y., D. C. Crane, J. M. Musser, S. Sreevatsan, and C. E. Barry III. 1997. MMAS-1, the branch point between *cis*- and *trans*-cyclopropane-containing oxygenated mycolates in *Mycobacterium tuberculosis*. *J. Biol. Chem.* **272:**10041–10049.

Yuan, Y., R. E. Lee, G. S. Besra, J. T. Belisle, and C. E. Barry III. 1995. Identification of a gene involved in the biosynthesis of cyclopropanated mycolic acids in *Mycobacterium tuberculosis*. *Proc. Natl. Acad. Sci. USA* **92:**6630–6634.

Yuan, Y., D. Mead, B. G. Schroeder, Y. Zhu, and C. E. Barry III. 1998a. The biosynthesis of mycolic acids in *Mycobacterium tuberculosis*. Enzymatic methyl(ene) transfer to acyl carrier protein bound meromycolic acid *in vitro*. *J. Biol. Chem.* **273:** 21282–21290.

Yuan, Y., Y. Zhu, D. D. Crane, and C. E. Barry III. 1998b. The effect of oxygenated mycolic acid composition on cell wall func-

tion and macrophage growth in *Mycobacterium tuberculosis*. *Mol. Microbiol.* **29:**1449–1458.

Zhang, Y., and J. E. Cronan, Jr. 1996. Polar allele duplication for transcriptional analysis of consecutive essential genes: application to a cluster of *Escherichia coli* fatty acid biosynthetic genes. *J. Bacteriol.* **178:**3614–3620.

Zhang, Y. M., H. Marrakchi, S. W. White, and C. O. Rock. 2003. The application of computational methods to explore the diversity and structure of bacterial fatty acid synthase. *J. Lipid Res.* **44:**1–10.

Zimhony, O., J. S. Cox, J. T. Welch, C. Vilcheze, and W. R. J. Jacobs. 2000. Pyrazinamide inhibits the eukaryotic-like fatty acid synthetase I (FASI) of *Mycobacterium tuberculosis*. *Nat. Med.* **9:**1043–1047.

The Mycobacterial Cell Envelope
Edited by M. Daffé and J.-M. Reyrat
© 2008 ASM Press, Washington, DC

Chapter 5

Cording, Cord Factors, and Trehalose Dimycolate

MICHAEL S. GLICKMAN

This chapter has two purposes. The first is to review the chemistry, biosynthesis, biologic activity, and pathogenic function for trehalose dimycolate (TDM). TDM is also known as "cord factor," a designation of great historical significance in mycobacteriology, but the source of some confusion and controversy. Therefore, the second purpose of this chapter is to review the history, genetic determinants, and biochemical basis of the cording morphology. The historical controversy about the relationship between cording and TDM has been partially clarified by recent studies, which we use as the basis for our attempts to summarize the importance of TDM and cording in mycobacterial pathogenesis.

THE CORDING MORPHOLOGY: HISTORY AND MORPHOLOGIC DEFINITION

Robert Koch, in his original description of *Mycobacterium tuberculosis*, noted the following morphology: "...these bacilli can be found in large numbers. They ordinarily form small groups of cells which are pressed together and arranged in bundles..."(Koch, 1982). Although we do not have direct micrographs of the structures Koch observed, his description is highly suggestive of the cording morphology. *Mycobacterium tuberculosis* cells, when grown in the absence of detergent, align along their long axes. This alignment produces a microscopic structure in which the bacteria form linear strands of tightly packed bacteria that resemble "cords" or strings. Cording must be distinguished from clumping, which is a general property of mycobacteria owing to their hydrophobic surface. Although cording could be seen as a specific pattern of highly ordered clumping, many mycobacteria clump in a random orientation and therefore do not form cords. A picture of *M. tuberculosis* Erdman forming characteristic cords is pictured in Fig. 1.

For decades after Koch's original description, the cording morphology was of mostly historic significance until a series of studies correlated the cording morphology of mycobacterial strains with full and attenuated virulence. These studies demonstrated a positive relationship between the cording morphology and virulence by comparing the strength of cording to the virulence of strains for experimental animals (Fahr, 1956; Middlebrook et al., 1947). Fully virulent strains of *M. tuberculosis* retained full cording strength, quantitated by the amount of detergent required to dissolve cording, whereas an attenuated strain (BCG) had intermediate cording strength and a more attenuated strain (H37Ra) lacked cording. Although these studies did not establish a causal relationship between the cording morphology and a specific virulence trait, as an assayable morphologic phenotype, cording segregated with virulence. Additionally, these studies fit with a larger literature that links colonial morphology of pathogenic bacteria, and its loss through in vitro passage, as a marker of virulence. Similar examples include rough mutants of *Salmonella* which have truncated LPS molecules (Qureshi et al., 1985) and rough variants of Pneumococcus which have lost capsule biosynthesis (Austrian, 1981).

GENETIC DETERMINANTS OF CORDING

Naturally Occurring Cording Mutants

Middlebrook's original description of the correlation between cording and virulence was derived from comparison of *M. tuberculosis* H37Rv, H37Ra, and BCG. BCG and H37Ra are attenuated strains of

Michael S. Glickman • Infectious Diseases Service, Immunology Program, Sloan Kettering Institute, 415 E. 68th Street, Z1605, New York, NY 10065.

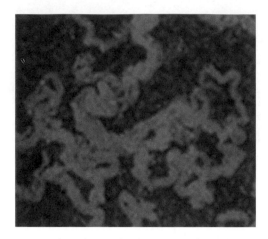

Figure 1. *M. tuberculosis* Erdman grown without detergent and stained with auramine rhodamine. Characteristic cording in which the bacteria align along their long axes to form linear strands is visible.

M. bovis and H37Rv respectively derived from prolonged in vitro passage. Whereas the genetic basis for BCG attenuation has been partially unraveled, the genetic basis for H37Ra severe attenuation is less clear. Nevertheless, the virulence attenuation in both these strains is accompanied by loss of cording, providing naturally occurring cording mutants for analysis. Comparative genomic hybridization of BCG substrains on an *M. tuberculosis* whole genome microarray identified multiple regions of difference (RD), some of which were shared between all BCG strains, and some of which allowed differentiation between BCG substrains (Behr et al., 1999). The RD1 region has received the most attention and is a major attenuating deletion of BCG (Hsu et al., 2003; Lewis et al., 2003), which in part encodes a secretion system for the immunodominant antigens CFP10 and ESAT-6 (Champion et al., 2006; Stanley et al., 2003) (see chapter 12). During the prolonged in vitro passage of *M. bovis* from which BCG was derived, a change in colonial morphology occurred that likely represents cording (see description in Pym et al., 2002). Reintroduction of the RD1 region into BCG-Pasteur altered colony morphology toward a flatter more spreading colony, consistent with restoration of the *M. tuberculosis* cording phenotype (Pym et al., 2002). As such, the RD1 region can be seen as a genetic determinant of cording that was lost during passage of *M. bovis*. However, the genes within the RD1 region responsible for this restoration of cording and whether this phenotype is due to restored synthesis of cell envelope lipids or a primary effect of loss of ESAT-6/CFP10 secretion has not been examined, and awaits further study.

H37Ra is an attenuated derivative of the virulent lab strain H37Rv whose noncording phenotype was included in early descriptions of the cording morphology. Microarray analysis of H37Ra transcriptome compared with H37Rv sought to deduce the transcriptional basis for the cording phenotype (Gao et al., 2004). The study compared transcriptional profiles of H37Rv and H37Ra in five different growth conditions. Genes found to be differentially regulated between H37Rv and H37Ra in all growth conditions were candidate "cording genes." 22 genes were underexpressed in H37Ra compared with H37Rv. Lipid metabolism genes were included in this list, along with many other classes of genes. However, the functional contribution of these genes to cording is unknown without further confirmation studies that delete these candidate genes from H37Rv and show loss of cording, with or without alteration in lipid biosynthesis.

GENETICALLY DEFINED MUTANTS

Despite its strong correlations and the existence of cording mutants derived by in vitro passage, the link between cording and virulence remained unexamined by modern techniques of microbial genetics. As originally suggested by Barksdale and Kim in their landmark review (Barksdale and Kim, 1977), techniques of mutagenesis could be applied to the cording morphology as a screenable phenotype on plates that enhance cording (Lorian, 1966, 1969). With the advent of efficient transposon mutagenesis for slow growing mycobacteria (Bardarov et al., 1997), such a screen was executed in *M. bovis* BCG Pasteur. A noncording mutant of BCG was identified on cord-reading agar and the noncording phenotype confirmed by microscopic examination. The transposon in this strain had interrupted a gene of unknown function (originally annotated at *umaA2*) with substantial amino acid similarity to cyclopropane fatty acid synthase of *E. coli* which was a member of a family of 8 S-adenosylmethionine dependent methyltransferases present in the *M. tuberculosis* chromosome.

Characterization of *umaA2* null mutants of BCG and *M. tuberculosis* revealed a strain that lacked the proximal cyclopropane ring of the alpha mycolic acid. Based on this phenotype, *umaA2* was renamed *pcaA*. This data demonstrated that one genetic determinant of the cording morphology is required for cyclopropanation of mycolic acids and raised the possibility that the fine structure of mycolic acids, rather than their presence or absence, was an important determinant of cording. The *M. tuberculosis pcaA* mutant was attenuated for virulence in mice, providing genetic validation of the link between cording and virulence. The *pcaA* mutant is discussed in more detail below, but this mutant provided the first link between a specific cell surface lipid modification present only in pathogenic mycobacteria and the cording morphology.

kasB encodes a Beta-ketoacyl-AcpM synthase of the multicomponent fatty acid synthase 2 of mycobacteria that is cotranscribed with *kasA* (please see chapter 4 for more details). An *M. marinum kasB* transposon mutant was isolated during a screen for mutants that were impaired for growth in J774 macrophages (Gao et al., 2003). *M. marinum* lacking KasB function had attenuated cording morphology by colonial and microscopic examination. In addition, loss of *kasB* diminished retention of Auramine-Rhodamine upon destaining in the acid fast staining procedure, indicating some relative loss of acid fastness. Lipid analysis of this mutant revealed shorter meromycolate chains by approximately 2 to 4 carbons (Gao et al., 2003), providing a second example of a noncording mutant with altered mycolic acid composition.

A *kasB* deletion mutant of *M. tuberculosis* was recently reported (Bhatt et al., 2007). Loss of *kasB* caused loss of colonial and microscopic cording as well as loss of acid fast staining. Although the *M. tuberculosis kasB* null mutant had shorter mycolic acids, this strain also lacked *trans* cyclopropane rings in the ketomycolates and reduced synthesis of *trans* cyclopropanated methoxymycolate, a partial phenotype of the complete absence of *trans* cyclopropane rings found in the *cmaA2* null mutant (Glickman et al., 2001). In contrast with the *cmaA2* strain, which displays hypervirulence (Rao et al., 2006), the *kasB* strain was highly attenuated in mice as measured by bacterial load and host morbidity (Bhatt et al., 2007). As lack of *trans* cyclopropane rings through loss of *cmaA2* or *mmaA1* does not lead to loss of cording (M. Glickman, unpublished), the cording defect of the *kasB* mutant can be ascribed to the alteration in mycolic chain length, either alone or in combination with trans cyclopropane defect. Thus, in both *M. marinum* and *M. tuberculosis*, loss of *kasB*, which shortens meromycolate chains by 2 to 6 carbons in both esterified and extractable mycolic acids, causes loss of the cording morphology.

All living organisms have regulatory mechanisms to control the composition of their cell membranes, often through the integration of membrane composition with gene expression. The unique structure and chemical composition of the mycobacterial cell envelope suggested that mycobacteria may have unique mechanisms to regulate the lipid composition of the cell wall, especially in response to signals during infection. Rv2869c, along with Rv0359 and Rv2625c, is a member of the M50 class of membrane-bound zinc metalloproteases (Rawlings et al., 2006), which are widely distributed in bacteria (Makinoshima and Glickman, 2006). The founding member of the M50 class is mammalian site two protease (S2P), first described as the second site protease of sterol regulatory element binding protein (SREBP) (Rawson, 1997). SREBPs activate transcription of fatty acid and cholesterol biosynthetic genes after proteolytic release from the membrane by the sequential action of site one protease (S1P) and S2P. Based on the hypothesis that regulated intramembrane proteolysis (RIP) mediated by M50 proteases might regulate mycobacterial cell envelope composition, an *M. tuberculosis* Rv2869c null mutant was constructed by specialized transduction. This mutant displayed an altered cording morphology in both BCG Pasteur and *M. tuberculosis* Erdman (Makinoshima and Glickman, 2005), providing an early confirmation the Rv2869c participated in cell wall biosynthesis. Lipid analysis of this mutant revealed multiple alterations in mycolic acid biosynthesis and phosphatidylinositol mannoside composition. Most striking was the inability of the Rv2869c mutant strain to maintain synthesis of the extractable mycolic acid classes upon transfer to detergent free media, conditions under which cording is measured (Makinoshima and Glickman, 2005). Arabinogalactan esterified mycolates were unaffected. Consistent with its noncording phenotype, the Rv2869c mutant was defective for initial growth in the lungs of mice after aerosol infection and also severely defective for persistence. This combined growth and persistence defect differed from previously described mutants that displayed either a growth or persistence phenotype, but not both (Hingley-Wilson et al., 2003). This data is consistent with a model in which the Rv2869c RIP protease participates in multiple lipid biosynthetic pathways possibly through cleavage of membrane bound transcriptional regulators, as has been described for other prokaryotic site two protease family members (Makinoshima and Glickman, 2006).

Further genetic analyses have revealed that cording can be affected by mutations in genes not directly involved in lipid biosynthesis. Deletion of *phoP*, part of the PhoP/PhoR two component system, caused loss of microscopic and colonial cording in *M. tuberculosis* MT103 (Perez et al., 2001). The *phoP* null mutant was highly attenuated in BALB/c mice and failed to replicate in murine bone marrow derived macrophages (Perez et al., 2001). Analysis of the cell envelope composition of the *phoP* mutant revealed lack of sulfolipid and acylated trehaloses, but preserved mycolic acid profiles (Gonzalo Asensio et al., 2006; Walters et al., 2006). These results indicate that cell wall products other than mycolic acids are necessary for cording even when mycolates are not altered. Therefore, the PhoP/PhoR system is genetically required for cording, likely through its multiple effects on cell envelope lipids, but this cording phenotype is apparently independent of mycolic acids and TDM.

GENETIC DETERMINANTS OF CORDING: SUMMARY

Genetic analyses of the cording morphology in *M. tuberculosis* have identified multiple genes that are necessary for the expression of the cording morphology. Whereas some of these genes are directly involved in lipid biosynthesis or modification, others are upstream regulators of lipid biosynthesis or have no clear relationship to cell envelope composition. Thus, on a genetic basis, cording is a complex polygenic phenotype which can be affected by alteration in multiple different cell envelope lipids, including mycolic acid cyclopropanation and chain length. However, this analysis does not support a model in which a single lipid species (a true "cord factor") is solely responsible for cording. Rather, multiple perturbations in cell envelope composition lead to lack of cording. Although one group of noncording mutant has alterations in mycolic acid biosynthesis which affect TDM, others do not. To date, all of the mutations that abolish cording also attenuate virulence, confirming that the phenotypic association observed by Middlebrook remains valid, although not attributable to a single lipid end product.

"Cord Factor" and its Relationship to the Morphologic Trait of Cording

The link between cording and virulence stimulated interest in identifying bacterial products responsible for cording. Hubert Bloch, working at the Public Health Research Institute in New York City, sought to identify such compounds by directly observing the dissolution of cords with various solvents, including paraffin oil and petroleum ether. He found that petroleum ether rapidly dissolved cording and a petroleum ether extract of cording bacteria contained a substance that inhibited leukocyte migration and was lethal to black mice (Bloch et al., 1953; Sorkin et al., 1952). Chemical characterization of the active compound revealed TDM (Noll et al., 1956; Noll and Bloch, 1955), a glycolipid whose common name became "cord factor" based on its derivation from Bloch's petroleum ether extracts. These studies earned TDM the designation as the first "virulence factor" of *M. tuberculosis* and stimu-

lated intense interest in this glycolipid, which continues to this day. However, enthusiasm for a unique role for TDM in *M. tuberculosis* pathogenesis and cording waned with the later discovery that the chemical motif of TDM was present in all mycobacteria (except for *M. leprae* in which only trehalose monomycolate (TMM) could be detected [Dhariwal et al., 1987]) that contain mycolic acids, including nonpathogenic/noncording strains.

Trehalose Dimycolate: Chemical Structure

TDM is a hydrophobic glycolipid produced by mycobacteria (except possibly *M. leprae*), corynebacteria, and *Nocardia*. The chemical structure of TDM is shown in Fig. 2. Trehalose forms the polar headgroup of TDM and consists of two glucose molecules linked in an α,α configuration at carbon 1. The 6 and 6′ hydroxyl groups of the two glucose molecules are esterified to mycolic acids with both positions being chemically equivalent. Mycolic acids in the mycobacterial cell wall exist in the TDM fraction and esterified to the arabinogalactan, but the chemical structure of mycolic acids on TDM is identical to that found in the esterified mycolic acid layer. The structurally related glycolipid trehalose monomycolate is ubiquitous in all mycobacteria and contains a single mycolic acid chain. Mycolic acid chemical structure has been extensively reviewed and is discussed in chapter 4 and therefore is not described here (Barry et al., 1998). Early studies of TDM isolated from a variety of mycobacteria revealed that the two mycolic acids in TDM can occur in multiple combinations and that the mycolic acid composition of TDM reflects all of the possible 6 and 6′ mycolate combinations using the mycolic acids produced by the mycobacterial strain (Strain et al., 1977). More sophisticated recent analysis using MALDI-TOF mass spectroscopy has contributed greatly to our understanding of the chemical diversity of TDM as it occurs in mycobacteria (Fujita et al., 2005). TDM from *Mycobacterium tuberculosis* H37Rv and Aoyama B contains a complex mixture of mycolic acid subtypes that encompasses all major mycolic acid subspecies and multiple chain length isomers within the major classes. In addition, within the limitation of the technique, intact TDM from these strains contains every combination of alpha, methoxy, and ketomycolate on the 6 and 6′ positions of trehalose. Taking the 5 major mycolic acids in *M. tuberculosis* (alpha, methoxy/*cis*, methoxy/*trans*, keto-*cis*, and keto-*trans*), there are 13 different combinations of TDM within the glycolipid fraction isolated from whole bacilli as TDM. Thus, the fraction TDM as isolated from intact bacteria is actually a complex

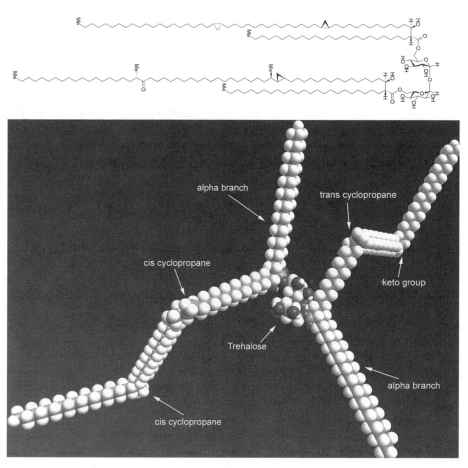

Figure 2. The chemical structure at the top is trehalose dimycolate. Esterified to the trehalose headgroup are two mycolic acids. In this example, *trans* cyclopropanated ketomycolate and alpha mycolate are pictured. The two mycolic acid positions can be occupied by any combination of mycolic acids, making TDM isolated from bacteria a complex mixture of compounds. The space-filling model is of the same TDM pictured on the top with the relevant functional groups indicated.

mixture of glycolipids with dozens of major molecular species. These individual TDM molecules are not available by purification and therefore have not been examined individually for biologic activity. As such, reference to TDM both here and in the literature refers to a complex mixture of compounds which are assayed as a mixture.

Trehalose Dimycolate: Biosynthesis and Physiologic Function: Localization in Envelope

Trehalose dimycolate is extractable from *M. tuberculosis* cells with organic solvents along with multiple other cell envelope lipids and glycolipids. Although this property confirms that TDM is not covalently attached to the cell wall, it does not determine the exact localization of TDM within the envelope. Relatively little information is available about where TDM localizes within the cell wall, although it is commonly stated that it resides in the outer "capsule" layer. One study used varying duration of Tween extraction to as-

sign possible locations to glycolipids within the cell envelope. This study concluded the TDM is not surface exposed but resides in a deeper layer of the capsule (Ortalo-Magne et al., 1996).

ANTIGEN 85 AND THE BIOGENESIS OF TDM

The antigen 85 (Ag85) complex is a group of 3 related proteins (Ag85A, -B, and -C; also known as fbpA, -B, and -C) that share substantial amino acid identity. These proteins were first defined as culture filtrate proteins, are antigens recognized by host immunity, and are included as components of TB vaccines in development (Huygen et al., 1996). A biochemical screen for the molecular basis for mycolic acid transfer from glucose monomycolate (GMM) to free trehalose identified antigen 85 A-C (Belisle et al., 1997). Each of these proteins transferred a mycolic acid from glucose monomycolate to free trehalose or GMM to produce GMM or TDM respectively. This activity was dependent on a conserved carboxylesterase motif serine. Structural analysis of the three antigen 85 isoforms showed a highly conserved active site, consistent with their common catalytic specificity (Ronning et al., 2000; Ronning et al., 2004). Genetic studies confirmed a role for the Ag85 complex in mycolic acid transfer as an *M. tuberculosis* antigen 85c null mutant had 40% less cell wall bound mycolates (Jackson et al., 1999) than wild type, a phenotype not found in the antigen 85A or B mutants (Puech et al., 2002). However, overexpression of Ag85A or B could restore wild type mycolate profiles to the Ag85c mutant (Puech et al., 2002). In addition, analysis of antigen85 in *Corynebacterium* suggested that these genes may act to transfer mycolate directly to the cell wall fraction without affecting TDM and TMM (Puech et al., 2000). In summary, the antigen85 complex possesses the in vitro biochemical activity for a key role in TDM synthesis. Genetic definition of their function has been difficult because of functional redundancy but generally supports a role for these proteins in TDM biogenesis. A null mutant for all three antigen85 genes in *M. tuberculosis* may help clarify the role of these genes in TDM and esterified mycolic acid biosynthesis, although such a mutant may be nonviable.

TREHALOSE DIMYCOLATE: ROLE IN PATHOGENESIS

Biophysical Models Which Link TDM to Cording

TDM is an amphipathic hydrophobic glycolipid which forms micelles in aqueous solution. Early studies of TDM noted that the biologic activity of the glycolipid was affected by the mode of administration. TDM emulsified in oil retained biologic activity, while micellar suspensions did not (Behling et al., 1993; Retzinger et al., 1982). Furthermore, TDM administered on the surface of beads was active, as was TDM presented as a monolayer dried from organic solvents. These observations suggested that the conformation of TDM was important for recognition by mammalian cells, either because of a specific receptor or because of a nonspecific detergent activity. Attempts to examine the physical conformation of TDM using scanning electron microscopy led to a model of the TDM monolayer in which periodically spaced trehalose head groups are separated by mycolic acids with a thickness of 29 angstroms. Beads coated with TDM aggregate to form a morphology reminiscent of cording (Behling et al., 1993).

Activities of Trehalose Dimycolate for Mammalian Cells

Since its discovery, TDM has generated interest due to its inflammatory activity for animal cells, adjuvant activity, and potential role as a virulence determinant of *Mycobacterium tuberculosis*. A plethora of studies have used purified TDM from various mycobacteria to stimulate macrophages, inject into mice, cause tumor regression, and induce immune responses. These studies are too numerous to be cited individually here, but we will highlight representative examples. For further discussion of this topic, the reader is directed to recent reviews (Hunter et al., 2006; Ryll et al., 2001).

Cytokine Release and Granuloma Formation

TDM from *M. tuberculosis* stimulates macrophages to produce TNF, IL-12, MCP-1, IL-6 and IL-1 (Behling et al., 1993; Rao et al., 2006; Silva, 1989; Rao et al., 2005; Indrigo et al., 2003). Induction of these inflammatory cytokines and chemokines produce granulomatous inflammation in mice (Bekierkunst, 1968; Yarkoni and Rapp, 1977; Perez et al., 2000; Hunter et al., 2006), rabbits (Hamasaki et al., 2000), and guinea pigs (Sugawara et al., 2002). The granuloma forming property of TDM was an early indicator of the importance of this glycolipid in *M. tuberculosis*-induced immunopathology. Multiple studies used TDM to immunize mice against challenge with virulent *M. tuberculosis*, either alone or in combination with specific antigens or other microbial products (Lima et al., 2003; Masihi et al., 1984; Ribi et al., 1982).

STRUCTURE FUNCTION STUDIES OF TREHALOSE DIMYCOLATE INFLAMMATORY ACTIVITY

TDM can be isolated from all mycobacteria that synthesize mycolic acids. However, the two mycolic acids on TDM reflect the mycolate composition of the mycobacterial strain from which the TDM is isolated. As such, the chemical structure of TDM differs substantially between strains because the mycolic acid composition of mycobacterial strains differs. These strain differences provide a natural source of chemically distinct TDM mixtures that can be tested for biologic activity. Some studies have compared the activity of TDM isolated from distinct mycobacteria. In a comparison of TDM from *M. tuberculosis*, *M. bovis* BCG (no substrain specified), and *M. kansasii*,

M. kansasii TDM was most potent for granuloma induction in mice followed by *M. tuberculosis* and BCG (Bekierkunst et al., 1969). In contrast, another study found BCG TDM to be more potent than *M. kansasii* TDM for lung inflammation as measured by organ weight (Baba et al., 1997). Using protection from infectious challenge as the readout for TDM potency, *M. tuberculosis* and *M. bovis* TDMs were significantly more potent than *M. avium* TDM in protecting mice against influenza and Toxoplasma challenge (Masihi et al., 1985). Taken together, these studies suggested that the mycolic acid composition of TDM could influence the inflammatory activity of the glycolipid, but the specific lipid structures responsible for these differences could not be deduced due to the complexity of the differences in mycolic acid structures between strains and the multiple combinations of mycolic acids in any TDM mixture.

Cyclopropanation of Trehalose Dimycolate and Inflammatory Activity

Recent analyses of *M. tuberculosis* strains deficient for mycolic acid cyclopropanation have provided strong evidence that the mycolic acid composition of TDM is a major determinant of its inflammatory activity and strongly implicated TDM as an immunomodulatory virulence determinant of *M. tuberculosis*. *M. tuberculosis* mycolic acids, along with those of *M. bovis*, *M. avium*, and *M. leprae*, are modified by cyclopropane rings in positions occupied by double bonds in nonpathogenic mycobacteria (see chapter 4). These cyclopropane rings are synthesized by a family of *S*-adenosylmethionine dependent methyltransferases encoded in the *M. tuberculosis* chromosome. Genetic analysis of this methyltransferase family has revealed their biosynthetic role in mycolic acid modification (Glickman et al., 2001; Glickman, 2003; Glickman et al., 2000; Dubnau et al., 2000). Deletion of the methyltransferase *pcaA* from *M. tuberculosis* Erdman attenuates *M. tuberculosis* in mice and causes a 50-fold reduction in bacterial titers 1 week after aerosol infection (Rao et al., 2005). The *pcaA* mutant also displays a mild persistence defect late in infection and attenuated immunopathology. In contrast, deletion of *cmaA2*, which is necessary for the *trans*-cyclopropane modification of methoxy and ketomycolates, causes hypervirulence in *M. tuberculosis* Erdman without a change in bacterial titers. These two mutants, both defective for cyclopropane modification, but differing in the stereochemistry and position of the cyclopropane ring affected, have in vivo phenotypes that are opposite in terms of virulence and severity of immunopathology. These results

led to the hypothesis that TDM from these strains was a direct mediator of pathogenesis and that the cyclopropane content of the glycolipid was important for inflammatory activity.

To test this hypothesis, TDM from the *pcaA* and *cmaA2* mutants was examined for potency in macrophages and mice. TDM isolated from the *pcaA* mutant strain was four-fold less potent than wild type or complemented mutant derived TDM for TNF release from macrophages (Rao, 2005). In contrast, *cmaA2* mutant-derived TDM was fivefold more potent than wild type or complemented mutant-derived TDM (Rao et al., 2006). Lipid transfer experiments demonstrated that the hypo- (*pcaA*) and hyper- (*cmaA2*) inflammatory phenotypes of the mutant strains could be transferred to delipidated wild-type cells with petroleum ether extractable lipids. These experiments provided direct evidence that the extractable lipids are sufficient for the early inflammatory activity for *M. tuberculosis* for macrophages and that the cyclopropane content of TDM has both stimulatory and suppressive effects on innate immune activation. These studies provide a new perspective on the role of TDM in pathogenesis and demonstrate that the specific chemical composition of TDM can be critical to its inflammatory activity. Nevertheless, these studies still examine TDM isolated as a mixture and do not identify the specific TDM molecule responsible for suppression or activation of innate immunity. One hypothesis that emerges from the *cmaA2* study is that the *trans* cyclopropane ring directly suppresses macrophage activation and its lack in the *cmaA2* mutant leads to hyperinflammation, whereas its overabundance in the *pcaA* mutant (in ketomycolate) leads to hypoinflammation. Alternatively, it is possible that the hypoinflammatory effect of *pcaA* TDM is attributable to loss of an inflammatory activity through the *cis* cyclopropane ring of the alpha mycolate. Definitive examination of these ideas awaits total synthesis of defined TDM molecules and construction of combined cyclopropane synthase mutant strains.

HOST RECOGNITION OF TREHALOSE DIMYCOLATE

The broad inflammatory activity of TDM and its granuloma forming activity indicate that this glycolipid may be recognized specifically by host receptors as foreign microbial product. Despite decades of interest in TDM and substantial recent advances in our understanding of the molecular recognition of microbial products (Beutler et al., 2006), very little information is available about the molecular basis for host cell recognition of TDM.

CD1

CD1 is an antigen-presenting molecule that presents glycolipid antigens to T cells (Dutronc and Porcelli, 2002). There are four CD1 isoforms in humans but only one in mice. Many CD1-presented lipid antigens are mycobacterial lipids, including mycolic acids, isoprenoids (Moody et al., 2000), and phosphatidylinositol mannosides (Fischer et al., 2004). Glucose monomycolate is presented by CD1b to T-cells, an activity that is exquisitely sensitive to changes in the sugar but insensitive to polymorphism in the lipid chain (Moody et al., 1997). The crystal structure of CD1b with bound lipids demonstrated hydrophobic channels which accommodate two alkyl chains such as are found in the meromycolate chain and alpha branch of mycolic acids. Under the present model of CD1 lipid binding, the 4 alkyl chains of TDM should not bind the CD1 grooves (Batuwangala et al., 2004; Gadola et al., 2002). CD1 does not present TDM and dendritic cells are unable to hydrolyze the alpha glycosidic bond of trehalose to generate GMM in vivo, at least under the conditions tested (Moody et al., 1999). To investigate whether CD1 has any role in murine response to TDM, CD1-deficient mice challenged with *M. smegmatis* TDM on beads develop pulmonary granulomas twice as large as wild type mice (Actor et al., 2001). Whether this effect represents a direct physical interaction between CD1 and TDM is unknown and awaits further study.

TLR/MyD88

The Toll-like receptors (TLRs) are a recently described family of proteins which are genetically required for cellular inflammatory response to a broad range of microbial products including LPS, flagellin, double stranded RNA, and lipoproteins, among others. Several studies have asked whether TLRs and their adapter protein MyD88 play a role in recognizing mycobacterial cell wall products, including TDM. Using intraperitoneal implantation of matrigel-containing macrophages and TDM on beads, Russell and colleagues analyzed cellular recruitment and cytokine production. TDM recruited a mixed inflammatory cell population with a prominent neutrophil component (Geisel et al., 2005). Deficiency of TLR2 or TLR4 had no effect on TDM stimulated cellular recruitment. In contrast, MyD88 deficiency abolished cellular recruitment as well as IL-1, IL-6, and TNF production in the matrix (Geisel et al., 2005). Other authors examined host immune requirements for TDM-induced granuloma formation in mice and found no effect of deficiency of Interferon gamma, but granuloma formation was abolished by coadmin-istration of an anti-TNF monoclonal antibody (Takimoto et al., 2006).

TDM is active as an adjuvant and is a component of the Ribi adjuvant system (RAS), named for Edgar Ribi, a pioneer in the investigation of the adjuvant properties of microbial products. The TDM in the Ribi adjuvant is a synthetic dicorynomycolate. In addition to TDM, Ribi adjuvant contains monophosphoryl lipid A (MPL), which is a truncated LPS derivative originally isolated from a rough mutant of *Salmonella minnesota* (Johnson et al., 1990; Qureshi et al., 1985). MPL has lost potency to induce endotoxemia but retains immunostimulant activity (Carpati et al., 1992), which is transduced by TLR4. A recent study examined the innate immune pathways required for RAS adjuvant activity. Ribi adjuvant retains its activity for stimulating antibody production in mice doubly deficient for MyD88 and TRIF (TIR domain-containing adapter inducing IFN-beta [Gavin et al., 2006]). Because MyD88/TRIF mutant mice lack all known TLR signaling and MPL signals through TLR4, the retained adjuvant activity of RAS in these mice is attributable to either TDM or the oil vehicle. This data suggests that the adjuvant activity of TDM, as measured by antibody production, is independent of TLR pathway signaling. Given the emerging role of TDM in the pathogenesis of *M. tuberculosis* infection, identification of the molecular basis for host cell TDM recognition is an area of fertile future investigation.

SUMMARY

Middlebrook's original observation that the degree of cording correlates with virulence has remained valid through recent genetic analyses of non-cording mutant strains. Cording remains a reliable predictor of attenuation and in many cases is due to loss of specific cell envelope lipids. TDM, the original cord factor, is still legitimately linked to cording, as alteration in mycolic acid structure in the extractable mycolic acids lead to loss of cording. In these examples, it is the fine structure of the mycolic acids on TDM which determine cording and TDM inflammatory activity, not the presence or absence of the glycolipid. However, TDM is not the only cording factor, as multiple alterations in cell envelope structure can lead to loss of cording. Thus, it is most accurate to say that TDM is a cording factor among many such factors but that TDM is emerging as a major determinant of *M. tuberculosis*-induced inflammatory pathology. Forty-eight years after Bloch's original description of TDM and 60 years after Middlebrook's description of the cording-virulence connection, both of these discoveries remain highly relevant to our present conceptions of cell envelope lipids in pathogenesis. Some of the major unanswered questions that arise from recent work which will be the subject of future investigations are: What are the properties of individual subspecies of TDM with defined mycolic acid combinations? Do these single molecule TDMs differ from bulk TDM isolated as a mixture? What is the molecular basis for host cell recognition of TDM? Is the molecular composition of TDM dynamic during infection in response to host signals? What is the full set of genetic determinants of the cording morphology?

With these major questions still unanswered, TDM and cording will continue to fascinate *M. tuberculosis* cell envelope investigators for decades.

REFERENCES

Actor, J. K., M. Olsen, R. L. Hunter, Jr., and Y. J. Geng. 2001. Dysregulated response to mycobacterial cord factor trehalose-6,6'-dimycolate in CD1D-/- mice. *J. Interferon Cytokine Res.* 21:1089–1096.

Austrian, R. 1981. Some observations on the pneumococcus and on the current status of pneumococcal disease and its prevention. *Rev. Infect. Dis.* 3(Suppl.):S1–17.

Baba, T., Y. Natsuhara, K. Kaneda, and I. Yano. 1997. Granuloma formation activity and mycolic acid composition of mycobacterial cord factor. *Cell. Mol. Life Sci.* 53:227–232.

Bardarov, S., J. Kriakov, C. Carriere, S. Yu, C. Vaamonde, R. A. McAdam, B. R. Bloom, G. F. Hatfull, and W. R. Jacobs, Jr. 1997. Conditionally replicating mycobacteriophages: a system for transposon delivery to *Mycobacterium tuberculosis*. *Proc. Natl. Acad. Sci. USA* 94:10961–10966.

Barksdale, L., and K. S. Kim. 1977. Mycobacterium. *Bacteriol. Rev.* 41:217–372.

Barry, C. E., 3rd, R. E. Lee, K. Mdluli, A. E. Sampson, B. G. Schroeder, R. A. Slayden, and Y. Yuan. 1998. Mycolic acids: structure, biosynthesis and physiological functions. *Prog. Lipid Res.* 37:143–179.

Batuwangala, T., D. Shepherd, S. D. Gadola, K. J. Gibson, N. R. Zaccai, A. R. Fersht, G. S. Besra, V. Cerundolo, and E. Y. Jones. 2004. The crystal structure of human CD1b with a bound bacterial glycolipid. *J. Immunol.* 172:2382–2388.

Behling, C. A., B. Bennett, K. Takayama, and R. L. Hunter. 1993. Development of a trehalose 6,6'-dimycolate model which explains cord formation by *Mycobacterium tuberculosis*. *Infect. Immun.* 61:2296–2303.

Behling, C. A., R. L. Perez, M. R. Kidd, G. W. Staton, Jr., and R. L. Hunter. 1993. Induction of pulmonary granulomas, macrophage procoagulant activity, and tumor necrosis factor-alpha by trehalose glycolipids. *Ann. Clin. Lab. Sci.* 23:256–266.

Behr, M. A., M. A. Wilson, W. P. Gill, H. Salamon, G. K. Schoolnik, S. Rane, and P. M. Small. 1999. Comparative genomics of BCG vaccines by whole-genome DNA microarray. *Science* 284:1520–1523.

Bekierkunst, A. 1968. Acute granulomatous response produced in mice by trehalose-6,6-dimycolate. *J. Bacteriol.* 96:958–961.

Bekierkunst, A., I. S. Levij, E. Yarkoni, E. Vilkas, A. Adam, and E. Lederer. 1969. Granuloma formation induced in mice by chemically defined mycobacterial fractions. *J. Bacteriol.* 100:95–102.

Belisle, J. T., V. D. Vissa, T. Sievert, K. Takayama, P. J. Brennan, and G. S. Besra. 1997. Role of the major antigen of *Mycobacterium tuberculosis* in cell wall biogenesis. *Science* 276:1420–1422.

Beutler, B., Z. Jiang, P. Georgel, K. Crozat, B. Croker, S. Rutschmann, X. Du, and K. Hoebe. 2006. Genetic analysis of host resistance: Toll-like receptor signaling and immunity at large. *Annu. Rev. Immunol.* 24:353–389.

Bhatt, A., N. Fujiwara, K. Bhatt, S. S. Gurcha, L. Kremer, B. Chen, J. Chan, S. A. Porcelli, K. Kobayashi, G. S. Besra, and W. R. Jacobs, Jr. 2007. Deletion of kasB in *Mycobacterium tuberculosis* causes loss of acid-fastness and subclinical latent tuberculosis in immunocompetent mice. *Proc. Natl. Acad. Sci. USA* 104:5157–5162.

Bloch, H., E. Sorkin, and H. Erlenmeyer. 1953. A toxic lipid component of the tubercle bacillus (cord factor). I. Isolation from petroleum ether extracts of young bacterial cultures. *Am. Rev. Tuberc.* 67:629–643.

Carpati, C. M., M. E. Astiz, E. C. Rackow, J. W. Kim, Y. B. Kim, and M. H. Weil. 1992. Monophosphoryl lipid A attenuates the effects of endotoxic shock in pigs. *J. Lab. Clin. Med.* 119:346–353.

Champion, P. A., S. A. Stanley, M. M. Champion, E. J. Brown, and J. S. Cox. 2006. C-terminal signal sequence promotes virulence factor secretion in *Mycobacterium tuberculosis*. *Science* 313:1632–1636.

Darzins, E., and G. Fahr. 1956. Cord-forming property, lethality and pathogenicity of mycobacteria. *Dis. Chest* 30:642–648.

Dhariwal, K. R., Y. M. Yang, H. M. Fales, and M. B. Goren. 1987. Detection of trehalose monomycolate in *Mycobacterium leprae* grown in armadillo tissues. *J. Gen. Microbiol.* 133:201–209.

Dubnau, E., J. Chan, C. Raynaud, V. P. Mohan, M. A. Laneelle, K. Yu, A. Quemard, I. Smith, and M. Daffe. 2000. Oxygenated mycolic acids are necessary for virulence of *Mycobacterium tuberculosis* in mice. *Mol. Microbiol.* 36:630–637.

Dutronc, Y., and S. A. Porcelli. 2002. The CD1 family and T cell recognition of lipid antigens. *Tissue Antigens* 60:337–353.

Fischer, K., E. Scotet, M. Niemeyer, H. Koebernick, J. Zerrahn, S. Maillet, R. Hurwitz, M. Kursar, M. Bonneville, S. H. Kaufmann, and U. E. Schaible. 2004. Mycobacterial phosphatidylinositol mannoside is a natural antigen for CD1d-restricted T cells. *Proc. Natl. Acad. Sci. USA* 101:10685–10690.

Fujita, Y., T. Naka, M. R. McNeil, and I. Yano. 2005. Intact molecular characterization of cord factor (trehalose 6,6'-dimycolate) from nine species of mycobacteria by MALDI-TOF mass spectrometry. *Microbiology* 151:3403–3416.

Gadola, S. D., N. R. Zaccai, K. Harlos, D. Shepherd, J. C. Castro-Palomino, G. Ritter, R. R. Schmidt, E. Y. Jones, and V. Cerundolo. 2002. Structure of human CD1b with bound ligands at 2.3 A, a maze for alkyl chains. *Nat. Immunol.* 3:721–726.

Gao, L. Y., F. Laval, E. H. Lawson, R. K. Groger, A. Woodruff, J. H. Morisaki, J. S. Cox, M. Daffe, and E. J. Brown. 2003. Requirement for kasB in Mycobacterium mycolic acid biosynthesis, cell wall impermeability and intracellular survival: implications for therapy. *Mol. Microbiol.* 49:1547–1563.

Gao, Q., K. Kripke, Z. Arinc, M. Voskuil, and P. Small. 2004. Comparative expression studies of a complex phenotype: cord formation in *Mycobacterium tuberculosis*. *Tuberculosis* (Edinburgh) 84:188–196.

Gavin, A. L., K. Hoebe, B. Duong, T. Ota, C. Martin, B. Beutler, and D. Nemazee. 2006. Adjuvant-enhanced antibody responses in the absence of toll-like receptor signaling. *Science* 314:1936–1938.

Geisel, R. E., K. Sakamoto, D. G. Russell, and E. R. Rhoades. 2005. In vivo activity of released cell wall lipids of *Mycobacterium bovis* bacillus Calmette-Guerin is due principally to trehalose mycolates. *J. Immunol.* **174**:5007–5015.

Glickman, M. S. 2003. The mmaA2 gene of *Mycobacterium tuberculosis* encodes the distal cyclopropane synthase of the alpha-mycolic acid. *J. Biol. Chem.* **278**:7844–7849.

Glickman, M. S., S. M. Cahill, and W. R. Jacobs, Jr. 2001. The *Mycobacterium tuberculosis* cmaA2 gene encodes a mycolic acid *trans* cyclopropane synthetase. *J. Biol. Chem.* **276**:2228–2233.

Glickman, M. S., J. S. Cox, and W. R. Jacobs, Jr. 2000. A novel mycolic acid cyclopropane synthetase is required for cording, persistence, and virulence of *Mycobacterium tuberculosis*. *Mol. Cell* **5**:717–727.

Gonzalo Asensio, J., C. Maia, N. L. Ferrer, N. Barilone, F. Laval, C. Y. Soto, N. Winter, M. Daffe, B. Gicquel, C. Martin, and M. Jackson. 2006. The virulence-associated two-component PhoP-PhoR system controls the biosynthesis of polyketide-derived lipids in *Mycobacterium tuberculosis*. *J. Biol. Chem.* **281**:1313–1316.

Hamasaki, N., K. Isowa, K. Kamada, Y. Terano, T. Matsumoto, T. Arakawa, K. Kobayashi, and I. Yano. 2000. In vivo administration of mycobacterial cord factor (Trehalose 6, 6'-dimycolate) can induce lung and liver granulomas and thymic atrophy in rabbits. *Infect. Immun.* **68**:3704–3709.

Hingley-Wilson, S. M., V. K. Sambandamurthy, and W. R. Jacobs, Jr. 2003. Survival perspectives from the world's most successful pathogen, *Mycobacterium tuberculosis*. *Nat. Immunol.* **4**:949–955.

Hsu, T., S. M. Hingley-Wilson, B. Chen, M. Chen, A. Z. Dai, P. M. Morin, C. B. Marks, J. Padiyar, C. Goulding, M. Gingery, D. Eisenberg, R. G. Russell, S. C. Derrick, F. M. Collins, S. L. Morris, C. H. King, and W. R. Jacobs, Jr. 2003. The primary mechanism of attenuation of bacillus Calmette-Guerin is a loss of secreted lytic function required for invasion of lung interstitial tissue. *Proc. Natl. Acad. Sci. USA* **100**:12420–12425.

Hunter, R. L., M. Olsen, C. Jagannath, and J. K. Actor. 2006. Trehalose 6,6'-dimycolate and lipid in the pathogenesis of caseating granulomas of tuberculosis in mice. *Am. J. Pathol.* **168**:1249–1261.

Hunter, R. L., M. R. Olsen, C. Jagannath, and J. K. Actor. 2006. Multiple roles of cord factor in the pathogenesis of primary, secondary, and cavitary tuberculosis, including a revised description of the pathology of secondary disease. *Ann. Clin. Lab. Sci.* **36**:371–386.

Huygen, K., J. Content, O. Denis, D. L. Montgomery, A. M. Yawman, R. R. Deck, C. M. DeWitt, I. M. Orme, S. Baldwin, C. D'Souza, A. Drowart, E. Lozes, P. Vandenbussche, J. P. Van Vooren, M. A. Liu, and J. B. Ulmer. 1996. Immunogenicity and protective efficacy of a tuberculosis DNA vaccine. *Nat. Med.* **2**:893–898.

Indrigo, J., R. L. Hunter, Jr., and J. K. Actor. 2003. Cord factor trehalose 6,6'-dimycolate (TDM) mediates trafficking events during mycobacterial infection of murine macrophages. *Microbiology* **149**:2049–2059.

Jackson, M., C. Raynaud, M. A. Laneelle, C. Guilhot, C. Laurent-Winter, D. Ensergueix, B. Gicquel, and M. Daffe. 1999. Inactivation of the antigen 85C gene profoundly affects the mycolate content and alters the permeability of the *Mycobacterium tuberculosis* cell envelope. *Mol. Microbiol.* **31**:1573–1587.

Johnson, R. S., G. R. Her, J. Grabarek, J. Hawiger, and V. N. Reinhold. 1990. Structural characterization of monophosphoryl lipid A homologs obtained from Salmonella minnesota Re595 lipopolysaccharide. *J. Biol. Chem.* **265**:8108–8116.

Koch, R. 1982. Classics in infectious diseases. The etiology of tuberculosis: Robert Koch. Berlin, Germany 1882. *Rev. Infect. Dis.* **4**:1270–1274.

Lewis, K. N., R. Liao, K. M. Guinn, M. J. Hickey, S. Smith, M. A. Behr, and D. R. Sherman. 2003. Deletion of RD1 from *Mycobacterium tuberculosis* mimics bacille Calmette-Guerin attenuation. *J. Infect. Dis.* **187**:117–123.

Lima, K. M., S. A. Santos, V. M. Lima, A. A. Coelho-Castelo, J. M. Rodrigues, Jr., and C. L. Silva. 2003. Single dose of a vaccine based on DNA encoding mycobacterial hsp65 protein plus TDM-loaded PLGA microspheres protects mice against a virulent strain of *Mycobacterium tuberculosis*. *Gene Ther.* **10**:678–685.

Lorian, V. 1969. Direct cord reading agar in routine mycobacteriology. *Appl. Microbiol.* **17**:559–562.

Lorian, V. 1966. Direct cord reading medium for isolation of mycobacteria. *Appl. Microbiol.* **14**:603–607.

Makinoshima, H., and M. S. Glickman. 2005. Regulation of *Mycobacterium tuberculosis* cell envelope composition and virulence by intramembrane proteolysis. *Nature* **436**:406–409.

Makinoshima, H., and M. S. Glickman. 2006. Site-2 proteases in prokaryotes: regulated intramembrane proteolysis expands to microbial pathogenesis. *Microbes Infect.* **8**:1882–1888.

Masihi, K. N., W. Brehmer, W. Lange, E. Ribi, and S. Schwartzman. 1984. Protective effect of muramyl dipeptide analogs in combination with trehalose dimycolate against aerogenic influenza virus and *Mycobacterium tuberculosis* infections in mice. *J. Biol. Response Mod.* **3**:663–671.

Masihi, K. N., W. Brehmer, W. Lange, H. Werner, and E. Ribi. 1985. Trehalose dimycolate from various mycobacterial species induces differing anti-infectious activities in combination with muramyl dipeptide. *Infect. Immun.* **50**:938–940.

Middlebrook, G., R.J. Dobos, and C. Pierce. 1947. Virulence and morphological characteristics of mammalian tubercle bacilli. *J. Exp. Med.* **86**:175–184.

Moody, D. B., B. B. Reinhold, M. R. Guy, E. M. Beckman, D. E. Frederique, S. T. Furlong, S. Ye, V. N. Reinhold, P. A. Sieling, R. L. Modlin, G. S. Besra, and S. A. Porcelli. 1997. Structural requirements for glycolipid antigen recognition by CD1b-restricted T cells. *Science* **278**:283–286.

Moody, D. B., B. B. Reinhold, V. N. Reinhold, G. S. Besra, and S. A. Porcelli. 1999. Uptake and processing of glycosylated mycolates for presentation to CD1b-restricted T cells. *Immunol. Lett.* **65**:85–91.

Moody, D. B., T. Ulrichs, W. Muhlecker, D. C. Young, S. S. Gurcha, E. Grant, J. P. Rosat, M. B. Brenner, C. E. Costello, G. S. Besra, and S. A. Porcelli. 2000. CD1c-mediated T-cell recognition of isoprenoid glycolipids in *Mycobacterium tuberculosis* infection. *Nature* **404**:884–888.

Noll, H., and H. Bloch. 1955. Studies on the chemistry of the cord factor of *Mycobacterium tuberculosis*. *J. Biol. Chem.* **214**:251–265.

Noll, H., H. Bloch, J. Asselineau, and E. Lederer. 1956. The chemical structure of the cord factor of *Mycobacterium tuberculosis*. *Biochim. Biophys. Acta* **20**:299–309.

Ortalo-Magne, A., A. Lemassu, M. A. Laneelle, F. Bardou, G. Silve, P. Gounon, G. Marchal, and M. Daffe. 1996. Identification of the surface-exposed lipids on the cell envelopes of *Mycobacterium tuberculosis* and other mycobacterial species. *J. Bacteriol.* **178**:456–461.

Perez, E., S. Samper, Y. Bordas, C. Guilhot, B. Gicquel, and C. Martin. 2001. An essential role for phoP in *Mycobacterium tuberculosis* virulence. *Mol. Microbiol.* **41**:179–187.

Perez, R. L., J. Roman, S. Roser, C. Little, M. Olsen, J. Indrigo, R. L. Hunter, and J. K. Actor. 2000. Cytokine message and protein expression during lung granuloma formation and resolution induced by the mycobacterial cord factor trehalose-6,6'-dimycolate. *J. Interferon Cytokine Res.* **20**:795–804.

Puech, V., N. Bayan, K. Salim, G. Leblon, and M. Daffe. 2000. Characterization of the in vivo acceptors of the mycoloyl residues transferred by the corynebacterial PS1 and the related mycobacterial antigens 85. *Mol. Microbiol.* **35**:1026–1041.

Puech, V., C. Guilhot, E. Perez, M. Tropis, L. Y. Armitige, B. Gicquel, and M. Daffe. 2002. Evidence for a partial redundancy of the fibronectin-binding proteins for the transfer of mycoloyl residues onto the cell wall arabinogalactan termini of *Mycobacterium tuberculosis*. *Mol. Microbiol.* **44**:1109–1122.

Pym, A. S., P. Brodin, R. Brosch, M. Huerre, and S. T. Cole. 2002. Loss of RD1 contributed to the attenuation of the live tuberculosis vaccines *Mycobacterium bovis* BCG and *Mycobacterium microti*. *Mol. Microbiol.* **46**:709–717.

Qureshi, N., P. Mascagni, E. Ribi, and K. Takayama. 1985. Monophosphoryl lipid A obtained from lipopolysaccharides of Salmonella minnesota R595. Purification of the dimethyl derivative by high performance liquid chromatography and complete structural determination. *J. Biol. Chem.* **260**:5271–5278.

Rao, V., N. Fujiwara, S. A. Porcelli, and M. S. Glickman. 2005. *Mycobacterium tuberculosis* controls host innate immune activation through cyclopropane modification of a glycolipid effector molecule. *J. Exp. Med.* **201**:535–543.

Rao, V., F. Gao, B. Chen, W. R. Jacobs, Jr., and M. S. Glickman. 2006. *Trans*-cyclopropanation of mycolic acids on trehalose dimycolate suppresses *Mycobacterium tuberculosis*-induced inflammation and virulence. *J. Clin. Invest.* **116**:1660–1667.

Rawlings, N. D., F. R. Morton, and A. J. Barrett. 2006. MEROPS: the peptidase database. *Nucleic Acids Res.* **34**:D270–272.

Rawson, R. B., N. G. Zelenski, D. Nijhawan, J. Ye, J. Sakai, M. T. Hasan, T. Y. Chang, M. S. Brown, and J. L. Goldstein. 1997. Complementation cloning of S2P, a gene encoding a putative metalloprotease required for intramembrane cleavage of SREBPs. *Mol. Cell* **1**:47–57.

Retzinger, G. S., S. C. Meredith, R. L. Hunter, K. Takayama, and F. J. Kezdy. 1982. Identification of the physiologically active state of the mycobacterial glycolipid trehalose 6,6'-dimycolate and the role of fibrinogen in the biologic activities of trehalose 6,6'-dimycolate monolayers. *J. Immunol.* **129**:735–744.

Ribi, E., D. L. Granger, K. C. Milner, K. Yamamoto, S. M. Strain, R. Parker, R. W. Smith, W. Brehmer, and I. Azuma. 1982. Induction of resistance to tuberculosis in mice with defined components of mycobacteria and with some unrelated materials. *Immunology* **46**:297–305.

Ronning, D. R., T. Klabunde, G. S. Besra, V. D. Vissa, J. T. Belisle, and J. C. Sacchettini. 2000. Crystal structure of the secreted form of antigen 85C reveals potential targets for mycobacterial drugs and vaccines. *Nat. Struct. Biol.* **7**:141–146.

Ronning, D. R., V. Vissa, G. S. Besra, J. T. Belisle, and J. C. Sacchettini. 2004. *Mycobacterium tuberculosis* antigen 85A and 85C structures confirm binding orientation and conserved substrate specificity. *J. Biol. Chem.* **279**:36771–36777.

Ryll, R., Y. Kumazawa, and I. Yano. 2001. Immunological properties of trehalose dimycolate (cord factor) and other mycolic acid-containing glycolipids—a review. *Microbiol. Immunol.* **45**: 801–811.

Silva, C. L. 1989. Participation of tumor necrosis factor in the antitumor activity of mycobacterial trehalose dimycolate (cord factor). *Braz. J. Med. Biol. Res.* **22**:341–344.

Sorkin, E., H. Erlenmeyer, and H. Bloch. 1952. Purification of a lipid material ('cord factor') obtained from young cultures of tubercle bacilli. *Nature* **170**:124.

Stanley, S. A., S. Raghavan, W. W. Hwang, and J. S. Cox. 2003. Acute infection and macrophage subversion by *Mycobacterium tuberculosis* require a specialized secretion system. *Proc. Natl. Acad. Sci. USA* **100**:13001–13006.

Strain, S. M., R. Toubiana, E. Ribi, and R. Parker. 1977. Separation of the mixture of trehalose 6,6'-dimycolates comprising the mycobacterial glycolipid fraction, "P3." *Biochem. Biophys. Res. Commun.* **77**:449–456.

Sugawara, I., T. Udagawa, S. C. Hua, M. Reza-Gholizadeh, K. Otomo, Y. Saito, and H. Yamada. 2002. Pulmonary granulomas of guinea pigs induced by inhalation exposure of heat-treated BCG Pasteur, purified trehalose dimycolate and methyl ketomycolate. *J. Med. Microbiol.* **51**:131–137.

Takimoto, H., H. Maruyama, K. I. Shimada, R. Yakabe, I. Yano, and Y. Kumazawa. 2006. Interferon-gamma independent formation of pulmonary granuloma in mice by injections with trehalose dimycolate (cord factor), lipoarabinomannan and phosphatidylinositol mannosides isolated from *Mycobacterium tuberculosis*. *Clin. Exp. Immunol.* **144**:134–141.

Walters, S. B., E. Dubnau, I. Kolesnikova, F. Laval, M. Daffe, and I. Smith. 2006. The *Mycobacterium tuberculosis* PhoPR two-component system regulates genes essential for virulence and complex lipid biosynthesis. *Mol. Microbiol.* **60**:312–330.

Yarkoni, E., and H. J. Rapp. 1977. Granuloma formation in lungs of mice after intravenous administration of emulsified trehalose-6,6'-dimycolate (cord factor): reaction intensity depends on size distribution of the oil droplets. *Infect. Immun.* **18**:552–554.

Chapter 6

Structure, Biosynthesis, and Activities of the Phosphatidyl-*myo*-Inositol-Based Lipoglycans

Martine Gilleron, Mary Jackson, Jérôme Nigou, and Germain Puzo

Mycobacterium tuberculosis bacilli are phagocytozed mostly by alveolar macrophages following entry into the lung. Paradoxically, these phagocytic cells constitute a niche that protects *M. tuberculosis* from rapid eradication by host defense mechanisms, particularly antibodies. Indeed, at the beginning of the infection, *M. tuberculosis* is an intracellular parasite and, consequently, the stimulation of both Th1 and cytotoxic T cells is required for it to be killed. Both macrophages and dendritic cells (DCs) contribute to innate immunity by secreting proinflammatory cytokines through the recognition by the Toll-like receptors (TLRs) of *M. tuberculosis*-associated molecular patterns, thereby inducing a Th1 polarization. Cytokines, including tumor necrosis factor alpha (TNF-α) produced by phagocytes, are essential to granuloma formation and containment of the infection.

DCs appear to play a key role in the immune responses to *M. tuberculosis*, by carrying mycobacterial antigens to lymph nodes where *M. tuberculosis*-specific conventional and unconventional T cells are stimulated. Conventional CD4$^+$ and CD8$^+$ αβ T cells recognize peptide antigens in the context of MHC II or MHC I proteins on antigen-presenting cells (APCs) respectively. Unconventional T cells include γδ T cells stimulated by phosphorylated ligands and CD1-restricted αβ T cells, which recognize lipid antigens (De Libero and Mori, 2005; Moody et al., 2005).

M. tuberculosis interactions with phagocytes are central to both host protective immunity and tuberculosis pathogenesis. It is still not clearly understood how *M. tuberculosis* interferes and manipulates host immunomodulatory networks. Nevertheless, recent findings reveal some of the mechanisms by which the tubercle bacilli can survive and replicate inside

phagocytes, and work on the development of subunit vaccines has made progress in the characterization of protective mycobacterial antigens of conventional and unconventional αβT cells. Furthermore, the effector T-cell mechanisms involved in *M. tuberculosis* killing have been elucidated. These various recent findings reveal that lipids of the mycobacterial envelope are key determinants of the interactions between the pathogen and the host immune cells.

The pioneering works of Lederer (Petit and Lederer, 1984), Asselineau (Asselineau and Asselineau, 1984), Brennan (Brennan, 1984), Goren (Goren, 1984) and others established that the envelope of *M. tuberculosis* has an extraordinarily high lipid content and molecular diversity. Lipids constitute up to 60% of the dry weight of the bacilli but only about 20% for gram-negative organisms despite their lipid-rich envelopes (Brennan and Nikaido, 1995; Daffe and Draper, 1998; Puzo, 1993). Approximately 250 genes are involved in the lipid metabolism in *M. tuberculosis* (Cole et al., 1998).

The *M. tuberculosis* envelope lipids include phosphatidyl-*myo*-inositol mannosides (PIM) and their multiglycosylated counterparts, lipomannans (LM) and mannosylated lipoarabinomannans (ManLAM). These molecules are involved in the modulation of the host immune responses (for detailed references, see reviews by Briken et al., 2004, Chatterjee and Khoo, 1998, Nigou et al., 2002b, 2003). ManLAM reproduces several salient properties of *M. tuberculosis* in the phagosome, including reduced acquisition of late endosomal markers and inhibition of phagosomal acidification (Fratti et al., 2003). It is a potential ligand for the entry of *M. tuberculosis* into DCs through the C-type lectin DC-SIGN (DC-specific intercellular adhesion molecule 3-grabbing non-integrin receptor)

Martine Gilleron, Jérôme Nigou, and Germain Puzo • Department of Molecular Mechanisms of Mycobacterial Infections, Institut de Pharmacologie et de Biologie Structurale du Centre National de la Recherche Scientifique (CNRS) and Université Paul Sabatier, 205 route de Narbonne, 31077 Toulouse Cedex 4, France. **Mary Jackson** • Department of Microbiology, Immunology and Pathology, Colorado State University, Fort Collins, CO 80523.

(Geijtenbeek et al., 2003; Tailleux et al., 2003a). It alters the innate immunity response by interacting with the host C-type lectins, leading to the inhibition of the production of proinflammatory cytokines interleukin-12 (IL-12) and TNF-α by activated macrophages and DCs (Knutson et al., 1998; Nigou et al., 2001). ManLAM also inhibits both infection-induced apoptosis in macrophages (Rojas et al., 2000) and their activation by interferon-γ (IFN-γ) (Sibley et al., 1988).

In contrast to ManLAM, PIM and LM are antigens of innate immunity; they are agonists of TLR2 (Gilleron et al., 2003; Quesniaux et al., 2004), a receptor involved in host defense against invading pathogens. PIM stimulates unconventional αβ T cells in the context of CD1 proteins, and the mechanism of PIM processing by lysosomal mannosidase assisted by the protein CD1e was recently described (De la Salle et al., 2005).

The structure and the distribution of these lipoglycans in various mycobacterial species will be presented in this review, and the different hypotheses concerning their localization in the mycobacterial envelope will be discussed. Then our current knowledge of the lipoglycan biosynthetic pathways will be described. The study of these molecules started more than 40 years ago with the pioneering work of Ballou and Brennan and the characterization of the first two glycosylation steps involved in PIM biosynthesis (Brennan and Ballou, 1967; Hill and Ballou, 1966). The recent technical advances allowing the genetic manipulation of mycobacteria, the availability of a growing number of genome sequences, and the development of bioinformatic software have led to the discovery of key glycosyl-transferases involved in ManLAM biosynthesis. Finally, we will consider recent progress in deciphering the structure-function relationships of these lipoglycans in the context of *M. tuberculosis* pathogenesis or as antigens of host protective immune responses.

LIPOGLYCAN STRUCTURE AND LOCALIZATION

Structural Characterization

PIM and their multiglycosylated counterparts LM and LAM are ubiquitous mycobacterial envelope lipoglycans that are biosynthetically related. They all share a conserved mannosyl-phosphatidyl-*myo*-inositol anchor (MPI), based on a *sn*-glycero-3-phospho-(1-D-*myo*-inositol) unit with one α-D-Man$p$ unit linked at O-2 and another at O-6 of the *myo*-inositol (Ins) unit, suggesting that they are all metabolically related. LM is described as polymannosylated PIM$_4$,

whereas LAM seems to correspond to LM with an attached D-arabinan domain (Fig. 1).

The PIM family comprises phosphatidyl-*myo*-inositol di-, tri-, tetra-, penta-, and hexa-mannosides (PIM$_2$ to PIM$_6$) (Ballou et al., 1963; Lee and Ballou, 1965). PIM$_2$ and PIM$_6$ are the most abundant glycoforms found in *M. tuberculosis*, *Mycobacterium bovis* BCG, and *Mycobacterium smegmatis* (Gilleron et al., 2003). PIM$_2$ is composed of a phospho-*myo*-inositol glycosylated with a mannosyl unit at each of positions 2 and 6 of the Ins. PIM$_4$, corresponding to the extension of PIM$_2$ by the addition of the disaccharidyl unit t-α-Man$p$(1→6)-α-Man$p$(1→6) to the Man$p$ linked to O-6 of the Ins, appears to be the intermediate structure giving rise either to PIM$_6$ or to LM and LAM. Indeed, PIM$_6$ results from the elongation of PIM$_4$ by the disaccharidyl unit t-α-Man$p$ (1→2)-α-Man$p$(1→2), whereas LM results from the elongation of PIM$_4$ by (1→6)-linked α-Man$p$ residues to form first a linear D-mannan segment. Chemically synthesized PIM (Ainge et al., 2006; Dyer et al., 2007; Liu et al., 2006b; Stadelmaier and Schmidt, 2003) and LM (Jayaprakash et al., 2005) are now available. The D-mannan domain of LM and LAM is described from the data of Chatterjee and colleagues (1992a) to consist of an (α1→6)-linked Man$p$ core, which is substituted at O-2 by single α-Man$p$ units. The arabinan polymer of LAM consists of a linear α(1→5)-linked arabinofuranosyl backbone punctuated by branching at O-3 with linear tetra-arabinofuranosides (Ara$_4$) and branched hexa-arabinofuranosides (Ara$_6$) (Chatterjee et al., 1991, 1993).

Having defined the general structural features shared by all LAM, we will consider the specific aspects of different types of LAM. LAM in any strain is heterogeneous, with various acyl-forms, glyco-forms, and charge numbers. It also differs between species in terms of the capping motifs, the degree and the type of branching of the mannan core, the presence of only recently discovered substitutions (succinyl appendages, peculiar monosaccharides including 5-deoxy-5-methylthio-*xylo*-furanose [MTP]). Various LAM structures will be presented in chronological order of discovery.

The first study highlighting heterogeneity of LAM demonstrated the existence of different types of capping motifs on the nonreducing termini of the arabinosyl side chains (Chatterjee et al., 1992b). Consequently, LAM is now classified into three major structural families, according to the presence and type of capping motifs (Fig. 1). The arabinan termini of LAM from *M. bovis* BCG (Prinzis et al., 1993; Venisse et al., 1993), the pathogenic species *M. tuberculosis* (Chatterjee et al., 1992b), *Mycobacterium leprae* (Khoo et al., 1995a), and *Mycobacterium bovis*

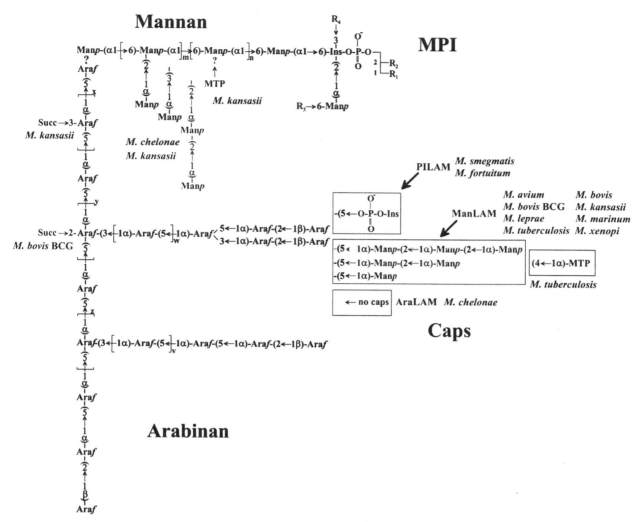

Figure 1. Schematic representation of mycobacterial LAM. Ara*f*, arabinofuranose; Ins, *myo*-inositol; Man*p*, mannopyranose; MPI, mannosyl-phosphatidyl-*myo*-inositol; R$_n$, fatty acyl residue. The mean molecular mass of *M. tuberculosis* and *M. bovis* BCG ManLAM is around 17 kDa, with a heterogeneity estimated at 6 kDa (Venisse et al., 1993). We estimate that these ManLAM contain approximately 60 Ara*f* and 50 Man*p* units. Man*p* units are distributed between the mannose caps and the mannan core (30 to 35 Man*p* units) (Nigou et al., 2000). The mannan domain of the *M. kansasii* ManLAM contains a low proportion of disaccharide side chains (Guerardel et al., 2003). 5-Methylthiopentose (MTP) was identified as 5-deoxy-5-methylthio-*xylo*-furanose (Joe et al., 2006) and has been described on the ManLAM of *M. tuberculosis* strains (Ludwiczak et al., 2002; Treumann et al., 2002) and ManLAM and LM of a *M. kansasii* clinical isolate (Guerardel et al., 2003). Succ indicates succinyl residues located on the arabinan domain of ManLAM of *M. bovis* BCG (Delmas et al., 1997) and of a *M. kansasii* clinical isolate (Guerardel et al., 2003). One to four succinyl groups, depending on the *M. bovis* BCG strain, esterify the 3,5-α-Ara*f* units at position O-2 (Delmas et al., 1997), and an average of two succinic acids per LAM was found in the case of *M. kansasii* ManLAM (Guerardel et al., 2003). *(See the color insert for the color version of this figure.)*

(Pitarque et al., 2005), and the opportunistic species *Mycobacterium avium* (Khoo et al., 2001; Pitarque et al., 2005), *Mycobacterium kansasii* (Guerardel et al., 2003; Pitarque et al., 2005), *Mycobacterium xenopi* (Pitarque et al., 2005), and *Mycobacterium marinum* (Pitarque et al., 2005) are modified with caps consisting of a single Man*p*, a dimannoside α-Man*p*(1→2)-α-Man*p* or a trimannoside α-Man*p* (1→2)-α-Man*p* (1→2)-α-Man*p*, with dimannosides almost always predominating. These molecules are designated ManLAM. However, the number and

type (mono-, di-, and trimannoside percentage) of mannoside caps vary according to the species and the strains (Nigou et al., 2003; Pitarque et al., 2005); as an example, single Man*p* is the most abundant cap motif in *M. marinum* (Pitarque et al., 2005). In the fast-growing species *M. smegmatis* (Gilleron et al., 1997) and *Mycobacterium fortuitum* (unpublished results) and in an unidentified fast growing *Mycobacterium* species (Khoo et al., 1995a), some branches of the terminal arabinan are ended by phospho-inositol caps, and in these cases, the LAM is designated as

PILAM. A third LAM family, devoid of both the manno-oligosaccharide and phospho-inositol caps, was recently identified in *Mycobacterium chelonae* and designated AraLAM (Guerardel et al., 2002).

A second source of LAM heterogeneity was revealed by studying the MPI anchor, variations occurring through the number, the location, and the nature of the fatty acids. Indeed, LAM, LM, and PIM exist in different acyl-forms, which have been clearly evidenced by ^{31}P nuclear magnetic resonance (NMR) analysis (Gilleron et al., 1999, 2000, 2001b; Nigou, 1999b). Four potential acylation sites have now been unambiguously defined on the anchor of these molecules: positions 1 and 2 of the glycerol (Gro) unit, position 6 of the Man*p* unit linked at O-2 of Ins, and position 3 of the Ins (Fig. 1). They are predominantly acylated by palmitic and tuberculostearic (10-methyl-octadecanoic) acids, and to a lesser extent by stearic acids. A larger diversity of fatty acyl substituents has been described for *M. chelonae* AraLAM (Guerardel, 2002). Characterization of the various PIM acyl-forms was easy because of their low molecular mass (<2,500 Da), such that these lipids can be completely defined by NMR (Gilleron et al., 2001b, 2003). The most abundant acyl-forms in *M. tuberculosis*, *M. smegmatis*, and *M. bovis* BCG are the tri- and tetra-acylated forms. The tri-acylated forms bear two fatty acyl groups on the Gro and one on the Man*p* unit. We have developed a strategy based on 1D ^{31}P and 2D ^{1}H-^{31}P experiments for differentiating between and determining the relative abundance of phosphate of the different acyl-forms, according to the acylation state of the units esterifying the phosphate, i.e. the Gro and Ins units (Fig. 1). Although Man*p* unit acylation cannot be investigated by this technique, it is nevertheless particularly useful for the characterization of the LAM and LM acyl-forms, because these high-molecular-mass molecules cannot be fully characterized by NMR. The relative abundance of the different acyl-forms of LAM differs between mycobacterial species (Nigou et al., 2003). For example, *M. smegmatis* and *M. fortuitum* have LAM that is significantly less acylated than that in other investigated species, including *M. avium* (unpublished results), *M. bovis* (Nigou et al., 2003), *M. bovis* BCG (Nigou et al., 1999b), *M. chelonae* (Guerardel et al., 2002), *M. kansasii* (Guerardel et al., 2003), *M. marinum* (unpublished results), *M. tuberculosis* H37Ra (Nigou et al., 1999b), and H37Rv (Gilleron et al., 2000); the LAM in *M. smegmatis* and *M. fortuitum* has predominantly monoacyl-glycerol rather than diacyl-glycerol (unpublished data). Recently, the different acyl-forms of *M. bovis* BCG LM have been purified and completely characterized by matrix-assisted laser desorption/ionization-time of flight-mass spec-

trometry (MALDI-TOF-MS) associated with 2D ^{1}H-^{31}P NMR (Gilleron et al., 2006). Four acyl-forms, mono- to tetra-acylated (Ac$_1$ to Ac$_4$) LM, were evidenced, and thus the existence of a tetra-acylated form of LM was definitively proved. Moreover, MALDI-TOF mass spectra revealed the presence of the different glyco-forms, and showed the molecular complexity of LM in terms of mass distribution and degree of mannosylation. Indeed, whatever the extent of acylation, the number of Man*p* units for each acyl-form was anywhere between 15 and 45, with a systematic preponderance of the glyco-form containing 26 Man*p* units (Gilleron et al., 2006).

There is also diversity in the structure and the degree of branching of the mannan core, which differs in different species (Khoo et al., 1996; Nigou et al., 2000). The mannan domain has been described as a core substituted at O-2 by single Man*p* units in the LAM of the majority of species, including *M. tuberculosis*, *M. leprae*, *M. bovis* BCG, and *M. smegmatis*, but at O-3 by single Man*p* units in *M. chelonae* (Guerardel et al., 2002) (Fig. 1). The size of branching chains also appears to differ between species. In particular, α1,2-dimannopyrannosyl side chains have been described on the LAM from a clinical isolate of *M. kansasii* (Guerardel et al., 2003).

Various substitutions have been found decorating LAM, and they include succinyl groups, evidenced by Brennan's group in the 80s (Hunter et al., 1986) and subsequently accurately mapped on the O-2 of the 3,5-Ara*f* units in ManLAM from *M. bovis* BCG (Delmas et al., 1997). These units also substitute some O-3 positions of 5-Ara*f* residues in *M. kansasii* ManLAM (Guerardel et al., 2003). The most recently discovered substituent is a 5-methylthiopentose (MTP) unit (Treumann et al., 2002), finally identified as 5-deoxy-5-methylthio-*xylo*furanose (Joe et al., 2006). MTP was first identified on *M. tuberculosis* ManLAM (Ludwiczak et al., 2002; Treumann et al., 2002) linked to a terminal mannosyl unit of the cap (Treumann et al., 2002; Turnbull et al., 2004), most probably at position O-4 (Joe et al., 2006). It has also been found on *M. avium* ManLAM (unpublished results) and both *M. kansasii* ManLAM and LM, probably substituting the mannan core (Guerardel et al., 2003).

Chatterjee and colleagues recently showed that LAM from all species could be resolved into discrete forms by isoelectric focusing. This separation is apparently based on their arabinan heterogeneity, and LM is always found as only a single isoform (Torrelles et al., 2004). This apparent heterogeneity of LAM, possibly the consequence of different numbers of acidic appendages on the arabinan domain, has still not been fully explained (Torrelles et al., 2004).

Localization in the Cell Envelope

LAM, LM, and PIM are not covalently attached to the mycobacterial cell envelope. PIM are found in the plasma membrane among other phospholipids and also in the capsule, where they seem to be randomly distributed from the cell surface to its innermost layers (Daffe and Draper, 1998; Ortalo-Magne et al., 1996b). The exact localizations of LAM and LM remain unclear and controversial. They have never been found in culture supernatants or in the surface-exposed material obtained by gentle mechanical treatment of cells with glass beads or detergent (Daffe and Draper, 1998; Ortalo-Magne et al., 1995), suggesting that they are embedded in the cell wall. The model of Minnikin (Minnikin, 1982) is now accepted for the physical organization of the cell wall: an asymmetric lipid bilayer in which the inner leaflet contains mycolic acids covalently linked to arabinogalactan and the outer leaflet contains the extractable lipids. Mycolic acids are believed to interact in a tightly packed, parallel arrangement perpendicular to the plane of the cytoplasma membrane, and a recent molecular modeling and simulation computational study of the *M. tuberculosis* cell wall supports this view (Hong and Hopfinger, 2004).

This model has led to two suggested LAM localizations: LAM molecules could be inserted, through their MPI anchor into the outer leaflet of the outer membrane among other different lipids (Rastogi, 1991) or into the plasma membrane (Gaylord and Brennan, 1987; McNeil and Brennan, 1991), or both. Chatterjee and Khoo (Chatterjee and Khoo, 1998) argued for the plasma membrane because of the relatively aggressive conditions (cell disruption and organic solvent extraction) needed to release LAM from mycobacteria. The recent elucidation of the biosynthetic pathways of LAM also supports a plasma membrane localization. Glycosyl-transferases involved in LAM biosynthesis are either cytosolic, for the early steps of PIM biosynthesis (Belanger and Inamine, 2000; Kordulakova et al., 2002), or membrane anchored, for the late steps (Berg et al., 2005; Escuyer et al., 2001), including the recently discovered mannosyl-transferase that catalyzes addition of the first mannose cap residue (Dinadayala et al., 2006) (see "Biogenesis of PIM, LM and LAM" below). There is now almost no doubt that mature LAM is produced while still anchored to the outer leaflet of the plasma membrane (Berg et al., 2005).

The main problem with this model is that it is incompatible with the presentation of terminal arabinose or mannose caps at the mycobacterial cell surface. Indeed, it raises two questions: First, how can this molecule cross the mycoloyl-arabinogalactan-peptidoglycan complex that forms a tight corset around the cell? Second, is LAM large enough to reach the cell surface while anchored in the plasma membrane?

The answer to the second question seems to be no. Electron microscopy analyses indicate that the thickness of the cell wall is approximately 25 to 30 nm and that of the capsule surrounding it is approximately 20 to 100 nm, according to the strain (Brennan and Nikaido, 1995; Daffe and Draper, 1998; Paul and Beveridge, 1992, 1994). Using dynamic light scattering and electron microscopy, Rivière and associates (2004) recently showed that *M. bovis* BCG ManLAM in solution form micelles of approximately 30 nm in diameter. This suggests that LAM, with an estimated "length" of 15 nm, is not large enough even to straddle the cell wall and consequently, when anchored in the plasma membrane, it is not accessible to the outside. Nevertheless, there is convincing evidence that some LAM and LM molecules are exposed at the mycobacterial cell surface, at least in *M. bovis* BCG, *M. xenopi,* and *M. smegmatis* (Pitarque et al., 2005; Pitarque et al., submitted for publication): after periodate oxidation of the cells and then labeling surface-exposed carbohydrates with biotin-hydrazide, purified LAM and LM were found to be tagged with biotin. So it seems likely that, in addition to the plasma membrane-anchored LAM and LM presumably located in the periplasmic space, there is a second pool of these lipoglycans accessible to the outside; this pool might be anchored in the outer leaflet of the outer membrane. This surface-exposed pool might be the molecules that play a role in the immune processes, including adhesins for C-type lectin-mediated phagocytosis or trafficking inside phagocytic cells after their release from the mycobacterial cell surface. The presence of LAM on the cell surface is supported by the fact that arabinomannan (AM), which has a structure identical to LAM but devoid of the MPI anchor (Nigou et al., 1999a; Ortalo-Magne et al., 1996a), has been found in culture supernatant and in the surface-exposed material (Ortalo-Magne et al., 1996a). AM is thought to be derived from LAM, and its production involves an endo-mannosidase. However, this raises the question of how these lipoglycans are transported from the plasma membrane to the cell surface. Braibant and associates (2000) have reported an inventory and described the assembly of the typical subunits of the ABC transporters encoded by the *M. tuberculosis* genome; they propose that *M. tuberculosis* expresses at least 26 complete ABC transporters. Homologies suggest that a putative transporter, encoded by genes *Rv3781* and *Rv3783*, is likely to be involved in the export of polysaccharides (Braibant et al., 2000). We have investigated its putative involvement in the transport of lipoglycans but have found that a *M. smegmatis* knockout mutant disrupted in the ortholog of *Rv3783* still expressed wild-type amounts of lipoglycans at its surface (Pitarque et al., submitted manuscript). Thus, if lipoglycan transporters exist, their identity remains to be discovered.

BIOGENESIS OF PIM, LM, AND LAM

Although the structures of PIM, LM, and LAM have been described, their biosynthetic pathways are still incomplete, and current knowledge of the involved catalytic steps mostly concerns the synthesis of their MPI anchor. The considerable progress in characterizing biosynthetic enzymes in the last 7 years is largely attributable to the advances in the possibilities for genetic manipulation of mycobacteria, together with the availability of a growing number of genome sequences of mycobacterial species and other LAM-like lipoglycan-producing actinomycetes. The exponentially growing use of genetic approaches, together with biochemical studies, are rapidly filling the gaps in our understanding of PIM/LM/LAM biosynthesis, and a clearer picture of these complex biosynthetic pathways is now beginning to emerge.

Structural similarities between PIM, LM, and LAM based on a conserved MPI anchor point to a metabolic relationship (Chatterjee et al., 1992a; Hunter and Brennan, 1990; Khoo et al., 1995b; Lee and Ballou, 1965). Studies initiated 40 years ago in Ballou's laboratory provide evidence that the early steps of PIM synthesis start with the transfer of a mannosyl residue (Man$p$) from GDP-Man$p$ to the 2-position of the Ins ring of PI to form phosphatidyl-*myo*-inositol monomannosides (PIM$_1$) (Hill and Ballou, 1966). This glycolipid or its acylated counterpart (Ac$_1$PIM$_1$) is then glycosylated by another Man$p$ residue at the 6-position of Ins to form phosphatidyl-*myo*-inositol dimannoside (PIM$_2$) or acylated phosphatidyl-*myo*-inositol dimannoside (Ac$_1$PIM$_2$), respectively (Brennan and Ballou, 1967, 1968; Hill and Ballou, 1966; Takayama and Goldman, 1969). It is thought that PIM$_2$ or Ac$_1$PIM$_2$ then undergo several glycosylation steps with Man$p$ to form higher PIM (PIM$_3$-PIM$_6$) and the highly branched LM, and then with Ara$f$ to form LAM (Besra et al., 1997). Recent evidence indicates that PIM$_4$ is at the branch point at which PIM and LM/LAM biosynthesis bifurcates (Kovacevic et al., 2006; Morita et al., 2004, 2006). This glycolipid can be further elongated with Man$p$, leading to the formation of the linear α(1→6)-linked mannan backbone of LM, or be mannosylated with two consecutive α(1→2)-linked mannosyl units to form PIM$_6$, generally considered to be terminal PIM products, not involved in LM and LAM synthesis. The two or three first mannosylation steps leading to the formation of PIM$_{1-3}$ occur on the cytosolic face of the plasma membrane, where the involved mannosyltransferases use the soluble sugar nucleotide GDP-Man$p$ as Man$p$ donor (Brennan and Ballou, 1967; Hill and Ballou, 1966; Kordulakova et al., 2002; Kremer et al., 2002; Schaeffer et al., 1999). Subsequent mannosylation and arabinosylation steps in polar PIM and LM/LAM biosynthesis are thought to take place on the periplasmic side of the plasma membrane and to require polyprenyl-phosphate-based sugar donors (Berg et al., 2005; Besra et al., 1997; Morita et al., 2004, 2005, 2006). Interestingly, many if not all of these later glycosylation steps seem to involve integral membrane enzymes that are evolutionarily related to eukaryotic glycosyl-transferases responsible for the final glycosylation steps of asparagine-linked glycans (*N*-glycans) or the mannosylation of the glycosylphosphatidylinositol (GPI) anchor of cell-surface proteins (Liu and Mushegian, 2003; Oriol et al., 2002).

This section and Fig. 2 summarize the current knowledge concerning the enzymes involved in lipoglycan synthesis.

Phosphatidyl-*myo*-Inositol Synthesis

Phosphatidyl-*myo*-inositol (PI) is made from CDP-diacylglycerol (CDP-DAG) and Ins (Salman et al., 1999) in a reaction catalyzed by the PI synthase PgsA1 (Rv2612c) (Jackson et al., 2000). PgsA1 is an essential enzyme for *M. smegmatis*, indicating that PI or metabolically derived molecules are required for mycobacterial growth.

The mycobacterial pathway for de novo Ins biosynthesis is apparently similar to that of eukaryotic systems and involves two catalytic steps. In the first step, a single enzyme, inositol-1-phosphate synthase (Ino1), converts glucose-6-phosphate to inositol-1-phosphate. In the second step, the dephosphorylation of inositol-1-phosphate by an inositol monophosphatase (IMPase) yields inositol (Nigou and Besra, 2002a). The *ino1* gene of *M. tuberculosis* H37Rv (*Rv0046c*) was originally identified by sequence homology searches and shown to complement functionally an *ino1* mutant of *Saccharomyces cerevisiae* (Bachhawat and Mande, 1999). *ino1* knockout mutants of *M. tuberculosis* and *M. smegmatis* have been generated by homologous recombination and are auxotrophic for inositol, indicating that *ino1* encodes the only inositol-1-phosphate synthase in these species (Haites et al., 2005; Movahedzadeh et al., 2004). The crystal structure of the *M. tuberculosis*

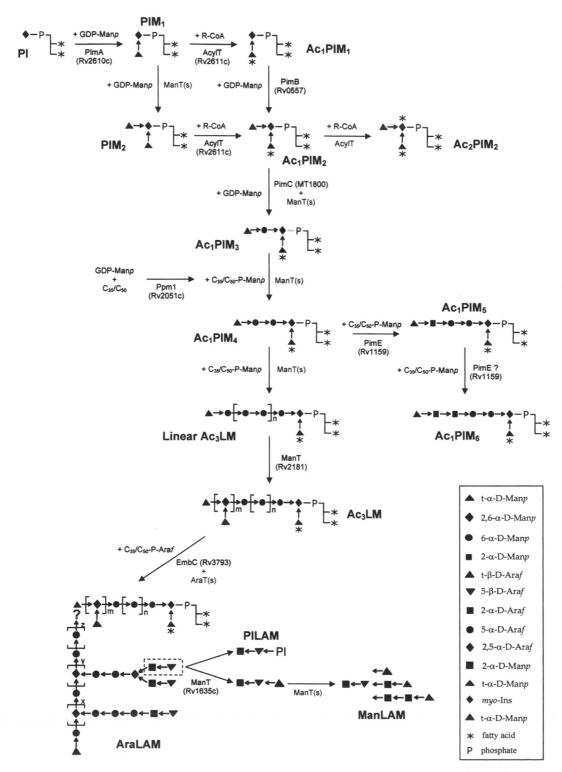

Figure 2. LAM biosynthesis schema. The biosynthesis of the triacylated forms of PIM and lipoglycans is shown. PimA is essential in *M. smegmatis* (Kordulakova et al., 2002). Rv2611c appears to be essential in *M. tuberculosis*, but not in *M. smegmatis* (Kordulakova et al., 2003; G. Stadthagen, M. Jackson, and B. Gicquel, unpublished results). AcylT, acyl-transferase; ManT(s), mannosyl-transferase(s); AraT(s), arabinosyl-transferase(s); C_{35}/C_{50}, polyprenol; C_{35}/C_{50}-P-Man, polyprenol-monophosphorylmannose; C_{35}/C_{50}-P-Ara*f*, polyprenol-monophosphoryl-β-D-Ara*f*. *(See the color insert for the color version of this figure.)*

enzyme was solved (Norman et al., 2002). The completion of the genome of *M. tuberculosis* has revealed at least four putative candidates for the dephosphorylation of inositol-1-phosphate (*Rv1604*, *Rv2701c*, *Rv3137*, and *Rv2131c*), suggesting the existence of redundant inositol monophosphatase activities in mycobacteria. The in vitro characterization of SuhB (Rv2701c) as a functional IMP (Nigou et al., 2002a) and the finding that a transposon insertion within the *impA* (*Rv1604*) gene of *M. smegmatis* only partially affects the PIM content of this mutant (Parish et al., 1997) supports this view.

CDP-DAG is the common precursor for the biosynthesis of phospholipids and is synthesized from phosphatidic acid and phosphatidate cytidyltransferase (CTP) by the CDP-DAG synthase. This enzymatic activity has been detected in *M. smegmatis* and is membrane associated (Nigou and Besra, 2002b). The structural gene for CDP-DAG synthase in the genome of *M. tuberculosis* H37Rv is predicted to be *cdsA* (*Rv2881c*).

The Biogenesis of PIM

The biosynthetic pathway of polar and apolar PIM, although incomplete, is by far the best documented aspect of the biosynthesis of PI-based lipoglycans. Using *M. smegmatis* recombinant strains and cell-free systems as experimental models, five glycosyltransferases and acyl-transferases have been characterized. Some of the enzymes required for the production of the Man donors used in the many mannosylation steps of these pathways have also been described.

GDP-Man$p$ is produced either through the conversion of exogenously acquired mannose to mannose-6-phosphate by a hexokinase or de novo through the conversion of fructose-6-phosphate to mannose-6-phosphate by a phosphomannose isomerase (PMI). Mannose-6-phosphate is then converted to mannose-1-phosphate by a phosphomannomutase and then to GDP-Man$p$ by a GDP-mannose pyrophosphorylase (GDPMP). Rv3264c has been identified as a GDPMP (Ma et al., 2001; Ning and Elbein, 1999) and ManA (Rv3255c) as a PMI (Patterson et al., 2003). Disruption of the *manA* gene in *M. smegmatis* yielded a mutant that is auxotrophic for mannose, indicating that it is the sole source of PMI activity in this species. In contrast, four genes in the genome of *M. tuberculosis* H37Rv share similarities with known PPM. Only one of those, *Rv3257c*, has been studied and was shown to display both phosphomannomutase and phosphoglucomutase activities in vivo and in vitro. Its overexpression in *M. smegmatis* resulted in increased production of PIM, LM and LAM (McCarthy et al., 2005).

The first step in PIM synthesis involves the transfer of a mannosyl residue from GDP-Man$p$ to the 2-position of the Ins ring of PI to form phosphatidyl-*myo*-inositol monomannoside, PIM$_1$ (Fig. 2). We have identified PimA (Rv2610c) as the α-mannosyltransferase responsible for this catalytic step and found it to be an essential enzyme in *M. smegmatis* (Gu et al., 2005; Kordulakova et al., 2002). Because it is only found in a few PIM-producing actinomycetes and is essential for viability, PimA is an attractive therapeutic target for the development of novel anti-TB drugs. Efforts have been made toward this goal and the recent determination of the three-dimensional structure of this protein (M. Guerin, A. Buschiazzo, P. Alzari, unpublished results) along with the development of a spectrophotometric assay for measuring PimA activity in a microtiter plate format (J. Korduláková, Z. Svetliková, P. Dianisková, M. McNeil, K. Mikusová, unpublished results) have paved the way for screening potential inhibitors of this enzyme both in vitro and in silico.

Interestingly, *pimA* belongs to a cluster of five open reading frames (ORFs) potentially organized as a single transcriptional unit that is conserved in all mycobacterial sequenced genomes. The function of the first ORF of this cluster, *Rv2613c*, has not been demonstrated but the presence of the conserved motif of Histidine Triad (HIT) superfamily proteins in the primary sequence of its product suggests that it may participate in the release of CMP from CDP-DAG in the PI synthase reaction catalyzed by PgsA1. The second ORF encodes the PI synthase PgsA1 described earlier (Jackson et al., 2000). The third ORF (*Rv2611c*) encodes a protein with similarities to bacterial acyl-transferases, and the fourth and fifth ORFs encode PimA and a putative GDP-Mannose hydrolase (Rv2609c), respectively. This genetic organization suggests that Rv2611c is involved in the acylation of the MPI anchor of PIM, LM and LAM, and this was confirmed when we showed that Rv2611c catalyzes the 6-O-acylation of the Man$p$ residue linked to 2-position of Ins in PIM$_1$ and PIM$_2$ (Kordulakova et al., 2003) (Fig. 2). An *M. smegmatis* mutant lacking this enzyme accumulates PIM$_1$ and PIM$_2$, and had very much lower Ac$_1$PIM$_1$ and Ac$_1$PIM$_2$ contents than the wild-type parent. These alterations in the PIM content of the mutant were responsible for severe growth defects that could be partially reversed by removing Tween 80 from the culture medium and increasing the concentration of NaCl. Although these results are suggestive of redundant MPI-acylating activities in *M. smegmatis*, such may not be the case in *M. tuberculosis*, in which we found *Rv2611c* to be an essential gene (G. Stadthagen, M. Jackson, B. Gicquel, unpublished results). Although Rv2613c and Rv2609c are also likely to participate in the formation or regulation of the synthesis of the early forms of PIM, their precise functions remain to be characterized. Likewise, the acyl-transferases responsible for the Ins acylation leading to the formation of Ac_2PIM_1 and Ac_2PIM_2 are not known. Based on recent biochemical evidence (Morita et al., 2004), Ac_2PIM_1 seems to be an end-product rather than an intermediate in the biosynthesis of PIM, LM, and LAM in *M. smegmatis*. It is not clear whether this is also the case for Ac_2PIM_2. It is also possible that these highly acylated PIM species serve as stable storage forms of apolar PIM that could reenter the PIM/LM/LAM pathways upon deacylation. The existence of such deacylating activities was suggested by pioneering studies (Brennan and Ballou, 1968).

PimB (Rv0557) was identified as the second mannosyl-transferase involved in the transfer of a Man*p* unit from GDP-Man*p* to the 6-position of the Ins ring of Ac_1PIM_1 to form Ac_1PIM_2 (Schaeffer et al., 1999). PimB is apparently a nonessential enzyme in *M. tuberculosis* (cited in Kremer et al., 2002 as L. DesJardin, G.S. Besra and L. Schlesinger, unpublished work) suggesting that more than one mannosyl-transferase participate in the formation of the di-mannosylated forms of PIM. Given that two distinct pathways apparently lead to the formation of the major apolar PIM species, Ac_1PIM_2, from PIM_1 (Kordulakova et al., 2003) (Fig. 2), it is expected that in the absence of PimB, an as yet unidentified mannosyl-transferase transfers Man*p* to the 6-position of the Ins ring of PIM_1 before the acylation of the resulting di-mannosylated product by Rv2611c.

PimC from *M. tuberculosis* CDC1551 is another enzyme that, when overproduced in *M. smegmatis*, displays GDP-Man-dependent α-mannosyl-transferase activity on apolar PIM in a cell-free assay (Kremer et al., 2002). PimC mediates the transfer of the third Man*p* residue from GDP-Man*p* to form trimannosylated PIM (Fig. 2). However, *pimC* is not present in all *M. tuberculosis* isolates, and its disruption in *M. bovis* BCG does not affect the production of PIM, LM or LAM (Kremer et al., 2002). PIM_3 is an obligate intermediate in the biosynthetic pathway of polar PIM, LM and LAM, so alternate routes must exist for its synthesis.

PimA, PimB, and PimC all belong to the GT-B structural superfamily of glycosyl-transferases (Liu and Mushegian, 2003) and also to the CAZy family 4 (http://afmb.cnrs-mrs.fr/CAZY/) of glycosyl-transferases, characterized by an α-retaining catalytic mechanism for glycosyl transfer and a conserved [D/E]-X_7-E motif involved in sugar-nucleotide binding (Abdian et al., 2000; Geremia et al., 1996) (M. Guerin, A. Buschiazzo, P. Alzari, unpublished re-

sults). They are all membrane-associated enzymes (Kordulakova et al., 2003; Kremer et al., 2002; Morita et al., 2005; Schaeffer et al., 1999) and catalyze glycosyl transfer on the cytosolic face of the plasma membrane, consistent with their use of GDP-Man*p* as Man*p* donor. The *M. tuberculosis* H37Rv genome is predicted to encode four more as yet uncharacterized CAZy family-4 glycosyl-transferases (Rv1212c, Rv3032, Rv0225, Rv2188c), all of which are candidate enzymes for the compensatory mannosyl-transferase activities mentioned earlier, with the exception of the probable glycogen synthase Rv1212c (M. Jackson, T. Sambou, A. Lemassu, M. Daffé, unpublished work).

Besra and coworkers (1997) and Morita and associates (2004) used cell-free assays in the presence of amphomycin, a lipopeptide antibiotic that specifically inhibits polyprenyl-P-requiring synthases. These studies established that most, if not all, of the mannosylation steps downstream from PIM_2 or PIM_3 are catalyzed by C_{35}/C_{50} polyprenyl monophosphomannose (C_{35}/C_{50}-P-Man*p*)-using mannosyl-transferases, most probably on the periplasmic face of the plasma membrane. The presence of amphomycin does not affect the production of PIM_1 and PIM_2 by membranes or cell lysates of *M. smegmatis*; it completely abolishes the synthesis of all forms of PIM_4, PIM_5, PIM_6, and an α(1→6)-linked "linear" form of LM. C_{35}/C_{50}-P-Man*p* is synthesized from GDP-Man*p* and polyprenyl phosphates by the polyprenol monophosphomannose (PPM) synthase, encoded by the *ppm1* (*Rv2051c*) gene (Gurcha et al., 2002). No attempts to disrupt the mycobacterial PPM synthase have been reported, but the disruption of the ortholog of *ppm1* in *Corynebacterium glutamicum* resulted in a mutant with an altered growth rate. The mutant produced wild-type apolar PIM (i.e., PIM_2) but failed to synthesize C_{55}-P-Man*p* and totally lacked mature LM- and LAM-like lipoglycans (Gibson et al., 2003c). Complementation of the mutant with the *M. tuberculosis ppm1* gene restored wild-type C_{55}-P-Man*p* and lipoglycan profiles, suggesting conserved biosynthetic machineries in the two bacilli. This finding also highlights the usefulness of other members of the *Corynebacteriaceae* family for studying the biogenesis of the mycobacterial cell wall. The specific (C_{35}/C_{50}-P-Man*p*)-using mannosyl-transferase responsible for the synthesis of PIM_4 has not been identified. In view of the central role of this PIM intermediate at the branch point at which PIM and LM/LAM biosynthesis bifurcates (Kovacevic et al., 2006; Morita et al., 2006), it is expected to be an essential enzyme.

Morita and coworkers (2006) performed sequence similarity searches with the eukaryotic dolichol-phosphate-mannose-using enzyme PIG-M and transmembrane topology predictions and thereby identified six putative glycosyl-transferases in the genome of *M. tuberculosis*, namely, Rv1159, Rv1459c, Rv2673, Rv2174, Rv2181, and Rv0051, likely to use polyprenyl-linked sugar donors. All of these enzymes belong to the recently described GT-C superfamily of integral membrane-inverting glycosyl-transferases, characterized by multiple (8 to 13) transmembrane domains and the presence of conserved charged residues in the amino-terminal extra-cytosolic loop that are involved in substrate binding or catalysis (Liu and Mushegian, 2003). Interestingly, representatives of GT-C are found in all sequenced eukaryotic genomes but not in archae; mycobacteria and related Actinomycetes are the only prokaryotes in which these enzymes are found (Liu and Mushegian, 2003). The function of Rv1459c has not been discovered, but preliminary results obtained in *Corynebacterium matruchotii* by Takayama's group suggest that it participates in mycolic acid synthesis rather than in the formation of PIM/LM/LAM (Wang et al., 2006).

Other *M. tuberculosis* proteins predicted to belong to the GT-C superfamily include Rv1002c, a mannosyl-transferase involved in the O-mannosylation of proteins (VanderVen et al., 2005), the arabinosyl-transferase AtfA (Rv3792) catalyzing for the first arabinosylation step in arabinogalactan synthesis (Alderwick et al., 2006b), the Man-capping enzyme Rv1635c (Dinadayala et al., 2006) (see later), the three Emb proteins (Rv3793 to Rv3795) (Berg et al., 2005) (see later), and three proteins of unknown function, Rv0236c, Rv1508c, and Rv3779 (Kaur et al., 2006). Targeted inactivation of the ortholog of *Rv1159* (*pimE*) in *M. smegmatis* resulted in a mutant with wild-type LM and LAM profiles but lacking all forms of PIM_6 and accumulating Ac_1PIM_4 and Ac_2PIM_4 in its place (Morita et al., 2006). Metabolic pulse chase labeling and cell-free assays further demonstrated that the mutant was defective in PIM_5 and PIM_6 synthesis, leading to the proposal that Rv 1159 is the $\alpha(1\rightarrow2)$-mannosyl-transferase required for the production of the terminal polar forms of PIM, PIM_5, and PIM_6. It is not clear whether Rv1159 catalyzes the transfer of only the fifth mannose onto Ac_1PIM_4 (and/or Ac_2PIM_4) or whether it catalyzes the transfer of both the fifth and sixth mannoses.

lpqW (*Rv1166*), a gene mapping close to *pimE* in the chromosome of *M. tuberculosis*, encodes a putative lipoprotein that was suggested to be involved in channeling early PIM intermediates into polar PIM or LM/LAM biosynthesis (Kovacevic et al., 2006). Disruption of the *lpqW* gene of *M. smegmatis* yielded a mutant with wild-type apolar and polar PIM profiles but that was unable to make LAM, al-

though it apparently retained the ability to produce LM. Interestingly, the mutant displayed an unstable phenotype and rapidly recovered the ability to synthesize LAM at the expense of apolar PIM. Like the *pimE* mutant, this "second generation-mutant" accumulated the branch-point intermediate Ac_1PIM_4. The precise determination of the function of LpqW is expected to provide key information about the signals and mechanisms underlying the regulation of PI-based glycolipid and lipoglycan synthesis. The crystal structure of the LpqW protein of *M. smegmatis* has recently been reported (Marland et al., 2005, 2006).

Another enzyme described as being involved in the biosynthesis of PIM is Rv2252 (Owens et al., 2006). Purified Rv2252, which harbors the consensus motif of diacylglycerol kinases, can phosphorylate a variety of host and bacterial amphipatic lipids in vitro, including *M. tuberculosis* diacylglycerol. Disruption of *Rv2252* in *M. tuberculosis* resulted in a mutant with reduced Ac_2PIM_2 and Ac_2PIM_6 contents, leading the authors to propose that Rv2252 is involved in the phosphorylation of DAG to yield phosphatidic acid, which is, in turn, used in the formation of CDP-DAG and PI. Given the nonessential character of this enzyme, mycobacteria presumably have other sources of phosphatidic acid.

Finally, PimF (Rv1500) was initially reported as a mannosyl-transferase involved in the synthesis of PIM_6/PIM_7, LM, and LAM (Alexander et al., 2004) but was more recently shown to be a glycosyl-transferase (now termed LosA) required for the final assembly of trehalose-based lipooligosaccharides (LOS) (Burguière et al., 2005).

An investigation of the subcellular localization of the enzymes involved in PIM synthesis demonstrated compartmentalization of PIM biosynthetic enzymes within the plasma membrane reflecting the nature of the mannose donor used in the different glycosylation steps (Morita et al., 2005, 2006). The early steps of PIM biosynthesis were localized in a subfraction of the plasma membrane, termed PM_f, devoid of cell wall components and carrying the two cytosolically oriented GDP-Man-dependent mannosyl-transferases, PimA and PimB. In contrast, the biosynthesis of polar PIM requiring (C_{35}/C_{50}-P-Man$p$)-using mannosyl-transferases and, thus, expected to occur on the external face of the plasma membrane, was exclusively associated with another subfraction, termed PM-CW, rich in cell wall fragments. Interestingly, these two subfractions also differed in their lipid compositions, and significantly more of the early PIM intermediate Ac_1PIM_2 was found in PM-CW than in PM_f. The biosynthesis of PIM is thus topologically complex and probably requires the active translocation of intermediates, most likely Ac_1PIM_2, from the cytosolic to the "periplasmic" side of the plasma membrane. None of the transporters involved has been identified.

Biogenesis of LM and LAM

The $\alpha(1\rightarrow6)$-mannosyl-transferases required for the polymerization steps leading to the linear mannan backbone of LM from PIM$_4$ have not been identified. From the various studies described earlier (Besra et al., 1997; Morita et al., 2004), it can be inferred that most if not all of these enzymes (as well as that responsible for the production of PIM$_4$) probably use polyprenyl-linked mannose as Man$p$ donor. Hence, good candidates for these functions are the GT-C proteins described in the previous section. Our laboratories, in collaboration with others, are currently testing for the involvement of these glycosyl-transferases in the biogenesis of cell wall glycoconjugates. It has been shown that Rv2181 is responsible for the $\alpha(1\rightarrow2)$-linked branching of the mannan core of LM and LAM (Kaur, 2006) (Fig. 2). The Colorado State University group also provided evidence that Rv1635c is the mannosyl-transferase catalyzing the transfer of the first Man$p$ residue of the Man-cap of LAM (Dinadayala et al., 2006) (Fig. 2). Disruption of *Rv1635c* in *M. tuberculosis* by transposon mutagenesis yielded a mutant devoid of Man-capping. Conversely, the expression of this gene, normally restricted to Man-capping species of mycobacteria, in *M. smegmatis* resulted in a recombinant strain producing a LAM capped with one single Man$p$ linked $\alpha(1\rightarrow5)$ to the terminal β-Ara$f$ residues. The *Rv1635c* mutant of *M. tuberculosis* will be invaluable for studying the contribution of ManLAM to the immunopathogenesis of tuberculosis.

Mature LM is thought to be glycosylated with arabinan to form LAM. However, relatively little is known about the arabinosylation steps involved. Polyprenol-monophosphoryl-β-D-arabinose (C_{35}/C_{50}-P-Ara$f$) is the only known Ara$f$ donor in mycobacteria (Wolucka et al., 1994). It is synthesized from 5-phosphoribose-pyrophosphate (Scherman et al., 1996) through four catalytic steps that have been characterized (Huang et al., 2005; Mikusova et al., 2005). The first committed step in this synthesis is catalyzed by the product of *Rv3806c*, a 5-phospho-α-D-ribose-1-diphosphate: decaprenyl-phosphate 5-phosphoribosyltransferase. Consistent with the central role of C_{35}/C_{50}-P-Ara$f$ in arabinan synthesis, disruption of the ortholog of *Rv3806c* (*ubiA*) in *C. glutamicum* by allelic replacement resulted in a mutant totally devoid of cell wall arabinan (Alderwick et al., 2005). The same enzyme is predicted to be essential in *M. tuberculosis*.

The only enzymes shown to participate in arabinan synthesis are AtfA (Rv3792) and the Emb proteins (Rv3793-Rv3795) involved in resistance to ethambutol (Belanger et al., 1996; Telenti et al., 1997). Evidence was recently provided that AtfA specifically catalyzes the first arabinosylation step in arabinogalactan synthesis (Alderwick et al., 2006b). The targeted disruption of *embA*, *embB*, and *embC* in *M. smegmatis* and chemical analyses of the mutants indicate that EmbA and EmbB participate in the formation of the Ara$_6$ motif characteristic of the arabinan domain of arabinogalactan, and that EmbC (Rv3793) is specifically required for the arabinosylation of LAM (Escuyer et al., 2001; Zhang et al., 2003). Inactivation of *embC* in *M. smegmatis* results in complete cessation of LAM synthesis without affecting the production of PIM, LM or arabinogalactan (Zhang et al., 2003). Our preliminary data indicate that *embC* is an essential gene in *M. tuberculosis* (G. Stadthagen, B. Gicquel, and M. Jackson, unpublished results). Although definitive proof that Emb proteins are arabinosyl-transferases is still lacking, the finding that these proteins share the topology and glycosyl-transferase motif of superfamily GT-C enzymes, and results from site-directed mutagenesis of the GT-C motif of EmbC tend to substantiate this assumption (Berg et al., 2005). In addition to the GT-C motif, all Emb proteins harbor a proline-rich motif homologous to that of other bacterial cytosolic membrane proteins involved in the regulation of polysaccharide biosynthesis, known as polysaccharide copolymerases (Morona et al., 2000). Amino acid substitutions within this motif lead to the synthesis of a LAM with a smaller arabinan structure than wild-type LAM (owing in part to a substantial decrease in the linear nonreducing Ara$_4$ termini of LAM) but in which relative linkage composition remained almost unchanged (Berg et al., 2005). It was proposed that, in addition to its arabinosyl-transferase activity, EmbC regulates the polymerization of the arabinan of LAM. It was subsequently shown that the C-terminal domain of EmbC is directly involved in the linear extension of LAM (Shi et al., 2006). It has also been suggested that the transmembrane segments of the Emb proteins are involved in translocating arabinan segments across the plasma membrane (Escuyer et al., 2001; Zhang et al., 2003). Emb proteins may thus be multifunctional proteins expressing catalytic, translocating and regulating activities. Although much remains to be done in describing the roles of EmbC in the formation of LAM, these findings and parallels with the polymerization of polysaccharides in other bacteria led Berg and coworkers (2005) to propose various models for LAM synthesis, dominated by the EmbC-dependent formation of a nonbranched chain of 5-linked α-Ara*f* sugars built on a polyprenyl diphosphoryl lipid, followed by polymerization/ligation events.

EmbR is a protein homologous to the OmpR class of transcriptional regulators implicated in the regulation of the *embCAB* operon (Belanger et al., 1996; Sharma et al., 2006b). EmbR is phosphorylated by the serine/threonine kinase PknH (Molle et al., 2003), and there is evidence that the phosphorylation of EmbR by PknH enhances its binding to the promoter region of *embCAB*, resulting in a positive regulatory effect on the transcription of the operon. The upregulation of the expression of *embC* via PknH results in an increased LAM/LM ratio in *M. smegmatis* (Sharma et al., 2006b). Recent in vitro evidence suggests that EmbR is also the substrate of other serine/threonine kinases (PknA and PknB) and of a serine/threonine phosphatase, Mstp (Sharma et al., 2006a). Whether these newly identified interactions affect the transcription of *embCAB* in vivo remains to be determined. The structure of EmbR has been solved (Alderwick et al., 2006a).

The biosynthetic origin of the 5-deoxy-5-methylthio-xylofuranose (MTX) motif linked to Man*p* residues in the LAM from *M. tuberculosis* and *M. kansasii* (Joe et al., 2006; Treumann et al., 2002; Turnbull et al., 2004; Ludwiczak et al., 2002) has not yet been investigated. Nevertheless, it has been suggested that MTX is derived from a byproduct of polyamine biosynthesis (Turnbull et al., 2004). The glycosyl-transferase subsequently involved in its α(1→4) linkage to Man*p* residues of LAM is not known.

Roles of PIM, LM, and LAM in the Physiology of Mycobacteria

Defective or deficient PIM/LM/LAM synthesis is associated with lethality or growth defects, and this raises the issue of the contribution of these complex molecules to the physiology of *Mycobacterium sp.* In *M. bovis* BCG, PI and PIM make up as much as 56% of all phospholipids in the cell wall and 37% of those in the plasma membrane. Therefore, they are considered to be important structural components acting as "cementing substances" for the cell wall skeleton (Goren, 1984). Supporting this notion, we and others have found that the PIM_2 content of the cell envelope of *M. smegmatis* determines the sensitivity of the bacterium to both hydrophobic and hydrophilic antibiotics (Kordulakova et al., 2002, Parish et al., 1997). Also, a decrease in PIM_2 content is associated with a decreased permeability to hydrophobic molecules including chenodeoxycholate (Parish et al., 1997).

Other possible clues as to the functions of PIM and metabolically related lipoglycans can be found in the morphotypes of the various viable PIM/LM/LAM mutants that have been generated, and by analysis of the environmental factors affecting their synthesis and examination of their production rate during growth. The rate of biosynthesis of PIM, LM, and LAM is apparently high in actively dividing cells and low during stationary phase (Haites et al., 2005; Morita et al., 2005). The available sources of carbon and nitrogen have considerable effects on the production of polar and apolar PIM (Dhariwal et al., 1977). Concentrations of polar and apolar PIM relative to other phospholipids drop with a decrease in growth temperature from 37°C to 27°C, suggesting that the regulation of PIM concentrations is important in maintaining membrane fluidity in response to temperature changes (Dhariwal et al., 1977; Taneja et al., 1979). Several reports also indicate that the phospholipid and PIM contents of *M. smegmatis* remain relatively constant during bacterial growth, whereas the levels of polar PIM increase with the age of the culture (Penumarti and Khuller, 1983). LM and LAM were reported to be detectable in *M. smegmatis* grown in rich medium when the cultures reached stationary phase (Morita et al., 2006). The rate of apolar PIM synthesis decreases when *M. smegmatis* enters stationary phase (Morita et al., 2005), so these observations suggest that apolar PIM are progressively chased into polar PIM, perhaps at the expense of LM and LAM, when cell division ceases. A study aimed at analyzing the effects of inositol starvation on the PI-based lipoglycan contents and viability of *M. smegmatis* demonstrated that inositol starvation in actively dividing cells resulted in the rapid depletion of PI and apolar PIM following their channeling into polar PIM and LAM synthesis (Haites et al., 2005). Interestingly, depletion of apolar PIM was not associated with loss of viability, indicating that the higher mannosylated types are sufficient to sustain cell viability. Conversely, the fact that polar PIM are inessential for the growth of *M. smegmatis* (Morita et al., 2006) indicates that their functions can be fulfilled, at least in part, by less mannosylated PIM species or LM/LAM. Morita and coworkers' findings for an *M. smegmatis pimE* mutant suggest that polar PIM contribute to the hydrophobicity of the cell surface and membrane integrity (Morita et al., 2006). Changes in the PIM_2 composition of *M. smegmatis* were shown to have much more dramatic effects on growth, consistent with PIM_2 being important early intermediates in the biosynthesis of polar PIM, LM, and LAM (Kordulakova et al., 2003; Parish et al., 1997). An *Rv2611c* mutant of *M. smegmatis* with defects in the acylation of PIM could only be grown in a medium devoid of detergent (Tween 80) and containing high concentration of

NaCl (Kordulakova et al., 2003). These findings further implicate certain forms of PIM or metabolically derived products in membrane stability. An *M. smegmatis* mannose auxotroph grown in the absence of exogenous mannose, and thus impaired in the synthesis of all mannose-containing glycoconjugates, was found to exhibit a hyperseptation phenotype (Patterson et al., 2003). It was proposed that the various PIM species may not be evenly distributed or synthesized in the plasma membrane and that the existence of such subdomains may influence septum formation and cell division (Morita et al., 2005).

In conclusion, although the diversity and complexity of the envelope compositions of the different mutants do not allow specific functions to be attributed to any given PIM, LM, or LAM molecule, common or recurrent phenotypic traits are beginning to emerge implicating PIM and related products in (1) the permeability barrier of the cell envelope, (2) cell membrane integrity or functions and, (3) regulation of cell septation and cell division.

ANTIGENS OF INNATE AND ADAPTIVE IMMUNITY

Lipoglycans Are Ligands of C-Type Lectins

ManLAM, LM and PIM are pathogen-associated molecular patterns (PAMPs) recognized by host cell pattern-recognition receptors (PRRs), the C-type lectins, mannose receptor (MR) (Schlesinger et al., 1994, 1996; Venisse et al., 1995), mannose-binding protein (MBP) (Hoppe et al., 1997; Polotsky et al., 1997), surfactant protein A (SP-A) (Sidobre et al., 2000, 2002), surfactant protein D (SP-D) (Ferguson et al., 1999), and DC-SIGN (Geijtenbeek et al., 2003; Maeda et al., 2002; Tailleux et al., 2003c). C-type lectins recognize the conserved carbohydrate recognition domains (CRDs) of mannose-containing glycoconjugates in a Ca^{2+}-dependent manner (for review, see Figdor et al., 2002). MR and DC-SIGN are transmembrane receptors (Figdor et al., 2002). MR mediates entry of mycobacteria into monocyte-derived macrophages (MDM) (Astarie-Dequeker et al., 1999; Schlesinger, 1993), whereas DC-SIGN specifically binds *M. tuberculosis* complex species on human monocyte-derived DCs (Geijtenbeek et al., 2003; Tailleux et al., 2003c) and alveolar macrophages (Tailleux et al., 2005). MBP, SP-A, and SP-D are soluble collectins possibly involved in the opsonic phagocytosis of mycobacteria by either increasing (MBP, SP-A) or reducing (SP-D) their uptake (Ferguson et al., 1999; Hoppe et al., 1997; Polotsky et al., 1997).

The role of lipoglycans as ligands of C-type lectins has been investigated using cells or lipoglycans from different mycobacterial species and different types of phagocytic cells or cell lines transfected with various receptors (for detailed reviews, see Ehlers and Daffe, 1998; Ernst, 1998; Fenton et al., 2005; Gilleron et al., 2001a; Tailleux et al., 2003a). These studies have established the broad outline of the molecular bases of lipoglycan binding to lectins. Two kinds of experiments have been performed: determining the ability of soluble or bead-coated molecules to (1) bind the purified receptor or the cell-surface expressed receptor, and (2) inhibit the binding of the whole bacterium to the purified receptor or the receptor on the cell surface. Binding of ManLAM requires both the mannosyl caps and fatty acyl groups. ManLAM devoid of mannosyl caps, following α-mannosidase treatment, and PILAM and AraLAM that both do not bear mannose caps do not or only weakly bind C-type lectins. Mannosyl caps are the epitopes recognized by the CRDs of C-lectins. Crystal structures of CRDs bound to oligosaccharides show that the interactions with the Ca^{2+} site involve equatorial OH-3 and OH-4 hydroxyl groups of the terminal (or the penultimate in the case of DC-SIGN) sugar (Feinberg et al., 2000, 2001; Weis et al., 1992). In contrast, fatty acyl residues are not involved in the interaction but rather contribute to ManLAM clustering (Riviere et al., 2004). This supramolecular organization results in a huge increase in ManLAM valency, thus allowing a high avidity for C-type lectins through multivalent interaction with the multiple CRD of the lectins (Sidobre et al., 2000, 2002). Mannosyl residues of the mannan domain are not involved in the binding, most probably because of steric hindrance involving the bulky arabinan domain (Sidobre et al., 2000). The K_D value has been estimated by surface-plasmon-resonance to be around 10^{-6} M for the interaction between *M. bovis* BCG ManLAM and SP-A (Riviere et al., 2004). Some subtle differences in MR binding by ManLAM from different *M. tuberculosis* species have been reported (Schlesinger et al., 1996), but they cannot be explained by the known structural differences between the molecules. LM and PIM binding to C-type lectins probably involves their terminal mannosyl residues that are readily accessible to the receptor. Recently, a fine discrimination has been evidenced in MR and DC-SIGN recognition of individual PIM species according to their acylation or mannosylation patterns (Torrelles et al., 2006).

The subtleties of structure/function relationships (Schlesinger et al., 1996) deserve to be studied in greater detail, but it is clear that soluble lipoglycans are ligands of C-type lectins, and lipoglycans may serve as ligands for the phagocytosis of mycobacteria through these receptors. However, great care must be taken before concluding about ligand specificity from experiments using purified molecules. The fact that a purified molecule can inhibit the binding of a bacterium to a receptor or can bind to the receptor itself does not necessarily demonstrate the involvement of this compound in the binding of the whole bacterium. An illustrative example is provided by work dedicated to decipher the molecular bases of *M. tuberculosis* binding to DC-SIGN. ManLAM was proposed by others and us (Geijtenbeek et al., 2003; Koppel et al., 2004; Maeda et al., 2002; Tailleux et al., 2003c) as a possible DC-SIGN ligand in the *M. tuberculosis* cell envelope because (1) it could bind to DC-SIGN-Fc chimeras in solution, (2) it could inhibit *M. tuberculosis* and *M. bovis* BCG binding to DC-SIGN-expressing recombinant cells, (3) removal of the mannose caps on ManLAM abolished the inhibitory effect of the molecule, (4) synthetic mannosyl caps can bind to DC-SIGN-Fc chimeras in solution, and (5) fast growers, such as *M. smegmatis*, bound only very weakly to the lectin, and PILAM isolated from this species could not inhibit *M. tuberculosis* binding to DC-SIGN. However, we later found that other slow-growing mycobacterial species of the *M. tuberculosis* complex, *M. avium*, *M. xenopi*, *M. marinum*, and *M. kansasii* could not bind DC-SIGN, although they produce ManLAM exposed on their cell surface (at least in *M. xenopi*) (Pitarque et al., 2005) (see "Localization in the Cell Envelope" above). Moreover, ManLAM purified from all these strains are able to inhibit *M. tuberculosis* binding to DC-SIGN (Pitarque et al., 2005). *M. tuberculosis* has an envelope exceptionally rich in glycoconjugates and most particularly mannoconjugates, and the epitopes recognized by the C-type lectins are borne by several cell surface glycoconjugates, including glycoproteins, that are all putative ligands. Moreover, cooperation between several ligands in the binding of *M. tuberculosis* to these receptors is also plausible. To demonstrate formally that lipoglycans indeed constitute ligands in the *M. tuberculosis* envelope during its phagocytosis, lipoglycan-deficient strains are required. Consequently, the very recent report of a *M. tuberculosis* mutant deficient in mannose cap biosynthesis is of particular interest (Dinadayala et al., 2006). Its ability, relative to that of the wild-type strain, to bind the various receptors should be very illuminating about the role of ManLAM as a mycobacterial adhesin and an immunomodulatory molecule (see later). In a similar way, the construction of a PIM_6-deficient *M. smegmatis* strain (Morita et al., 2006) should help clarify the role of these molecules.

Lipoglycan Immunomodulatory Activities

Alteration of phagocytic cell functions

The ability of soluble lipoglycans to bind C-type lectins and TLR2 is of particular interest because mycobacterial compounds, including lipoglycans and PIM, are delivered from infected macrophages, through exosomes or apoptotic vesicles, to non-infected bystander DCs (Beatty et al., 2000; Rhoades et al., 2003; Schaible et al., 2003). So these molecules can stimulate bystander cells, even though they are not necessarily receptor ligands on the whole bacterium. This "detour pathway" is thought to be critical for eliciting an efficient immune response, including the TLR-mediated production of proinflammatory cytokines [see "Lipoglycans as agonists of Toll-like receptor-2 (TLR2)" below] and the presentation of antigens through CD1 to T cells (see paragraph "Lipoglycans as ligands of CD1 proteins") (Beatty et al., 2000; Rhoades et al., 2003; Schaible et al., 2003).

ManLAM modulate the functions of macrophages and DCs. They are mostly involved in inhibition of cellular responses including (1) the IFN-γ-mediated activation of macrophages, (2) the LPS-induced production of the Th1 pro-inflammatory cytokines IL-12 and TNF-α by macrophages and DCs, (3) *M. tuberculosis*-induced macrophage apoptosis, and (4) phagolysosome biogenesis in macrophages (for detailed references, see the reviews by Briken et al., 2004; Nigou et al., 2002b, 2003) (Fig. 3). All of these activities that contribute to inhibiting the host defense response parallel those observed for whole *M. tuberculosis* bacteria (Briken et al., 2004; Koul et al., 2004) implicating ManLAM in the ability of the pathogen to survive inside the infected host. Cross-linking of ManLAM to the C-type lectins MR or DC-SIGN seems to be required, at least for two of these inhibitory activities. The body of data now available suggests that all of these activities may result from the activation of common signaling pathways triggered by ManLAM binding to these lectins: MR on macrophages or DC-SIGN on DCs. The evidence supporting this hypothesis is presented below for each of these four inhibitory activities.

The structural requirements concerning ManLAM for IL-12 inhibitory activity on LPS-stimulated DCs (i.e., the presence of both the mannosyl caps and fatty acyl residues) are the same as those for binding to C-type lectins (see earlier) (Nigou et al., 2001, 2002b). The inhibitory effect of ManLAM can be mimicked by an agonist anti-MR monoclonal antibody and by mannan from *Saccharomyces cerevisiae*, a known agonist of the MR. This led us to

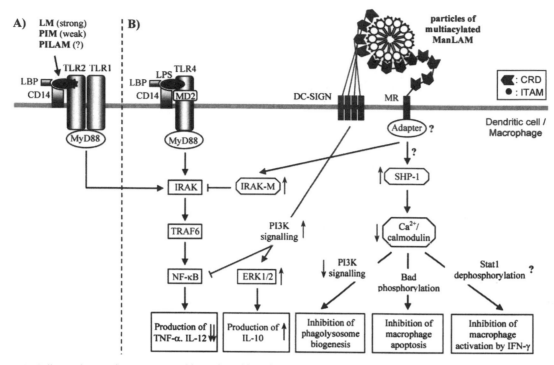

Figure 3. Cell signaling pathways triggered by PI-based lipoglycans. (A) LM (Quesniaux et al., 2004; Vignal et al., 2003), and to a lesser extent PIM (Gilleron et al., 2003), activate macrophages and DCs through a TLR2/TLR1-dependent but TLR6-independent pathway that requires MyD88 (Quesniaux et al., 2004). Only Ac$_3$ LM and Ac$_4$LM are active (Gilleron et al., 2006), whereas the residual PIM activity is independent of the acylation degree (from one to four fatty acids) (Gilleron et al., 2003). It is not known whether lipoglycans are presented to the receptor in a monomeric or multimeric form. ManLAM and AraLAM do not signal through TLR2 as a consequence of steric hindrance: the arabinan domain masks the lipomannan moiety of the molecule (Guerardel et al., 2003). The molecular bases of PILAM activity are not clear yet. (B) ManLAM inhibits IL-12 and TNF-α (Nigou et al., 2001) and induces IL-10 production by LPS-stimulated DCs through DC-SIGN ligation (Geijtenbeek et al., 2003). The signaling pathway involves activation of PI3K and ERK1/2 (Caparros et al., 2006). In macrophages, ManLAM inhibits the LPS-induced production of TNF-α and IL-12 (Knutson et al., 1998), independently of IL-10 production, through IRAK-M activation (Pathak et al., 2005). ManLAM exerts other inhibitory activities on macrophages including inhibition of IFN-γ-mediated activation (Sibley et al., 1988), *M. tuberculosis*-induced apoptosis (Rojas et al., 2000), and phagolysosome biogenesis (Fratti et al., 2003). Phagolysosome biogenesis is associated with ManLAM binding to MR (Kang et al., 2005) and requires inhibition of both the cytosolic Ca^{2+} rise/calmodulin pathway and PI3K signaling (Vergne et al., 2003). Inhibition of apoptosis (Rojas et al., 2000) and possibly IFN-γ-mediated activation (Briken et al., 2004) are also dependent on the alteration of Ca^{2+}-dependent intracellular events, suggesting that they could be also both mediated by MR. LM and PIM also bind MR and DC-SIGN (Pitarque et al., 2005; Torrelles et al., 2006); however, little is known about the functional consequences. LM induces a TLR2-dependent production of proinflammatory cytokines but concomitantly inhibits, most probably through C-type lectin binding, TLR4-mediated cytokine production (Quesniaux et al., 2004). The net cytokine response is dependent on the receptor equipment of the cells as well as the LM used and their acylation degree (Quesniaux et al., 2004). *(See the color insert for the color version of this figure.)*

propose that the inhibitory effect on ManLAM on DCs was mediated by MR (Nigou et al., 2001, 2002b). However, it was later shown that ManLAM binding to DCs is not inhibited by anti-MR but rather by anti-DC-SIGN antibodies, suggesting that DC-SIGN is the major receptor of ManLAM on DCs (Geijtenbeek et al., 2003) and, as a consequence, the mediator of IL-12 inhibition on DCs. This has been confirmed by data showing that a blocking anti-DC-SIGN antibody can inhibit ManLAM-induced IL-10 production by LPS-stimulated DCs (Geijtenbeek et al., 2003). Why ManLAM only binds DC-SIGN on

DCs although MR is expressed on these cells remains unexplained. Differences between the expression of these two receptors or different avidities for Man LAM are possible explanations. However, ManLAM not binding MR on DCs does not conflict with the ability of MR to trigger an anti-inflammatory signal. Indeed, this has been confirmed by other authors showing that cross-linking MR with various ligands (a mucin or a proteoglycan) as well as an anti-MR antibody, different from that used in our study, resulted in inhibition of IL-12 and TNF-α production by LPS-stimulated DCs (Chieppa et al., 2003).

Similarly, using small interfering RNA gene silencing, a negative regulatory role of MR on proinflammatory cytokine release by human alveolar macrophages has been demonstrated (Zhang et al., 2005). ManLAM has also been reported to inhibit IL-12 and TNF-α in human THP-1 (Knutson et al., 1998) and murine RAW (Pathak et al., 2005) macrophage cell lines although DC-SIGN is absent (Puig-Kroger et al., 2004), so MR is likely to be the mediator of the effects in these cells.

How DC-SIGN or MR signal into the cells and interfere with TLR4 signaling is not yet clear (Fig. 3). The presence of immunoreceptor tyrosine-based activation motifs (ITAMs) in its cytosolic tail suggests that DC-SIGN is capable of direct signaling (Engering et al., 2002). However, MR has no signaling motif in its cytosolic domain, raising the intriguing question as to whether it associates with adapter molecules to transduce signals (Chieppa et al., 2003). Recently, Pathak and colleagues (2005) demonstrated that ManLAM dampens IL-12 in a murine macrophage cell line, independently of IL-10 production, by directly acting on the TLR4 signaling cascade through an IRAK-M-mediated pathway. ManLAM induces the expression of IRAK-M, which can compete with IRAK1 for binding to TRAF6 and thus inhibit NF-κB activation (Liew et al., 2005) (Fig. 3). IRAK-M is a negative regulator of TLR4 signaling mostly produced by peripheral blood leukocytes (Liew et al., 2005). Modulation of cytokine production through DC-SIGN seems to proceed by another pathway. ManLAM binding to DC-SIGN increases IL-10 production by LPS-stimulated DCs (Geijtenbeek et al., 2003), an immunosuppressive cytokine that could be indirectly responsible for IL-12 and TNF-α inhibition (Nigou et al., 2001, 2002b). A very recent paper has thrown more light on the signaling pathway downstream from DC-SIGN: Caparros and coworkers (2006) report that DC-SIGN engagement by specific antibodies activates PI3 kinase (PI3K) and ERK1/2 in DCs. These results are of particular interest because activation of PI3K is involved in the negative regulation of IL-12 and TLR signaling (Fukao and Koyasu, 2003), whereas ERK1/2 activation favors IL-10 production (Lucas et al., 2005) (Fig. 3). The role of IL-10 is not clear and might be cell dependent because DC-SIGN ligation by ManLAM does not potentiate IL-10 production by LPS-stimulated alveolar macrophages (Tailleux et al., 2005).

Thus, MR and DC-SIGN trigger different intracellular pathways, but both lead to activation of well-known negative regulators of TLR signaling, IRAK-M and PI3K, respectively. These regulators are believed to help prevent excessive innate immune responses that are detrimental to the host (Fukao and Koyasu, 2003; Liew et al., 2005). They constitute the first identified pieces of the puzzle of the anti-inflammatory signaling cascade elicited by ManLAM binding to C-type lectins.

The second activity linked to ManLAM binding MR is the inhibition of phagolysosome biogenesis (Kang et al., 2005). The fusion of phagosomes containing ManLAM-coated microspheres with lysosomes is significantly reduced in human monocyte-derived macrophages and an MR-expressing cell line but not in monocytes that lack MR or in a DC-SIGN-expressing cell line (Kang et al., 2005). Moreover, a reversal of phagosome-lysosome fusion inhibition was observed following MR blockade by mannan or an anti-MR antibody (Kang et al., 2005).

The receptors for the two remaining activities of ManLAM, inhibition of IFN-γ-mediated macrophage activation or apoptosis have not been identified; however, there is some evidence indicating that the latter could be mediated by MR (Rojas et al., 2000; M. Rojas and L. F. Garcia, personal communication). As recently reviewed by Briken and associates (2004), ManLAM inhibition of Ca^{2+} cell signaling seems to be central to these two activities as well as to inhibition of phagosome maturation (for reviews, see Briken et al., 2004; Koul et al., 2004). ManLAM inhibits a cascade, which is essential for conversion of phagosomes into phagolysosomes, consisting of cytosolic Ca^{2+} transients, calmodulin, and PI3K (Deretic et al., 2006; Vergne et al., 2003). Similarly, Rojas and colleagues (2000) reported that ManLAM counteracts the alterations of Ca^{2+}-dependent intracellular events, including the increase in the permeability of the mitochondrial membrane and caspase activation, that occur during macrophage apoptosis. Inhibition of the cytosolic Ca^{2+} rise/calmodulin pathway may also contribute to the inhibition of IFN-γ signaling (Briken et al., 2004). ManLAM inhibitory activity on phagolysosome biogenesis is associated with its ability to bind the MR (Kang et al., 2005). It is plausible that all of these activities involving an inhibition of cytosolic Ca^{2+} rise are related, albeit by an unknown mechanism, to ManLAM binding to MR (Fig. 3). It is relevant that ManLAM has been shown to increase the activity of the tyrosine phosphatase SHP-1 (Knutson et al., 1998). This protein is required for ManLAM-mediated inhibition of macrophage apoptosis (Rojas et al., 2002) and might also be involved in inhibition of both IFN-γ and Ca^{2+} signaling (Briken et al., 2004). There are also data implicating PI3K in the signaling downstream from MR cross-linking (Chieppa et al., 2003).

It is hoped that future research will identify the links missing between the pieces of this puzzle and provide a better view of the signaling pathway triggered by ManLAM. However, it must not be forgotten that C-type lectin targeting by ManLAM directs them to specific endocytic/phagocytic pathways that may also determine their ability to modulate macrophage or DC functions (Kang et al., 2005).

From all of these data, it is tempting to suggest that pathogenic mycobacteria, through ManLAM binding to MR and DC-SIGN, have adapted so as to be able to exploit the immunosuppressive pathways triggered by these C-type lectins to their own advantage. However, this may also be in favor of the host to limit inflammation-induced immunopathology, for example (Tailleux et al., 2003a; Neyrolles et al., 2006). In addition, the ability of human DCs to control the growth of *M. tuberculosis* could result from the selective entry of the bacilli through DC-SIGN, a step that might direct their intracellular trafficking (Tailleux et al., 2003b). The recent finding that a rise of the level of DC-SIGN expression increases resistance to the development of tuberculosis supports the role played by this receptor in host immunity to this pathogen (Barreiro et al., 2006). However, further work is required to understand fully the functional consequences of DC-SIGN ligation in vivo (Neyrolles et al., 2006).

Lipoglycans as agonists of Toll-like receptor-2 (TLR2)

The successful eradication of *M. tuberculosis* from infected hosts requires a rapid stimulation of innate immunity responses. TLRs play a crucial role in innate immunity by the recognition of molecular patterns associated with mycobacteria. Stimulation of TLRs leads to the secretion by phagocytic cells of antimicrobial peptides and proinflammatory cytokines including TNF-α, IL-12 and IFN-γ. These proinflammatory cytokines favor the development of a Th1-type response, allowing the control of bacillary growth through macrophage activation and granuloma formation (Cook et al., 2004; Kaufmann, 2001, 2005; Stenger and Modlin, 2002).

The role of these cytokines in the host defense against *M. tuberculosis* is supported by the finding that human susceptibility to mycobacterial infections is increased in individuals functionally deficient either in IL-12 and INF-γ cytokines or in their receptors (Altare et al., 1998). Similarly, rheumatoid arthritis patients undergoing an anti-TNF-α therapy show an increased risk of tuberculosis reactivation (Ehlers, 2003). Polymorphisms in TLR2 that blunt NF-κB activation by *M. tuberculosis* seem to enhance susceptibility to tuberculosis in humans (Ben-Ali et al., 2004; Texereau et al., 2005; Yim et al., 2006). However, the role of TLR2 is still not clear. Stimulation of

APCs with mycobacterial lipoproteins through TLR2 results in a reduced MHC-II Ag presentation (Gehring et al., 2004; Noss et al., 2001; Pecora et al., 2006). This may be a negative feedback mechanism for control of inflammation. An elegant study supports a connection between TLR and vitamin D-mediated innate immunity, suggesting that differences in the ability of individuals to synthesize vitamin D, including polymorphism in the vitamin D receptor, may also contribute to differences in susceptibility to tuberculosis (Liu et al., 2006a).

Several mycobacterial components trigger TLRs. TLR2, TLR4, and more recently, TLR1, which heterodimerizes with TLR2, have been implicated in the recognition of *M. tuberculosis* antigens. An important role for TLR2 in immune recognition of *M. tuberculosis* has been demonstrated, and a TLR4-dependent activity has also been reported but only for *M. tuberculosis* (Fenton et al., 2005). Some TLR-activating mycobacterial compounds have been identified over the past 10 years: the 19 kDa (Brightbill et al., 1999), 38 kDa (Brightbill et al., 1999; Jung et al., 2006), LprG (Gehring et al., 2004) and LprA (Pecora et al., 2006) lipoproteins, PIM (Gilleron et al., 2003; Jones et al., 2001), LM (Quesniaux et al., 2004; Vignal et al., 2003), and PILAM (Chatterjee et al., 1992c; Gilleron et al., 1997; Khoo et al., 1995a). All of them are TLR2 agonists; TLR4 ligands still await identification.

PIM_2 and PIM_6, the most abundant PIM glycoforms (see paragraph "Structural characterization"), are weak inducers of TNF-α and IL-12 in murine and human macrophages (Gilleron et al., 2003), an activity mediated by TLR2. This induction requires the PIM fatty acyl appendages but is independent of the acylation degree.

In contrast, LM are strong TLR2 proinflammatory stimuli, but the activity is restricted to Ac_3LM and Ac_4LM acyl-forms (Gilleron et al., 2006). Ac_1LM and Ac_2LM failed to induce TNF-α and IL-12 production from human monocytic cells or primary murine macrophages (Gilleron et al., 2006). Lipopeptides of gram-positive bacteria are also TLR2 ligands. Interestingly, murine TLR2 is able to discriminate between di- and triacylated lipopeptides, depending on the heterodimerization with TLR6 or TLR1, respectively (Akira, 2003). LM activity is dependent on TLR2 and TLR1 but independent of TLR6 (Quesniaux et al., 2004; Gilleron et al., 2006). Glycosyl-phosphatidyl-inositol (GPI) anchors from *Plasmodium falciparum*, which have a structure similar to that of mycobacterial MPI-based lipoglycans, have also been reported to signal mainly through TLR2 (Krishnegowda et al., 2005). However, in contrast to LM, the activity of *sn*-2-lyso-GPIs, with a total of two fatty acids, is similar to that of intact GPIs containing three fatty acids, including two on the Gro. Thus, the activity is independent of whether there are two or three fatty acids, but it differs considerably according to the auxiliary receptor, TLR6 versus TLR1, as in the case of lipopeptides (Krishnegowda et al., 2005). So LM activity is different from that of these two families of TLR2 agonists in such a way that whatever the acylation degree, LM require TLR1 but not TLR6 and diacylated forms are not active.

The size of the oligomannan domain affects lipoglycan activity. LM are much more active than PIM (see earlier). Studying lipoglycan variants from different actinomycete species (Garton et al., 2002; Gibson et al., 2003a, 2003b, 2004, 2005; Gilleron et al., 2005), we have found that a linear ($\alpha1\rightarrow6$)-Man$p$ chain linked to a MPI anchor is sufficient for TLR2-dependent proinflammatory activity (Gibson, 2004). However, adding side chains, like those in mycobacterial LM, and increasing their size, as in *Saccharothrix aerocolonigenes* LM, potentiate this activity (Gibson et al., 2005). So, the D-mannan domain participates in optimal recognition of these lipoglycans by TLR2 by a mechanism that is not yet understood.

PILAM isolated from *M. smegmatis* and a fast-growing *Mycobacterium* sp. activate macrophages in a TLR2-dependent manner (Chatterjee et al., 1992c; Gilleron et al., 1997; Means et al., 1999); AraLAM (Guerardel et al., 2002) and ManLAM are not, or only poor, inducers of cytokine secretion. This difference of activity between AraLAM and ManLAM, on the one hand, and PILAM, on the other, has been tentatively explained by the specific presence of PI caps in PILAM. However, PILAM contains only one phosphoinositol cap per molecule, the majority of the arabinan side chains being ended by terminal β-D-Ara$f$ units as in AraLAM. So, we consider that the molecular bases of PILAM as a ligand of TLR2 remain unclear and more work is required to support this claim.

The nonactivity of AraLAM and ManLAM, despite the fact that both molecules contain a LM moiety in their structure, can be explained by steric hindrance: the arabinan domain masks the mannan core (Riviere et al., 2004), thus limiting its access to the TLR. This has been demonstrated in the case of *M. kansasii* ManLAM, in which chemical degradation of the arabinan domain yielding a lipomannan-like structure restores cytokine-inducing activity (Guerardel et al., 2003).

The LM/LAM ratio in the fast-growing mycobacteria *M. smegmatis* and *M. chelonae* is higher than that in *M. tuberculosis* (Guerardel et al., 2002, and unpublished results). They thus contain rela-

tively more proinflammatory stimuli than *M. tuberculosis*. Moreover, in contrast to PILAM or AraLAM, *M. tuberculosis*, ManLAM is anti-inflammatory. These observations might explain, at least in part, why *M. tuberculosis* induces less proinflammatory cytokines than *M. smegmatis*.

In conclusion, it can be suggested that lipoglycans, as immunoregulators of proinflammatory cytokine secretion, play a key role in determining the ability of *M. tuberculosis* to persist and replicate inside macrophages. In addition, by stimulating TLRs on APCs, lipoglycans not only modulate cytokine secretion, but also endogeneous lipid antigen synthesis and CD1 protein translation (Moody, 2006). Krutzik and colleagues (2005) have shown that CD1b$^+$ cells can be rapidly induced by activation of TLRs (TLR2, TLR4, and TLR5) on peripheral blood monocytes in vitro. Moreover, PILAM and PIM, but not ManLAM, have been shown to upregulate, in a TLR2-dependent manner, group 1 CD1 expression in DCs (Roura-Mir et al., 2005).

Lipoglycans as ligands of CD1 proteins

Recent studies have shown that lipid antigens are able to stimulate the immune system and may play a role in protective immunity. Conventional αβT lymphocytes are stimulated by specific recognition of peptides in the context of major histocompatibility complex class I and II molecules, whereas unconventional αβT cells are activated by lipid antigens. Lipid antigens are presented by a newly identified family of antigen-presenting molecules, the CD1 proteins, which is composed of several isoforms that are able to bind different lipid structures (Porcelli, 1995).

Indeed, human APCs express five CD1 proteins, CD1a to CD1e, with different structures and different localizations on the cell. These proteins have been classified into two groups (for reviews, see Brigl and Brenner, 2004; De Libero and Mori, 2005; Lawton and Kronenberg, 2004; Moody and Porcetti, 2003; Moody et al., 2005). The group 1 of CD1 molecules, CD1a, CD1b, and CD1c, restricts αβT cells. These T cells primarily secrete IFN-γ and have cytolytic functions, characterizing them as prime effector T cells in infections including tuberculosis. The sole member of group 2 CD1 proteins, CD1d, also present in mice, restricts NKT cells. These T cells bear an evolutionarily conserved T-cell receptor using the invariant α-chains: Vα14Jα18 in mice and Vα24JαQ in humans. NKT cells probably fulfill regulatory as well as effector functions consistent with their dual secretion of IFN-γ and IL-4 and their cytolytic activity.

Crystallographic studies have revealed that the grooves of CD1a, CD1b and CD1d differ in the number (two or four), shape, and connectivity of their antigen-binding pockets. The crystal structures of CD1b proteins show that their antigen-binding groove is composed of four pockets (A', C', F' and T') and two antigen portals (C' and F') (Batuwangala et al., 2004; Gadola et al., 2002). This architecture allows anchoring of long fatty acids (up to C_{80}), accommodating A' passing through T' and coming out by F'. The CD1a (Zajonc et al., 2003, 2005b) and CD1d (Giabbai, 2005; Koch et al., 2005; Zajonc et al., 2005a, 2005c; Zeng et al., 1997) binding grooves have only two pockets (A' and F') that are more or less connected, which can accommodate lipids with an overall length of up to $\sim C_{32}$ and $\sim C_{40}$, respectively. They differ from one another in the overall volume of their hydrophobic groove (hCD1a, 1,900 $Å^3$; hCD1b, 2,600 $Å^3$; mCD1d1, 2,300 $Å^3$) (De la Salle et al., 2005). Because immunogenic lipids could have very different acyl chain lengths, it is now clear that the pockets of CD1 antigen-binding grooves influence ligand specificity and facilitate the presentation of various kinds of antigenic lipids. The apparent mismatch between the CD1 groove volume and lipid length is greatest for CD1b, which presents both glycolipids with short-chain fatty acids (PIM and LAM with C_{16}/C_{19}) and those with long-chain fatty acids (mycolic acids, glucose monomycolate, and sulfoglycolipids). CD1c presents bacterial mannophosphoisoprenoids, and CD1a binds mycobacterial lipopeptides (for a review, see Young et al., 2006). The existence of CD1-restricted T cells specific for several of the most abundant mycobacterial lipids strongly suggests that CD1 proteins are important in the host response to *M. tuberculosis*.

PIM and CD1d

The microbial antigens for NKT cells include mycobacterial PIM, sphingolipids from *Ehrlichia* and *Sphingomonas* species, and lipophosphoglycan from *Leishmania* parasites (for a review, see Moody et al., 2005). The first indication that PIM is a ligand for NKT cells came from a study by Gachelin's group (Apostolou et al., 1999), showing that a lipid fraction of *M. tuberculosis* injected into mice induced a granuloma and recruitment of NKT cells in the lesions. The granuloma formation was associated with a crude fraction containing PIM and was described to be dependent on CD1d-restricted T cells. The involvement of PIM in this phenomenon was demonstrated subsequently with highly purified molecules (Gilleron et al., 2001b). The minimal structure required for this activity was investigated with the aim of demonstrating a CD1d-dependent phenomenon. We found that there is an absolute requirement for

the molecules to possess at least one fatty acyl chain, but the number, localization and size of the acyl chains was irrelevant. Moreover, increasing the complexity of the carbohydrate moiety ($PI/PIM_2/PIM_6$) did not lead to significant differences in the biological responses. This finding indicates the minor role of the carbohydrate part in NKT cell recruitment and a lack of specificity of the TCR toward sugars, a feature independently suggested by analysis of the TCR repertoire showing that it remained invariant whatever the stimuli used (Apostolou et al.; 2000; Ronet et al., 2001). The absence of a specific recognition by the TCR suggested that the TCR-CD1 axis does not contribute to NKT cell recruitment.

Fischer and colleagues (2004) identified PIM as the first bacterial antigens for both human and murine NKT cells. Although the lipid was initially claimed to be PIM_4, recent re-examination of the active fraction revealed that it did not contain PIM_4, but rather a triacylated form of PIM_2 (Ac_1PIM_2) (Schaible, Niemeyer and Gilleron, unpublished results). For the authors, NKT cell activation may result from a direct CD1d-resticted recognition of PIM (Fischer et al., 2004), but purified PIM acyl-forms (Gilleron et al., 2001b, 2003) were unable to activate NKT cells using the same bioassay (unpublished data). Alternatively, upregulation of an endogenous ligand through other receptors, such as TLRs could lead to its presentation by CD1d and subsequent NKT cell activation (Tsuji, 2006). In conclusion, the data from Fischer and associates reveal a mycobacterial product able to activate NKT cells in the context of CD1 proteins, but the precise structure of this antigen remains to be established.

Lipoglycans and CD1 proteins of group 1

PIM and LAM were also found to stimulate unconventional CD4 $\alpha\beta$ T lymphocytes in the context of CD1b proteins (De la Salle et al., 2005; Sieling et al., 1995). The study conducted by Sieling and associates was one of the first reports showing that nonpeptide antigens could be presented to T cells by molecules that are not encoded by the major histocompatibility complex (MHC). To characterize the antigens presented by CD1 molecules to T cells, mycobacterium-reactive double negative (DN) $\alpha\beta$ T cell lines were established: one cell line from the skin lesion of a leprosy patient (line LDN4) and another from the peripheral blood mononuclear cell (PBMC) of a normal donor (line BDN2). Both cell lines were CD1b restricted and stimulated by *M. leprae* ManLAM. Cell line BDN2 but not cell line LDN4 was also stimulated by *M. tuberculosis* ManLAM. A mixture of PIM_2 and PIM_6 induced strong proliferation of

BDN2 cells but only weak stimulation of LDN4 cells. PIM$_2$ did not induce proliferation in any of the DN T cells. These findings reveal as yet uncharacterized differences in the specificity of recognition of LAM and PIM.

Presentation of lipid antigens requires a series of events including their uptake by APCs, loading into CD1, and presentation to T cells (for reviews, see De Libero and Mori, 2006a, 2006b). *M. tuberculosis* infects phagocytic cells of the lung, using DC-SIGN to enter DCs, and the complement receptor 3 (CR3) and the MR to enter macrophages (Fig. 4) (Herrmann et al., 2006; Tailleux et al., 2003c; Villeneuve et al., 2005). After infection, mycobacteria reside in early phagosomes and escape the immune system by inhibiting phagosome maturation and fusion with lysosomes (Russell, 2003). Then, mycobacterial antigens are segregated in phagosomes of APCs. Because CD1 proteins of group 1 are expressed prominently by myeloic lineage DCs (Porcelli, 1995) and because they survey endocytic compartments of DCs, they obviously must encounter antigens from extracellular sources. Various mechanisms underlying antigen delivery to APC have been described recently: (1) release of surface glycoconjugates (like PIM) into the endocytic network of the host macrophage in exosomes (Beatty et al., 2000; Rhoades et al., 2003), and (2) apoptosis induced by mycobacteria in macrophages causing the release of apoptotic vesicles that carry mycobacterial antigens to noninfected bystander cells, such as DCs (Schaible et al., 2003). Once released from infected cells, the antigens are delivered by serum apolipoproteins (ApoE) to activate T cells (van den Elzen et al., 2005). ApoE binds lipid antigens and delivers them by receptor-mediated uptake into endosomal compartments containing CD1 in APCs. Moreover, receptor-mediated endocytosis of ManLAM by the MR and subsequent colocalization of ManLAM and CD1b in late endosomes have been reported (Prigozy et al., 1997). Moreover, Langerhans cells, a subset of DCs that initiate an immune response in skin, use langerin to capture and present antigens to CD1a-restricted T cells (Hunger et al., 2004).

Sieling and coworkers (1995) demonstrated that the presentation of ManLAM required a transit to endosomal compartments. Indeed, the loading of lipid antigens onto CD1 proteins is mostly intracellular in the endosomal network. The length of lipid acyl chains influences the endosomal compartments

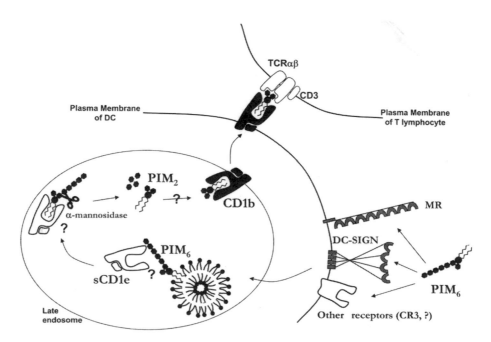

Figure 4. Pathway of PIM presentation to T lymphocytes via CD1b. PIM are normally assembled in aqueous biological solutions in micelles or integrated into biological membranes. They have been provided to antigen-presenting cells (APCs) as membrane fragments (exosomes) (Beatty et al., 2000; Rhoades et al., 2003), apoptotic bodies (Schaible et al., 2003) or lipoprotein complexes (van den Elzen et al., 2005). The uptake of PIM was shown to be mediated by host cell C-type lectins: the mannose receptor (MR), the dendritic-cell-specific intercellular adhesion molecule 3-grabbing nonintegrin receptor (DC-SIGN), and the complement receptor 3 (CR3). PIM are then segregated in late endosomes, where they meet CD1b, saposins and enzymes. PIM$_6$ must be processed by a α-mannosidase to generate a structure (PIM$_2$ in the diagram, but which could be even simpler than PIM$_2$), that is presented by CD1b to stimulate the T lymphocytes. This phenomenon is CD1e-assisted (De la Salle et al., 2005), but the exact role of the soluble CD1e protein (sCD1e) is still unknown (see text).

where the lipid is loaded into CD1 (Mukherjee et al., 1999). In the case of glucose monomycolate (GMM), the recognition of GMM with long (C_{80}) alkyl chains involves traffic to the late endosomes, where it is efficiently loaded onto CD1b molecules, whereas a less immunogenic form of GMM with shorter alkyl chains (C_{32}) might bind to CD1b molecules at the cell surface (Moody et al., 2002). This preferential transport to late endosomes where the pH is acidic might facilitate the loading of lipids with long alkyl chains, which are more frequently of microbial origin. Indeed, endosomal acidification seems to act directly to insert lipids into CD1b by inducing a partial unfolding (Cheng et al., 2006). The lipid part of ManLAM is constituted of short alkyl chains (C_{16} and C_{19}), and therefore this does not explain transit to endosomal compartment.

However, antigen loading requires a lipid transfer process from membranes to CD1, a phenomenon described as being assisted by lipid transfer proteins (LTP) including saposins (or SAP for sphingolipid activator proteins) and GM2-activator protein (GM2A). These proteins are localized in late (not early) endosomes. The intervention of this type of accessory molecule has been described in the case of CD1d- (Kang and Cresswell, 2004; Yuan et al., 2007; Zhou et al., 2004) and CD1b-restricted lipids (Winau et al., 2004). Winau and colleagues (2004) have demonstrated the involvement of SAP-C as an accessory molecule required for LAM loading onto CD1b. They have shown that fibroblasts deficient in SAP and transfected with CD1b failed to activate lipid-specific T cells. The T-cell response was restored when fibroblasts were reconstituted with SAP-C but not with other SAPs. SAP-C appears to extract LAM efficiently from liposomes, and the study demonstrates a direct molecular interaction between SAP-C and CD1b. These results led the authors to propose a model in which SAP-C is a previously missing link in the antigen presentation of lipids through CD1b to human T cells, serving to expose lipid antigens from intralysosomal membranes for loading onto CD1b.

Lipoglycans Processing

It is not yet clear how antigens with large polysaccharide moieties, such as ManLAM, can be positioned between the CD1 and the TCR without sterically blocking the contact of the TCR with isoform-specific determinants on the outer surface of CD1 and thereby loosing the CD1 restriction. Several mechanisms have been proposed to explain the recognition of large carbohydrate structures (Young and Moody, 2006). The first possibility is that large, branched carbohydrates extend laterally along the plane of

CD1–TCR contact, so that only a few carbohydrate units are positioned directly between the TCR and the CD1. This seems highly improbable in the case of a huge molecule such as ManLAM (molecular mass of approximately 17 kDa). The second possibility is that the large carbohydrate moieties are enzymatically degraded in the APC before their contact with the TCR. Indeed, glycolipids, like proteins, are subjected to partial degradation generating smaller molecules compatible with TCR interaction. The previously described transit of ManLAM through late endosomal compartments allows their colocalization with enzymes potentially involved in glycolipid processing.

The first study that looked for an enzymatic processing using trehalose dimycolate (or cord factor) was unable to find any evidence of such processing generating the antigenic GMM (Moody et al., 1999). The first report of glycolipid processing was from a model system of the CD1d-restricted recognition of α-galactosylceramide (α-GalCer) in which a synthetic Galα1→2GalCer requires the removal of the termini galactose by α-galactosidase A to stimulate the lymphocytes (Prigozy et al., 2001).

The second report of the existence of a glycolipid processing was the demonstration that the mycobacterial antigen PIM_6 stimulated CD1b-restricted T cells only after partial digestion of the oligomannose moiety by lysosomal α-mannosidase (De la Salle et al., 2005). This processing required the assistance of the soluble CD1e protein. CD1e molecules are exclusively localized in the lysosomes of mature DCs (Angenieux et al., 2005) in which they are cleaved into a stable soluble form (sCD1e). Remarkably, CD1e never transits through the cell surface of DC, a characteristic that excludes a direct interaction with T cells. However, the accumulation of CD1e in the $CD1b^+$ lysosomes of mature DCs suggests that CD1e may have a role in processing or presentation of CD1b-restricted antigens. Recombinant CD1e binds glycolipids and assists in vitro in the digestion of PIM_6 to PIM_2 by α-mannosidase (De la Salle et al., 2005). Therefore, CD1e appears to be directly involved in the editing and the expansion of the repertoire of glycolipid T-cell antigens. However, the exact role of CD1e is still unknown (Fig. 4): it may perturb the membrane, allowing the release and exposure of the lipid substrate to its enzyme; it may play a more active role, extracting PIM from the membrane; it may interact with PIM to transfer it from CD1b to α-mannosidase; and it may interact directly with the α-mannosidase or be required only for mannosidase activity. The true antigenic structure, that is, the minimal PIM-derived structure presented by CD1b and able to activate T cells, is not known. The influence of the different PIM fatty acids should be investigated, because surprisingly in view of the shape of CD1b binding groove, tetra-acylated PIM activate T cells (Gilleron, Puzo and De Libero, unpublished results). Also, lipases may be involved in PIM processing, and if so, they should be identified.

CONCLUSIONS AND OPEN QUESTIONS

Studies using purified molecules in vitro indicate that ManLAM is a major virulence factor of *M. tuberculosis*, despite it not being strictly restricted to pathogenic species. LAM, LM, and PIM, thanks to their MPI anchor, traffic by different mechanisms from the mycobacterial phagosome to different cell compartments or to bystander phagocytic cells by which they are endocytozed mainly by C-type lectins. This intra- or intercellular traffic is a clue to their ability to modulate the host immune responses. It seems likely that lipoglycan-mediated *M. tuberculosis* binding to phagocytic cells through C-type lectins, and lipoglycan migration from the mycobacterial phagosome to the cytosol, requires their presence in the outermost layers of the mycobacterial envelope. This implies that there are cargo proteins allowing their translocation from the plasma membrane, where they are biosynthesized, to the cell surface. The characterization of the genes encoding these putative lipoglycan transporter proteins, the mechanisms of their assemblage with lipoglycans, and the traffic through the mycobacterial cell wall are all important issues that need to be addressed. One protein, LppX, involved in the translocation of phthiocerol dimycocerosates (DIM) from the plasma membrane to the envelope surface was recently reported (Sulzenbacher, 2006). Interestingly, abolition of DIM transport following disruption of the *lppX* gene was accompanied by a large attenuation of *M. tuberculosis* virulence.

The biosynthesis pathway of the MPI anchor, shared by PIM, LM, and LAM, is now well understood, except the acyl-transferase involved in the acylation of Ins that remains to be discovered. Several glycosyl-transferases involved in the biogenesis of the polysaccharide backbone of LM and LAM have been evidenced, but most of them, particularly arabinosyl-transferases, have not been discovered and their characterization remains a major challenge. Identifying the genes encoding these enzymes and generating selective knockout mutants when possible should help in determining the contribution of ManLAM, as well as LM and PIM, to *M. tuberculosis* virulence. These enzymes are attractive therapeutic targets because lipoglycan deficiencies are associated with lethality or growth defects. The identification of

biosynthetic enzymes would be greatly facilitated by high-throughput screening of lipoglycan structures in mutants generated by either homologous recombination or transposition mutagenesis. To make this approach feasible, we are currently trying to develop analytical procedures, referred to as "glycomics" and based on those used for proteomics. They include the separation of lipoglycans by 1D or 2D gel electrophoresis, followed by in-gel or ex-gel digests and structural analysis by capillary electrophoresis, mass spectrometry and NMR.

ManLAM binding to the C-type lectins, MR and DC-SIGN elicits cell signaling pathways. This finding and the demonstration that C-type lectins can trigger an intracellular signal have opened a new area of research: deciphering the signaling pathways downstream from these receptors. Some pieces of these complex puzzles have already been identified, but many remain mysterious, and more work is required to connect all the pieces together.

PIM and LM stimulate non-conventional $\alpha\beta T$ cells restricted by the CD1 proteins and innate immunity through TLR2 binding. Considerable progress has been made in characterizing mycobacterial lipid antigens relevant to CD1 proteins, but the molecular mechanisms of their processing have not been elucidated. We are beginning to understand the structure/function relationships underlying lipoglycan signaling through TLR2, although nothing is known about the molecular bases of their binding.

The finding that CD1 proteins present (glyco) lipids means that the number of potential microbial antigens recognized by T cells is very much greater than previously suspected. These molecules have the potential to be highly antigenic; are able to stimulate CD4$^+$, CD8$^+$ or CD4-CD8$^-$ $\alpha\beta$ T lymphocytes; and their structure is often sufficiently different from their eukaryotic counterparts that they may be considered as foreign (Skold and Behar, 2005).

There is now diverse evidence that these lipid antigens could be suitable for the development of subunit vaccines: (1) *M. tuberculosis* has an extraordinary repertoire of lipids (Puzo, 1993); (2) they can traffic more easily than proteins between phagocyte intracellular compartments or between phagocytic cells by cross priming processes (Schaible et al., 2003); (3) they are the products of multistep biosynthetic pathways, and therefore rapid structural variation of these antigens is unlikely, in contrast to microbial protein antigens; (4) CD1 proteins are less polymorphic than MHC proteins; and importantly (5) these molecules generate a memory effect, as CD1-restricted T cells contribute to the human immune response to *M. tuberculosis* (Gilleron et al., 2004; Moody et al., 2000). Thus, lipid antigens should be tested in animal models either as single components or in combination with mycobacterial protein antigens (Martin, 2005; Skeiky and Sadoff, 2006). A first study has shown that immunization with a mycobacterial lipid vaccine improves pulmonary pathology in the guinea pig model of tuberculosis (Dascher et al., 2003). Indeed, guinea pigs are currently the best model for investigating the contribution of CD1-restricted T cells activated by lipids in the protective response (Orme, 2005). However, the possibility of expressing human group 1 CD1 genes in transgenic mice is another viable approach to study the role of CD1-restricted T cells in microbial immunity.

Acknowledgments. Both of our laboratories are supported by the National Institute of Allergy and Infectious Diseases/National Institutes of Health Grant AI64798 (of which Patrick J. Brennan of Colorado State University in Fort Collins is the principal investigator) and the European Union (Cluster for a Tuberculosis Vaccine, LSHP-CT-2003-503367). M.G., J.N., and G.P. were funded by the Agence Nationale de la Recherche (ANR-05-MIIM-006-02 and ANR-05-MIIM-038-04) and by the Centre National de la Recherche Scientifique (CNRS). M.J. is supported by the Institut Pasteur GPH-05 (Tuberculose) research program and the European Union project "New Medicines for Tuberculosis" (LSHP-CT-2005-018923).

REFERENCES

Abdian, P. L., A. C. Lellouch, C. Gautier, L. Ielpi, and R. A. Geremia. 2000. Identification of essential amino acids in the bacterial alpha-mannosyltransferase aceA. *J. Biol. Chem.* **275:**40568–40575.

Ainge, G. D., J. Hudson, D. S. Larsen, G. F. Painter, G. S. Gill, and J. L. Harper. 2006. Phosphatidylinositol mannosides: Synthesis and suppression of allergic airway disease. *Bioorg. Med. Chem.* **14:**5632–5642.

Akira, S. 2003. Mammalian Toll-like receptors. *Curr. Opin. Immunol.* **15:**5–11.

Alderwick, L. J., V. Molle, L. Kremer, A. J. Cozzone, T. R. Dafforn, G. S. Besra, and K. Futterer. 2006a. Molecular structure of EmbR, a response element of Ser/Thr kinase signaling in *Mycobacterium tuberculosis. Proc. Natl. Acad. Sci. USA* **103:**2558–2563.

Alderwick, L. J., E. Radmacher, M. Seidel, R. Gande, P. G. Hitchen, H. R. Morris, A. Dell, H. Sahm, L. Eggeling, and G. S. Besra. 2005. Deletion of Cg-*emb* in Corynebacterianeae leads to a novel truncated cell wall arabinogalactan, whereas inactivation of Cg-*ubiA* results in an arabinan-deficient mutant with a cell wall galactan core. *J. Biol. Chem.* **280:**32362–32371.

Alderwick, L. J., M. Seidel, H. Sahm, G. S. Besra, and L. Eggeling. 2006b. Identification of a novel arabinosyl transferase (AtfA) involved in cell wall arabinan biosynthesis in *Mycobacterium tuberculosis. J. Biol. Chem.* **281:**15653–15661.

Alexander, D. C., J. R. W. Jones, T. Tan, J. M. Chen, and J. Liu. 2004. PimF, a mannosyltransferase of Mycobacteria, is involved in the biosynthesis of phosphatidylinositol mannosides and lipoarabinomannan. *J. Biol. Chem.* **279:**18824–18833.

Altare, F., A. Durandy, D. Lammas, J. F. Emile, S. Lamhamedi, F. Le Deist, P. Drysdale, E. Jouanguy, R. Doffinger, F. Bernaudin, O. Jeppsson, J. A. Gollob, E. Meinl, A. W. Segal, A. Fischer, D. Kumararatne, and J. L. Casanova. 1998. Impairment of mycobacterial immunity in human interleukin-12 receptor deficiency. *Science* **280:**1432–1435.

Angenieux, C., V. Fraisier, B. Maitre, V. Racine, N. van der Wel, D. Fricker, F. Proamer, M. Sachse, J. P. Cazenave, P. Peters, B. Goud, D. Hanau, J. B. Sibarita, J. Salamero, and H. de la Salle. 2005. The cellular pathway of CD1e in immature and maturing dendritic cells. *Traffic* **6:**286–302.

Apostolou, I., A. Cumano, G. Gachelin, and P. Kourilsky. 2000. Evidence for two subgroups of CD4-CD8- NKT cells with distinct TCR alpha beta repertoires and differential distribution in lymphoid tissues. *J. Immunol.* **165:**2481–2490.

Apostolou, I., Y. Takahama, C. Belmant, T. Kawano, M. Huerre, G. Marchal, J. Cui, M. Taniguchi, H. Nakauchi, J. J. Fournie, P. Kourilsky, and G. Gachelin. 1999. Murine natural killer T(NKT) cells [correction of natural killer cells] contribute to the granulomatous reaction caused by mycobacterial cell walls. *Proc. Natl. Acad. Sci. USA* **96:**5141–5146.

Asselineau, C., and J. Asselineau. 1984. Waxes, mycosides and related compounds, p. 345–360. *In* G. P. Kubica and L. G. Wayne (ed.), *The Mycobacteria—a Source Book,* vol. 15. Marcel Dekker, Inc., New York, NY.

Astarie-Dequeker, C., E. N. N'Diaye, V. Le Cabec, M. G. Rittig, J. Prandi, and I. Maridonneau-Parini. 1999. The mannose receptor mediates uptake of pathogenic and nonpathogenic mycobacteria and bypasses bactericidal responses in human macrophages. *Infect. Immun.* **67:**469–477.

Bachhawat, N., and S. C. Mande. 1999. Identification of the *INO1* gene of *Mycobacterium tuberculosis* H37Rv reveals a novel class of inositol-1-phosphate synthase enzyme. *J. Mol. Biol.* **291:**531–536.

Ballou, C. E., E. Vilkas, and E. Lederer. 1963. Structural studies on the *myo*-inositol phospholipids of *Mycobacterium tuberculosis* (var. *bovis,* strain BCG). *J. Biol. Chem.* **238:**69–76.

Barreiro, L. B., O. Neyrolles, C. L. Babb, L. Tailleux, H. Quach, K. McElreavey, P. D. Helden, E. G. Hoal, B. Gicquel, and L. Quintana-Murci. 2006. Promoter variation in the DC-SIGN-encoding gene CD209 is associated with tuberculosis. *PLoS Med.* **3:**e20.

Batuwangala, T., D. Shepherd, S. D. Gadola, K. J. Gibson, N. R. Zaccai, A. R. Fersht, G. S. Besra, V. Cerundolo, and E. Y. Jones. 2004. The crystal structure of human CD1b with a bound bacterial glycolipid. *J. Immunol.* **172:**2382–2388.

Beatty, W. L., E. R. Rhoades, H. J. Ullrich, D. Chatterjee, J. E. Heuser, and D. G. Russell. 2000. Trafficking and release of mycobacterial lipids from infected macrophages. *Traffic* **1:**235–247.

Belanger, A. E., G. S. Besra, M. E. Ford, K. Mikusova, J. T. Belisle, P. J. Brennan, and J. M. Inamine. 1996. The *embAB* genes of *Mycobacterium avium* encode an arabinosyl transferase involved in cell wall arabinan biosynthesis that is the target for the antimycobacterial drug ethambutol. *Proc. Natl. Acad. Sci. USA* **93:**11919–11924.

Belanger, A. E., and J. M. Inamine. 2000. Genetics of cell wall biosynthesis, p. 191–202. *In* G. F. Hatfull and W. R. Jacobs (ed.), *Molecular Genetics of Mycobacteria.* ASM Press, Washington, DC.

Ben-Ali, M., M. R. Barbouche, S. Bousnina, A. Chabbou, and K. Dellagi. 2004. Toll-like receptor 2 Arg677Trp polymorphism is associated with susceptibility to tuberculosis in Tunisian patients. *Clin. Diagn. Lab. Immunol.* **11:**625–626.

Berg, S., J. Starbuck, J. B. Torrelles, V. D. Vissa, D. C. Crick, D. Chatterjee, and P. J. Brennan. 2005. Roles of conserved proline and glycosyltransferase motifs of EmbC in biosynthesis of lipoarabinomannan. *J. Biol. Chem.* **280:**5651–5663.

Besra, G. S., C. B. Morehouse, C. M. Rittner, C. J. Waechter and P. J. Brennan. 1997. Biosynthesis of mycobacterial lipoarabinomannan. *J. Biol. Chem.* **272:**18460–18466.

Braibant, M., P. Gilot, and J. Content. 2000. The ATP binding cassette (ABC) transport systems of *Mycobacterium tuberculosis. FEMS Microbiol. Rev.* **24:**449–467.

Brennan, P. J. 1984. Antigenic peptidoglycolipids, phospholipids and glycolipids, p. 467–490. *In* G. P. Kubica and L. G. Wayne (ed.), *The Mycobacteria—a Source Book*, vol. 15. Marcel Dekker, Inc., New York, NY.

Brennan, P. J., and C. E. Ballou. 1968. Biosynthesis of mannophosphoinositides by *Mycobacterium phlei*. Enzymatic acylation of the dimannophosphoinositides. *J. Biol. Chem.* 243:2975–2984.

Brennan, P. J., and C. E. Ballou. 1967. Biosynthesis of mannophosphoinositides by *Mycobacterium phlei*. The family of dimannophosphoinositides. *J. Biol. Chem.* 242:3046–3056.

Brennan, P. J., and H. Nikaido. 1995. The envelope of mycobacteria. *Annu. Rev. Biochem.* 64:29–63.

Brightbill, H. D., D. H. Libraty, S. R. Krutzik, R. B. Yang, J. T. Belisle, J. R. Bleharski, M. Maitland, M. V. Norgard, S. E. Plevy, S. T. Smale, P. J. Brennan, B. R. Bloom, P. J. Godowski, and R. L. Modlin. 1999. Host defense mechanisms triggered by microbial lipoproteins through toll-like receptors. *Science* 285:732–736.

Brigl, M., and M. B. Brenner. 2004. CD1: antigen presentation and T cell function. *Annu. Rev. Immunol.* 22:817–890.

Briken, V., S. A. Porcelli, G. S. Besra, and L. Kremer. 2004. Mycobacterial lipoarabinomannan and related lipoglycans: from biogenesis to modulation of the immune response. *Mol. Microbiol.* 53:391–403.

Burguière, A., P. G. Hitchen, L. G. Dover, L. Kremer, M. Ridell, D. C. Alexander, J. Liu, H. R. Morris, D. E. Minnikin, A. Dell, and G. S. Besra. 2005. LosA, a key glycosyltransferase involved in the biosynthesis of a novel family of glycosylated acyltrehalose lipooligosaccharides from *Mycobacterium marinum*. *J. Biol. Chem.* 280:42124–42133.

Caparros, E., P. Munoz, E. Sierra-Filardi, D. Serrano-Gomez, A. Puig-Kroger, J. L. Rodriguez-Fernandez, M. Mellado, J. Sancho, M. Zubiaur, and A. L. Corbi. 2006. DC-SIGN ligation on dendritic cells results in ERK and PI3K activation and modulates cytokine production. *Blood* 107:3950–3958.

Chatterjee, D., C. M. Bozic, M. McNeil, and P. J. Brennan. 1991. Structural features of the arabinan component of the lipoarabinomannan of *Mycobacterium tuberculosis*. *J. Biol. Chem.* 266:9652–9660.

Chatterjee, D., S. W. Hunter, M. McNeil, and P. J. Brennan. 1992a. Lipoarabinomannan. Multiglycosylated form of the mycobacterial mannosylphosphatidylinositols. *J. Biol. Chem.* 267:6228–6233.

Chatterjee, D., and K. H. Khoo. 1998. Mycobacterial lipoarabinomannan: an extraordinary lipoheteroglycan with profound physiological effects. *Glycobiology* 8:113–120.

Chatterjee, D., K. H. Khoo, M. R. McNeil, A. Dell, H. R. Morris, and P. J. Brennan. 1993. Structural definition of the non-reducing termini of mannose-capped LAM from *Mycobacterium tuberculosis* through selective enzymatic degradation and fast atom bombardment-mass spectrometry. *Glycobiology* 3:497–506.

Chatterjee, D., K. Lowell, B. Rivoire, M. R. McNeil, and P. J. Brennan. 1992b. Lipoarabinomannan of *Mycobacterium tuberculosis*. Capping with mannosyl residues in some strains. *J. Biol. Chem.* 267:6234–6239.

Chatterjee, D., A. D. Roberts, K. Lowell, P. J. Brennan, and I. M. Orme. 1992c. Structural basis of capacity of lipoarabinomannan to induce secretion of tumor necrosis factor. *Infect. Immun.* 60:1249–1253.

Cheng, T. Y., M. Relloso, I. Van Rhijn, D. C. Young, G. S. Besra, V. Briken, D. M. Zajonc, I. A. Wilson, S. Porcelli, and D. B. Moody. 2006. Role of lipid trimming and CD1 groove size in cellular antigen presentation. *EMBO J.* 25:2989–2999.

Chieppa, M., G. Bianchi, A. Doni, A. Del Prete, M. Sironi, G. Laskarin, P. Monti, L. Piemonti, A. Biondi, A. Mantovani, M. Introna, and P. Allavena. 2003. Cross-linking of the man-

nose receptor on monocyte-derived dendritic cells activates an anti-inflammatory immunosuppressive program. *J. Immunol.* 171:4552–4560.

Cole, S. T., R. Brosch, J. Parkhill, T. Garnier, C. Churcher, D. Harris, S. V. Gordon, K. Eiglmeier, S. Gas, C. E. Barry, 3rd, F. Tekaia, K. Badcock, D. Basham, D. Brown, T. Chillingworth, R. Connor, R. Davies, K. Devlin, T. Feltwell, S. Gentles, N. Hamlin, S. Holroyd, T. Hornsby, K. Jagels, A. Krogh, J. McLean, S. Moule, L. Murphy, K. Oliver, J. Osborne, M. A. Quail, M. A. Rajandream, J. Rogers, S. Rutter, K. Seeger, J. Skelton, R. Squares, S. Squares, J. E. Sulston, K. Taylor, S. Whitehead, and B. G. Barrell. 1998. Deciphering the biology of *Mycobacterium tuberculosis* from the complete genome sequence. *Nature* 393:537–544.

Cook, D. N., D. S. Pisetsky, and D. A. Schwartz. 2004. Toll-like receptors in the pathogenesis of human disease. *Nat. Immunol.* 5:975–979.

Daffe, M., and P. Draper. 1998. The envelope layers of mycobacteria with reference to their pathogenicity. *Adv. Microb. Physiol.* 39:131–203.

Dascher, C. C., K. Hiromatsu, X. Xiong, C. Morehouse, G. Watts, G. Liu, D. N. McMurray, K. P. LeClair, S. A. Porcelli, and M. B. Brenner. 2003. Immunization with a mycobacterial lipid vaccine improves pulmonary pathology in the guinea pig model of tuberculosis. *Int. Immunol.* 15:915–925.

De la Salle, H., S. Mariotti, C. Angenieux, M. Gilleron, L. F. Garcia-Alles, D. Malm, T. Berg, S. Paoletti, B. Maitre, L. Mourey, J. Salamero, J. P. Cazenave, D. Hanau, L. Mori, G. Puzo, and G. De Libero. 2005. Assistance of microbial glycolipid antigen processing by CD1e. *Science* 310:1321–1324.

De Libero, G., and L. Mori. 2006a. How T lymphocytes recognize lipid antigens. *FEBS Lett.* 580:5580–5587.

De Libero, G., and L. Mori. 2006b. Mechanisms of lipid-antigen generation and presentation to T cells. *Trends Immunol.* 27:485–492.

De Libero, G., and L. Mori. 2005. Recognition of lipid antigens by T cells. *Nat. Rev. Immunol.* 5:485–496.

Delmas, C., M. Gilleron, T. Brando, A. Vercellone, M. Gheorghui, M. Rivière, and G. Puzo. 1997. Comparative structural study of the mannosylated-lipoarabinomannans from *Mycobacterium bovis* BCG vaccine strains: characterization and localization of succinates. *Glycobiology* 7:811–817.

Deretic, V., S. Singh, S. Master, J. Harris, E. Roberts, G. Kyei, A. Davis, S. de Haro, J. Naylor, H. H. Lee, and I. Vergne. 2006. *Mycobacterium tuberculosis* inhibition of phagolysosome biogenesis and autophagy as a host defence mechanism. *Cell. Microbiol.* 8:719–727.

Dhariwal, K. R., A. Chander, and T. A. Venkitasubramanian. 1977. Environmental effects on lipids of *Mycobacterium phlei* ATCC 354. *Can. J. Microbiol.* 23:7–19.

Dinadayala, P., D. Kaur, S. Berg, A. G. Amin, V. D. Vissa, D. Chatterjee, P. J. Brennan, and D. C. Crick. 2006. Genetic basis for the synthesis of the immunomodulatory mannose caps of lipoarabinomannan in *Mycobacterium tuberculosis*. *J. Biol. Chem.* 281:20027–20035.

Dyer, B. S., J. D. Jones, G. D. Ainge, M. Denis, D. S. Larsen, and G. F. Painter. 2007. Synthesis and structure of phosphatidylinositol dimannoside. *J. Org. Chem.* 72:3282–3288.

Ehlers, M. R., and M. Daffe. 1998. Interactions between *Mycobacterium tuberculosis* and host cells: are mycobacterial sugars the key? *Trends Microbiol.* 6:328–335.

Ehlers, S. 2003. Role of tumour necrosis factor (TNF) in host defence against tuberculosis: implications for immunotherapies targeting TNF. *Ann. Rheum. Dis.* 62(Suppl 2):ii37–ii42.

Engering, A., T. B. Geijtenbeek, and Y. van Kooyk. 2002. Immune escape through C-type lectins on dendritic cells. *Trends Immunol.* 23:480–485.

Ernst, J. D. 1998. Macrophage receptors for *Mycobacterium tuberculosis*. *Infect. Immun.* **66**:1277–1281.

Escuyer, V. E., M.-A. Lety, J. B. Torrelles, K.-H. Khoo, J.-B. Tang, C. D. Rithner, C. Frehel, M. R. McNeil, P. J. Brennan, and D. Chatterjee. 2001. The role of the *embA* and *embB* gene products in the biosynthesis of the terminal hexaarabinofuranosyl motif of *Mycobacterium smegmatis* arabinogalactan. *J. Biol. Chem.* **276**:48854–48862.

Feinberg, H., D. A. Mitchell, K. Drickamer, and W. I. Weis. 2001. Structural basis for selective recognition of oligosaccharides by DC-SIGN and DC-SIGNR. *Science* **294**:2163–2166.

Feinberg, H., S. Park-Snyder, A. R. Kolatkar, C. T. Heise, M. E. Taylor, and W. I. Weis. 2000. Structure of a C-type carbohydrate recognition domain from the macrophage mannose receptor. *J. Biol. Chem.* **275**:21539–21548.

Fenton, M. J., L. W. Riley, and L. S. Schlesinger. 2005. Receptor-mediated recognition of *Mycobacterium tuberculosis* by host cells, p. 405–426. *In* S. T. Cole, K. D. Eisenach, D. N. McMurray and W. R. Jacobs, Jr. (ed.), *Tuberculosis and the Tubercle Bacillus.* ASM Press, Washington, DC.

Ferguson, J. S., D. R. Voelker, F. X. McCormack and L. S. Schlesinger. 1999. Surfactant protein D binds to *Mycobacterium tuberculosis* bacilli and lipoarabinomannan via carbohydrate-lectin interactions resulting in reduced phagocytosis of the bacteria by macrophages. *J. Immunol.* **163**:312–321.

Figdor, C. G., Y. van Kooyk, and G. J. Adema. 2002. C-type lectin receptors on dendritic cells and Langerhans cells. *Nat. Rev. Immunol.* **2**:77–84.

Fischer, K., E. Scotet, M. Niemeyer, H. Koebernick, J. Zerrahn, S. Maillet, R. Hurwitz, M. Kursar, M. Bonneville, S. H. Kaufmann, and U. E. Schaible. 2004. Mycobacterial phosphatidylinositol mannoside is a natural antigen for CD1d-restricted T cells. *Proc. Natl. Acad. Sci. USA* **101**:10685–10690.

Fratti, R. A., J. Chua, I. Vergne, and V. Deretic. 2003. *Mycobacterium tuberculosis* glycosylated phosphatidylinositol causes phagosome maturation arrest. *Proc. Natl. Acad. Sci. USA* **100**:5437–5442.

Fukao, T., and S. Koyasu. 2003. PI3K and negative regulation of TLR signaling. *Trends Immunol.* **24**:358–363.

Gadola, S. D., N. R. Zaccai, K. Harlos, D. Shepherd, J. C. Castro-Palomino, G. Ritter, R. R. Schmidt, E. Y. Jones, and V. Cerundolo. 2002. Structure of human CD1b with bound ligands at 2.3 A, a maze for alkyl chains. *Nat. Immunol.* **3**:721–726.

Garton, N. J., M. Gilleron, T. Brando, H. H. Dan, S. Giguere, G. Puzo, J. F. Prescott and I. C. Sutcliffe. 2002. A novel lipoarabinomannan from the equine pathogen *Rhodococcus equi*: structure and effect on macrophage cytokine production. *J. Biol. Chem.* **277**:31722–31733.

Gaylord, H., and P. J. Brennan. 1987. Leprosy and the leprosy bacillus: recent developments in characterization of antigens and immunology of the disease. *Annu. Rev. Microbiol.* **41**:645–675.

Gehring, A. J., K. M. Dobos, J. T. Belisle, C. V. Harding, and W. H. Boom. 2004. *Mycobacterium tuberculosis* LprG (Rv1411c): a novel TLR-2 ligand that inhibits human macrophage class II MHC antigen processing. *J. Immunol.* **173**:2660–2668.

Geijtenbeek, T. B., S. J. Van Vliet, E. A. Koppel, M. Sanchez-Hernandez, C. M. Vandenbroucke-Grauls, B. Appelmelk, and Y. Van Kooyk. 2003. Mycobacteria target DC-SIGN to suppress dendritic cell function. *J. Exp. Med.* **197**:7–17.

Geremia, R. A., E. A. Petroni, L. Ielpi, and B. Henrissat. 1996. Towards a classification of glycosyltransferases based on amino acid sequence similarities: prokaryotic alpha-mannosyltransferases. *Biochem. J.* **318**:133–138.

Giabbai, B., S. Sidobre, M. D. Crispin, Y. Sanchez-Ruiz, A. Bachi, M. Kronenberg, I. A. Wilson, and M. Degano. 2005. Crystal structure of mouse CD1d bound to the self ligand phosphatidylcholine: a molecular basis for NKT cell activation. *J. Immunol.* **175**:977–984.

Gibson, K. J., M. Gilleron, P. Constant, T. Brando, G. Puzo, G. S. Besra, and J. Nigou. 2004. *Tsukamurella paurometabola* lipoglycan: a new lipoarabinomannan variant with pro-inflammatory activity. *J. Biol. Chem.* **279**:22973–22982.

Gibson, K. J., M. Gilleron, P. Constant, G. Puzo, J. Nigou, and G. S. Besra. 2003a. Structural and functional features of *Rhodococcus ruber* lipoarabinomannan. *Microbiology* **149**:1437–1445.

Gibson, K. J., M. Gilleron, P. Constant, G. Puzo, J. Nigou, and G. S. Besra. 2003b. Identification of a novel mannose-capped lipoarabinomannan from *Amycolatopsis sulphurea*. *Biochem. J.* **372**:821–829.

Gibson, K. J., M. Gilleron, P. Constant, B. Sichi, G. Puzo, G. S. Besra, and J. Nigou. 2005. A lipomannan variant with strong TLR-2-dependent pro-inflammatory activity in *Saccharothrix aerocolonigenes*. *J. Biol. Chem.* **280**:28347–28356.

Gibson, K. J. C., L. Eggeling, W. N. Maughan, K. Krumbach, S. S. Gurcha, J. Nigou, G. Puzo, H. Sahm, and G. S. Besra. 2003c. Disruption of *Cg*-Ppm1, a polyprenyl monophosphomannose synthase, and the generation of lipoglycan-less mutants in *Corynebacterium glutamicum*. *J. Biol. Chem.* **278**:40842–40850.

Gilleron, M., L. Bala, T. Brando, A. Vercellone, and G. Puzo. 2000. *Mycobacterium tuberculosis* H37Rv parietal and cellular lipoarabinomannans. Characterization of the acyl- and glycoforms. *J. Biol. Chem.* **275**:677–684.

Gilleron, M., N. J. Garton, J. Nigou, T. Brando, G. Puzo, and I. C. Sutcliffe. 2005. Characterization of a truncated lipoarabinomannan from the actinomycete *Turicella otitidis*. *J. Bacteriol.* **187**:854–861.

Gilleron, M., N. Himoudi, O. Adam, P. Constant, A. Venisse, M. Riviere, and G. Puzo. 1997. *Mycobacterium smegmatis* phosphoinositols-glyceroarabinomannans. Structure and localization of alkali-labile and alkali-stable phosphoinositides. *J. Biol. Chem.* **272**:117–124.

Gilleron, M., J. Nigou, B. Cahuzac, and G. Puzo. 1999. Structural study of the lipomannans from *Mycobacterium bovis* BCG: characterisation of multiacylated forms of the phosphatidyl-*myo*-inositol anchor. *J. Mol. Biol.* **285**:2147–2160.

Gilleron, M., J. Nigou, D. Nicolle, V. Quesniaux, and G. Puzo. 2006. The acylation state of mycobacterial lipomannans modulates innate immunity response through toll-like receptor 2. *Chem. Biol.* **13**:39–47.

Gilleron, M., V. F. Quesniaux, and G. Puzo. 2003. Acylation state of the phosphatidylinositol hexamannosides from *Mycobacterium bovis* bacillus Calmette Guerin and mycobacterium tuberculosis H37Rv and its implication in Toll-like receptor response. *J. Biol. Chem.* **278**:29880–29889.

Gilleron, M., M. Riviere, and G. Puzo. 2001a. Role of glycans in the bacterial infections: interaction host-mycobacteria, p. 113–140. *In* M. Aubery (ed.), *Glycans in Cell Interaction and Recognition: Therapeutic Aspects.* Harwood Academic, Amsterdam, The Netherlands.

Gilleron, M., C. Ronet, M. Mempel, B. Monsarrat, G. Gachelin and G. Puzo. 2001b. Acylation state of the phosphatidylinositol mannosides from *Mycobacterium bovis* bacillus Calmette Guerin and ability to induce granuloma and recruit natural killer T cells. *J. Biol. Chem.* **276**:34896–34904.

Gilleron, M., S. Stenger, Z. Mazorra, F. Wittke, S. Mariotti, G. Bohmer, J. Prandi, L. Mori, G. Puzo, and G. De Libero. 2004. Diacylated sulfoglycolipids are novel mycobacterial antigens stimulating CD1-restricted T Cells during infection with *Mycobacterium tuberculosis*. *J. Exp. Med.* **199**:649–659.

Goren, M. B. 1984. Biosynthesis and structures of phospholipids and sulfatides, p. 379–415. *In* G. P. Kubica and L. G. Wayne (ed.), *The Mycobacteria—a Source Book*, vol. 1. Marcel Dekker, Inc., New York, NY.

Gu, X., M. Chen, Q. Wang, M. Zhang, B. Wang, and H. Wang. 2005. Expression and purification of a functionally active recombinant GDP-mannosyltransferase (PimA) from *Mycobacterium tuberculosis* H37Rv. *Protein Expr. Purif.* **42**:47–53.

Guerardel, Y., E. Maes, V. Briken, F. Chirat, Y. Leroy, C. Locht, G. Strecker, and L. Kremer. 2003. Lipomannan and lipoarabinomannan from a clinical isolate of *Mycobacterium kansasii*: novel structural features and apoptosis-inducing properties. *J. Biol. Chem.* **278**:36637–36651.

Guerardel, Y., E. Maes, E. Elass, Y. Leroy, P. Timmerman, G. S. Besra, C. Locht, G. Strecker, and L. Kremer. 2002. Structural study of lipomannan and lipoarabinomannan from *Mycobacterium chelonae*. Presence of unusual components with alpha 1,3-mannopyranose side chains. *J. Biol. Chem.* **277**:30635–30648.

Gurcha, S. S., A. R. Baulard, L. Kremer, C. Locht, D. B. Moody, W. Muhlecker, C. E. Costellos, D. C. Crick, P. J. Brennan, and G. S. Besra. 2002. Ppm1, a novel polyprenol monophosphomannose synthase from *Mycobacterium tuberculosis*. *Biochem. J.* **365**:441–450.

Haites, R. E., Y. S. Morita, M. J. McConville, and H. Billman-Jacobe. 2005. Function of phosphatidylinositol in mycobacteria. *J. Biol. Chem.* **280**:10981–10987.

Herrmann, J. L., L. Tailleux, J. Nigou, B. Giquel, G. Puzo, P. H. Lagrange, and O. Neyrolles. 2006. The role of human dendritic cells in tuberculosis: protector or non-protector? *Rev. Mal. Respir.* **23**:21–28.

Hill, D. L., and C. E. Ballou. 1966. Biosynthesis of mannophospholipids by *Mycobacterium phlei*. *J. Biol. Chem.* **241**:895–902.

Hong, X., and A. J. Hopfinger. 2004. Construction, molecular modeling, and simulation of *Mycobacterium tuberculosis* cell walls. *Biomacromolecules* **5**:1052–1065.

Hoppe, H. C., B. J. de Wet, C. Cywes, M. Daffe, and M. R. Ehlers. 1997. Identification of phosphatidylinositol mannoside as a mycobacterial adhesin mediating both direct and opsonic binding to nonphagocytic mammalian cells. *Infect. Immun.* **65**:3896–3905.

Huang, H., M. S. Scherman, W. D'Haeze, D. Vereecke, M. Holsters, D. C. Crick, and M. R. McNeil. 2005. Identification and active expression of the *Mycobacterium tuberculosis* gene encoding 5-phospho-α-D-ribose-1-diphosphate: decaprenyl-phosphate 5-phosphoribosyltransferase, the first enzyme committed to decaprenylphosphoryl-D-arabinose synthesis. *J. Biol. Chem.* **280**:24539–24543.

Hunger, R. E., P. A. Sieling, M. T. Ochoa, M. Sugaya, A. E. Burdick, T. H. Rea, P. J. Brennan, J. T. Belisle, A. Blauvelt, S. A. Porcelli, and R. L. Modlin. 2004. Langerhans cells utilize CD1a and langerin to efficiently present nonpeptide antigens to T cells. *J. Clin. Investig.* **113**:701–708.

Hunter, S. W., and P. J. Brennan. 1990. Evidence for the presence of a phosphatidylinositol anchor on the lipoarabinomannan and lipomannan of *Mycobacterium tuberculosis*. *J. Biol. Chem.* **265**:9272–9279.

Hunter, S. W., H. Gaylord, and P. J. Brennan. 1986. Structure and antigenicity of the phosphorylated lipopolysaccharide antigens from the leprosy and tubercle bacilli. *J. Biol. Chem.* **261**:12345–12351.

Jackson, M., D. C. Crick, and P. J. Brennan. 2000. Phosphatidylinositol is an essential phospholipid of mycobacteria. *J. Biol. Chem.* **275**:30092–30099.

Jayaprakash, K. N., J. Lu, and B. Fraser-Reid. 2005. Synthesis of a lipomannan component of the cell-wall complex of *Mycobacterium tuberculosis* is based on Paulsen's concept of donor/acceptor "match." *Angew. Chem. Int. Ed. Engl.* **44**:5894–5898.

Joe, M., D. Sun, H. Taha, G. C. Completo, J. E. Croudace, D. A. Lammas, G. S. Besra, and T. L. Lowary. 2006. The 5-deoxy-5-methylthio-xylofuranose residue in mycobacterial lipoarabinomannan. Absolute stereochemistry, linkage position, conformation, and immunomodulatory activity. *J. Am. Chem. Soc.* **128**:5059–5072.

Jones, B. W., T. K. Means, K. A. Heldwein, M. A. Keen, P. J. Hill, J. T. Belisle, and M. J. Fenton. 2001. Different Toll-like receptor agonists induce distinct macrophage responses. *J. Leukoc. Biol.* **69**:1036–1044.

Jung, S. B., C. S. Yang, J. S. Lee, A. R. Shin, S. S. Jung, J. W. Son, C. V. Harding, H. J. Kim, J. K. Park, T. H. Paik, C. H. Song, and E. K. Jo. 2006. The mycobacterial 38-kilodalton glycolipoprotein antigen activates the mitogen-activated protein kinase pathway and release of proinflammatory cytokines through Toll-like receptors 2 and 4 in human monocytes. *Infect. Immun.* **74**:2686–2696.

Kang, P. B., A. K. Azad, J. B. Torrelles, T. M. Kaufman, A. Beharka, E. Tibesar, L. E. DesJardin, and L. S. Schlesinger. 2005. The human macrophage mannose receptor directs *Mycobacterium tuberculosis* lipoarabinomannan-mediated phagosome biogenesis. *J. Exp. Med.* **202**:987–999.

Kang, S. J., and P. Cresswell. 2004. Saposins facilitate CD1d-restricted presentation of an exogenous lipid antigen to T cells. *Nat. Immunol.* **5**:175–181.

Kaufmann, S. H. 2001. How can immunology contribute to the control of tuberculosis? *Nat. Rev. Immunol.* **1**:20–30.

Kaufmann, S. H. 2005. Recent findings in immunology give tuberculosis vaccines a new boost. *Trends Immunol.* **26**:660–667.

Kaur, D., S. Berg, P. Dinadayala, B. Gicquel, D. Chatterjee, M. R. McNeil, V. D. Vissa, D. C. Crick, M. Jackson, and P. J. Brennan. 2006. Biosynthesis of mycobacterial lipoarabinomannan: role of a branching mannosyltransferase. *Proc. Natl Acad. Sci. USA* **103**:13664–13669.

Khoo, K. H., A. Dell, H. R. Morris, P. J. Brennan. and D. Chatterjee. 1995a. Inositol phosphate capping of the nonreducing termini of lipoarabinomannan from rapidly growing strains of *Mycobacterium*. *J. Biol. Chem.* **270**:12380–12389.

Khoo, K.-H., A. Dell, H. R. Morris, P. J. Brennan, and D. Chatterjee. 1995b. Structural definition of acylated phosphatidylinositol mannosides from *Mycobacterium tuberculosis*: definition of a common anchor for lipomannan and lipoarabinomannan. *Glycobiology* **5**:117–127.

Khoo, K. H., E. Douglas, P. Azadi, J. M. Inamine, G. S. Besra, K. Mikusova, P. J. Brennan, and D. Chatterjee. 1996. Truncated structural variants of lipoarabinomannan in ethambutol drug-resistant strains of *Mycobacterium smegmatis*. Inhibition of arabinan biosynthesis by ethambutol. *J. Biol. Chem.* **271**:28682–28690.

Khoo, K. H., J. B. Tang, and D. Chatterjee. 2001. Variation in mannose-capped terminal arabinan motifs of lipoarabinomannans from clinical isolates of *Mycobacterium tuberculosis* and *Mycobacterium avium* complex. *J. Biol. Chem.* **276**:3863–3871.

Knutson, K. L., Z. Hmama, P. Herrera-Velit, R. Rochford, and N. E. Reiner. 1998. Lipoarabinomannan of *Mycobacterium tuberculosis* promotes protein tyrosine dephosphorylation and inhibition of mitogen-activated protein kinase in human mononuclear phagocytes. Role of the Src homology 2 containing tyrosine phosphatase 1. *J. Biol. Chem.* **273**:645–652.

Koch, M., V. S. Stronge, D. Shepherd, S. D. Gadola, B. Mathew, G. Ritter, A. R. Fersht, G. S. Besra, R. R. Schmidt, E. Y. Jones, and V. Cerundolo. 2005. The crystal structure of human CD1d with and without alpha-galactosylceramide. *Nat. Immunol.* **6**:819–826.

Koppel, E. A., I. S. Ludwig, M. S. Hernandez, T. L. Lowary, R. R. Gadikota, A. B. Tuzikov, C. M. Vandenbroucke-Grauls, Y. van

Kooyk, B. J. Appelmelk, and T. B. Geijtenbeek. 2004. Identification of the mycobacterial carbohydrate structure that binds the C-type lectins DC-SIGN, L-SIGN and SIGNR1. *Immunobiology* 209:117–127.

Kordulakova, J., M. Gilleron, K. Mikusova, G. Puzo, P. J. Brennan, B. Gicquel and M. Jackson. 2002. Definition of the first mannosylation step in phosphatidylinositol synthesis: PimA is essential for growth of mycobacteria. *J. Biol. Chem.* 277:31335–31344.

Kordulakova, J., M. Gilleron, G. Puzo, P. J. Brennan, B. Gicquel, K. Mikusova, and M. Jackson. 2003. Identification of the required acyltransferase step in the biosynthesis of the phosphatidylinositol mannosides of *Mycobacterium* species. *J. Biol. Chem.* 278:36285–36295.

Koul, A., T. Herget, B. Klebl, and A. Ullrich. 2004. Interplay between mycobacteria and host signalling pathways. *Nat. Rev. Microbiol.* 2:189–202.

Kovacevic, S., D. Anderson, Y. S. Morita, J. H. Patterson, R. E. Haites, B. N. I. McMillan, R. Coppel, M. J. McConville, and H. Billman-Jacobe. 2006. Identification of a novel protein with a role in lipoarabinomannan biosynthesis in mycobacteria. *J. Biol. Chem.* 281:9011–9017.

Kremer, L., S. S. Gurcha, P. Bifani, P. G. Hitchen, A. Baulard, H. R. Morris, A. Dell, P. J. Brennan, and G. S. Besra. 2002. Characterization of a putative α-mannosyltransferase involved in phosphatidylinositol trimannoside biosynthesis in *Mycobacterium tuberculosis*. *Biochem. J.* 363:437–447.

Krishnegowda, G., A. M. Hajjar, J. Zhu, E. J. Douglass, S. Uematsu, S. Akira, A. S. Woods, and D. C. Gowda. 2005. Induction of proinflammatory responses in macrophages by the glycosylphosphatidylinositols of *Plasmodium falciparum*: cell signaling receptors, glycosylphosphatidylinositol (GPI) structural requirement, and regulation of GPI activity. *J. Biol. Chem.* 280:8606–8616.

Krutzik, S. R., B. Tan, H. Li, M. T. Ochoa, P. T. Liu, S. E. Sharfstein, T. G. Graeber, P. A. Sieling, Y. J. Liu, T. H. Rea, B. R. Bloom, and R. L. Modlin. 2005. TLR activation triggers the rapid differentiation of monocytes into macrophages and dendritic cells. *Nat. Med.* 11:653–660.

Lawton, A. P., and M. Kronenberg. 2004. The Third Way: Progress on pathways of antigen processing and presentation by CD1. *Immunol. Cell. Biol.* 82:295–306.

Lee, Y. C., and C. E. Ballou. 1965. Complete structures of the glycophospholipids of mycobacteria. *Biochemistry* 4:1395–1404.

Liew, F. Y., D. Xu, E. K. Brint, and L. A. O'Neill. 2005. Negative regulation of toll-like receptor-mediated immune responses. *Nat. Rev. Immunol.* 5:446–458.

Liu, J., and A. Mushegian. 2003. Three monophyletic superfamilies account for the majority of the known glycosyltransferases. *Protein Sci.* 12:1418–1431.

Liu, P. T., S. Stenger, H. Li, L. Wenzel, B. H. Tan, S. R. Krutzik, M. T. Ochoa, J. Schauber, K. Wu, C. Meinken, D. L. Kamen, M. Wagner, R. Bals, A. Steinmeyer, U. Zugel, R. L. Gallo, D. Eisenberg, M. Hewison, B. W. Hollis, J. S. Adams, B. R. Bloom, and R. L. Modlin. 2006a. Toll-like receptor triggering of a vitamin D-mediated human antimicrobial response. *Science* 311:1770–1773.

Liu, X., B. L. Stocker, and P. H. Seeberger. 2006b. Total synthesis of phosphatidylinositol mannosides of *Mycobacterium tuberculosis*. *J. Am. Chem. Soc.* 128:3638–3648.

Lucas, M., X. Zhang, V. Prasanna, and D. M. Mosser. 2005. ERK activation following macrophage FcgammaR ligation leads to chromatin modifications at the IL-10 locus. *J. Immunol.* 175:469–477.

Ludwiczak, P., M. Gilleron, Y. Bordat, C. Martin, B. Gicquel, and G. Puzo. 2002. *Mycobacterium tuberculosis* phoP mutant: lipo-arabinomannan molecular structure. *Microbiology* 148:3029–3037.

Ma, Y., R. J. Stern, M. S. Scherman, V. D. Vissa, W. Yan, V. C. Jones, F. Zhang, S. G. Franzblau, W. H. Lewis, and M. R. McNeil. 2001. Drug targeting *Mycobacterium tuberculosis* cell wall synthesis: genetics of dTDP-rhamnose synthetic enzymes and development of a microtiter plate-based screen for inhibitors of conversion of dTDP-glucose to dTDP-rhamnose. *Antimicrob. Agents Chemother.* 45:1407–1416.

Maeda, N., J. Nigou, J. L. Herrmann, M. Jackson, A. Amara, P. H. Lagrange, G. Puzo, B. Gicquel, and O. Neyrolles. 2002. The cell surface receptor DC-SIGN discriminates between *Mycobacterium* species through selective recognition of the mannose caps on lipoarabinomannan. *J. Biol. Chem.* 278:5513–5516.

Marland, Z., T. Beddoe, L. Zaker-Tabrizi, R. L. Coppel, P. K. Crellin, and J. Rossjohn. 2005. Expression, purification, crystallization and preliminary X-ray diffraction analysis of an essential lipoprotein implicated in cell-wall biosynthesis in mycobacteria. *Acta Crystallogr. Sect. F* 61:1081–1083.

Marland, Z., T. Beddoe, L. Zaker-Tabrizi, I. S. Lucet, R. Brammananth, J. C. Whisstock, M. C. J. Wilce, R. L. Coppel, P. K. Crellin, and J. Rossjohn. 2006. Hijacking of a substrate-binding protein scaffold for use in mycobacterial cell wall biosynthesis. *J. Mol. Biol.* 359:983–997.

Martin, C. 2005. The dream of a vaccine against tuberculosis; new vaccines improving or replacing BCG? *Eur. Respir. J.* 26:162–167.

McCarthy, T. R., J. B. Torrelles, A. S. MacFarlane, M. Katawczik, B. Kutzbach, L. E. DesJardin, S. Clegg, J. B. Goldberg, and L. S. Schlesinger. 2005. Overexpression of *Mycobacterium tuberculosis manB*, a phosphomannomutase that increases phosphatidylinositol mannoside biosynthesis in *Mycobacterium smegmatis* and mycobacterial association with human macrophages. *Mol. Microbiol.* 58:774–790.

McNeil, M. R., and P. J. Brennan. 1991. Structure, function and biogenesis of the cell envelope of mycobacteria in relation to bacterial physiology, pathogenesis and drug resistance; some thoughts and possibilities arising from recent structural information. *Res. Microbiol.* 142:451–463.

Means, T. K., E. Lien, A. Yoshimura, S. Wang, D. T. Golenbock, and M. J. Fenton. 1999. The CD14 ligands lipoarabinomannan and lipopolysaccharide differ in their requirement for Toll-like receptors. *J. Immunol.* 163:6748–6755.

Mikusova, K., H. Huang, T. Yagi, M. Holsters, D. Vereecke, W. D'Haeze, M. S. Scherman, P. J. Brennan, M. R. McNeil, and D. C. Crick. 2005. Decaprenylphosphoryl arabinofuranose, the donor of the D-arabinofuranosyl residues of mycobacterial arabinan, is formed via a two-step epimerization of decaprenyl-phosphoryl ribose. *J. Bacteriol.* 187:8020–8025.

Minnikin, D. E. 1982. Lipids: complex lipids, their chemistry, biosynthesis and role, p. 95–184. *In* C. Ratledge and J. Standford (ed.), *The Biology of the Mycobacteria*, vol. 1. Academic Press Ltd., London, United Kingdom.

Molle, V., L. Kremer, C. Girard-Blanc, G. S. Besra, A. J. Cozzone, and J. F. Prost. 2003. An FHA phosphoprotein recognition domain mediates EmbR phosphorylation by PknH, a Ser/Thr protein kinase from *Mycobacterium tuberculosis*. *Biochemistry* 42:15300–15309.

Moody, D. B. 2006. TLR gateways to CD1 function. *Nat. Immunol.* 7:811–817.

Moody, D. B., V. Briken, T. Y. Cheng, C. Roura-Mir, M. R. Guy, D. H. Geho, M. L. Tykocinski, G. S. Besra, and S. A. Porcelli. 2002. Lipid length controls antigen entry into endosomal and nonendosomal pathways for CD1b presentation. *Nat. Immunol.* 3:435–442.

Moody, D. B., and S. A. Porcelli. 2003. Intracellular pathways of CD1 antigen presentation. *Nat. Rev. Immunol.* 3:11–22.

Moody, D. B., B. B. Reinhold, V. N. Reinhold, G. S. Besra, and S. A. Porcelli. 1999. Uptake and processing of glycosylated mycolates for presentation to CD1b-restricted T cells. *Immunol. Lett.* 65:85–91.

Moody, D. B., T. Ulrichs, W. Muhlecker, D. C. Young, S. S. Gurcha, E. Grant, J. P. Rosat, M. B. Brenner, C. E. Costello, G. S. Besra, and S. A. Porcelli. 2000. CD1c-mediated T-cell recognition of isoprenoid glycolipids in *Mycobacterium tuberculosis* infection. *Nature* 404:884–888.

Moody, D. B., D. M. Zajonc, and I. A. Wilson. 2005. Anatomy of CD1-lipid antigen complexes. *Nat. Rev. Immunol.* 5:387–399.

Morita, Y. S., J. H. Patterson, H. Billman-Jacobe, and M. J. McConville. 2004. Biosynthesis of mycobacterial phosphatidylinositol mannosides. *Biochem. J.* 378:589–597.

Morita, Y. S., C. B. C. Sena, R. F. Waller, K. Kurokawa, M. F. Sernee, F. Nakatani, R. E. Haites, H. Billman-Jacobe, M. J. McConville, Y. Maeda, and T. Kinoshita. 2006. PimE is a polyprenol-phosphate-mannose-dependent mannosyltransferase that transfers the fifth mannose of phosphatidylinositol mannoside in mycobacteria. *J. Biol. Chem.* 281:25143–25155.

Morita, Y. S., R. Velasquez, E. Taig, R. F. Waller, J. H. Patterson, D. Tull, S. J. Williams, H. Billman-Jacobe, and M. J. McConville. 2005. Compartmentalization of lipid biosynthesis in mycobacteria. *J. Biol. Chem.* 280:21645–21652.

Morona, R., L. Van Den Bosch, and C. Daniels. 2000. Evaluation of Wzz/MPA1/MPA2 proteins based on the presence of coiled coil regions. *Microbiology* 146:1–4.

Movahedzadeh, F., D. A. Smith, R. A. Norman, P. Dinadayala, J. Murray-Rust, D. G. Russell, S. L. Kendall, S. C. G. Rison, M. S. B. McAlister, G. J. Bancroft, N. Q. McDonald, M. Daffé, Y. Av-Gay, and N. G. Stocker. 2004. The *Mycobacterium tuberculosis ino1* gene is essential for growth and virulence. *Mol. Microbiol.* 51:1003–1014.

Mukherjee, S., T. T. Soe, and F. R. Maxfield. 1999. Endocytic sorting of lipid analogues differing solely in the chemistry of their hydrophobic tails. *J. Cell. Biol.* 144:1271–1284.

Neyrolles, O., B. Gicquel, and L. Quintana-Murci. 2006. Towards a crucial role for DC-SIGN in tuberculosis and beyond. *Trends Microbiol.* 14:383–387.

Nigou, J., and G. S. Besra. 2002a. Characterization and regulation of inositol monophosphatase activity in *Mycobacterium smegmatis*. *Biochem. J.* 361:385–390.

Nigou, J., and G. S. Besra. 2002b. Cytidine diphosphate-diacylglycerol synthesis in *Mycobacterium smegmatis*. *Biochem. J.* 367:157–162.

Nigou, J., L. G. Dover, and G. S. Besra. 2002a. Purification and biochemical characterization of *Mycobacterium tuberculosis* SuhB, an inositol monophosphatase involved in inositol biosynthesis. *Biochemistry* 41:4392–4398.

Nigou, J., M. Gilleron, T. Brando, A. Vercellone, and G. Puzo. 1999a. Structural definition of arabinomannans from *Mycobacterium bovis* BCG. *Glycoconj. J.* 16:257–264.

Nigou, J., M. Gilleron, and G. Puzo. 1999b. Lipoarabinomannans: characterization of the multiacylated forms of the phosphatidyl-myo-inositol anchor by NMR spectroscopy. *Biochem. J.* 337:453–460.

Nigou, J., M. Gilleron, and G. Puzo. 2003. Lipoarabinomannans: from structure to biosynthesis. *Biochimie* 85:153–166.

Nigou, J., M. Gilleron, M. Rojas, L. F. Garcia, M. Thurnher, and G. Puzo. 2002b. Mycobacterial lipoarabinomannans: modulators of dendritic cell function and the apoptotic response. *Microbes Infect.* 4:945–953.

Nigou, J., A. Vercellone, and G. Puzo. 2000. New structural insights into the molecular deciphering of mycobacterial lipogly-can binding to C-type lectins: lipoarabinomannan glycoform characterization and quantification by capillary electrophoresis at the subnanomole level. *J. Mol. Biol.* 299:1353–1362.

Nigou, J., C. Zelle-Rieser, M. Gilleron, M. Thurnher, and G. Puzo. 2001. Mannosylated lipoarabinomannans inhibit IL-12 production by human dendritic cells: evidence for a negative signal delivered through the mannose receptor. *J. Immunol.* 166:7477–7485.

Ning, B., and A. D. Elbein. 1999. Purification and properties of mycobacterial GDP-mannose pyrophosphorylase. *Arch. Biochem. Biophys.* 362:339–345.

Norman, R. A., M. S. McAlister, J. Murray-Rust, F. Movahedzadeh, N. G. Stocker, and N. Q. McDonald. 2002. Crystal structure of inositol 1-phosphate synthase from *Mycobacterium tuberculosis*, a key enzyme in phosphatidylinositol synthesis. *Structure* 10:393–402.

Noss, E. H., R. K. Pai, T. J. Sellati, J. D. Radolf, J. Belisle, D. T. Golenbock, W. H. Boom, and C. V. Harding. 2001. Toll-like receptor 2-dependent inhibition of macrophage class II MHC expression and antigen processing by 19-kDa lipoprotein of *Mycobacterium tuberculosis*. *J. Immunol.* 167:910–918.

Oriol, R., I. Martinez-Duncker, I. Chantret, R. Mollicone, and P. Codogno. 2002. Common origin and evolution of glycosyltransferases using Dol-P-monosaccharides as donor substrate. *Mol. Biol. Evol.* 19:1451–1463.

Orme, I. M. 2005. The use of animal models to guide rational vaccine design. *Microbes Infect.* 7:905–910.

Ortalo-Magne, A., A. B. Andersen, and M. Daffe. 1996a. The outermost capsular arabinomannans and other mannoconjugates of virulent and avirulent tubercle bacilli. *Microbiology* 142:927–935.

Ortalo-Magne, A., M. A. Dupont, A. Lemassu, A. B. Andersen, P. Gounon, and M. Daffe. 1995. Molecular composition of the outermost capsular material of the tubercle bacillus. *Microbiology* 141:1609–1620.

Ortalo-Magne, A., A. Lemassu, M. A. Laneelle, F. Bardou, G. Silve, P. Gounon, G. Marchal, and M. Daffe. 1996b. Identification of the surface-exposed lipids on the cell envelopes of *Mycobacterium tuberculosis* and other mycobacterial species. *J. Bacteriol.* 178:456–461.

Owens, R. M., F. F. Hsu, B. C. VanderVen, G. E. Purdy, E. Hesteande, P. Giannakas, J. C. Sacchettini, J. D. McKinney, P. J. Hill, J. T. Belisle, B. A. Butcher, K. Pethe, and D. G. Russell. 2006. *M. tuberculosis* Rv2252 encodes a diacylglycerol kinase involved in the biosynthesis of phosphatidylinositol mannosides (PIMs). *Mol. Microbiol.* 60:1152–1163.

Parish, T., J. Liu, H. Nikaido, and N. G. Stoker. 1997. A *Mycobacterium smegmatis* mutant with a defective inositol monophosphate phosphatase gene homolog has altered cell envelope permeability. *J. Bacteriol.* 179:7827–7833.

Pathak, S. K., S. Basu, A. Bhattacharyya, S. Pathak, M. Kundu, and J. Basu. 2005. *Mycobacterium tuberculosis* lipoarabinomannan-mediated IRAK-M induction negatively regulates Toll-like receptor-dependent interleukin-12 p40 production in macrophages. *J. Biol. Chem.* 280:42794–42800.

Patterson, J. H., R. F. Waller, D. Jeevarajah, H. Billman-Jacobe, and M. J. McConville. 2003. Mannose metabolism is required for mycobacterial growth. *Biochem. J.* 372:77–86.

Paul, T. R., and T. J. Beveridge. 1994. Preservation of surface lipids and determination of ultrastructure of *Mycobacterium kansasii* by freeze-substitution. *Infect. Immun.* 62:1542–1550.

Paul, T. R., and T. J. Beveridge. 1992. Reevaluation of envelope profiles and cytoplasmic ultrastructure of mycobacteria processed by conventional embedding and freeze-substitution protocols. *J. Bacteriol.* 174:6508–6517.

Pecora, N. D., A. J. Gehring, D. H. Canaday, W. H. Boom, and C. V. Harding. 2006. *Mycobacterium tuberculosis* LprA is a lipoprotein agonist of TLR2 that regulates innate immunity and APC function. *J. Immunol.* 177:422–429.

Penumarti, N., and G. K. Khuller. 1983. Subcellular distribution of mannophosphoinositides in *Mycobacterium smegmatis* during growth. *Experientia* 39:882–884.

Petit, J. F., and E. Lederer. 1984. The structure of the mycobacterial cell wall, p. 301–314. *In* G. P. Kubica and L. G. Wayne (ed.), *The Mycobacteria—a Source Book*, vol. 15. Marcel Dekker, Inc., New York, NY.

Pitarque, S., J. L. Herrmann, J. L. Duteyrat, M. Jackson, G. R. Stewart, F. Lecointe, B. Payre, O. Schwartz, D. B. Young, G. Marchal, P. H. Lagrange, G. Puzo, B. Gicquel, J. Nigou, and O. Neyrolles. 2005. Deciphering the molecular bases of *Mycobacterium tuberculosis* binding to the lectin DC-SIGN reveals an underestimated complexity. *Biochem. J.* 392:615–624.

Polotsky, V. Y., J. T. Belisle, K. Mikusova, R. A. Ezekowitz, and K. A. Joiner. 1997. Interaction of human mannose-binding protein with *Mycobacterium avium*. *J. Infect. Dis.* 175:1159–1168.

Porcelli, S. A. 1995. The CD1 family: a third lineage of antigen-presenting molecules. *Adv. Immunol.* 59:1–98.

Prigozy, T. I., O. Naidenko, P. Qasba, D. Elewaut, L. Brossay, A. Khurana, T. Natori, Y. Koezuka, A. Kulkarni, and M. Kronenberg. 2001. Glycolipid antigen processing for presentation by CD1d molecules. *Science* 291:664–667.

Prigozy, T. I., P. A. Sieling, D. Clemens, P. L. Stewart, S. M. Behar, S. A. Porcelli, M. B. Brenner, R. L. Modlin, and M. Kronenberg. 1997. The mannose receptor delivers lipoglycan antigens to endosomes for presentation to T cells by CD1b molecules. *Immunity* 6:187–197.

Prinzis, S., D. Chatterjee, and P. J. Brennan. 1993. Structure and antigenicity of lipoarabinomannan from *Mycobacterium bovis* BCG. *J. Gen. Microbiol.* 139:2649–2658.

Puig-Kroger, A., D. Serrano-Gomez, E. Caparros, A. Dominguez-Soto, M. Relloso, M. Colmenares, L. Martinez-Munoz, N. Longo, N. Sanchez-Sanchez, M. Rincon, L. Rivas, P. Sanchez-Mateos, E. Fernandez-Ruiz, and A. L. Corbi. 2004. Regulated expression of the pathogen receptor dendritic cell-specific intercellular adhesion molecule 3 (ICAM-3)-grabbing nonintegrin in THP-1 human leukemic cells, monocytes, and macrophages. *J. Biol. Chem* 279:25680–25688.

Puzo, G. 1993. La paroi mycobactérienne: structure et organisation des glycoconjugués majeurs. *Ann. Inst. Pasteur* 4:225–238.

Quesniaux, V. J., D. M. Nicolle, D. Torres, L. Kremer, Y. Guerardel, J. Nigou, G. Puzo, F. Erard, and B. Ryffel. 2004. Toll-like receptor 2 (TLR2)-dependent-positive and TLR2-independent-negative regulation of proinflammatory cytokines by mycobacterial lipomannans. *J. Immunol.* 172:4425–4434.

Rastogi, N. 1991. Recent observations concerning structure and function relationships in the mycobacterial cell envelope: elaboration of a model in terms of mycobacterial pathogenicity, virulence and drug-resistance. *Res. Microbiol.* 142:464–476.

Rhoades, E., F. Hsu, J. B. Torrelles, J. Turk, D. Chatterjee, and D. G. Russell. 2003. Identification and macrophage-activating activity of glycolipids released from intracellular *Mycobacterium bovis* BCG. *Mol. Microbiol.* 48:875–888.

Riviere, M., A. Moisand, A. Lopez, and G. Puzo. 2004. Highly ordered supra-molecular organization of the mycobacterial lipoarabinomannans in solution. Evidence of a relationship between supra-molecular organization and biological activity. *J. Mol. Biol.* 344:907–918.

Rojas, M., L. F. Garcia, J. Nigou, G. Puzo, and M. Olivier. 2000. Mannosylated lipoarabinomannan antagonizes *Mycobacterium tuberculosis*-induced macrophage apoptosis by altering Ca+2-dependent cell signaling. *J. Infect. Dis.* 182:240–251.

Rojas, M., M. Olivier, and L. F. Garcia. 2002. Activation of JAK2/STAT1-alpha-dependent signaling events during *Mycobacterium tuberculosis*-induced macrophage apoptosis. *Cell. Immunol.* 217:58–66.

Ronet, C., M. Mempel, N. Thieblemont, A. Lehuen, P. Kourilsky, and G. Gachelin. 2001. Role of the complementarity-determining region 3 (CDR3) of the TCR-beta chains associated with the Valpha14 semi-invariant TCR alpha-chain in the selection of CD4(+) NK T cells. *J. Immunol.* 166:1755–1762.

Roura-Mir, C., L. Wang, T. Y. Cheng, I. Matsunaga, C. C. Dascher, S. L. Peng, M. J. Fenton, C. Kirschning, and D. B. Moody. 2005. *Mycobacterium tuberculosis* regulates CD1 antigen presentation pathways through TLR-2. *J. Immunol.* 175:1758–1766.

Russell, D. G. 2003. Phagosomes, fatty acids and tuberculosis. *Nat. Cell. Biol.* 5:776–778.

Salman, M., J. T. Lonsdale, G. S. Besra, and P. J. Brennan. 1999. Phosphatidylinositol synthesis in mycobacteria. *Biochim. Biophys. Acta* 1436:437–450.

Schaeffer, M. L., K.-H. Khoo, G. S. Besra, D. Chatterjee, P. J. Brennan, J. T. Belisle, and J. M. Inamine. 1999. The *pimB* gene of *Mycobacterium tuberculosis* encodes a mannosyltransferase involved in lipoarabinomannan biosynthesis. *J. Biol. Chem.* 274:31625–31631.

Schaible, U. E., F. Winau, P. A. Sieling, K. Fischer, H. L. Collins, K. Hagens, R. L. Modlin, V. Brinkmann, and S. H. Kaufmann. 2003. Apoptosis facilitates antigen presentation to T lymphocytes through MHC-I and CD1 in tuberculosis. *Nat. Med.* 9:1039–1046.

Scherman, M. S., L. Kalbe-Bournonville, D. Bush, Y. Xin, L. Deng, and M. R. McNeil. 1996. Polyprenylphosphate-pentoses in mycobacteria are synthesized from 5-phosphoribose pyrophosphate. *J. Biol. Chem.* 271:29652–29658.

Schlesinger, L. S. 1993. Macrophage phagocytosis of virulent but not attenuated strains of *Mycobacterium tuberculosis* is mediated by mannose receptors in addition to complement receptors. *J. Immunol.* 150:2920–2930.

Schlesinger, L. S., S. R. Hull, and T. M. Kaufman. 1994. Binding of the terminal mannosyl units of lipoarabinomannan from a virulent strain of *Mycobacterium tuberculosis* to human macrophages. *J. Immunol.* 152:4070–4079.

Schlesinger, L. S., T. M. Kaufman, S. Iyer, S. R. Hull, and L. K. Marchiando. 1996. Differences in mannose receptor-mediated uptake of lipoarabinomannan from virulent and attenuated strains of *Mycobacterium tuberculosis* by human macrophages. *J. Immunol.* 157:4568–4575.

Sharma, K., M. Gupta, A. Krupa, N. Srinivasan, and Y. Singh. 2006a. EmbR, a regulatory protein with ATPase activity, is a substrate of multiple serine/threonine kinases and phosphatase in *Mycobacterium tuberculosis*. *FEBS J.* 273:2711–2721.

Sharma, K., M. Gupta, M. Pathak, N. Gupta, A. Koul, S. Sarangi, R. Baweja, and Y. Singh. 2006b. Transcriptional control of the mycobacterial *embCAB* operon by PknH through a regulatory protein, EmbR, *in vivo*. *J. Bacteriol.* 188:2936–2944.

Shi, L., S. Berg, A. Lee, J. S. Spencer, J. Zhang, V. Vissa, M. R. McNeil, K.-H. Khoo, and D. Chatterjee. 2006. The carboxy terminus of EmbC from *Mycobacterium smegmatis* mediates chain length extension of the arabinan in lipoarabinomannan. *J. Biol. Chem.* 281:19542–19546.

Sibley, L. D., S. W. Hunter, P. J. Brennan, and J. L. Krahenbuhl. 1988. Mycobacterial lipoarabinomannan inhibits gamma interferon-mediated activation of macrophages. *Infect. Immun.* 56:1232–1236.

Sidobre, S., J. Nigou, G. Puzo, and M. Riviere. 2000. Lipoglycans are putative ligands for the human pulmonary surfactant protein A attachment to mycobacteria. Critical role of the lipids for lectin-carbohydrate recognition. *J. Biol. Chem.* **275**:2415–2422.

Sidobre, S., G. Puzo, and M. Riviere. 2002. Lipid-restricted recognition of mycobacterial lipoglycans by human pulmonary surfactant protein A: a surface-plasmon-resonance study. *Biochem. J.* **365**:89–97.

Sieling, P. A., D. Chatterjee, S. A. Porcelli, T. I. Prigozy, R. J. Mazzaccaro, T. Soriano, B. R. Bloom, M. B. Brenner, M. Kronenberg, P. J. Brennan, et al. 1995. CD1-restricted T cell recognition of microbial lipoglycan antigens. *Science* **269**:227–230.

Skeiky, Y. A., and J. C. Sadoff. 2006. Advances in tuberculosis vaccine strategies. *Nat. Rev. Microbiol.* **4**:469–476.

Skold, M., and S. M. Behar. 2005. The role of group 1 and group 2 CD1-restricted T cells in microbial immunity. *Microbes Infect.* **7**:544–551.

Stadelmaier, A., and R. R. Schmidt. 2003. Synthesis of phosphatidylinositol mannosides (PIMs). *Carbohydr. Res.* **338**:2557–2569.

Stenger, S., and R. L. Modlin. 2002. Control of *Mycobacterium tuberculosis* through mammalian Toll-like receptors. *Curr. Opin. Immunol.* **14**:452–427.

Sulzenbacher, G., S. Canaan, Y. Bordat, O. Neyrolles, G. Stadthagen, V. Roig-Zamboni, J. Rauzier, D. Maurin, F. Laval, M. Daffe, C. Cambillau, B. Gicquel, Y. Bourne, and M. Jackson. 2006. LppX is a lipoprotein required for the translocation of phthiocerol dimycocerosates to the surface of *Mycobacterium tuberculosis*. *Embo J.* **25**:1436–1444.

Tailleux, L., N. Maeda, J. Nigou, B. Gicquel, and O. Neyrolles. 2003a. How is the phagocyte lectin keyboard played? Master class lesson by *Mycobacterium tuberculosis*. *Trends Microbiol.* **11**:259–263.

Tailleux, L., O. Neyrolles, S. Honore-Bouakline, E. Perret, F. Sanchez, J. P. Abastado, P. H. Lagrange, J. C. Gluckman, M. Rosenzwajg, and J. L. Herrmann. 2003b. Constrained intracellular survival of *Mycobacterium tuberculosis* in human dendritic cells. *J. Immunol.* **170**:1939–1948.

Tailleux, L., N. Pham-Thi, A. Bergeron-Lafaurie, J. L. Herrmann, P. Charles, O. Schwartz, P. Scheinmann, P. H. Lagrange, J. de Blic, A. Tazi, B. Gicquel, and O. Neyrolles. 2005. DC-SIGN induction in alveolar macrophages defines privileged target host cells for mycobacteria in patients with tuberculosis. *PLoS Med.* **2**:e381.

Tailleux, L., O. Schwartz, J. L. Herrmann, E. Pivert, M. Jackson, A. Amara, L. Legres, D. Dreher, L. P. Nicod, J. C. Gluckman, P. H. Lagrange, B. Gicquel and O. Neyrolles. 2003c. DC-SIGN is the major *Mycobacterium tuberculosis* receptor on human dendritic cells. *J. Exp. Med.* **197**:121–127.

Takayama, K., and D. S. Goldman. 1969. Pathway for the synthesis of mannophospholipids in *Mycobacterium tuberculosis*. *Biochim. Biophys. Acta* **176**:196–198.

Taneja, R., U. Malik, and G. K. Khuller. 1979. Effect of growth temperature on the lipid composition of *Mycobacterium smegmatis* ATCC 607. *J. Gen. Microbiol.* **113**:413–416.

Telenti, A., W. J. Philipp, S. Sreevatsan, C. Bernasconi, K. E. Stockbauer, B. Wieles, J. M. Musser, and W. R. Jacobs Jr. 1997. The *emb* operon, a gene cluster of *Mycobacterium tuberculosis* involved in resistance to ethambutol. *Nat. Med.* **3**:567–570.

Texereau, J., J. D. Chiche, W. Taylor, G. Choukroun, B. Comba, and J. P. Mira. 2005. The importance of Toll-like receptor 2 polymorphisms in severe infections. *Clin. Infect. Dis.* **41**:S408–415.

Torrelles, J. B., A. K. Azad, and L. S. Schlesinger. 2006. Fine discrimination in the recognition of individual species of phosphatidyl-myo-inositol mannosides from *Mycobacterium tuberculosis* by C-type lectin pattern recognition receptors. *J. Immunol.* **177**:1805–1816.

Torrelles, J. B., K. H. Khoo, P. A. Sieling, R. L. Modlin, N. Zhang, A. M. Marques, A. Treumann, C. D. Rithner, P. J. Brennan, and D. Chatterjee. 2004. Truncated structural variants of lipoarabinomannan in *Mycobacterium leprae* and an ethambutol-resistant strain of *Mycobacterium tuberculosis*. *J. Biol. Chem.* **279**:41227–42139.

Treumann, A., F. Xidong, L. McDonnell, P. J. Derrick, A. E. Ashcroft, D. Chatterjee, and S. W. Homans. 2002. 5-methylthiopentose: a new substituent on lipoarabinomannan in *Mycobacterium tuberculosis*. *J. Mol. Biol.* **316**:89–100.

Tsuji, M. 2006. Glycolipids and phospholipids as natural CD1d-binding NKT cell ligands. *Cell. Mol. Life Sci.* **63**:1889–1898.

Turnbull, W. B., K. H. Shimizu, D. Chatterjee, S. W. Homans, and A. Treumann. 2004. Identification of the 5-methylthiopentosyl substituent in *Mycobacterium tuberculosis* lipoarabinomannan. *Angew. Chem. Int. Ed.* **43**:3918–3922.

van den Elzen, P., S. Garg, L. Leon, M. Brigl, E. A. Leadbetter, J. E. Gumperz, C. C. Dascher, T. Y. Cheng, F. M. Sacks, P. A. Illarionov, G. S. Besra, S. C. Kent, D. B. Moody, and M. B. Brenner. 2005. Apolipoprotein-mediated pathways of lipid antigen presentation. *Nature* **437**:906–910.

VanderVen, B. C., J. D. Harder, D. C. Crick and J. T. Belisle. 2005. Export-mediated assembly of mycobacterial glycoproteins parallels eukaryotic pathways. *Science* **309**:941–943.

Venisse, A., J. M. Berjeaud, P. Chaurand, M. Gilleron and G. Puzo. 1993. Structural features of lipoarabinomannan from *Mycobacterium bovis* BCG. Determination of molecular mass by laser desorption mass spectrometry. *J. Biol. Chem.* **268**:12401–12411.

Venisse, A., J. J. Fournie, and G. Puzo. 1995. Mannosylated lipoarabinomannan interacts with phagocytes. *Eur. J. Biochem.* **231**:440–447.

Vergne, I., J. Chua, and V. Deretic. 2003. Tuberculosis toxin blocking phagosome maturation inhibits a novel Ca2+/calmodulin-PI3K hVPS34 cascade. *J. Exp. Med.* **198**:653–659.

Vignal, C., Y. Guerardel, L. Kremer, M. Masson, D. Legrand, J. Mazurier, and E. Elass. 2003. Lipomannans, but not lipoarabinomannans, purified from *Mycobacterium chelonae* and *Mycobacterium kansasii* induce TNF-alpha and IL-8 secretion by a CD14-toll-like receptor 2-dependent mechanism. *J. Immunol.* **171**:2014–2023.

Villeneuve, C., M. Gilleron, I. Maridonneau-Parini, M. Daffe, C. Astarie-Dequeker, and G. Etienne. 2005. Mycobacteria use their surface-exposed glycolipids to infect human macrophages through a receptor-dependent process. *J. Lipid Res.* **46**:475–483.

Wang, C., B. Hayes, M. M. Vestling, and K. Takayama. 2006. Transposome mutagenesis of an integral membrane transporter in *Corynebacterium matruchotii*. *Biochem. Biophys. Res. Commun.* **340**:953–960.

Weis, W. I., K. Drickamer, and W. A. Hendrickson. 1992. Structure of a C-type mannose-binding protein complexed with an oligosaccharide. *Nature* **360**:127–134.

Winau, F., V. Schwierzeck, R. Hurwitz, N. Remmel, P. A. Sieling, R. L. Modlin, S. A. Porcelli, V. Brinkmann, M. Sugita, K. Sandhoff, S. H. Kaufmann, and U. E. Schaible. 2004. Saposin C is required for lipid presentation by human CD1b. *Nat. Immunol.* **5**:169–174.

Wolucka, B. A., M. R. McNeil, E. de Hoffmann, T. Chojnacki, and P. J. Brennan. 1994. Recognition of the lipid intermediate for arabinogalactan/arabinomannan biosynthesis and its relation to the mode of action of ethambutol on mycobacteria. *J. Biol. Chem.* **269**:23328–23335.

Yim, J. J., H. W. Lee, H. S. Lee, Y. W. Kim, S. K. Han, Y. S. Shim, and S. M. Holland. 2006. The association between microsatellite polymorphisms in intron II of the human Toll-like receptor 2 gene and tuberculosis among Koreans. *Genes Immun.* **7**:150–155.

Young, D. C., and D. B. Moody. 2006. T-cell recognition of glyco-lipids presented by CD1 proteins. *Glycobiology* **16**:103R–112R.

Yuan, W., X. Qi, P. Tsang, S. J. Kang, P. A. Illarionov, G. S. Besra, J. Gumperz, and P. Cresswell. 2007. Saposin B is the dominant saposin that facilitates lipid binding to human CD1d molecules. *Proc. Natl. Acad. Sci. USA* **104**:5551–5556.

Zajonc, D. M., C. Cantu, 3rd, J. Mattner, D. Zhou, P. B. Savage, A. Bendelac, I. A. Wilson, and L. Teyton. 2005a. Structure and function of a potent agonist for the semi-invariant natural killer T cell receptor. *Nat. Immunol.* **6**:810–818.

Zajonc, D. M., M. D. Crispin, T. A. Bowden, D. C. Young, T. Y. Cheng, J. Hu, C. E. Costello, P. M. Rudd, R. A. Dwek, M. J. Miller, M. B. Brenner, D. B. Moody, and I. A. Wilson. 2005b. Molecular mechanism of lipopeptide presentation by CD1a. *Immunity* **22**:209–219.

Zajonc, D. M., M. A. Elsliger, L. Teyton, and I. A. Wilson. 2003. Crystal structure of CD1a in complex with a sulfatide self anti-gen at a resolution of 2.15 A. *Nat. Immunol.* **4**:808–815.

Zajonc, D. M., I. Maricic, D. Wu, R. Halder, K. Roy, C. H. Wong, V. Kumar, and I. A. Wilson. 2005c. Structural basis for CD1d presentation of a sulfatide derived from myelin and its implica-tions for autoimmunity. *J. Exp. Med.* **202**:1517–1526.

Zeng, Z.-H., A. R. Castano, B. W. Segelke, E. A. Stura, P. A. Peterson, and I. A. Wilson. 1997. Crystal structure of mouse CD1: an MHC-like fold with a large hydrophobic binding groove. *Science* **277**:339–345.

Zhang, J., S. D. Tachado, N. Patel, J. Zhu, A. Imrich, P. Man-fruelli, M. Cushion, T. B. Kinane, and H. Koziel. 2005. Negative regulatory role of mannose receptors on human alveo-lar macrophage proinflammatory cytokine release in vitro. *J. Leukoc. Biol.* **78**:665–674.

Zhang, N., J. B. Torrelles, M. R. McNeil, V. E. Escuyer, K.-H. Khoo, P. J. Brennan, and D. Chatterjee. 2003. The Emb proteins of my-cobacteria direct arabinosylation of lipoarabinomannan and ara-binogalactan via an N-terminal recognition region and a C-termi-nal synthetic region. *Mol. Microbiol.* **50**:69–76.

Zhou, D., J. Mattner, C. Cantu, 3rd, N. Schrantz, N. Yin, Y. Gao, Y. Sagiv, K. Hudspeth, Y. P. Wu, T. Yamashita, S. Teneberg, D. Wang, R. L. Proia, S. B. Levery, P. B. Savage, L. Teyton, and A. Bendelac. 2004. Lysosomal glycosphingolipid recognition by NKT cells. *Science* **306**:1786–1789.

The Mycobacterial Cell Envelope
Edited by M. Daffé and J.-M. Reyrat
© 2008 ASM Press, Washington, DC

Chapter 7

Antibiotics and New Inhibitors of the Cell Wall

LYNN G. DOVER, LUKE ALDERWICK, VEEMAL BHOWRUTH, ALISTAIR K. BROWN,
LAURENT KREMER, AND GURDYAL S. BESRA

A "golden age" of tuberculosis (TB) chemotherapy was heralded by the discovery of streptomycin in 1944 (Schatz et al., 1944). Over the next few years, the potent antitubercular activity of *p*-aminosalicylic acid was discovered (Lehmann, 1946), and the investigation of nicotinamide activity on *Mycobacterium tuberculosis* (Chorine, 1945) led to the simultaneous discovery of isoniazid (INH) (Bernstein et al., 1952; Fox, 1952; Offe et al., 1952) and pyrazinamide (PZA) (Malone et al., 1952). Further development of the same lead then gave rise to ethionamide (ETH) and prothionamide (PTH) (Liebermann et al., 1956). Investigation of the anti-TB activity of polyamines and diamines spawned a series of diamine analogues that included ethambutol (EMB) (Thomas et al., 1961). The anti-TB activity of substituted thioureas (Eisman et al., 1954) were refined to produce the potent isoxyl (ISO) (Crowle et al., 1963; Urbancik et al., 1964). Several other anti-TB agents, including D-cycloserine (Kurosawa, 1952), kanamycin (Umezawa et al., 1957), viomycin (Bartz et al., 1951), capreomycin (Herr et al., 1960), rifamycin (Sensi et al., 1959), and its derivative rifampin (RIF) (Maggi et al., 1966), were discovered through screening of compounds produced by soil isolates. Modern TB therapy was developed essentially using this array of agents; with the broad-range quinolones representing the only significant addition to the anti-TB arsenal in the late 1980s (Tsunekawa et al., 1987; Yew et al., 1990).

At present, the recommended standard chemotherapeutic regimen for TB treatment is prescribed under directly observed treatment, short course (DOTS). Initially, this acronym described the chemotherapy regimen but has since become used to de-scribe a broader public health strategy with five principal elements: political commitment; case detection by sputum smear microscopy; standard short-course chemotherapy with supportive patient management, including DOT; a system to ensure regular drug supplies; and a standard recording and reporting system, including the evaluation of treatment outcomes (Dye et al., 2005). The chemotherapeutic regimen consists of an initial 2-month phase of treatment with INH, RIF, PZA, and EMB, followed by a continuation phase of treatment lasting four months with INH and RIF.

The need for this long and relatively complex regimen stems from several contributing factors. Bacterial dormancy and phenotypic resistance to antimicrobials are certainly important, as is the issue of exposure to drugs, which might be limited by the potential intracellular location of the bacteria or the variable structure and pathology of necrotic tuberculous lesions. Such heterogeneity in TB lesions effectively define the metabolic status of their bacterial inhabitants and give rise to at least four subpopulations; active growers, that can be killed by INH, those with sporadic metabolic bursts that can be killed by RIF, a population with low metabolic activity that are considered likely to experience acidic surroundings and hypoxia killed by PZA, and finally dormant bacilli that are not killed by any current agents (Mitchison, 1979).

Poor patient compliance can promote the emergence of drug resistance, and this is particularly true in TB chemotherapy. The complex nature and length of the treatment certainly contributed to unacceptable levels of noncompliance. Likewise, logistical problems concerning drug supply and the tendency

Lynn G. Dover • Biomedical Sciences Division, School of Applied Sciences, Ellison Building, Northumbria University, Newcastle upon Tyne NE1 8ST, United Kingdom. **Luke Alderwick, Veemal Bhowruth, Alistair K. Brown, and Gurdyal S. Besra** • School of Biosciences, University of Birmingham, Edgbaston, Birmingham B15 2TT, United Kingdom. **Laurent Kremer** • CNRS UMR 5235, Dynamique des Interactions, Membranaires Normales et Pathologiques, Université de Montpellier II, Case 107, Place Eugène Bataillon, 34095 Montpellier Cedex 5, France.

for patients to feel well long before safe completion of the prescribed course compounded the problem. This degree of noncompliance promoted drug resistance and prompted the development of DOTS (Sharma and Mohan, 2004; Spigelman and Gillespie, 2006).

INH and RIF represent the cornerstones of anti-TB therapy; INH is the most powerful mycobactericidal drug available, normally ensuring early sputum conversion and thereby decreasing TB transmission as a consequence. RIF, by its mycobactericidal activity is crucial for preventing relapses. Strains of *M. tuberculosis* resistant to both INH and RIF, regardless of profiles of sensitivity and resistance to other drugs, have been termed multidrug-resistant strains. Multidrug-resistant TB (MDR-TB) is a major concern because TB patients in whom treatment has failed have a high risk of death. Although resistance to either drug may be managed with other first-line drugs, MDR-TB requires treatment with second-line drugs under DOTS-Plus. These agents possess limited sterilizing capacity and are not suitable for short-term treatment necessitating prolonged treatment with drugs that are less effective and more toxic (Moore-Gillon, 2001; Sharma and Mohan, 2004; Wright et al., 2006). Recently, the emergence of extensively drug-resistant TB (XDR-TB) strains with resistance to at least three of the six classes of second-line drugs (aminoglycosides, polypeptides, fluoroquinolones, thioamides, cycloserine, and *p*-aminosalicylic acid) has been reported (Wright et al., 2006). In some regions, approximately 20% of MDR-TB cases were classified as XDR-TB, raising concerns over a future epidemic of virtually untreatable TB (Wright et al., 2006).

Given this backdrop, the need for rapid and continued progress in the understanding of current anti-TB agents, development of new lead compounds, and the identification and characterization of novel drug targets to engage medicinal chemists is clearly evident. Important considerations for new agents include enhancement of penetration of infection sites, such as lung cavities, and long biological half-lives; achieving either might represent a significant advance toward shortening therapy and lead to simpler treatment regimens with improved patient compliance (Spigelman and Gillespie, 2006). Another crucial consideration is their effectiveness in coadministration with antivirals used to treat acquired immunodeficiency syndrome (AIDS) (Kwara et al., 2005; Harries et al., 2006; Spigelman and Gillespie, 2006); RIF activates cytochrome P450 enzymes that metabolize antivirals, significantly reducing their plasma concentrations (Kwara et al., 2005).

THE CELL WALL IS AN ESTABLISHED DRUG TARGET

Several of the major anti-TB agents disrupt the biosynthesis of cell wall components (Fig. 1). Both INH and ETH inhibit the enoyl-acyl carrier protein (ACP) reductase component of the dissociable fatty acid synthase (FAS-II) devoted to the synthesis of meromycolic acid precursors (Quémard et al., 1991; Banerjee et al., 1994; Kremer et al., 2003; Vilchèze et al., 2006) (see chapter 4). Likewise, EMB disrupts

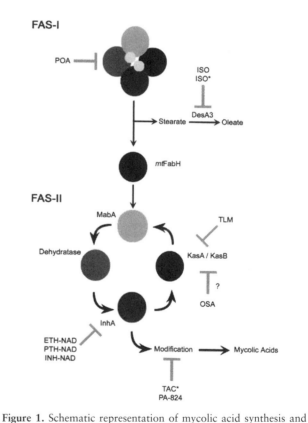

Figure 1. Schematic representation of mycolic acid synthesis and its inhibition. De novo fatty acid biosynthesis is carried out by fatty acid synthase (FAS-I), a function that appears to be inhibited by pyrazinoic acid (POA) and analogues. Stearate is transformed to the monounsaturated oleate by the Δ9 desaturase DesA3, the target of ISO or ISO*. Medium-chain-length fatty acyl primers are extended to form meromycolic acid precursors by the enzymes of FAS-II (one of which is likely a target of ISO*). All enzymes and domains of FAS-I are shaded to signify function; from lightest to darkest, these are β-ketoacyl-ACP reductase, β-hydroxyacyl-ACP dehydratase, enoyl-ACP reductase (inhibited by INH/ETH/PTH NAD adducts), β-ketoacyl-ACP synthase (inhibited by TLM and possibly OSA). During elongation, meromycolic acids are variously modified. The methyltransferases responsible are likely the targets for TAC* and PA-824. Drugs: POA, pyrazinoic acid; ISO, isoxyl; ISO*, activated ISO; ETH, ethionamide; PTH, prothionamide; INH, isoniazid; -NAD signifies an adduct with nicotinamide adenine dinucleotide, TAC*, activated thiacetazone; OSA, *n*-octylsulfonylacetamide; TLM, thiolactomycin.

the deposition of the arabinan domain of the cell wall polysaccharide arabinogalactan (AG) that tethers the inner mycolic acid-based leaflet of the outer wall permeability barrier to the underlying peptidoglycan (Takayama and Kilburn, 1989) (see chapter 1). The following sections relate current understanding of the modes of action of cell wall-active drugs and interesting new lead compounds. INH and the thiocarbonyl-containing ETH, PTH, ISO, and thiacetazone (TAC) are all prodrugs; we examine their activation and potential for augmenting their use with additional agents in the future.

ETHAMBUTOL INHIBITS CELL WALL ARABINAN DEPOSITION

EMB (Fig. 2) is one of the front-line drugs used to treat infections with most strains of *M. tuberculosis*. The pleiotropic effects induced in EMB-treated mycobacteria have led to the proposal of several hypotheses concerning its mode of action (Forbes et al., 1965; Kilburn et al., 1981; Poso et al., 1983; Silve et al., 1993), but the widely accepted view is that its primary effect is the disruption of cell wall arabinan biosynthesis. Initial studies focused on the early events after exposure of mycobacteria to the drug in order to define its mode of action. Exposure of *Mycobacterium smegmatis* to EMB led to disaggregation of bacillary clusters during the first 4 hours of exposure, producing the illusion of growth stimulation (Kilburn and Greenberg, 1977), the accumulation of trehalose-mycolate and free mycolic acids, reduced deposition of mycolates in mycolyl-AG (Takayama et al., 1979; Kilburn and Takayama, 1981),

and disruption of phospholipid metabolism (Kilburn et al., 1981). The accumulation of trehalose-mycolate conjugates occurs with alacrity and can be observed after as little as 1 minute (Kilburn and Takayama, 1981) after treatment of *M. smegmatis* with EMB, and after around 10 minutes, significant decreases in mycolate deposition in the wall are evident (Kilburn and Takayama, 1981). Trehalose conjugates, free mycolic acids and phosphatidylethanolamine (PE) all begin to leak from treated cells after 30 minutes (Kilburn and Takayama, 1981; Kilburn et al., 1981). Along with this leakage of PE, the synthesis of both cardiolipin and phosphatidylinositol dimannoside declines (Kilburn et al., 1981). Although this leakage of mycolates and PE from the cell is not associated with significant cell lysis (Kilburn et al., 1981), there are morphological disturbances that might indicate a structural flaw sufficient to promote this lipid efflux (Kilburn and Greenberg, 1977).

Prelabeling of mycolates in the cell wall with ^{14}C and inhibition of further mycolate biosynthesis by treatment with INH revealed that the ^{14}C content of the cell wall did not decline on EMB treatment, ruling out that the observed inhibition of mycolyl transfer was mediated through a drug-enhanced hydrolysis of AG-bound mycolates (Takayama et al., 1979). Two candidate sites for the lesion in mycolyl transfer were thus apparent; either the inhibition of mycolyl-transfer from trehalose conjugates to AG or the synthesis of the AG acceptor itself (Takayama et al., 1979).

Focus moved to the latter when in vivo pulse-chase labeling experiments using *M. smegmatis* revealed immediate inhibition of [^{14}C]Ara incorporation into both AG and lipoarabinomannan (LAM) on EMB treatment, an effect that required an accordingly higher dosage in an EMB-resistant strain (Takayama and Kilburn, 1989). Consistent with EMB causing a lesion in arabinan polymerisation, administration of the drug was also shown to cause accumulation of decaprenyl-monophosphoryl-Ara*f* (DPA) (Takayama and Kilburn, 1989), the activated sugar donor involved in the transfer of Ara*f* residues in AG biosynthesis (Wolucka et al., 1994; Xin et al., 1997). EMB has also been suggested to strongly inhibit the synthesis of cell wall galactan (Deng et al., 1995), probably through dysregulation of cell wall synthesis as a consequence of the lesion in arabinan biosynthesis (Deng et al., 1995); however, subsequent studies have not confirmed this secondary effect (Mikusová et al., 1995).

Interestingly and somewhat at odds with a previous study that showed no release of ^{14}C mycolates from cell walls after treatment with 3 μg of EMB/ml

EMB

SQ109

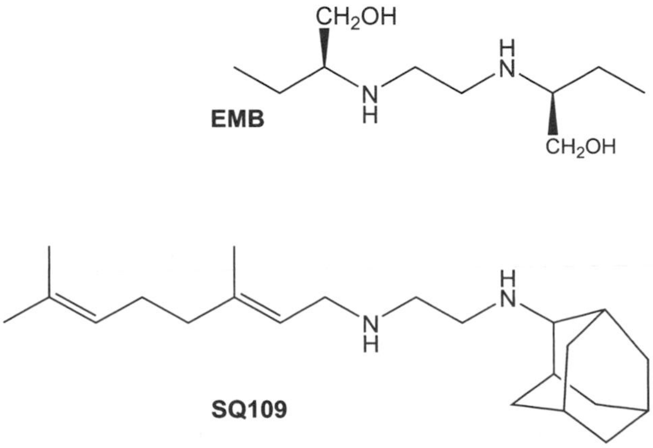

Figure 2. Structures of ethambutol and SQ109. Ethambutol inhibits arabinan deposition, whereas SQ109 appears to interfere with lipid biosynthesis.

(Takayama and Kilburn, 1989), administration of the drug at 10 μg/ml promoted the cleavage of Ara*f* residues from the cell wall; more than 50% of the arabinan in the cell wall core was removed from the wall 1 hour after addition of the drug to growing mycobacterial cultures (Deng et al., 1995). Significant quantities of the released arabinan were apparently lipidated, presumably mycolylated, but water-soluble arabinan was also detected, suggesting that nonmycolylated AG and LAM could represent significant substrates of this arabinase. Although the identity and natural function of the hydrolytic enzyme responsible for this effect is not yet known, a role in cell wall remodeling during growth and cell division appears likely. Its action, once AG synthesis is inhibited by EMB, results in severe disruption of the mycobacterial cell wall. Accordingly, EMB-induced damage to the cell wall provides a molecular explanation for the known synergistic effects of EMB with other drugs (Cavalieri et al., 1995; Jagannath et al., 1995; Rastogi et al., 1998; De Logu et al., 2002), presumably through facilitating their penetration of the damaged cell wall (Deng et al., 1995). In vitro analyses of this arabinase activity have recently determined that its activity is not enhanced by EMB treatment (Dong et al., 2006).

In contrast to AG arabinan, synthesis of the arabinan of LAM was less severely affected by EMB, demonstrating a differential effect on the synthesis of these two arabinans (Deng et al., 1995; Mikusová et al., 1995) and leading to the postulate that these two distinct assemblies are synthesized through different arabinosyltransferases with varying sensitivity to the drug (Mikusová et al., 1995).

Molecular Targets and Prediction of Clinical EMB Resistance

The products of the *emb* locus of *Mycobacterium avium* were identified as the targets for EMB using a strategy of target overexpression. The locus contains three genes, *embR*, *embA*, and *embB*; the former encodes a putative regulator of *embA* and *embB* and is expendable for the resistant phenotype, which is copy number dependent (Belanger et al., 1996). Consistent with these findings, overexpression of *M. avium embA* and *embB* conferred EMB resistance in *M. smegmatis*. The inhibitory effect of EMB on the in vitro incorporation of Ara*f* from DPA into arabinan was partially recovered in extracts from cells overexpressing *embAB* (Belanger et al., 1996). Together, these data suggested that the highly homologous EmbA and EmbB were arabinosyltransferases involved in arabinan polymerisation and were the cellular targets of EMB, the former awaits formal proof.

M. tuberculosis possesses three closely related *emb* genes, clustered *embCAB* with *embR* present but occupying a distant locus. A similar *emb* cluster occurs in *M. smegmatis* mc^{2}155 and recent gene knockout studies in this organism have shed light on this apparent redundancy in arabinosyltransferases (Escuyer et al., 2001). Individual mutants inactivated in *embC*, *embA*, and *embB* were characterized; all three strains were viable, but of them, the *embB* mutant was most profoundly affected. Significant effects on cell wall integrity were indicated by morphological changes, and the cells displayed increased sensitivity to hydrophobic drugs and detergents. The arabinose content of the AG was diminished for both the *embA* and *embB* strains. Nuclear magnetic resonance studies showed that these two mutations resulted in considerable effects on the formation of the terminal hexa-arabinofuranosyl motifs of AG, specifically the addition of the β-D-Ara*f*-1→2-β-D-Ara*f* disaccharide to the 3 position of the 3,5-linked Ara*f* residue, resulting in a linear terminal motif. The AG formation of the *embC* strain, however, seemed unaffected, yet arabinan deposition in LAM was abolished. This finding supported the earlier hypothesis that AG and LAM were synthesized through different arabinosyltransferases (Mikusová et al., 1995); EmbA and EmbB are critical to the formation of the hexa-arabinofuranosyl motifs of AG that are crucial for the esterification of mycolic acids (Escuyer et al., 2001), whereas EmbC is dedicated to LAM biogenesis.

The topological decoding of the *emb* operon led to the identification of regions within this gene cluster that were implicated in EMB resistance (Sreevatsan et al., 1997; Telenti et al., 1997; Ramaswamy et al., 2000, 2004). Indeed, the region conferring resistance to EMB was termed the EMB resistance-determining region and was shown to reside on the second intracellular loop of EmbB (Telenti et al., 1997). Automated DNA sequencing and single-stranded conformational polymorphism analysis of EMB-resistant *M. tuberculosis* strains highlighted mutations at codon 306, with Met306 being replaced by Leu, Ile, Val, or Thr residues (Lety et al., 1997; Sreevatsan et al., 1997; Telenti et al., 1997; Escalante et al., 1998; Van Rie et al., 2001; Lee et al., 2002, 2004; Mokrousov et al., 2004). An equivalent single nucleotide polymorphism was also reported in EMB resistant strains of *M. smegmatis* (Lety et al., 1997; Telenti et al., 1997). Because the polymorphism occurs in approximately 60% of all EMB-resistant isolates, then screening for this mutation has been widely suggested to form the basis for clinical tests to predict EMB resistance (Escuyer et al., 2001; Rinder et al., 2001; Van Rie et al., 2001; Lee et al., 2002; Mokrousov et al., 2004; Wada et al., 2004; Isola et al., 2005; Parsons et al., 2005; Plinke et al., 2006). Recently, however, Hazbón et al. (2005) carried out an extensive study of *embB*306 mutations prompted by reports of these polymorphisms in EMB-susceptible strains (Mokrousov et al., 2002; Post et al., 2004; Tracevska et al., 2004); here, the mutation was hypothesized to be associated with development of broad multidrug resistance (Hazbón et al., 2005). A more recent study designed to test this hypothesis and incorporating appropriate EMB-sensitive MDR control strains has contradicted these findings, with *embB*306 mutations being detected in EMB-resistant strains only (Hazbón et al., 2005; Plinke et al., 2006). Undoubtedly, further studies are required to clarify this matter and to determine the utility of this polymorphism in drug sensitivity prediction.

Other polymorphisms in *embB* that are associated with resistance to EMB include those that cause Ser297Ala, Asp328Gly/Tyr, Phe330Val, and Phe334 His substitutions (all of which are positioned on intracellular loop 2 of EmbB) and Gly745Asp, Asp 959Ala, Met1000Arg, and Asp1024Asn, which form part of the large C-terminal globular region of EmbB (Ramaswamy et al., 2000). However, other single nucleotide polymorphisms correlated with EMB resistance occur elsewhere within the *M. tuberculosis* genome, suggesting that other mechanisms for resistance might exist (Ramaswamy et al., 2000).

SQ109, an EMB Analogue with a Distinct Mode of Action

Recently, screening of a combinatorial library founded on an EMB pharmacophore resulted in the selection of *N*-geranyl-*N'*-(2-adamantyl)ethane (SQ109) (Fig. 2) on the basis of improved activity against *M. tuberculosis* compared with EMB (Lee et al., 2003); sensitive strains included those singly resistant to RIF, INH, and, encouragingly, EMB (Lee et al., 2003; Jia et al., 2005b, 2006; Protopopova et al., 2005). Interestingly, SQ109 is also more potent than EMB against mycobacteria within macrophages and mice but is less potent than INH (Protopopova et al., 2005). SQ109 demonstrated strong synergistic activity with INH and RIF in inhibition of *M. tuberculosis* growth. The marked synergy between SQ109 and RIF was also evident when using RIF-resistant *M. tuberculosis* strains (Chen et al., 2006). Additive effects were observed between SQ109 and streptomycin, but neither synergy nor additive effects were observed with the combination of SQ109 with EMB or PZA (Chen et al., 2006).

Pharmacoproteomic studies with *M. tuberculosis* H37Rv revealed that similar protein profiles were catalogued after both EMB and SQ109 treatments

(Jia et al., 2005a). However, transcriptional profiling of *M. tuberculosis* H37Rv after several different drug treatments revealed that, despite many similarities, SQ109 and EMB differentially affected a regulon containing genes associated with FAS-II (Rv2241-Rv2247), as well as genes implicated in fatty acid modification (Boshoff et al., 2004). This potential mechanistic divergence was confirmed by the observation that, whereas *M. tuberculosis* cells treated with EMB rapidly lost acid fastness, those treated with SQ109 retained the property. Furthermore, unlike cells treated with EMB, walls of those treated with SQ109 did not contain low arabinose levels (Deng et al., 1995; Boshoff et al., 2004).

Despite its low oral bioavailability in dogs and rodents, SQ109 is effectively transferred to the lungs (Jia et al., 2005b; Jia et al., 2006) and is currently in clinical trials. The in vitro and in vivo data derived from study of SQ109 identify it as an attractive compound for further drug development, and fortuitously, the mechanistic divergence from the parent compound EMB appears to avoid cross-resistance problems. Encouragingly, the synergy with INH and RIF, two of the most important front-line TB drugs, promises the development of new combination therapies incorporating SQ109.

AGENTS THAT INHIBIT MYCOLIC ACID BIOSYNTHESIS

Isoniazid is a Prodrug Requiring Peroxidative Activation

Evidence to date suggests that normal mycolic acid metabolism is crucial to the survival of *M. tuberculosis*. The strongest evidence for this comes from work on the action of the bactericidal agent INH (Mackaness and Smith, 1953; Singh and Mitchison, 1954; Barclay and Winberg, 1964) (Fig. 3A), which interferes, in mycolic acid biosynthesis (Winder and Collins, 1970; Winder et al., 1970, 1971; Takayama et al., 1972; Wang and Takayama, 1972; Takayama and Qureshi, 1979). Initial insight into a mechanism of drug resistance was gained from the observation of low catalase activity in INH-resistant strains (Cohn et al., 1954; Middlebrook, 1954; Middlebrook et al., 1954) and an inverse relationship between catalase-peroxidase activity and the minimal inhibitory concentration (MIC) of INH (Hedgecock and Faucher, 1957). In the early 1990s, Cole and coworkers cloned and characterized the *M. tuberculosis* catalase peroxidase encoded by *katG* (Zhang et al., 1992; Heym et al., 1993). The multifunctional enzyme also carries peroxynitritase (Wengenack et al., 1999) and NADH oxidase activities (Singh et al., 2004), and it appears

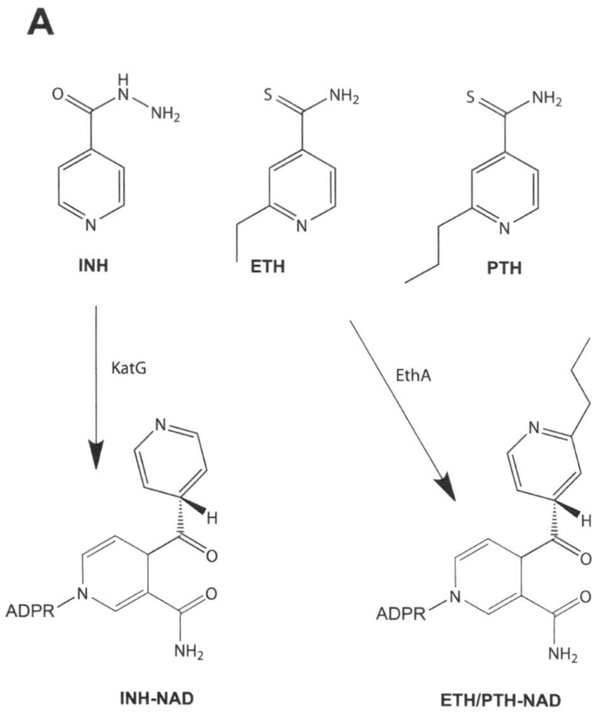

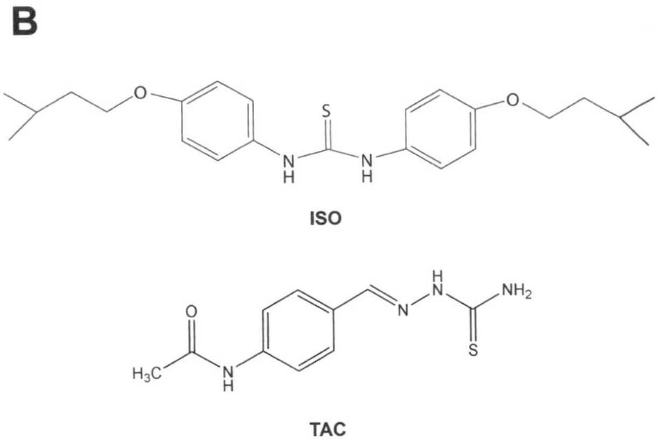

A

B

Figure 3. Antimycobacterial prodrug inhibitors of mycolic acid biosynthesis. (A) Formation of isonicotinoyl adducts with isoniazid (INH), ethionamide (ETH), and prothionamide (PTH). INH is peroxidatively activated through the peroxidase activity of KatG and reacts with NAD$^+$ to form its adduct. ADPR represents adenosine diphosphate ribose. Similar products form after the oxidative activation of ETH by EthA. All three adducts represent tight-binding slow inhibitors of InhA. The adduct depicted is PTH-NAD, which has an extended alkyl branch over ETH-NAD. (B) Structures of the EthA-activable prodrugs isoxyl (ISO) and thiacetazone (TAC).

that the major role of KatG in pathogenesis is to catabolize the peroxides generated by phagocyte NADPH oxidase. In the absence of this host antimicrobial mechanism, that is, in mutant phagosomes lacking a functional NADPH oxidase, KatG is dispensable (Ng et al., 2004). However, there does appear to be strong selection for INH resistance during treatment (Zhang et al., 1992; Heym et al., 1995; Barry et al., 1998; Slayden and Barry, 2000). Loss of *katG* had been correlated with INH resistance in *M. tuberculosis* (Zhang et al., 1992), and its participation in INH activity was confirmed by transformation of resistant strains of *M. tuberculosis* (Zhang et al., 1993) and *M. smegmatis* (Zhang et al., 1992) to susceptibility with a plasmid bearing wild-type *katG*. Many INH-resistant strains, however, do not bear deletions in *katG* but have acquired point mutations (Musser et al., 1996; Rouse et al., 1996; Marttila et al., 1998). One important point mutation (Ser315Thr) is found in more than 50% of INH-resistant clinical isolates; this *katG* allele encodes a competent catalase-peroxidase with reduced activity toward INH (Wengenack et al., 1997). Recent structural determinations suggest that a narrowing of the channel leading to the heme prosthetic group (Bertrand et al., 2004; Pierattelli et al., 2004) in the Ser315Thr variant might limit the access of INH to the active site (Zhao et al., 2006), consistent with the findings of previous biochemical studies (Wengenack et al., 1998; Todorovic et al., 1999; Lukat-Rodgers et al., 2000; Yu et al., 2003).

Complex Debate over the Primary Cellular Target for Isoniazid

The identity of the primary target for INH in *M. tuberculosis* has been a contentious issue in recent years. We provide a historical perspective on the debate in which we attempt to faithfully reflect the evidence provided in support of both KasA and InhA in this respect.

A spontaneous mutant of *M. smegmatis* designated mc^{2}651 is resistant to INH, but retains wild-type KatG activity. This strain is also resistant to ETH, but exhibits normal susceptibility to other antibiotics. Transfer of resistance to sensitive *M. smegmatis* was mediated by transformation with a cosmid library derived from genomic DNA of the resistant strain. Subcloning of cosmid DNA imparting INH resistance revealed two open reading frames (ORFs) designated *orf1*(*mabA*) and *inhA* (Banerjee et al., 1994). InhA and MabA have since been shown to be the enoyl-ACP reductase (Quémard et al., 1995) and β-ketoacyl-ACP reductase components of FAS-II (Banerjee et al., 1998; Marrakchi et al., 2002). The wild-type *inhA* gene conferred INH and ETH resistance when transferred on a multicopy plasmid vector to *M. smegmatis* and *M. bovis* BCG (Banerjee et al., 1994), whereas similar experiments using *mabA* have shown that the other FAS-II reductase does not confer INH/ETH resistance (Banerjee et al., 1998). InhA is able to catalyze the NADH-specific reduction of medium-chain (C_{12}-C_{24}) 2-*trans*-enoyl-thioesters, which is consistent with its involvement in the early stages of meromycolic acid biosynthesis. However INH does not bind directly to the enzyme (Quémard et al., 1995) consistent with its need for prior peroxidative activation through KatG (Quémard et al., 1996).

Sequencing of the *inhA* locus in resistant strains unmasked an A→G transversion at position 280, resulting in a Ser94Ala substitution (Banerjee et al., 1994). This mutation was suggested to alter the binding affinity of InhA to NADH, thus resulting in increased resistance against INH (Dessen et al., 1995). Other resistance-conferring mutations have also been found in the promoter region leading to overexpression of *inhA*. Thus, increased resistance to INH in *M. smegmatis* may also result from overproduction of the InhA target. In some laboratories, however, overexpression of *inhA* in *M. tuberculosis* was not seen to impart an increased resistance to INH (Mdluli et al., 1996; Barry et al., 1998), casting some doubt on the identity of the primary target of INH.

Mdluli et al. (1998) suggested that InhA is not the major target for activated INH in *M. tuberculosis* and reiterated that INH might specifically inhibit the insertion of a Δ^5-double bond into a C_{24} fatty acid. Earlier studies conducted by Takayama and coworkers suggested that INH acted by inhibiting synthesis of meromycolate precursors of mycolic acids. By purifying and analyzing intermediates in fatty acid biosynthesis that accumulated on exposure to INH, it appeared that INH specifically inhibited the desaturation of C_{24} and C_{26} fatty acids, and that the molecular target of INH was most likely a C_{24} desaturase (Takayama et al., 1972; Takayama et al., 1974; Davidson and Takayama, 1979). These investigators also suggested that INH may be involved in some aspects of the elongation of C_{30}-C_{56} meromycolates based on the inhibitory effects of INH on the synthesis of fatty acids and hydroxy lipids in a cell-free preparation of *M. tuberculosis* H37Ra (Takayama and Qureshi, 1979). However, the addition of NADH and NADPH appeared to neutralize the inhibitory effects of INH in this cell-free system, consistent with the nature of the InhA protein (Quémard et al., 1995) and possibly MabA inhibition by INH, which has been observed in vitro (Ducasse-Cabanot et al., 2004). A similar study using extracts of *Mycobacterium aurum* A^+ identified 24:1 *cis*-5 elongase as the primary INH target rather than the C_{24} desaturase as 24:1 *cis*-5 did not restore INH-inhibited mycolic acid biosynthetic activity, suggesting that the drug affects biosyntheses occurring after the formation of the Δ-5 double bond. Consistently, mycolic acid synthesis was completely inhibited by INH in a 24:1 *cis*-5 elongase assay. Finally, the only intermedi-

ates that accumulated as a result of the addition of INH were C_{24} fatty acids. Both 24:0 and 24:1 fatty acids were observed in a similar ratio whether INH was present or not, even though concomitant mycolic acid biosynthesis was inhibited by INH (Wheeler and Anderson, 1996), an observation that does not support a C_{24} desaturase as the INH target. Together, these findings were totally consistent with a FAS-II component providing the primary target for the drug.

In accord with the implication of FAS-II in INH sensitivity, Mdluli et al. (1998) identified a complex formed by three covalently-linked partners, which were identified as INH, acyl carrier protein (AcpM), and KasA, a β-ketoacyl-AcpM synthase component of FAS-II. INH-treated mycobacteria accumulate C_{26} fatty acid, and inhibition of KasA would expect to stop fatty acid elongation and result in the accumulation of long chain-saturated fatty acids. Moreover, AcpM and KasA, unlike InhA, were substantially upregulated by INH treatment, suggesting that transcription of *acpM* and *kasA* is sensitive to mycolic acid synthesis. The potential involvement of *kasA* mutations in the development of INH-resistance was also examined. Four *kasA* mutations were identified among 28 INH-resistant isolates, whereas none were found in 43 INH-susceptible counterparts. In two of these strains, no alteration was found in other loci involved in INH resistance (*katG*, *inhA*, *ahpC*), whereas in the other two strains, a *katG* alteration was found (Mdluli et al., 1998).

Despite this evidence implicating KasA as the primary target of INH, Vilchèze et al. (2000) examined whether inactivation of InhA produced an identical effect to INH treatment of mycobacteria. A temperature-sensitive mutation in *M. smegmatis inhA* rendered the reductase inactive at 42°C. Thermal inactivation of InhA in *M. smegmatis* resulted in the inhibition of mycolic acid biosynthesis, a decrease in $C_{16:0}$, and a concomitant increase of $C_{24:0}$ in a manner equivalent to that seen in INH-treated cells. Furthermore, the InhA-inactivated cells, like INH-treated cells, underwent drastic morphological changes, leading to cell lysis (Vilchèze et al., 2000). At this time, a molecular basis for direct INH-dependent effects on InhA became clear when Rozwarski et al. (1998) published their structural characterization of a complex formed between an INH-NAD adduct and the enoyl-ACP reductase (see later).

Slayden et al. (2000) compared the consequences of INH treatment of *M. tuberculosis* with two inhibitors having well-defined targets: triclosan (TRC), which inhibits InhA, and thiolactomycin (TLM), which inhibits KasA. INH and TLM, but not TRC, caused an increase in the expression of an operon (Rv2241-Rv2247) containing genes encoding five FAS-II-related components, including *kasA* and *acpM*. Although all three compounds inhibit mycolic acid synthesis, treatment with INH and TLM, but not with TRC, resulted in the accumulation of ACP-bound fully saturated C_{26} mycolic acid precursors. TLM-resistant mutants of *M. tuberculosis* showed more cross-resistance to INH than TRC-resistant mutants. Here, overexpression of *kasA* conferred more resistance to TLM and INH than to TRC, whereas overexpression of InhA conferred more resistance to TRC than to INH and TLM. Co-overexpression of both InhA and KasA resulted in strong INH resistance, in addition to cross-resistance to both TLM and TRC. These authors suggested that the components of FAS-II are not independently regulated and that alterations in *inhA* expression affect expression of *kasA*. Nonetheless, INH appeared to resemble TLM more closely overall, and KasA levels appeared to be tightly correlated with INH sensitivity (Slayden et al., 2000). Furthermore, these authors used an in vitro meromycolic acid synthase assay to analyze the anabolic roles of the FAS-II β-ketoacyl-ACP synthases KasA and KasB (Slayden and Barry, 2002). Overproduction of KasA and KasB individually and together in *M. smegmatis* enabled cell-free incorporation of ^{14}C from [^{14}C]malonyl-CoA into lipids whose chain length was dependent on the *M. tuberculosis* elongating enzyme used. KasA specifically elongated C_{16}-CoA to monounsaturated fatty acids that averaged 40 carbons in length, whereas KasB and KasA together produced longer chain multiunsaturated chains averaging 54 carbons in length. These products comigrated with a synthetic standard of meromycolic acid and their production was sensitive to INH, TLM, and TRC. Mutations in *kasA* associated with INH resistance diminished overall catalytic activity but conferred resistance to INH. In vivo analyses confirmed that overexpression of each of the four mutant *kasA* alleles enhanced INH resistance when compared with overexpression of wild-type KasA (Slayden and Barry, 2002).

Because the previous studies investigating whether overexpressed *inhA* or *kasA* could confer resistance to INH yielded disparate results, Larsen et al. (2002) transformed *M. smegmatis*, *M. bovis* BCG, and three different *M. tuberculosis* strains with multicopy plasmids bearing either *inhA* or *kasA*. These transformants, as well as previously published *M. tuberculosis* strains with multicopy *inhA* or *kasAB* plasmids, were tested for resistance to INH, ETH, or TLM. Mycobacteria overexpressing *inhA* uniformly exhibited at least 20-fold increased resistance to INH and at least 10-fold increased resistance to ETH. In contrast, the *kasA* merodiploids exhibited no increased resistance to INH or ETH, but resistance to the

known KasA inhibitor TLM was observed. Because upregulation of *kasA* had been described in susceptible *M. tuberculosis* strains exposed to INH (Slayden et al., 2000), *inhA* and *kasA* mRNA levels were quantified. Increased *inhA* mRNA correlated with INH resistance, whereas *kasA* mRNA levels did not. These authors concluded that overexpressed *inhA*, but not *kasA*, conferred INH and ETH resistance to *M. smegmatis*, *M. bovis* BCG, and *M. tuberculosis* and thus, InhA is the primary target of action of INH and ETH in all three mycobacteria (Larsen et al., 2002).

These groups followed this study with a molecular analysis of a KasA-containing complex similar to that reported by Mdluli et al. (1998) in INH-treated mycobacteria using anti-KasA sera (Kremer et al., 2003). The KasA-containing complex was readily detected in immunoblots of proteins from INH-treated *M. smegmatis*, along with similar amounts of nonsequestered KasA to that seen in untreated cells, where they were presumably sufficient to ensure mycolic acid biosynthesis. Interestingly, a *furA*-lacking strain, which would overexpress *katG* (Pym et al., 2001), induced the complex at lower concentrations of INH compared with the control strain, whereas higher INH concentrations were necessary to induce the complex in a strain that lacked *katG*, consistent with a need for INH activation by KatG to induce the KasA-containing complex. However, ETH and the InhA inhibitor diazaborine also induced the complex, whereas the β-ketoacyl ACP synthase inhibitors cerulenin and TLM did not. Thus, formation of this complex was related to InhA inhibition rather than as a specific sequel to INH toxicity. Interestingly, the KasA-containing complex was found in several thermosensitive *M. smegmatis inhA* mutant strains without drug treatment, suggesting that this complex may be a normal component of FAS-II that might well be stabilized when InhA activity is compromised. In addition, in vitro assays using purified InhA and KasA demonstrated that KatG-activated INH, TRC, and diazaborine inhibited InhA but not KasA activity. These authors concluded that only inhibition of InhA, but not KasA, induces this KasA-containing complex, INH is not a component of this complex, and that INH does not target KasA, consistent with InhA being the primary target (Kremer et al., 2003). Recently, the introduction of a single point mutation in *inhA*, resulting in the Ser94Ala substitution, into the chromosome of *M. tuberculosis* correlated with increased resistance to INH providing the strongest evidence to date that INH targets the enoyl-ACP-reductase of FAS-II (Vilchèze et al., 2006).

The matter of the primary cellular target for INH has been controversial recently. The disparity in conclusion between laboratories must have a (possibly important) mechanistic basis. Future analyses may illuminate this point of discrepancy. An avenue of investigation might stem from the possibility that the *kasA* mutations implicated by some in INH resistance might impact upon the formation of multiprotein FAS-II complexes (Veyron-Churlet et al., 2004, 2005) and provide some alleviation of the symptoms of INH inhibition of InhA experienced by the bacterium.

Isonicotinoyl Radicals React with NAD(P)$^+$ to Form Various Enzyme Inhibitors

A major advance in understanding its mode of action was the discovery that the oxidation of INH in the presence of NAD led to the formation of an inhibitory substance (Johnsson et al., 1995). The resolution of the crystal structure of the inhibitor-enzyme complex revealed the inhibitor to be a covalent INH-NAD adduct (in the S-configuration at the generated stereochemical center [Fig. 3A]) formed through reaction of a putative isonicotinoyl radical with NAD$^+$, supposedly within the active site of InhA (Rozwarski et al., 1998). This adduct represents a slow, tight-binding inhibitor of the enzyme at subnanomolar concentrations (Rawat et al., 2003). This and similar adducts have since been formed in free solution by the fast addition of isonicotinoyl radicals to electron-deficient heterocycles such as NAD$^+$ (Wilming and Johnsson, 1999). Mutations that reduce NADH dehydrogenase activity (Ndh; type II) cause multiple phenotypes, including coresistance to INH and ethionamide; the resulting in vivo increase of NADH/NAD$^+$ ratios is consistent with the suppression of this suggested role for NAD$^+$ in INH and ETH toxicity in *M. smegmatis* and *M. bovis* BCG (Miesel et al., 1998; Vilchèze et al., 2005).

Initial ideas regarding altered NADH affinity for resistant mutants, which appeared consistent with the prebinding of NADH to the enzyme before adduct formation, were correct but analysis of inhibition parameters for Ile21Val and Ile47Thr, two InhA mutations detected in INH-resistant clinical isolates, and the Ser94Ala substitution mutant highlighted earlier, are similar to those for the wild-type enzyme. Although the adduct is obviously a potent inhibitor of InhA, these kinetic and thermodynamic studies indicate that resistance cannot be explained simply in terms of a decrease in the ability of purified InhA mutants to bind the adduct (Rawat et al., 2003). FAS-II enzymes appear to associate in heterotypic complexes (Veyron-Churlet et al., 2004, 2005); interactions of InhA with the condensing enzymes KasA and KasB, consistent with channeling of

substrates to these enzymes for a subsequent round of chain elongation, have been reported (Veyron-Churlet et al., 2004). Such interactions could be the basis not only of the apparent similarity in the kinetics of INH-NAD adduct binding to purified wild type InhA and resistant mutants (Rawat et al., 2003) but may also be a contributing factor at the heart of the debate regarding the identity of the primary target of INH (Veyron-Churlet et al., 2004).

Similarly, INH-NADP adducts have been shown to inhibit MabA (Ducasse-Cabanot et al., 2004) and dihydrofolate reductase (Argyrou et al., 2006b), which contributes to nucleotide biosynthesis, in vitro. In these artificial systems, however, nonphysiological concentrations of NADP are present in a vastly simplified milieu compared with the interior of *M. tuberculosis*. An affinity profiling study using both immobilized INH-NAD and INH-NADP adducts has highlighted that a further 16 components of the *M. tuberculosis* proteome may also bind, directly or indirectly, to these adducts with high affinity (Argyrou et al., 2006a). A total of 18 proteins were identified from 11 protein bands excised from SDS-PAGE gels. The majority of these are predicted to be NAD(P)-dependent dehydrogenases/reductases involved in diverse cellular processes. The validity of these hits now requires further investigation but suggest that the spectrum of inhibition of INH through NAD(P) adducts alone might be more significant than previously thought.

ETH: a Lack of INH Cross-Resistance Indicates Discrete Routes to Prodrug Activation

Patients infected with MDR-TB are committed to a regimen of second-line agents that are generally more expensive and more toxic than first-line agents, such as INH and RIF. The second-line drug ETH (Fig. 3A) is associated with important gastrointestinal disturbances and hepatotoxicity, and thus its prescription is limited mainly to retreatment of patients relapsing with MDR-TB (Jenner and Smith, 1987).

Like its structural analogue INH, ETH has been proposed to target the enoyl-acyl carrier protein reductase InhA (Banerjee et al., 1994). Both INH and ETH are prodrugs that require activation by specific mycobacterial enzymes; however, the lack of cross-resistance in clinical isolates (Fattorini et al., 1999) as well as *katG* mutants suggested a different route to activation. Recently, two groups independently identified two neighboring *M. tuberculosis* genes, *ethA* and *ethR*, that govern activation of ETH (Baulard et al., 2000; DeBarber et al., 2000). EthA is a FAD-containing Baeyer-Villiger monooxygenase of unknown metabolic significance that is crucial to ETH activation (Vannelli et al., 2002; Fraaije et al., 2004) through a non-Baeyer-Villiger reaction mechanism. EthR regulates the expression of *ethA* in *M. tuberculosis*; overexpression of *ethR* leads to ETH resistance, whereas chromosomal inactivation of *ethR* promotes ETH hypersensitivity (Baulard et al., 2000). EthR was found to bind directly and specifically to DNA sequences corresponding to the *ethRA* intergenic region (Engohang-Ndong et al., 2004).

The nature of the activated ETH metabolite has since been sought through a series of studies. Vannelli et al. (2002) showed that, in vitro, EthA can transform ETH to its S-oxide and further to its amide derivative, 2-ethyl-4-amidopyridine suggesting that this proceeded *via* an unstable sulfinic acid intermediate. In an elegant alternative approach, Hanoulle et al. (2005) followed ETH metabolism within dense cell pastes prepared from *M. smegmatis* overexpressing *ethA* using ^{1}H high-resolution magic-angle spinning nuclear magnetic resonance. Using this novel technique, these workers described an undefined metabolite, ETH*, that differed from the prodrug itself and, after comparison with standards through their addition to ETH-treated *M. smegmatis* cell lysates, all other proposed metabolites. The previously identified ETH metabolite 2-ethyl-4-hydroxymethylpyridine is produced in substantial amounts by the ETH-treated mycobacteria and is detected exclusively outside of the bacterium. Interestingly, ETH* was the only ETH derivative detected after the application of a diffusion filter to the nuclear magnetic resonance data. Furthermore, it was also not detected in the culture medium, suggesting that it was unable to cross the bacterial envelope and consequently accumulates within the cytoplasm of the treated bacteria (Hanoulle et al., 2006). As might be expected, nonrecombinant bacteria produced 34-fold less ETH* than the recombinant, but interestingly, no trace of ETH could be found in cells overproducing the *ethA* repressor EthR, suggesting that the drug is rapidly expelled from the cell in the absence of activation.

Very recently, the nature of a metabolite that fulfils the criteria for identity with ETH* has been described using a cell-based activation method in which EthA was overproduced alongside either *M. tuberculosis* or *M. leprae* InhA in *Escherichia coli* in the presence of ETH (Wang et al., 2007). InhA purified directly from the ETH-treated system retained only 1% activity of the protein isolated from untreated bacteria. Through analysis of crystal structures of the inhibited *M. leprae* and *M. tuberculosis* InhA complexes, Wang et al. (2007) provided definitive evidence that both ETH and PTH form covalent adducts with NAD in an S configuration akin to the INH-NAD adduct (Fig. 3A). Indeed, the atoms common to both INH and ETH adducts occupy almost identical positions within *M. tuberculosis* InhA. Affinity purification of these adducts facilitated kinetic studies that determined them to be tight-binding inhibitors of the enzymes expressing K_i values in the low nanomolar range. This striking similarity in the modes of action of both INH and ETH is surprising given that their oxidative activation occurs by a wholly different enzymic route. Nevertheless, the potency of these minor side products of mycobacterial metabolism not only underscores the importance of InhA as drug target but probably highlights the most efficient manner through which toxicity can be expressed therein.

Other EthA-Activated Thiocarbonyl Prodrugs

ETH is only one example of a thiocarbonyl-containing antituberculosis drug approved in clinical use. Among the second-line TB therapeutics, there are two other such molecules, thiacetazone (TAC) and isoxyl (ISO, thiocarlide) (Fig. 3B) that might similarly be activated by EthA-catalyzed S-oxidation (DeBarber et al., 2000).

Eisman and coworkers (1954) first published in vitro and in vivo data on the anti-TB activity of substituted thioureas tested in mice and guinea pigs. Subsequent modification and biological testing produced a library of compounds in which the thiourea derivative 4,4'-diisoamythio-carbanilide (isoxyl, thiocarlide, ISO) was shown to have considerable antimycobacterial activity in mice, guinea pigs, and rabbits (Crowle et al., 1963; Urbancik et al., 1964). ISO was initially used in the 1960s for the treatment of TB (Picone et al., 1965; Urbancik, 1966, 1970; Lambelin, 1970). Furthermore, ISO was shown to be effective against various multidrug-resistant *M. tuberculosis* isolates (Phetsuksiri et al., 1999). In an attempt to understand the biological basis for the activity of ISO, Phetsuksiri et al. (2003) investigated its mode of action and reported that ISO, like INH and ETH, strongly inhibited the synthesis of mycolic acids. The specific targets in mycolic acid biosynthesis remain unknown but presumably play a role in meromycolic acid extension because the synthesis of all classes of mycolic acid is affected by treatment with the drug. Moreover, ISO also affected the synthesis of oleic acid, a ubiquitous constituent of mycobacterial membrane phospholipids (Okuyama et al., 1967; Walker et al., 1970). Oleate is derived through fatty acid synthase I (FAS-I) in mycobacteria, which synthesizes medium-chain length fatty acids, including stearic acid, which is subsequently desaturated to oleic acid. Overexpression of the putative *M. tuberculosis* fatty acid desaturases in *M. bovis* BCG implicated *desA3*, which encodes the membrane-bound Δ9-acyl-CoA desaturase, in oleic acid synthesis. Overexpression of *desA3* in *M. bovis* BCG also resulted in increased resistance to ISO, identifying *desA3* as a novel target (Phetsuksiri et al., 2003).

TAC is a thiosemicarbazone antimicrobial that has been widely used for TB treatment in Africa and South America as a cheap and effective substitute for *p*-aminosalicylic acid (Davidson and Le, 1992). This drug is bacteriostatic against *M. tuberculosis* but is useful in combination with INH (Houston and Fanning, 1994). Because TAC is often associated with dermatological side effects and Stevens-Johnson syndrome in AIDS patients, it is not available in the United States (Houston and Fanning, 1994). EMB replaced TAC in many low-income countries during the AIDS epidemic, when patients infected with human immunodeficiency virus (HIV) were found to experience a high rate of serious and sometimes fatal skin reactions to TAC. It has also been shown to be very active against *M. avium* in vitro (Heifets et al., 1990) and in mice (Bermudez et al., 2003). However, the mechanism of action of TAC remains poorly understood.

Since EthA has been implied in the activation of ETH, a role in the activation of other thiocarbonyl-based drugs, including ISO, has been suggested on the basis of cross-resistance. DeBarber et al. (2000) characterized 14 multidrug-resistant isolates selected on the basis of TAC resistance, which were then characterized with respect to ETH resistance. Eleven of these isolates were ETH cross-resistant and, although none of the patients had been treated with ISO, all but one of the isolates showed ISO cross-resistance. Recently, two studies provided supportive in vitro evidence, confirming the oxidation of TAC, thiobenzamide, and isothionicotinamide by EthA (Vannelli et al., 2002; Qian and Ortiz de Montellano, 2006). This broad specificity in thiocarbonyl oxidation is in total accord with the structural diversity tolerated by the enzyme in terms of its Baeyer-Villiger substrates, which range from phenylacetone to long-chain ketones (Fraaije et al., 2004). Although in vitro oxidation of these drugs by EthA was certainly possible, formal proof of their status as prodrugs required the demonstration of an associated increase in their toxicity. A recent study has now confirmed that the MICs for TAC, ISO and derivatives thereof are altered through modulation of *ethA* expression in vivo (Dover et al., 2007). Moreover, this study also demonstrated that although TAC, like ETH and ISO, inhibited mycolic acid synthesis, the generation of a novel and likely unsaturated fatty acid that appeared to represent a meromycolic acid precursor suggested that this drug created a different lesion in mycolate biosynthesis.

It now appears that the possession of this enigmatic oxygenase, for which the native physiological role is not understood, can be envisaged as an Achilles' heel through which mycolic acid biosynthesis can be assaulted on many sides. Different strands of medicinal chemistry research effort have thus far unknowingly exploited the broad-range thiocarbonyl oxidation activity of EthA to great effect. The challenge now is to determine the structural constraints defining efficient activation and apply these to the development of the next generation of antimycobacterial thiocarbonyl-based drugs. A more complete understanding of the modes of action for these individual drugs through the identification and structural definition of their molecular targets will undoubtedly inform the development of these promising drugs. Effective combination of these data will ultimately require a compromise in design between efficient activation and optimized binding of the target enzyme active site.

Recent Development of the ISO Lead

Recently, two studies have determined the antimicrobial activities of modified thioureas based on ISO. In both cases, the thiourea core was not modified because early studies had proven it essential for activity (Bhowruth et al., 2006). In the hope of generating more effective variants, randomly chosen substitutions were made in the side chains attached to this key structure (Phetsuksiri et al., 1999), resulting in an array of new ISO derivatives with variations in the symmetry and asymmetry of the side chains. Some thioureas substituted in the *para* and *para'* positions by alkyl, alkoxy, or sulfur functional groups were transformed from inactive thiocarbanilides into substances with considerable antimycobacterial activity. For instance, the butyl derivative of ISO (compound B01) possessed low MICs (0.1 to 0.5 μg/ml) and was chosen for further evaluation of its effects on mycobacteria. Replacement of the oxygen with sulfur in the side chains provided extremely potent antibacterial activity against *M. tuberculosis*. Several of the ISO derivatives with various side chains of allyl, alkoxy, and alkylthio units were superior to the parent compound in their activities against *M. tuberculosis*. Other slow growers, including *M. avium*, were also susceptible to ISO and its derivatives within a narrow range of low MICs. These results suggested that a concerted approach to chemical modification of the basic thiourea nucleus of ISO would lead to even more powerful inhibitors of *M. tuberculosis* and *M. avium* (Phetsuksiri et al., 1999).

A later study demonstrated that the incorporation of a biphenyl moiety at one or both of the *para* positions was also tolerated. Although the incorporation of biphenyl to both positions produced poorer in vivo activity in comparison to ISO, inhibition against both mycolate and oleate production was observed. Interestingly the unsymmetrical biphenyl ISO failed to inhibit oleate production, suggesting that this compound does not inhibit DesA3 desaturase activity but some other enzyme specifically involved in the extension of meromycolic acid precursors (Bhowruth et al., 2006).

Interestingly, the activities of the oxygen-containing analogues produced in this study indicate that oxygen is required for a more effective inhibitor of oleate biosynthesis. Nearly all of the compounds that do not have two oxygen-containing R groups tend to be poor inhibitors of oleate biosynthesis, whereas compounds containing two oxygens inhibit oleate biosynthesis to the same degree if not better than ISO (Bhowruth et al., 2006).

Regulation of EthA: Scope for Potentiation of Prodrug Activation?

Recently two groups determined the X-ray crystal structure of the *ethA*-associated regulatory protein EthR (Dover et al., 2004, 2007; Frenois et al., 2004). The interesting feature of the structure is the possession of a long ligand-binding pocket in the C-terminal domain of the protein. Indeed, in one of these structures this pocket carried a hexadecyl octanoate ligand, presumably derived from the bacteria in which the recombinant protein had been generated, whether this be *M. smegmatis* or *E. coli* (Frenois et al., 2004). Encouragingly, when administered to an EthR-regulated reporter system, this wax ester appeared to derepress gene expression directed by the *ethA* promoter (Frenois et al., 2004). The physiological relevance of this ligand is not yet clear, but the ester could represent a product of the Baeyer-Villiger oxidation of tetracosan-8-one by EthA. Although this would be an unusual reaction, a FAD-dependent Baeyer-Villiger monooxygenase of *Pseudomonas cepacia* has previously been characterized as a ketone oxygenase and, consistent with the structure of this unusual ligand, appeared to prefer tridecan-7-one to tridecan-2-one (Britton and Markovetz, 1977). Thus, a regulatory mechanism could be envisaged in which the sensing of EthA-derived wax ester products by EthR indicated the availability of a suitable substrate and alleviated its partial inhibition of *ethA* transcription. Efficiency in such a system would be evident only if the ester product represented a superior EthR ligand than the ketone substrate. Another avenue that appears worthy of investigation is the coadminstration of a synthetic EthR ligand that might derepress *ethA* tran-

scription in vivo and thus promote more efficient oxidation of the prodrugs. Because all of these thiocarbonyl-containing second-line prodrug agents have associated toxicity problems, progress in any of the approaches to optimization outlined here may facilitate a useful reduction in effective dosage.

Thiolactomycin Inhibits Condensing Enzymes of FAS-II

Thiolactomycin (TLM) (Fig. 4) (Sasaki et al., 1982) is a broad-range thiolactone antibiotic discovered through screening *Nocardia* spp soil isolates (Oishi et al., 1982) and exhibits potent in vitro activity against many pathogenic bacteria from both gram-negative (Hayashi et al., 1983, 1984; Hamada et al., 1990) and gram-positive (Noto et al., 1982) genera including *M. tuberculosis*. It has also shown significant activity against *Plasmodium falciparum* (Waller et al., 1998) and trypanosomes (Jones et al., 2004); however, stability and synthetic issues have limited significant progress of this drug. The total synthesis of TLM was first reported by Wang and Salvino (1984) and was recently improved to yield the active 5R stereoisomer (McFadden et al., 2002).

TLM reversibly inhibits β-ketoacyl-ACP synthase (KAS) condensing enzymes (Hayashi et al., 1983, 1984; Nishida et al., 1986) associated with the dissociable fatty acid synthase FAS-II, and its mode of action has been studied extensively in *E. coli*. At low concentrations, incorporation of acetate in fatty acids and phospholipids is markedly inhibited (Hayashi et al., 1983, 1984). The primary target for TLM appears to be FabB, which is broadly analogous to mycobacterial KasA; screening a recombinant *E. coli* DNA library for genes that confer resistance to TLM identified only *fabB* clones (Tsay et al., 1992).

Although the presence of a thiolactone moiety within TLM suggested that a covalent adduct may form with the condensing enzymes through a thioester exchange reaction, analysis of the crystal structure of an *E. coli* FabB-TLM complex revealed that the drug makes a number of specific but noncovalent interactions with the enzyme active site. The C-9 and C-10 methyl groups nestled within hydrophobic pockets bounded by Phe229, Phe392, Pro272, and Phe390. The exocyclic oxygens O1 and O2 are involved in hydrogen-bonding interactions. O1 interacts with the Nε-2 nitrogen of two active site histidine residues (His298 and His333). The O2 oxygen binds to the carbonyl oxygen of residue 270 and the amide nitrogen of residue 305 through the lattice of water molecules at the base of the active site tunnel. Finally, the isoprenoid moiety is sandwiched between

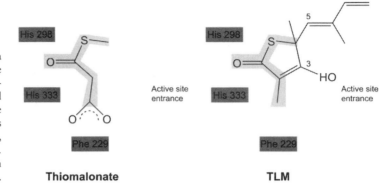

Figure 4. Thiolactomycin may mimic malonyl-ACP in the active site of β-ketoacyl-ACP synthases. The figure illustrates the perceived similarity (shaded area) between the structure of thiolactomycin (TLM) (right) and the thiomalonate moiety of malonyl-ACP (left). The amino acid residues highlighted interact with TLM in its complex with *E. coli* β-ketoacyl-ACP synthase FabB, which is broadly analogous to KasA of *M. tuberculosis*. The numerals 3 and 5 indicate carbon atoms through which analogues of TLM have been generated (see text).

three delocalized systems. The ring system of TLM remains intact within the active site and appears to mimic the bent conformation that thiomalonate is thought to adopt within the active site (Fig. 4) (Price et al., 2001).

Like *E. coli*, *M. tuberculosis* possesses three KAS enzymes, *mt*FabH, KasA, and KasB (Slayden et al., 1996; Choi et al., 2000; Kremer et al., 2000; Schaeffer et al., 2001); however, in *Mycobacterium* these are components of FAS-II involved in the synthesis of meromycolic acid. Overexpression of *kasA* and *kasB*, both individually and simultaneously, increased the resistance of *M. bovis* BCG to TLM. TLM affected the synthesis of all classes of mycolates in *M. bovis* BCG, but mycolate biosynthesis in those strains over-producing the condensing enzymes was less sensitive to inhibition by the drug (Kremer et al., 2000). Although mtFabH, the remaining KAS enzyme of *M. tuberculosis* that initiates FAS-II by condensing FAS-I products with malonyl AcpM, is sensitive to TLM in vitro, its overproduction in *M. bovis* BCG did not increase the resistance of the host strain, suggesting that this was not a primary target for the drug (Choi et al., 2000).

Several analogues have since been designed and tested against *M. tuberculosis* (Slayden et al., 1996; Douglas et al., 2002; Senior et al., 2003, 2004; Kamal et al., 2005; Kim et al., 2006). The development of TLM analogues with C-5 derivatives stemmed from the understanding that extending the length of the chain might occlude the hydrophobic pocket of the KAS enzymes, thus inhibiting the elongation of fatty acyl chains. Indeed, preliminary studies showed that modifications to the thiolactone core in the C-5 position with a 5-tetrahydrogeranyl substituent gave an MIC_{90} of 29 μM and 92% inhibition in extracts of *M. smegmatis*, as compared with 125 μM and 54%, respectively for TLM (Douglas et al., 2002). A series of C-5 substituted biphenyl and acetylene analogues were developed, and two compounds were markedly more potent against *mt*FabH in vitro (Senior et al.,

2003, 2004). 5-[3-(4-acetyl-phenyl)-propyl-2-ynyl]-4-hydroxy-3,5-dimethyl-5H-thiophen-2-one and 5-(4'-benzyloxy-biphenyl-4-yl-methyl)-4-hydroxy-3,5-dimethyl-5H-thiophen-2-one gave an 18-fold and 4-fold increase in activity with an IC_{50} value of 17 μM and 4 μM respectively, compared with 74.9 μM for TLM. In contrast, others have suggested that C5 substitutions of the TLM core render the analogues inactive against *M. tuberculosis* (Kim et al., 2006); this matter awaits clarification. However, the recent determination of the crystal structure of *M. tuberculosis* KasB and subsequent homology modeling of KasA, using the KasB structure as a template, supports the potential for C5-derivatization of the TLM scaffold toward the design of improved antimycobacterial KAS inhibitors (Sridharan et al., 2007).

Interestingly, Douglas et al. (2002) also produced 3-alkyl-3,5-dimethyl-thiophene-2,4-dione analogues of TLM. These had no effect upon fatty acid or mycolic acid biosynthesis, but a C-3 geranyl derivative did show encouraging activity against *M. tuberculosis* with an MIC of 60 μM. The inactivity of this derivative against the primary target of TLM suggests the possibility of a useful novel antimycobacterial drug target that can be explored using existing leads.

Is Pyrazinamide a Prodrug Active against De Novo Fatty Acid Biosynthesis?

Pyrazinamide (PZA) (Fig. 5) is an extremely useful anti-TB drug that plays a unique role in shortening TB therapy from a period of 9 to 12 months down to 6 months. Unusually, PZA displays remarkable activity in vivo (McCune et al., 1956) yet has no activity against *M. tuberculosis* grown on normal culture medium (Tarshis and Weed, 1953) except under acidic conditions (McDermott and Tompsett, 1954); indeed an eight-fold decrease in MIC for *M. tuberculosis* occurs between pH 5.95 and pH 5.5 (Salfinger and Heifets, 1988). Thus, it is widely considered to kill semidormant bacteria exposed to a

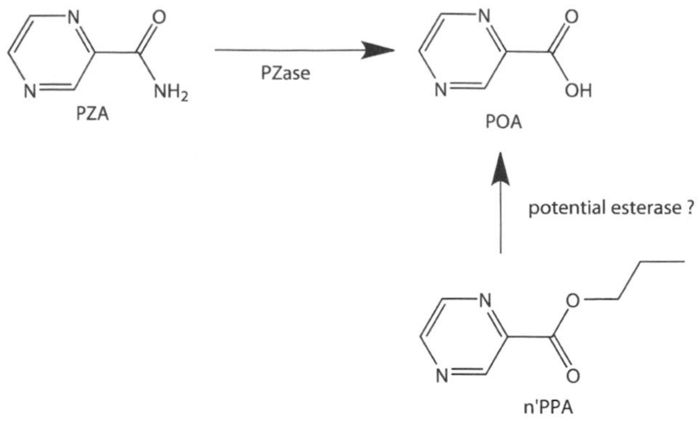

Figure 5. Pyrazinamide and analogues, proposed inhibitors of fatty acid synthase (FAS-I). Pyrazinamide (PZA) is deamidated to pyrazinoic acid (POA) through pyrazinamidase (PZase). This processing at least promotes its retention in the *M. tuberculosis* cytoplasm but may unmask its toxicity. Pyrazinoic acid esters, represented here by *n*-propyl pyrazinoate (*n*'PPA) can be hydrolyzed to produce POA. It is not known whether this modification, which presumably could be carried out by a mycobacterial esterase, is required to activate the drug.

supposed acidic environment that are not killed by other drugs (Heifets and Lindholm-Levy, 1992). However, an interesting recent study using an ex vivo culture system to evaluate drug susceptibility in *M. tuberculosis* demonstrated a window of PZA susceptibility that coincides with the onset of the T-cell-mediated immune response in mice. It is proposed that PZA acts early on a population of bacteria that might be exposed to an acidic environment as a result of immune activation (Turner et al., 2002).

M. tuberculosis is uniquely sensitive to PZA; other bacteria are completely insensitive to the drug. In *M. tuberculosis*, sensitivity to PZA correlates with the presence of a single enzyme bearing nicotinamidase and pyrazinamidase (PZase) activities (Konno et al., 1967). Strains of *M. tuberculosis* that are resistant to PZA are often defective in PZase activity (Konno et al., 1967; McClatchy et al., 1981; Trivedi and Desai, 1987) and PZA resistance in some strains of *M. tuberculosis* and the inherently-resistant *M. bovis* is related to mutations in the PZase-encoding gene, *pncA*, that abolish the amidase activity of its product (Scorpio and Zhang, 1996; Scorpio et al., 1997a, 1997b). Therefore, like INH, PZA appears to represent another antituberculous nicotinamide-based (Malone et al., 1952) prodrug that has a requirement for activation by a mycobacterial enzyme to express its toxicity (Konno et al., 1967). However, the correlation between PZase activity and PZA susceptibility does not extend to nontuberculous mycobacteria; these species possess significant PZase activity but are, nevertheless, intrinsically resistant to PZA (Tarnok and Rohrscheidt, 1976).

The product of PZA deamidation is pyrazinoic acid (POA, Fig. 5) (Konno et al., 1967; Scorpio and Zhang, 1996). However, when given orally to infected mice, POA was less effective than PZA, presumably due to inferior pharmacokinetics (Konno et al., 1967). Recently, Zhang et al. (1999) clarified the significance of acidic conditions to the efficacy of PZA/POA against *M. tuberculosis*; an acidic medium promotes the accumulation of protonated POA for which *M. tuberculosis* has only a weak efflux mechanism unlike *M. smegmatis*, which rapidly exports the activated drug.

Like the situation with its analogue INH, the search for the primary cellular target for POA has been subject to robust debate. Because POA-resistant mutants, that is, those arising in the cellular target, had not been isolated (Scorpio et al., 1997b; Zimhony et al., 2000), it was thought that this target might either represent an essential function with limited ability to express resistance-conferring mutations or that the drug might exert its toxicity through multiple targets. Interestingly and unlike PZA, which is active only against *M. tuberculosis*, various PZA analogues are active against other mycobacteria, some of which lack PZase activity (Liu et al., 1991; Cynamon et al., 1995; Speirs et al., 1995; Cynamon et al., 1998). Zimhony et al. (2000) reasoned that the targets of PZA and these analogues were likely the same and that their differential activities might be due to changes in transport, efflux, or activation and attempted to define a cellular target for 5-chloro-PZA (5-Cl-PZA) using a target overexpression strategy. This approach was unsuccessful in *M. tuberculosis* but demonstrated that overexpression of *fas* from *M. avium*, *M. bovis* BCG, or *M. tuberculosis* in *M. smegmatis* conferred resistance to the PZA analogue (Zimhony et al., 2000). The reason for failure of this approach in *M. tuberculosis* was apparent when it became clear that the bacterium cannot tolerate overexpression of *fas*.

Analysis of extracted lipids from [^{14}C]acetate-labeled *M. smegmatis* cells revealed that administration of 5-Cl-PZA inhibited fatty acid biosynthesis, an inhibition that was relieved by overexpression of *fas*. Likewise, treatment of *M. tuberculosis* cells with PZA and POA under acidic culture conditions produced a similar inhibition of fatty acid biosynthesis, implicating FAS-I as the target of PZA in *M. tuberculosis*. Another recent study, employed the pyrazinoic acid ester (PAE) *n*-propyl pyrazinoate (*n*'PPA, Fig. 5), which is assumed to release POA through an esterase activity, in a study of fatty acid synthesis in replicating *M. tuberculosis* (Zimhony et al., 2007). Although inhibition of palmitate synthesis in vivo was apparent on administration of POA and its ester, it is yet to be established whether lysis of the ester bond is required for activation of n'PPA (Zimhony et al., 2007). Both studies demonstrated that only at acidic pH did activity against tubercule bacilli implicate FAS-I inhibition; this was not apparent when bacilli were treated at neutral pH with the same dose of PZA or POA (Zimhony et al., 2000, 2007).

However, others contest that FAS-I represents the primary target of PZA-derived POA. Boshoff et al. (2002) argued that POA inhibition of FAS-I was difficult to reconcile with their previous study in which complementation of a pncA knockout mutant of M. tuberculosis with the broader spectrum PzaA amidase from M. smegmatis conferred sensitivity to the other aromatic amides, benzamide and nicotinamide, as well as PZA (Boshoff and Mizrahi, 2000). Rather than exerting toxicity through a specific binding event with a single key enzyme target, they suggested that the aromatic acids formed on activation of these three compounds exert broad inhibitory effects on the bacterium (Boshoff and Mizrahi, 2000; Boshoff et al., 2002); specifically, that intracellular accumulation of aromatic acids places stress on the pH homeostasis mechanisms of the bacterium in concert with metabolic disturbances caused by anion accumulation (Boshoff et al., 2002).

Treatment with all three aromatic amides resulted in a profile of newly synthesized fatty acids that was similar to that generated by the relevant low pH control and reflecting biosynthesis in response to extracellular acid stress. Under these conditions, however, 5-Cl-PZA clearly inhibited fatty acid synthesis (Boshoff et al., 2002). As PZase-deficient PZA-resistant mutants retain sensitivity to 5-Cl-PZA, it has been suggested that PZA and 5-Cl-PZA diverge mechanistically (Cynamon et al., 1998). However, this observation could be explained equally by a lack of POA accumulation. The lack of inhibition of fatty acid biosynthesis by PZA and the aromatic amides benzamide and nicotinamide in vivo was corroborated by a lack of inhibition of [^{14}C]malonyl-CoA incorporation into fatty acids by nicotinic acid and POA in assays using purified M. smegmatis FAS-I. However, similar assays unequivocally confirmed that FAS-I is irreversibly inhibited by 5-Cl-PZA, leading to their suggestion that the 5-chloro substitution is a prerequisite for FAS-I inhibition (Boshoff et al., 2002).

Using low-pH environments, Zhang et al. (2003) observed a general decrease in metabolism alongside inhibition of serine uptake on PZA treatment and proposed that accumulation of POA might deenergize the plasma membrane by collapsing its pH gradient under these highly acidic conditions. An observed synergy between PZA and inhibitors of membrane energetics targeting cytochrome C oxidase and the F_1F_o ATPase were presented in support of this hypothesis. Although these data provide support for a mechanism in which POA deenergises the plasma membrane, they must be viewed with some caution as an earlier study also demonstrated that energy poisons can inhibit POA extrusion from the tubercle bacillus (Zhang et al., 1999), which allows the argument that elevated intracellular POA concentrations could promote inhibition of FAS-I.

Interestingly, hypoxia enhanced the activity of PZA against M. tuberculosis, with anoxia providing a greater enhancing effect than microaerobic conditions, thereby contributing an explanation for the higher sterilizing activity of PZA in vivo, that is, within lesions with poor oxygenation, than under in vitro drug susceptibility testing conditions with ambient oxygen levels (Wade and Zhang, 2004). Although F_1F_o ATPase and respiratory chain enzyme inhibitors did not enhance PZA activity under these hypoxic conditions (Wade and Zhang, 2004), the provision of nitrate as an alternative electron acceptor reduced PZA activity under anaerobic conditions. In support of a deenergizing mode of action for POA, this nitrate was proposed to facilitate energy production and thus compensate for the depletion of energy caused by POA.

It is not beyond reason that the toxicity of POA might combine synergistic contributions from inhibition of de novo fatty acid synthesis and compromised fatty acid uptake due to the deenergizing of membranes. A complete understanding of the mode of action of this intriguing drug requires resolution of several matters. First, are the related issues of environmental acidity and the activity of PZase relevant other than to the accumulation of protonated POA? Indeed, the most frequently cited evidence that M. tuberculosis occupies acidic compartments arose from the perceived necessity for medium acidification to achieve PZA activity (Crowle et al., 1991). Interestingly, Crowle and May (1990) found that chloroquine, which raises the pH of macrophage phagosomes, actually enhances the activity of PZA against intracellular tubercle bacilli. A subsequent study determined that live virulent M. tuberculosis occupy nonacidified compartments in human macrophages (Crowle et al., 1991); an assertion that would ultimately be confirmed by the discovery that mycobacteria promote the exclusion of the vesicular proton ATPase from phagosomes, which consequently fail to acidify maintaining pH around 6.5 (Sturgill-Koszycki et al., 1994). Likewise, the precise physical microenvironment experienced by extracellular tubercle bacilli in the granuloma has not been characterized in detail. It remains a possibility that the enhancing effect of the hypoxia encountered by *M. tuberculosis* in the granuloma is more significant than the potential for acidification of the lesion. Because there is no correlation between the hydrolytic stability of PAEs and antimycobacterial potency (Cynamon et al., 1995), the intrinsic activity of PAEs may be expressed independently of hydrolysis to POA. If the activity of PZA analogues against *M. avium* was solely dependent on conversion and accumulation of POA, *M. avium* would be resistant to the PAEs as well as POA (Cynamon et al., 1995). Taking these considerations into account, it seems that conversion of PZA to POA and the requirement of acidic pH are conditions for the accumulation of POA, rather than being the sole mechanism of action and that description of PZA as an amidase-activated prodrug may ultimately be misleading.

Next we must consider whether these related drugs act in the same manner. We await publication of direct enzymological and structural analyses of *M. tuberculosis* FAS-I required to define the interaction and mode of inhibition of 5-Cl-PZA, which appears to be a bona fide FAS-I inhibitor. Because PZase activation is not required for 5-Cl-PZA (Cynamon et al., 1998) and an analogous esterase activation may not be required for PAE toxicity (Cynamon et al., 1995), then the investigation of unmodified PAEs and *N*-acyl PZA analogues may also be enlightening and ultimately informative in defining a role for all of these nicotinamide-based inhibitors in the inhibition of de novo fatty acid synthesis. It should not yet be ruled out that PZA-derived POA might ultimately represent a weakly active member of a class of FAS-I inhibitors that requires highly specific conditions to express toxicity and that, fortuitously, these conditions are particularly relevant to the pathobiology of *M. tuberculosis* infection.

Ultimately, the role of FAS-I as the target of these drugs can, and must, be investigated further. McCarthy (1971) and Weir et al. (1972) observed uptake of long-chain fatty acids from culture medium by mycobacteria and their incorporation into triacylglycerols. In *M. tuberculosis* infection, these internalized C_{16} and C_{18} fatty acids might also enter more complex and essential lipid syntheses, for instance FAS-II and be used for the synthesis of mycolic acids. Interestingly and despite massive decay in its genome, *M. leprae* has retained a functional FAS-I and is competent in de novo fatty acid biosynthesis (Wheeler et al., 1990). Taking into consideration the relatively large size of the *fas* gene (Cole et al., 2001), the bacterium's privileged intracellular niche and capacity to scavenge host lipids (Wheeler et al., 1990), it is tempting to suggest that FAS-I might be essential in vivo despite the availability of host fatty acids. *M. bovis* BCG FAS-I is recognized to produce a bimodal

distribution (C_{16} and C_{26}) of fatty acids (Kikuchi et al., 1992), and it may be this balance of fatty acids that is crucial to the survival of the bacterium. Interestingly, the larger modal acyl chain length is equivalent to that of the α-chain of mycolic acids. Could inhibition of the extension of shorter fatty acids to this length result in the inhibition of mycolate production or lead to the production of mycolates that are not fit for function in vivo?

NOVEL DRUGS TARGETING MYCOLATE BIOSYNTHESIS

n-Octanesulfonylacetamide (OSA)

β-Sulfonylcarboxamide compounds were designed as potential inhibitors of β-ketoacyl-ACP synthases (KAS) of pathogenic mycobacteria by acting as mimics (both geometrically and electronically) of the putative transition state (Fig. 6) in the Claisen-type condensation reaction catalyzed by these enzymes or domains involved in the biosynthesis of fatty acids, the mycolic acid components of the cell wall and various polyketides in *M. tuberculosis* (Jones et al., 2000). The most active compounds discovered through this strategy were amide derivatives of 3-sulfonyl fatty acids bearing alkyl chains of between 8 and 10 car-

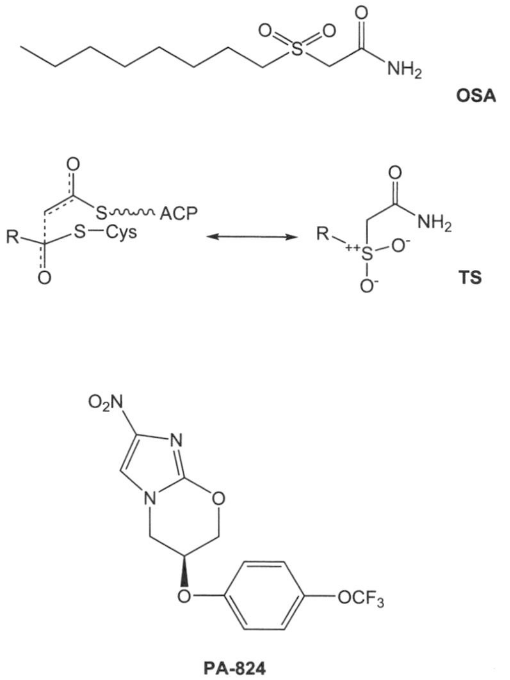

Figure 6. The structures of novel agents targeting mycolic acid biosynthesis. The structures of *n*-octylsulfonylacetamide (OSA) and nitroimidazopyran PA-824, which affect mycolic acid biosynthesis, are illustrated. OSA mimics the proposed transition state (TS) generated during the Claisen-like condensation reaction.

bons in length with MICs as low as 0.75 to 1.5 μg/ml and comparable with front-line anti-TB agents. Surprisingly, the compounds were particularly species specific, showing no activity against other bacteria, even nonpathogenic fast-growing mycobacteria (Jones et al., 2000). One of these, n-octanesulfonylacetamide (OSA) (Fig. 6), inhibited the growth of a range of slow-growing pathogenic and drug-resistant mycobacteria including MDR-TB strains (Parrish et al., 2001). Furthermore, analysis of mycobacterial lipids revealed a marked inhibition of all subtypes of mycolic acids in M. bovis BCG without affecting the panoply of polar and nonpolar extractable lipids generated by the bacterium; mycolic acid biosynthesis in the relatively insensitive M. smegmatis was unaffected (Parrish et al., 2001). Analyses of treated sensitive bacteria using electron microscopy revealed dysfunction in cell wall biosynthesis and incomplete septation (Parrish et al., 2001). Further proteomic analyses of the effects of OSA revealed the overproduction of two small proteins (~18 kDa) in M. bovis BCG, the β-subunit of ATP synthase (F_1F_o ATPase) and a small heat shock protein encoded by atpF and hsp (Rv0251c), respectively (Parrish et al., 2004). The overexpression of the atpF suggested the possible involvement of ATP synthase, either in a direct or indirect manner. Consistent with this idea, cellular ATP levels were shown to decrease upon treatment with OSA (Parrish et al., 2004) mimicking the effect of the known ATP synthase inhibitor dicyclohexylcarbodiimide (Zhang et al., 2003). None of the other antimycobacterial agents tested elicited the same response. Reminiscent of the diarylquinoline R207910 (Andries et al., 2005), this OSA-mediated decrease in cellular ATP may inhibit the operation of the F_1F_o complex or may target other unidentified regulatory components involved in the energy production.

The Nitroimidazopyran PA-824

The nitroimidazopyran PA-824 (Fig. 6) is a promising new compound for the treatment of TB that is currently undergoing clinical trials. PA-824 is a prodrug of the nitroimidazole class, requiring bioreductive activation of an aromatic nitro group to exert an antitubercular effect (Stover et al., 2000). PA-824 is highly active, with a MIC as low as 0.015 to 0.025 μg/ml against M. tuberculosis and MDR-TB. The bioreductive activation of PA-824 requires the bacterial F240-dependent glucose-6-phosphate dehydrogenase (FGD1) and nitroreductase to activate components that then inhibit bacterial mycolic acid and protein synthesis. Initial data suggest that PA-824 induces an accumulation of hydroxymycolic acids, biosynthetic precursors of organic extractable and cell wall-bound methoxymycolic and ketomy-

colic acids. However, it is important to consider that a mutant in mmaA4, which is devoid of oxygenated mycolates, is viable although attenuated in mice (Dubnau et al., 2000).

In preclinical testing against a broad panel of multidrug resistant clinical isolates in vitro, PA-824 was found to be highly active against all isolates with MICs lower than 1 μg/ml (Lenaerts et al., 2005). In a short-course mouse infection model, the efficacy of PA-824 was comparable with those of INH, RIF, and moxifloxacin. PA-824 also demonstrated potent activity during the continuation phase of therapy, during which it targeted bacilli that had persisted through an initial 2-month intensive phase of treatment with RIF, INH, and PZA (Tyagi et al., 2005). More recently, a novel nitroimidazo-oxazine specific protein has been identified and shown to be involved in PA-824 resistance in M. tuberculosis (Manjunatha et al., 2006). Resistance to PA-824 is most commonly mediated by a loss of the FGD1 or its deazaflavin cofactor F240 involved in reductive activation of this class of molecules. Although FGD1 and F240 are necessary, they are not sufficient for activation of PA-824, which requires the involvement of additional accessory proteins that directly interact with the nitroimidazole. To fully understand the mechanism of activation, an extensive panel of PA-824 resistant mutants were characterized and revealed a small class of mutants with lesions in Rv3547 that possessed normal FGD1 and F240 capabilities. Complementation with intact Rv3547 fully restored sensitivity to nitroimidazo-oxazines and restored the ability of M. tuberculosis to metabolize PA-824 (Manjunatha et al., 2006). The function of this novel protein has yet to be determined.

UNEXPLORED TARGETS IN CELL WALL BIOSYNTHESIS

The crucial AG component of the M. tuberculosis cell wall core underpins the covalent tethering of the mycolic acid-based "pseudo" outer membrane to its peptidoglycan sacculus and is referred to as the mycolyl-arabinogalactan-peptidoglycan (mAGP) complex (see chapter 3). The mAGP structure is essential and provides the bacterium with a formidable protective barrier against toxic insult, such as antibiotics and components of the macrophage's bactericidal arsenal (Gao et al., 2003).

Based on the findings that wbbL (Rv3265c), the likely rhamnosyltransferase involved in the biosynthesis of the rhamnosyl-N-acetylglucosamine disaccharide linking peptidoglycan and galactan, is an essential enzyme (Mills et al., 2004), and that EMB and INH target cell wall biosynthesis at later steps, we suggest that the intermediate steps in mAGP synthesis, notably galactan polymerisation, also represent extremely attractive novel drug targets.

A slightly controversial issue is the discovery of new drugs that inhibit *M. tuberculosis* cell wall synthesis and whether they might have a major impact on TB control. It is believed that adding new drugs that act on actively growing bacilli will have utility only in treating multidrug resistant organisms (admittedly an increasing concern) but may do little to crack the main problem, which is the length of time of treatment. For this, drugs that kill nonmultiplying persistent organisms are needed. Some argue that cell wall synthesis may be already defective or unnecessary in these organisms (they are not stainable or otherwise detectable microscopically). We would strongly argue in the favor for the need for new drug targets and inhibitors that target the mycobacterial cell wall. Since their discovery in the 1950s, inhibitors of cell wall synthesis have been the mainstay of many chemotherapeutic regimens. Although drugs that act on nonreplicating forms appear desirable on theoretical grounds, we are not aware of any direct evidence that their use would actually accelerate TB therapy. We simply do not know how new drugs will impact on time to cure, and we need to await animal and clinical trial data before drawing any such conclusions.

The view that cell wall synthesis in persistent organisms may be already defective or unnecessary based on staining or microscopy is not entirely correct. For instance, Stover et al. (2000) reported that nitroimidazopyran PA-824 was effective against static forms (i.e., cultures grown under anaerobic conditions) of *M. tuberculosis* by inhibiting both lipid and protein synthesis, and hydroxymycolic acids accumulated. As a result, the study reports one of the first drugs to be active against persistent organisms and the targeting of the synthesis of mycolic acids and the cell wall. Therefore, a basal cell wall core, mAGP, is always maintained, and cell wall turnover possibly modulated through autolytic activity as found for other bacilli that enter a dormant or persistent state.

The increased abundance of CmaA2 (Glickman et al., 2001) and KasB (Gao et al., 2003), involved in mycolic acid biosynthesis under anaerobic conditions (Starck et al., 2004) suggests a level of metabolic activity related to mycolic acid biosynthesis under conditions usually associated with a transition to dormancy that may be linked, resulting in modulation of mycolic acid chain length during a dormant or persistent anaerobic state.

Acknowledgments. G.S.B. acknowledges support from James Bardrick in the form of a Personal Chair, the Lister Institute as a former Jenner Research Fellow, the Medical Research Council (United Kingdom) and the Wellcome Trust.

REFERENCES

Andries, K., P. Verhasselt, J. Guillemont, H. W. Gohlmann, J. M. Neefs, H. Winkler, J. Van Gestel, P. Timmerman, M. Zhu, E. Lee, P. Williams, D. de Chaffoy, E. Huitric, S. Hoffner, E. Cambau, C. Truffot-Pernot, N. Lounis, and V. Jarlier. 2005. A diarylquinoline drug active on the ATP synthase of *Mycobacterium tuberculosis. Science* 307:223–227.

Argyrou, A., L. Jin, L. Siconilfi-Baez, R. H. Angeletti, and J. S. Blanchard. 2006a. Proteome-wide profiling of isoniazid targets in *Mycobacterium tuberculosis. Biochemistry* 45:13947–13953.

Argyrou, A., M. W. Vetting, B. Aladegbami, and J. S. Blanchard. 2006b. *Mycobacterium tuberculosis* dihydrofolate reductase is a target for isoniazid. *Nat. Struct. Mol. Biol.* 13:408–413.

Banerjee, A., E. Dubnau, A. Quémard, V. Balasubramanian, K. S. Um, T. Wilson, D. Collins, G. de Lisle, and W. R. Jacobs, Jr. 1994. *inhA*, a gene encoding a target for isoniazid and ethionamide in *Mycobacterium tuberculosis. Science* 263:227–230.

Banerjee, A., M. Sugantino, J. C. Sacchettini, and W. R. Jacobs, Jr. 1998. The *mabA* gene from the *inhA* operon of *Mycobacterium tuberculosis* encodes a 3-ketoacyl reductase that fails to confer isoniazid resistance. *Microbiology* 144:2697–2704.

Barclay, W. R., and E. Winberg. 1964. Bactericidal effect of isoniazid as a function of time. *Am. Rev. Respir. Dis.* 90:749–753.

Barry, C. E., R. A. Slayden, and K. Mludli. 1998. Mechanisms of isoniazid resistance in *Mycobacterium tuberculosis. Drug Resist. Update* 1:128–134.

Bartz, Q. R., J. Ehrlich, J. D. Mold, M. A. Penner, and R. M. Smith. 1951. Viomycin, a new tuberculostatic antibiotic. *Am. Rev. Tuberc.* 63:4–6.

Baulard, A. R., J. C. Betts, J. Engohang-Ndong, S. Quan, R. A. McAdam, P. J. Brennan, C. Locht, and G. S. Besra. 2000. Activation of the pro-drug ethionamide is regulated in mycobacteria. *J. Biol. Chem.* 275:28326–28331.

Belanger, A. E., G. S. Besra, M. E. Ford, K. Mikusová, J. T. Belisle, P. J. Brennan, and J. M. Inamine. 1996. The *embAB* genes of *Mycobacterium avium* encode an arabinosyl transferase involved in cell wall arabinan biosynthesis that is the target for the antimycobacterial drug ethambutol. *Proc. Natl. Acad. Sci. USA* 93:11919–11924.

Bermudez, L. E., R. Reynolds, P. Kolonoski, P. Aralar, C. B. Inderlied, and L. S. Young. 2003. Thiosemicarbazole (thiacetazone-like) compound with activity against *Mycobacterium avium* in mice. *Antimicrob. Agents Chemother.* 47:2685–2687.

Bernstein, J., W. A. Lott, B. A. Steinberg, and H. L. Yale. 1952. Chemotherapy of experimental tuberculosis. V. Isonicotinic acid hydrazide (nydrazid) and related compounds. *Am. Rev. Tuberc.* 65:357–364.

Bertrand, T., N. A. J. Eady, J. N. Jones, J. M. Nagy, B. Jamart-Gregoire, E. L. Raven, and K. A. Brown. 2004. Crystal structure of *Mycobacterium tuberculosis* catalase-peroxidase. *J. Biol. Chem.* 279:38991–38999.

Bhowruth, V., A. K. Brown, R. C. Reynolds, G. D. Coxon, S. P. Mackay, D. E. Minnikin, and G. S. Besra. 2006. Symmetrical and unsymmetrical analogues of isoxyl; active agents against *Mycobacterium tuberculosis. Bioorg. Med. Chem. Lett.* 16:4743–4747.

Boshoff, H. I., and V. Mizrahi. 2000. Expression of *Mycobacterium smegmatis* pyrazinamidase in *Mycobacterium tuberculosis* confers hypersensitivity to pyrazinamide and related amides. *J. Bacteriol.* 182:5479-5485.

Boshoff, H. I., V. Mizrahi, and C. E. Barry III. 2002. Effects of pyrazinamide on fatty acid synthesis by whole mycobacterial cells and purified fatty acid synthase I. *J. Bacteriol.* 184:2167–2172.

Boshoff, H. I., T. G. Myers, B. R. Copp, M. R. McNeil, M. A. Wilson, and C. E. Barry III. 2004. The transcriptional responses of *Mycobacterium tuberculosis* to inhibitors of metabolism:

novel insights into drug mechanisms of action. *J. Biol. Chem.* 279:40174–40184.

Britton, L. N., and A. J. Markovetz. 1977. A novel ketone monooxygenase from *Pseudomonas cepacia*. Purification and properties. *J. Biol. Chem.* 252:8561–8566.

Cavalieri, S. J., J. R. Biehle, and W. E. Sanders, Jr. 1995. Synergistic activities of clarithromycin and antituberculous drugs against multidrug-resistant *Mycobacterium tuberculosis*. *Antimicrob. Agents Chemother.* 39:1542–1545.

Chen, P., J. Gearhart, M. Protopopova, L. Einck, and C. A. Nacy. 2006. Synergistic interactions of SQ109, a new ethylene diamine, with front-line antitubercular drugs in vitro. *J. Antimicrob. Chemother.* 58:332–337.

Choi, K. H., L. Kremer, G. S. Besra, and C. O. Rock. 2000. Identification and substrate specificity of beta-ketoacyl (acyl carrier protein) synthase III (mtFabH) from *Mycobacterium tuberculosis*. *J. Biol. Chem.* 275:28201–28207.

Chorine, V. 1945. Action de l'amide nicotinique sur les bacilless du genre *Mycobacterium*. *C. R. Acad. Sci.* 220:150–151.

Cohn, M. L., C. Kovitz, U. Oda, and G. Middlebrook. 1954. Studies on isoniazid and tubercle bacilli. 2. The growth requirements, catalase activities, and pathogenic properties of isoniazid-resistant mutants. *Am. Rev. Tuberc.* 70:641–664.

Cole, S. T., K. Eiglmeier, J. Parkhill, K. D. James, N. R. Thomson, P. R. Wheeler, N. Honore, T. Garnier, C. Churcher, D. Harris, K. Mungall, D. Basham, D. Brown, T. Chillingworth, R. Connor, R. M. Davies, K. Devlin, S. Duthoy, T. Feltwell, A. Fraser, N. Hamlin, S. Holroyd, T. Hornsby, K. Jagels, C. Lacroix, J. Maclean, S. Moule, L. Murphy, K. Oliver, M. A. Quail, M. A. Rajandream, K. M. Rutherford, S. Rutter, K. Seeger, S. Simon, M. Simmonds, J. Skelton, R. Squares, S. Squares, K. Stevens, K. Taylor, S. Whitehead, J. R. Woodward, and B. G. Barrell. 2001. Massive gene decay in the leprosy bacillus. *Nature* 409:1007–1011.

Crowle, A. J., R. S. Mitchell, and T. L. Petty. 1963. The effectiveness of a thiocarbanilide (Isoxyl) as a therapeutic drug in mouse tuberculosis. *Am. Rev. Respir. Dis.* 88:716–717.

Crowle, A. J., and M. H. May. 1990. Inhibition of tubercle bacilli in cultured human macrophages by chloroquine used alone and in combination with streptomycin, isoniazid, pyrazinamide, and two metabolites of vitamin D3. *Antimicrob. Agents Chemother.* 34:2217–2222.

Crowle, A. J., R. Dahl, E. Ross, and M. H. May. 1991. Evidence that vesicles containing living, virulent *Mycobacterium tuberculosis* or *Mycobacterium avium* in cultured human macrophages are not acidic. *Infect. Immun.* 59:1823–1831.

Cynamon, M. H., R. Gimi, F. Gyenes, C. A. Sharpe, K. E. Bergmann, H. J. Han, L. B. Gregor, R. Rapolu, G. Luciano, and J. T. Welch. 1995. Pyrazinoic acid esters with broad spectrum in vitro antimycobacterial activity. *J. Med. Chem.* 38:3902–3907.

Cynamon, M. H., R. J. Speirs, and J. T. Welch. 1998. In vitro antimycobacterial activity of 5-chloropyrazinamide. *Antimicrob. Agents Chemother.* 42:462–463.

Davidson, L. A., and K. Takayama. 1979. Isoniazid inhibition of the synthesis of monounsaturated long-chain fatty acids in *Mycobacterium tuberculosis* H37Ra. *Antimicrob. Agents Chemother.* 16:104–105.

Davidson, P. T., and H. Q. Le. 1992. Drug treatment of tuberculosis-1992. *Drugs* 43:651–673.

De Logu, A., V. Onnis, B. Saddi, C. Congiu, M. L. Schivo, and M. T. Cocco. 2002. Activity of a new class of isonicotinoylhydrazones used alone and in combination with isoniazid, rifampicin, ethambutol, para-aminosalicylic acid and clofazimine against *Mycobacterium tuberculosis*. *J. Antimicrob. Chemother.* 49:275–282.

DeBarber, A. E., K. Mdluli, M. Bosman, L. G. Bekker, and C. E. Barry III. 2000. Ethionamide activation and sensitivity in

multidrug-resistant *Mycobacterium tuberculosis*. *Proc. Natl. Acad. Sci. USA* 97:9677–9682.

Deng, L., K. Mikusová, K. G. Robuck, M. Scherman, P. J. Brennan, and M. R. McNeil. 1995. Recognition of multiple effects of ethambutol on metabolism of mycobacterial cell envelope. *Antimicrob. Agents Chemother.* 39:694–701.

Dessen, A., A. Quémard, J. S. Blanchard, W. R. Jacobs, Jr., and J. C. Sacchettini. 1995. Crystal structure and function of the isoniazid target of *Mycobacterium tuberculosis*. *Science* 267:1638–1641.

Dong, X., S. Bhamidi, M. Scherman, Y. Xin, and M. R. McNeil. 2006. Development of a quantitative assay for mycobacterial endogenous arabinase and ensuing studies of arabinase levels and arabinan metabolism in *Mycobacterium smegmatis*. *Appl. Environ. Microbiol.* 72:2601–2605.

Douglas, J. D., S. J. Senior, C. Morehouse, B. Phetsukiri, I. B. Campbell, G. S. Besra, and D. E. Minnikin. 2002. Analogues of thiolactomycin: potential drugs with enhanced anti-mycobacterial activity. *Microbiology* 148:3101–3109.

Dover, L. G., P. E. Corsino, I. R. Daniels, S. L. Cocklin, V. Tatituri, G. S. Besra, and K. Futterer. 2004. Crystal structure of the TetR/CamR family repressor *Mycobacterium tuberculosis* EthR implicated in ethionamide resistance. *J. Mol. Biol.* 340:1095–1105.

Dover, L. G., A. Alahari, P. Gratraud, J. M. Gomes, V. Bhowruth, R. C. Reynolds, G. S. Besra, and L. Kremer. 2007. EthA, a common activator of thiocarbamide-containing drugs acting on different mycobacterial targets. *Antimicrob. Agents Chemother.* 51:1055–1063.

Dubnau, E., J. Chan, C. Raynaud, V. P. Mohan, M. A. Laneelle, K. Yu, A. Quemard, I. Smith, and M. Daffe. 2000. Oxygenated mycolic acids are necessary for virulence of *Mycobacterium tuberculosis* in mice. *Mol. Microbiol.* 36:630–637.

Ducasse-Cabanot, S., M. Cohen-Gonsaud, H. Marrakchi, M. Nguyen, D. Zerbib, J. Bernadou, M. Daffé, G. Labesse, and A. Quémard. 2004. In vitro inhibition of the *Mycobacterium tuberculosis* beta-ketoacyl-acyl carrier protein reductase MabA by isoniazid. *Antimicrob. Agents Chemother.* 48:242–249.

Dye, C., C. J. Watt, D. M. Bleed, S. M. Hosseini, and M. C. Raviglione. 2005. Evolution of tuberculosis control and prospects for reducing tuberculosis incidence, prevalence, and deaths globally. *J. A. M. A.* 293:2767–2775.

Eisman, P. C., E. A. Konopka, and R. L. Mayer. 1954. Antituberculous activity of substituted thioureas. II. Activity in mice. *Am. Rev. Tuberc.* 70:121–129.

Engohang-Ndong, J., D. Baillat, M. Aumercier, F. Bellefontaine, G. S. Besra, C. Locht, and A. R. Baulard. 2004. EthR, a repressor of the TetR/CamR family implicated in ethionamide resistance in mycobacteria, octamerizes cooperatively on its operator. *Mol. Microbiol.* 51:175–188.

Escalante, P., S. Ramaswamy, H. Sanabria, H. Soini, X. Pan, O. Valiente-Castillo, and J. M. Musser. 1998. Genotypic characterization of drug-resistant *Mycobacterium tuberculosis* isolates from Peru. *Tuber. Lung Dis.* 79:111–118.

Escuyer, V. E., M. A. Lety, J. B. Torrelles, K. H. Khoo, J. B. Tang, C. D. Rithner, C. Frehel, M. R. McNeil, P. J. Brennan, and D. Chatterjee. 2001. The role of the *embA* and *embB* gene products in the biosynthesis of the terminal hexaarabinofuranosyl motif of *Mycobacterium smegmatis* arabinogalactan. *J. Biol. Chem.* 276:48854–48862.

Fattorini, L., E. Iona, M. L. Ricci, O. F. Thoresen, G. Orru, M. R. Oggioni, C. Piersimoni, P. Chiaradonna, M. Tronci, G. Pozzi, and G. Orefici. 1999. Activity of 16 antimicrobial agents against drug-resistant strains of *Mycobacterium tuberculosis*. *Microb. Drug Resist.* 5:265–270.

Forbes, M., N. A. Kuck, and E. A. Peets. 1965. Effect of ethambutol on nucleic acid metabolism in *Mycobacterium smegmatis* and its reversal bypolyamines and divalent cations. *J. Bacteriol.* 89:1299–1305.

Fox, H. H. 1952. The chemical approach to the control of tuberculosis. *Science* 115:129–134.

Fraaije, M. W., N. M. Kamerbeek, A. J. Heidekamp, R. Fortin, and D. B. Janssen. 2004. The prodrug activator EtaA from *Mycobacterium tuberculosis* is a Baeyer-Villiger monooxygenase. *J. Biol. Chem.* 279:3354–3360.

Frenois, F., J. Engohang-Ndong, C. Locht, A. R. Baulard, and V. Villeret. 2004. Structure of EthR in a ligand bound conformation reveals therapeutic perspectives against tuberculosis. *Mol. Cell* 16:301–307.

Gao, L. Y., F. Laval, E. H. Lawson, R. K. Groger, A. Woodruff, J. H. Morisaki, J. S. Cox, M. Daffé, and E. J. Brown. 2003. Requirement for *kasB* in *Mycobacterium* mycolic acid biosynthesis, cell wall impermeability and intracellular survival: implications for therapy. *Mol. Microbiol.* 49:1547–1563.

Glickman, M. S., S. M. Cahill, and W. R. Jacobs, Jr. 2001. The *Mycobacterium tuberculosis* cmaA2 gene encodes a mycolic acid trans-cyclopropane synthetase. *J. Biol. Chem.* 276:2228–2233.

Hamada, S., T. Fujiwara, H. Shimauchi, T. Ogawa, T. Nishihara, T. Koga, T. Nehashi, and T. Matsuno. 1990. Antimicrobial activities of thiolactomycin against gram-negative anaerobes associated with periodontal disease. *Oral Microbiol. Immunol.* 5:340–345.

Hanoulle, X., J. M. Wieruszeski, P. Rousselot-Pailley, I. Landrieu, A. R. Baulard, and G. Lippens. 2005. Monitoring of the ethionamide pro-drug activation in mycobacteria by ^{1}H high resolution magic angle spinning NMR. *Biochem. Biophys. Res. Commun.* 331:452–458.

Hanoulle, X., J. M. Wieruszeski, P. Rousselot-Pailley, I. Landrieu, C. Locht, G. Lippens, and A. R. Baulard. 2006. Selective intracellular accumulation of the major metabolite issued from the activation of the prodrug ethionamide in mycobacteria. *J. Antimicrob. Chemother.* 58:768–772.

Harries, A. D., R. Chimzizi, and R. Zachariah. 2006. Safety, effectiveness, and outcomes of concomitant use of highly active antiretroviral therapy with drugs for tuberculosis in resource-poor settings. *Lancet* 367:944–945.

Hayashi, T., O. Yamamoto, H. Sasaki, A. Kawaguchi, and H. Okazaki. 1983. Mechanism of action of the antibiotic thiolactomycin inhibition of fatty acid synthesis of *Escherichia coli*. *Biochem. Biophys. Res. Commun.* 115:1108–1113.

Hayashi, T., O. Yamamoto, H. Sasaki, H. Okazaki, and A. Kawaguchi. 1984. Inhibition of fatty acid synthesis by the antibiotic thiolactomycin. *J. Antibiot.* (Tokyo) 37:1456–1461.

Hazbón, M. H., M. Bobadilla del Valle, M. I. Guerrero, M. Varma-Basil, I. Filliol, M. Cavatore, R. Colangeli, H. Safi, H. Billman-Jacobe, C. Lavender, J. Fyfe, L. Garcia-Garcia, A. Davidow, M. Brimacombe, C. I. Leon, T. Porras, M. Bose, F. Chaves, K. D. Eisenach, J. Sifuentes-Osornio, A. Ponce de Leon, M. D. Cave, and D. Alland. 2005. Role of *embB* codon 306 mutations in *Mycobacterium tuberculosis* revisited: a novel association with broad drug resistance and IS6110 clustering rather than ethambutol resistance. *Antimicrob. Agents Chemother.* 49:3794–3802.

Hedgecock, L. W., and I. O. Faucher. 1957. Relation of pyrogallol-peroxidative activity to isoniazid resistance in *Mycobacterium tuberculosis*. *Am. Rev. Tuberc. Pulmon. Dis.* 75:670–674.

Heifets, L., and P. Lindholm-Levy. 1992. Pyrazinamide sterilizing activity in vitro against semidormant *Mycobacterium tuberculosis* bacterial populations. *Am. Rev. Respir. Dis.* 145:1223–1225.

Heifets, L. B., P. J. Lindholm-Levy, and M. Flory. 1990. Thiacetazone: in vitro activity against *Mycobacterium avium* and *M. tuberculosis*. *Tubercle* 71:287–291.

Herr, E. B., M. E. Haney, G. E. Pittenger, and C. E. Higgens. 1960. Isolation and characterisation of a new peptide antibiotic. *Proc. Indiana Acad. Sci.* 69:134.

Heym, B., Y. Zhang, S. Poulet, D. Young, and S. T. Cole. 1993. Characterization of the *katG* gene encoding a catalase-peroxidase required for the isoniazid susceptibility of *Mycobacterium tuberculosis*. *J. Bacteriol.* 175:4255–4259.

Heym, B., P. M. Alzari, N. Honoré, and S. T. Cole. 1995. Missense mutations in the catalase-peroxidase gene, KatG, are associated with isoniazid resistance in *Mycobacterium tuberculosis*. *Mol. Microbiol.* 15:235–245.

Houston, S., and A. Fanning. 1994. Current and potential treatment of tuberculosis. *Drugs* 48:689–708.

Isola, D., M. Pardini, F. Varaine, S. Niemann, S. Rusch-Gerdes, L. Fattorini, G. Orefici, F. Meacci, C. Trappetti, M. Rinaldo Oggioni, and G. Orru. 2005. A pyrosequencing assay for rapid recognition of SNPs in *Mycobacterium tuberculosis* embB306 region. *J. Microbiol. Methods* 62:113–120.

Jagannath, C., V. M. Reddy, and P. R. Gangadharam. 1995. Enhancement of drug susceptibility of multi-drug resistant strains of *Mycobacterium tuberculosis* by ethambutol and dimethyl sulphoxide. *J. Antimicrob. Chemother.* 35:381–390.

Jenner, P. J., and S. E. Smith. 1987. Plasma levels of ethionamide and prothionamide in a volunteer following intravenous and oral dosages. *Lepr. Rev.* 58:31–37.

Jia, L., L. Coward, G. S. Gorman, P. E. Noker, and J. E. Tomaszewski. 2005a. Pharmacoproteomic effects of isoniazid, ethambutol, and N-geranyl-N'-(2-adamantyl)ethane-1,2-diamine (SQ109) on *Mycobacterium tuberculosis* H37Rv. *J. Pharmacol. Exp. Ther.* 315:905–911.

Jia, L., J. E. Tomaszewski, C. Hanrahan, L. Coward, P. Noker, G. Gorman, B. Nikonenko, and M. Protopopova. 2005b. Pharmacodynamics and pharmacokinetics of SQ109, a new diamine-based antitubercular drug. *Br. J. Pharmacol.* 144:80–87.

Jia, L., P. E. Noker, L. Coward, G. S. Gorman, M. Protopopova, and J. E. Tomaszewski. 2006. Interspecies pharmacokinetics and in vitro metabolism of SQ109. *Br. J. Pharmacol.* 147:476–485.

Johnsson, K., D. S. King, and P. G. Schultz. 1995. Studies on the mechanism of action of isoniazid and ethionamide in the chemotherapy of tuberculosis. *JACS* 117:5009–5010.

Jones, P. B., N. M. Parrish, T. A. Houston, A. Stapon, N. P. Bansal, J. D. Dick, and C. A. Townsend. 2000. A new class of antituberculosis agents. *J. Med. Chem.* 43:3304–3314.

Jones, S. M., J. E. Urch, R. Brun, J. L. Harwood, C. Berry, and I. H. Gilbert. 2004. Analogues of thiolactomycin as potential antimalarial and anti-trypanosomal agents. *Bioorg. Med. Chem.* 12:683–692.

Kamal, A., A. A. Shaik, R. Sinha, J. S. Yadav, and S. K. Arora. 2005. Antitubercular agents. Part 2: New thiolactomycin analogues active against *Mycobacterium tuberculosis*. *Bioorg. Med. Chem. Lett.* 15:1927–1929.

Kikuchi, S., D. L. Rainwater, and P. E. Kolattukudy. 1992. Purification and characterization of an unusually large fatty acid synthase from *Mycobacterium tuberculosis* var. *bovis* BCG. *Arch. Biochem. Biophys.* 295:318–326.

Kilburn, J. O., and J. Greenberg. 1977. Effect of ethambutol on the viable cell count in *Mycobacterium smegmatis*. *Antimicrob. Agents Chemother.* 11:534–540.

Kilburn, J. O., and K. Takayama. 1981. Effects of ethambutol on accumulation and secretion of trehalose mycolates and free mycolic acid in *Mycobacterium smegmatis*. *Antimicrob. Agents Chemother.* 20:401–404.

Kilburn, J. O., K. Takayama, E. L. Armstrong, and J. Greenberg. 1981. Effects of ethambutol on phospholipid metabolism in *Mycobacterium smegmatis*. *Antimicrob. Agents Chemother.* 19:346–348.

Kim, P., Y. M. Zhang, G. Shenoy, Q. A. Nguyen, H. I. Boshoff, U. H. Manjunatha, M. B. Goodwin, J. Lonsdale, A. C. Price, D. J. Miller, K. Duncan, S. W. White, C. O. Rock, C. E. Barry, and C. S. Dowd. 2006. Structure-activity relationships at the 5-position of thiolactomycin: An intact (5R)-isoprene unit is required for activity against the condensing enzymes from *Mycobacterium tuberculosis* and *Escherichia coli*. *J. Med. Chem.* **49:** 159–171.

Konno, K., F. M. Feldmann, and W. McDermott. 1967. Pyrazinamide susceptibility and amidase activity of tubercle bacilli. *Am. Rev. Respir. Dis.* **95:**461–469.

Kremer, L., J. D. Douglas, A. R. Baulard, C. Morehouse, M. R. Guy, D. Alland, L. G. Dover, J. H. Lakey, W. R. Jacobs, P. J. Brennan, D. E. Minnikin, and G. S. Besra. 2000. Thiolactomycin and related analogues as novel anti-mycobacterial agents targeting KasA and KasB condensing enzymes in *Mycobacterium tuberculosis*. *J. Biol. Chem.* **275:**16857–16864.

Kremer, L., L. G. Dover, H. R. Morbidoni, C. Vilchèze, W. N. Maughan, A. Baulard, S. C. Tu, N. Honoré, V. Deretic, J. C. Sacchettini, C. Locht, W. R. Jacobs, Jr., and G. S. Besra. 2003. Inhibition of InhA activity, but not KasA activity, induces formation of a KasA-containing complex in mycobacteria. *J. Biol. Chem.* **278:**20547–20554.

Kurosawa, H. 1952. Studies on the antibiotic substances from actinomycetes. XXIII. The isolation of an antibiotic produced by strain of *Streptomyces* "K 30." *J. Antibiot. Ser. B* **5:**682–688.

Kwara, A., T. P. Flanigan, and E. J. Carter. 2005. Highly active antiretroviral therapy (HAART) in adults with tuberculosis: current status. *Int. J. Tuberc. Lung Dis.* **9:**248–257.

Lambelin, G. 1970. Pharmacology and toxicology of Isoxyl. *Antibiot. Chemother.* **16:**84–95.

Larsen, M. H., C. Vilchèze, L. Kremer, G. S. Besra, L. Parsons, M. Salfinger, L. Heifets, M. H. Hazbon, D. Alland, J. C. Sacchettini, and W. R. Jacobs, Jr. 2002. Overexpression of *inhA*, but not *kasA*, confers resistance to isoniazid and ethionamide in *Mycobacterium smegmatis*, *M. bovis* BCG and *M. tuberculosis*. *Mol. Microbiol.* **46:**453–466.

Lee, A. S., S. N. Othman, Y. M. Ho, and S. Y. Wong. 2004. Novel mutations within the *embB* gene in ethambutol-susceptible clinical isolates of *Mycobacterium tuberculosis*. *Antimicrob. Agents Chemother.* **48:**4447–4449.

Lee, H. Y., H. J. Myoung, H. E. Bang, G. H. Bai, S. J. Kim, J. D. Kim, and S. N. Cho. 2002. Mutations in the *embB* locus among Korean clinical isolates of *Mycobacterium tuberculosis* resistant to ethambutol. *Yonsei Med. J.* **43:**59–64.

Lee, R. E., M. Protopopova, E. Crooks, R. A. Slayden, M. Terrot, and C. E. Barry III. 2003. Combinatorial lead optimization of [1,2]-diamines based on ethambutol as potential antituberculosis preclinical candidates. *J. Comb. Chem.* **5:**172–187.

Lehmann, J. 1946. *p*-aminosalicylic acid in the treatment of tuberculosis. *Lancet* **1:**15–16.

Lenaerts, A. J., V. Gruppo, K. S. Marietta, C. M. Johnson, D. K. Driscoll, N. M. Tompkins, J. D. Rose, R. C. Reynolds, and I. M. Orme. 2005. Preclinical testing of the nitroimidazopyran PA-824 for activity against *Mycobacterium tuberculosis* in a series of in vitro and in vivo models. *Antimicrob. Agents Chemother.* **49:**2294–2301.

Lety, M. A., S. Nair, P. Berche, and V. Escuyer. 1997. A single point mutation in the *embB* gene is responsible for resistance to ethambutol in *Mycobacterium smegmatis*. *Antimicrob. Agents Chemother.* **41:**2629–2633.

Liebermann, D., M. Moyeux, N. Rist, and F. Grumbach. 1956. Sur le preparation de nouveaux thioamides pyridiniques acitfs dans la tuberculose exerimentale. *C. R. Acad. Sci.* **242:**2409–2412.

Liu, Z. Z., X. D. Guo, L. E. Straub, G. Erdos, R. J. Prankerd, R. J. Gonzalez-Rothi, and H. Schreier. 1991. Lipophilic N-acylpyraz-inamide derivatives: synthesis, physicochemical characterization, liposome incorporation, and in vitro activity against *Mycobacterium avium-intracellulare*. *Drug Des. Discov.* **8:**57–67.

Lukat-Rodgers, G. S., N. L. Wengenack, F. Rusnak, and K. R. Rodgers. 2000. Spectroscopic comparison of the heme active sites in WT KatG and its S315T mutant. *Biochemistry* **39:**9984–9993.

Mackaness, G. B., and N. Smith. 1953. The bactericidal action of isoniazid, streptomycin, and terramycin on extracellular and intracellular tubercle bacilli. *Am. Rev. Tuberc.* **67:**322–340.

Maggi, N., C. R. Pasqualucci, R. Ballotta, and P. Sensi. 1966. Rifampicin: a new orally active rifamycin. *Chemotherapy* **11:** 285–292.

Malone, L., A. Schurr, H. Lindh, D. McKenzie, J. S. Kiser, and J. H. Williams. 1952. The effect of pyrazinamide (aldinamide) on experimental tuberculosis in mice. *Am. Rev. Tuberc.* **65:**511–518.

Manjunatha, U. H., H. Boshoff, C. S. Dowd, L. Zhang, T. J. Albert, J. E. Norton, L. Daniels, T. Dick, S. S. Pang, and C. E. Barry III. 2006. Identification of a nitroimidazo-oxazine-specific protein involved in PA-824 resistance in *Mycobacterium tuberculosis*. *Proc. Natl. Acad. Sci. USA* **103:**431–436.

Marrakchi, H., S. Ducasse, G. Labesse, H. Montrozier, E. Margeat, L. Emorine, X. Charpentier, M. Daffé, and A. Quémard. 2002. MabA (FabG1), a *Mycobacterium tuberculosis* protein involved in the long-chain fatty acid elongation system FAS-II. *Microbiology* **148:**951–960.

Marttila, H. J., H. Soini, E. Eerola, E. Vyshnevskaya, B. I. Vyshnevskiy, T. F. Otten, A. V. Vasilyef, and M. K. Viljanen. 1998. A Ser315Thr substitution in KatG is predominant in genetically heterogeneous multidrug-resistant *Mycobacterium tuberculosis* isolates originating from the St. Petersburg area in Russia. *Antimicrob. Agents Chemother.* **42:**2443–2445.

McCarthy, C. 1971. Utilization of palmitic acid by *Mycobacterium avium*. *Infect. Immun.* **4:**199–204.

McClatchy, J. K., A. Y. Tsang, and M. S. Cernich. 1981. Use of pyrazinamidase activity on *Mycobacterium tuberculosis* as a rapid method for determination of pyrazinamide susceptibility. *Antimicrob. Agents Chemother.* **20:**556–557.

McCune, R. M., Jr., W. McDermott, and R. Tompsett. 1956. The fate of *Mycobacterium tuberculosis* in mouse tissues as determined by the microbial enumeration technique. II. The conversion of tuberculous infection to the latent state by the administration of pyrazinamide and a companion drug. *J. Exp. Med.* **104:**763–802.

McDermott, W., and R. Tompsett. 1954. Activation of pyrazinamide and nicotinamide in acidic environments in vitro. *Am. Rev. Tuberc.* **70:**748–754.

McFadden, J. M., G. L. Frehywot, and C. A. Townsend. 2002. A flexible route to (5R)-thiolactomycin, a naturally occurring inhibitor of fatty acid synthesis. *Org. Lett.* **4:**3859–3862.

Mdluli, K., D. R. Sherman, M. J. Hickey, B. N. Kreiswirth, S. Morris, C. K. Stover, and C. E. Barry III. 1996. Biochemical and genetic data suggest that InhA is not the primary target for activated isoniazid in *Mycobacterium tuberculosis*. *J. Infect. Dis.* **174:**1085–1090.

Mdluli, K., R. A. Slayden, Y. Zhu, S. Ramaswamy, X. Pan, D. Mead, D. D. Crane, J. M. Musser, and C. E. Barry III. 1998. Inhibition of a *Mycobacterium tuberculosis* beta-ketoacyl ACP synthase by isoniazid. *Science* **280:**1607–1610.

Middlebrook, G. 1954. Isoniazid-resistance and catalase activity of tubercle bacilli-a preliminary report. *Am. Rev. Tuberc.* **69:** 471–472.

Middlebrook, G., M. L. Cohn, and W. B. Schaefer. 1954. Studies on isoniazid and tubercle bacilli. 3. The isolation, drug-susceptibility, and catalase-testing of tubercle bacilli from isoniazid-treated patients. *Am. Rev. Tuberc.* **70:**852–872.

Miesel, L., T. R. Weisbrod, J. A. Marcinkeviciene, R. Bittman, and W. R. Jacobs, Jr. 1998. NADH dehydrogenase defects confer isoniazid resistance and conditional lethality in *Mycobacterium smegmatis*. *J. Bacteriol.* 180:2459–2467.

Mikusová, K., R. A. Slayden, G. S. Besra, and P. J. Brennan. 1995. Biogenesis of the mycobacterial cell wall and the site of action of ethambutol. *Antimicrob. Agents Chemother.* 39:2484–2489.

Mills, J. A., K. Motichka, M. Jucker, H. P. Wu, B. C. Uhlik, R. J. Stern, M. S. Scherman, V. D. Vissa, F. Pan, M. Kundu, Y. F. Ma, and M. McNeil. 2004. Inactivation of the mycobacterial rhamnosyltransferase, which is needed for the formation of the arabinogalactan-peptidoglycan linker, leads to irreversible loss of viability. *J. Biol. Chem.* 279:43540–43546.

Mitchison, D. A. 1979. Basic mechanisms of chemotherapy. *Chest* 76:771–781.

Mokrousov, I., T. Otten, B. Vyshnevskiy, and O. Narvskaya. 2002. Detection of *embB306* mutations in ethambutol-susceptible clinical isolates of *Mycobacterium tuberculosis* from Northwestern Russia: implications for genotypic resistance testing. *J. Clin. Microbiol.* 40:3810–3813.

Mokrousov, I., N. V. Bhanu, P. N. Suffys, G. V. Kadival, S. F. Yap, S. N. Cho, A. M. Jordaan, O. Narvskaya, U. B. Singh, H. M. Gomes, H. Lee, S. P. Kulkarni, K. C. Lim, B. K. Khan, D. van Soolingen, T. C. Victor, and L. M. Schouls. 2004. Multicenter evaluation of reverse line blot assay for detection of drug resistance in *Mycobacterium tuberculosis* clinical isolates. *J. Microbiol. Methods* 57:323–335.

Moore-Gillon, J. 2001. Multidrug-resistant tuberculosis: this is the cost. *Ann. N. Y. Acad. Sci.* 953:233–240.

Musser, J. M., V. Kapur, D. L. Williams, B. N. Kreiswirth, D. vanSoolingen, and J. D. A. vanEmbden. 1996. Characterization of the catalase-peroxidase gene (*katG*) and *inhA* locus in isoniazid-resistant and -susceptible strains of *Mycobacterium tuberculosis* by automated DNA sequencing: Restricted array of mutations associated with drug resistance. *J. Infect. Dis.* 173:196–202.

Ng, V. H., J. S. Cox, A. O. Sousa, J. D. MacMicking, and J. D. McKinney. 2004. Role of KatG catalase-peroxidase in mycobacterial pathogenesis: countering the phagocyte oxidative burst. *Mol. Microbiol.* 52:1291–1302.

Nishida, I., A. Kawaguchi, and M. Yamada. 1986. Effect of thiolactomycin on the individual enzymes of the fatty acid synthase system in *Escherichia coli*. *J. Biochem.* (Tokyo) 99:1447–1454.

Noto, T., S. Miyakawa, H. Oishi, H. Endo, and H. Okazaki. 1982. Thiolactomycin, a new antibiotic. III. In vitro antibacterial activity. *J. Antibiot.* (Tokyo) 35:401–410.

Offe, H. A., W. Sieken, and G. Domagk. 1952. The tuberculostatic activity of hydrazine derivatives from pyridine carboxylic acids and carbonyl compounds. *Z. Naturforsch.* 7:462–468.

Oishi, H., T. Noto, H. Sasaki, K. Suzuki, T. Hayashi, H. Okazaki, K. Ando, and M. Sawada. 1982. Thiolactomycin, a new antibiotic. I. Taxonomy of the producing organism, fermentation and biological properties. *J. Antibiot.* 35:391–395.

Okuyama, H., T. Kankura, and S. Nojima. 1967. Positional distribution of fatty acids in phospholipids from Mycobacteria. *J. Biochem.* (Tokyo) 61:732–737.

Parrish, N. M., T. Houston, P. B. Jones, C. Townsend, and J. D. Dick. 2001. In vitro activity of a novel antimycobacterial compound, N-octanesulfonylacetamide, and its effects on lipid and mycolic acid synthesis. *Antimicrob. Agents Chemother.* 45:1143–1150.

Parrish, N. M., C. G. Ko, M. A. Hughes, C. A. Townsend, and J. D. Dick. 2004. Effect of N-octanesulphonylacetamide (OSA) on ATP and protein expression in *Mycobacterium bovis* BCG. *J. Antimicrob. Chemother.* 54:722–729.

Parsons, L. M., M. Salfinger, A. Clobridge, J. Dormandy, L. Mirabello, V. L. Polletta, A. Sanic, O. Sinyavskiy, S. C. Larsen, J. Driscoll, G. Zickas, and H. W. Taber. 2005. Phenotypic and molecular characterization of *Mycobacterium tuberculosis* isolates resistant to both isoniazid and ethambutol. *Antimicrob. Agents Chemother.* 49:2218–2225.

Phetsuksiri, B., A. R. Baulard, A. M. Cooper, D. E. Minnikin, J. D. Douglas, G. S. Besra, and P. J. Brennan. 1999. Antimycobacterial activities of isoxyl and new derivatives through the inhibition of mycolic acid synthesis. *Antimicrob. Agents Chemother.* 43:1042–1051.

Phetsuksiri, B., M. Jackson, H. Scherman, M. McNeil, G. S. Besra, A. R. Baulard, R. A. Slayden, A. E. DeBarber, C. E. Barry III, M. S. Baird, D. C. Crick, and P. J. Brennan. 2003. Unique mechanism of action of the thiourea drug isoxyl on *Mycobacterium tuberculosis*. *J. Biol. Chem.* 278:53123–53130.

Picone, A., M. Di Vincenzo, and C. Russo. 1965. 4,4′Diisoamyloxythiocarbanilide (isoxyl) in the therapy of chronic pulmonary tuberculosis resistant to antibiotics and the usual chemotherapeutic agents. *G. Ital. Chemioter.* 12:99–106. (In Italian.)

Pierattelli, R., L. Banci, N. A. J. Eady, J. Bodiguel, J. N. Jones, P. C. E. Moody, E. L. Raven, B. Jamart-Gregoire, and K. A. Brown. 2004. Enzyme-catalyzed mechanism of isoniazid activation in class I and class III peroxidases. *J. Biol. Chem.* 279:39000–39009.

Plinke, C., S. Rusch-Gerdes, and S. Niemann. 2006. Significance of mutations in *embB* codon 306 for prediction of ethambutol resistance in clinical *Mycobacterium tuberculosis* isolates. *Antimicrob. Agents Chemother.* 50:1900–1902.

Poso, H., L. Paulin, and E. Brander. 1983. Specific inhibition of spermidine synthase from mycobacteria by ethambutol. *Lancet* 2:1418.

Post, F. A., P. A. Willcox, B. Mathema, L. M. Steyn, K. Shean, S. V. Ramaswamy, E. A. Graviss, E. Shashkina, B. N. Kreiswirth, and G. Kaplan. 2004. Genetic polymorphism in *Mycobacterium tuberculosis* isolates from patients with chronic multidrug-resistant tuberculosis. *J. Infect. Dis.* 190:99–106.

Price, A. C., K. H. Choi, R. J. Heath, Z. Li, S. W. White, and C. O. Rock. 2001. Inhibition of beta-ketoacyl-acyl carrier protein synthases by thiolactomycin and cerulenin. Structure and mechanism. *J. Biol. Chem.* 276:6551–6559.

Protopopova, M., C. Hanrahan, B. Nikonenko, R. Samala, P. Chen, J. Gearhart, L. Einck, and C. A. Nacy. 2005. Identification of a new antitubercular drug candidate, SQ109, from a combinatorial library of 1,2-ethylenediamines. *J. Antimicrob. Chemother.* 56:968–974.

Pym, A. S., P. Domenech, N. Honoré, J. Song, V. Deretic, and S. T. Cole. 2001. Regulation of catalase-peroxidase (KatG) expression, isoniazid sensitivity and virulence by *furA* of *Mycobacterium tuberculosis*. *Mol. Microbiol.* 40:879–889.

Qian, L., and P. R. Ortiz de Montellano. 2006. Oxidative activation of thiacetazone by the *Mycobacterium tuberculosis* flavin monooxygenase EtaA and human FMO1 and FMO3. *Chem. Res. Toxicol.* 19:443–449.

Quémard, A., C. Lacave, and G. Lanéelle. 1991. Isoniazid inhibition of mycolic acid synthesis by cell extracts of sensitive and resistant strains of *Mycobacterium aurum*. *Antimicrob. Agents Chemother.* 35:1035–1039.

Quémard, A., J. C. Sacchettini, A. Dessen, C. Vilchèze, R. Bittman, W. R. Jacobs, Jr., and J. S. Blanchard. 1995. Enzymatic characterization of the target for isoniazid in *Mycobacterium tuberculosis*. *Biochemistry* 34:8235–8241.

Quémard, A., A. Dessen, M. Sugantino, W. R. Jacobs, J. C. Sacchettini, and J. S. Blanchard. 1996. Binding of catalase-peroxidase-activated isoniazid to wild-type and mutant Myco-

bacterium tuberculosis enoyl-ACP reductases. *JACS* 118:1561-1562.

Ramaswamy, S. V., A. G. Amin, S. Goksel, C. E. Stager, S. J. Dou, H. El Sahly, S. L. Moghazeh, B. N. Kreiswirth, and J. M. Musser. 2000. Molecular genetic analysis of nucleotide polymorphisms associated with ethambutol resistance in human isolates of *Mycobacterium tuberculosis*. *Antimicrob. Agents Chemother.* 44:326–336.

Ramaswamy, S. V., S. J. Dou, A. Rendon, Z. Yang, M. D. Cave, and E. A. Graviss. 2004. Genotypic analysis of multidrug-resistant *Mycobacterium tuberculosis* isolates from Monterrey, Mexico. *J. Med. Microbiol.* 53:107–113.

Rastogi, N., K. S. Goh, L. Horgen, and W. W. Barrow. 1998. Synergistic activities of antituberculous drugs with cerulenin and trans-cinnamic acid against *Mycobacterium tuberculosis*. *FEMS Immunol. Med. Microbiol.* 21:149–157.

Rawat, R., A. Whitty, and P. J. Tonge. 2003. The isoniazid-NAD adduct is a slow, tight-binding inhibitor of InhA, the *Mycobacterium tuberculosis* enoyl reductase: adduct affinity and drug resistance. *Proc. Natl. Acad. Sci. USA* 100:13881–13886.

Rinder, H., K. T. Mieskes, E. Tortoli, E. Richter, M. Casal, M. Vaquero, E. Cambau, K. Feldmann, and T. Loscher. 2001. Detection of *embB* codon 306 mutations in ethambutol resistant *Mycobacterium tuberculosis* directly from sputum samples: a low-cost, rapid approach. *Mol. Cell Probes* 15:37–42.

Rouse, D. A., J. A. DeVito, Z. M. Li, H. Byer, and S. L. Morris. 1996. Site-directed mutagenesis of the *katG* gene of *Mycobacterium tuberculosis*: Effects on catalase-peroxidase activities and isoniazid resistance. *Mol. Microbiol.* 22:583–592.

Rozwarski, D. A., G. A. Grant, D. H. Barton, W. R. Jacobs, Jr., and J. C. Sacchettini. 1998. Modification of the NADH of the isoniazid target (InhA) from *Mycobacterium tuberculosis*. *Science* 279:98–102.

Salfinger, M., and L. B. Heifets. 1988. Determination of pyrazinamide MICs for *Mycobacterium tuberculosis* at different pHs by the radiometric method. *Antimicrob. Agents Chemother.* 32:1002–1004.

Sasaki, H., H. Oishi, T. Hayashi, I. Matsuura, K. Ando, and M. Sawada. 1982. Thiolactomycin, a new antibiotic. II. Structure elucidation. *J. Antibiot.* (Tokyo) 35:396–400.

Schaeffer, M. L., G. Agnihotri, C. Volker, H. Kallender, P. J. Brennan, and J. T. Lonsdale. 2001. Purification and biochemical characterization of the *Mycobacterium tuberculosis* beta-ketoacyl-acyl carrier protein synthases KasA and KasB. *J. Biol. Chem.* 276:47029–47037.

Schatz, A., E. Bugie, and S. A. Waksman. 1944. Streptomycin, a substance exhibiting antibiotic activity against gram-positive and gram-negative bacteria. *Proc. Soc. Exp. Biol. Med.* 55:66–69.

Scorpio, A., and Y. Zhang. 1996. Mutations in *pncA*, a gene encoding pyrazinamidase/nicotinamidase, cause resistance to the antituberculous drug pyrazinamide in tubercle bacillus. *Nat. Med.* 2:662–667.

Scorpio, A., D. Collins, D. Whipple, D. Cave, J. Bates, and Y. Zhang. 1997a. Rapid differentiation of bovine and human tubercle bacilli based on a characteristic mutation in the bovine pyrazinamidase gene. *J. Clin. Microbiol.* 35:106–110.

Scorpio, A., P. Lindholm-Levy, L. Heifets, R. Gilman, S. Siddiqi, M. Cynamon, and Y. Zhang. 1997b. Characterization of *pncA* mutations in pyrazinamide-resistant *Mycobacterium tuberculosis*. *Antimicrob. Agents Chemother.* 41:540–543.

Senior, S. J., P. A. Illarionova, S. S. Gurcha, I. B. Campbell, M. L. Schaeffer, D. E. Minnikin, and G. S. Besra. 2003. Biphenyl-based analogues of thiolactomycin, active against *Mycobacterium tuberculosis* mtFabH fatty acid condensing enzyme. *Bioorg. Med. Chem. Lett.* 13:3685–3688.

Senior, S. J., P. A. Illarionov, S. S. Gurcha, I. B. Campbell, M. L. Schaeffer, D. E. Minnikin, and G. S. Besra. 2004. Acetylene-based analogues of thiolactomycin, active against *Mycobacterium tuberculosis* mtFabH fatty acid condensing enzyme. *Bioorg. Med. Chem. Lett.* 14:373–376.

Sensi, P., P. Margalith, and M. T. Timbal. 1959. Rifomycin, a new antibiotic; preliminary report. *Farmaco [Sci]* 14:146–147.

Sharma, S. K., and A. Mohan. 2004. Multidrug-resistant tuberculosis. *Indian J. Med. Res.* 120:354–376.

Silve, G., P. Valero-Guillen, A. Quémard, M. A. Dupont, M. Daffé, and G. Lanéelle. 1993. Ethambutol inhibition of glucose metabolism in mycobacteria: a possible target of the drug. *Antimicrob. Agents Chemother.* 37:1536–1538.

Singh, B., and D. A. Mitchison. 1954. Bactericidal activity of streptomycin and isoniazid against tubercle bacilli. *BMJ* i:130–132.

Singh, R., B. Wiseman, T. Deemagarn, L. J. Donald, H. W. Duckworth, X. Carpena, I. Fita, and P. C. Loewen. 2004. Catalase-peroxidases (KatG) exhibit NADH oxidase activity. *J. Biol. Chem.* 279:43098–43106.

Slayden, R. A., R. E. Lee, J. W. Armour, A. M. Cooper, I. M. Orme, P. J. Brennan, and G. S. Besra. 1996. Antimycobacterial action of thiolactomycin: an inhibitor of fatty acid and mycolic acid synthesis. *Antimicrob. Agents Chemother.* 40:2813–2819.

Slayden, R. A., and C. E. Barry. 2000. The genetics and biochemistry of isoniazid resistance in *Mycobacterium tuberculosis*. *Microb. Infect.* 2:659–669.

Slayden, R. A., R. E. Lee, and C. E. Barry III. 2000. Isoniazid affects multiple components of the type II fatty acid synthase system of *Mycobacterium tuberculosis*. *Mol. Microbiol.* 38:514–525.

Slayden, R. A., and C. E. Barry III. 2002. The role of KasA and KasB in the biosynthesis of meromycolic acids and isoniazid resistance in *Mycobacterium tuberculosis*. *Tuberculosis* 82:149–160.

Speirs, R. J., J. T. Welch, and M. H. Cynamon. 1995. Activity of n-propyl pyrazinoate against pyrazinamide-resistant *Mycobacterium tuberculosis*: investigations into mechanism of action of and mechanism of resistance to pyrazinamide. *Antimicrob. Agents Chemother.* 39:1269–1271.

Spigelman, M., and S. Gillespie. 2006. Tuberculosis drug development pipeline: progress and hope. *Lancet* 367:945–947.

Sreevatsan, S., K. E. Stockbauer, X. Pan, B. N. Kreiswirth, S. L. Moghazeh, W. R. Jacobs, Jr., A. Telenti, and J. M. Musser. 1997. Ethambutol resistance in *Mycobacterium tuberculosis*: critical role of *embB* mutations. *Antimicrob. Agents Chemother.* 41:1677–1681.

Sridharan, S., L. Wang, A. K. Brown, L. G. Dover, L. Kremer, G. S. Besra, and J. C. Sacchettini. 2007. X-Ray crystal structure of *Mycobacterium tuberculosis* beta-ketoacyl acyl carrier protein synthase II (mtKasB). *J. Mol. Biol.* 366:469–480.

Starck, J., G. Kallenius, B. I. Marklund, D. I. Andersson, and T. Akerlund. 2004. Comparative proteome analysis of *Mycobacterium tuberculosis* grown under aerobic and anaerobic conditions. *Microbiology* 150:3821–3829.

Stover, C. K., P. Warrener, D. R. VanDevanter, D. R. Sherman, T. M. Arain, M. H. Langhorne, S. W. Anderson, J. A. Towell, Y. Yuan, D. N. McMurray, B. N. Kreiswirth, C. E. Barry, and W. R. Baker. 2000. A small-molecule nitroimidazopyran drug candidate for the treatment of tuberculosis. *Nature* 405:962–966.

Sturgill-Koszycki, S., P. H. Schlesinger, P. Chakraborty, P. L. Haddix, H. L. Collins, A. K. Fok, R. D. Allen, S. L. Gluck, J. Heuser, and D. G. Russell. 1994. Lack of acidification in *Mycobacterium* phagosomes produced by exclusion of the vesicular proton-ATPase. *Science* 263:678–681.

Takayama, K., L. Wang, and H. L. David. 1972. Effect on isoniazid on in-vivo mycolic acid synthesis, cell-growth, and viability of *Mycobacterium tuberculosis*. *Antimicrob. Agents Chemother.* 2:29–35.

Takayama, K., E. L. Armstrong, and H. L. David. 1974. Restoration of mycolate synthetase activity in *Mycobacterium tuberculosis* exposed to isoniazid. *Am. Rev. Respir. Dis.* 110:43–48.

Takayama, K., E. L. Armstrong, K. A. Kunugi, and J. O. Kilburn. 1979. Inhibition by ethambutol of mycolic acid transfer into the cell wall of *Mycobacterium smegmatis*. *Antimicrob. Agents Chemother.* 16:240–242.

Takayama, K., and N. Qureshi. 1979. Presented at the 14th US-Japan Tuberculosis Research Conference.

Takayama, K., and J. O. Kilburn. 1989. Inhibition of synthesis of arabinogalactan by ethambutol in *Mycobacterium smegmatis*. *Antimicrob. Agents Chemother.* 33:1493–1499.

Tarnok, I., and E. Rohrscheidt. 1976. Biochemical background of some enzymatic tests used for the differentiation of mycobacteria. *Tubercle* 57:145–150.

Tarshis, M. S., and W. A. Weed, Jr. 1953. Lack of significant in vitro sensitivity of *Mycobacterium tuberculosis* to pyrazinamide on three different solid media. *Am. Rev. Tuberc.* 67:391–395.

Telenti, A., W. J. Philipp, S. Sreevatsan, C. Bernasconi, K. E. Stockbauer, B. Wieles, J. M. Musser, and W. R. Jacobs, Jr. 1997. The *emb* operon, a gene cluster of *Mycobacterium tuberculosis* involved in resistance to ethambutol. *Nat. Med.* 3:567–570.

Thomas, J. P., C. O. Baughn, R. G. Wilkinson, and R. G. Shepherd. 1961. A new synthetic compound with antituberculous activity in mice: ethambutol (dextro-2,2'-(ethylenediimino)-di-l-butanol). *Am. Rev. Respir. Dis.* 83:891–893.

Todorovic, S., N. Juranic, S. Macura, and F. Rusnak. 1999. Binding of N-15-labeled isoniazid to KatG and KatG(S315T): Use of two-spin [zz]-order relaxation rate for N-15-Fe distance determination. *JACS* 121:10962–10966.

Tracevska, T., I. Jansone, A. Nodieva, O. Marga, G. Skenders, and V. Baumanis. 2004. Characterisation of *rpsL*, *rrs* and *embB* mutations associated with streptomycin and ethambutol resistance in *Mycobacterium tuberculosis*. *Res. Microbiol.* 155:830–834.

Trivedi, S. S., and S. G. Desai. 1987. Pyrazinamidase activity of *Mycobacterium tuberculosis*-a test of sensitivity to pyrazinamide. *Tubercle* 68:221–224.

Tsay, J. T., C. O. Rock, and S. Jackowski. 1992. Overproduction of beta-ketoacyl-acyl carrier protein synthase I imparts thiolactomycin resistance to *Escherichia coli* K-12. *J. Bacteriol.* 174:508–513.

Tsunekawa, H., T. Miyachi, E. Nakamura, M. Tsukamura, and H. Amano. 1987. Therapeutic effect of ofloxacin on 'treatment-failure' pulmonary tuberculosis. *Kekkaku* 62:435–439. (In Japanese.)

Turner, D. J., S. L. Hoyle, V. A. Snewin, M. P. Gares, I. N. Brown, and D. B. Young. 2002. An ex vivo culture model for screening drug activity against in vivo phenotypes of *Mycobacterium tuberculosis*. *Microbiology* 148:2929–2936.

Tyagi, S., E. Nuermberger, T. Yoshimatsu, K. Williams, I. Rosenthal, N. Lounis, W. Bishai, and J. Grosset. 2005. Bactericidal activity of the nitroimidazopyran PA-824 in a murine model of tuberculosis. *Antimicrob. Agents Chemother.* 49:2289–2293.

Umezawa, H., M. Ueda, K. Maeda, K. Yagishita, S. Kondo, Y. Okami, R. Utahara, Y. Osato, K. Nitta, and T. Takeuchi. 1957. Production and isolation of a new antibiotic: kanamycin. *J. Antibiot.* (Tokyo) 10:181–188.

Urbancik, B. 1966. A clinical trial of thiocarlide (4-4′ diisoamyloxythiocarbanilide). *Tubercle* 47:283–288.

Urbancik, B. 1970. Clinical experience with thiocarlide (Isoxyl). *Antibiot. Chemother.* 16:117–123.

Urbancik, R., L. Trnka, J. Kruml, and H. Polenska. 1964. Antimycobacterial activity of isoxyl. II. Experiments with guinea pigs and rabbits. *Pathol. Microbiol.* 27:79–87. (In German.)

Van Rie, A., R. Warren, I. Mshanga, A. M. Jordaan, G. D. van der Spuy, M. Richardson, J. Simpson, R. P. Gie, D. A. Enarson, N. Beyers, P. D. van Helden, and T. C. Victor. 2001. Analysis for a limited number of gene codons can predict drug resistance of *Mycobacterium tuberculosis* in a high-incidence community. *J. Clin. Microbiol.* 39:636–641.

Vannelli, T. A., A. Dykman, and P. R. Ortiz de Montellano. 2002. The antituberculosis drug ethionamide is activated by a flavoprotein monooxygenase. *J. Biol. Chem.* 277:12824–12829.

Veyron-Churlet, R., O. Guerrini, L. Mourey, M. Daffé, and D. Zerbib. 2004. Protein-protein interactions within the Fatty Acid Synthase-II system of *Mycobacterium tuberculosis* are essential for mycobacterial viability. *Mol. Microbiol.* 54:1161–1172.

Veyron-Churlet, R., S. Bigot, O. Guerrini, S. Verdoux, W. Malaga, M. Daffé, and D. Zerbib. 2005. The biosynthesis of mycolic acids in *Mycobacterium tuberculosis* relies on multiple specialized elongation complexes interconnected by specific protein-protein interactions. *J. Mol. Biol.* 353:847–858.

Vilchèze, C., H. R. Morbidoni, T. R. Weisbrod, H. Iwamoto, M. Kuo, J. C. Sacchettini, and W. R. Jacobs, Jr. 2000. Inactivation of the *inhA*-encoded fatty acid synthase II (FASII) enoyl-acyl carrier protein reductase induces accumulation of the FASI end products and cell lysis of *Mycobacterium smegmatis*. *J. Bacteriol.* 182:4059–4067.

Vilchèze, C., T. R. Weisbrod, B. Chen, L. Kremer, M. H. Hazbon, F. Wang, D. Alland, J. C. Sacchettini, and W. R. Jacobs, Jr. 2005. Altered NADH/NAD+ ratio mediates coresistance to isoniazid and ethionamide in mycobacteria. *Antimicrob. Agents Chemother.* 49:708–720.

Vilchèze, C., F. Wang, M. Arai, M. H. Hazbon, R. Colangeli, L. Kremer, T. R. Weisbrod, D. Alland, J. C. Sacchettini, and W. R. Jacobs, Jr. 2006. Transfer of a point mutation in *Mycobacterium tuberculosis inhA* resolves the target of isoniazid. *Nat. Med.* 12:1027–1029.

Wada, T., S. Maeda, A. Tamaru, S. Imai, A. Hase, and K. Kobayashi. 2004. Dual-probe assay for rapid detection of drug-resistant *Mycobacterium tuberculosis* by real-time PCR. *J. Clin. Microbiol.* 42:5277–5285.

Wade, M. M., and Y. Zhang. 2004. Anaerobic incubation conditions enhance pyrazinamide activity against *Mycobacterium tuberculosis*. *J. Med. Microbiol.* 53:769–773.

Walker, R. W., H. Barakat, and J. G. Hung. 1970. The positional distribution of fatty acids in the phospholipids and triglycerides of *Mycobacterium smegmatis* and *M. bovis* BCG. *Lipids* 5:684–691.

Waller, R. F., P. J. Keeling, R. G. Donald, B. Striepen, E. Handman, N. Lang-Unnasch, A. F. Cowman, G. S. Besra, D. S. Roos, and G. I. McFadden. 1998. Nuclear-encoded proteins target to the plastid in *Toxoplasma gondii* and *Plasmodium falciparum*. *Proc. Natl. Acad. Sci. USA* 95:12352–12357.

Wang, C.-L. J., and J. M. Salvino. 1984. Total synthesis of (±) thiolactomycin. *Tetrahedron Lett.* 25:5243–5246.

Wang, F., R. Langley, G. Gulten, L. G. Dover, G. S. Besra, W. R. Jacobs, Jr., and J. C. Sacchettini. 2007. Mechanism of thioamide drug action against tuberculosis and leprosy. *J. Exp. Med.* 204:73-78.

Wang, L., and K. Takayama. 1972. Relationship between uptake of isoniazid and its action on in vivo mycolic acid synthesis in *Mycobacterium tuberculosis*. *Antimicrob. Agents Chemother.* 2:438-441.

Weir, M. P., W. H. Langridge III, and R. W. Walker. 1972. Relationships between oleic acid uptake and lipid metabolism in *Mycobacterium smegmatis*. *Am. Rev. Respir. Dis.* 106:450–457.

Wengenack, N. L., J. R. Uhl, A. L. S. Amand, A. J. Tomlinson, L. M. Benson, S. Naylor, B. C. Kline, F. R. Cockerill, and F. Rusnak. 1997. Recombinant *Mycobacterium tuberculosis* KatG(S315T) is a competent catalase-peroxidase with reduced activity toward isoniazid. *J. Infect. Dis.* 176:722–727.

Wengenack, N. L., S. Todorovic, L. Yu, and F. Rusnak. 1998. Evidence for differential binding of isoniazid by *Mycobacterium tuberculosis* KatG and the isoniazid-resistant mutant KatG (S315T). *Biochemistry* 37:15825–15834.

Wengenack, N. L., M. P. Jensen, F. Rusnak, and M. K. Stern. 1999. *Mycobacterium tuberculosis* KatG is a peroxynitritase. *Biochem. Biophys. Res. Commun.* 256:485–487.

Wheeler, P. R., K. Bulmer, and C. Ratledge. 1990. Enzymes for biosynthesis de novo and elongation of fatty acids in mycobacteria grown in host cells: is *Mycobacterium leprae* competent in fatty acid biosynthesis? *J. Gen. Microbiol.* 136:211–217.

Wheeler, P. R., and P. M. Anderson. 1996. Determination of the primary target for isoniazid in mycobacterial mycolic acid biosynthesis with *Mycobacterium aurum* A+. *Biochem. J.* 318:451–457.

Wilming, M., and K. Johnsson. 1999. Spontaneous formation of the bioactive form of the tuberculosis drug isoniazid. *Angew. Chem. Int. Ed. Engl.* 38:2588–2590.

Winder, F. G., and P. B. Collins. 1970. Inhibition by isoniazid of synthesis of mycolic acids in *Mycobacterium tuberculosis*. *J. Gen. Microbiol.* 63:41–48.

Winder, F. G., P. J. Brennan, and D. Whelan. 1971. Investigations into mechanism of interference by isoniazid in mycolic acid synthesis. *J. Gen. Microbiol.* 68:R1.

Winder, F. G. A., P. Collins, and S. A. Rooney. 1970. Effects of isoniazid on mycolic acid synthesis in *Mycobacterium tuberculosis* and on its cell envelope. *Biochem. J.* 117:27p.

Wolucka, B. A., M. R. McNeil, E. de Hoffmann, T. Chojnacki, and P. J. Brennan. 1994. Recognition of the lipid intermediate for arabinogalactan/arabinomannan biosynthesis and its relation to the mode of action of ethambutol on mycobacteria. *J. Biol. Chem.* 269:23328–23335.

Wright, A., G. Bai, L. Barrera, F. Boulahbal, N. Martín-Casabona, C. Gilpin, F. Drobniewski, M. Havelková, R. Lepe, R. Lumb, B. Metchock, F. Portaels, M. Rodrigues, S. Rüsch-Gerdes, A. Van Deun, V. Vincent, V. Leimane, V. Riekstina, G. Skenders, T. Holtz, R. Pratt, K. Laserson, C. Wells, P. Cegielski, and N. S.

Shah. 2006. Emergence of *Mycobacterium tuberculosis* with extensive resistance to second-line drugs-worldwide, 2000-2004. *Morb. Mortal. Wkly. Rep.* 55:301–305.

Xin, Y., R. E. Lee, M. S. Scherman, K. H. Khoo, G. S. Besra, P. J. Brennan, and M. McNeil. 1997. Characterization of the in vitro synthesized arabinan of mycobacterial cell walls. *Biochim. Biophys. Acta* 1335:231–234.

Yew, W. W., S. Y. Kwan, W. K. Ma, M. A. Khin, and P. Y. Chau. 1990. In-vitro activity of ofloxacin against *Mycobacterium tuberculosis* and its clinical efficacy in multiply resistant pulmonary tuberculosis. *J. Antimicrob. Chemother.* 26:227–236.

Yu, S. W., S. Girotto, C. Lee, and R. S. Magliozzo. 2003. Reduced affinity for isoniazid in the S315T mutant of *Mycobacterium tuberculosis* KatG is a key factor in antibiotic resistance. *J. Biol. Chem.* 278:14769–14775.

Zhang, Y., B. Heym, B. Allen, D. Young, and S. Cole. 1992. The catalase-peroxidase gene and isoniazid resistance of *Mycobacterium tuberculosis*. *Nature* 358:591–593.

Zhang, Y., T. Garbe, and D. Young. 1993. Transformation with *katG* restores isoniazid-sensitivity in *Mycobacterium tuberculosis* isolates resistant to a range of drug concentrations. *Mol. Microbiol.* 8:521–524.

Zhang, Y., A. Scorpio, H. Nikaido, and Z. Sun. 1999. Role of acid pH and deficient efflux of pyrazinoic acid in unique susceptibility of *Mycobacterium tuberculosis* to pyrazinamide. *J. Bacteriol.* 181:2044–2049.

Zhang, Y., M. M. Wade, A. Scorpio, H. Zhang, and Z. Sun. 2003. Mode of action of pyrazinamide: disruption of *Mycobacterium tuberculosis* membrane transport and energetics by pyrazinoic acid. *J. Antimicrob. Chemother.* 52:790–795.

Zhao, X. B., H. Yu, S. W. Yu, F. Wang, J. C. Sacchettini, and R. S. Magliozzo. 2006. Hydrogen peroxide-mediated isoniazid activation catalyzed by *Mycobacterium tuberculosis* catalase-peroxidase (KatG) and its S315T mutant. *Biochemistry* 45:4131–4140.

Zimhony, O., J. S. Cox, J. T. Welch, C. Vilchèze, and W. R. Jacobs, Jr. 2000. Pyrazinamide inhibits the eukaryotic-like fatty acid synthetase I (FASI) of *Mycobacterium tuberculosis*. *Nat. Med.* 6:1043–1047.

Zimhony, O., C. Vilchèze, M. Arai, J. T. Welch, and W. R. Jacobs, Jr. 2007. Pyrazinoic acid and its n-propyl ester inhibit fatty acid synthase type I in replicating tubercle bacilli. *Antimicrob. Agents Chemother.* 51:752–754.

The Mycobacterial Cell Envelope
Edited by M. Daffé and J.-M. Reyrat
© 2008 ASM Press, Washington, DC

Chapter 8

Enigmatic Proteins from the Surface: the Erp, PE, and PPE Protein Families

Giovanni Delogu, Fabiana Bigi, Seyed E. Hasnain, and Angel Cataldi

THE ERP FAMILY OF EXPORTED PROTEINS

The Discovery of Erp Proteins

Erp is a cell wall-associated exported protein, also known as P36, encoded by *Mycobacterium tuberculosis* complex (MTBC) gene *pirG* (Rv3810) (Cole et al., 1998), that was identified in *Mycobacterium bovis* by Bigi et al. in 1995 (1995) after screening an *M. bovis* expression library with TB-infected bovine sera. Simultaneously, Berthet et al. (1995) at the Institut Pasteur of Paris identified Erp protein in a *phoA* fusion library of *M. tuberculosis*. However, the first member was later defined as Erp family was a 28-kDa protein from *M. leprae* described by Cherayil and Young (1998). Although the exact biochemical function of Erp has remained very elusive, conclusive experiments have demonstrated its crucial role in MTBC virulence.

Sequence and Biochemical Characteristics

Erp is a 27,669-Da protein that comprises 284 amino acids and possesses a classic signal sequence. Three different protein domains can be identified within Erp. The amino-terminal domain encompasses the region from amino acids 1 to 80. This domain includes the signal sequence for exportation composed by first four charged amino acids, followed by 14 nonpolar amino acids, and then a probable cleavage site for the signal peptidase. The central segment has 11 repeated peptides, whose consensus is formed by the aminoacidic residues PGLTS. Four pentapeptides match exactly the consensus, whereas the others are degenerated. Blast searches have shown no significant clues about the biochemical role of repeats. The

amino acid sequence of the carboxy-terminal domain shows a glycine-rich stretch, followed by a hydrophobic domain rich in proline and alanine.

Erp protein has been detected in culture supernatants or cell wall preparations but not in cell extracts. The association with the cell wall has been well demonstrated by the above-mentioned French group by means of immunoelectron microscopy (Berthet et al., 1995).

There is a strong divergence between the theoretical molecular weight of Erp, which is 28 kDa, and that observed in Western blots, because serum raised against Erp recombinant antigen reacts with a protein of 36 kDa in *M. bovis* supernatants. Also, expressed in *Escherichia coli*, Erp has a molecular mass of 38 kDa. The discrepancy among the molecular masses deduced from the sequence and that from the native protein in MTBC (36 kDa) could be attributed either to the high proline content (14.8%) or to posttranslational modifications that may alter the migration properties of the protein (Romain et al., 1993). Another intriguing aspect of the native Erp protein is that a double band is recognized by anti Erp antiserum in the cultured supernatants of MTBC complex bacteria (Bigi et al., 1995). Either the smaller polypeptide is a degradation or processing product of the larger one, or the two proteins represent different stages of a posttranslational modification, as it happens with MPB70 (Fifis et al., 1991).

Erp has been detected in protein samples of several mycobacterial species by Western blot. However, it has not been directly identified by Coomassie blue or silver staining techniques. These observations indicate either that Erp is produced in very low amounts or that the protein is resistant to the staining methods

---

Giovanni Delogu • Institute of Microbiology, Catholic University of the Sacred Heart, Largo A. Gemelli, 8, 00168 Rome, Italy.
Fabiana Bigi and Angel Cataldi • Institute of Biotechnology, National Institute of Agricultural Research, Los Reseros y Las Cabañas, 1712, Castelar, Argentina. Seyed E. Hasnain • Vice-Chancellor's Office, University of Hyderabad, Central University P.O., Hyderabad 500 046, India.

tested. This may also explain why the protein is not detected in the reference 2D gels database (http://www.mpiib-berlin.mpg.de/2D-PAGE/).

The overall content of GC in the *erp* (*pirG*) gene is 68%, slightly over that in the *M. tuberculosis* genome. The transcription start point (TSP) was located 93 base pairs upstream from the GTG start codon. Both a −10 promoter element, similar to other present in *E. coli* and *B. subtillis,* and a less typical −35 element have been associated with this TSP (Berthet et al., 1995). A typical ribosomal binding site motif (GAGG) occurs 12 base pairs upstream from the GTG start codon. Erp and its neighbor gene downstream, *csp*, do not seem to form an operon because 205 base pairs separate each gene from each other.

Ubiquity in Mycobacteria and Sequence Polymorphism

The Erp family is an expanding group of proteins originally identified and characterized from mycobacteria causing tuberculosis (TB) and leprosy. Subsequently, de Mendonça-Lima et al. (2003) demonstrated that genes homologous to the *erp* gene are also present in saprophytic, opportunistic pathogens and environmental mycobacteria species such as *Mycobacterium smegmatis, Mycobacterium avium, Mycobacterium marinum,* and *Mycobacterium xenopi,*

and in extracellular toxin-producing mycobacteria such as *Mycobacterium ulcerans.*

The protein alignment of the sequenced members of this family indicates that the main divergence is in the central domain of the proteins. As described earlier, this domain is composed of variable numbers of repeats conforming to the consensus sequence PGLTS (Fig. 1). The length of the central domain varies considerably among species: *M. leprae*, for example, has four repeats, whereas *M. smegmatis* has 21 repeats. On the other hand, the sequences of the repeats are variable, so that in some species, such as *M. xenopi* or *M. smegmatis*, half the repeats contain two mismatches. Some of the patterns of repeats are conserved among different species, for example, *M. marinum, M. ulcerans, M. shottsii* and *M. tuberculosis*. The sequence variation in the number of repeats has helped to distinguish *Mycobacterium pseudoshottsii* from *Mycobacterium shottsii* (Rhodes et al., 2005). The sequence similarity among amino-terminal part of the proteins, which includes the canonical signal sequence, is more than 70%, whereas the carboxyl-terminal of the proteins exhibits more than 50% identity among species. Interestingly, despite being a highly variable region among mycobacterial species, the central domain displays a perfect degree of conservation in five clinical isolates of *M. tuberculosis* (with different spoligotypes and geographical origins), because it

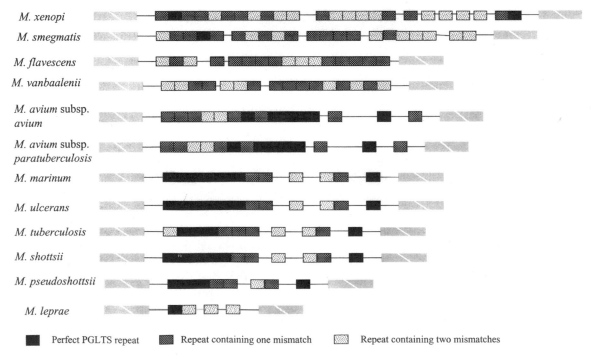

Figure 1. Schematic representation of the Erp orthologues identified in all mycobacterial genomes in the public domain by January, 2007. The conserved amino- and carboxyterminal domains are shown as gray bars. This figure is adapted from de Mendonça-Lima et al. (2001).

has been demonstrated by sequencing the genes, suggesting that the repeated region was not subject to allelic variation (de Mendonça-Lima et al., 2001). To date, the availability of new genome sequences has enabled new members of the family to be identified in other mycobacterial species, for example *M. avium* subsp *paratuberculosis*.

A striking feature of this family is that it has no orthologous sequences from outside *Mycobacterium* genus. Indeed, none of the three domains is significantly similar to any protein sequence in public database. These findings have allowed the definition of a new *Mycobacterium*-specific family of secreted proteins.

Genomic Localization of the *pirG* Gene

The genome sequence of *M. tuberculosis* shows that *pirG* lies between *glf* (Rv3809) and Rv3811. Glf is an UDP-galactopyranose mutase that is essential for bacterial growth and involved in the synthesis of galactofuran moiety of the cell wall (Pan et al., 2001). In turn, Csp encodes for a protein with similarity to peptidoglycan recognition protein from mammals (http://genolist.pasteur.fr/TuberculList/). Interestingly, the genetic context of the above-mentioned *erp* locus is conserved in gene content and order in all mycobacteria tested, suggesting a strong selective pressure that has maintained this locus unchanged from saprophytic to pathogenic mycobacteria. Remarkably, that genomic conservation is extensive to the close relative *Corynebacterium diphtheriae*, in which the *glf* and *csp* genes are present but in which the *erp* gene is replaced by the orthologues of *Rv0836c* and *Rv0837c*. However, in *M. leprae*, numerous point mutations in the *csp* gene probably lead to the absence of an active product, suggesting that, unlike *glp*, *csp* may not be required for a strictly intracellular lifestyle.

Role of Erp in Virulence

Genetic manipulation of pathogenic mycobacteria species has always been complicated because of the limitations to transform them efficiently and to obtain allelic replacement. As a result, for a long time it has been very difficult to create mutant strains for the analysis of their virulence properties. One of the first reports describing the generation of site-specific knockout mutant in virulence-associated genes from slow-growing mycobacterial species was published before delivering of *M. tuberculosis* H37Rv genome sequence, in 1998 (Berthet et al., 1998). In that work, Berthet et al. obtained *erp*-deficient mutants of *M. tuberculosis* and *M. bovis* BCG, and demonstrated that the disruption of *erp* negatively affects the multiplication of both strains in mammalian hosts. Therefore, the initial evidence supporting a role of the Erp protein in mycobacterial pathogenesis came from this study. By comparing the number of colony-forming units (CFU) of *erp*-deficient mutants and parental strains from infected cultured bone marrow-derived macrophages and mice, the authors found that the multiplication of mutant strains was significantly reduced. Reintroduction of the *erp* gene recovered the multiplication of *erp* mutants, revealing that *erp* may play a pathogenic role in *M. tuberculosis* and *M. bovis* BCG. Remarkably, although both *erp*-deficient mutant strains exhibited a similar degree of CFU reduction inside cultured macrophages compared with their respective parental strains, their kinetics of replication in mice was completely different. The BCG *erp* mutant was rapidly cleared from the lungs and spleen of infected animals, whereas the H37Rv *erp* mutant survived but multiplied very slowly in the lungs and the spleen as compared with the parental strains (Berthet et al., 1998). Consistent with those previous results, there are observations demonstrating that the disruption of the *M. bovis erp* gene impairs the growth of an *M. bovis* virulent strain, in vivo (Bigi et al., 2005). Unlike the *erp*-deficient BCG mutant strain, the *erp*-deficient *M. bovis* strain does not present a dramatic decrease in its bacterial load either in spleens or lungs from mice, indicating that the mutant strain is able to survive in mice in spite of its inability to multiply. Here again, suppression of attenuation is obtained after overexpression of Erp from a multicopy plasmid. The role of Erp in *M. tuberculosis* complex virulence is also supported by the findings of Sassetti and Rubin (2003), who, by using a high-density library of *M. tuberculosis* insertional mutants, have defined this gene as required for in vivo growth.

Based on the time course of *M. tuberculosis* murine lung infection, Glickman and Jacobs (2001) have divided attenuated mutant strains of *M. tuberculosis* in three types of theoretical phenotypes: (1) growth in vivo (giv) mutants, which are defective for replication in the initial stage of infection; (2) severe giv (sgiv) mutants, which do not replicate at all in vivo; and (3) persistence mutants (per), which replicate normally but decline in titer after the onset of host immunity. Because the above-mentioned features regarding growth and persistence in vivo of *erp*-deficient mutant strains match perfectly with those of giv mutants, Erp mutants should be included in this group.

The persistence without multiplication in mice observed in *erp*-deficient *M. tuberculosis* and *M. bovis* could be an important attribute to achieve the optimal balance between attenuation and immunogenicity in a live vaccine. Indeed, *erp*-deficient BCG induces a delayed-type hypersensitivity response in guinea pigs sensitized with mycobacterial purified protein derivative (PPD) comparable to that induced by the parental BCG, suggesting that the mutant strain persists long enough to stimulate antimycobacterial immunity (Berthet et al., 1998). Thus, these findings might indicate that *Erp*-mutant strains represent good candidates to be evaluated as a potential live vaccine against TB.

de Mendonça-Lima et al. (2001) have studied the role of central repeats of Erp in the virulence of *M. tuberculosis*. They observed that the *erp*-deficient *M. tuberculosis* strain could be fully complemented by the *erp* orthologues from *M. leprae* and *M. smegmatis* in terms of bacterial load in spleen and lungs from mice at the late stage of infection. However, the *M. smegmatis* allele has shown to complement less efficiently and the *M. leprae* allele more efficiently than the *M. tuberculosis* allele in the lungs at the early stages of infection. In addition, bacterial growth in early infection correlates directly with lung damage at later stages, in terms of granuloma and infiltration. Thus, the higher the bacterial load during early infection, the more severe the damage occurring later in the infection, even on day 80 after infection, when the bacterial loads of all complemented strains arc similar. From these data, it may be concluded that at early time-points of infection in lungs, the *erp* allele from nonpathogenic *M. smegmatis* is not able to restore the wild-type virulence of the *erp*-deficient *M. tuberculosis* strain, whereas the *erp* allele from pathogenic *M. leprae* induces a hypervirulent phenotype. Because both *erp* alleles are rather conserved in both amino and carboxyl terminal domains but differ considerably in the central domain, it is likely that initial bacterial growth and, in turn, the histological outcome of the infection are conditioned by the nature and number of repeats in Erp proteins.

The Elusive Biological Function of Erp

Although the precise roles of Erp proteins have remained evasive, the number of reports demonstrating that Erp is a crucial factor for survival and multiplication in vitro in cell culture assays is increasing. It has been very recently published that the growth of *erp*-deficient *M. marinum* is attenuated in cultured macrophage monolayers and during chronic granulomatous infection of leopard frogs, its natural host species, suggesting that Erp function is similarly required for the virulence of mycobacterial species other than the species belonging to the *M. tuberculosis* complex. In addition, microinjection of zebrafish embryos with the *erp*-deficient *M. marinum* strain

has resulted in low-burden infection and a paucity of macrophage aggregates. The attenuation characteristics, in terms of survival and bacterial counts, of *erp*-deficient *M. marinum* strain are indistinguishable from those of *M. marinum* strain carrying a deletion in RD1 locus. This RD1 region encodes for Esat-6 that is one of the most widely studied proteins because of its antigenicity and virulence properties. Thus, this last finding reinforces the role of Erp as virulent factor of pathogenic mycobacteria. Closer examination using fluorescence and differential interference contrast microscopy in infected zebrafish embryos revealed that macrophages of embryos infected with the ΔRD1 mutant typically harbor large numbers of bacteria, whereas those from embryos infected with the erp::aph mutant, frequently harbor few bacteria. Authors propose that *erp*-deficient bacteria are attenuated primarily because of reduced intracellular growth and/or survival, whereas the ΔRD1 mutant fails to induce macrophage aggregation and intercellular spread (Cosma et al., 2006).

More than 10 years after the discovery of the Erp protein, little information regarding its biochemical function has been discovered. Speculations about this function might be made based on the finding that Erp expression is strongly induced in *M. tuberculosis* upon nutrient starvation (Betts et al., 2002). However, these results are dispersed and more research is necessary to deal with this issue. Erp has no orthologues in other bacteria and a search in PROSITE or Pfam websites has not revealed any functional or family consensus. Consequently, the function of Erp may be related with exclusive properties of the bacillus such as those involved in the synthesis, degradation, stability, or assembly of a specific mycobacterial membrane or cell wall component. Differently, the role of Erp during in vivo infection and disease is clearer.

Immune Response against Erp

The Erp protein has not been generally included in high throughput assessments of mycobacterial antigens (Wilsher et al., 1999; Landowski et al., 2001; Aagaard et al., 2003; Ben et al., 2005; Kulshrestha et al., 2005; Weldingh et al., 2005), and only some individual evaluations of Erp antigenicity have been performed. This scarcity is striking in light of the fact that members of the Erp family, such as the 28-kDa protein from *M. leprae*, Erp from MTBC, and the 34-kDa antigen from *M. avium* subsp. *paratuberculosis*, have all been identified on screening expression libraries with antibodies. In addition to these preliminary evidences that underscore the relevance of Erp to induce humoral responses in the host, further studies have confirmed the humoral properties of the Erp antigen. Sera from cattle infected with *M. bovis* or from *M. avium* subsp. *paratuberculosis* recognize their respective Erp recombinant allele when tested in western blot or enzyme-linked immunosorbent assays (ELISAs) (Bigi et al., 1995; Weldingh et al., 2005). Furthermore, sera from animals with both clinical and subclinical paratuberculosis were reactive to Erp (Willemsen et al., 2006).

During human TB, the humoral recognition of Erp seems to happen at advanced TB stages. Although neither sera from healthy PPD-positive patients nor those from noncavitary TB patients detect the protein, sera from cavitary TB patients strongly recognize Erp (Singh et al., 2001). Several evidences suggest that the central PGLTS domain contains all the antigenic determinants of Erp. For instance, mouse sera absorbed with recombinant PGLTS do not recognize the full-length Erp. Also, rabbit sera raised against Erp do not recognize a truncated version of Erp lacking the central portion (unpublished results). Interestingly, sera from leprosy patients recognize recombinant Erp from *M. bovis* (Bigi et al., 1995), whereas Erp protein from *M. leprae* is not recognized by a serum raised against *M. tuberculosis* Erp (de Mendonça-Lima et al., 2001).

With regard to the cellular immune response, it has been preliminarily observed that bovine T cells from cattle experimentally infected with *M. bovis* frequently recognize Erp (M. Govaerts, personal communication).

Influence of Erp in Cell Morphology

The contribution of each Erp protein domain to mycobacterial pathogenicity and colony morphology has been addressed by Reyrat and collaborators. It has been demonstrated that the absence of Erp in both BCG and *M. smegmatis* strains transforms the flat and spreading wild-type colony phenotype toward a smaller and raised colony morphology (Fig. 2), when bacteria are plated on solid medium from liquid culture inoculums (Kocincova et al., 2004a) and animal organs (Berthet et al., 1998). Surprisingly, *erp* alleles from *M. tuberculosis*, *M. leprae*, and *M. smegmatis* strains complement identically the wild-type phenotype, suggesting that neither the nature nor the number of PGLTS repeats affect the colony morphology of *erp*-deficient *M. smegmatis* (Kocincova et al., 2004a). As well as what has already been mentioned, these findings clearly show that members of the Erp protein family share a similar function with regard to colony morphology, and might either directly or indirectly contribute to cell wall structure. Finally, the role of the conserved carboxyl-terminal domain of Erp in *M. tuberculosis* and *M. smegmatis* has been addressed in two reports. By complementing an *erp*-deficient *M. tuberculosis* strain with an Erp version lacking the most hydrophobic region of the carboxyl-terminal domain, it has been demonstrated that this domain of the Erp protein is not essential for either organ multiplication or persistence, or even induction of lung damage (Kocincova et al., 2004b). In addition, by using a similar experimental approach, the carboxyl terminal domain of Erp from *M. tuberculosis* has proved to be necessary for the complementation of the altered detergent resistance of the *erp*-deficient *M. smegmatis* (Kocincova et al., 2004a) but not for restoring wild-type colony morphology. Therefore, based on these results, it seems that loss of wild-type colony morphology correlates positively in some circumstances, with loss of residual virulence, but the characterization of the *erp*-deficient *M. tuberculosis* strain complemented with the carboxyl-terminal deleted Erp allele is essential to address this issue. However, it is important to note that despite the fact that different Erp orthologues suppress the abnormal colony morphology in *erp*-deficient *M. smegmatis*, their capacity to restore the virulence of *erp*-deficient *M. tuberculosis* is dissimilar.

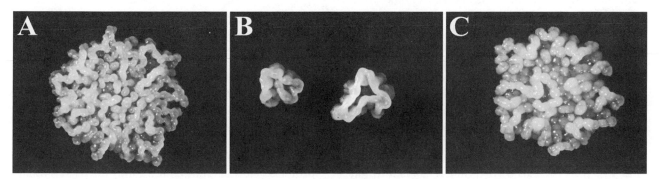

Figure 2. Photomicrographs of colonies of *M. smegmatis*. (A) *M. smegmatis* wild-type; (B) *M. smegmatis* erp::aph; (C) *M. smegmatis* erp::aph complemented with a wild-type gene of *M. tuberculosis*.

Although it has been conclusively demonstrated that the suppression of Erp does not affect the growth of *M. bovis*, *M. tuberculosis*, and *M. marinum* in liquid synthetic media, several factors indicate that in some particular conditions of culture, the loss of the *erp* gene can be of detriment to the growth of mycobacterial species, for example, (1) the growth of *erp*-deficient BCG, *M. bovis* and *M. marinum* on 7H10 agar is compromised because mutant colonies appear later than wild-type ones (Berthet et al., 1998; Bigi et al., 2005; Cosma et al., 2006); (2) when bacterial strains are cultured from frozen cell stocks, the *erp*-deficient *M. bovis* strain exhibits a reproducible increase in lag phase (20 days) as compared with the parental and the complemented strains (unpublished results).

The reports presented here converge to the conclusion that, probably, a major reason for the attenuation observed in the *erp*-deficient mutant strains is a difficulty to establish the initial infection due to impaired multiplication, as indicated by (1) reduced intracellular growth in individual macrophages infected with *erp*-deficient *M. marinum* (Cosma et al., 2006); (2) lower bacterial count in lungs of mice at the early stage of infection with a partially complemented (by the *M. smegmatis* allele) *erp*-deficient *M. tuberculosis* strain compared with bacterial count of fully complemented strain (de Mendonça-Lima et al., 2001); and (3), as discussed above, delayed in vitro growth, observed when *erp*-deficient *M. tuberculosis*, *M. bovis*, BCG, and *M. marinum* strains are cultured in solid 7H10 media from frozen stocks or animal organs (Berthet et al., 1998; Bigi et al., 2005; Cosma et al., 2006).

Interactomics of Erp

Two studies have provided indirect evidence demonstrating the role of Erp proteins in maintaining integrity of the cell wall and outer lipid layer. The recent work of Cosma et al. (2006) shows that Erp-deficient *M. marinum* is more susceptible to the lipophilic antibiotics rifampin, erythromycin, and chloramphenicol than the wild-type and complemented strains, and they postulated that this is due to cell wall defects in the mutant strain. Consistent with these findings is the fact that the loss of Erp, in particular the carboxyl-terminal domain of the protein (as mentioned above), renders the bacterial more susceptible to sodium dodecyl sulphate (SDS), probably as a consequence of that permeability defect. Additionally, Kocíncova et al. (2004b) has proposed that Erp uses the carboxyl-terminal domain to interact with some other molecules of the cell wall to achieve its correct structure. In this regard, by screening a *M. tuberculosis* DNA library with full-length

Erp using a bacterial two-hybrid system (Karimova et al., 1998), it has been found that Erp interacts with the proteins encoded in Rv1417 and Rv2617c, a possible conserved membrane protein and a probable transmembrane protein, respectively. Furthermore, all three proteins have been revealed to interact together and self-interaction has also been detected in Rv1417 and Rv2617c. However, it has been found that the carboxyl-terminal domain of Erp does not have a role in those interactions and the region of interaction has been mapped in the amino-terminal and central repeated domain of Erp (L. Klepp, M. P. Santangelo, F. Blanco, A. Cataldi, and F. Bigi, unpublished data). If these data are taken together, it is possible to speculate that Erp proteins, which have extracellular localization, may have an important role in cell wall structure by interaction with membrane-associated proteins. Although this might be attractive, it is almost certainly too simplistic, because Erp proteins are found mostly secreted rather than cell wall associated (Bigi et al., 1995). One possible explanation formulated by Kocíncova et al. (2004a) is that Erp proteins are probably released from the surface by Tween treatment, rather than actively exported into the supernatant. However, small vesicles near phagosomes have revealed that they contain free Erp protein after infection of a murine macrophage cell line with wild-type *M. tuberculosis* (Berthet et al., 1995). More evidence is necessary to clarify this controversial point.

The availability of *M. tuberculosis* genome sequence and its comparison with other mycobacterial genomes has accelerated the study of pathogenesis as never before and has raised a number of questions concerning the roles and functions of a large group of putative unknown proteins, in which Erp is included. Therefore, the application of postgenomic tools in the near future will eventually contribute to our understanding in this relevant area of mycobacterial investigation.

THE PE AND PPE FAMILIES OF PROTEINS

Discovery of the PE and PPE Families

The genome of *M. tuberculosis* H37Rv is made up of 4,411,529 nucleotides and encodes about 3,986 proteins (Cole et al., 1998). Subsequently, when the *M. tuberculosis* CDC1551 strain was sequenced, the genome size was found to be more or less similar (4,403,836 nucleotides) encoding 4,187 proteins (Fleischmann et al., 2002) (http://www.nabi.nlm.nih.gov/cgi-bin/Entrez). One of the major surprises of the genome sequence was the presence of two large unrelated families of proteins with no known functions. These protein families that make up about 10% of the coding capacity were referred to as PE/PPE family of genes and are primarily glycine and alanine rich. The name PE is derived from the signature motif Pro-Glu, whereas PPE represents the Pro-Pro-Glu motif. A general overview of these families is shown in Fig. 3. The 110 or approximately 180 amino acid residue-long N-terminal end of PE or PPE, respectively, is conserved, whereas the C terminal end is highly variable and thought to confer antigenic variation among different strains of *M. tuberculosis* (Cole et al., 1998). The existence of these families was reported even before the genome was sequenced (Poulet and Cole, 1995; Abdallah et al., 2006). These families of proteins have now been shown to be present in other mycobacteria as well. The PE family has approximately 100 members, whereas PPE includes 68 members. Based on the domain structure, the *M. tuberculosis* H37Rv PE family has been subdivided into three main classes. In one class are grouped proteins that contain just the PE domain (<110 aa), and in H37Rv have been detected 20 of these genes (Fig. 4). To a second class, which comprises 18 members, belong proteins that carry the PE domain followed by a C-terminal comprising unique sequences. The largest class, with 63 members, carries the PE domain followed by multiple tandem repeats of Gly-Gly-Ala or Gly-Gly-Asn. This PE subfamily is referred to as polymorphic GC-rich repetitive sequences (PE_PGRS) and has been discussed separately in this chapter. Based on the sequence analysis, the PPE protein family has also been classified into three, or more recently four, categories (Gey van Pittius et al., 2006). The first group of 12 PPEs contains proteins, with unrelated sequences downstream from the PPE domain. The second group

is referred as PPE-PPW; it shows the consensus motif PxxPxxW and includes 10 members. Proteins with major polymorphic tandem repeats (PPE-MPTR), represented by the amino acid sequence motif NxGWxG, is the major subfamily with 23 proteins. The third subfamily has the conserved GxxSVPxxW motif present around position 350, is referred as PPE-SVP, and comprises 24 proteins in *M. tuberculosis* H37Rv.

Amino acid sequence analysis of PPE proteins of *M. tuberculosis* H37Rv as well as CDC1551 strains revealed a previously uncharacterized 225 amino acid-residue common region in 22 proteins (Adindla and Guruprasad, 2004). At position 15, there was a conservation of amino acid residues and the secondary structure corresponding to this region was predicted to be a mixture of alpha helices and β-strands. Another group of 20 PPE proteins had GFxGT and PxxPxxW motifs within the 44 amino acid residues at the conserved C-terminal region. Analysis of the C-terminal variable regions of the PE proteins similarly identified the presence of 41 to 43 amino acid-long tandem repeats in two PE proteins, namely Rv0978 and Rv0980. These repeats correspond to AB repeats, which were first identified in some proteins of the *Methanosarcina mazei* genome and were demonstrated as surface antigens (Mayerhofer et al., 1995). Therefore, it has been suggested that such AB repeat-carrying PE proteins of *M. tuberculosis* may represent surface antigens (Adindla and Guruprasad, 2003).

Intraspecies Polymorphism

Very few reports have pointed to the functional importance of these proteins, leading to speculations about their role in generating antigenic variation, virulence, vaccine antigens, or as diagnostic tools

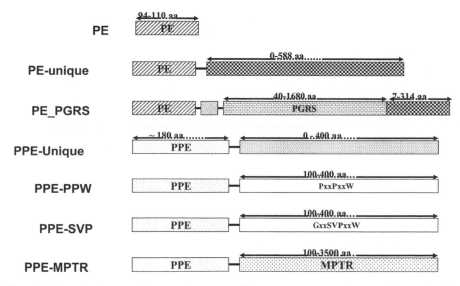

Figure 3. Schematic showing the general structure of the PE, PE_PGRS, and PPE multigene families.

Mycobacterial Genome

Gene Family and subfamily		M. tuberculosis H37Rv		M. ulcerans		M. leprae TN		M. avium paratuberculosis K10		M. smegmatis mc2155	
Size (Kb)		4411		5631		3268		4829		6988	
Genes		3993		4281		1614		4350		6776	
PE		37	20	21 (6)[1]	9	14 (5)[1]	5	10 (1)[1]	10 (1)[1]	2 (1)[1]	2
PE-Unique			17	pseud[2] 6	6	pseud 9	0		0		
PE_PGRS		61 pseud 9	51	121 (104)[1] pseud 66	43-54	7 (4)[1] pseud 7	0	0	0	0	0
PPE- Unique		69	12	81 (55)[1] pseud 34	6	27 (20)[1] pseud 19	2	37 (18)[1]	1	2	
PPE-PPW			10		8		3		3		2
PPE-SVP			24		22		2		13		
PPE- MPTR			23		11		0		1		

[1] Numbers in parentheses indicate species-specific genes.

[2] pseud: indicates the number of pseudogenes of that given category

Figure 4. Number of genes belonging to the PE and PPE multigene families found in different mycobacterial genomes.

(Chakhaiyar et al., 2004). Their possible localization to the mycobacteria envelope and the consequent functional significance has resulted in the PE multigene family being compared with a molecular mantra for mycobacteria (Brennan and Delogu, 2002). Nucleotide sequence polymorphism vis-à-vis genome evolution and plasticity has encouraged prediction of genomic microsatellite locations, which could be affected by frameshift mutations. Many such regions also include the PPE, PE_PGRS family of proteins (Sreenu et al., 2006). Comparison of clinical isolates with reference strains indeed showed evidence of such polymorphism (Chakhaiyar et al., 2004; Talarico et al., 2005). The PE and PPE family of proteins have been shown to be differentially expressed in *M. tuberculosis* during in vitro growth (Flores and Espitia, 2003) and, at the same time, exhibit gene expression diversity in different clinical isolates (Gao et al., 2005). Whole-genome comparison of *M. tuberculosis* clinical and laboratory strains revealed that polymorphism among the clinical strains within these regions is more extensive than was initially thought, and furthermore, such genetic variations may play an important role in disease pathogenesis and immunity (Fleischmann et al., 2002).

Genomic Organization of the PE/PPE Family

The genomic organization of the PE/PPE family of genes within the *M. tuberculosis* genome is not random but follows a distinct pattern such that most PE members are upstream of PPE members (Tundup et al., 2006). Of 41 such operonic arrangements, 28 operons contained only the PE/PPE genes, whereas the remaining were clusters of either PE or PPE with non-PE/PPE members such as Rv0915c, Rv3872, ESAT-6 like secretory proteins, and so on. This definite pattern of operonic arrangement in which PE is separated from PPE or non-PE/PPE genes by less than 90 bp enables the two genes to be transcribed as a single mRNA which is then translated independently to give rise to two different gene products. In heterologous *E. coli* system, the two proteins when expressed alone go into inclusion bodies but when coexpressed interact with each other and solubilize each other. A classic example of such an operonic organization is Rv2430c encoding an immunodominant antigen (Choudhary et al., 2003), earlier identified as one of the genes overexpressed in microarray analysis in an IdeR mutant (Rodriguez et al., 2002), and PE 2431c. This interaction leads to heterooligomerization of the PPE Rv2430c as well as the PE Rv2431c proteins. This was evident from protein cross-linking as well as immunoprecipitation and Western blot analysis. The coexpressed recombinant proteins were found to be highly soluble in vitro at room temperature compared with individually expressed proteins pointing to the importance of this interaction. Although both these proteins have been shown to elicit immunodominant B-cell response (Choudhary et al., 2003), it has been speculated that such cooperativity between in vivo coexpressed

PE/PPE family of proteins is a prelude to likely immune sensing role within the cell for these proteins (Tundup et al., 2006).

Such evidence of in vitro cooperativity was elegantly supported by structural data of a complex of Rv2430c and Rv2431c (Tundup et al., 2006). When present as a complex, the PE-PPE proteins mate along an extended apolar interface to form a four alpha helical bundle. For this structural arrangement, the two alpha helices are contributed each by the two proteins, PE and PPE (Fig. 5). The structural evidence also points to the possibility that some of the other PE/PPE complexes may be involved in signal transduction, either as membrane tethered proteins or as soluble proteins.

Role of PE and PPE in Virulence

Macrophages, which are the first line of defense against *M. tuberculosis*, harbor a large repertoire of immunological "artillery," which includes the interleukins, tumor necrosis factor alpha (TNF-α), interferon gamma (IFN-γ), reactive nitrogen intermediates, reactive oxygen species, and so on (Schluger and Rom, 1998). Despite the powerful host defense mechanisms *M. tuberculosis* is able to mount a successful infection or stay dormant in the host by adopting various strategies (Ellner, 1997; Keane et

al., 2000; Noss et al., 2000). It is not clear whether the PE/PPE proteins are directly involved in promoting intracellular survival and infection by perturbing macrophage-immune function. Being facultative intracellular pathogens, they could likely adopt a strategy to selectively express specific genes in macrophages to facilitate replication and persistence in vivo. This line of reasoning is supported by the observation that some of the PE_PGRS proteins are selectively induced inside macrophages to facilitate intracellular survival of the bacilli (Triccas et al., 1999). The amino acid composition predicts the PE/PPE proteins to be surface localized (Cole et al., 1998), thereby pointing to the likelihood of the bacilli interacting directly with the macrophages through these proteins. In addition to PE/PPE proteins, many members of the PE_PGRS family (discussed elsewhere in this chapter) have been reported to be involved in virulence. Inactivation of *PPE46 (Rv3018c)* by signature tagged mutagenesis resulted in the attenuation of *M. tuberculosis* in an animal model (Camacho et al., 1999). Transposon-mediated interruption of an *M. avium* PPE gene (which is 52% homologous to *M. tuberculosis* Rv1887) showed that the mutant mycobacterium did not prevent phagosome-lysosome fusion and was comparable to the wild type in terms of macrophage entry but not growth in 7H9 broth. Challenge of mice with this mutant resulted in severely reduced

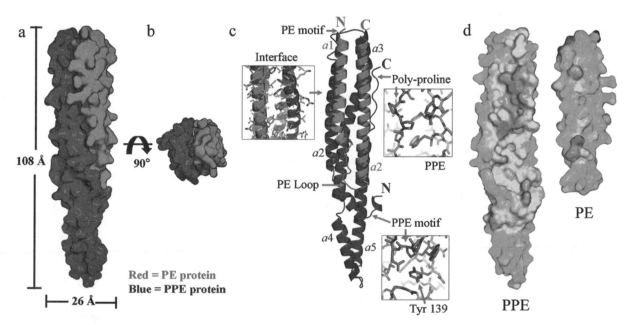

Figure 5. Crystal structure of the *M.tb.* PE/PPE protein complex. (A) Surface representation of the PE/PPE protein complex. (B) The PE/PPE protein complex viewed down its longitudinal axis. (C) Ribbon diagram of the PE/PPE protein complex. The complex is composed of seven helices. Two helices of the PE protein interact with two helices of the PPE protein to form a four-helix bundle. Regions of high sequence conservation are indicated by arrows and discussed in the text. (D) Interface hydrophobicity of the PPE and PE proteins. Notice the extensive apolar regions that are shielded from solvent as the complex forms. Adapted from Strong et al. (2006). *(See the color insert for the color version of this figure.)*

bacterial count in the spleen pointing to its likely role in virulence (Li et al., 2005).

The attenuation of *M. bovis* BCG vaccine strain is a result of serial passage leading to deletion of certain regions in the ancestral *M. bovis* genome. The most important of these is the RD1 region, whose importance in virulence has been established experimentally though the mechanism is unknown. RD1 region contains the *ESAT-6* and *CFP-10* genes along with *PE (Rv3872)* and *PPE (Rv3873)* genes, which work together to form a single virulence determinant (Guinn et al., 2005). Other RD regions, which are deleted among BCG strains, also include *PE/PPE* genes (Gordon et al., 1999). Various PE/PPE family genes such as *Rv0279, Rv0285*, and *Rv2123* are controlled by Iron dependent Regulator (IdeR), which is a regulator of various virulence factors and iron acquisition systems. The fact that iron acquisition is necessary for virulence in the host (Rodriguez et al., 2002) further implicates *PE/PPE* genes in virulence.

Additional studies further support the view that PE/PPE proteins may play an important role in assisting the mycobacteria in better survival. Using microarray and proteome analysis, the response of *M. tuberculosis* during nutrient starvation was investigated (Betts et al., 2002). Upregulation in the induction of four *PPE* genes was apparent at 24-hour time point. These genes were found to be important for maintaining long-term survival of the bacteria during nutrient starvation condition. In another study (Dillon et al., 1999), the PPE protein Rv1206 was found to be expressed during infection, and mice immunized with Rv1196 DNA showed increased protection from *M tb* challenge, as indicated by a reduction of bacterial load compared with the nonimmunized mice. A very recent report by Abdallah et al. (2006) elegantly demonstrated that PPE41 in *Mycobacterium marinum* is secreted through a process involving the ESX-5 locus, which is present in various pathogenic mycobacteria but not in saprophytic species such as *Mycobacterium smegmatis*. The authors were able to show that the secretion of PPE41 affects spreading of the mycobacteria to uninfected macrophages (Abdallah et al., 2006). These findings clearly suggest that the PE/PPE proteins could act as virulence factors, helping the bacteria to establish infection inside the host. However, the molecular mechanisms by which the PE/PPE proteins support bacterial survival inside the host is not known. It is likely that the bacilli abrogate macrophage-immune function by directly or indirectly regulating macrophage-signaling cascades. That *M. tuberculosis* makes an effort to shift the Th1 response to Th2 kind (unpublished) is also supported by the observation that as compared with the healthy control group, the anti-PPD T-cell response in infected patients is biased toward the Th2 type (Baliko et al., 1998; Vekemans et al., 2004). Therefore, a decision of protection versus survival is very much dependent on the production of IL-12 and IL-10 cytokines from the macrophages.

Antigenicity of PE/PPE Proteins

When the complete genome sequence of *M. tuberculosis* was described, the PE/PPE genes, the nucleotide sequence of which resembled the hot spots for bacterial recombinations, were predicted to be responsible for creating antigenic variability (Brosch et al., 1998). A recombinant member of the PE_PGRS family (Rv1859c) was shown to recognize patient serum samples. This protein was earlier known to have fibronectin binding activity (Espitia et al., 1999). Another surface localized PPE protein (Rv1918c) was considered as a possible diagnostic target (Sampson et al., 2001).

In silico-based analysis has been used to predict the antigenicity of *M. tuberculosis* putative antigens (Mayerhofer et al., 1995; Choudhary et al., 2003; Chakhaiyar and Hasnain, 2004; Mustafa, 2005). Using a panel of human sera representing healthy controls, patients reporting with fresh infections with mycobacterium TB (category 1), patients with relapsed TB (category 2), and extra pulmonary TB (category 3), the recombinant protein corresponding to a hypothetical PPE ORF Rv2430c was used in an ELISA-based assay. Well known antigens such as HSP10 as well as the (PPDs) were used as control for comparison of the B-cell response. This study (Choudhary et al., 2003) revealed that a high percentage of patients belonging to category 1 showed stronger antibody response against this protein when compared with anti-immunoglobulin M (IgM) or anti-immunoglobin G (IgG) response. This study demonstrated that the PPE protein Rv2430c, which is likely cytosolic or secretory in localization, has apparent diagnostic potential, along with possible use in a cocktail vaccination with other immunodominant antigens (Dhar et al., 2000). A biophysical structural feature analyses of PPE Rv2430c suggested a similar role for other members of this family (Ghosh et al., 2004). In another study that was similar in silico approach was used to identify immunodominant antigens from the PPE major polymorphic tandem repeat (MPTR) subfamily (Chakhaiyar et al., 2004). The recombinant protein corresponding to another hypothetical ORF Rv2608, a member of the PPE MPTR protein subfamily, was used to study the humoral as well as cellular immune response. Once again, well characterized sera from patients belonging to three different categories, as described earlier, were used. About 16%

of the clinical isolates showed a deviation of the PCR profile of Rv2608 from *M. tuberculosis* H37Rv, which provided support to the hypothesis that these genes are a likely source of antigenic variability. Rv2608, which has a conserved N-terminal regional and a "C"-terminal region with MPTRs of Gly-X-Gly-Asn-X-Gly repeat motifs, was used in ELISA to screen human sera. In parallel, synthetic peptides corresponding to regions of antigenicity were used to map the antigenic potential of the repeat motifs per se. Results indicated that category 2 patients showed evidence of strong humoral immune response when synthetic peptides were used, and this was also more or less similar to what was obtained when the total protein was used. When T-cell proliferation assays were performed to evaluate the response to different synthetic peptides, there was no evidence of their ability to distinguish patients from various categories. The high antibody response to the peptides and a low T-cell response therefore easily explains the relapse of infection in category 2 patients (relapse TB patients). It was argued that in vivo, the responsive T cells are not able to expand because of the classic Gly-x-Gly-Asn-x-Gly repeat motif, which somehow prevents antigen processing in a manner somewhat analogous to Epstein-Barr virus nuclear antigen, in which the Gly-Ala repeat similarly prevents antigen processing (Levitskaya et al., 1995). Complementing this line of reasoning was the evidence that regions within the repeat motifs that had fewer Gly-Asn repeats showed a comparatively higher T-cell response. In both of these studies (Choudhary et al., 2003; Chakhaiyar et al., 2004), it was very clear that PE/PPE family protein members mounted an equal or higher response than the well known secretory protein HSP10.

The Elusive Biological Function of PE_PPE Proteins

The possible function of the PE/PPE protein family also became apparent. It was hypothesized that these antigens play a role in evading the host immune response, thereby preventing the establishment of an effective cellular response that is required to contain the disease. Although this study (Chakhaiyar et al., 2004) demonstrated for the first time in a clinical setting, the ability of repeat sequences to elicit a high humoral immune response and a low T-cell response it was suggested that other members of the same PPE-MPTR family would possibly display a similar differential response.

A critical step in host-pathogen interaction in the course of bacterial infection is the entry of the pathogen into the host cells and tissues. Many *M. tuberculosis* genes, such as *mce, eis,* and *erp,* which encode proteins involved in enhancing mycobacterial entry have been identified in *M. tuberculosis* (Arruda et al., 1993; Berthet et al., 1998; Lefevre et al., 2000). Several PE/PPE proteins have putative transmembrane domains, which are an indication of their membrane localization. PE/PPE proteins are also reported as cell surface components and are involved in cell entry. It has been reported that individual RD1-region genes are required for export of ESAT-6/CFP-10 and for virulence of *M. tuberculosis*. The RD1 region also includes a pair of *PE (Rv3871)* and *PPE (Rv3872)* genes, and it has been argued that these RD1 region genes are possibly a part of a transport secretion system for ESAT-6 and CFP-10 proteins, which are secretory antigens (Guinn et al., 2004). Being part of this probable secretory apparatus, this PE/PPE protein pair may have cell wall localization.

THE PE_PGRS MULTIGENE SUBFAMILY

Main Features of the PE_PGRS Multigene Subfamily

The PE_PGRS proteins belong to a subgroup of the PE protein family that have the distinct feature of containing a highly repetitive domain rich in the amino acids glycine and alanine (Cole et al., 1998; Brennan and Delogu, 2002). The term PGRS comes from earlier epidemiological studies that used these repetitive DNA sequences in DNA fingerprinting (Poulet and Cole, 1995; van Soolingen et al., 1993). All the PE_PGRS proteins share a common molecular architecture (Fig. 6), with the PGRS domain being linked to the highly conserved PE domain (approximately 94 aa) that constitutes the N-terminal portion of the protein. The PE and PGRS domains are linked by a region of 35 to 40 aa containing a conserved GRPLIGNG motif that is found in all but one (PE_PGRS39) PE_PGRS protein (AC, anchoring domain). It has been proposed that this domain forms a putative transmembrane helix involved in the cellular localization of PE_PGRS proteins (Brennan and Delogu, 2002; Strong and Goulding, 2006). Alternatively, this domain may serve to anchor PE_PGRS proteins to a specific, yet unidentified, portion of the mycobacterial cell wall. The PGRS domain contains multiple tandem repetitions of Gly-Gly-Ala and Gly-Gly-Asn motifs and may vary in number from a few (tens) to several hundred amino acids (Cole et al., 2001; Brennan and Delogu, 2002). Downstream from the PGRS region is usually found a sequence that is not conserved between the different proteins and that may vary in size.

In the PGRS domain, the Gly-Gly-X repeats are intercalated by short sequences of diverse sequence

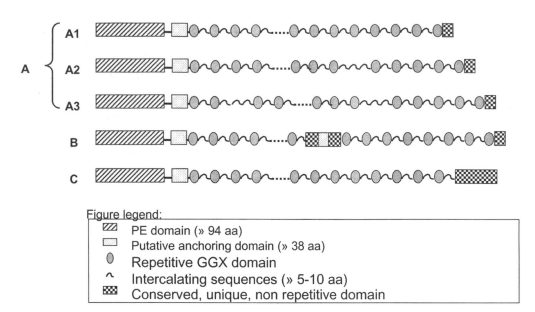

Figure legend:

▨	PE domain (» 94 aa)
▢	Putative anchoring domain (» 38 aa)
◍	Repetitive GGX domain
⌃	Intercalating sequences (» 5-10 aa)
▩	Conserved, unique, non repetitive domain

Figure 6. Schematic showing the molecular organization of the different PE_PGRS subgroups.

composition and size. In most proteins, these intercalating sequences are systematically short in size (4 to 10 aa) throughout the domain, and only the sequence found at the C-terminus may extend up to 15 to 27 aa (group A1). In several PE_PGRS proteins, one or more of these intercalating sequences may extend up to 20 aa. In some proteins only one or two of these longer sequences are found (group A2), whereas other proteins contain a higher number of these variable intercalating sequences (group A3). A group of four PE_PGRS proteins present within the PGRS region, a domain varying in size from 25 to 59 aa and containing the conserved sequence GRPLIGNG, that we refer here as AC domain (group B). It is possible that this sequence provides an additional anchor to the mycobacterial cell wall or cell membrane or mediates the localization of these PE_PGRS proteins in a specific micro compartment. In some of the PE_PGRS proteins, a domain that can be as large as 300 aa is found downstream from the PGRS region (group C). It is difficult at this point to grasp the meaning of this peculiar molecular organization, because the structural and functional role of the different domains (PGRS, intercalating sequences, and C-terminal) are ignored. For these reasons, the assignment to these groups is arbitrary and serves only to point out that the PGRS domain is not simply a repetitive and redundant domain but rather contains variable sequences interspersed among conserved sequence patterns that makes the PGRS domain unique for each PE_PGRS protein.

In the genome of *M. tuberculosis* H37Rv 63 PE_PGRS, genes have been identified (http://pasteur.fr/TubercuList) (Brennan et al., 2004), but a number of them have frameshift mutations that prevent the synthesis of a complete gene product (PE_PGRS6, *-12, -13, -36, -37, -49, -56, -60, -61*). Two genes annotated as PE_PGRS (PE_PGRS62 and *-63*) lack both the typical PGRS domain and the conserved region that links the PE and PGRS domains, and for these reasons, they should not be considered true PE_PGRS genes. Hence, in *M. tuberculosis* H37Rv, there are 51 potentially functional PE_PGRS genes (Cole et al., 1998; Fleischmann et al., 2002; Garnier et al., 2003). These genes are scattered throughout the genome, appear to be transcribed mostly as single ORFs, and only three PE_PGRS genes are separated by less than 90 bp by another gene and may be part of multigene operons (PE_PGRS9, *-10, -34* and *-48*). This genomic organization is different from what is observed for the PE_PPE gene family, and suggests an autonomous and independent regulation of gene expression for most of the PE_PGRS proteins.

Cellular Localization of PE_PGRS Proteins

The possibility that the PE_PGRS proteins might be available at the mycobacterial surface was proposed by Cole et al. (1998). Following the observation that a PE_PGRS protein (Rv0749c, wag22) (Abou-Zeid et al., 1991) shows fibronectin-binding properties, Espitia et al. (1999) suggested the possibility that these proteins may localize in the mycobacterial surface and interact with host-components. In agreement with these findings, it was later shown that PE_PGRS33 localized at the cell surface of *M. bovis* BCG, where it influences interactions with other bacterial cells and host cells (Brennan et al., 2001). Further experimental evidences were provided by Banu et al. (2002), using immunoelectron microscopy studies and subcellular fractionation experiments, to show that PE_PGRS proteins localize in the cell wall and cell membrane of *M. tuberculosis*. Using a trypsin sensitivity assay, it has been possible to demonstrate that PE_PGRS33 is surface exposed and available for the interaction with external components (Delogu et al., 2004). Also, overexpression of PE_PGRS33 in *M. tuberculosis* influenced colony morphology and bacterial cell structure, implicating the localization of the protein in the mycobacterial cell wall (Delogu et al., 2004). Much of the published evidence has been obtained using the PE_PGRS33 as a model, and it remains to be seen whether other PE_PGRS proteins show a similar cellular localization and impact on cell morphology.

The highly conserved PE domain may provide the PE_PGRS proteins a common structural and functional feature. A first attempt to dissect the role of the PE and PGRS domains was made using a set of in frame derivatives and GFP-tagged proteins of the Rv1818c gene (PE_PGRS33) (Delogu et al., 2004). It was shown that both the PE_PGRS-GFP and the PE-GFP fusion proteins localize in the mycobacterial cell wall, mostly at the bacterial cell poles. The results obtained led the authors to conclude that the PE domain is necessary for the subcellular localization of PE_PGRS33 and the PE domain may contain the information sufficient to transport or translocate the PE_PGRS proteins to the mycobacterial cell wall (Delogu et al., 2004). However, because the PE-GFP recombinant strains used in this experiment lacked the PGRS domain but contained the conserved AC domain, it remains to be seen whether the PE domain lacking the AC shows a similar pattern. Given the fact that the PE plus AC domain are highly conserved in all PE_PGRS proteins, it is likely that all members of this family share the same cellular localization in the mycobacterial cell wall. This hypothesis on the role of the PE domain has recently found solid experimental support in studies that addressed the association between PE and PPE genes that are discussed elsewhere in this chapter (Tundup et al., 2006).

Expression Patterns of PE_PGRS Proteins

The first evidence that some PE_PGRS genes are actually expressed by *M. tuberculosis* came from studies that demonstrated the presence of antibodies against PE_PGRS in sera of TB patients (Espitia et al., 1999; Talarico et al., 2005) or in mice experimentally infected with *M. tuberculosis* (Delogu and Brennan, 2001). Expression of some PE_PGRS genes in vitro was also demonstrated by immunoblot

analysis of crude *M. tuberculosis* cell lysates (Brennan et al., 2001; Banu et al., 2002). The interpretation of these results was complicated by the low specificity of the sera used that targeted the PGRS domain and could detect more than one protein. Unfortunately, the lack of specific and reliable immunological tools remains a major obstacle to the study of the PE_PGRS protein family.

Detection of PE_PGRS transcripts by RT-PCR both in in vitro and in vivo models provided a better understanding of the expression profile of these genes. In a study by Banu et al. (2002), 10 PE_PGRS genes selected were shown to be expressed by *M. tuberculosis* in culture, and in another report by Dheenadhayalan et al. (2006b), evaluation of the expression of 16 PE_PGRS genes was carried out under various growth conditions. The data that have been accumulating in the last few years, especially expression analysis data obtained by microarray analysis, are confirming the complexity of PE_PGRS gene expression regulation (Voskuil et al., 2004). For instance, PE_PGRS4 has been shown to be iron induced, whereas PE_PGRS51 is iron repressed (Rodriguez et al., 2002) and PE_PGRS15 and PE_PGRS51 are upregulated under nutrient starvation conditions (Betts et al., 2002). More recent data have demonstrated that expression of some PE_PGRS genes is controlled by the SigL ECF sigma factor (Hahn et al., 2005) and others by the PhoPR two-component system (Walters et al., 2006) that control expression of membrane-associated proteins or proteins involved in the pathogenesis of TB.

Quantitative analysis of PE_PGRS transcripts carried out in in vitro and in in vivo models of *M. tuberculosis* infection indicate that the level of expression may vary from gene to gene and that the relative amount of transcripts are in some cases (for example PE_PGRS33) similar to what has been observed for genes known to play an important role in the biology of *M. tuberculosis*, like *fbp*ABC, and *sod*A (Shi et al., 2003; Delogu et al., 2006). It is remarkable that none of the PE_PGRS proteins have, so far, been identified in *M. tuberculosis* by proteomic techniques (Jungblut et al., 1999; Pleissner et al., 2004), because the level of expression should warrant detection (Gu et al., 2003; Mawuenyega et al., 2005).

The quantitative analysis of transcripts is very useful to assess the regulation of gene expression during in vivo growth and persistence. Results obtained using the mouse model of infection have shown that *M. tuberculosis* constitutively expresses PE_PGRS33 in the lung and in the spleen tissues. Upregulation of the PE_PGRS9, PE_PGRS16, and PE_PGRS30 genes was observed in *M. tuberculosis* infecting macrophages and in host tissues during the chronic steps of infection, and upregulation of these three PE_PGRS genes was higher in the spleen compared with the lung (Dheenadhayalan et al., 2006b). Genes that are upregulated during the chronic steps of infection may play a role in protecting *M. tuberculosis* from host insults (Cunningham and Spreadbury, 1998; Timm et al., 2003), and the spleen tissue, which contains a high number of activated macrophages, may represent a harsher environment compared with that of the lung tissue. The current body of evidence suggests that PE_PGRS genes are differentially expressed by *M. tuberculosis* in host tissues, and regulation of these genes is finely tuned by the bacilli to sense different environmental signals encountered during the complex steps of the infectious process.

Immune Responses to PE_PGRS Proteins

It has been proposed that PE_PGRS proteins may be actively involved in the host-pathogen interaction process in different ways (Flynn and Chan, 2001). The strong homology found between the PGRS domain and the Epstein-Barr virus nuclear antigen (EBNA1) has suggested an active role for the PE_PGRS in helping *M. tuberculosis* evade the host immune response by inhibiting the class I antigen processing and presentation pathway (Cole et al., 1998; Brennan and Delogu, 2002). Recently, a study by Dheenadhayalan et al. (2006a) has shown that overexpression of PE_PGRS33 in *M. smegmatis* enhances the virulent properties of this nonpathogenic mycobacteria, both in vitro and in vivo, through a process that involves an increased secretion of TNF and IL-10, and a reduction of NO production by infected macrophages that results in cell necrosis and enhanced mycobacterial persistence. This effect was not observed when a strain of *M. smegmatis* expressing only the PE domain of PE_PGRS33 was used, suggesting that the process is mediated by the PGRS domain (Dheenadhayalan et al. 2006a). Because TNF production is dependent on Toll-like receptor (TLR) signaling (Underhill et al., 1999), PE_PGRS proteins may induce TNF by interactions of the repetitive Gly-Gly-X motifs found in PGRS domain with TLRs. Indeed, a recent report by Basu et al. (2006) demonstrated that PE_PGRS33 triggers apoptosis in macrophages in a TLR2-dependent process that results in the secretion of TNF. These authors were also able to show that this process is mediated by the PGRS domain of PE_PGRS33 and that polymorphisms in this domain affect the ability of the protein to trigger TNF secretion. It remains to be determined whether these properties are specific for PE_PGRS33, because two other PE_PGRS proteins could not trigger apoptosis in macrophages (Basu et al. 2006). Nevertheless, these results suggest that PE_PGRS proteins may be part of the long list of molecules and proteins with immunoregulatory activities found in the mycobacterial cell wall that play an active role in the immunopathogenetic process of TB.

Polymorphisms of PE_PGRS Proteins

Comparative analysis of the genomes of *M. tuberculosis* H37Rv, *M. tuberculosis* CDC1551 and *M. bovis* demonstrated that PE_PGRS genes are among the few *M. tuberculosis* complex genes that can support extensive sequence variation, and for these reasons, it has been proposed that they may be a source of genetic variability (Cole et al., 1998; Fleischmann et al., 2002; Garnier et al., 2003). Variability mostly occurs in the repetitive PGRS domain and is probably the result of intergenic and intragenic recombinational events or results from strand slippage during replication (Cole et al., 2001). The selective pressure that acts upon these sequences is not known, but the identification of hot spots within each gene provides important information that is useful to infer the relevance of a given domain. Talarico et al. (2005) addressed this issue by analyzing the gene sequence of PE_PGRS33 in 123 *M. tuberculosis* clinical strains. These authors found that variations in the PE_PGRS33 gene are extensive compared with the limited genetic variability of the whole *M. tuberculosis* genome. Sequence variation resulted from insertions, deletions, and single nucleotide polymorphisms. With only one exception, insertions and deletions did not change the reading frame, suggesting that PE_PGRS33 is important for the survival of *M. tuberculosis*. The PGRS domain showed the highest polymorphism, whereas the PE and the C-terminal domains were highly conserved (Talarico et al., 2005). The significance of these observations are not yet clear, although it has been hypothesized that polymorphism at the PGRS region may be actively involved in modulating the innate host immune response through a direct interaction with TLR2 receptor (Basu et al., 2006). Sequence comparison of the PE_PGRS30 gene among more than 20 *M. tuberculosis* and *M. bovis* clinical isolates indicated that both the PE and C-terminal domains (more than 300 aa) are highly conserved (X. Sali and G. Delogu, unpublished results). These results suggest that both the PE and C-terminal domains are under genetic pressure to remain intact and therefore likely have an important functional role. It is possible that the PE domain is highly conserved because a polymorphism would interfere with the ability of the PE domain to direct the cellular localization of the PE_PGRS proteins. The role of the C-terminal domain is less clear, but the location of a major C terminal domain in the *M. marinum mag24* gene indicates that this region may be directly implicated in pathogenesis (Ramakrishnan et al., 2000).

Distribution among Mycobacteria

Molecular studies carried out in the pregenomic era and involving the use of genetic probes had suggested that PE_PGRS genes exist not only in the *M. tuberculosis* complex but also in a few other pathogenic mycobacteria, such as *M. marinum*, *M. kansasii*, and *M. gordonae* (Ross et al., 1992; Bigi et al., 1995; Poulet and Cole, 1995). The availability of the complete genomes of several other mycobacterial species provided new insights on the role of this protein sub-family. The complete sequencing of the *M. avium* sub. *paratuberculosis* genome confirmed the absence of PE_PGRS genes in this species (Li et al., 2005), and all of the seven PE_PGRS genes detected in the *M. leprae* genome were classified as pseudogenes (Cole et al., 2001). Recently, Gey van Pittius et al. (2006) highlighted the impact of these proteins in the evolution of *M. tuberculosis* complex and closely related mycobacteria by comparing the presence of PE_PGRS genes in different genomes. Fig. 7 summarizes the number of PE_PGRS detected in selected mycobacterial genomes.

A

	PE_PGRS #	Gene # (Mtb H37Rv)	Length (aa)	C-ter (aa)	PE_PGRS #	Gene # (Mtb H37Rv)	Length (aa)	C-ter (aa)
A1	*1*	*Rv0109*	496	12	*25*	*Rv1396c*	576	18
	2	*Rv0124*	487	27	*29*	*Rv1468c*	370	10
	5	*Rv0297*	591	8	*33*	*Rv1818c*	498	13
	8	*Rv0742*	175	0	*38*	*Rv2162c*	532	16
	19	*Rv1067c*	667	14	*41*	*Rv2396*	361	11
	20	*Rv1068c*	463	13	*44*	*Rv2591*	543	13
	22	*Rv1091*	853	13	*47*	*Rv2741*	525	13
	24	*Rv1325c*	603	13	*52*	*Rv3388*	731	4
A2	*7*	*Rv0578c*	1306	0	*45*	*Rv2615c*	461	16
	14	*Rv0834c*	882	14	*46*	*Rv2634c*	778	18
	15	*Rv0872c*	606	20	*48*	*Rv2853*	615	25
	21	*Rv1087*	767	13	*51*	*Rv3367*	588	13
	23	*Rv1243c*	562	29	*54*	*Rv3508*	1901	0
	26	*Rv1441c*	491	9	*55*	*Rv3511*	714	13
	28	*Rv1452c*	741	3	*57*	*Rv3514*	1489	0
	31	*Rv1768*	618	11	*58*	*Rv3590c*	584	23
	32	*Rv1803c*	639	11	*59*	*Rv3595c*	439	39
	34	*Rv1840c*	515	11				
A3	*27*	*Rv1450c*	1329	0	*50*	*Rv3345c*	1538	94
	42	*R2487c*	694	21	*53*	*Rv3507*	1381	0
	43	*Rv2490c*	1660	8				

B

PE_PGRS #	Gene # (Mtb H37Rv)	Length (aa)	PGRS domain (aa)	Internal domain (aa)	C-ter (aa)
3	*Rv0278c*	957	720	55	77
4	*Rv0279c*	837	700	40	13
9	*Rv0746*	783	630	59	12
10	*Rv0747*	801	660	25	11

C

PE_PGRS #	Gene # (Mtb H37Rv)	Length (aa)	PGRS domain (aa)	C-ter (aa)
11	*Rv0754*	584	104	314
16	*Rv0977*	923	570	273
17	*Rv0978c*	331	78	123
18	*Rv0980c*	457	150	177
30	*Rv1651c*	1011	560	306
35	*Rv1983*	558	130	292
39	*Rv2340*	413	165	130

Figure 7. Major features of the PE_PGRS proteins of *M. tuberculosis* H37Rv.

Both *M. smegmatis* and *M. avium* lack PE_PGRS genes, and *M. leprae* does not have functional PE_PGRS genes. In the *M. ulcerans* genome (see BuruList at http://genolist.pasteur.fr/BuruList) 121 PE_PGRS genes have been detected, 104 of which are *M. ulcerans* specific and 66 are pseudogenes (59 *M. ulcerans* specific). Although the *M. marinum* genome has not yet been annotated, a comparative analysis with the *M. tuberculosis* genome suggests that the number of genes and phylogenetic trees generated are very similar to what was seen for *M. ulcerans* (Gey van Pittius et al., 2006). Of course, it remains to be determined what is the role and impact of PE_PGRS genes in these two closely related mycobacteria and it will be important to ascertain the presence of the PGRS domain in proteins included in the PE_PGRS subfamily. For instance, *mag24-1*, the *M. tuberculosis* PE_PGRS30 orthologue in *M. marinum* does not have PGRS domain, and the *mag24-2* gene showing homology with the PGRS region was not expressed either constitutively or in macrophages by *M. marinum* (Rhodes et al., 2005). Annotation of the complete genome sequence of *M. marinum* will certainly provide a more precise indication of the presence of PE_PGRS genes in this mycobacterium (http://www.sanger.uk/Projects/M_marinum). It has been proposed that PE_PGRS genes have originated after the divergence of *M. avium*, and these genes expanded before the divergence of *M. leprae* under an evolutionary pressure that affected the evolution of other PE and PPE genes and the Esat-6 clusters as well. These observations certainly have functional implications that at this point cannot be fully appreciated.

CONCLUDING REMARKS

The role of Erp, PPE, PE, and PE_PGRS proteins in the biology and mechanism of pathogenesis of *M. tuberculosis* still remains elusive, and we have a long way ahead before the "enigma" of these gene families will be solved. The paucity of experimental data, the lack of well characterized orthologues, and generally, the complexity of the mycobacterial cell wall make this task even more puzzling. At this point in time, it is possible only to hint at some function or propose some hypothesis that may be useful to focus future research activities.

An aspect that deserves attention is the molecular structure of these proteins. Erp, PE, PPE, and PE_PGRS proteins have a highly conserved N-terminal domain that, though different for Erp, PE, and PPE, appears to serve the purpose of exporting these proteins to the surface. Another interesting common pattern found in these three protein families is the presence of internal repeats. The Erp proteins contain repeats with the PGLTS consensus motif, whose number may vary among the different orthologues and are polymorphic. A number of PE and PPE proteins show AB repeats that have been associated in *M. mazei* with surface-exposed antigens and several PPE proteins show the polymorphic MPTR repeats with the NxGWxG motif. Then, PE_PGRS proteins with the highly repetitive GGX-GGX motifs that vary in size among the different paralogs and is polymorphic among different strains. The C-terminal domain of Erp protein is less conserved among orthologues and seems to be involved in the interaction with other cellular components. The C-terminal domain of PE_PGRS proteins varies among paralogs but is highly conserved for a given gene among different strains. A few questions arise: Why do these proteins have repeat sequences? What is their role in protein localization or function? Why do these sequences seems to be more prone to undergo mutations, or as in the case of the PGRS domains, insertions and deletions?

The mycobacterial cell wall needs a complex and sophisticated set of proteins for its synthesis and assembly. With the emergence of new and powerful tools of functional genomics, proteomics, and confocal microscopy, and their use in addressing classical biological questions will throw more light on the function of these protein families in the coming years.

Acknowledgment. We thank Laura Klepp for the critical and careful reading of the chapter.

REFERENCES

Aagaard, C., M. Govaerts, O. L. Meng, P. Andersen, and J. M. Pollock. 2003. Genomic approach to identification of *Mycobacterium bovis* diagnostic antigens in cattle. *J. Clin. Microbiol.* 41:3719–3728.

Abdallah, A. M., T. Verboom, F. Hannes, M. Safi, M. Strong, D. Eisenberg, R. J. Musters, C. M. Vandenbroucke-Grauls, B. J. Appelmelk, J. Luirink, and W. Bitter. 2006. A specific secretion system mediates PPE41 transport in pathogenic mycobacteria. *Mol. Microbiol.* 62:667–679.

Abou-Zeid, C., T. Garbe, R. Lathigra, H. G. Wiker, M. Harboe, G. A. Rook, and D. B. Young. 1991. Genetic and immunological analysis of *Mycobacterium tuberculosis* fibronectin-binding proteins. *Infect. Immun.* 59:2712–2718.

Adindla, S., and L. Guruprasad. 2003. Sequence analysis corresponding to the PPE and PE proteins in *Mycobacterium tuberculosis* and other genomes. *J. Biosci.* 28:169–179.

Arruda, S., G. Bomfim, R. Knights, T. Huima-Byron, and L. W. Riley. 1993. Cloning of an M. tuberculosis DNA fragment associated with entry and survival inside cells. *Science* 261:1454–1457.

Baliko, Z., L. Szereday, and J. Szekeres-Bartho. 1998. Th2 biased immune response in cases with active *Mycobacterium tuberculosis* infection and tuberculin anergy. *FEMS Immunol. Med. Microbiol.* 22:199–204.

Banu, S., N. Honore, B. Saint-Joanis, D. Philpott, M. C. Prevost, and S. T. Cole. 2002. Are the PE-PGRS proteins of *Mycobacte-*

rium tuberculosis variable surface antigens? *Mol. Microbiol.* 44:9–19.

Basu, S., S. K. Pathak, A. Banerjee, S. Pathak, A. Bhattacharyya, Z. Yang, S. Talarico, M. Kundu, and J. Basu. 2006. Execution of macrophage apoptosis by PE_PGRS33 of *Mycobacterium tuberculosis* is mediated by toll-like receptor 2-dependent release of tumor necrosis factor-alpha. *J. Biol. Chem.* 282:1039–1050

Ben, A. Y., E. Shashkina, S. Johnson, P. J. Bifani, N. Kurepina, B. Kreiswirth, S. Bhattacharya, J. Spencer, A. Rendon, A. Catanzaro, and M. L. Gennaro. 2005. Immunological characterization of novel secreted antigens of *Mycobacterium tuberculosis*. *Scand. J. Immunol.* 61:139–146.

Berthet, F. X., M. Lagranderie, P. Gounon, C. Laurent-Winter, D. Ensergueix, P. Chavarot, F. Thouron, E. Maranghi, V. Pelicic, D. Portnoi, G. Marchal, and B. Gicquel. 1998. Attenuation of virulence by disruption of the *Mycobacterium tuberculosis erp* gene. *Science* 282:759–762.

Berthet, F. X., J. Rauzier, E. M. Lim, W. Philipp, B. Gicquel, and D. Portnoi. 1995. Characterization of the *Mycobacterium tuberculosis erp* gene encoding a potential cell surface protein with repetitive structures. *Microbiology* 141(Pt. 9):2123–2130.

Betts, J. C., P. T. Lukey, L. C. Robb, R. A. McAdam, and K. Duncan. 2002. Evaluation of a nutrient starvation model of *Mycobacterium tuberculosis* persistence by gene and protein expression profiling. *Mol. Microbiol.* 43:717–731.

Bigi, F., A. Alito, J. C. Fisanotti, M. I. Romano, and A. Cataldi. 1995. Characterization of a novel *Mycobacterium bovis* secreted antigen containing PGLTS repeats. *Infect. Immun.* 63:2581–2586.

Bigi, F., A. Gioffre, L. Klepp, M. P. Santangelo, C. A. Velicovsky, G. H. Giambartolomei, C. A. Fossati, M. I. Romano, T. Mendum, J. J. McFadden, and A. Cataldi. 2005. Mutation in the P36 gene of *Mycobacterium bovis* provokes attenuation of the bacillus in a mouse model. *Tuberculosis* (Edinburgh) 85:221–226.

Bigi, F., M. I. Romano, A. Alito, and A. Cataldi. 1995. Cloning of a novel polymorphic GC-rich repetitive DNA from *Mycobacterium bovis*. *Res. Microbiol.* 146:341–348.

Brennan, M. J., and G. Delogu. 2002. The PE multigene family: a 'molecular mantra' for mycobacteria. *Trends Microbiol.* 10:246–249.

Brennan, M. J., G. Delogu, Y. Chen, S. Bardarov, J. Kriakov, M. Alavi, and W. R. Jacobs, Jr. 2001. Evidence that mycobacterial PE_PGRS proteins are cell surface constituents that influence interactions with other cells. *Infect. Immun.* 69:7326–7333.

Brennan, M. J., N. C. Gey van Pittius, and C. Espitia. 2005. The PE and PPE multigene families of mycobacteria, p. 513–525. *In* S. T. Cole, K. D. Eisenach, D. N. McMurray, and W. R. Jacobs, Jr. (ed.), *Tuberculosis and the Tubercle Bacillus*. ASM Press, Washington, DC.

Brosch, R., S. V. Gordon, A. Billault, T. Garnier, K. Eiglmeier, C. Soravito, B. G. Barrell, and S. T. Cole. 1998. Use of a *Mycobacterium tuberculosis* H37Rv bacterial artificial chromosome library for genome mapping, sequencing, and comparative genomics. *Infect. Immun.* 66:2221–2229.

Camacho, L. R., D. Ensergueix, E. Perez, B. Gicquel, and C. Guilhot. 1999. Identification of a virulence gene cluster of *Mycobacterium tuberculosis* by signature-tagged transposon mutagenesis. *Mol. Microbiol.* 34:257–267.

Chakhaiyar, P., and S. E. Hasnain. 2004. Defining the mandate of tuberculosis research in a postgenomic era. *Med. Princ. Pract.* 13:177–184.

Chakhaiyar, P., Y. Nagalakshmi, B. Aruna, K. J. Murthy, V. M. Katoch, and S. E. Hasnain. 2004. Regions of high antigenicity within the hypothetical PPE major polymorphic tandem repeat open-reading frame, Rv2608, show a differential humoral response and a low T cell response in various categories of patients with tuberculosis. *J. Infect. Dis.* 190:1237–1244.

Cherayil, B. J., and R. A. Young. 1988. A 28-kDa protein from *Mycobacterium leprae* is a target of the human antibody response in lepromatous leprosy. *J. Immunol.* 141:4370–4375.

Choudhary, R. K., S. Mukhopadhyay, P. Chakhaiyar, N. Sharma, K. J. Murthy, V. M. Katoch, and S. E. Hasnain. 2003. PPE antigen Rv2430c of *Mycobacterium tuberculosis* induces a strong B-cell response. *Infect. Immun.* 71:6338–6343.

Cole, S. T., K. Eiglmeier, J. Parkhill, K. D. James, N. R. Thomson, P. R. Wheeler, N. Honore, T. Garnier, C. Churcher, D. Harris, K. Mungall, D. Basham, D. Brown, T. Chillingworth, R. Connor, R. M. Davies, K. Devlin, S. Duthoy, T. Feltwell, A. Fraser, N. Hamlin, S. Holroyd, T. Hornsby, K. Jagels, C. Lacroix, J. Maclean, S. Moule, L. Murphy, K. Oliver, M. A. Quail, M. A. Rajandream, K. M. Rutherford, S. Rutter, K. Seeger, S. Simon, M. Simmonds, J. Skelton, R. Squares, S. Squares, K. Stevens, K. Taylor, S. Whitehead, J. R. Woodward, and B. G. Barrell. 2001. Massive gene decay in the leprosy bacillus. *Nature* 409:1007–1011.

Cole, S. T., R. Brosch, J. Parkhill, T. Garnier, C. Churcher, D. Harris, S. V. Gordon, K. Eiglmeier, S. Gas, C. E. Barry, III, F. Tekaia, K. Badcock, D. Basham, D. Brown, T. Chillingworth, R. Connor, R. Davies, K. Devlin, T. Feltwell, S. Gentles, N. Hamlin, S. Holroyd, T. Hornsby, K. Jagels, and B. G. Barrell. 1998. Deciphering the biology of *Mycobacterium tuberculosis* from the complete genome sequence. *Nature* 393:537–544. (Erratum, 396:190.)

Cosma, C. L., K. Klein, R. Kim, D. Beery, and L. Ramakrishnan. 2006. *Mycobacterium marinum* Erp is a virulence determinant required for cell wall integrity and intracellular survival. *Infect. Immun.* 74:3125–3133.

Cunningham, A. F., and C. L. Spreadbury. 1998. Mycobacterial stationary phase induced by low oxygen tension: cell wall thickening and localization of the 16-kilodalton alpha-crystallin homolog. *J. Bacteriol.* 180:801–808.

de Mendonça-Lima, L., M. Picardeau, C. Raynaud, J. Rauzier, Y. O. de la Salmoniere, L. Barker, F. Bigi, A. Cataldi, B. Gicquel, and J. M. Reyrat. 2001. Erp, an extracellular protein family specific to mycobacteria. *Microbiology* 147:2315–2320.

de Mendonça-Lima, L., Y. Bordat, E. Pivert, C. Recchi, O. Neyrolles, A. Maitournam, B. Gicquel, and J. M. Reyrat. 2003. The allele encoding the mycobacterial Erp protein affects lung disease in mice. *Cell Microbiol.* 5:65–73.

Delogu, G., and M. J. Brennan. 2001. Comparative immune response to PE and PE_PGRS antigens of *Mycobacterium tuberculosis*. *Infect. Immun.* 69:5606–5611.

Delogu, G., C. Pusceddu, A. Bua, G. Fadda, M. J. Brennan, and S. Zanetti. 2004. Rv1818c-encoded PE_PGRS protein of *Mycobacterium tuberculosis* is surface exposed and influences bacterial cell structure. *Mol. Microbiol.* 52:725–733.

Delogu, G., M. Sanguinetti, C. Pusceddu, A. Bua, M. J. Brennan, S. Zanetti, and G. Fadda. 2006. PE_PGRS proteins are differentially expressed by *Mycobacterium tuberculosis* in host tissues. *Microbes. Infect.* 8:2061–2067.

Dhar, N., V. Rao, and A. K. Tyagi. 2000. Recombinant BCG approach for development of vaccines: cloning and expression of immunodominant antigens of M. tuberculosis. *FEMS Microbiol. Lett.* 190:309–316.

Dheenadhayalan, V., G. Delogu, and M. J. Brennan. 2006a. Expression of the PE_PGRS 33 protein in *Mycobacterium smegmatis* triggers necrosis in macrophages and enhanced mycobacterial survival. *Microbes. Infect.* 8:262–272.

Dheenadhayalan, V., G. Delogu, M. Sanguinetti, G. Fadda, and M. J. Brennan. 2006b. Variable expression patterns of *Mycobacterium tuberculosis* PE_PGRS genes: evidence that PE_PGRS16 and PE_PGRS26 are inversely regulated in vivo. *J. Bacteriol.* 188:3721–3725.

Dillon, D. C., M. R. Alderson, C. H. Day, D. M. Lewinsohn, R. Coler, T. Bement, A. Campos-Neto, Y. A. Skeiky, I. M. Orme, A. Roberts, S. Steen, W. Dalemans, R. Badaro, and S. G. Reed. 1999. Molecular characterization and human T-cell responses to a member of a novel *Mycobacterium tuberculosis* mtb39 gene family. *Infect. Immun.* 67:2941–2950.

Ellner, J. J. 1997. Regulation of the human immune response during tuberculosis. *J. Lab Clin. Med.* 130:469–475.

Espitia, C., J. P. Laclette, M. Mondragon-Palomino, A. Amador, J. Campuzano, A. Martens, M. Singh, R. Cicero, Y. Zhang, and C. Moreno. 1999. The PE-PGRS glycine-rich proteins of *Mycobacterium tuberculosis*: a new family of fibronectin-binding proteins? *Microbiology* 145(Pt 12):3487–3495.

Fifis, T., C. Costopoulos, A. J. Radford, A. Bacic, and P. R. Wood. 1991. Purification and characterization of major antigens from a *Mycobacterium bovis* culture filtrate. *Infect. Immun.* 59:800–807.

Fleischmann, R. D., D. Alland, J. A. Eisen, L. Carpenter, O. White, J. Peterson, R. DeBoy, R. Dodson, M. Gwinn, D. Haft, E. Hickey, J. F. Kolonay, W. C. Nelson, L. A. Umayam, M. Ermolaeva, S. L. Salzberg, A. Delcher, T. Utterback, J. Weidman, H. Khouri, J. Gill, A. Mikula, W. Bishai, J. W. Jacobs, Jr., J. C. Venter, and C. M. Fraser. 2002. Whole-genome comparison of *Mycobacterium tuberculosis* clinical and laboratory strains. *J. Bacteriol.* 184:5479–5490.

Flores, J., and C. Espitia. 2003. Differential expression of PE and PE_PGRS genes in *Mycobacterium tuberculosis* strains. *Gene* 318:75–81.

Flynn, J. L., and J. Chan. 2001. Immunology of tuberculosis. *Annu. Rev. Immunol.* 19:93–129.

Gao, Q., K. E. Kripke, A. J. Saldanha, W. Yan, S. Holmes, and P. M. Small. 2005. Gene expression diversity among *Mycobacterium tuberculosis* clinical isolates. *Microbiology* 151:5–14.

Garnier, T., K. Eiglmeier, J. C. Camus, N. Medina, H. Mansoor, M. Pryor, S. Duthoy, S. Grondin, C. Lacroix, C. Monsempe, S. Simon, B. Harris, R. Atkin, J. Doggett, R. Mayes, L. Keating, P. R. Wheeler, J. Parkhill, B. G. Barrell, S. T. Cole, S. V. Gordon, and R. G. Hewinson. 2003. The complete genome sequence of *Mycobacterium bovis*. *Proc. Natl. Acad. Sci. USA* 100:7877–7882.

Gey van Pittius, N. C., S. L. Sampson, H. Lee, Y. Kim, P. D. van Helden, and R. M. Warren. 2006. Evolution and expansion of the *Mycobacterium tuberculosis* PE and PPE multigene families and their association with the duplication of the ESAT-6 (esx) gene cluster regions. *BMC Evol. Biol.* 6:95.

Ghosh, S., S. Rasheedi, S. S. Rahim, S. Banerjee, R. K. Choudhary, P. Chakhaiyar, N. Z. Ehtesham, S. Mukhopadhyay, and S. E. Hasnain. 2004. Method for enhancing solubility of the expressed recombinant proteins in Escherichia coli. *BioTechniques* 37:418, 420, 422–423.

Glickman, M. S., and W. R. Jacobs, Jr. 2001. Microbial pathogenesis of *Mycobacterium tuberculosis*: dawn of a discipline. *Cell* 104:485.

Gordon, S. V., R. Brosch, A. Billault, T. Garnier, K. Eiglmeier, and S. T. Cole. 1999. Identification of variable regions in the genomes of tubercle bacilli using bacterial artificial chromosome arrays. *Mol. Microbiol.* 32:643–655.

Gu, S., J. Chen, K. M. Dobos, E. M. Bradbury, J. T. Belisle, and X. Chen. 2003. Comprehensive proteomic profiling of the membrane constituents of a *Mycobacterium tuberculosis* strain. *Mol. Cell. Proteomics* 2:1284–1296.

Guinn, K. M., M. J. Hickey, S. K. Mathur, K. L. Zakel, J. E. Grotzke, D. M. Lewinsohn, S. Smith, and D. R. Sherman. 2004. Individual RD1-region genes are required for export of ESAT-6/CFP-10 and for virulence of *Mycobacterium tuberculosis*. *Mol. Microbiol.* 51:359–370.

Hahn, M. Y., S. Raman, M. Anaya, and R. N. Husson. 2005. The *Mycobacterium tuberculosis* extracytoplasmic-function sigma factor SigL regulates polyketide synthases and secreted or membrane proteins and is required for virulence. *J. Bacteriol.* 187:7062–7071.

Jungblut, P. R., U. E. Schaible, H. J. Mollenkopf, U. Zimny-Arndt, B. Raupach, J. Mattow, P. Halada, S. Lamer, K. Hagens, and S. H. Kaufmann. 1999. Comparative proteome analysis of *Mycobacterium tuberculosis* and *Mycobacterium bovis* BCG strains: towards functional genomics of microbial pathogens. *Mol. Microbiol.* 33:1103–1117.

Karimova, G., J. Pidoux, A. Ullmann, and D. Ladant. 1998. A bacterial two-hybrid system based on a reconstituted signal transduction pathway. *Proc. Natl. Acad. Sci. USA* 95:5752–5756.

Keane, J., H. G. Remold, and H. Kornfeld. 2000. Virulent *Mycobacterium tuberculosis* strains evade apoptosis of infected alveolar macrophages. *J. Immunol.* 164:2016–2020.

Kocincova, D., B. Sonden, Y. Bordat, E. Pivert, L. de Mendonca-Lima, B. Gicquel, and J. M. Reyrat. 2004b. The hydrophobic domain of the Mycobacterial Erp protein is not essential for the virulence of *Mycobacterium tuberculosis*. *Infect. Immun.* 72:2379–2382.

Kocincova, D., B. Sonden, L. de Mendonca-Lima, B. Gicquel, and J. M. Reyrat. 2004a. The Erp protein is anchored at the surface by a carboxy-terminal hydrophobic domain and is important for cell-wall structure in *Mycobacterium smegmatis*. *FEMS Microbiol. Lett.* 231:191–196.

Kulshrestha, A., A. Gupta, N. Verma, S. K. Sharma, A. K. Tyagi, and V. K. Chaudhary. 2005. Expression and purification of recombinant antigens of *Mycobacterium tuberculosis* for application in serodiagnosis. *Protein Expr. Purif.* 44:75–85.

Landowski, C. P., H. P. Godfrey, S. I. Bentley-Hibbert, X. Liu, Z. Huang, R. Sepulveda, K. Huygen, M. L. Gennaro, F. H. Moy, S. A. Lesley, and M. Haak-Frendscho. 2001. Combinatorial use of antibodies to secreted mycobacterial proteins in a host immune system-independent test for tuberculosis. *J. Clin. Microbiol.* 39:2418–2424.

Lefevre, P., O. Denis, L. De Wit, A. Tanghe, P. Vandenbussche, J. Content, and K. Huygen. 2000. Cloning of the gene encoding a 22-kilodalton cell surface antigen of *Mycobacterium bovis* BCG and analysis of its potential for DNA vaccination against tuberculosis. *Infect. Immun.* 68:1040–1047.

Levitskaya, J., M. Coram, V. Levitsky, S. Imreh, P. M. Steigerwald-Mullen, G. Klein, M. G. Kurilla, and M. G. Masucci. 1995. Inhibition of antigen processing by the internal repeat region of the Epstein-Barr virus nuclear antigen-1. *Nature* 375:685–688.

Li, L., J. P. Bannantine, Q. Zhang, A. Amonsin, B. J. May, D. Alt, N. Banerji, S. Kanjilal, and V. Kapur. 2005. The complete genome sequence of *Mycobacterium avium* subspecies *paratuberculosis*. *Proc. Natl. Acad. Sci. USA* 102:12344–12349.

Mawuenyega, K. G., C. V. Forst, K. M. Dobos, J. T. Belisle, J. Chen, E. M. Bradbury, A. M. Bradbury, and X. Chen. 2005. *Mycobacterium tuberculosis* functional network analysis by global subcellular protein profiling. *Mol. Biol. Cell* 16:396–404.

Mayerhofer, L. E., E. Conway de Macario, and A. J. Macario. 1995. Conservation and variability in Archaea: protein antigens with tandem repeats encoded by a cluster of genes with common motifs in *Methanosarcina mazei* S-6. *Gene* 165:87–91

Mustafa, A. S. 2005. Recombinant and synthetic peptides to identify *Mycobacterium tuberculosis* antigens and epitopes of diagnostic and vaccine relevance. *Tuberculosis* (Edinburgh) 85:367–376.

Noss, E. H., C. V. Harding, and W. H. Boom. 2000. *Mycobacterium tuberculosis* inhibits MHC class II antigen processing in murine bone marrow macrophages. *Cell Immunol.* 201:63–74.

Pan, F., M. Jackson, Y. Ma, and M. McNeil. 2001. Cell wall core galactofuran synthesis is essential for growth of mycobacteria. *J. Bacteriol.* 183:3991–3998.

Pleissner, K. P., T. Eifert, S. Buettner, F. Schmidt, M. Boehme, T. F. Meyer, S. H. Kaufmann, and P. R. Jungblut. 2004. Web-accessible proteome databases for microbial research. *Proteomics* 4:1305–1313.

Poulet, S., and S. T. Cole. 1995. Characterization of the highly abundant polymorphic GC-rich-repetitive sequence (PGRS) present in *Mycobacterium tuberculosis. Arch. Microbiol.* 163:87–95.

Ramakrishnan, L., N. A. Federspiel, and S. Falkow. 2000. Granuloma-specific expression of mycobacterium virulence proteins from the glycine-rich PE-PGRS family. *Science* 288:1436–1439.

Rhodes, M. W., H. Kator, A. McNabb, C. Deshayes, J. M. Reyrat, B. A. Brown-Elliott, R. Wallace Jr, K. A. Trott, J. M. Parker, B. Lifland, G. Osterhout, I. Kaattari, K. Reece, W. Vogelbein, and C. A. Ottinger. 2005. *Mycobacterium pseudoshottsii* sp. nov., a slowly growing chromogenic species isolated from Chesapeake Bay striped bass (Morone saxatilis). *Int. J. Syst. Evol. Microbiol.* 55:1139–1147.

Rodriguez, G. M., M. I. Voskuil, B. Gold, G. K. Schoolnik, and I. Smith. 2002. ideR, An essential gene in *Mycobacterium tuberculosis*: role of IdeR in iron-dependent gene expression, iron metabolism, and oxidative stress response. *Infect. Immun.* 70:3371–3381.

Romain, F., J. Augier, P. Pescher, and G. Marchal. 1993. Isolation of a proline-rich mycobacterial protein eliciting delayed-type hypersensitivity reactions only in guinea pigs immunized with living mycobacteria. *Proc. Natl. Acad. Sci. USA* 90:5322–5326.

Ross, B. C., K. Raios, K. Jackson, and B. Dwyer. 1992. Molecular cloning of a highly repeated DNA element from *Mycobacterium tuberculosis* and its use as an epidemiological tool. *J. Clin. Microbiol.* 30:942–946.

Sampson, S. L., P. Lukey, R. M. Warren, P. D. van Helden, M. Richardson, and M. J. Everett. 2001. Expression, characterization and subcellular localization of the *Mycobacterium tuberculosis* PPE gene Rv1917c. *Tuberculosis* (Edinburgh) 81:305–317.

Sassetti, C. M., and E. J. Rubin. 2003. Genetic requirements for mycobacterial survival during infection. *Proc. Natl. Acad. Sci. USA* 100:12989–12994.

Schluger, N. W., and W. N. Rom. 1998. The host immune response to tuberculosis. *Am. J. Respir. Crit. Care Med.* 157:679–691.

Shi, L., Y. Jung, S. Tyagi, M. L. Gennaro, and R. North. 2003. Expression of Th1-mediated immunity in mouse lungs induces a *Mycobacterium tuberculosis* transcription pattern characteristic of nonreplicating persistence. *Proc. Natl. Acad. Sci. USA* 100:241–246.

Singh, K. K., X. Zhang, A. S. Patibandla, P. Chien, Jr., and S. Laal. 2001. Antigens of *Mycobacterium tuberculosis* expressed during preclinical tuberculosis: serological immunodominance of proteins with repetitive amino acid sequences. *Infect. Immun.* 69:4185–4191.

Sreenu, V. B., P. Kumar, J. Nagaraju, and H. A. Nagarajaram. 2006. Microsatellite polymorphism across the M. tuberculosis and M. bovis genomes: implications on genome evolution and plasticity. BMC. *Genomics* 7:78.

Strong, M., and C. W. Goulding. 2006. Structural proteomics and computational analysis of a deadly pathogen: combating *Myco-bacterium tuberculosis* from multiple fronts. *Methods Biochem. Anal.* 49:245–269.

Strong, M., M. R. Sawaya, S. Wang, M. Phillips, D. Cascio, and D. Eisenberg. 2006. Toward the structural genomics of complexes: crystal structure of a PE/PPE protein complex from *Mycobacterium tuberculosis. Proc. Natl. Acad. Sci. USA* 103:8060–8065.

Talarico, S., M. D. Cave, C. F. Marrs, B. Foxman, L. Zhang, and Z. Yang. 2005. Variation of the *Mycobacterium tuberculosis* PE_PGRS 33 gene among clinical isolates. *J. Clin. Microbiol.* 43:4954–4960.

Timm, J., F. A. Post, L. G. Bekker, G. B. Walther, H. C. Wainwright, R. Manganelli, W. Chan, L. Tsenova, B. Gold, I. Smith, G. Kaplan, and J. D. McKinney. 2003. Differential expression of iron-, carbon-, and oxygen- responsive mycobacterial genes in the lungs of chronically infected mice and tuberculosis patients. *Proc. Natl. Acad. Sci. USA* 100:14321–14326.

Triccas, J. A., F. X. Berthet, V. Pelicic, and B. Gicquel. 1999. Use of fluorescence induction and sucrose counterselection to identify *Mycobacterium tuberculosis* genes expressed within host cells. *Microbiology* 145(Pt. 10):2923–2930.

Tundup, S., Y. Akhter, D. Thiagarajan, and S. E. Hasnain. 2006. Clusters of PE and PPE genes of *Mycobacterium tuberculosis* are organized in operons: evidence that PE Rv2431c is co-transcribed with PPE Rv2430c and their gene products interact with each other. *FEBS Lett.* 580:1285–1293.

Underhill, D. M., A. Ozinsky, K. D. Smith, and A. Aderem. 1999. Toll-like receptor-2 mediates mycobacteria-induced proinflammatory signaling in macrophages. *Proc. Natl. Acad. Sci. USA* 96:14459–14463.

van Soolingen, D., P. E. de Haas, P. W. Hermans, P. M. Groenen, and J. D. van Embden. 1993. Comparison of various repetitive DNA elements as genetic markers for strain differentiation and epidemiology of *Mycobacterium tuberculosis. J. Clin. Microbiol.* 31:1987–1995.

Vekemans, J., M. O. Ota, J. Sillah, K. Fielding, M. R. Alderson, Y. A. Skeiky, W. Dalemans, K. P. McAdam, C. Lienhardt, and A. Marchant. 2004. Immune responses to mycobacterial antigens in the Gambian population: implications for vaccines and immunodiagnostic test design. *Infect. Immun.* 72:381–388.

Voskuil, M. I., D. Schnappinger, R. Rutherford, Y. Liu, and G. K. Schoolnik. 2004. Regulation of the *Mycobacterium tuberculosis* PE/PPE genes. *Tuberculosis* (Edinburgh) 84:256–262.

Walters, S. B., E. Dubnau, I. Kolesnikova, F. Laval, M. Daffe, and I. Smith. 2006. The *Mycobacterium tuberculosis* PhoPR two-component system regulates genes essential for virulence and complex lipid biosynthesis. *Mol. Microbiol.* 60:312–330.

Weldingh, K., I. Rosenkrands, L. M. Okkels, T. M. Doherty, and P. Andersen. 2005. Assessing the serodiagnostic potential of 35 *Mycobacterium tuberculosis* proteins and identification of four novel serological antigens. *J. Clin. Microbiol.* 43:57–65.

Willemsen, P. T., J. Westerveen, A. Dinkla, D. Bakker, F. G. van Zijderveld, and J. E. Thole. 2006. Secreted antigens of *Mycobacterium avium* subspecies *paratuberculosis* as prominent immune targets. *Vet. Microbiol.* 114:337–344.

Wilsher, M. L., C. Hagan, R. Prestidge, A. U. Wells, and G. Murison. 1999. Human in vitro immune responses to *Mycobacterium tuberculosis. Tuber. Lung Dis.* 79:371–377.

The Mycobacterial Cell Envelope
Edited by M. Daffé and J.-M. Reyrat
© 2008 ASM Press, Washington, DC

Chapter 9

Mycobacterial Porins

MICHAEL NIEDERWEIS

GENERAL INTEREST IN MYCOBACTERIA

Scientific interest in mycobacteria has not only been sparked by the paramount medical importance of *Mycobacterium tuberculosis* but also by properties that distinguish mycobacteria from other bacteria. Mycobacteria devote a large part of the coding capacity of its genome to fatty acid biosynthesis. For example, 225 genes of *M. tuberculosis* are involved in lipid metabolism in contrast to 50 genes of *E. coli*, whereas both bacteria have a similar genome size (Cole and Barrell, 1998). Mycobacteria use these genes to synthesize a fascinating diversity of lipids, many of which are unique (Brennan and Nikaido, 1995). Among those are exceptionally long fatty acids called mycolic acids, which account for 30% to 40% of the dry weight of the cell envelope (Barry, 2001; Rastogi et al., 2001). Most of these lipids are used by mycobacteria to produce a unique cell envelope (Fig. 1), which establishes an efficient permeability barrier and plays a crucial role in the high intrinsic drug resistance and in extracellular and intracellular survival of mycobacteria under harsh conditions (Brennan and Nikaido, 1995). The architecture of the cell envelope has been reviewed in chapters 1 and 2.

FUNCTIONS OF BACTERIAL PORINS

Porins are water-filled channel proteins that mediate diffusion of small and hydrophilic nutrient molecules across bacterial outer membranes. This function has been reviewed for gram-negative bacteria (Nikaido, 2003) and for mycobacteria (Niederweis, 2003). In addition, extracellular loops of porins are potential binding sites for antibodies, phages, and bacterial or mammalian cells. Therefore, porins of gram-negative bacteria have multiple roles and are often involved in the virulence of pathogenic bacteria

(Fig. 2). The porins of *Neisseria meningitidis* and *N. gonorrhoeae* serve as examples. Neisserial porins function as pores and are essential for bacterial survival because they modulate the exchange of ions between the bacteria and the surrounding environment (Massari et al., 2003). Porins of *N. meningitidis* induce actin nucleation in the host cell, suggesting a role in cell actin reorganization (Wen et al., 2000), which might influence the invasive ability of the bacteria. It has been proposed that neisserial porins bind or insert into the membrane of epithelial cells at the sites of close contact between the bacteria and host cells during infection (Merz and So, 2000), which might contribute to bacterial pathogenicity (Rudel et al., 1996). The *N. gonorrhoeae* porin PorB1A has recently been shown to be associated with increased bacterial invasiveness (Bauer et al., 1999). Gonococcal strains bearing PorB1A, which are responsible for disseminated disease, are more resistant to killing by normal human serum and invade cells in vitro to a greater degree than gonococci carrying the PorB1B porin (van Putten et al., 1998). This suggests that a difference in the porin subtype influences the ability of the bacteria to invade eukaryotic cells. In a particularly interesting case, it was shown that the porin OprF of *Pseudomonas aeruginosa* binds interferon-γ (IFN-γ) to recognize the status of host immune activation and to induce expression of virulence factors such as the type I lectin (PA-I) and the cytotoxin pyocyanin through activation of the quorum sensing system (Wagner et al., 2006; Wu et al., 2005). The impact of *Salmonella* sp. porins on infection and progress of the diseases salmonellosis or typhoid fever is documented for all stages of pathogenesis. OmpC and OmpD of *S. enterica* serovar Typhimurium are involved in the initial host-cell recognition by macrophages (Negm and Pistole, 1998, 1999) and both $\Delta ompC/\Delta ompF$ and $\Delta ompS1/\Delta ompS2$ mutant strains are attenuated for their virulence in mice (Negm and Pistole, 1999; Rodriguez-Morales et al.,

Michael Niederweis • Department of Microbiology, University of Alabama at Birmingham, Birmingham, AL 35294.

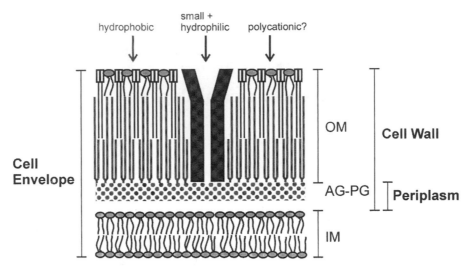

Figure 1. Transport processes across the mycobacterial cell envelope. A simplified schematic representation of the mycobacterial cell envelope is shown. Adapted from Niederweis, (2003). This representation is based on the model proposed by Minnikin (1982). According to this model, the inner leaflet of the asymmetric outer membrane (OM) is composed of mycolic acids (MA), which are covalently linked to the arabinogalactan (AG)-peptidoglycan (PG) copolymer. A variety of extractable lipids presumably form the outer leaflet of the outer membrane. Surface layers such as a capsule (Daffé and Draper, 1998; Daffé and Etienne, 1999; Lemassu et al., 1996) were omitted from this figure for clarity.

2006). Furthermore, *Salmonella* porins induce cellular activation (Galdiero et al., 2003), a sustained long-term antibody response (Secundino et al., 2006), cytokine release, and the activation of protein kinases A and C (Galdiero et al., 2003). The nucleoside-specific porin TsX from *S. enterica* serovar Typhi is suggested to participate in membrane assembly (Bucarey et al., 2005). Porins from several gram-negative bacteria, including *Salmonella* (Galdiero et al., 1990), *Escherichia coli* (Vordermeier and Bessler, 1987; Vordermeier et al., 1987) and neisserial species (Massari et al., 2003), induce inflammation in mice, activation of human and murine B cells with a mitogenic effect, release of cytokines from human cells (Galdiero et al., 1993), and stimulation of the production of platelet-activating factor by human endothelial cells

(Tufano et al., 1993). Because mycobacterial porins have been shown to be surface-accessible both in *M. smegmatis* (Faller et al., 2004; Stahl et al., 2001) and in *M. tuberculosis* (Mailaender et al., 2004), it is tempting to speculate that porins of pathogenic mycobacteria are also involved in host-pathogen interactions. However, none of these potential additional functions have been unveiled for mycobacterial porins yet.

NUTRIENT UPTAKE BY MYCOBACTERIA

The growth and nutritional requirements of mycobacteria have been intensely studied since the discovery of *Mycobacterium tuberculosis* (Koch, 1882).

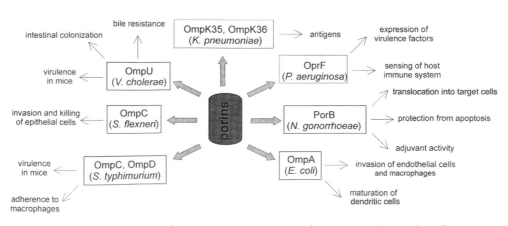

Figure 2. Functions of porins of gram-negative bacteria in pathogenesis. See the text for references.

These studies have resulted in an overwhelming body of literature on the physiology of mycobacterial metabolism in the years before the dawn of molecular biology (Edson, 1951; Ramakrishnan et al., 1972; Ratledge, 1982). The carbon metabolism of mycobacteria has attracted renewed interest since the discovery that *M. tuberculosis* relies on the glyoxylate cycle for survival in mice (McKinney et al., 2000; Munoz-Elias and McKinney, 2005). This observation indicates that *M. tuberculosis* uses lipids as the main carbon source during infection. On the other side, genes that encode a putative sugar transporter were essential for *M. tuberculosis* during the first week of infection (Sassetti and Rubin, 2003), indicating that *M. tuberculosis* may switch its main carbon source from carbohydrates to lipids with the onset of the adaptive immune response. Obviously, this requires that *M. tuberculosis* has the capacity to take up these particular carbohydrates and lipids. However, both the nutrients and the corresponding uptake proteins are unknown for *M. tuberculosis* inside the human host. Surprisingly, this is also true for *M. tuberculosis* growing in vitro and for *M. smegmatis*. which is often used as a fast-growing, nonpathogenic model organism to learn more about basic mycobacterial physiology. There is no doubt that the uptake pathways have been adapted to the human body as habitat of *M. tuberculosis* and the soil as habitat of *M. smegmatis*. Thus, much can be learned about the lifestyles of both organisms by a comparison of the complement of specific nutrient uptake proteins. In this review, I focus primarily on the porin-mediated uptake of solutes across mycobacterial outer membranes.

THE CELL ENVELOPE OF MYCOBACTERIA

Permeability Barriers in Mycobacterial Cell Envelopes

Nutrient uptake mechanisms obviously depend on the permeability barriers imposed by the cell envelope. Therefore, it is necessary to summarize the current status of the research about the cell envelope of mycobacteria. In order to avoid confusion about the terminology, I want to refer to the definitions put forward primarily by Terry Beveridge and coworkers (Beveridge, 1995; Beveridge and Kadurugamuwa, 1996; Graham et al., 1991): The cell wall consists of the periplasm, the peptidoglycan layer, and for gram-negative bacteria, the outer membrane. The periplasmic space represents an extracytoplasmic compartment confined between the plasma membrane and an outer structure (outer membrane versus a peptidoglycan-teichoic acid-protein network for gram-negative and gram-positive bacteria, respectively). The constituent

periplasm is composed mostly of highly deformable, low-density, soluble components. The cell envelope comprises the inner membrane and the cell wall. The components of the cell envelopes of both gram-positive and gram-negative bacteria have been beautifully visualized recently by cryoelectron microscopy (Graham et al., 1991; Matias et al., 2003; Matias and Beveridge, 2005, 2006).

Mycobacteria have evolved a complex cell wall, comprising a peptidoglycan-arabinogalactan polymer with covalently bound mycolic acids of considerable size (up to 90 carbon atoms), a variety of extractable lipids (Barry et al., 1998; Daffé and Draper, 1998), and pore-forming proteins (Niederweis, 2003). Most of the mycobacterial lipids are constituents of the cell envelope, which provides an extraordinarily efficient permeability barrier to noxious compounds, rendering mycobacteria intrinsically resistant to many drugs (Brennan and Nikaido, 1995). Owing to the paramount medical importance of *Mycobacterium tuberculosis*, the ultrastructure of mycobacterial cell envelopes has been intensively studied for decades by electron microscopy. A thick, electron-transparent zone has been observed in the cell wall in stained thin sections of many mycobacterial species (Mineda et al., 1998; Paul and Beveridge, 1992) and was shown to comprise lipids (Paul and Beveridge, 1994). To account for the extraordinary efficiency of the mycobacterial cell wall as a permeability barrier, Minnikin (1982) originally proposed an ingenious model (Fig. 1) in which the mycolic acids are covalently bound to the peptidoglycan-arabinogalactan copolymer and form the inner leaflet of an asymmetrical bilayer. Other lipids extractable by organic solvents were thought to form the outer leaflet of this outer bilayer. Experimental evidence for this model was provided by X-ray diffraction studies, which showed that the mycolic acids are oriented parallel to each other and perpendicular to the plane of the cell envelope (Nikaido et al., 1993). Mutants and treatments affecting mycolic acid biosynthesis and the production of extractable lipids showed an increase of cell wall permeability and a drastic decrease of virulence, underscoring the importance of the integrity of the cell wall for intracellular survival of *M. tuberculosis* (Barry et al., 1998). These indirect structural, biochemical, and genetic data are consistent with the existence of an outer lipid bilayer as proposed by Minnikin (1982). However, this model (Fig. 1) faced criticism mainly because electron microscopy of mycobacteria, in particular thin sections thereof, never showed evidence for an additional, outer lipid bilayer (Daffé and Draper, 1998; Draper, 1998). A reason may be that such a lipid membrane is notoriously poorly preserved during chemical fixation and plastic embedding (Beveridge, 1999).

Thus, direct evidence for an outer lipid bilayer in mycobacteria, which are classified as gram-positive bacteria, is lacking. Consequently, a periplasmic space as defined above has been regarded as hypothetical for mycobacteria (Daffé and Draper, 1998; Etienne et al., 2005). Recently, cryoelectron tomography of intact *M. smegmatis* and *M. bovis* BCG cells revealed an architecture of the cell envelope very similar to that of *E. coli*. The cells are covered by two main layers resembling the inner and outer membranes of *E. coli* in appearance and dimensions. This is the first visualization of native outer membranes in mycobacteria (C. Hoffmann, M. Niederweis, and H. Engelhardt, submitted for publication). This discovery is likely to trigger research efforts to examine the roles of outer membrane proteins in general and outer membrane proteins in particular in nutrient uptake, drug resistance, and host-pathogen interactions.

Principal Pathways across the Mycobacterial Cell Envelope

The unique mycolic acid bilayer is an extremely efficient permeability barrier protecting the cell from toxic compounds and is generally thought to be the major determinant of the intrinsic resistance of mycobacteria to most common antibiotics, chemotherapeutic agents and chemical disinfectants (Brennan and Nikaido, 1995). The mycolic acid bilayer is functionally and structurally analogous to the outer membrane of gram-negative bacteria, a similarity that merits a more detailed comparison. Three general and several specific pathways for transport across the outer membrane of gram-negative bacteria exist. (1) Hydrophobic compounds penetrate the membrane through the lipid pathway by temporarily dissolving in the lipid bilayer. (2) Small and hydrophilic compounds diffuse through water-filled protein channels, the porins (Nikaido, 1994). Some of the diffusion channels show specificity toward certain classes of compounds, such as maltodextrin (Dumas et al., 2000) or nucleosides (Maier et al., 1988). (3) Polycationic compounds are thought to disorganize the outer membrane, thereby mediating their own uptake in a process, which was termed "self-promoted uptake" (Hancock et al., 1991). (4) Certain compounds are specifically taken up by transporter proteins such as FhuA and FepA, which transport iron-loaded siderophores in an energy-dependent process across the outer membrane of *E. coli* (Braun and Killmann, 1999). It is likely that these pathways across the outer membrane also exist in mycobacteria.

PORINS OF FAST-GROWING MYCOBACTERIA

MspA, the Major Porin of *M. smegmatis*

Physiological functions of the Msp porins of *M. smegmatis*

Porins are defined as nonspecific protein channels in bacterial outer membranes that enable the influx of hydrophilic solutes (Nikaido, 2003). The discovery of the existence of porins in mycobacteria was a surprise (Trias et al., 1992), considering the classification of mycobacteria as gram-positive bacteria (Stackebrandt et al., 1997), which do not have a second membrane and, therefore, do not possess outer membrane proteins such as porins. However, genome-wide analyses find a closer relationship of mycobacteria to gram-negative bacteria (Fu and Fu-Liu, 2002a, 2002b) and also confirm the previous classifications that were based on 16S rRNA phylogenetic trees (Gao et al., 2006).

MspA was discovered as the first porin of *M. smegmatis* (Niederweis et al., 1999). Deletion of *mspA* reduced the outer membrane permeability of *M. smegmatis* toward cephaloridine and glucose nine- and fourfold, respectively (Stahl et al., 2001). These results show that MspA is the major general diffusion pathway for these solutes in *M. smegmatis*. Consecutive deletions of the two porin genes *mspA* and *mspC* reduced the number of pores by 15-fold compared with wild-type *M. smegmatis*. The loss of these porins lowered the permeability for glucose by 75-fold, and the growth rate of *M. smegmatis* on plates and in full and minimal liquid medium dropped drastically (Stephan et al., 2005) (Table 1). This showed for the first time that the porin-mediated influx of hydrophilic nutrients limited the growth rate of this porin mutant. However, it is not clear which nutrient was in low supply for the porin mutants and whether the influx of nutrients limits the growth rate of wild-type *M. smegmatis*. The fact that the lack of the MspA and MspC porins also caused a reduced uptake of phosphates and a slower growth on low-phosphate plates (Wolschendorf et al., 2007) indeed suggested that the slow uptake of essential hydrophilic nutrients other than the carbon source may also contribute to the slow growth of *M. smegmatis* porin mutants.

Structure and pore-forming properties of MspA

The crystal structure of MspA revealed an octameric goblet-like conformation with a single central channel of 10-nm length and constitutes the first structure of a mycobacterial outer membrane protein (Faller et al., 2004). This structure is different from

Table 1. Mycobacterial porins of the MspA family[a]

Protein	Organism	Identity/similarity to MspA (%)	Length (aa)	Reference
MspA/Msmeg0965	*M. smegmatis*	100/100	211	gb|ABK74363.1| (Stahl et al., 2001)
MspB/Msmeg0520	*M. smegmatis*	94/95	215	gb|ABK73437.1| (Stahl et al., 2001)
MspC/Msmeg5483	*M. smegmatis*	93/95	215	gb|ABK74976.1| (Stahl et al., 2001)
MspD/Msmeg6057	*M. smegmatis*	82/89	207	gb|ABK72453.1| (Stahl et al., 2001)
MppA	*M. phlei*	100/100	211	AJ812030 (Dorner et al., 2004)
PorM1	*M. fortuitum*	95/96	211	emb|CAI54228.1|
PorM2	*M. fortuitum*	91/93	215	emb|CAL29811.1|
PorM1	*M. peregrinum*	94/96	211	emb|CAI54230.1|
Mmcs4296	*Mycobacterium* sp. strain MCS	85/91	216	gb|ABG10401.1|
Mmcs4297	*Mycobacterium* sp. strain MCS	85/91	216	gb|ABG10402.1|
Mmcs3857	*Mycobacterium* sp. strain MCS	30/44	235	gb|ABG09962.1|
Mmcs4382	*Mycobacterium* sp. strain MCS	85/91	216	gb|ABL93573.1|
Mmcs4383	*Mycobacterium* sp. strain MCS	85/91	216	gb|ABL93574.1|
Mjls3843	*Mycobacterium* sp. strain JLS	26/40	235	gb|ABN99619.1|
Mjls3857	*Mycobacterium* sp. strain JLS	26/40	235	gb|ABG09962.1|
Mjls3931	*Mycobacterium* sp. strain JLS	26/40	235	gb|ABL93123.1|
Mjls4674	*Mycobacterium* sp. strain JLS	85/89	216	gb|ABO00440.1|
Mjls4675	*Mycobacterium* sp. strain JLS	83/89	216	gb|ABO00441.1|
Mjls4677	*Mycobacterium* sp. strain JLS	84/89	216	gb|ABO00443.1|
Map3123c	*M. avium* subsp. *paratuberculosis*	24/39	220	gb|AAS05671.1|
Mav3943	*M. avium*	24/39	227	gb|ABK66660.1|
Mvan1836	*M. vanbaalenii* PYR-1	83/89	209	gb|ABM12657.1|
Mvan4117	*M. vanbaalenii* PYR-1	33/45	239	gb|ABM14894.1|
Mvan4839	*M. vanbaalenii* PYR-1	84/89	209	gb|ABM15612.1|
Mvan4840	*M. vanbaalenii* PYR-1	84/90	209	gb|ABM15613.1|
Mvan5016	*M. vanbaalenii* PYR-1	27/40	238	gb|ABM15788.1|
Mvan5017	*M. vanbaalenii* PYR-1	25/35	227	gb|ABM15789.1|
Mvan5768	*M. vanbaalenii* PYR-1	21/32	216	gb|ABM16533.1|
MUL_2391	*M. ulcerans* Agy99	21/34	233	gb|ABL04749.1|
Mflv1734	*M. gilvum* PYR-GCK	21/32	225	gb|ABP44214.1|
Mflv1735	*M. gilvum* PYR-GCK	32/41	226	gb|ABP44215.1|
Mflv2295	*M. gilvum* PYR-GCK	25/40	250	gb|ABP44773.1|
ND[b]	*M. gilvum* PYR-GCK	84/90	217	gb|ABP44371.1|

[a] Only proteins with amino acid similarities over the full length of the protein were included. Data were obtained by PSI-Blast using the NIH GenBank database at http://www.ncbi.nlm.nih.gov/blast/Blast.cgi.
[b] ND, not determined

that of the trimeric porins of gram-negative bacteria, which have one pore per monomer and are approximately 5 nm long (Koebnik et al., 2000) and defines MspA as the founding member of a new class of outer membrane proteins (Fig. 3). The crystal structure also revealed that the constriction zone of MspA consists of 16 aspartate residues (D90/D91). This creates a zone of highly negative charges and likely explains the preference of MspA for cations (Niederweis et al., 1999). The high number of negative charges in the vestibule and the channel interior may also contribute to the cation preference of MspA (Fig. 3B). The MspA pore provides an example how outer membrane transport proteins can contribute

to the selectivity of mycobacteria toward particular nutrients.

Topology of MspA in the outer membrane of *M. smegmatis*

MspA consists of two consecutive hydrophobic β-barrels of 3.7 nm total length and a more hydrophilic, globular rim domain (Faller et al., 2004). Thus, the length of the hydrophobic domain does not match the thickness of mycobacterial outer membranes (OMs) of 5 to 12 nm as derived from ultrathin sections of mycobacterial cells by electron microscopy (Paul and Beveridge, 1992, 1993, 1994).

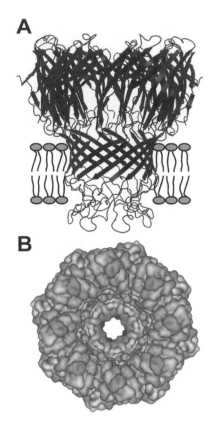

Figure 3. MspA, a general porin of *M. smegmatis*. (A) Side view of MspA integrated into a lipid bilayer. (B) Electrostatic potential of MspA in top view. The electrostatic potential is represented by the Gasteiger charges for the atoms in the surface of MspA. These figures are based on the crystal structure of MspA (Faller et al., 2004). Panels A and B were created with the visualization software PyMol (DeLano Scientific LLC) and ViewerLight (Accelrys), respectively. (*See color insert for the color version of this figure.*)

There are two alternative explanations for this hydrophobic mismatch: Either the OM of *M. smegmatis* is not thicker than 3.7 nm, or the large and hydrophilic rim domain of MspA is partially or completely embedded into the membrane. To solve this puzzle, the topology of MspA in the OM of *M. smegmatis* was examined by cysteine scanning mutagenesis (Mahfoud et al., 2006). The accessibility of 47 single cysteine mutants of MspA to a membrane-impermeable biotinylation reagent was compared in vitro and in vivo. Seventeen mutants covered the surface of the rim domain and were efficiently biotinylated in vitro. Six of these mutants were protected from biotinylation in *M. smegmatis* cells. These residues reside above a plane, which is defined by tryptophan 21. The eight phenylalanines at position 99 form a ring at the periplasmic end of the hydrophobic β-barrel domain. These results indicated that the membrane boundaries of MspA are defined by aromatic girdles. This may be a general feature of

mycobacterial outer membrane proteins as in gram-negative bacteria. Second, a 7.1-nm long stretch of MspA is not accessible to hydrophilic membrane-impermeable reagents. This fits very nicely with the thickness of 6.9 ± 1.0 nm as determined for the outer membrane of *M. smegmatis* from cryothin sections by electron microscopy. These results strongly indicated that a 7.1-nm-long stretch of MspA is inserted into the outer membrane of *M. smegmatis* (Faller et al., 2004). However, the chemical interactions between the hydrophilic rim domain of MspA and the constituents of the outer membrane are still unclear.

MspA is the prototype of a new family of porins of mycolic acid-containing bacteria

Based on phylogenetic trees derived from 16S rRNA sequences, Stackebrandt and colleagues (1997) proposed a new classification system for actinomycetes. According to this classification, mycobacteria belong to the suborder *Corynebacterineae* in the class *Actinobacteria,* which comprises among others the bacterial families *Corynebacteriaceae, Nocardiaceae,* and *Mycobacteriaceae.* Species of these bacterial families contain mycolic acids and are thought to have a similar cell wall architecture. Indirect experimental evidence that this assumption is correct comes from the increasing list of MspA-like porins in those bacteria (Dorner et al., 2004; Riess and Benz, 2000; Riess et al., 2001, 2003). Thus, MspA is now established as the prototype of a new family of porins of mycolic acid-containing bacteria. In particular, 33 MspA-like proteins have been identied in fast-growing mycobacteria (Table 2), as initially proposed based on hybridization of chromosomal DNA with the *mspA* gene (Niederweis et al., 1999). The existence of MspA-like porins in *M. ulcerans* and *M. avium* is of immediate medical interest considering the importance of MspA for drug susceptibility of *M. smegmatis* (see earlier in this chapter) and the difficulties of drug treatments of infections by *M. avium* (Rastogi et al., 1981; Tomioka, 2004) and *M. ulcerans* (Marsollier et al., 2003).

Porin-Mediated Outer Membrane Permeability and Slow Growth of Mycobacteria

The number of porins per cell determines the influx rate of hydrophilic solutes across the OM and was calculated for wild-type *M. smegmatis* and the porin mutants MN01 ($\Delta mspA$) and ML10 ($\Delta mspA$, $\Delta mspC$) to 2400, 800, and 150 pores, respectively, using a surface area of 2.4 μm^2 for an exponentially growing cell (Stephan et al., 2005). A striking observation was that the growth rate of *M. smegmatis*,

Table 2. Kinetic parameters and permeability coefficients P of mycobacteria and *E. coli* for glucose

Uptake parameter[a]	M. smegmatis			M. chelonae[d]	E. coli[e]
	Wild type[b]	ΔmspA[b]	ΔmspA, ΔmspC[c]		
V_{max} (nmol × min^{-1} × mg^{-1})	2.6	1.4	0.5	2.34	107
K_M (μM)	9.2	19.1	132	550	3
P (cm × s^{-1})	1.8×10^{-5}	4.6×10^{-6}	2.4×10^{-7}	2.7×10^{-7}	1.4×10^{-2}

[a]The V_{max} and K_M values are measured with whole cells and reflect the overall transport across both inner and outer membranes.
[b]Data from Stahl et al. (2001).
[c]Data from Stephan et al. (2005).
[d]Data from Jarlier and Nikaido (1990).
[e]Data from Bavoil et al. (1977).

both in rich and minimal medium and on agar plates, dropped drastically with its porin-mediated OM permeability. This is in contrast to porin mutants of *E. coli* and *Salmonella typhimurium*, which did not show a growth defect in rich media (Henning and Haller, 1975; Nurminen et al., 1976). Only under glucose limitation and in competition experiments, a growth advantage of wild-type *E. coli* over an *ompF* mutant was observed (Liu and Ferenci, 1998). Loss of both the OmpF and OmpC porins increased the doubling time of *E. coli* by 15% (Bavoil et al., 1977).

Why do porin mutants of *M. smegmatis* show a severe growth defect, but porin mutants of *E. coli* don't? Bavoil and Nikaido (1977) found that a 1,000-fold reduced residual porin level is sufficient to support growth of *E. coli* at maximal rate with 0.2% glucose. By contrast, in *M. smegmatis*, a modest 3- and 15-fold reduced number of porins increased the generation time by 30% and 80%, respectively. In minimal medium, the growth rate of *M. smegmatis* declined more rapidly with the number of porins than in rich medium (Stephan et al., 2005), underscoring the greater importance of porins at low nutrient concentration. This has been observed very early for *E. coli* (Bavoil et al., 1977) and is a consequence of the first Fick's law of diffusion. Higher influx rates are achieved at higher nutrient concentrations, and less porins are needed to maintain a constant nutrient influx into the cell under those conditions. These data fit perfectly to a straight line demonstrating that porin-mediated nutrient influx across the OM is the sole determinant of the growth rate of *M. smegmatis* porin mutants at low carbon concentrations. This was confirmed by complementation experiments, which restored both the OM permeability to hydrophilic solutes and the growth rate to wild-type levels. However, it is not clear whether the low efficiency of the porin pathway limits the growth rate of wild-type *M. smegmatis*. Experimental proof of this hypothesis would need to show that a larger number of porins in the wild-type strain would increase the growth rate. However, different *mspA* expression

vectors using strong mycobacterial promoters only increased the initial amount of *mspA* transcript but not the number of MspA porins in the OM (Hillmann and Niederweis, unpublished), indicating that unknown mechanisms limit integration of MspA into the OM. Acceleration of growth was observed after heterologous expression of *mspA* in *M. bovis* BCG (Mailaender et al., 2004; Sharbati-Tehrani et al., 2004). Taken together, these results establish the slow porin-mediated uptake of hydrophilic nutrients as an important determinant of the slow growth of mycobacteria as proposed earlier by Nikaido and coworkers (Jarlier and Nikaido, 1990). It is likely that slow nutrient uptake across the OM is not the only growth rate-limiting step and that the rates of other cellular processes have been adapted. Indeed, it was observed for *E. coli* that the effective permeabilities of the inner and the outer membranes are well matched for galactosides for concentrations in the range of 100 to 200 μM (West and Page, 1984). One obvious advantage of slow growth would be that less porins mean a lower permeability of the outer membrane for small toxic compounds such as the reactive oxygen and nitrogen intermediates that are generated by activated macrophages (Bogdan et al., 2000). Indeed, it was shown that porins contribute to a rapid elimination of *M. smegmatis* from macrophages, probably by increasing the permeability of toxic compounds generated by immune cells (Sharbati-Tehrani et al., 2005). These observations underscore the importance for *M. smegmatis* to balance its porin-mediated OM permeability between two mutually exclusive needs of rapid uptake of nutrients and protection from toxic compounds.

Role of Porins for Survival of Mycobacteria in Macrophages

Porin mutants are not yet available for pathogenic mycobacteria. Therefore, two approaches were used to examine the role of porins in survival of mycobacteria in macrophages. The first approach was motivated by the apparent lack of Msp-like porins in slowly growing mycobacteria. Expression of the major porin MspA of *M. smegmatis* increased the rate of glucose uptake (Mailaender et al., 2004) and the growth rate of *M. bovis* BCG (Mailaender et al., 2004; Sharbati-Tehrani et al., 2004). The faster growth of this recombinant *M. bovis* BCG strain in mouse macrophages indicated that nutrient uptake limits the growth of wild-type *M. bovis* BCG inside the phagosomal compartment. In apparent contrast, survival in macrophages of a double-porin mutant of *M. smegmatis* lacking MspA and MspC was enhanced compared with *M. smegmatis* (Sharbati-Tehrani et al., 2005). This indicated that nutrient uptake by *M. smegmatis* in phagosomes was not growth rate limiting under those conditions. Fast-growing mycobacteria such as *M. smegmatis* reside in phagosomes, representing a very different environment. They cannot prevent the maturation of the phagosomes (Cotter and Hill, 2003) and are accordingly heavily attacked by lysosomal compounds. The increased permeability of *M. smegmatis* for these solutes apparently outweighs the favorable effect of porins on nutrient supply. Taken together, these results underscore the pivotal importance of the outer membrane permeability for intracellular persistence of mycobacteria. The different effects of porin expression indicate that nutrient uptake represents a limiting factor for slow-growing mycobacteria, whereas less porins are beneficial for the fast-growing *M. smegmatis,* which are vulnerable to toxic lysosomal compounds.

Role of Porins for Antibiotic Sensitivity of Mycobacteria

Mycobacteria contain an outer membrane of unusually low permeability that contributes to their intrinsic resistance to many agents. It was assumed that small and hydrophilic antibiotics cross the outer membrane through porins, whereas hydrophobic antibiotics may diffuse through the membrane directly. First, a mutant of *M. smegmatis* lacking the major porin MspA was used to examine the role of the porin pathway in antibiotic sensitivity. Deletion of the *mspA* gene caused high-level resistance of *M. smegmatis* to 256 μg of ampicillin/ml by increasing the MIC 16-fold. The permeation of cephaloridine in the mspA mutant was reduced ninefold, and the resistance increased eightfold. This established a clear relationship between the activity and the outer membrane permeation of cephaloridine (Stephan et al., 2004). Recently, we showed that deletion of the porins MspC and MspD in addition to MspA rendered *M. smegmatis* completely resistant to β-lactam

antibiotics, whereas the β-lactamase activity remained unchanged (Danilchanka et al., submitted). The resistance to rifampicin, erythromycin, vancomycin, kanamycin, tetracycline, and some fluoroquinolones was increased by twofold for the triple-porin mutant, indicating minor or indirect contributions of the porins to uptake of these antibiotics.

The susceptibility of both *M. bovis* BCG and *M. tuberculosis* to zwitterionic β-lactam antibiotics was substantially enhanced by MspA, decreasing the minimal inhibitory concentration up to 16-fold. Furthermore, *M. tuberculosis* became significantly more susceptible to isoniazid, ethambutol, and streptomycin (Mailaender et al., 2004). These results highlight the prominent role of outer membrane permeability for the sensitivity of mycobacteria to antibiotics. An understanding of the pathways across the outer membrane is essential to the successful design of chemotherapeutic agents with activities against mycobacteria.

PORINS OF *M. TUBERCULOSIS*

Evidence for the Existence of Porins in Slow-Growing Mycobacteria

Direct evidence for the existence of proteins with channel-forming activity was obtained both for *M. tuberculosis* (Kartmann et al., 1999; Senaratne et al., 1998) and for *M. bovis* BCG (Lichtinger et al., 1999) by lipid bilayer experiments. In *M. tuberculosis,* protein channels with a rather small conductance of 0.7 nS in 1 M KCl were observed (Table 1). Kartmann et al. (1999) also detected channels with a larger conductance of 3 nS in 1 M KCl. Apparently consistent with these results was the detection of proteins that produced channels of 0.8 and 4 nS in detergent extracts of *M. bovis* BCG (Lichtinger et al., 1998), a species that is genetically almost identical to *M. tuberculosis* except for a limited number of chromosomal deletions (Behr et al., 1999; Brosch et al., 2001). However, the low-conductance channel of *M. bovis* BCG was found to be anion selective (Lichtinger et al., 1998) in contrast to the low-conductance channel of *M. tuberculosis* (Kartmann et al., 1999). This indicated that the proteins producing these channels are probably not identical. These studies also showed that *M. tuberculosis* and *M. bovis* BCG contained at least two different types of porins. Unfortunately, the amount of porins in detergent extracts was too low to allow purification and identification of these proteins.

There is also considerable evidence that *M. tuberculosis* takes up hydrophilic nutrients both in vitro and in vivo. *M. tuberculosis* can grow in vitro on small and hydrophilic nutrients including glucose, glycerol, and lactate as the sole carbon and energy sources (Edson, 1951; Schaefer et al., 1949). In addition, it has been shown that carbohydrate (Sassetti and Rubin, 2003), magnesium (Buchmeier et al., 2000) and phosphate transporters (Sassetti and Rubin, 2003) in the inner membrane are virulence factors of *M. tuberculosis*. This means that these nutrients also must cross the outer membrane of *M. tuberculosis*. Because direct diffusion of phosphates through model lipid membranes is extremely slow (permeability coefficient of the monoanion: 5×10^{-12} cm/s [Chakrabarti and Deamer, 1992]), it appears likely that mycobacteria use outer membrane pore proteins for uptake of phosphate. Evidence for this assumption was recently provided for *M. smegmatis* (Wolschendorf et al., 2007). A similar argument can be made also for the uptake of carbohydrates and divalent cations. However, uptake of these solutes across the outer membrane of *M. tuberculosis* may also be mediated by solute-specific proteins instead of porins.

OmpATb, a Pore-Forming Protein of *M. tuberculosis*

OmpATb (Rv0899) was discovered as a channel-forming protein of *M. tuberculosis* due to its (weak) similarity to OmpA proteins of gram-negative bacteria (Senaratne et al., 1998). To examine the physiological function of this protein, a mutant in *M. tuberculosis* was constructed. The suprising finding was that growth of the *ompATb* mutant was delayed by between 8 and 15 days after reinoculation into an acidified medium (pH 5.5) compared with medium with a pH of 7.2. Wild-type *M. tuberculosis*, on the other hand, needed only 2 days (Raynaud et al., 2002). Importantly, the organ burden established by the *ompATb* mutant in the lungs and the spleen of Balb/c mice was up to 1,000-fold reduced in intravenously infected mice (Raynaud et al., 2002). The failure of the *ompATb* mutant to grow at low pH is likely the underlying mechanism of its virulence defect in mice because activated macrophages are able to override the block of phagosome acidification exerted by pathogenic mycobacteria and to lower the pH to 5.2 inside phagocytic vacuoles (Schaible et al., 1998). The strong induction of *ompATb* transcription at pH 5.5 in vitro (30-fold) and in murine bone marrow-derived macrophages (fivefold) (Raynaud et al., 2002) suggests that acidification of the phagosome is the signal which triggers an OmpATb-depending defense mechanism of *M. tuberculosis* in macrophages to cope with growth-limiting proton concentrations. This fascinating observation has been reviewed recently (Niederweis, 2003; Smith,

2003). The molecular mechanism by which OmpATb enables *M. tuberculosis* to overcome the block of growth at low pH is unknown. It was concluded that OmpATb may be a porin of *M. tuberculosis* based on its channel-forming properties (Senaratne et al., 1998). Apparently consistent with this assumption was the observation that uptake of serine but not of glycine was reduced in the *ompATb* mutant compared with that of wild-type *M. tuberculosis* (Raynaud et al., 2002). However, the overall permeability of the outer membrane of *M. tuberculosis* was reduced at pH 5.5 compared with pH 7.2, although the levels of OmpATb in the outer membrane were strongly increased (Raynaud et al., 2002). Considering these contradicting results, it is doubtful that the major function of OmpATb in *M. tuberculosis* is that of a porin (Niederweis, 2003). The recent observation that a central domain of approximately 150 amino acids is sufficient for the channel activity of OmpATb in vitro confirmed the initial findings that OmpATb is indeed a channel protein but did not contribute to the understanding of its biological functions (Molle et al., 2006). Pore-forming proteins of *M. tuberculosis* other than OmpATb are currently not known.

How to Demonstrate that a Protein has the Physiological Function of a Porin

The long-standing discussion about the putative physiological function of OmpA as a porin of *E. coli* (Arora et al., 2001; Pautsch and Schulz, 2000; Sugawara and Nikaido, 1992, 1994; Sugawara et al., 2006) motivated me to write a note of caution. Channel activity of a purified protein is a necessary but not sufficient requirement of porins because many outer membrane proteins form channels but do not have the physiological function of porins. The function of porins is to form (transiently) open channels that allow the influx of small and hydrophilic nutrient molecules. The classification of a pore protein as a porin requires to demonstrate that a mutant lacking this protein shows a reduced permeability to small and hydrophilic solutes. This has been demonstrated for many gram-negative bacteria including *E. coli*, *P. aeruginosa*, and others (Hancock and Brinkman, 2002; Nikaido, 2003). This requirement imposes an experimental problem: most bacteria possess several porins with complete or partially redundant functions. For example, the genome of *E. coli* encodes for at least seven porins including the main porins OmpC and OmpF (Nikaido, 2003). Thus, often single porin mutants would not show a permeability defect. As a consequence, it may require the construction of multiple deletions in the same strain to demonstrate the physiological function of a particular protein as a porin. Using this strategy we demonstrated that MspA and the other Msp proteins function as general porins that mediate the uptake of very diverse small and hydrophilic solutes, such as glucose, serine, phosphate, and β-lactam antibiotics (Stahl et al., 2001; Stephan et al., 2004; Stephan et al., 2005; Wolschendorf et al., 2007). It is still unclear whether *M. tuberculosis* has proteins similar to the Msp porins.

Role of the Porin Pathway for *M. tuberculosis*

To assess the role of the porin pathway in slowly growing mycobacteria, the porin MspA of *M. smegmatis* was expressed in *M. bovis* BCG and *M. tuberculosis* (Mailaender et al., 2004; Sharbati-Tehrani et al., 2004). Only 20 and 35 MspA molecules per mm^2 of cell wall were expressed in *M. tuberculosis* and *M. bovis* BCG, respectively, compared with approximately 2,400 from the same expression vector in *M. smegmatis* (Mailaender et al., 2004). It was surprising that this low number of MspA pores caused a twofold faster glucose uptake by *M. bovis* BCG, indicating that the outer membrane permeability to small and hydrophilic solutes was increased. The accelerated growth of *M. bovis* BCG on expression of MspA identified slow nutrient uptake as one of the determinants of slow growth in mycobacteria. Both *M. bovis* BCG and *M. tuberculosis* became significantly more susceptible to zwitter-ionic β-lactam antibiotics, isoniazid, ethambutol, and streptomycin. These results demonstrated that MspA enhanced both nutrient uptake, growth rate, and drug susceptibility. This study provided the first experimental evidence that porins are important for drug susceptibility of *M. tuberculosis* (Mailaender et al., 2004).

Acknowledgments. I thank Jason Huff for critically reading the manuscript and Claudia Mailaender for an initial version of Fig. 2.

This work was supported by grant AI063432 from the National Institutes of Health.

REFERENCES

Arora, A., F. Abildgaard, J. H. Bushweller, and L. K. Tamm. 2001. Structure of outer membrane protein A transmembrane domain by NMR spectroscopy. *Nat. Struct. Biol.* **8:**334–338.

Barry, C. E. 2001. Interpreting cell wall 'virulence factors' of *Mycobacterium tuberculosis. Trends Microbiol.* **9:**237–241.

Barry, C. E., III, R. E. Lee, K. Mdluli, A. E. Sampson, B. G. Schroeder, R. A. Slayden, and Y. Yuan. 1998. Mycolic acids: structure, biosynthesis and physiological functions. *Prog. Lipid Res.* **37:**143–179.

Bauer, F. J., T. Rudel, M. Stein, and T. F. Meyer. 1999. Mutagenesis of the *Neisseria gonorrhoeae* porin reduces invasion in epithelial cells and enhances phagocyte responsiveness. *Mol. Microbiol.* **31:**903–913.

Bavoil, P., H. Nikaido, and K. von Meyenburg. 1977. Pleiotropic transport mutants of *Escherichia coli* lack porin, a major outer membrane protein. *Mol. Gen. Genet.* **158:**23–33.

Behr, M. A., M. A. Wilson, W. P. Gill, H. Salamon, G. K. Schoolnik, S. Rane, and P. M Small. 1999. Comparative genomics of BCG vaccines by whole-genome DNA microarray. *Science* **284:**1520–1523.

Beveridge, T. J. 1995. The periplasmic space and the periplasm in gram-positive and gram-negative bacteria. *ASM News* **61:**125–130.

Beveridge, T. J. 1999. Structures of gram-negative cell walls and their derived membrane vesicles. *J. Bacteriol.* **181:**4725–4733.

Beveridge, T. J., and J. L. Kadurugamuwa. 1996. Periplasm, periplasmic spaces, and their relation to bacterial wall structure: novel secretion of selected periplasmic proteins from *Pseudomonas aeruginosa. Microb. Drug Resist.* **2:**1–8.

Bogdan, C., M. Röllinghoff, and A. Diefenbach. 2000. Reactive oxygen and reactive nitrogen intermediates in innate and specific immunity. *Curr. Opin. Immunol.* **12:**64–76.

Braun, V., and H. Killmann. 1999. Bacterial solutions to the iron-supply problem. *Trends Biochem. Sci.* **24:**104–109.

Brennan, P. J., and H. Nikaido. 1995. The envelope of mycobacteria. *Annu. Rev. Biochem.* **64:**29–63.

Brosch, R., A. S. Pym, S. V. Gordon, and S. T. Cole. 2001. The evolution of mycobacterial pathogenicity: clues from comparative genomics. *Trends Microbiol.* **9:**452–458.

Bucarey, S. A., N. A. Villagra, M. P. Martinic, A. N. Trombert, C. A. Santiviago, N. P. Maulen, P. Youderian, and G. C. Mora. 2005. The *Salmonella enterica* serovar Typhi *tsx* gene, encoding a nucleoside-specific porin, is essential for prototrophic growth in the absence of nucleosides. *Infect. Immun.* **73:**6210–6219.

Buchmeier, N., A. Blanc-Potard, S. Ehrt, D. Piddington, L. Riley, and E. A. Groisman. 2000. A parallel intraphagosomal survival strategy shared by *Mycobacterium tuberculosis* and *Salmonella enterica. Mol. Microbiol.* **35:**1375–1382.

Chakrabarti, A. C., and D. W. Deamer. 1992. Permeability of lipid bilayers to amino acids and phosphate. *Biochim. Biophys. Acta* **1111:**171–177.

Cole, S. T., and B. G. Barrell. 1998. Analysis of the genome of Mycobacterium tuberculosis H37Rv. *Novartis Found. Symp.* **217:**160–172; discussion 172–177.

Cotter, P. D., and C. Hill. 2003. Surviving the acid test: responses of gram-positive bacteria to low pH. *Microbiol. Mol. Biol. Rev.* **67:**429–453.

Daffé, M., and P. Draper. 1998. The envelope layers of mycobacteria with reference to their pathogenicity. *Adv. Microb. Physiol.* **39:**131–203.

Daffé, M., and G. Etienne. 1999. The capsule of *Mycobacterium tuberculosis* and its implications for pathogenicity. *Tuber. Lung Dis.* **79:**153–169.

Dorner, U., E. Maier, and R. Benz. 2004. Identification of a cation-specific channel (TipA) in the cell wall of the gram-positive mycolata Tsukamurella inchonensis: the gene of the channel-forming protein is identical to mspA of Mycobacterium smegmatis and mppA of Mycobacterium phlei. *Biochim. Biophys. Acta* **1667:**47–55.

Draper, P. 1998. The outer parts of the mycobacterial envelope as permeability barriers. *Front. Biosci.* **3:**1253–1261.

Dumas, F., R. Koebnik, M. Winterhalter, and P. van Gelder. 2000. Sugar transport through maltoporin of *Escherichia coli.* Role of polar tracks. *J. Biol. Chem.* **275:**19747–19751.

Edson, N. L. 1951. The intermediary metabolism of the mycobacteria. *Bacteriol. Rev.* **15:**147–182.

Etienne, G., F. Laval, C. Villeneuve, P. Dinadayala, A. Abouwarda, D. Zerbib, A. Galamba, and M. Daffe. 2005. The cell envelope structure and properties of *Mycobacterium smegmatis* mc(2)155: is there a clue for the unique transformability of the strain? *Microbiology* **151:**2075–2086.

Faller, M., M. Niederweis, and G. E. Schulz. 2004. The structure of a mycobacterial outer-membrane channel. *Science* 303:1189–1192.

Fu, L. M., and C. S. Fu-Liu. 2002a. Is *Mycobacterium tuberculosis* a closer relative to Gram-positive or Gram-negative bacterial pathogens? *Tuberculosis* 82:85–90.

Fu, L. M., and C. S. Fu-Liu. 2002b. Genome comparison of *Mycobacterium tuberculosis* and other bacteria. *Omics* 6:199–206.

Galdiero, F., G. C. de L'ero, N. Benedetto, M. Galdiero, and M. A. Tufano. 1993. Release of cytokines induced by Salmonella typhimurium porins. *Infect. Immun.* 61:155–161.

Galdiero, M., M. G. Pisciotta, E. Galdiero, and C. R. Carratelli. 2003. Porins and lipopolysaccharide from *Salmonella typhimurium* regulate the expression of CD80 and CD86 molecules on B cells and macrophages but not CD28 and CD152 on T cells. *Clin. Microbiol. Infect.* 9:1104–1111.

Galdiero, F., M. A. Tufano, M. Galdiero, S. Masiello, and M. Di Rosa. 1990. Inflammatory effects of Salmonella typhimurium porins. *Infect. Immun.* 58:3183–3186.

Gao, B., R. Paramanathan, and R. S. Gupta. 2006. Signature proteins that are distinctive characteristics of *Actinobacteria* and their subgroups. *Antonie Leeuwenhoek* 90:69–91.

Graham, L. L., T. J. Beveridge, and N. Nanninga. 1991. Periplasmic space and the concept of the periplasm. *Trends Biochem. Sci.* 16:328–329.

Hancock, R. E., and F. S. Brinkman. 2002. Function of *Pseudomonas* porins in uptake and efflux. *Annu. Rev. Microbiol.* 56:17–38.

Hancock, R. E., S. W. Farmer, Z. S. Li, and K. Poole. 1991. Interaction of aminoglycosides with the outer membranes and purified lipopolysaccharide and OmpF porin of *Escherichia coli*. *Antimicrob. Agents Chemother.* 35:1309–1314.

Henning, U., and I. Haller. 1975. Mutants of *Escherichia coli* K12 lacking all 'major' proteins of the outer cell envelope membrane. *FEBS Lett.* 55:161–164.

Jarlier, V., and H. Nikaido. 1990. Permeability barrier to hydrophilic solutes in *Mycobacterium chelonei*. *J. Bacteriol.* 172:1418–1423.

Kartmann, B., S. Stenger, and M. Niederweis. 1999. Porins in the cell wall of *Mycobacterium tuberculosis*. *J. Bacteriol.* 181:6543–6546. (Authors' correction appeared in *J. Bacteriol.* 181:7650).

Koch, R. 1882. Die Aetiologie der Tuberculose. *Berl. Klin. Wochenzeitsch.* 19:18.

Koebnik, R., K. P. Locher, and P. van Gelder. 2000. Structure and function of bacterial outer membrane proteins: barrels in a nutshell. *Mol. Microbiol.* 37:239–253.

Lemassu, A., A. Ortalo-Magne, F. Bardou, G. Silve, M. A. Laneelle, and M. Daffé. 1996. Extracellular and surface-exposed polysaccharides of non-tuberculous mycobacteria. *Microbiology* 142:1513–1520.

Lichtinger, T., A. Burkovski, M. Niederweis, R. Kramer, and R. Benz. 1998. Biochemical and biophysical characterization of the cell wall porin of *Corynebacterium glutamicum*: the channel is formed by a low molecular mass polypeptide. *Biochemistry* 37:15024–15032.

Lichtinger, T., B. Heym, E. Maier, H. Eichner, S. T. Cole, and R. Benz. 1999. Evidence for a small anion-selective channel in the cell wall of *Mycobacterium bovis* BCG besides a wide cation-selective pore. *FEBS Lett.* 454:349–355.

Liu, X., and T. Ferenci. 1998. Regulation of porin-mediated outer membrane permeability by nutrient limitation in *Escherichia coli*. *J. Bacteriol.* 180:3917–3922.

Mahfoud, M., S. Sukumaran, P. Hülsmann, K. Grieger, and M. Niederweis. 2006. Topology of the porin MspA in the outer membrane of *Mycobacterium smegmatis*. *J. Biol. Chem.* 281:5908–5915.

Maier, C., E. Bremer, A. Schmid, and R. Benz. 1988. Pore-forming activity of the Tsx protein from the outer membrane of *Escherichia coli*. Demonstration of a nucleoside-specific binding site. *J. Biol. Chem.* 263:2493–2499.

Mailaender, C., N. Reiling, H. Engelhardt, S. Bossmann, S. Ehlers, and M. Niederweis. 2004. The MspA porin promotes growth and increases antibiotic susceptibility of both *Mycobacterium bovis* BCG and *Mycobacterium tuberculosis*. *Microbiology* 150:853–864.

Marsollier, L., G. Prevot, N. Honore, P. Legras, A. L. Manceau, C. Payan, H. Kouakou, and B. Carbonnelle. 2003. Susceptibility of *Mycobacterium ulcerans* to a combination of amikacin/rifampicin. *Int. J. Antimicrob. Agents* 22:562–566.

Massari, P., S. Ram, H. Macleod, and L. M. Wetzler. 2003. The role of porins in neisserial pathogenesis and immunity. *Trends Microbiol.* 11:87–93.

Matias, V. R., A. Al-Amoudi, J. Dubochet, and T. J. Beveridge. 2003. Cryo-transmission electron microscopy of frozen-hydrated sections of *Escherichia coli* and *Pseudomonas aeruginosa*. *J. Bacteriol.* 185:6112–6118.

Matias, V. R., and T. J. Beveridge. 2005. Cryo-electron microscopy reveals native polymeric cell wall structure in *Bacillus subtilis 168* and the existence of a periplasmic space. *Mol. Microbiol.* 56:240–251.

Matias, V. R., and T. J. Beveridge. 2006. Native cell wall organization shown by cryo-electron microscopy confirms the existence of a periplasmic space in *Staphylococcus aureus*. *J. Bacteriol.* 188:1011–1021.

McKinney, J. D., K. Honer zu Bentrup, E. J. Munoz-Elias, A. Miczak, B. Chen, W. T. Chan, D. Swenson, J. C. Sacchettini, W. R. Jacobs, Jr., and D. G. Russell. 2000. Persistence of *Mycobacterium tuberculosis* in macrophages and mice requires the glyoxylate shunt enzyme isocitrate lyase. *Nature* 406:735–738.

Merz, A. J., and M. So. 2000. Interactions of pathogenic *Neisseriae* with epithelial cell membranes. *Annu. Rev. Cell Dev. Biol.* 16:423–457.

Mineda, T., N. Ohara, H. Yukitake, and T. Yamada. 1998. The ribosomes contents of mycobacteria. *New Microbiol.* 21:1–7.

Minnikin, D. E. 1982. Lipids: complex lipids, their chemistry, biosynthesis and roles, p. 95–184. In C. Ratledge and J. Stanford (ed.), *The Biology of the Mycobacteria: Physiology, Identification and Classification*, vol. I. Academic Press, London, United Kingdom.

Molle, V., N. Saint, S. Campagna, L. Kremer, E. Lea, P. Draper, and G. Molle. 2006. pH-dependent pore-forming activity of OmpATb from *Mycobacterium tuberculosis* and characterization of the channel by peptidic dissection. *Mol. Microbiol.* 61:826–837.

Munoz-Elias, E. J., and J. D. McKinney. 2005. *Mycobacterium tuberculosis* isocitrate lyases 1 and 2 are jointly required for in vivo growth and virulence. *Nat. Med.* 11:638–644.

Negm, R. S., and T. G. Pistole. 1998. Macrophages recognize and adhere to an OmpD-like protein of *Salmonella typhimurium*. *FEMS Immunol. Med. Microbiol.* 20:191–199.

Negm, R. S., and T. G. Pistole. 1999. The porin OmpC of *Salmonella typhimurium* mediates adherence to macrophages. *Can. J. Microbiol.* 45:658–669.

Niederweis, M. 2003. Mycobacterial porins-new channel proteins in unique outer membranes. *Mol. Microbiol.* 49:1167–1177.

Niederweis, M., S. Ehrt, C. Heinz, U. Klöcker, S. Karosi, K. M. Swiderek, L. W. Riley, and R. Benz. 1999. Cloning of the *mspA* gene encoding a porin from *Mycobacterium smegmatis*. *Mol. Microbiol.* 33:933–945.

Nikaido, H. 1994. Porins and specific diffusion channels in bacterial outer membranes. *J. Biol. Chem.* 269:3905–3908.

Nikaido, H. 2003. Molecular basis of bacterial outer membrane permeability revisited. *Microbiol. Mol. Biol. Rev.* 67:593–656.

Nikaido, H., S. H. Kim, and E. Y. Rosenberg. 1993. Physical organization of lipids in the cell wall of *Mycobacterium chelonae*. *Mol. Microbiol.* **8:**1025–1030.

Nurminen, M., K. Lounatmaa, M. Sarvas, P. H. Makela, and T. Nakae. 1976. Bacteriophage-resistant mutants of *Salmonella typhimurium* deficient in two major outer membrane proteins. *J. Bacteriol.* **127:**941–955.

Paul, T. R., and T. J. Beveridge. 1992. Reevaluation of envelope profiles and cytoplasmic ultrastructure of mycobacteria processed by conventional embedding and freeze-substitution protocols. *J. Bacteriol.* **174:**6508–6517.

Paul, T. R., and T. J. Beveridge. 1993. Ultrastructure of mycobacterial surfaces by freeze-substitution. *Zentbl. Bakteriol.* **279:**450–457.

Paul, T. R., and T. J. Beveridge. 1994. Preservation of surface lipids and determination of ultrastructure of *Mycobacterium kansasii* by freeze-substitution. *Infect. Immun.* **62:**1542–1550.

Pautsch, A., and G. E. Schulz. 2000. High-resolution structure of the OmpA membrane domain. *J. Mol. Biol.* **298:**273–282.

Ramakrishnan, T., P. S. Murthy, and K. P. Gopinathan. 1972. Intermediary metabolism of mycobacteria. *Bacteriol. Rev.* **36:**65–108.

Rastogi, N., C. Frehel, A. Ryter, H. Ohayon, M. Lesourd, and H. L. David. 1981. Multiple drug resistance in *Mycobacterium avium*: is the wall architecture responsible for exclusion of antimicrobial agents? *Antimicrob. Agents Chemother.* **20:**666–677.

Rastogi, N., E. Legrand, and C. Sola. 2001. The mycobacteria: an introduction to nomenclature and pathogenesis. *Rev. Sci. Tech.* **20:**21–54.

Ratledge, C. 1982. Nutrition, growth and metabolism, p. 186–212. *In* C. Ratledge and J. Stanford (ed.), *The Biology of the Mycobacteria*. Academic Press, Ltd., London, United Kingdom.

Raynaud, C., K. G. Papavinasasundaram, R. A. Speight, B. Springer, P. Sander, E. C. Böttger, M. J. Colston, and P. Draper. 2002. The functions of OmpATb, a pore-forming protein of *Mycobacterium tuberculosis*. *Mol. Microbiol.* **46:**191–201.

Riess, F. G., and R. Benz. 2000. Discovery of a novel channel-forming protein in the cell wall of the non-pathogenic *Nocardia corynebacteroides*. *Biochim. Biophys. Acta* **1509:**485–495.

Riess, F. G., U. Dorner, B. Schiffler, and R. Benz. 2001. Study of the properties of a channel-forming protein of the cell wall of the Gram-positive bacterium *Mycobacterium phlei*. *J. Membr. Biol.* **182:**147–157.

Riess, F. G., M. Elflein, M. Benk, B. Schiffler, R. Benz, N. Garton, and I. Sutcliffe. 2003. The cell wall of the pathogenic bacterium *Rhodococcus equi* contains two channel-forming proteins with different properties. *J. Bacteriol.* **185:**2952–2960.

Rodriguez-Morales, O., M. Fernandez-Mora, I. Hernandez-Lucas, A. Vazquez, J. L. Puente, and E. Calva. 2006. *Salmonella* enterica serovar Typhimurium ompS1 and ompS2 mutants are attenuated for virulence in mice. *Infect. Immun.* **74:**1398–1402.

Rudel, T., A. Schmid, R. Benz, H. A. Kolb, F. Lang, and T. F. Meyer. 1996. Modulation of Neisseria porin (PorB) by cytosolic ATP/GTP of target cells: parallels between pathogen accommodation and mitochondrial endosymbiosis. *Cell* **85:**391–402.

Sassetti, C. M., and E. J. Rubin. 2003. Genetic requirements for mycobacterial survival during infection. *Proc. Natl. Acad. Sci. USA* **100:**12989–12994.

Schaefer, W. B., A. Marshak, and B. Burkhart. 1949. The growth of *Mycobacterium tuberculosis* as a function of its nutrients. *J. Bacteriol.* **58:**549–563.

Schaible, U. E., S. Sturgill-Koszycki, P. H. Schlesinger, and D. G. Russell. 1998. Cytokine activation leads to acidification and increases maturation of *Mycobacterium avium*-containing phagosomes in murine macrophages. *J. Immunol.* **160:**1290–1296.

Secundino, I., C. Lopez-Macias, L. Cervantes-Barragan, C. Gil-Cruz, N. Rios-Sarabia, R. Pastelin-Palacios, M. A. Villasis-Keever, I. Becker, J. L. Puente, E. Calva, and A. Isibasi. 2006. *Salmonella* porins induce a sustained, lifelong specific bactericidal antibody memory response. *Immunology* **117:**59–70.

Senaratne, R. H., H. Mobasheri, K. G. Papavinasasundaram, P. Jenner, E. J. Lea, and P. Draper. 1998. Expression of a gene for a porin-like protein of the OmpA family from *Mycobacterium tuberculosis* H37Rv. *J. Bacteriol.* **180:**3541–3547.

Sharbati-Tehrani, S., B. Meister, B. Appel, and A. Lewin. 2004. The porin MspA from *Mycobacterium smegmatis* improves growth of *Mycobacterium bovis* BCG. *Int. J. Med. Microbiol.* **294:**235–245.

Sharbati-Tehrani, S., J. Stephan, G. Holland, B. Appel, M. Niederweis, and A. Lewin. 2005. Porins limit the intracellular persistence of *Mycobacterium smegmatis*. *Microbiology* **151:**2403–2410.

Smith, I. 2003. *Mycobacterium tuberculosis* pathogenesis and molecular determinants of virulence. *Clin. Microbiol. Rev.* **16:**463–496.

Stackebrandt, E., F. A. Rainey, and N. L. Ward-Rainey. 1997. Proposal for a new hierarchic classification system, *Actinobacteria* classis nov. *Int. J. Syst. Bacteriol.* **47:**479–491.

Stahl, C., S. Kubetzko, I. Kaps, S. Seeber, H. Engelhardt, and M. Niederweis. 2001. MspA provides the main hydrophilic pathway through the cell wall of *Mycobacterium smegmatis*. *Mol. Microbiol.* **40:**451–464. (Authors' correction, **57:**1509, 2005.)

Stephan, J., J. Bender, F. Wolschendorf, C. Hoffmann, E. Roth, C. Mailänder, H. Engelhardt, and M. Niederweis. 2005. The growth rate of *Mycobacterium smegmatis* depends on sufficient porin-mediated influx of nutrients. *Mol. Microbiol.* **58:**714–730.

Stephan, J., C. Mailaender, G. Etienne, M. Daffe, and M. Niederweis. 2004. Multidrug resistance of a porin deletion mutant of *Mycobacterium smegmatis*. *Antimicrob. Agents Chemother.* **48:**4163–4170.

Sugawara, E., E. M. Nestorovich, S. M. Bezrukov, and H. Nikaido. 2006. *Pseudomonas aeruginosa* porin OprF exists in two different conformations. *J. Biol. Chem.* **281:**16220–16229.

Sugawara, E., and H. Nikaido. 1992. Pore-forming activity of OmpA protein of *Escherichia coli*. *J. Biol. Chem.* **267:**2507–2511.

Sugawara, E., and H. Nikaido. 1994. OmpA protein of *Escherichia coli* outer membrane occurs in open and closed channel forms. *J. Biol. Chem.* **269:**17981–17987.

Tomioka, H. 2004. Present status and future prospects of chemotherapeutics for intractable infections due to *Mycobacterium avium* complex. *Curr. Drug Discov. Technol.* **1:**255–268.

Trias, J., V. Jarlier, and R. Benz. 1992. Porins in the cell wall of mycobacteria. *Science* **258:**1479–1481.

Tufano, M. A., L. Biancone, F. Rossano, C. Capasso, A. Baroni, A. De Martino, E. L. Iorio, L. Silvestro, and G. Camussi. 1993. Outer-membrane porins from gram-negative bacteria stimulate platelet-activating-factor biosynthesis by cultured human endothelial cells. *Eur. J. Biochem.* **214:**685–693.

van Putten, J. P., T. D. Duensing, and J. Carlson. 1998. Gonococcal invasion of epithelial cells driven by P.IA, a bacterial ion channel with GTP binding properties. *J. Exp. Med.* **188:**941–952.

Vordermeier, H. M., and W. G. Bessler. 1987. Polyclonal activation of murine B lymphocytes in vitro by Salmonella typhimurium porins. *Immunobiology* **175:**245–251.

Vordermeier, H. M., H. Drexler, and W. G. Bessler. 1987. Polyclonal activation of human peripheral blood lymphocytes by bacterial porins and defined porin fragments. *Immunol. Lett.* 15:121–126.

Wagner, V. E., J. G. Frelinger, R. K. Barth, and B. II. Iglewski. 2006. Quorum sensing: dynamic response of *Pseudomonas aeruginosa* to external signals. *Trends Microbiol.* 14:55–58.

Wen, K. K., P. C. Giardina, M. S. Blake, J. Edwards, M. A. Apicella, and P. A. Rubenstein. 2000. Interaction of the gonococcal porin P.IB with G- and F-actin. *Biochemistry* 39:8638–8647.

West, I. C., and M. G. Page. 1984. When is the outer membrane of *Escherichia coli* rate-limiting for uptake of galactosides? *J. Theor. Biol.* 110:11–19.

Wolschendorf, F., M. Mahfoud, and M. Niederweis. 2007. Porins are required for uptake of phosphates by *Mycobacterium smegmatis. J. Bacteriol.* 189:2435–2442.

Wu, L., O. Estrada, O. Zaborina, M. Bains, L. Shen, J. E. Kohler, N. Patel, M. W. Musch, E. B. Chang, Y. X. Fu, M. A. Jacobs, M. I. Nishimura, R. E. Hancock, J. R. Turner, and J. C. Alverdy. 2005. Recognition of host immune activation by *Pseudomonas aeruginosa. Science* 309:774–777.

The Mycobacterial Cell Envelope
Edited by M. Daffé and J.-M. Reyrat
© 2008 ASM Press, Washington, DC

Chapter 10

Iron Uptake in Mycobacteria

LUIS E. N. QUADRI

Iron plays central roles in vital metabolic functions and is an essential element for most bacteria, which require a constant iron supply that they must obtain from their environment (Braun, 2001; Jurado, 1997). Typically, iron is found in its ferric form (Fe^{3+}) and as virtually insoluble ferric hydroxides under aerobic conditions in the absence of iron-solubilizing agents. The total concentration of soluble iron at pH $\approx$ 7 is believed to be in the order of 10^{-10} M to 10^{-9} M (Boukhalfa and Crumbliss, 2002; Chipperfield and Ratledge, 2000). This concentration was revised from previous lower estimates, but it is still several orders of magnitude below the minimal concentration required by bacteria, which need 10^{-7} to 10^{-6} M iron in their environment for proper multiplication (Braun, 2001; Jurado, 1997). Thus, to guarantee a suitable iron supply, both pathogenic and nonpathogenic bacteria necessitate efficient iron acquisition systems. Among these systems are those based on small (<1,000 Da) iron-chelating compounds, called siderophores, that have very high affinity for Fe^{3+} (e.g., K_d <10^{-25} M) (Drechsel and Winkelmann, 1997; Neilands, 1995). Siderophores act as Fe^{3+} solubilizing and chelating agents and are, in most cases, secreted into the extracellular environment and transported back into the cell after chelating extracellular Fe^{3+} (Neilands, 1995; Ratledge and Dover, 2000). Inside the cell, the iron can be recovered from the Fe^{3+}-siderophore complex (ferrisiderophore) for metabolic use or temporary storage (Drechsel and Winkelmann, 1997; Neilands, 1995; Ratledge and Dover, 2000). It is important to note that this general model may not be fully applicable to all siderophores. For example, as discussed in this chapter, it is not yet clear whether some mycobacterial siderophores enter the cell's cytoplasm or simply act as ionophores to facilitate passage of iron across the plasma membrane.

In the case of pathogenic bacteria, the organisms must obtain iron from the host. In this situation, bacteria face conditions that make iron acquisition particularly challenging, and thus pathogens have evolved sophisticated mechanisms that allow them to capture iron from host sources (Ratledge and Dover, 2000; Wandersman and Delepelaire, 2004; Wooldridge and Williams, 1993). In mammals, for example, iron is bound to intracellular and extracellular components, thus rendering the free iron concentration (10^{-15} to 10^{-24} M) well below what is required for bacterial multiplication (Jurado, 1997; Ward et al., 1996; Weinberg, 1978; Weinberg, 1993). During infection, the level of free iron is further reduced by host iron withholding mechanisms that lead to hypoferremia. This host defense strategy is intended to restrict the multiplication of the invading pathogens by limiting the bioavailability of iron (Jurado, 1997; Ward et al., 1996; Weinberg, 1993). Notably, recent reports indicate that siderocalin (lipocalin 2), a member of the lipocalin family of binding proteins, binds ferrisiderophores, including ferricarboxymycobactins (Goetz et al., 2002; Holmes et al., 2005b). Carboxymycobactins are a type of mycobacterial siderophores that will be discussed in the ensuing text. Siderocalin is found in neutrophil granules and uterine secretions, and its levels are dramatically elevated in serum and synovial membrane during bacterial infection (Kjeldsen et al., 2000; Xu and Venge, 2000). This ferrisiderophore-binding protein is also secreted by epithelial cells in response to tumorigenesis or inflammation (Kjeldsen et al., 2000). Siderocalin has strong bacteriostatic activity in vitro under iron-limiting conditions, and siderocalin knockout mice have increased susceptibility to bacterial infection (Flo et al., 2004; Goetz et al., 2002). These observations have led to the hypothesis that sequestration of ferrisiderophore complexes by siderocalin complements the iron-withholding defense mechanisms of the mammalian innate immune system (Goetz et al., 2002). Notably, the importance of iron availability for the human pathogen *Mycobacterium tuberculosis* is consistent with the fact that iron overload intensifies

Luis E. N. Quadri • Department of Microbiology and Immunology, Weill Medical College of Cornell University, New York, NY 10021.

disease susceptibility and progression in animal models and humans (Cronje and Bornman, 2005; Gangaidzo et al., 2001; Lounis et al., 2001; Schaible et al., 2002).

It has been widely recognized for quite some time that iron limitation triggers up-regulation of siderophore systems to facilitate an increase in the sequestration of environmental iron via siderophore-dependent Fe^{3+} uptake (Crosa, 1997; Ratledge and Dover, 2000). Nevertheless, although iron uptake is an essential process, it must be tightly controlled to maintain proper iron homeostasis and prevent uptake of an excessive amount of iron. Excess of iron results in cellular toxicity due to the promotion of reactive oxygen intermediate production through the Fenton reaction (Andrews et al., 2003; Touati, 2000). Thus, upon iron sufficiency, the siderophore systems are actively repressed at the level of transcription. In fact, the upregulation of siderophore systems during the iron-deficient state of the cell is due to the cessation of transcriptional repression. Recent studies conducted in the laboratory of the author (Palma et al., 2003) and in other laboratories addressing global analysis of gene and protein expression changes induced in pathogenic and nonpathogenic bacteria growing under low iron conditions suggest that iron restriction not only leads to an upregulation of siderophore production but also to a complex bacterial adaptive response (Baichoo et al., 2002; Bjarnason et al., 2003; Ducey et al., 2005; Ernst et al., 2005; Grifantini et al., 2003; Holmes et al., 2005a; Merrell et al., 2003; Miethke et al., 2006; Ochsner et al., 2002; Palyada et al., 2004; Paustian et al., 2001; Rodriguez et al., 2002; Singh et al., 2003; Todd et al., 2005; Wan et al., 2004; Wehrl et al., 2004). This response includes changes in metabolism to tailor the bacterial physiology to the condition of low iron availability. Gene expression analyses of various bacterial pathogens grown in cellular or animal models of infection have revealed that bacteria upregulate the genes of the siderophore systems and of other iron uptake systems while in the host (Boyce et al., 2002; Fernandez et al., 2004; Gold et al., 2001; Hou et al., 2002; Schnappinger et al., 2003; Talaat et al., 2004; Timm et al., 2003; Williams et al., 2004). These observations are consistent with the expectation that bacteria experience iron-limiting conditions in the host and with the recognized relevance of siderophores for efficient in vivo multiplication and establishment of successful infections by many pathogens (Bearden et al., 1997; Carniel et al., 1992; Cendrowski et al., 2004; Dale et al., 2004; De Voss et al., 2000; Der Vartanian et al., 1992; Fernandez et al., 2004; Fetherston et al., 1995; Heesemann, 1987; Henderson and Payne, 1994; Litwin et al., 1996;

Rakin et al., 1994; Register et al., 2001; Sokol et al., 1999; Takase et al., 2000; Yancey et al., 1979). An important practical implication of these observations is that it should be possible to find anti-infectives that target siderophore production or other functions essential to the physiology of *iron-starved* bacteria. This new paradigm impels research efforts in the laboratory of the author toward development and identification of antimicrobials. In fact, the first siderophore biosynthesis inhibitor was recently reported by Ferreras et al. (2005) and shown to have antimicrobial activity against *M. tuberculosis* and *Yersinia pestis* (the etiologic agent of plague) growing in iron-deficient media.

This chapter highlights both the well established and the yet poorly understood aspects of siderophore-mediated iron acquisition in pathogenic and nonpathogenic mycobacteria, with a particular emphasis in the siderophore system of *M. tuberculosis*. The *M. tuberculosis* siderophore system is believed to play a crucial role in the procurement of a suitable iron supply to support bacterial multiplication in vivo and to be a key factor in the ability of this human pathogen to produce successful infections.

GENERAL FEATURES OF THE MYCOBACTERIAL SIDEROPHORE SYSTEMS

Iron is an essential element to both pathogenic and nonpathogenic mycobacteria. The mycobacteria examined for iron-acquisition systems appear to rely on siderophores with high affinity for the ferric ion as the primary mechanism for iron acquisition. One intriguing exception is *Mycobacterium leprae*, the causative agent of leprosy, which is incapable of producing siderophores and, therefore, must have an alternative mechanism (probably based on iron transporters) for capturing iron from the host (Agranoff and Krishna, 2004; Cole et al., 2001; Kato, 1985; Morrison, 1995). Analysis of the genome of *M. leprae* has conclusively shown that this pathogen lacks the gene clusters that are required for biosynthesis of the siderophores found in other *Mycobacterium* species (Cole et al., 2001). Interestingly, *M. leprae* appears to be capable of using at least one mycobacterial siderophore, exochelin MN from *Mycobacterium neoaurum* (Hall et al., 1987). However, this phenomenon is not expected to have implications in the context of the pathogenesis of *M. leprae* because this bacterium does not encounter exochelin MN in the human host.

Two different siderophore systems have been found in mycobacteria, and they are typically referred to as the exochelin (EXC) and the mycobactin

(MBT)/carboxymycobactin (CMBT) systems. Each of these two systems is discussed in detail in specific sections of this chapter. For additional information, the reader is referred to Quadri and Ratledge (2005), which presents the latest comprehensive review on the subject. Salicylic acid, which is not only a precursor in MBT/CMBT biosynthesis (Quadri et al., 1998a) but is also produced and secreted as an independent entity by mycobacteria growing under iron-limiting conditions (Ratledge and Winder, 1962), has been occasionally considered a siderophore in the past (Meyer et al., 1992; Sokol et al., 1999). Today, salicylic acid is no longer thought of as a siderophore due to its relatively poor iron-chelating properties (Chipperfield

and Ratledge, 2000). Nevertheless, it has been recognized that this relatively simple molecule plays a key, yet unclear, role in mycobacterial physiology (Adilakshmi et al., 2000).

Significant structural diversity exists between the EXCs and the MBTs/CMBTs, yet both of these siderophore types provide the metal-ion coordination sites required for high-affinity binding of the Fe^{3+} ion. From a structural point of view, the core scaffolds of the EXCs are peptide based, whereas the core scaffolds of the MBTs and CMBTs are salicyl capped peptide polyketide based (De Voss et al., 1999; Quadri, 2000; Quadri and Ratledge, 2005) (Fig. 1A). The biosynthetic strategy by which these types of core scaffolds are assembled is often referred

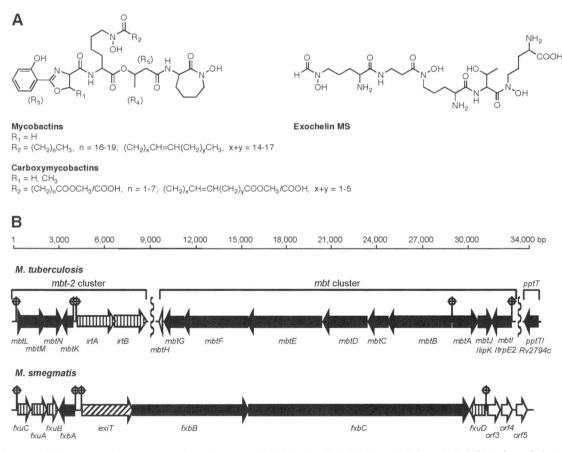

Mycobactins
R_1 = H
R_2 = $(CH_2)_nCH_3$, n = 16-19; $(CH_2)_xCH=CH(CH_2)_yCH_3$, x+y = 14-17

Carboxymycobactins
R_1 = H, CH_3
R_2 = $(CH_2)_nCOOCH_3/COOH$, n = 1-7; $(CH_2)_xCH=CH(CH_2)_yCOOCH_3/COOH$, x+y = 1-5

Exochelin MS

Figure 1. (A) Representative structures of mycobactins and carboxymycobactins from *M. tuberculosis* (left) and exochelin MS from *M. smegmatis* (right). The primary difference between mycobactins and carboxymycobactins is in the substituent at R_2. Alternative alkyl substituents at R_3, R_4, and R_5 are found in the siderophores of other *Mycobacterium* species. (B) *M. tuberculosis* chromosomal loci containing the genes of the mycobactin/carboxymycobactin system (upper scheme) and *M. smegmatis* chromosomal locus containing the genes of the exochelin MS system (bottom scheme). The *M. tuberculosis* gene map represents three distinct chromosomal loci (*mbt-2*, *mbt*, and *pptT* from left to right) that are shown in their relative orientation and separated from each other by a wavy line. The functions of most of the genes depicted have been predicted but not experimentally confirmed. Genes encoding proteins involved in siderophore biosynthesis, ferrisiderophore uptake or receptor functions, and siderophore secretion are shown in black, vertical-line hatched, and diagonal-line hatched arrows, respectively. Genes of unknown functions are shown in white. IdeR binding sequences (confirmed or predicted) are marked with the symbol ⊕. Gene *leuW* (84 bp), encoding tRNA-Leu and located between *mbtK* and *irtA*, is not depicted. See text for information on individual genes.

to as the thiotemplate mechanism, and it involves systems of multifunctional nonribosomal peptide synthetases, polyketide synthases, and nonribosomal peptide-polyketide hybrid synthetases (Du and Shen, 2001; Marahiel et al., 1997; Quadri, 2000; Walsh, 2004). Interestingly, the thiotemplate mechanism is also used by *M. tuberculosis*, *M. leprae*, and other mycobacteria to synthesize other secreted compounds involved in mycobacterial virulence, including various forms of dimycocerosate esters (e.g., phthiocerol dimycocerosates and phenolic glycolipids; reviewed by Onwueme et al. [2005]).

The current working models for the biosynthesis of mycobacterial siderophores are primarily derived from (1) the knowledge gained from studies on the biosyntheses of various types of nonribosomal peptide and polyketide natural products, including several investigations on biosynthesis of siderophores from other bacterial genera (e.g., yersiniabactin and pyochelin [Gehring et al., 1998a; Quadri et al., 1999]), and (2) in silico analyses of the biosynthetic capacities that are encoded in the siderophore biosynthesis gene loci found in pathogenic and saprophytic *Mycobacterium* species (De Voss et al., 1999; Quadri, 2000; Quadri and Ratledge, 2005; Quadri et al., 1998a; Yu et al., 1998; Zhu et al., 1998). In general terms, these working models illustrate the notion that the nonribosomal peptide synthetases involved in biosynthesis of siderophores assemble peptide moieties from proteogenic and nonproteogenic amino acids, whereas the polyketide synthase components assemble polyketide moieties from small carboxylic acids (preactivated as acyl-coenzyme A thioesters). The core scaffolds synthesized by the action of these enzymes are often modified by other enzymes that add substituents (e.g., acyl chains and hydroxyl groups) to yield the final products, that is, the siderophores.

Posttranslational modification of selected domains (carrier protein domains) found in the siderophore biosynthesis protein machinery is essential for the siderophore biosynthesis process in mycobacteria and other microorganisms. This posttranslational modification is the addition of a phosphopantetheinyl prosthetic group onto the carrier protein domains, a reaction that is mediated by specific enzymes called phosphopantetheinyl transferases (Quadri et al., 1998a, 1998b, 1999; Walsh et al., 1997). Phosphopantetheinyl transferases catalyze the transfer of the phosphopantetheinyl moiety from coenzyme A to the hydroxyl group in the side chain of a conserved serine residue that is present in each carrier protein domain of the system. The phosphopantetheinyl prosthetic groups act as tethers to which the biosynthetic intermediates are bound to the siderophore

biosynthesis protein machinery during synthesis (Walsh et al., 1997). The biosynthetic intermediates are linked by a thioester bond to the thiol end of the cysteamine in the prosthetic group. Thus, phosphopantetheinylation of unmodified, inactive, carrier protein domains (*apo* domains) converts them into active domains (*holo* domains) competent for holding the biosynthetic intermediates.

The initial covalent loading of amino acid monomers onto holo peptidyl carrier protein domains of nonribosomal peptide synthetases, and, in the case of MBT/CMBT biosynthesis, of a salicylic acid monomer onto a holo aroyl carrier protein domain as well, requires preactivation of these basic building blocks as aminoacyl or aroyl adenylates. The monomer activation and loading steps are catalyzed by monomer specific adenylation domains of the siderophore biosynthesis pathways. Similarly, covalent loading of small acyl chains onto holo *acyl* carrier protein domains of the polyketide synthases involved in MBT/CMBT biosynthesis requires previous activation of the carboxylic acid building blocks as acyl-coenzyme A thioesters. This activation is catalyzed by acyl-coenzyme A synthetases, and the monomer loading is catalyzed by acyltransferase functions. The protein-bound aminoacyl, aroyl, and acyl monomers are subsequently condensed in a programmed fashion by condensation catalysts of the biosynthesis protein machinery. Additional catalytic activities (e.g., acyltransferase and oxygenase) are often required to complete the biosynthesis of the siderophore. The preceding text outlines the most characteristic and well-accepted features of the current models for siderophore biosynthesis (Crosa and Walsh, 2002; Quadri, 2000). Considerable research efforts are needed to validate different aspects of these models and to gain a detailed understanding of these biosynthetic pathways, which represent archetypal examples of complex natural product biosyntheses.

As noted in the preceding section, maintenance of proper iron homeostasis is essential for cellular well-being (Andrews et al., 2003; Touati, 2000). Like in other prokaryotes, the mycobacteria respond to the intracellular concentration of iron by altering gene expression (Prakash et al., 2005; Quadri and Ratledge, 2005; Rodriguez, 2006; Rodriguez and Smith 2003; Yellaboina et al., 2006). The ultimate goal of this iron-dependent regulation of gene expression is the adaptation to oscillating high and low iron states and the fine-tuning of iron transport and storage processes toward achieving optimal iron homeostasis. Transcription of genes of the EXC and MBT/CMBT systems mentioned earlier is derepressed when the bacterium experiences iron limitations (i.e., it senses that the intracellular iron concentration is below optimal), thus leading to siderophore biosynthesis and siderophore-mediated iron uptake (Dussurget et al., 1996, 1999; Gold et al., 2001; Raghu et al., 1993; Rodriguez et al., 2002). The resulting iron accumulation in the cell triggers repression of the transcription of the genes of the siderophore systems. This repression results in a reduction of iron uptake, a key event to prevent accumulation of excess iron that would result in cellular toxicity. Other mycobacterial genes are also regulated (upregulated or downregulated) in response to the intracellular concentration of iron. Although some of these genes are involved in iron metabolism (e.g., bacterioferritin for iron storage), others encode functions that are regulated to afford a better metabolic fitness by customizing the cellular physiology to the iron levels in the cell (Quadri and Ratledge, 2005; Rodriguez, 2006; Rodriguez and Smith, 2003).

The iron-dependent regulatory protein called IdeR (Rodriguez, 2006; Rodriguez and Smith, 2003), an essential protein in *M. tuberculosis* (Rodriguez et al., 2002) but nonessential in *Mycobacterium smegmatis* (Dussurget et al., 1998), is the key player at the center of the mechanism of iron-dependent regulation of gene expression in mycobacteria. Among the genes that IdeR regulates are those encoding functions of the mycobacterial siderophore systems. When intracellular iron is abundant, an Fe^{2+}-IdeR complex is easily formed. In complex with Fe^{2+}, IdeR is competent to bind a specific sequence motif (the IdeR binding site) present in the promoter region of each IdeR-regulated gene transcript. Typically, this binding represses gene transcription. Conversely, when intracellular iron is in short supply, IdeR is found predominantly free, that is, not in complex with Fe^{2+}. The free IdeR is not competent to bind its target sequence, thus allowing the genes under IdeR repression control to express. Using this mechanism, IdeR represses transcription of key genes of the siderophore systems under iron-rich conditions, but it also permits transcription of these genes under iron-limiting conditions, thus leading to increased siderophore-mediated iron uptake. See references (Quadri and Ratledge, 2005; Rodriguez, 2006; Rodriguez and Smith, 2003) for recent reviews on iron-dependent regulation of gene expression in mycobacteria.

THE MYCOBACTIN/CARBOXYMYCOBACTIN SYSTEM

Several *Mycobacterium* species produce two structurally related families of high-affinity Fe^{3+}-binding siderophores, the MBTs and the CMBTs (De Voss et al., 1999; Quadri and Ratledge, 2005; Ratledge and Dover, 2000; Snow, 1970) (Fig. 1A). Particular efforts have been made to characterize the MBT/CMBT siderophore system because it is considered a mycobacterial virulence factor. The system is required for proper multiplication of *M. tuberculosis* inside human macrophages and in iron-deficient media in vitro (De Voss et al., 2000).

The MBTs are water-insoluble, lipophilic siderophores that are associated with the mycobacterial cell envelope (Snow, 1970). On the other hand, the CMBTs have better solubility in aqueous environments and are secreted into the extracellular milieu, where they are likely to serve as the primary agents for solubilizing/chelating the extracellular Fe^{3+} (Gobin et al., 1995; Gobin et al., 1999; Lane et al., 1998; Lane et al., 1995; Ratledge and Ewing, 1996; Wong et al., 1996). Isolation of cell-associated MBTs began in the mid-1960s, and by 1970, close to a dozen MBTs had been purified and characterized from several *Mycobacterium* species (Merkal et al., 1981; Snow, 1970). The first CMBTs were isolated from supernatants of mycobacterial cultures and characterized in the mid-1990s. CMBTs have been characterized from *M. smegmatis* (Lane et al., 1998; Ratledge and Ewing, 1996), *Mycobacterium avium* (Lane et al., 1995; Wong et al., 1996), *M. tuberculosis* (Gobin et al., 1995), and *Mycobacterium bovis* (Gobin et al., 1999). The CMBTs characterized from *M. tuberculosis* and *M. bovis* were initially called EXCs, a name that was also given to another siderophore type isolated from the supernatants of cultures of saprophytic mycobacteria. The EXCs from these saprophytic mycobacteria are pentapeptide or hexapeptide derivatives and are discussed in a subsequent section (Sharman et al., 1995a, 1995b) (Fig. 1A). To prevent nomenclature confusion, the term CMBTs has been primarily used in current literature to refer to the MBT-related (salicylate-derived) siderophores recovered for *M. tuberculosis* and *M. bovis* culture supernatants. The term CMBT alludes to the fact that some of these siderophores have a carboxylic acid group, which is not present in the cell-associated MBTs. However, it is worth noting that the methyl ester form of these water-soluble siderophores has also been isolated. Although the final nomenclature is still under debate, in this chapter the term CMBT refers to both the carboxylic acid and the methyl ester forms of the MBT-related siderophores recovered from the supernatants of *M. smegmatis*, *M. avium*, *M. tuberculosis*, and *M. bovis* cultures (Gobin et al., 1995, 1999; Lane et al., 1995, 1998; Ratledge and Ewing, 1996; Wong et al., 1996).

Both MBTs and CMBTs have an almost invariable salicyl-capped nonribosomal peptide-polyketide core scaffold with variable substitutions (De Voss et al., 1999; Quadri, 2000; Quadri and Ratledge, 2005; Ratledge and Dover, 2000). The representative core scaffold contains a 2-hydroxyphenyl(methyl)oxazoline moiety connected to an acylated N^ε-hydroxylysine residue, which is linked to a terminal cyclo-N^ε-hydroxylysine by a four-carbon insert. In particular, the nature of the substituent on the internal N^ε-hydroxylysine residue provides the base for the structural and functional differences between MBTs and CMBTs (De Voss et al., 1999; Quadri, 2000; Quadri and Ratledge, 2005; Ratledge and Dover, 2000). The acyl substituents in the CMBTs are carbon chains with a terminal carboxylic acid group or its methyl ester, whereas the acyl substituents in the MBTs are longer alkyl chains with no carboxylic acid group (Fig. 1A). The nitrogen and the phenol oxygen of the 2-hydroxyphenyl(methyl)oxazoline moiety and the two hydroxamic acid groups provide the sites for Fe^{3+} ion binding in the ferri-MBT and ferri-CMBT complexes. Based on studies done with the MBTs from *M. smegmatis*, the ferri-MBT complexes have a predicted formation constant of approximately 10^{26} M^{-1} at neutral pH (MacCordick et al., 1985). Notably, the conserved core scaffold of these mycobacterial siderophores is remarkably similar to that of the siderophores formobactin and nocobactin NA from *Nocardia* spp. (Murakami et al., 1996; Ratledge and Snow, 1974).

In 1998, more than 30 years after the first MBT structure was reported, the first gene cluster encoding the core enzymatic machinery for MBT/CMBT biosynthesis was identified in the genome of *M. tuberculosis* (Cole et al., 1998; Quadri et al., 1998a) (Fig. 1B). The gene cluster (named the *mbt* cluster) was initially assigned to MBT/CMBT biosynthesis through sequence and biochemical analyses (Cole et al., 1998; Quadri et al., 1998a). As expected, gene expression studies have revealed that various *mbt* genes are upregulated (derepressed) under iron-limiting conditions in vitro (Rodriguez et al., 2002), during *M. tuberculosis* infection of macrophages (Gold et al., 2001; Schnappinger et al., 2003), and in *M. tuberculosis* recovered from mouse lung (Talaat et al., 2004; Timm et al., 2003). Upregulation of *mbt* genes has also been documented in *M. avium* during growth in human macrophages (Hou et al., 2002).

The *mbt* cluster of *M. tuberculosis* encompasses approximately 24 kb and contains 10 genes, which have been named *mbtA* through *mbtJ* (Cole et al., 1998; Quadri et al., 1998a). Based on sequence analysis, the cluster is predicted to encode a bifunctional salicyl-AMP ligase/salicyl-S-aroyl carrier protein domain synthetase (MbtA, 565 amino acid residues), three nonribosomal peptide synthetases (MbtB, MbtE,

and MbtF, with 1,413, 1,682, and 1,462 amino acid residues, respectively), and two polyketide synthases (MbtC and MbtD, with 444 and 1,004 amino acid residues, respectively). The functions and domain organizations of these enzymes were primarily assigned based on their sequence homology with other nonribosomal peptide synthetases and polyketide synthases (Quadri et al., 1998a). Other proteins encoded in the cluster are: MbtG, a predicted lysine-N-oxygenase required for hydroxylation of lysine residues; MbtJ (LipK), a putative esterase/acetyl hydrolase of unclear function; MbtI (TrpE2), required for salicylic acid biosynthesis; and MbtH, with no homology to proteins of known activity or function (Cole et al., 1998; Quadri et al., 1998a) (Fig. 1B).

Mutational analysis has conclusively linked the *mbt* gene cluster to production of both MBTs and CMBTs. It has been demonstrated that the transposon insertion-based disruption of *mbtF* and the antibiotic marker insertion-based disruption of *mbtB* rendered *M. tuberculosis* mutants incapable of producing MBTs or CMBTs (De Voss et al., 2000; Ferreras et al., 2005). The siderophore-deficient *mbtB* mutant has been shown to display marginal growth in iron-deficient medium and to be impaired for multiplication inside human macrophages cultured in vitro (De Voss et al., 2000). Similar *mbt* gene clusters have been recently identified in the chromosomes of other MBT/CMBT-producing *Mycobacterium* species (Quadri et al., 2005). Among these species are *M. smegmatis*, in which a transposon-based disruption of the orthologue of *M. tuberculosis mbtE* abrogates siderophore production (LaMarca et al., 2004), and *M. avium*, in which expression of *mbt* genes has been shown to be upregulated during growth in human macrophages (Hou et al., 2002). The *mbt* gene clusters of mycobacteria other than *M. tuberculosis* have relatively minor variation in gene organization compared with the cluster of *M. tuberculosis* and have been extensively described by Quadri and Ratledge (2005).

The strategy for the biosynthesis of the MBT/CMBT salicyl-capped peptide-polyketide core scaffold by the genes encoded in the *mbt* cluster has been proposed within the framework of the current models for nonribosomal peptide and polyketide biosynthesis and based on the results of the in silico analysis of the catalytic functions encoded by the *mbt* genes (De Voss et al., 1999; Quadri et al., 1998a). In line with these models, the MbtA-MbtB-MbtC-MbtD-MbtE-MbtF system would catalyze the biosynthesis of the MBT/CMBT salicyl-capped peptide-polyketide core scaffold from salicylic acid, serine or threonine, (hydroxy)lysine, and acyl-coenzyme A thioesters (e.g., acetyl- and malonyl-coenzyme A as donors for synthesis of the polyketide moiety). In vitro studies have provided biochemical evidence for the involvement of MbtA and MbtB in MBT/CMBT acyl chain initiation (Quadri et al., 1998a). These studies have been conducted with purified proteins produced recombinantly in *Escherichia coli*. MbtA has been shown to adenylate salicylic acid (salicyl-AMP ligase activity), the first building block of the MBT/CMBT acyl chain (Quadri et al., 1998a). The predicted salicyl-S-aroyl carrier protein domain synthetase activity of MbtA was also characterized (Quadri et al., 1998a). MbtA is able to catalyze the transfer of the salicyl group from the salicyl-AMP intermediate to the phosphopantetheinyl group on the aroyl carrier protein domain of MbtB (Quadri et al., 1998a). The activity of the cyclization domain-serine/threonine adenylation domain-peptidyl carrier protein domain multifunctional module of MbtB (Quadri et al., 1998a) has also been demonstrated in vitro. These studies were carried out with a recombinant nonribosomal peptide synthetase hybrid constructed by fusing the MbtB module and functional domains from nonribosomal peptide synthetases involved in the biosynthesis of bacitracin and tyrocidine, and they demonstrated the formation of the predicted products: heterocyclic Ile-Thr$_{oxazoline}$ and Ile-Ser$_{oxazoline}$ dipeptides (Duerfahrt et al., 2004). The predicted lysine-N-oxygenase activity of MbtG, required for hydroxylation of lysine residues (Quadri et al., 1998a), has been validated using surrogate substrates (N6-lauroyl Z-lysine-OMe, L-lysine, and acetylated lysine) (Krithika et al., 2006). Last, in vitro activity and crystallographic studies on MbtI have demonstrated that this protein, earlier proposed to be required for salicylic acid synthesis (Quadri et al., 1998a), is indeed a salicylic acid synthase that catalyzes the formation of salicylic acid from chorismate (Harrison et al., 2006).

In addition to the *mbt* gene cluster mentioned earlier, two other gene loci are involved in MBT/CMBT assembly and/or ferrisiderophore uptake in *M. tuberculosis* and possibly in other *Mycobacterium* species. One of these loci contains a single gene, designated *pptT* (*Rv2794c*) (Quadri et al., 1998a). The gene *pptT* encodes a member of a subclass of phosphopantetheinyl transferases represented by the archetypical phosphopantetheinyl transferase Sfp (Quadri et al., 1998b). PptT is predicted to catalyze the posttranslational modification of the carrier protein domains of the nonribosomal peptide synthetases and a polyketide synthase of the MBT/CMBT pathway (Chalut et al., 2006; Quadri et al., 1998a). The activity of PptT has been validated and characterized in vitro. The protein has been shown to phosphopantetheinylate the N-terminal aroyl carrier pro-

tein domain of MbtB and a carrier protein domain from MbtE (Quadri et al., 1998a).

The third locus has a cluster with six genes, *Rv1344* to *Rv1349* (Card et al., 2005; Krithika et al., 2006; Rodriguez and Smith, 2006). The *Rv1344* to *Rv1347c* group has been named *mbt-2* gene cluster (Krithika et al., 2006); however, the genes *Rv1348* and *Rv1349* should be considered as part of the *mbt-2* cluster as well due to their recently reported functional relation with the MBT/CMBT system (see later) (Fig. 1B). As is the case for the genes of the *mbt* cluster mentioned earlier, expression of the genes in the *mbt-2* cluster is derepressed under iron-limiting conditions (Rodriguez et al., 2002) and probably repressed by IdeR during iron abundance (Rodriguez, 2006; Rodriguez and Smith, 2006; Rodriguez et al., 2002). The annotations *mbtK*, *mbtL*, *mbtM*, and *mbtN* have been recently proposed for the *Rv1347c*, *Rv1344*, *Rv1345* (*fadD33*), and *Rv1346* (*fadE14*) genes, respectively, found in the *mbt-2* cluster (Krithika et al., 2006). These annotations will be used in the ensuing text. These four genes are proposed to be involved in the biosynthesis and attachment of the acyl substituent to the N$^\varepsilon$ in the side chain of the internal hydroxylysine residue in the core scaffold of the MBT/CMBT molecules (Krithika et al., 2006). The product of *mbtK* was predicted to catalyze the addition of the acyl substituent to the hydroxylysine side chain in the siderophore's core scaffold (Card et al., 2005). The function of MbtK was proposed based on the analysis of the three-dimensional structure of the protein, which showed that MbtK has the fold of the GCN5-related N-acetyltransferase family of enzymes, the molecular modeling of MbtK-ligand binding, and the demonstration of MbtK N-acylation activity in vitro using substrate mimics (Card et al., 2005; Krithika et al., 2006). The acyl donor in the reaction was initially suggested to be an acyl-coenzyme A thioester (Card et al., 2005), but recent studies strongly suggest that the acyl donor is an acyl-acyl carrier protein domain thioester (Krithika et al., 2006). The protein encoded by *mbtM* has been shown to have acyl-adenylate (acyl-AMP) synthetase activity and to transfer the acyl moiety of the acyl-adenylate onto the acyl carrier protein MbtL to generate an acyl-acyl carrier protein domain thioester, the predicted acyl donor used by MbtK (Krithika et al., 2006). Notably, transposon-based disruption of the orthologue of *mbtM* found in *M. smegmatis* generates a mutant that shows reduced production of MBTs, with the produced siderophores apparently having altered fatty acid side chains (LaMarca et al., 2004). The *mbtN* gene product has detectable fatty acyl-acyl carrier protein dehydrogenase activity in vitro, and it has been proposed that MbtN catalyzes the formation of the double bond in the acyl substituent on the N^ε-hydroxylysine side chain of the internal lysine residue of the siderophores (Krithika et al., 2006).

The other two genes in the *mbt-2* cluster are *Rv1348* and *Rv1349*, which have been named *irtA* and *irtB*, respectively, and encode putative transporter components of the ATP binding cassette transporter family (Rodriguez and Smith, 2006). Mutational analysis indicates that these genes are required for bacterial growth in iron-deficient medium, but not needed for CMBT biosynthesis or secretion (Rodriguez and Smith, 2006). However, the mutants are compromised in their ability to use ferri-CMBTs supplemented in the growth medium. Based on these results and the sequence similarity observed between the IrtA and IrtB proteins and the proteins required for transport of the siderophore yersiniabactin, it has been hypothesized that the IrtAB system is involved in ferrisiderophore uptake (Rodriguez and Smith, 2006). Notably, during the study of the IrtAB system, the *mbtB* mutant of *M. tuberculosis* deficient in production of both MBTs and CMBTs was shown to use supplemented ferri-CMBTs as an iron source (Rodriguez and Smith, 2006). This observation suggests that, at least ex vivo, the lipophilic MBTs are not essential for uptake of iron from ferri-CMBTs. It is also noteworthy that the residual, yet significant, growth of the IrtAB system mutant in iron-deficient conditions suggests that the bacterium acquires iron from ferri-MBTs/CMBTs through an alternative mechanism with lower efficiency than that based on the IrtAB system. Importantly, the IrtAB system mutant is markedly deficient for multiplication in human macrophages and in mouse lung, thus implicating this system in virulence (Rodriguez and Smith, 2006).

Although mutational analysis keeps facilitating identification of mycobacterial genes involved in iron acquisition and enzyme biochemistry continues to afford insight into siderophore biosynthesis, progress toward a better understanding of various mechanistic aspects of the iron acquisition process has proven more challenging. Most conspicuously, the pathway of (ferri-)MBT and (ferri-)CMBT trafficking and the mechanism by which the ferric ion bound to MBTs or CMBTs finds its way into the bacterial cytoplasm remain poorly understood. Fig. 2A outlines different aspects of the MBT/CMBT-mediated iron acquisition process by an intraphagosomal bacterium such as *M. tuberculosis*, including possible alternatives for (ferri-)MBT/(ferri-)CMBT trafficking and delivery of iron to the bacterial cytoplasm. Although some of these alternatives are directly or indirectly supported or suggested by experimental observations, others are sensible possibilities that cannot be ruled out as yet, and warrant consideration until future investigation

provides a more conclusive mechanistic view of the MBT/CMBT-mediated iron acquisition process.

As stated above, MBTs are lipophilic siderophores found associated with the mycobacterial cell envelope (Snow, 1970). The CMBTs have better solubility in aqueous environments and are thought to be the Fe^{3+}-solubilizing/chelating siderophores that are secreted into the extracellular milieu (Gobin et al., 1995, 1999; Lane et al., 1995, 1998; Ratledge and Ewing, 1996; Wong et al., 1996). Studies conducted in vitro demonstrated that the CMBTs of *M. tuberculosis* can indeed take up Fe^{3+} from human transferrin, lactoferrin, and ferritin (Gobin and Horwitz, 1996). Furthermore, it has been demonstrated that ferri-CMBT can donate the Fe^{3+} ion to cell-associated MBTs (Gobin and Horwitz, 1996). Unlike CMBTs, cell-associated MBTs are not competent for acquisition of Fe^{3+} from transferrin in vitro (Gobin and Horwitz, 1996). Recent cross-feeding experiments indicate that the *mbtB* mutant of *M. tuberculosis* deficient in production of both MBTs and CMBTs can use supplemented ferri-CMBTs, thus suggesting that, at least ex vivo, the lipophilic MBTs are not essential for uptake of iron from ferri-CMBTs (Rodriguez and Smith, 2006). However, recent studies discussed later suggest that the lipophilic MBTs can sequester iron from host sources (Luo et al., 2005). These and other notions discussed later are represented in Fig. 2A.

Based on electron microscopy studies of whole *M. smegmatis* cells stained with vanadate, the lipophilic MBTs are believed to be at the cell envelope, where they appear to localize to a discrete region close to, or perhaps within, the plasma membrane (Ratledge et al., 1982). These studies led to the proposal that MBTs provide a place at the cell surface for short-term iron storage or act as ionophores of the mobile carrier type (non-channel-forming ionophores) (Ratledge et al., 1982). As ionophores, the lipophilic MBT molecules would bind Fe^{3+}, shielding its charge from the membrane environment, and thus facilitating Fe^{3+} mobilization across or into the bacterial membrane. As with other ionophores, this process would not necessitate coupling to energy sources and would instead be driven by concentration gradients. In connection with the MBT ionophore hypothesis, it is pertinent to mention that early studies revealed that the mobilization of iron from externally added ferri-CMBTs to a cell-associated state is insensitive to electron transport inhibitors and uncouplers of ATP generation, thus suggesting an energy independent process (Stephenson and Ratledge, 1979, 1980). However, these studies do not formally differentiate between iron incorporated into the cytoplasm and iron accumulated at the cell envelope, perhaps bound to MBTs, nor do they provide information as to

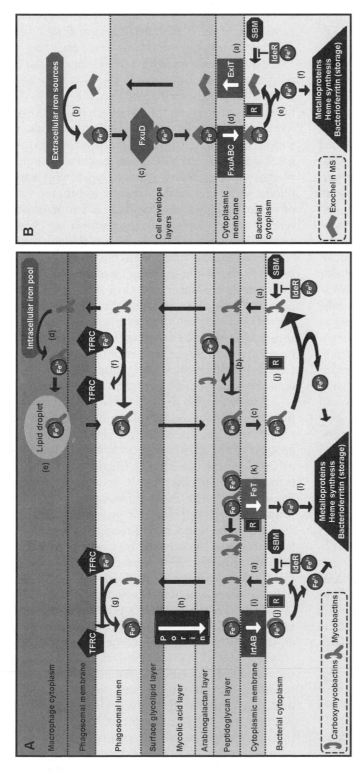

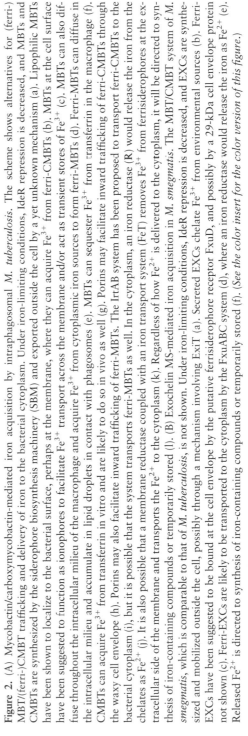

Figure 2. (A) Mycobactin/carboxymycobactin-mediated iron acquisition by intraphagosomal *M. tuberculosis*. The scheme shows alternatives for (ferri-)MBT/(ferri-)CMBT trafficking and delivery of iron to the bacterial cytoplasm. Under iron-limiting conditions, IdeR repression is decreased, and MBTs and CMBTs are synthesized by the siderophore biosynthesis machinery (SBM) and exported outside the cell by a yet unknown mechanism (a). Lipophilic MBTs have been shown to localize to the bacterial surface, perhaps at the membrane, where they can acquire Fe³⁺ from ferri-CMBTs (b). MBTs at the cell surface have been suggested to function as ionophores to facilitate Fe³⁺ transport across the membrane and/or act as transient stores of Fe³⁺ (c). MBTs can also diffuse throughout the intracellular milieu of the macrophage and acquire Fe³⁺ from cytoplasmic iron sources to form ferri-MBTs (d). Ferri-MBTs can diffuse in the intracellular milieu and accumulate in lipid droplets in contact with phagosomes (e). MBTs can sequester Fe³⁺ from transferrin in the macrophage (f). CMBTs can acquire Fe³⁺ from transferrin in vitro and are likely to do so in vivo as well (g). Porins may facilitate inward trafficking of ferri-CMBTs through the waxy cell envelope (h). Porins may also facilitate inward trafficking of ferri-MBTs. The IrtAB system has been proposed to transport ferri-CMBTs to the bacterial cytoplasm (i), but it is possible that the system transports ferri-MBTs as well. In the cytoplasm, an iron reductase (R) would release the iron from the chelates as Fe²⁺ (j). It is also possible that a membrane reductase coupled with an iron transport system (FeT) removes Fe³⁺ from ferrisiderophores at the extracellular side of the membrane and transports the Fe²⁺ to the cytoplasm (k). Regardless of how Fe²⁺ is delivered to the cytoplasm, it will be directed to synthesis of iron-containing compounds or temporarily stored (l). (B) Exochelin MS-mediated iron acquisition in *M. smegmatis*. The MBT/CMBT system of *M. smegmatis*, which is comparable to that of *M. tuberculosis*, is not shown. Under iron-limiting conditions, IdeR repression is decreased, and EXCs are synthesized and mobilized outside the cell, possibly through a mechanism involving ExiT (a). Secreted EXCs chelate Fe³⁺ from environmental sources (b). Ferri-EXCs have been suggested to be bound at the cell envelope by the putative ferrisiderophore receptor FxuD, and possibly by a 29-kDa cell envelope protein not shown (c). Ferri-EXCs are likely to be transported to the cytoplasm by the FxuABC system (d), where an iron reductase would release the iron as Fe²⁺ (e). Released Fe²⁺ is directed to synthesis of iron-containing compounds or temporarily stored (f). (*See the color insert for the color version of this figure.*)

whether the ferrisiderophores enter the cytoplasm or only move between the outer and inner membrane leaflets. Last, the recently suggested involvement of the predicted ATP-dependent IrtAB transport system mentioned earlier in MBT/CMBT-dependent iron uptake would suggest that iron uptake is an ATP-dependent process (Rodriguez and Smith, 2006). Clearly, further experimental exploration is required to gain a better insight into ferrisiderophore transport and the energetic requirement of iron uptake.

It has been demonstrated that intraphagosomal *M. tuberculosis* acquires iron from both internalized transferrin and intracellular pools that are created from iron supplied by extracellular transferrin (Olakanmi et al., 2002) (Fig. 2A). The *Mycobacterium*-containing phagosome has been proposed to acquire extracellular Fe^{3+}-transferrin complexes through receptor-mediated endocytosis to form early endosomes that subsequently fuse with the phagosomes (Clemens and Horwitz, 1996). Intraphagosomal *M. tuberculosis* can also acquire iron from extracellular Fe^{3+}-lactoferrin and Fe^{3+}-citrate complexes, two extracellular chelates from which macrophages can take up iron by as yet unclear mechanisms (Olakanmi et al., 2004). The ability of intraphagosomal *M. tuberculosis* to acquire iron from extracellular lactoferrin could be of importance because the Fe^{3+}-lactoferrin chelate is particularly abundant in lung airways. It is very likely that both MBTs and CMBTs are involved in the mobilization of iron from various iron host sources to the bacterial cell. Very recent studies suggest that the lipophilic MBTs could be directly involved in uptake of iron from host sources (Luo et al., 2005). These studies were conducted with MBT J, the lipophilic MBT from *M. paratuberculosis*, and revealed that the externally added iron-free MBT J diffuses into macrophages cultured in vitro and extracts iron from intracellular iron pools to form the Fe^{3+}-MBT J complex (Luo et al., 2005). Moreover, this complex also diffuses through the macrophage's intracellular milieu and accumulates preferentially in lipid droplets that are part of the lipid storage and lipid sorting system of the cell. Intriguingly, these ferri-MBT-containing lipid droplets appear to be in contact with the phagosomal membrane (Fig. 2A). This arrangement would leave the Fe^{3+}-MBT complex conveniently positioned to access the phagosome, where pathogenic mycobacteria such as *M. tuberculosis* reside. These studies suggest that the lipophilic MBTs (and perhaps the secreted CMBTs as well) from *M. tuberculosis* residing in the phagosome could diffuse into the membrane environments and cytoplasmic milieu of the macrophage, capture Fe^{3+} from cellular sources, and then deliver the Fe^{3+} to bacteria residing in the phagosome (Luo et al., 2005) (Fig. 2A). Recent studies have shown that phagosomes containing a mutant of *M. tuberculosis* deficient in MBT and CMBT production accumulate (or retain) less iron than those containing the wild-type bacterium (Wagner et al., 2005). This observation is in agreement with the proposed role of the MBT/CMBT system in mobilization of iron to intraphagosomal mycobacteria.

The studies with MBT J would suggest the possibility that, unlike the situation found in mycobacteria cultured in growth media, the lipophilic MBTs do not localize exclusively to the mycobacterial cell envelope when the bacteria are found intraphagosomally. Unlike in the case of MBTs of mycobacteria propagated ex vivo, the MBTs of mycobacteria in the phagosome could have access to extracellular hydrophobic vehicles (i.e., the dynamic membrane systems of the macrophage) that facilitate diffusion of the lipophilic siderophores from the bacteria into the host cell milieu. Once in the host cell, the siderophores could come in contact with host iron pools from which iron would be sequestered to form ferrisiderophores, which could then diffuse back to the bacterial cell. Thus, it is tempting to speculate that (ferri-)MBT trafficking takes place, or is facilitated, through (ferri-)MBT partitioning into the host membrane systems (or through membrane-cytoplasm interfaces). Additional investigation is needed to validate these mechanistic hypotheses.

Another aspect of the mycobacterial iron acquisition process that needs exploration is the delivery of the iron to the bacterial cytoplasm. Although the analysis of the IrtAB transport system mutants of *M. tuberculosis* discussed earlier suggests that this system is involved in ferrisiderophore uptake (Rodriguez and Smith, 2006), there are no direct experimental data to support or rule out a circuit of inward transport of ferri-MBTs or ferri-CMBTs for intracellular iron release followed by outward transport of the desferrisiderophores (Fig. 2A). A predicted two-component metal transport system (involving Irp10 and Mta22) induced under iron limiting conditions in *M. tuberculosis* has been suggested as a possible system for uptake of iron brought by the siderophores to, or stored at, the cell surface or plasma membrane (Calder and Horwitz, 1998; Quadri and Ratledge, 2005). However, there is no experimental confirmation that this system is indeed transporting iron or any other metal. To date, there is no conclusive evidence to ascertain whether ferri-MBT and/or ferri-CMBT (1) enter into the cell cytoplasm to deliver iron, (2) function as ionophores that do not leave the membrane, or (3) remain extracellular and act to concentrate iron at the cell surface that on release from the siderophores, probably as Fe^{2+}, is mobilized to the cytoplasm by metal transporters or other means.

Regardless of whether the ferri-MBT and ferri-CMBT complexes enter the cell's cytoplasm, remain in the plasma membrane, or accumulate at the cell surface, it is clear that a mechanism to release the tightly bound iron in the ferrisiderophore complexes will be required to make the iron available to the bacterial cell. An iron reductase activity has been hypothesized to mediate iron release from the ferri-siderophores by reducing the Fe^{3+} to Fe^{2+}, the iron ion for which the siderophores have no significant affinity. Indeed, a soluble iron reductase activity capable of promoting iron release from ferri-MBTs has been detected in supernatant fractions of *M. smegmatis* cell lysates after high-speed centrifugation (Brown and Ratledge, 1975). More recently, a protein with iron reductase activity was purified from the supernatant of a *M. paratuberculosis* culture (Homuth et al., 1998). It should be noted that it would be challenging to the cell to provide an adequate supply of reducing power to be used by an extracellular iron reductase. Additional studies are needed to consolidate the reductase-dependent iron release hypothesis and decipher the reductase identity and precise subcellular localization.

THE EXOCHELIN SYSTEM

Several *Mycobacterium* species that are normally found as environmental saprophytes release EXCs, the nonribosomally synthesized pentapeptide-based or hexapeptide-based siderophores mentioned earlier, into the extracellular environment. These siderophores are, from a structural point of view, different from the MBTs and CMBTs presented in the preceding section. EXCs have been purified from the supernatant of cultures of *M. smegmatis*, *M. neoaurum*, *Mycobacterium vaccae*, and other saprophytic mycobacteria (Messenger et al., 1986; Quadri and Ratledge, 2005; Sharman et al., 1995a, 1995b) and shown to have remarkably high affinity for Fe^{3+} (Dhungana et al., 2003, 2004). Detailed structural characterization has been reported for one of the EXCs of *M. smegmatis* (EXC MS) and an EXC of *M. neoaurum* (EXC MN) (Sharman et al., 1995a, 1995b). EXC MS is a formylated pentapeptide [*N*-(δ-*N*-formyl, δ-*N*-hydroxy-*R*-ornithyl)-β-alaninyl-δ-*N*-hydroxy-*R*-ornithinyl-*R*-allo-threoninyl-δ-*N*-hydroxy-*S*-ornithine] (Sharman et al., 1995b) (Fig. 1A), whereas EXC MN is a hexapeptide [L-*threo*-β-hydroxy-histidine-β-alanine-β-alanine-L-α-methyl ornithine-L-ornithine-L-(cyclo)ornithine] (Sharman et al., 1995a). It is noteworthy that Ratledge and coworkers reported that EXC MS is the most abundant of several EXC variants produced by *M. smegmatis* (Macham

et al., 1977; Sharman et al., 1995b). Preliminary examination of the structures of four of the minor EXCs of *M. smegmatis* indicated that three of them are hexapeptide variants (Quadri and Ratledge, 2005).

Interestingly, *M. smegmatis* and *M. neoaurum* also produce and secrete CMBTs into the extracellular environment (Ratledge and Ewing, 1996). In *M. smegmatis*, EXCs are the predominant siderophores, representing 90% of the total siderophores produced in vitro by bacteria propagated under iron-limiting conditions (Ratledge and Ewing, 1996). No information regarding the relative proportion of EXCs to CMBTs has been reported for *M. neoaurum*. The adaptive advantage of having both EXCs and CMBTs remains unclear. Furthermore, the contribution of each siderophore system to bacterial iron uptake in natural environments is unknown. More important, *M. smegmatis*, *M. neoaurum*, and other saprophytic mycobacteria are known to produce opportunistic infections in humans with different degrees of severity (Katoch, 2004). The relevance of the EXC and MBT/CMBT systems for the ability of these mycobacteria to cause infection remains to be investigated.

The *M. smegmatis* gene cluster involved in the biosynthesis and transport of EXCs has been identified (Fiss et al., 1994; Yu et al., 1998; Zhu et al., 1998) (Fig. 1B). The gene cluster encompasses approximately 34 kb and includes (1) *fxbA*, which encodes a protein with homology to formyl-transferases; (2) *fxbB* and *fxbC*, encoding proteins with sequence similarity to non-ribosomal peptide synthetases; (3) *fxuA*, *fxuB*, and *fxuC*, coding for proteins with homology to the FepG, FepC, and FepD subunits, respectively, of the ferrienterobactin permease complex of *E. coli*; (4) *fxuD*, encoding a protein with homology to periplasmic siderophore receptors; (5) *exiT*, with a gene product with sequence similarity to ABC transporters; (6) *orf3*, encoding a protein with an ATP-binding domain; and (7) *orf4* and *orf5*, encoding proteins with multiple membrane spanning segments and no significant sequence similarity to proteins of known function. Based on sequence similarity, FxbA would be involved in the formylation step of EXC biosynthesis and the FxbB-FxbC nonribosomal synthetase system is predicted to assemble the peptide backbone of the siderophores (Fiss et al., 1994; Yu et al., 1998). In silico analysis of functional domains in the FxbB-FxbC synthetase system revealed a total of six adenylation domains (Yu et al., 1998). This arrangement would, in principle, afford synthetic capacity for both the pentapeptide (Sharman et al., 1995b) and hexapeptide (Quadri and Ratedge, 2005) EXC variants produced by *M. smegmatis*. FxuA, FxuB, and FxuC are suggested to form a transport complex for ferri-EXCs, whereas ExiT is proposed to be involved in EXC secretion (Fiss et al., 1994; Yu et al., 1998; Zhu et al., 1998). Mutagenesis analysis has demonstrated that FxbA, the FxbB-FxbC synthetase system, and ExiT are indeed required for EXC production (Fiss et al., 1994; Yu et al., 1998; Zhu et al., 1998). Furthermore, because the *exiT* mutant does not accumulate EXCs, it has been proposed that EXC synthesis may be tightly coupled to EXC secretion (Zhu et al., 1998). The involvement of *orf3*, *orf4*, and *orf5* in EXC biosynthesis, export, uptake, or other aspects of the EXC system cannot be predicted from sequence inspection. Mutational analysis will be required to determine whether these gene loci are indeed part of the EXC system. A schematic model for EXC dependent-iron uptake is shown in Fig. 2B. The model incorporates notions derived from the studies with the EXC system of *M. smegmatis* outlined earlier and from other studies presented in the ensuing text. Some aspects of the model are supported or suggested (directly or indirectly) by experimental observations. Others are more speculative, yet sensible possibilities based on our general knowledge of siderophore-dependent iron uptake in other bacteria.

As shown in Fig. 2B, the uptake of ferri-EXC MS is proposed to involve the predicted ATP-dependent FxuA-FxuB-FxuC ferri-EXC permease systems. This hypothesis is consistent with the early observation that the mobilization of iron from externally added ferri-EXC MS to a *M. smegmatis* cell-associated state is a cellular process sensitive to inhibitors of energy generation (Stephenson and Ratledge, 1979). Conversely, studies with ferri-EXC MN from *M. neoaurum* revealed that mobilization of iron from externally added ferri-EXC MN to a *M. neoaurum* cell-associated state is insensitive to inhibitors of ATP generation (Hall and Ratledge, 1987; Hall et al., 1983). However, these early studies did not formally differentiate between iron incorporated into the cytoplasm and iron accumulated at the cell envelope, perhaps bound to MBTs; neither did they demonstrate that ferri-EXCs entered the cytoplasm.

Interestingly, the transfer of iron from ferri-EXC MS to iron-depleted cells of *M. smegmatis* appears to require a 29-kDa envelope protein that is upregulated under iron deficient conditions (Hall and Ratledge, 1987). It has been shown that the mobilization of iron from ferri-EXC MS to iron-depleted *M. smegmatis* cells is blocked in the presence of an antibody against this 29-kDa envelope protein (Hall and Ratledge, 1987). Notably, the antibody did not block transfer to iron-replete *M. smegmatis* cells, in which expression of the 29-kDa envelope protein is downregulated (Hall and Ratledge, 1987). More re-

cently, evidence of direct binding between ferri-EXC MS and this 29-kDa protein and another *M. smegmatis* envelope protein of 25 kDa was reported (Dover and Ratledge, 1996). It has been suggested that the 29-kDa protein is a ferri-EXC-binding protein and that the 25-kDa protein has a role in assembly or stability of the 29-kDa protein-ferri-EXC complex (Dover and Ratledge, 1996). The observed molecular weight of neither of these envelope proteins is close to that of FxuD, the putative siderophore receptor encoded in the EXC MS gene cluster mentioned earlier. The sequence of each of these two envelope proteins remains unknown.

Paralleling the iron-dependent regulation of the MBT/CMBT system of *M. smegmatis* (Adilakshmi et al., 2000; Dussurget et al., 1996), the EXC MS system is also upregulated under iron-limiting conditions, and the IdeR protein is required for the iron-dependent regulation of EXC MS production (Dussurget et al., 1996). As expected, the genetically engineered *ideR* mutant of *M. smegmatis* displays constitutive siderophore production (Dussurget et al., 1996). Intriguingly, however, the amount of EXC produced by this mutant is less than that produced by the wild type under iron-limiting conditions (Dussurget et al., 1996). This observation may suggest the existence of a regulatory layer beyond IdeR-dependent repression.

MYCOBACTERIAL IRON ACQUISITION PATHWAYS: NOVEL POTENTIAL TARGETS FOR DRUG DEVELOPMENT

Mycobacterial pathogens require iron to multiply in the host and produce disease. Thus, drugs that block the ability of these pathogens to acquire iron from the host are anticipated to be useful, alone or as part of combination chemotherapies, in the treatment and prophylaxis of tuberculosis and other *Mycobacterium*-produced diseases. Such new drugs might, for example, find a therapeutic niche in multidrug treatments against multidrug-resistant *M. tuberculosis* (Quadri, 2007).

The MBT/CMBT biosynthesis pathway of *M. tuberculosis* contains enzymes that do not have homologues in humans. These enzymes represent attractive targets for new drugs with the potential to restrict the growth of *M. tuberculosis* in the iron-limiting environments of the human host by blocking siderophore-mediated iron acquisition. The reported construction of mutants deficient in MBT/CMBT biosynthesis is, ipso facto, an indication that the MBT/CMBT system is not essential ex vivo (Darwin et al., 2003; De Voss et al., 2000; Ferreras et al.,

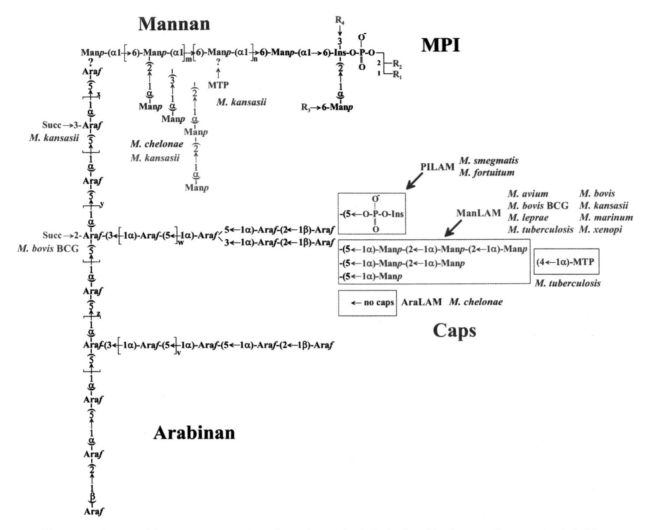

Chapter 6, Figure 1. Schematic representation of mycobacterial LAM. Ara*f*, arabinofuranose; Ins, *myo*-inositol; Man*p*, mannopyranose; MPI, mannosyl-phosphatidyl-*myo*-inositol; R_n, fatty acyl residue. The mean molecular mass of *M. tuberculosis* and *M. bovis* BCG ManLAM is around 17 kDa, with a heterogeneity estimated at 6 kDa (Venisse et al., 1993). We estimate that these ManLAM contain approximately 60 Ara*f* and 50 Man*p* units. Man*p* units are distributed between the mannose caps and the mannan core (30 to 35 Man*p* units) (Nigou et al., 2000). The mannan domain of the *M. kansasii* ManLAM contains a low proportion of disaccharide side chains (Guerardel et al., 2003). 5-Methylthiopentose (MTP) was identified as 5-deoxy-5-methylthio-*xylo*-furanose (Joe et al., 2006) and has been described on the ManLAM of *M. tuberculosis* strains (Ludwiczak et al., 2002; Treumann et al., 2002) and ManLAM and LM of a *M. kansasii* clinical isolate (Guerardel et al., 2003). Succ indicates succinyl residues located on the arabinan domain of ManLAM of *M. bovis* BCG (Delmas et al., 1997) and of a *M. kansasii* clinical isolate (Guerardel et al., 2003). One to four succinyl groups, depending on the *M. bovis* BCG strain, esterify the 3,5-α-Ara*f* units at position O-2 (Delmas et al., 1997), and an average of two succinic acids per LAM was found in the case of *M. kansasii* ManLAM (Guerardel et al., 2003).

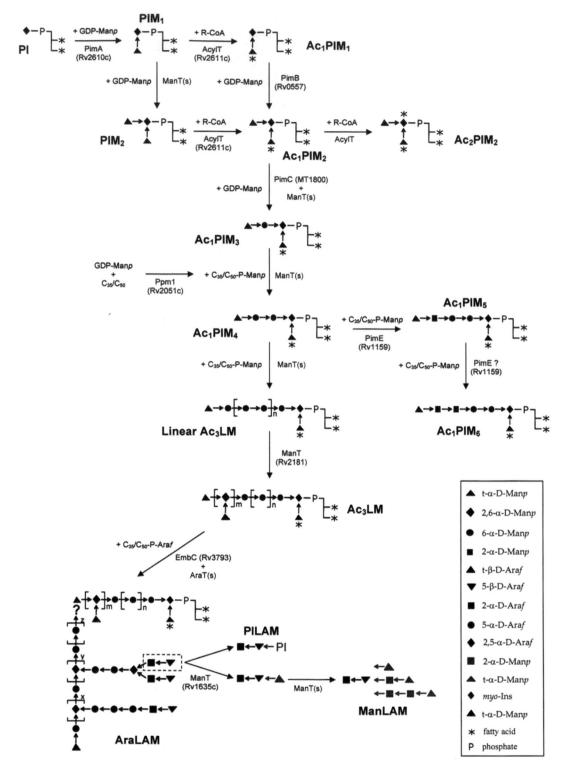

Chapter 6, Figure 2. LAM biosynthesis schema. The biosynthesis of the triacylated forms of PIM and lipoglycans is shown. PimA is essential in *M. smegmatis* (Kordulakova et al., 2002). Rv2611c appears to be essential in *M. tuberculosis*, but not in *M. smegmatis* (Kordulakova et al., 2003; G. Stadthagen, M. Jackson, and B. Gicquel, unpublished results). Sugar donors are in blue and identified glycosyl-transferases in red. AcylT, acyl-transferase; ManT(s), mannosyl-transferase(s); AraT(s), arabinosyl-transferase(s); C_{35}/C_{50}, polyprenol; C_{35}/C_{50}-P-Man, polyprenol-monophosphorylmannose; C_{35}/C_{50}-P-Araf, polyprenol-monophosphoryl-β-D-Araf.

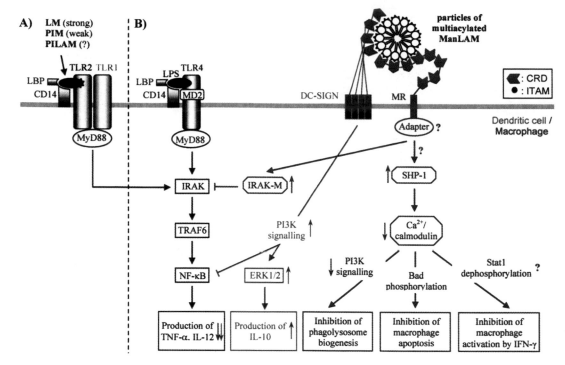

Chapter 6, Figure 3. Cell signaling pathways triggered by PI-based lipoglycans. (A) LM (Quesniaux et al., 2004; Vignal et al., 2003), and to a lesser extent PIM (Gilleron et al., 2003), activate macrophages and DCs through a TLR2/TLR1-dependent but TLR6-independent pathway that requires MyD88 (Quesniaux et al., 2004). Only Ac_3 LM and Ac_4LM are active (Gilleron et al., 2006), whereas the residual PIM activity is independent of the acylation degree (from one to four fatty acids) (Gilleron et al., 2003). It is not known whether lipoglycans are presented to the receptor in a monomeric or multimeric form. ManLAM and AraLAM do not signal through TLR2 as a consequence of steric hindrance: the arabinan domain masks the lipomannan moiety of the molecule (Guerardel et al., 2003). The molecular bases of PILAM activity are not clear yet. (B) ManLAM inhibits IL-12 and TNF-α (Nigou et al., 2001) and induces IL-10 production by LPS-stimulated DCs through DC-SIGN ligation (Geijtenbeek et al., 2003). The signaling pathway involves activation of PI3K and ERK1/2 (Caparros et al., 2006). In macrophages, ManLAM inhibits the LPS-induced production of TNF-α and IL-12 (Knutson et al., 1998), independently of IL-10 production, through IRAK-M activation (Pathak et al., 2005). ManLAM exerts other inhibitory activities on macrophages including inhibition of IFN-γ-mediated activation (Sibley et al., 1988), *M. tuberculosis*-induced apoptosis (Rojas et al., 2000), and phagolysosome biogenesis (Fratti et al., 2003). Phagolysosome biogenesis is associated with ManLAM binding to MR (Kang et al., 2005) and requires inhibition of both the cytosolic Ca^{2+} rise/calmodulin pathway and PI3K signaling (Vergne et al., 2003). Inhibition of apoptosis (Rojas et al., 2000) and possibly IFN-γ-mediated activation (Briken et al., 2004) are also dependent on the alteration of Ca^{2+}-dependent intracellular events, suggesting that they could be also both mediated by MR. LM and PIM also bind MR and DC-SIGN (Pitarque et al., 2005; Torrelles et al., 2006); however, little is known about the functional consequences. LM induces a TLR2-dependent production of proinflammatory cytokines but concomitantly inhibits, most probably through C-type lectin binding, TLR4-mediated cytokine production (Quesniaux et al., 2004). The net cytokine response is dependent on the receptor equipment of the cells as well as the LM used and their acylation degree (Quesniaux et al., 2004).

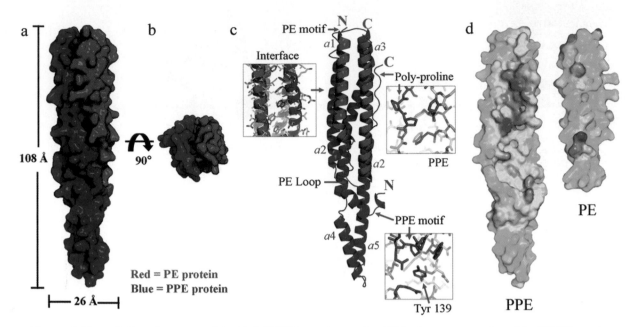

Chapter 8, Figure 5. Crystal structure of the *M.tb.* PE/PPE protein complex. (A) Surface representation of the PE/PPE protein complex. The PE protein Rv2431c is shown in red, and the PPE protein Rv2430c is in blue. (B) The PE/PPE protein complex viewed down its longitudinal axis. (C) Ribbon diagram of the PE/PPE protein complex. The complex is composed of seven helices. Two helices of the PE protein interact with two helices of the PPE protein to form a four-helix bundle. Regions of high sequence conservation are indicated by arrows and discussed in the text. (D) Interface hydrophobicity of the PPE and PE proteins. The hydrophobicity of the interaction interface between the PPE and PE protein is color coded: the most apolar regions are indicated in red, orange, and yellow, and the most polar regions are indicated in blue. Notice the extensive apolar regions that are shielded from solvent as the complex forms. Adapted from Strong et al. (2006).

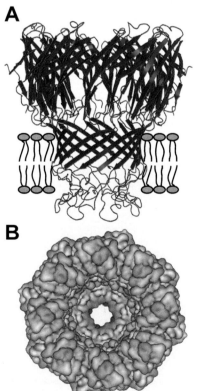

Chapter 9, Figure 3. MspA, a general porin of *M. smegmatis*. (A) Side view of MspA integrated into a lipid bilayer. (B) Electrostatic potential of MspA in top view. The electrostatic potential is represented by the Gasteiger charges for the atoms in the surface of MspA. Negative charges are shown in red, positive charges are shown in blue. These figures are based on the crystal structure of MspA (Faller et al., 2004). Panels A and B were created with the visualization software PyMol (DeLano Scientific LLC) and ViewerLight (Accelrys), respectively.

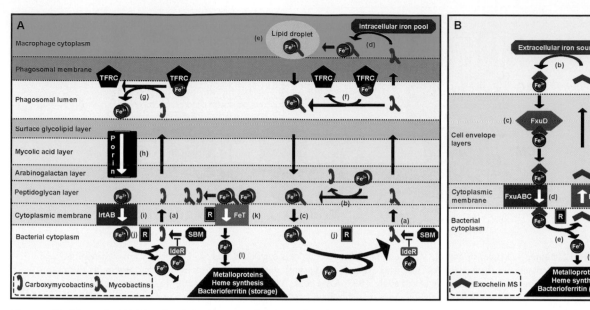

Chapter 10, Figure 2. (A) Mycobactin/carboxymycobactin-mediated iron acquisition by intraphagosomal *M. tuberculosis*. The scheme shows alternatives for (ferri-)MBT/(ferri-)CMBT trafficking and delivery of iron to the bacterial cytoplasm. Under iron-limiting conditions, IdeR repression is decreased, and MBTs and CMBTs are synthesized by the siderophore biosynthesis machinery (SBM) and exported outside the cell by a yet unknown mechanism (a). Lipophilic MBTs have been shown to localize to the bacterial surface, perhaps at the membrane, where they can acquire Fe^{3+} from ferri-CMBTs (b). MBTs at the cell surface have been suggested to function as ionophores to facilitate Fe^{3+} transport across the membrane and/or act as transient stores of Fe^{3+} (c). MBTs can also diffuse throughout the intracellular milieu of the macrophage and acquire Fe^{3+} from cytoplasmic iron sources to form ferri-MBTs (d). Ferri-MBTs can diffuse in the intracellular milieu and accumulate in lipid droplets in contact with phagosomes (e). MBTs can sequester Fe^{3+} from transferrin in the macrophage (f). CMBTs can acquire Fe^{3+} from transferrin in vitro and are likely to do so in vivo as well (g). Porins may facilitate inward trafficking of ferri-CMBTs through the waxy cell envelope (h). Porins may also facilitate inward trafficking of ferri-MBTs. The IrtAB system has been proposed to transport ferri-CMBTs to the bacterial cytoplasm (i), but it is possible that the system transports ferri-MBTs as well. In the cytoplasm, an iron reductase (R) would release the iron from the chelates as Fe^{2+} (j). It is also possible that a membrane reductase coupled with an iron transport system (FeT) removes Fe^{3+} from ferrisiderophores at the extracellular side of the membrane and transports the Fe^{2+} to the cytoplasm (k). Regardless of how Fe^{2+} is delivered to the cytoplasm, it will be directed to synthesis of iron-containing compounds or temporarily stored (l). (B) Exochelin MS-mediated iron acquisition in *M. smegmatis*. The MBT/CMBT system of *M. smegmatis*, which is comparable to that of *M. tuberculosis*, is not shown. Under iron-limiting conditions, IdeR repression is decreased, and EXCs are synthesized and mobilized outside the cell, possibly through a mechanism involving ExiT (a). Secreted EXCs chelate Fe^{3+} from environmental sources (b). Ferri-EXCs have been suggested to be bound at the cell envelope by the putative ferrisiderophore receptor FxuD, and possibly by a 29-kDa cell envelope protein not shown (c). Ferri-EXCs are likely to be transported to the cytoplasm by the FxuABC system (d), where an iron reductase would release the iron as Fe^{2+} (e). Released Fe^{2+} is directed to synthesis of iron-containing compounds or temporarily stored (f).

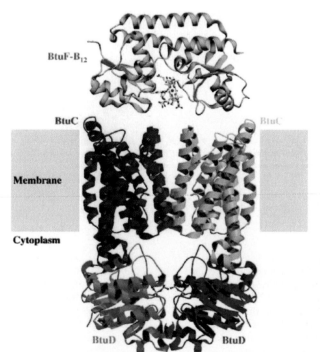

Chapter 11, Figure 1. Ribbon diagram of the *E. coli* vitamin B_{12}BtuCDF, ABC permease protein structure. The transporter is assembled from two membrane-spanning BtuC subunits (red and yellow) and two ABC cassettes BtuD (green and blue). At the ATP binding sites, cyclotetravanadate molecules are bound to the transporter (ball-and-stick models at the BtuD interface). Vitamin B_{12} is delivered to the periplasmic side of the transporter by a binding protein (BtuF, light blue), then translocated through a pathway provided at the interface of the two membrane-spanning BtuC subunits. It finally exits into the cytoplasm at the large gap between the four subunits. This transport cycle is powered by the hydrolysis of ATP by the ABC cassettes BtuD. Reprinted from Locher and Borths (2004), with permission of the authors.

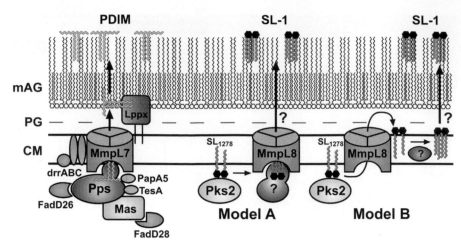

Chapter 12, Figure 4. Model of MmpL secretion. The proposed models for PDIM and SL-1 secretion through MmpL7 and MmpL8, respectively, are shown. PpsA-E and Mas are polyketide synthases that extend straight chain fatty acids to phthiocerol and mycocerosic acid, respectively (Azad et al., 1997; Azad et al., 1996; Trivedi et al., 2005). The enzymes FadD26 and FadD28 are thought to be AMP ligases that activate straight-chain fatty acids for transfer to the Pps and Mas enzymes (Trivedi et al., 2004). The thioesterase TesA is also required for the synthesis of PDIM and interacts with PpsE (Rao and Ranganathan, 2004). PapA5 is able to catalyze the esterification of mycocerosic acids to phthiocerol to form PDIM (Onwueme et al., 2004). MmpL7 and DrrABC are required to transport PDIM across the cell membrane (Camacho et al., 1999; Cox et al., 1999). Finally, LppX is a signal sequence containing protein that is thought to transport PDIM across the periplasm to the cell wall (Sulzenbacher et al., 2006). Two models for SL-1 secretion are shown. Pks2 is required for the synthesis of SL_{1278} (Sirakova et al., 2001). In model A, MmpL8 recruits an as yet unidentified biosynthetic factor to complete the synthesis of SL-1 from SL_{1278}, before transport across the cell membrane (Jain and Cox, 2005). In model B, MmpL8 exports SL_{1278} across the cell membrane, after which it is converted to SL-1 by an as yet unidentified enzyme (Converse et al., 2003). CM, cytoplasmic membrane; PG, peptidoglycan; mAG, mycolyl-arabinogalactan.

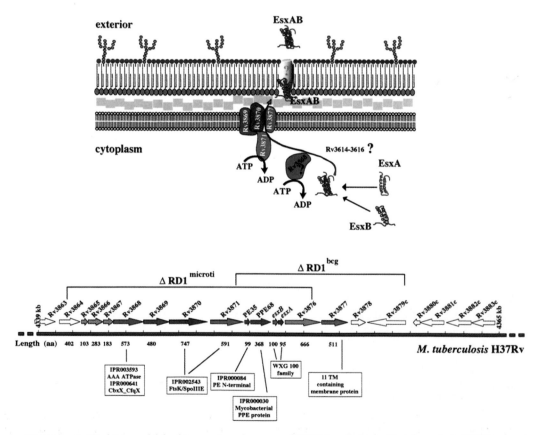

Chapter 13, Figure 3. Working model for biogenesis and export of ESAT-6 proteins in *M. tuberculosis* H37Rv plus the positions of various genes and deletions (Brodin et al., 2004b). The upper part presents a possible functional model indicating predicted subcellular localization and known or potential protein-protein interactions. Rosetta stone analysis indicates direct interaction between proteins Rv3870 and Rv3871; Rv3868 is an AAA-ATPase that is likely to act as a chaperone. ESAT-6 and CFP-10 are predicted to be exported through a transmembrane channel, consisting of at least Rv3870, Rv3871, and Rv3877 and possibly Rv3869 in a process catalyzed by ATP hydrolysis. The lower part shows key genes, the various proteins from the RD1 region, their sizes (number of amino acid residues), and the protein families.

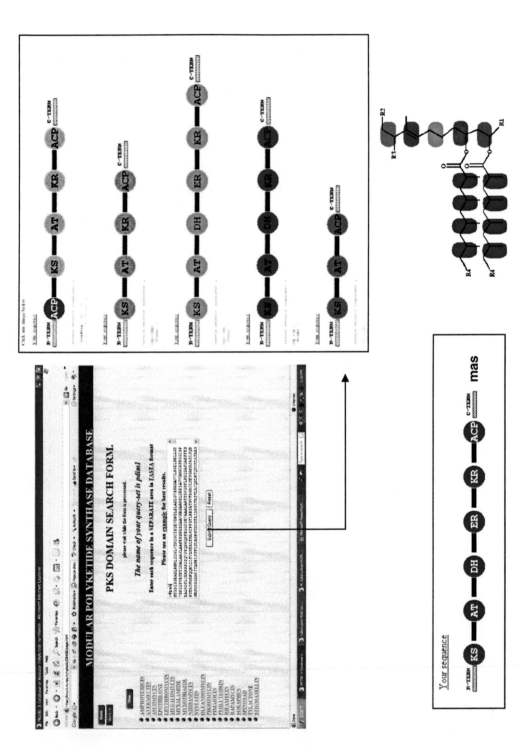

Chapter 15, Figure 3. Pictorial depiction of domains predicted in *PpsA-E* genes and Mas, by PKSDB and ITERDB. The ketide units in the final PDIM structure are indicated by colors similar to the synthesizing modules.

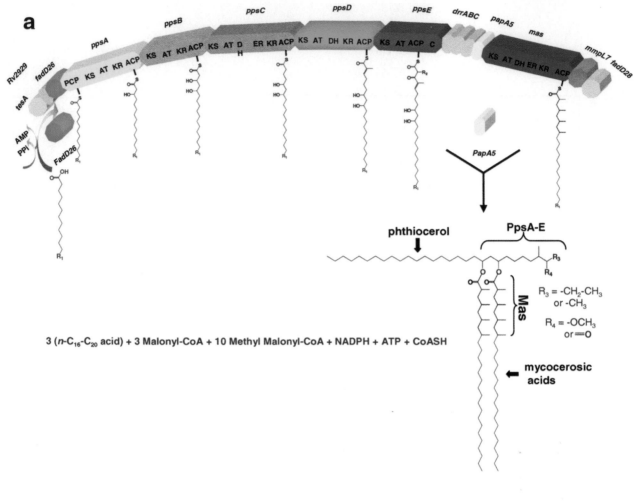

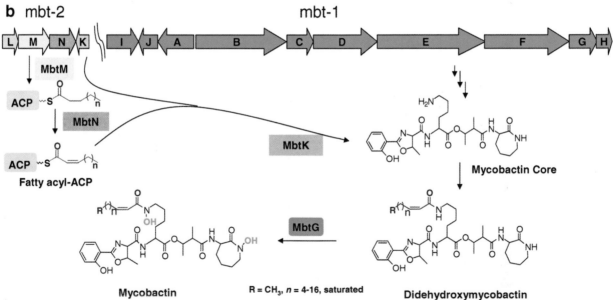

Chapter 15, Figure 4. (A) Biosynthesis of PDIM demonstrates thiotemplate-based assembly line enzymology carried out by PpsA-E, FadD26, Mas, and PapA5 proteins. (B) Mycobactin synthesis by mbt-1 and -2 clusters. mbt-1 cluster proteins synthesize the mycobactin core, and mbt-2 cluster proteins along with MbtG modify the core structure of mycobactin.

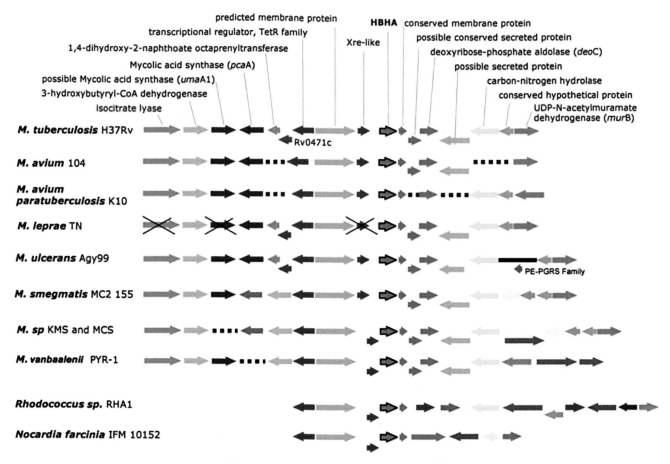

Chapter 19, Figure 2. Structure of the *hbhA* locus in different bacterial species. The *hbhA* gene of the different species indicated is highlighted as the blue arrow in bold. Conserved open reading frames between the different species are shown by the same color. The corresponding putative proteins encoded by the genes are indicated on the top. Chromosomal deletions or absence of genes are indicated by the dotted lines, and pseudogenes in *M. leprae* are depicted by the crosses.

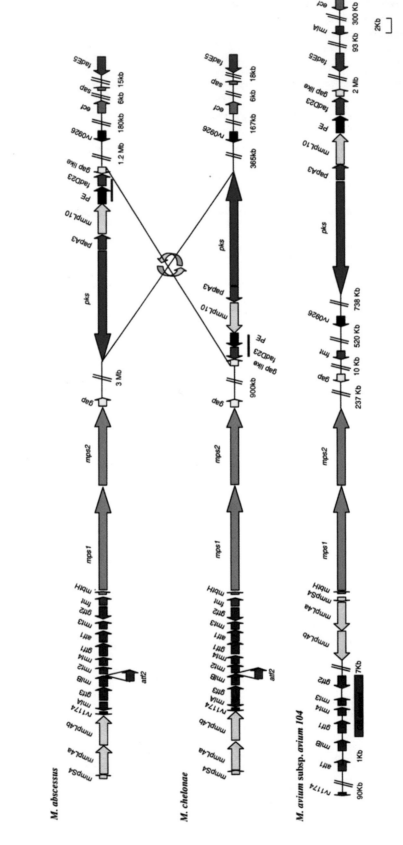

Chapter 21, Figure 3. Genetic organization of the GPL locus in various mycobacterial species. The ORFs are depicted as arrows and have been drawn proportionally to their size (Ripoll et al., 2007). Color code: light blue, *mmpL* family; black, unknown; purple, sugar biosynthesis, activation, transfer and modifications; red, lipid biosynthesis, activation, transfer and modifications; green, pseudopeptide biosynthesis; yellow, required for GPL transport to the surface; gray, regulation.

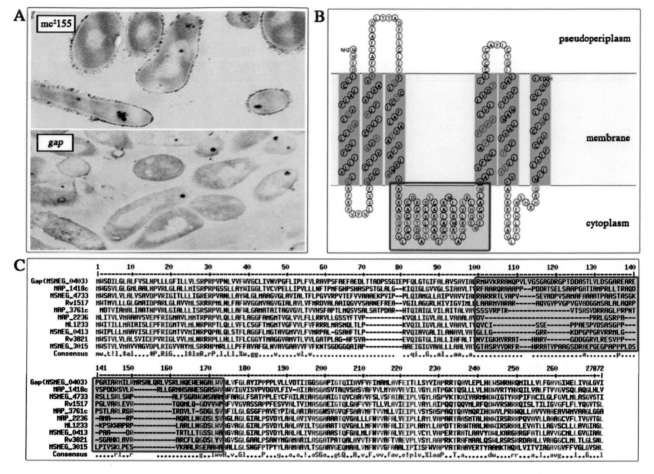

Chapter 21, Figure 4. (A) Transmission electron microscopy analysis of the *M. smegmatis* wild-type strain mc²155 and the gap mutant strain after ruthenium red staining (Sonden et al., 2005). (B) Topology of Gap protein predicted by Sosui program. (C) Amino acid alignment of various Gap orthologues found in sequenced mycobacterial genomes: *M. smegmatis* (MSMEG), *M. tuberculosis* (Rv), *M. avium* subsp. *paratuberculosis* (MAP) and *M. leprae* (ML). Color code: red, high consensus (>90%); blue, low consensus (>50%). The central parts bordered in orange correspond to the predicted cytoplasmic portion and may be the region of selectivity of Gap proteins.

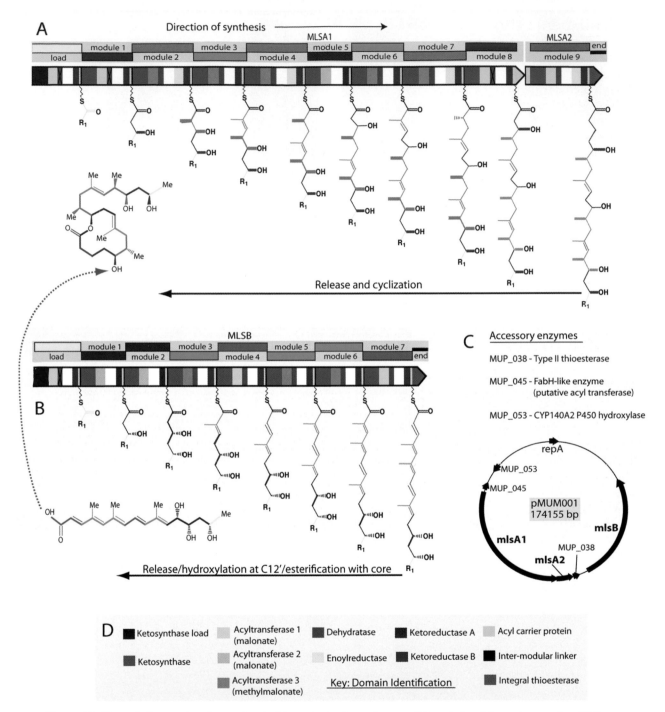

Chapter 22, Figure 3. Proposed pathways for the biosynthesis of mycolactone A/B as deduced from the complete DNA sequence of pMUM001 and transposon mutagenesis (Stinear et al., 2004). (A) Module and domain arrangement for synthesis of the core. (B) Module and domain arrangement for synthesis of the side chain. (C) Map of the pMUM001 plasmid showing the relative positions of the *mls* genes and accessory genes. (D) Key for domain function. Identical color indicates both same function and sequence (97% to 100% aa identity).

2005; Sassetti et al., 2003). However, the essential nature of this siderophore system is conditional to an environmental iron restriction, a situation that is encountered by *M. tuberculosis* in its human host. Although there are no published data on the capacity of MBT/CMBT deficient mutants of *M. tuberculosis* to multiply and cause disease in animal models of infection, several lines of indirect evidence support the view of an important role for the MBT/CMBT system in the maintenance of proper iron homeostasis during *M. tuberculosis* infections, namely (1) the MBT/CMBT system-deficient mutant is severely restricted for growth in iron-limiting medium and in human macrophages (De Voss et al., 2000); (2) the MBT/CMBT system-deficient mutant displays a growth defect even in iron-rich medium (Sassetti et al., 2003); (3) the expression of genes in the *mbt* cluster is upregulated in *M. tuberculosis* growing inside macrophages (Gold et al., 2001; Schnappinger et al., 2003) and mouse lung (Talaat et al., 2004; Timm et al., 2003), a fact suggesting the need for an iron starvation adaptive response during infection; (4) proper iron homeostasis is essential for *M. tuberculosis* growth (Rodriguez et al., 2002); and (5) MBT J (from *M. paratuberculosis*) efficiently extracts intracellular Fe^{3+} in macrophages (Luo et al., 2005). These observations warrant efforts to develop MBT/CMBT biosynthesis inhibitors and to explore the possible therapeutic value of the chemical inhibition of the MBT/CMBT pathway.

With the above-mentioned considerations in mind, the identification of inhibitors of MBT/CMBT

biosynthesis in *M. tuberculosis* is viewed as an important step toward evaluating the feasibility of developing drug candidates acting as siderophore biosynthesis blockers. In early 2005, the design, synthesis, and characterization of the first such inhibitor (a nucleoside antibiotic named salicyl-AMS) was reported by Ferreras et al. (2005). During the biosynthesis of mycobacterial MBTs and CMBTs, the enzyme MbtA catalyzes (1) salicylic acid adenylation to form salicyl-AMP and (2) transesterification of the salicyl moiety of the adenylate onto the prosthetic group of an aroyl carrier protein domain target in the peptide synthetase MbtB (Fig. 3A) (Quadri et al., 1998a). Salicyl-AMS is a potent inhibitor of the salicyl-AMP synthetase activity of MbtA's adenylation domain and the salicylic acid adenylation domains required for biosynthesis of salicylic acid-derived siderophores produced by other bacteria (Ferreras et al., 2005). Salicyl-AMS [5'-O-(N-salicylsulfamoyl)-adenosine] is a nonhydrolyzable mimic of salicyl-AMP (Fig. 3B), and was rationally conceived primarily based on the fact that nonhydrolyzable acyl-AMP analogues (e.g., acyl sulfamoyl adenosines) have been proven to be potent inhibitors of adenylate-forming enzymes (Finking et al., 2003; Lee et al., 2003) and the information afforded by a salicyl-AMP-enzyme binding model (Ferreras et al., 2005).

Salicyl-AMS inhibits MbtA in vitro and, most importantly, it inhibits MBT and CMBT biosynthesis and growth of *M. tuberculosis* under iron-limiting conditions (Ferreras et al., 2005). Salicyl-AMS is 20-fold less active against *M. tuberculosis* in iron-rich medium,

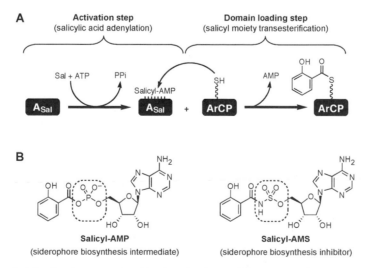

Figure 3. (A) Salicylic acid (Sal) adenylation catalyzed by the adenylation domain (A_{sal}) of *M. tuberculosis* MbtA and MbtA-dependent transesterification of the salicyl moiety onto the prosthetic group of the aroyl carrier protein domain (ArCP) of the multifunctional peptide synthetase MbtB. (B) Salicyl-AMP and its nonhydrolyzable mimic salicyl-AMS. The difference between these two molecules is highlighted.

where siderophores are not critical for growth (Ferreras et al., 2005). Salicyl-AMS also inhibits the salicylic acid adenylation activity of YbtE and PchD (Ferreras et al., 2005), the MbtA homologues required for biosynthesis of the salicylic acid-derived siderophores yersiniabactin (from *Yersinia pestis*) and pyochelin (from *Pseudomonas aeruginosa*) (Gehring et al., 1998b; Quadri et al., 1999), respectively. Thus, salicyl-AMS is considered a pan-inhibitor of salicylic acid-specific adenylation domains and represents a first lead compound for exploring the development of novel anti-infectives that act by blocking siderophore-mediated iron acquisition. The studies on salicyl-AMS by Ferreras et al. (2005) provide the first demonstration of the ability to target salicylic acid adenylation domains and the druggability of the MBT/CMBT biosynthesis pathway. More recently, analogues of salicyl-AMS have also been demonstrated to block MBT/CMBT biosynthesis and *M. tuberculosis* multiplication in iron-limiting conditions (Somu et al., 2006).

CONCLUDING REMARKS

Our current understanding of the siderophore-dependent iron acquisition systems of the mycobacteria is significant yet far from comprehensive. Much has been learned about the structures of the mycobacterial siderophores and many genes involved in their biosynthesis and transport have been identified by genetic analyses and bioinformatics. Remarkable advances have also been made towards unraveling the molecular mechanism underlying iron-dependent regulation of siderophore production. Although more progress is clearly under way in these areas, our knowledge of several other aspects of mycobacterial siderophore-dependent iron acquisition is lagging behind. In particular and of utmost importance, we need to expand our understanding of the function and relevance of the siderophore systems in the context of the mycobacterial infection. This is true not only for *M. tuberculosis* but also for *Mycobacterium* species that cause opportunistic infection in humans. This knowledge is important, because it will permit us to assess whether siderophore-dependent iron acquisition is, as hypothesized, a valid target for development of antimycobacterial drugs. Research efforts are also needed to validate different aspects of the working models of siderophore biosynthesis and characterize the enzymes involved in this process. Finally, new and creative multidisciplinary experimental approaches are required to gain a better insight into (ferri)siderophore trafficking and delivery of iron to the cytoplasm of mycobacteria in the host environment.

REFERENCES

Adilakshmi, T., P. D. Ayling, and C. Ratledge. 2000. Mutational analysis of a role for salicylic acid in iron metabolism of *Mycobacterium smegmatis*. *J. Bacteriol.* 182:264–271.

Agranoff, D., and S. Krishna. 2004. Metal ion transport and regulation in *Mycobacterium tuberculosis*. *Front. Biosci.* 9:2996–3006.

Andrews, S. C., A. K. Robinson, and F. Rodriguez-Quinones. 2003. Bacterial iron homeostasis. *FEMS Microbiol. Rev.* 27:215–237.

Baichoo, N., T. Wang, R. Ye, and J. D. Helmann. 2002. Global analysis of the *Bacillus subtilis* Fur regulon and the iron starvation stimulon. *Mol. Microbiol.* 45:1613–1629.

Bearden, S. W., J. D. Fetherston, and R. D. Perry. 1997. Genetic organization of the yersiniabactin biosynthetic region and construction of avirulent mutants in *Yersinia pestis*. *Infect. Immun.* 65:1659–1668.

Bjarnason, J., C. M. Southward, and M. G. Surette. 2003. Genomic profiling of iron-responsive genes in *Salmonella enterica* serovar *typhimurium* by high-throughput screening of a random promoter library. *J. Bacteriol.* 185:4973–4982.

Boukhalfa, H., and A. L. Crumbliss. 2002. Chemical aspects of siderophore mediated iron transport. *Biometals* 15:325–339.

Boyce, J. D., I. Wilkie, M. Harper, M. L. Paustian, V. Kapur, and B. Adler. 2002. Genomic scale analysis of *Pasteurella multocida* gene expression during growth within the natural chicken host. *Infect. Immun.* 70:6871–6879.

Braun, V. 2001. Iron uptake mechanisms and their regulation in pathogenic bacteria. *Int. J. Med. Microbiol.* 291:67–79.

Brown, K. A., and C. Ratledge. 1975. Iron transport in *Mycobacterium smegmatis*: ferrimycobactin reductase (NAD(P)H:ferrimycobactin oxidoreductase), the enzyme releasing iron from its carrier. *FEBS Lett.* 53:262–266.

Calder, K. M., and M. A. Horwitz. 1998. Identification of iron-regulated proteins of *Mycobacterium tuberculosis* and cloning of tandem genes encoding a low iron-induced protein and a metal transporting ATPase with similarities to two-component metal transport systems. *Microb. Pathog.* 24:133–143.

Card, G. L., N. A. Peterson, C. A. Smith, B. Rupp, B. M. Schick, and E. N. Baker. 2005. The crystal structure of Rv1347c, a putative antibiotic resistance protein from *Mycobacterium tuberculosis*, reveals a GCN5-related fold and suggests an alternative function in siderophore biosynthesis. *J. Biol. Chem.* 280:13978–13986.

Carniel, E., A. Guiyoule, I. Guilvout, and O. Mercereau-Puijalon. 1992. Molecular cloning, iron-regulation and mutagenesis of the *irp2* gene encoding HMWP2, a protein specific for the highly pathogenic *Yersinia*. *Mol. Microbiol.* 6:379–388.

Cendrowski, S., W. MacArthur, and P. Hanna. 2004. *Bacillus anthracis* requires siderophore biosynthesis for growth in macrophages and mouse virulence. *Mol. Microbiol.* 51:407–417.

Chalut, C., L. Botella, C. de Sousa-D'Auria, C. Houssin, and C. Guilhot. 2006. The nonredundant roles of two 4'-phosphopantetheinyl transferases in vital processes of Mycobacteria. *Proc. Natl. Acad. Sci. USA* 103:8511–8516.

Chipperfield, J. R., and C. Ratledge. 2000. Salicylic acid is not a bacterial siderophore: a theoretical study. *Biometals* 13:165–168.

Clemens, D. L., and M. A. Horwitz. 1996. The *Mycobacterium tuberculosis* phagosome interacts with early endosomes and is accessible to exogenously administered transferrin. *J. Exp. Med.* 184:1349–1355.

Cole, S. T., R. Brosch, J. Parkhill, T. Garnier, C. Churcher, D. Harris, S. V. Gordon, K. Eiglmeier, S. Gas, C. E. Barry III, F. Tekaia, K. Badcock, D. Basham, D. Brown, T. Chillingworth, R. Connor, R. Davies, K. Devlin, T. Feltwell, S. Gentles, N. Hamlin,

S. Holroyd, T. Hornsby, K. Jagels, A. J. McLean, S. Moule, L. Murphy, K. Oliver, J. Osborne, M. A. Quail, M. A. Rajandream, J. Rogers, S. Rutter, K. Seeger, J. Skelton, R. Squares, S. Squares, J. E. Sulston, K. Taylor, S. Whitehead, and B. G. Barrell. 1998. Deciphering the biology of *Mycobacterium tuberculosis* from the complete genome sequence. *Nature* 393:537–544.

Cole, S. T., K. Eiglmeier, J. Parkhill, K. D. James, N. R. Thomson, P. R. Wheeler, N. Honore, T. Garnier, C. Churcher, D. Harris, K. Mungall, D. Basham, D. Brown, T. Chillingworth, R. Connor, R. M. Davies, K. Devlin, S. Duthoy, T. Feltwell, A. Fraser, N. Hamlin, S. Holroyd, T. Hornsby, K. Jagels, C. Lacroix, J. Maclean, S. Moule, L. Murphy, K. Oliver, M. A. Quail, M. A. Rajandream, K. M. Rutherford, S. Rutter, K. Seeger, S. Simon, M. Simmonds, J. Skelton, R. Squares, S. Squares, K. Stevens, K. Taylor, S. Whitehead, J. R. Woodward, and B. G. Barrell. 2001. Massive gene decay in the leprosy bacillus. *Nature* 409:1007–1011.

Cronje, L., and L. Bornman. 2005. Iron overload and tuberculosis: a case for iron chelation therapy. *Int. J. Tuberc. Lung Dis.* 9:2–9.

Crosa, J. H. 1997. Signal transduction and transcriptional and posttranscriptional control of iron-regulated genes in bacteria. *Microbiol. Mol. Biol. Rev.* 61:319–336.

Crosa, J. H., and C. T. Walsh. 2002. Genetics and assembly line enzymology of siderophore biosynthesis in bacteria. *Microbiol. Mol. Biol. Rev.* 66:223–249.

Dale, S. E., A. Doherty-Kirby, G. Lajoie, and D. E. Heinrichs. 2004. Role of siderophore biosynthesis in virulence of *Staphylococcus aureus*: identification and characterization of genes involved in production of a siderophore. *Infect. Immun.* 72:29–37.

Darwin, K. H., S. Ehrt, J. C. Gutierrez-Ramos, N. Weich, and C. F. Nathan. 2003. The proteasome of *Mycobacterium tuberculosis* is required for resistance to nitric oxide. *Science* 302:1963–1966.

De Voss, J. J., K. Rutter, B. G. Schroeder, and C. E. Barry III. 1999. Iron acquisition and metabolism by mycobacteria. *J. Bacteriol.* 181:4443–4451.

De Voss, J. J., K. Rutter, B. G. Schroeder, H. Su, Y. Zhu, and C. E. Barry III. 2000. The salicylate-derived mycobactin siderophores of *Mycobacterium tuberculosis* are essential for growth in macrophages. *Proc. Natl. Acad. Sci. USA* 97:1252–1257.

Der Vartanian, M., B. Jaffeux, M. Contrepois, M. Chavarot, J. P. Girardeau, Y. Bertin, and C. Martin. 1992. Role of aerobactin in systemic spread of an opportunistic strain of *Escherichia coli* from the intestinal tract of gnotobiotic lambs. *Infect. Immun.* 60:2800–2807.

Dhungana, S., M. J. Miller, L. Dong, C. Ratledge, and A. L. Crumbliss. 2003. Iron chelation properties of an extracellular siderophore exochelin MN. *J. Am. Chem. Soc.* 125:7654–7663.

Dhungana, S., C. Ratledge, and A. L. Crumbliss. 2004. Iron chelation properties of an extracellular siderophore exochelin MS. *Inorg. Chem.* 43:6274–6283.

Dover, L. G., and C. Ratledge. 1996. Identification of a 29-kDa protein in the envelope of *Mycobacterium smegmatis* as a putative ferri-exochelin receptor. *Microbiology* 142:1521–1530.

Drechsel, H., and G. Winkelmann. 1997. Iron chelation and siderophores, p. 1–9. *In* G. Winkelmann and C. J. Carrano (ed.), *Transition Metals in Microbial Metabolism*. Harwood Acad., Amsterdam, The Netherlands.

Du, L., and B. Shen. 2001. Biosynthesis of hybrid peptide-polyketide natural products. *Curr. Opin. Drug Discov. Devel.* 4:215–228.

Ducey, T. F., M. B. Carson, J. Orvis, A. P. Stintzi, and D. W. Dyer. 2005. Identification of the iron-responsive genes of *Neisseria gonorrhoeae* by microarray analysis in defined medium. *J. Bacteriol.* 187:4865–4874.

Duerfahrt, T., K. Eppelmann, R. Muller, and M. A. Marahiel. 2004. Rational design of a bimodular model system for the investigation of heterocyclization in nonribosomal peptide biosynthesis. *Chem. Biol.* 11:261–271.

Dussurget, O., M. Rodriguez, and I. Smith. 1996. An *ideR* mutant of *Mycobacterium smegmatis* has derepressed siderophore production and an altered oxidative-stress response. *Mol. Microbiol.* 22:535–544.

Dussurget, O., M. Rodriguez, and I. Smith. 1998. Protective role of the *Mycobacterium smegmatis* IdeR against reactive oxygen species and isoniazid toxicity. *Tuber. Lung Dis.* 79:99–106.

Dussurget, O., J. Timm, M. Gomez, B. Gold, S. Yu, S. Z. Sabol, R. K. Holmes, W. R. Jacobs, Jr., and I. Smith. 1999. Transcriptional control of the iron-responsive *fxbA* gene by the mycobacterial regulator IdeR. *J. Bacteriol.* 181:3402–3408.

Ernst, F. D., S. Bereswill, B. Waidner, J. Stoof, U. Mader, J. G. Kusters, E. J. Kuipers, M. Kist, A. H. van Vliet, and G. Homuth. 2005. Transcriptional profiling of *Helicobacter pylori* Fur- and iron-regulated gene expression. *Microbiology* 151:533–546.

Fernandez, L., I. Marquez, and J. A. Guijarro. 2004. Identification of specific in vivo-induced (*ivi*) genes in *Yersinia ruckeri* and analysis of ruckerbactin, a catecholate siderophore iron acquisition system. *Appl. Environ. Microbiol.* 70:5199–5207.

Ferreras, J. A., J.-S. Ryu, F. Di Lello, D. S. Tan, and L. E. N. Quadri. 2005. Small molecule inhibition of siderophore biosynthesis in *Mycobacterium tuberculosis* and *Yersinia pestis*. *Nat. Chem. Biol.* 1:29–32.

Fetherston, J. D., J. W. Lillard, Jr., and R. D. Perry. 1995. Analysis of the pesticin receptor from *Yersinia pestis*: role in iron-deficient growth and possible regulation by its siderophore. *J. Bacteriol.* 177:1824–1833.

Finking, R., A. Neumueller, J. Solsbacher, D. Konz, G. Kretzschmar, M. Schweitzer, T. Krumm, and M. A. Marahiel. 2003. Aminoacyl adenylate substrate analogues for the inhibition of adenylation domains of nonribosomal peptide synthetases. *Chembiochem* 4:903–906.

Fiss, E. H., S. Yu, and W. R. Jacobs, Jr. 1994. Identification of genes involved in the sequestration of iron in mycobacteria: the ferric exochelin biosynthetic and uptake pathways. *Mol. Microbiol.* 14:557–569.

Flo, T. H., K. D. Smith, S. Sato, D. J. Rodriguez, M. A. Holmes, R. K. Strong, S. Akira, and A. Aderem. 2004. Lipocalin 2 mediates an innate immune response to bacterial infection by sequestrating iron. *Nature* 432:917–921.

Gangaidzo, I. T., V. M. Moyo, E. Mvundura, G. Aggrey, N. L. Murphree, H. Khumalo, T. Saungweme, I. Kasvosve, Z. A. Gomo, T. Rouault, J. R. Boelaert, and V. R. Gordeuk. 2001. Association of pulmonary tuberculosis with increased dietary iron. *J. Infect. Dis.* 184:936–939.

Gehring, A. M., E. DeMoll, J. D. Fetherston, I. Mori, G. F. Mayhew, F. R. Blattner, C. T. Walsh, and R. D. Perry. 1998a. Iron acquisition in plague: modular logic in enzymatic biogenesis of yersiniabactin by *Yersinia pestis*. *Chem. Biol.* 5:573–586.

Gehring, A. M., I. I. Mori, R. D. Perry, and C. T. Walsh. 1998b. The nonribosomal peptide synthetase HMWP2 forms a thiazoline ring during biogenesis of yersiniabactin, an iron-chelating virulence factor of *Yersinia pestis*. *Biochemistry* 37:11637–11650.

Gobin, J., and M. A. Horwitz. 1996. Exochelins of *Mycobacterium tuberculosis* remove iron from human iron-binding proteins and donate iron to mycobactins in the *M. tuberculosis* cell wall. *J. Exp. Med.* 183:1527–1532.

Gobin, J., C. H. Moore, J. R. Reeve, Jr., D. K. Wong, B. W. Gibson, and M. A. Horwitz. 1995. Iron acquisition by *Mycobacterium tuberculosis*: isolation and characterization of a family of iron-binding exochelins. *Proc. Natl. Acad. Sci. USA* **92:** 5189–5193.

Gobin, J., D. K. Wong, B. W. Gibson, and M. A. Horwitz. 1999. Characterization of exochelins of the *Mycobacterium bovis* type strain and BCG substrains. *Infect. Immun.* **67:**2035–2039.

Goetz, D. H., M. A. Holmes, N. Borregaard, M. E. Bluhm, K. N. Raymond, and R. K. Strong. 2002. The neutrophil lipocalin NGAL is a bacteriostatic agent that interferes with siderophore-mediated iron acquisition. *Mol. Cell.* **10:**1033–1043.

Gold, B., G. M. Rodriguez, S. A. Marras, M. Pentecost, and I. Smith. 2001. The *Mycobacterium tuberculosis* IdeR is a dual functional regulator that controls transcription of genes involved in iron acquisition, iron storage and survival in macrophages. *Mol. Microbiol.* **42:**851–865.

Grifantini, R., S. Sebastian, E. Frigimelica, M. Draghi, E. Bartolini, A. Muzzi, R. Rappuoli, G. Grandi, and C. A. Genco. 2003. Identification of iron-activated and -repressed Fur-dependent genes by transcriptome analysis of *Neisseria meningitidis* group B. *Proc. Natl. Acad. Sci. USA* **100:**9542–9547.

Hall, R. M., and C. Ratledge. 1987. Exochelin-mediated iron acquisition by the leprosy bacillus, *Mycobacterium leprae*. *J. Gen. Microbiol.* **133:**193–199.

Hall, R. M., M. Sritharan, A. J. Messenger, and C. Ratledge. 1987. Iron transport in *Mycobacterium smegmatis*: occurrence of iron-regulated envelope proteins as potential receptors for iron uptake. *J. Gen. Microbiol.* **133:**2107–2114.

Hall, R. M., P. R. Wheeler, and C. Ratledge. 1983. Exochelin-mediated iron uptake into *Mycobacterium leprae*. *Int. J. Lepr. Other Mycobact. Dis.* **51:**490–494.

Harrison, A. J., M. Yu, T. Gardenborg, M. Middleditch, R. J. Ramsay, E. N. Baker, and J. S. Lott. 2006. The structure of MbtI from *Mycobacterium tuberculosis*, the first enzyme in the biosynthesis of the siderophore mycobactin, reveals it to be a salicylate synthase. *J. Bacteriol.* **188:**6081–6091.

Heesemann, J. 1987. Chromosomal-encoded siderophores are required for mouse virulence of enteropathogenic *Yersinia* species. *FEMS Microbiol. Lett.* **48:**229–233.

Henderson, D. P., and S. M. Payne. 1994. *Vibrio cholerae* iron transport systems: roles of heme and siderophore iron transport in virulence and identification of a gene associated with multiple iron transport systems. *Infect. Immun.* **62:**5120–5125.

Holmes, K., F. Mulholland, B. M. Pearson, C. Pin, J. McNicholl-Kennedy, J. M. Ketley, and J. M. Wells. 2005a. *Campylobacter jejuni* gene expression in response to iron limitation and the role of Fur. *Microbiology* **151:**243–257.

Holmes, M. A., W. Paulsene, X. Jide, C. Ratledge, and R. K. Strong. 2005b. Siderocalin (Lcn 2) also binds carboxymycobactins, potentially defending against mycobacterial infections through iron sequestration. *Structure* **13:**29–41.

Homuth, M., P. Valentin-Weigand, M. Rohde, and G. F. Gerlach. 1998. Identification and characterization of a novel extracellular ferric reductase from *Mycobacterium paratuberculosis*. *Infect. Immun.* **66:**710–716.

Hou, J. Y., J. E. Graham, and J. E. Clark-Curtiss. 2002. *Mycobacterium avium* genes expressed during growth in human macrophages detected by selective capture of transcribed sequences (SCOTS). *Infect. Immun.* **70:**3714–3726.

Jurado, R. L. 1997. Iron, infections, and anemia of inflammation. *Clin. Infect. Dis.* **25:**888–895.

Kato, L. 1985. Absence of mycobactin in *Mycobacterium leprae*; probably a microbe dependent microorganism implications. *Indian J. Lepr.* **57:**58–70.

Katoch, V. M. 2004. Infections due to non-tuberculous mycobacteria (NTM). *Indian J. Med. Res.* **120:**290–304.

Kjeldsen, L., J. B. Cowland, and N. Borregaard. 2000. Human neutrophil gelatinase-associated lipocalin and homologous proteins in rat and mouse. *Biochim. Biophys. Acta* **1482:**272–283.

Krithika, R., U. Marathe, P. Saxena, M. Z. Ansari, D. Mohanty, and R. S. Gokhale. 2006. A genetic locus required for iron acquisition in *Mycobacterium tuberculosis*. *Proc. Natl. Acad. Sci. USA* **103:**2069–2074.

LaMarca, B. B., W. Zhu, J. E. Arceneaux, B. R. Byers, and M. D. Lundrigan. 2004. Participation of *fad* and *mbt* genes in synthesis of mycobactin in *Mycobacterium smegmatis*. *J. Bacteriol.* **186:**374–382.

Lane, S. J., P. S. Marshall, R. J. Upton, and C. Ratledge. 1998. Isolation and characterization of carboxymycobactins as the second extracellular siderophores in *Mycobacterium smegmatis*. *Biometals* **11:**13–20.

Lane, S. J., P. S. Marshall, R. J. Upton, C. Ratledge, and M. Ewing. 1995. Novel extracellular mycobactins, the carboxymycobactins from *Mycobacterium avium*. *Tetrahedron Lett.* **36:**4129–4132.

Lee, J., S. E. Kim, J. Y. Lee, S. Y. Kim, S. U. Kang, S. H. Seo, M. W. Chun, T. Kang, S. Y. Choi, and H. O. Kim. 2003. N-Alkoxysulfamide, N-hydroxysulfamide, and sulfamate analogues of methionyl and isoleucyl adenylates as inhibitors of methionyl-tRNA and isoleucyl-tRNA synthetases. *Bioorg. Med. Chem. Lett.* **13:**1087–1092.

Litwin, C. M., T. W. Rayback, and J. Skinner. 1996. Role of catechol siderophore synthesis in *Vibrio vulnificus* virulence. *Infect. Immun.* **64:**2834–2838.

Lounis, N., C. Truffot-Pernot, J. Grosset, V. R. Gordeuk, and J. R. Boelaert. 2001. Iron and *Mycobacterium tuberculosis* infection. *J. Clin. Virol.* **20:**123–126.

Luo, M., E. A. Fadeev, and J. T. Groves. 2005. Mycobactin-mediated iron acquisition within macrophages. *Nat. Chem. Biol.* **1:**149–153.

MacCordick, H. J., J. J. Schleiffer, and G. Duplatre. 1985. Radiochemical studies of iron binding and stability in ferrimycobactin S. *Radiochim. Acta* **38:**43–47.

Macham, L. P., M. C. Stephenson, and C. Ratledge. 1977. Iron transport in *Mycobacterium smegmatis*: the isolation, purification and function of exochelin MS. *J. Gen. Microbiol.* **101:**41–49.

Marahiel, M. A., T. Stachelhaus, and H. D. Mootz. 1997. Modular peptide synthetases involved in nonribosomal peptide synthesis. *Chem. Rev.* **97:**2651–2674.

Merkal, R. S., W. G. McCullough, and K. Takayama. 1981. Mycobactins, the state of the art. *Bull. Inst. Pasteur* **79:**251–259.

Merrell, D. S., L. J. Thompson, C. C. Kim, H. Mitchell, L. S. Tompkins, A. Lee, and S. Falkow. 2003. Growth phase-dependent response of *Helicobacter pylori* to iron starvation. *Infect. Immun.* **71:**6510–6525.

Messenger, A. J., R. M. Hall, and C. Ratledge. 1986. Iron uptake processes in *Mycobacterium vaccae* R877R, a mycobacterium lacking mycobactin. *J. Gen. Microbiol.* **132:**845–852.

Meyer, J. M., P. Azelvandre, and C. Georges. 1992. Iron metabolism in *Pseudomonas*: salicylic acid, a siderophore of *Pseudomonas fluorescens* CHAO. *Biofactors* **4:**23–27.

Miethke, M., H. Westers, E. J. Blom, O. P. Kuipers, and M. A. Marahiel. 2006. Iron starvation triggers the stringent response and induces amino acid biosynthesis for bacillibactin production in *Bacillus subtilis*. *J. Bacteriol.* **188:**8655–8657.

Morrison, N. E. 1995. *Mycobacterium leprae* iron nutrition: bacterioferritin, mycobactin, exochelin and intracellular growth. *Int. J. Lepr. Other Mycobact. Dis.* **63:**86–91.

Murakami, Y., S. Kato, M. Nakajima, M. Matsuoka, H. Kawai, K. Shin-Ya, and H. Seto. 1996. Formobactin, a novel free radical scavenging and neuronal cell protecting substance from *Nocardia* sp. *J. Antibiot.* (Tokyo) 49:839–845.

Neilands, J. B. 1995. Siderophores: structure and function of microbial iron transport compounds. *J. Biol. Chem.* 270:26723–26726.

Ochsner, U. A., P. J. Wilderman, A. I. Vasil, and M. L. Vasil. 2002. GeneChip expression analysis of the iron starvation response in *Pseudomonas aeruginosa*: identification of novel pyoverdine biosynthesis genes. *Mol. Microbiol.* 45:1277–1287.

Olakanmi, O., L. S. Schlesinger, A. Ahmed, and B. E. Britigan. 2002. Intraphagosomal *Mycobacterium tuberculosis* acquires iron from both extracellular transferrin and intracellular iron pools. Impact of interferon-gamma and hemochromatosis. *J. Biol. Chem.* 277:49727–49734.

Olakanmi, O., L. S. Schlesinger, A. Ahmed, and B. E. Britigan. 2004. The nature of extracellular iron influences iron acquisition by *Mycobacterium tuberculosis* residing within human macrophages. *Infect. Immun.* 72:2022–2028.

Onwueme, K. C., C. J. Vos, J. Zurita, J. A. Ferreras, and L. E. Quadri. 2005. The dimycocerosate ester polyketide virulence factors of mycobacteria. *Prog. Lipid Res.* 44:259–302.

Palma, M., S. Worgall, and L. E. Quadri. 2003. Transcriptome analysis of the *Pseudomonas aeruginosa* response to iron. *Arch. Microbiol.* 180:374–379.

Palyada, K., D. Threadgill, and A. Stintzi. 2004. Iron acquisition and regulation in *Campylobacter jejuni*. *J. Bacteriol.* 186:4714–4729.

Paustian, M. L., B. J. May, and V. Kapur. 2001. *Pasteurella multocida* gene expression in response to iron limitation. *Infect. Immun.* 69:4109–4115.

Prakash, P., S. Yellaboina, A. Ranjan, and S. E. Hasnain. 2005. Computational prediction and experimental verification of novel IdeR binding sites in the upstream sequences of *Mycobacterium tuberculosis* open reading frames. *Bioinformatics* 21:2161–2166.

Quadri, L. E. N. 2000. Assembly of aryl-capped siderophores by modular peptide synthetases and polyketide synthases. *Mol. Microbiol.* 37:1–12.

Quadri, L. E. N. 2007. Strategic paradigm shifts in the antimicrobial drug discovery process of the 21st century. *Infect. Disord. Drug Targets* 7:230–237.

Quadri, L. E. N., T. A. Keating, H. M. Patel, and C. T. Walsh. 1999. Assembly of the *Pseudomonas aeruginosa* nonribosomal peptide siderophore pyochelin: In vitro reconstitution of aryl-4,2-bisthiazoline synthetase activity from PchD, PchE, and PchF. *Biochemistry* 38:14941–14954.

Quadri, L. E. N., and C. Ratledge. 2005. Iron metabolism in the tubercle bacillus and other mycobacteria, p. 341-357. *In* S. T. Cole, K. D. Eisenach, D. N. McMurray, and W. R. Jacobs, Jr. (ed.), *Tuberculosis and the Tubercle Bacillus.* ASM Press, Washington, DC.

Quadri, L. E. N., J. Sello, T. A. Keating, P. H. Weinreb, and C. T. Walsh. 1998a. Identification of a *Mycobacterium tuberculosis* gene cluster encoding the biosynthetic enzymes for assembly of the virulence-conferring siderophore mycobactin. *Chem. Biol.* 5:631–645.

Quadri, L. E. N., P. H. Weinreb, M. Lei, M. M. Nakano, P. Zuber, and C. T. Walsh. 1998b. Characterization of Sfp, a *Bacillus subtilis* phosphopantetheinyl transferase for peptidyl carrier protein domains in peptide synthetases. *Biochemistry* 37:1585–1595.

Raghu, B., G. R. Sarma, and P. Venkatesan. 1993. Effect of iron on the growth and siderophore production of mycobacteria. *Biochem. Mol. Biol. Int.* 31:341–348.

Rakin, A., E. Saken, D. Harmsen, and J. Heesemann. 1994. The pesticin receptor of *Yersinia enterocolitica*: a novel virulence factor with dual function. *Mol. Microbiol.* 13:253–263.

Ratledge, C., and L. G. Dover. 2000. Iron metabolism in pathogenic bacteria. *Annu. Rev. Microbiol.* 54:881–941.

Ratledge, C., and M. Ewing. 1996. The occurrence of carboxymycobactin, the siderophore of pathogenic mycobacteria, as a second extracellular siderophore in *Mycobacterium smegmatis*. *Microbiology* 142:2207–2212.

Ratledge, C., P. V. Patel, and J. Mundy. 1982. Iron transport in *Mycobacterium smegmatis*: the location of mycobactin by electron microscopy. *J. Gen. Microbiol.* 128:1559–1565.

Ratledge, C., and G. A. Snow. 1974. Isolation and structure of nocobactin NA, a lipid-soluble iron-binding compound from *Nocardia asteroides*. *Biochem. J.* 139:407–413.

Ratledge, C., and F. G. Winder. 1962. The accumulation of salicylic acid by mycobacteria during growth on an iron-deficient medium. *Biochem. J.* 84:501–506.

Register, K. B., T. F. Ducey, S. L. Brockmeier, and D. W. Dyer. 2001. Reduced virulence of a *Bordetella bronchiseptica* siderophore mutant in neonatal swine. *Infect. Immun.* 69:2137–2143.

Rodriguez, G. M. 2006. Control of iron metabolism in *Mycobacterium tuberculosis*. *Trends Microbiol.* 14:320–327.

Rodriguez, G. M., and I. Smith. 2003. Mechanisms of iron regulation in mycobacteria: role in physiology and virulence. *Mol. Microbiol.* 47:1485–1494.

Rodriguez, G. M., and I. Smith. 2006. Identification of an ABC transporter required for iron acquisition and virulence in *Mycobacterium tuberculosis*. *J. Bacteriol.* 188:424–430.

Rodriguez, G. M., M. I. Voskuil, B. Gold, G. K. Schoolnik, and I. Smith. 2002. *ideR*, an essential gene in *Mycobacterium tuberculosis*: role of IdeR in iron-dependent gene expression, iron metabolism, and oxidative stress response. *Infect. Immun.* 70:3371–3381.

Sassetti, C. M., D. H. Boyd, and E. J. Rubin. 2003. Genes required for mycobacterial growth defined by high density mutagenesis. *Mol. Microbiol.* 48:77–84.

Schaible, U. E., H. L. Collins, F. Priem, and S. H. Kaufmann. 2002. Correction of the iron overload defect in β-2-microglobulin knockout mice by lactoferrin abolishes their increased susceptibility to tuberculosis. *J. Exp. Med.* 196:1507–1513.

Schnappinger, D., S. Ehrt, M. I. Voskuil, Y. Liu, J. A. Mangan, I. M. Monahan, G. Dolganov, B. Efron, P. D. Butcher, C. Nathan, and G. K. Schoolnik. 2003. Transcriptional adaptation of *Mycobacterium tuberculosis* within macrophages: Insights into the phagosomal environment. *J. Exp. Med.* 198:693–704.

Sharman, G. J., D. H. Williams, D. F. Ewing, and C. Ratledge. 1995a. Determination of the structure of exochelin MN, the extracellular siderophore from *Mycobacterium neoaurum*. *Chem. Biol.* 2:553–561.

Sharman, G. J., D. H. Williams, D. F. Ewing, and C. Ratledge. 1995b. Isolation, purification and structure of exochelin MS, the extracellular siderophore from *Mycobacterium smegmatis*. *Biochem. J.* 305:187–196.

Singh, A. K., L. M. McIntyre, and L. A. Sherman. 2003. Microarray analysis of the genome-wide response to iron deficiency and iron reconstitution in the cyanobacterium *Synechocystis* sp. PCC 6803. *Plant Physiol.* 132:1825–1839.

Snow, G. A. 1970. Mycobactins: iron-chelating growth factors from mycobacteria. *Bacteriol. Rev.* 34:99–125.

Sokol, P. A., P. Darling, D. E. Woods, E. Mahenthiralingam, and C. Kooi. 1999. Role of ornibactin biosynthesis in the virulence of *Burkholderia cepacia*: characterization of *pvdA*, the gene encoding L-ornithine N(5)-oxygenase. *Infect. Immun.* 67:4443–4455.

Somu, R. V., H. Boshoff, C. Qiao, E. M. Bennett, C. E. Barry III, and C. C. Aldrich. 2006. Rationally designed nucleoside antibiotics that inhibit siderophore biosynthesis of *Mycobacterium tuberculosis*. *J. Med. Chem.* **49:**31–34.

Stephenson, M. C., and C. Ratledge. 1979. Iron transport in *Mycobacterium smegmatis*: uptake of iron from ferriexochelin. *J. Gen. Microbiol.* **110:**193–202.

Stephenson, M. C., and C. Ratledge. 1980. Specificity of exochelins for iron transport in three species of mycobacteria. *J. Gen. Microbiol.* **116:**521–523.

Takase, H., H. Nitanai, K. Hoshino, and T. Otani. 2000. Impact of siderophore production on *Pseudomonas aeruginosa* infections in immunosuppressed mice. *Infect. Immun.* **68:**1834–1839.

Talaat, A. M., R. Lyons, S. T. Howard, and S. A. Johnston. 2004. The temporal expression profile of *Mycobacterium tuberculosis* infection in mice. *Proc. Natl. Acad. Sci. USA* **101:**4602–4607.

Timm, J., F. A. Post, L. G. Bekker, G. B. Walther, H. C. Wainwright, R. Manganelli, W. T. Chan, L. Tsenova, B. Gold, I. Smith, G. Kaplan, and J. D. McKinney. 2003. Differential expression of iron-, carbon-, and oxygen-responsive mycobacterial genes in the lungs of chronically infected mice and tuberculosis patients. *Proc. Natl. Acad. Sci. USA* **100:**14321–14326.

Todd, J. D., G. Sawers, and A. W. Johnston. 2005. Proteomic analysis reveals the wide-ranging effects of the novel, iron-responsive regulator RirA in *Rhizobium leguminosarum* bv. viciae. *Mol. Genet. Genomics* **273:**197–206.

Touati, D. 2000. Iron and oxidative stress in bacteria. *Arch. Biochem. Biophys.* **373:**1–6.

Wagner, D., J. Maser, B. Lai, Z. Cai, C. E. Barry III, K. Honer Zu Bentrup, D. G. Russell, and L. E. Bermudez. 2005. Elemental analysis of *Mycobacterium avium*-, *Mycobacterium tuberculosis*-, and *Mycobacterium smegmatis*-containing phagosomes indicates pathogen-induced microenvironments within the host cell's endosomal system. *J. Immunol.* **174:**1491–1500.

Walsh, C. T. 2004. Polyketide and nonribosomal peptide antibiotics: modularity and versatility. *Science* **303:**1805–1810.

Walsh, C. T., A. M. Gehring, P. H. Weinreb, L. E. Quadri, and R. S. Flugel. 1997. Post-translational modification of polyketide and nonribosomal peptide synthases. *Curr. Opin. Chem. Biol.* **1:**309–315.

Wan, X. F., N. C. Verberkmoes, L. A. McCue, D. Stanek, H. Connelly, L. J. Hauser, L. Wu, X. Liu, T. Yan, A. Leaphart, R. L. Hettich, J. Zhou, and D. K. Thompson. 2004. Transcriptomic and proteomic characterization of the Fur modulon in the metal-reducing bacterium *Shewanella oneidensis*. *J. Bacteriol.* **186:**8385–8400.

Wandersman, C., and P. Delepelaire. 2004. Bacterial iron sources: from siderophores to hemophores. *Annu. Rev. Microbiol.* **58:**611–647.

Ward, C. G., J. J. Bullen, and H. J. Rogers. 1996. Iron and infection: new developments and their implications. *J. Trauma* **41:**356–364.

Wehrl, W., T. F. Meyer, P. R. Jungblut, E. C. Muller, and A. J. Szczepek. 2004. Action and reaction: *Chlamydophila pneumoniae* proteome alteration in a persistent infection induced by iron deficiency. *Proteomics* **4:**2969–2981.

Weinberg, E. D. 1978. Iron and infection. *Microbiol. Rev.* **42:**45–66.

Weinberg, E. D. 1993. The development of awareness of iron-withholding defense. *Perspect. Biol. Med.* **36:**215–221.

Williams, D. L., M. Torrero, P. R. Wheeler, R. W. Truman, M. Yoder, N. Morrison, W. R. Bishai, and T. P. Gillis. 2004. Biological implications of *Mycobacterium leprae* gene expression during infection. *J. Mol. Microbiol. Biotechnol.* **8:**58–72.

Wong, D. K., J. Gobin, M. A. Horwitz, and B. W. Gibson. 1996. Characterization of exochelins of *Mycobacterium avium*: evidence for saturated and unsaturated and for acid and ester forms. *J. Bacteriol.* **178:**6394–6398.

Wooldridge, K. G., and P. H. Williams. 1993. Iron uptake mechanisms of pathogenic bacteria. *FEMS Microbiol. Rev.* **12:**325–348.

Xu, S., and P. Venge. 2000. Lipocalins as biochemical markers of disease. *Biochim. Biophys. Acta* **1482:**298–307.

Yancey, R. J., S. A. Breeding, and C. E. Lankford. 1979. Enterochelin (enterobactin): virulence factor for *Salmonella typhimurium*. *Infect. Immun.* **24:**174–180.

Yellaboina, S., S. Ranjan, V. Vindal, and A. Ranjan. 2006. Comparative analysis of iron regulated genes in mycobacteria. *FEBS Lett.* **580:**2567–2576.

Yu, S., E. Fiss, and W. R. Jacobs, Jr. 1998. Analysis of the exochelin locus in *Mycobacterium smegmatis*: biosynthesis genes have homology with genes of the peptide synthetase family. *J. Bacteriol.* **180:**4676–4685.

Zhu, W., J. E. Arceneaux, M. L. Beggs, B. R. Byers, K. D. Eisenach, and M. D. Lundrigan. 1998. Exochelin genes in *Mycobacterium smegmatis*: identification of an ABC transporter and two non-ribosomal peptide synthetase genes. *Mol. Microbiol.* **29:**629–639.

The Mycobacterial Cell Envelope
Edited by M. Daffé and J.-M. Reyrat
© 2008 ASM Press, Washington, DC

Chapter 11

The ABC Transporter Systems

Jean Content and Priska Peirs

When we unexpectedly cloned a gene belonging to a first mycobacterial ABC phosphate transporter in 1994 (Braibant ct al., 1994) and discovered their complexity (Lefèvre et al., 1997), we hoped to learn soon about their function, organization and topology, regulation by sensing systems, cellular localization, differential expression in various physiological situations, and finally, their role and significance in the control of mycobacterial virulence. Since then, the entire genome of *Mycobacterium tuberculosis* (Cole et al., 1998) and several other mycobacteria have been sequenced, and we have made abundant predictions on the function of potential mycobacterial ABC transporters on the basis of in silico analysis (Braibant et al., 2000).

Detailed information and review of the pertinent literature including an exhaustive inventory of *M. tuberculosis* ABC transporters have been published by us recently (Braibant et al., 2000; Content et al., 2005). Therefore, in this chapter, we summarize mainly the knowledge that has accumulated after 2000 on the basis of recently published experimental data.

The ATP-binding cassette (ABC) transporter systems are a conserved superfamily of multisubunit permeases that are found in all living organisms. They all share a structural organization consisting of four core domains: two transmembrane domains forming the path through which the transported molecules are translocated and two cytosolic ATP-binding domains (Fig. 1). They consist typically of two membrane-spanning domains (MSDs), also sometimes called integral membrane proteins (IMPs) associated with two nucleotide-binding domains (NBDs) (Braibant et al., 2000) (Fig. 2). These complexes function to transport specifically a large variety of molecules (ions, amino acids, peptides, antibiotics, polysaccharides, proteins, lipids, and so on) across biological membranes. In bacteria, the genes encoding the various subunits of each permease are usually clustered

and often organized in operons. Bacterial ABC transporters can be either importers or exporters. Bacterial importers are usually associated with an additional subunit called substrate-binding protein (SBP), which is a high-affinity extracytoplasmic substrate-binding protein either periplasmic and free in gram-negative bacteria or anchored by a lipid tail in the cytoplasmic membrane, as a lipoprotein, in gram-positive bacteria (Sutcliffe and Harrington, 2004) (Fig. 2).

The ABC proteins form the largest paralogous family of proteins in *Escherichia coli*. There are 57 systems encoded by the *E. coli* genome, of which 44 are periplasmic-binding protein-dependent uptake systems and 13 are presumed exporters. The genes encoding these ABC transporters occupy almost 5% of the genome. A phylogenetic analysis suggests that the majority of ABC proteins can be assigned to 10 subfamilies (Linton and Higgins, 1998).

MSDs usually contain six transmembrane α-helices constituting the channel that translocates the substrate through the cytoplasmic membrane, whereas NBDs bind and hydrolyze ATP, to activate the translocation, probably by articulating the two MSD subunits (Fig. 3). In this way, the permease generates the energy necessary to overcome the concentration gradient (for instance, when the nutriments are absorbed from a very diluted environment) between the outside and the intracellular medium.

In mycobacteria, those four distinct domains that constitute each transporter may be fused in different ways (Braibant et al., 2000) (Table 1). Structural characteristics including sequence signatures of both MSDs and NBDs have been reviewed in detail previously. In mycobacteria, they represent 2.5% of the *M. tuberculosis* genome, and correspond to more than 37 complete and incomplete ABC transporters (Braibant et al., 2000). Another argument that justifies the continuous study of these transporters in mycobacteria is that *M. tuberculosis* is undoubtedly a successful pathogen, and this can be

Jean Content and Priska Peirs • Molecular Microbiology Unit, Pasteur Institute of Brussels, 642, rue Engeland, B-1180 Brussels, Belgium.

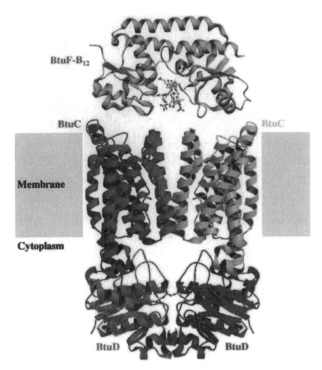

Figure 1. Ribbon diagram of the *E. coli* vitamin B_{12}BtuCDF, ABC permease protein structure. The transporter is assembled from two membrane-spanning BtuC subunits and two ABC cassettes BtuD. At the ATP binding sites, cyclotetravanadate molecules are bound to the transporter (ball-and-stick models at the BtuD interface). Vitamin B_{12} is delivered to the periplasmic side of the transporter by a binding protein (BtuF), then translocated through a pathway provided at the interface of the two membrane-spanning BtuC subunits. It finally exits into the cytoplasm at the large gap between the four subunits. This transport cycle is powered by the hydrolysis of ATP by the ABC cassettes BtuD. Reprinted from Locher and Borths (2004), with permission of the authors. *(See the color insert for the color version of this figure.)*

partly explained by (i) its ability to survive inside macrophage phagosomes, where it uses nutriments from the host cells; (ii) its capacity to alternate be-

tween proliferating and dormant conditions of growth; and (iii) its capacity to develop (multi)drug resistance. All these characteristics are potentially mediated through the very diverse functions of multiple ABC transporters (Mir et al., 2006). Their better functional knowledge might constitute interesting and specific targets for new drugs.

VALIDATED FUNCTIONAL SYSTEMS

In Table 1, we summarize the putative function and putative substrates of around 40 complete and incomplete ABC transporters including their classification as proposed earlier on the basis of in silico analysis. We shall now review briefly the transporter systems that have been studied recently and experimentally validated.

The DrrABC Transporter

It was initially predicted as a daunorubicin exporter on the basis of its similarity to the *drrAB* locus from *Streptomyces peucetius* (Guilfoile and Hutchinson, 1991). A strongly attenuated *M. tuberculosis* *drrC* (*Rv2938*) mutant (*drrC⁻*) was found during a signature tagged transposon mutagenesis experiment (Camacho et al., 1999). Biochemical analysis demonstrated that this *drrC⁻* mutant is unable to properly localize phthiocerol dimycocerosate (PDIM) in the cell envelope, a defect that can be corrected by expressing an intact copy of the *drrC* gene within the mutant strain (Camacho et al., 2001). In another study, the role of the *drrB* (*Rv2937*) gene in the complex lipid biosynthesis was investigated by analyzing a *M. tuberculosis* *drrB*-mutant strain (Waddell et al., 2005). It was reported that the disruption of *drrB* resulted in the loss of PDIM in *M. tuberculosis*.

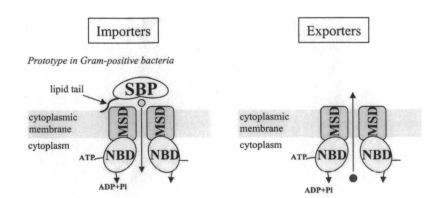

Figure 2. Topological organization of the prototypical ABC transporters. ABC transporters are classified as importers and exporters, depending on the direction of translocation of their substrate (indicated by an arrow). The prokaryotic prototype is composed of two membrane-spanning domains (MSDs) and two nucleotide-binding domains (NBDs) expressed as independent polypeptides. SBP indicates the presence of a substrate binding protein, usually present in importers. This figure was adapted from Braibant et al. (2000).

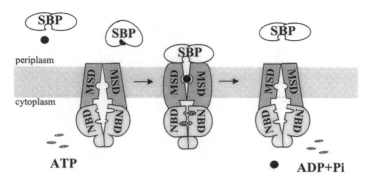

Figure 3. Maltose transport in *E. coli*. The maltose-binding protein (MBP) undergoes a conformational change from an open to a closed conformation on binding maltose in the periplasm. The transport complex (FGK$_2$) consists of two membrane-spanning subunits (MSD), MalF and MalG, likely consisting of bundles of eight and six α-helices, respectively, and two copies of the cytoplasmic NBD subunit, MalK. In either the open or the closed conformation, MBP binds to nucleotide-free FGK$_2$, in which the MalK NBDs are in an open conformation, and the periplasmic entrance to the translocation pathway is closed (P-closed state). MBP in the open conformation can interact with FGK$_2$, but only upon binding maltose and closing is it competent to initiate the transport cycle. ATP binding to MalK triggers NBD association, coinciding with simultaneous opening of both MBP and the periplasmic entrance to the translocation pathway (P-open state), allowing the transfer of sugar to FGK. ATP hydrolysis results in disruption of the MalK dimer interface and reorientation of the transmembrane helices to the starting conformation. Modified from Chen et al. (2003), with permission of the publisher.

Therefore, the *drrABC* transporter is necessary for DIM translocation, but it might not be sufficient because it was found that the transport of this complex lipid also requires the MmpL7 protein, which belongs to the resistance-nodulation-cell division superfamily (Onwueme et al., 2005) (see chapter 17). How these two transporters cooperate to export this complex branched lipid is still a matter of speculation. Moreover, a third component, the lipoprotein Lppx (Lefèvre et al., 2000) was found recently to be also required for the translocation of PDIM to the outer membrane of *M. tuberculosis* (Sulzenbacher et al., 2006).

This *drrABC* gene cluster has also been identified as being required for in vivo survival by Rubin and colleagues. They mutated virtually every nonessential gene of *M. tuberculosis* and determined the effect disrupting each gene on the growth rate of this pathogen during infection. Within the 194 genes identified, which are required for bacterial survival during infection, eight genes are predicted to encode for a component of an ABC transporter (Sassetti and Rubin, 2003). Additionally, another group found that *drrB* is one of the virulence genes that have been identified by isolating *M. tuberculosis* mutants attenuated in their ability to parasitize human macrophages (Rosas-Magallanes et al., 2007).

Interestingly, the *drrABC* cluster was directly demonstrated to function also as a doxorubicin export pump, at least as far as the DrrA and DrrB protein components are concerned. *drrA* (*Rv2936*) encodes an NBD, and *drrB* (*Rv2937*) and *drrC* both encode an MBD. Simultaneously, expression of DrrA and DrrB in *E. coli* or *Mycobacterium smegmatis* confers an in-

creased resistance of these cells to several structurally unrelated antibiotics (Choudhuri et al., 2002). This resistance phenotype could be reversed by verapamil and reserpine, two potent inhibitors of ABC transporters. Surprisingly, the authors did not study the participation of DrrC (Rv2938) in the transport. Because ABC transporters require two hydrophobic membrane-spanning domains and two hydrophilic NBDs, the reported data suggest that the transporter is able to function as a multidrug efflux pump composed of a homodimer of DrrA and a homodimer of DrrB as described in *Streptomyces* (Guilfoile and Hutchinson, 1991).

In conclusion, the *drrABC* cluster is an interesting example of a predicted antibiotic exporter that effectively exports doxorubicin in *M. smegmatis* and, on the other hand, strongly contributes to the virulence of *M. tuberculosis* by exporting the complex lipid PDIM to the bacterial envelope.

Carbohydrate Transport

Borich et al. (2000) cloned a putative *malE* gene encoding a maltose-binding protein from *Mycobacterium bovis*, similar to the *M. tuberculosis lpq*Y (*Rv1235*) gene and to the periplasmic maltose-binding protein MalE of *E. coli* and reported the genomic organization of the putative operon *sugC-sugB-sugA-lpqY* involved in maltose transport. No functional study of this permease has been reported. However, this maltose transporter, the sugar importer *Rv2041c-Rv2040c-Rv2039c-Rv2038c* and the glycerol-3-phosphate transporter *ugpA-ugpE-ugpB-ugpC* (*Rv2835c-Rv2834c-Rv2833c-Rv2832c*) have been recognized as

Table 1. Reconstituted *M. tuberculosis* ABC transporters[a]

ABC transporter-encoding gene(s)	Rv no.	Transporter organization	Putative function and/or substrate	Demonstrated function	Biochemically	Genetically	Subfamily	Reference(s)
Rv0072 Rv0073		(MSD)2/(NBD)2MSD/NBD	Glutamine importer			*Rv0072*[-]	4	
Rv0194		MSD-NBD-MSD-NBD	Multidrug resistance				6	
fecB2	*Rv0265c*	**SBP**	Iron (III) dicitrate import				8	Sassetti et al., 2003
glnH	*Rv0411c*	**SBP**	Glutamine import				4	
mkl(mceG)	*Rv0655*	NBD	Translocation of lipid substrates	Binding with YrbE1A and YrbE1B proteins		*mkl[-] mceG[-]*	12	Joshi et al., 2006; Lamichhane et al., 2005; Rengarajan et al., 2005; Sassetti and Rubin, 2003
phoT	*Rv0820*	NBD	Inorganic phosphate import	Virulence-restoring gene, phosphate uptake	X	*phoT[-]*	11	Collins et al., 2003; Rengarajan et al., 2005
pstS3 pstC2 pstA1 pknD pstB pstS2 pstS1 pstC1 pstA2	*Rv0928, Rv0929, Rv0930, Rv0931c, Rv0932c, Rv0933, Rv0934, Rv0935, Rv0936*	**SBP**/MSD/ MSD/PknD(NBD) **SBP**(NBD) 2/**SBP**/MSD/ MSD	Inorganic phosphate import	Phosphate uptake	X	*pstS1[-] pstS2[-] pstA1*	11	Braibant et al., 1994; Braibant et al., 1996a; Braibant et al., 1996b; Chang et al., 1994; Choudhary et al., 1994; Lefèvre et al., 1997; Peirs et al., 2005; Rengarajan et al., 2005; Sarin et al.,2001
Rv0986 Rv0987		(NBD)2/MSD MSD	Attachment/ export adhesion component			*Rv0986[-] Rv0987[-]*		Pethe et al., 2004; Rosas-Magallanes et al., 2007
lpqW *Rv1218c* *Rv1217c*	*Rv1166*	**SBP** (NBD)2/ MSD-MSD	Peptide import Antibiotic (tetronasin) resistance				2 6	

Gene(s)	Rv number(s)	Domain organization	Function	Notes	Mutant	No.	Reference(s)
sugC sugB sugA lpqY	*Rv1238, Rv1237, Rv1236, Rv1235*	(NBD)2/MSD/MSD/**SBP**	Sugar (maltose) import		*sugC*$^-$	5	Borich et al., 2000; Lamichhane et al., 2005; Rengarajan et al., 2005; Sassetti and Rubin, 2003
lpqZ *Rv1273c Rv1272c*	*Rv1244*	**SBP** MSD-NBD/MSD-NBD	Peptide import Lipid exporter			2	Sassetti and Rubin, 2003; Schnappinger et al., 2003
oppB oppC oppD **oppA**	*Rv1283c, Rv1282c, Rv1281c, Rv1280c*	MSD/MSD/NBD-NBD/**SBP**	Peptide import	Peptide uptake	*oppD*$^-$	2	Green et al., 2000
irtA irtB	*Rv1348, Rv1349*	MSD-NBD/MSD-NBD	Iron transporter	Fe-carboxy-mycobactin uptake	*irtA*$^-$ *irtB*$^-$	8	Rodriguez and Smith, 2006
Rv1458c Rv1457c Rv1456c		(NBD)2/MSD/MSD	Mycolic acid transport system	In *Corynebacterium*		12	Wang et al., 2006
Rv1463		NBD	Unknown	Part of the SUF machinery (Fe-S) cluster			
Rv1473		(NBD-NBD)	Macrolide antibiotics resistance			6	
cydD cydC	*Rv1621c, Rv1620c*	MSD-NBD/MSD-NBD	Involved in cytochrome biogenesis				
Rv1668c Rv1667c		(NBD/NBD)	Macrolide antibiotic resistance			6	Schnappinger et al., 2003
Rv1687c Rv1686c		(NBD)2/(MSD)2	Unknown			6	Curry et al., 2005
Rv1747		(MSD-NBD)2	Unknown	Phosphorylated by Ser/Thr PK			
Rv1819c		(MSD-NBD)2	Multidrug resistance		*Rv1819c*$^-$	6	Pethe et al., 2004
modA *modB modC*	*Rv1857, Rv1858, Rv1859*	**SBP**/(MSD)2/(NBD)2	Molybdate import		*modA*$^-$	11	Camacho et al., 2001
Rv2041c Rv2040c Rv2039c Rv2038c		**SBP**/MSD/MSD/(NBD)2	Sugar import			5	Sassetti et al., 2003; Sassetti and Rubin, 2003; Schnappinger et al., 2003

Continued on following page

Table 1. Reconstituted *M. tuberculosis* ABC transporters[a] *(continued)*

ABC transporter-encoding gene(s)	Rv no.	Transporter organization	Putative function and/or substrate	Demonstrated function	Biochemically	Genetically	Subfamily	Reference(s)
uspA uspE **uspC**	Rv2316, Rv2317, Rv2318	(MSD/MSD/**SBP**)	Sugar import				5	McAdam et al, 1995; Sassetti et al, 2003; Sassetti and Rubin, 2003; Wooff et al, 2002
Rv2326c		(MSD-NBD-)NBD	Unknown					
subI cysT cysW cysA	Rv2400c, Rv2399c, Rv2398c, Rv2397c	**SBP**/MSD/MSD/(NBD)2	Sulfate import	Sulfate uptake	X	subI⁻ cysA⁻	11	
Rv2477c		(NBD-NBD)	Macrolide antibiotic resistance				6	
Rv2563 glnQ	Rv2563, Rv2564	(MSD)2/(NBD)2	Glutamine importer				4	
Rv2585c		**SBP**					2	Schnappinger et al, 2003
Rv2688c Rv2687c Rv2686c		(NBD)2/MSD/MSD	Peptide import Antibiotic resistance				6	
ugpA ugpE **ugpB** ugpC	Rv2835c, Rv2834c, Rv2833c, Rv2832c	MSD/MSD/**SBP**/(NBD)2	Glycerol-3-phosphate import				5	Sassetti et al, 2003; Schnappinger et al, 2003
drrA drrB drrC	Rv2936c, Rv2937c, Rv2938c	(NBD)2/MSD/MSD	Lipid transporter	Doxorubicin export, DIM translocation	X	drrC⁻ drrB⁻	12	Camacho et al, 2001; Camacho et al, 1999; Choudhuri et al, 2002; Cox et al, 1999; Rosas-Magallanes et al, 2007; Waddell et al, 2005
Rv3041c **fecB**	Rv3044	NBD **SBP**	Unknown Iron (III) dicitrate import				8	
ftsX ftsE	Rv3101c, Rv3102c	(MSD)2/(NBD)2	Implicated in cell division	Complementation of the *E. coli* ftsE (Ts) phenotype				Mir et al, 2006

Gene	Rv number	Subfamily	SBP/MSD/NBD	Function	Reference
dppA dppB dppC dppD	Rv3666c, Rv3665c, Rv3664c, Rv3663c	2	SBP/MSD/MSD/NBD-NBD	Peptide import	Sassetti et al., 2003
proX proV proW proZ	Rv3759c, Rv3758c, Rv3757c, Rv3756c	4	SBP/(NBD)2/MSD/MSD	Glycine-betaine/L-proline import-osmoprotection	Sassetti and Rubin, 2003; Talaat et al., 2004
rfbE rfbD	Rv3781, Rv3783	5	(NBD)2/(MSD)2	Polysaccharide export	

[a]The ABC transporters are presented in the order of their genomic location (Rv number) and classified in subfamilies, indicated by numbers: 2: peptide transporter, 4: amino acid transporter, 5: carbohydrate transporter, 6: antibiotic transporter, 8: iron transporter, 11: anion transporter, 12: lipid exporter. Bold gene names indicate the SBP subunits.

being virulence factors in macrophages and in mice (Lamichhane et al., 2005; Rengarajan et al., 2005; Sassetti and Rubin, 2003; Schnappinger et al., 2003).

By screening a transposon library from *M. tuberculosis*, the genes *lpqY* and *Rv2038c* have been identified as required for bacterial survival during mice spleen infection (Sassetti and Rubin, 2003). Similarly, Rengarajan et al. (2005), using the same transposon library, identified all the genes from the putative maltose transporter operon as important for intracellular survival in primary macrophages. Furthermore, the sugC gene has been identified as essential for survival of *M. tuberculosis* in mouse lung (Lamichhane et al., 2005).

Transcriptome analysis of *M. tuberculosis* expression in macrophages (Schnappinger et al., 2003) showed that 454 genes were induced and 147 genes were repressed in activated or resting macrophages at 24 hours compared with broth culture. This set of 601 genes was called "the differential intraphagosome transcriptome." Within this set, there were 11 intraphagosomal induced ABC transporter genes that included five genes belonging to the two above-mentioned carbohydrate transporters, the sugar transporter and the glycerol-3-phosphate transporter.

Peptide Transport Systems

In several bacteria, peptide transport can be important for nutrition of the cell, signaling processes such as regulation of gene expression, sporulation, chemotaxis, competence, and virulence (Doeven et al., 2005). However in mycobacteria, the physiological role of peptide transport has not yet been established, but it is likely that acquisition of oligopeptides is mainly important as a source of essential nutriments, playing a crucial role in the maintenance of bacterial metabolism when intracellular survival of bacteria is limited during some stages of the infection.

The protein *oppD(Rv1281c)* encoded by the oligopeptide permease operon *oppBCDA* from *M. bovis* BCG shares similarity to the ATP-binding component of the oligopeptide permeases from other prokaryotes. Disruption of this operon by homologous recombination reduced the ability of the *M. bovis* mutant to grow using peptides as the sole carbon or nitrogen source. Furthermore, the mutant was more resistant to the toxic effects of glutathione and *S*-nitrosoglutathione, and displayed a reduced uptake of glutathione as compared with that observed in the wild-type strain or in the *oppD* mutant complemented with a wild copy of the operon (Green et al., 2000). Three additional SBP lipoproteins have been proposed (Rv1166, Rv1244, and Rv2585) as peptide transporters (Sutcliffe and Harrington, 2004).

Iron Transport and Metabolism

This is a vast and complex topic that was reviewed recently by others (Rodriguez, 2006) (see also chapter 10). In mycobacteria, several transporters coexist and must be tightly regulated because iron is both essential for several biochemical processes but also toxic if accumulated in excess within the bacteria. One of the *M. tuberculosis* iron transporters was discovered very recently (Rodriguez and Smith, 2006) on the basis of sequence similarity to the YbtPQ iron ABC transporter previously described in *Yersinia pestis* (Fetherston et al., 1999). This *M. tuberculosis* transporter consists of two similar proteins encoded by *irtA* and *irtB* (respectively, *Rv1348* and *Rv1349*) both containing a NBD fused with a typical six transmembrane helices MSD. Both genes are regulated by iron and the transcriptional regulator IdeR, and directly surrounded with genes probably implicated in the biosynthesis of mycobactin siderophores (Rodriguez, 2006). Knocking out the two genes has shown that the *irtAB* system is important for *M. tuberculosis* replication in human macrophages and mouse lung, and also in vitro under iron-deficient conditions. Thus, a direct correlation was observed between the ability to efficiently acquire iron and the capacity to replicate in vivo in the mouse model of tuberculosis infection (Rodriguez, 2006). Because the Fe^{3+} ion is not soluble under bacterial growth conditions (Braun and Killmann, 1999; Koster, 2005), it is usually combined to a bacterially produced and secreted siderophore, in the present case, probably Fe^{3+}-carboxymycobactin, and presented in this form to the bacterial cell. Finally, an unusual feature of this iron importer is that its SBP is not encoded by a neighboring gene and, therefore, is not precisely known: it could correspond to the 292 amino acids N terminal domain of the IrtA, or it could be located somewhere else in the genome (Rodriguez and Smith, 2006).

It is interesting that an independent study based on representational difference analysis between *Mycobacterium avium* subsp. *paratuberculosis* and *M. avium* subsp. *avium* identified an *mpt* operon-containing gene orthologous to the *M. tuberculosis irtAB* genes, within a 38-kb pathogenicity island. On the basis of sequence similarity and the presence of several *fur* regulatory boxes, it was assumed that this putative ABC transporter was involved in iron uptake in *M. avium* subsp. *paratuberculosis* (Stratmann et al., 2004).

In addition to acquisition of siderophore-solubilized iron, *M. tuberculosis* might be able to transport other forms of iron, like ferric citrate, because two mycobacterial genes *fecB* (*Rv3044*) and *fecB2* (*Rv0265c*) show homology with *fecB* from *E. coli*. In *E. coli*, FecB is a periplasmic solute-binding protein involved in the transport of iron from ferric citrate. The *fecB* gene from *M. avium* has been characterized, and it has been shown that the expression of this gene is influenced by the iron concentration in the medium: the lower the iron concentration, the higher the expression (Wagner et al., 2005).

Sulfate Transport

A methionine auxotrophic mutant in *M. bovis* BCG was generated by transposon mutagenesis and found to contain a transposon insertion within the *subI* (*Rv2400c*) gene, which encodes the predicted high-affinity sulfate binding subunit of the transporter (McAdam et al., 1995). Later, by insertional mutagenesis in *M. bovis* BCG, other auxotrophs for amino acids containing sulfur have been isolated (Wooff et al., 2002). One of these auxotrophs was found to be disrupted within the *cysA* (*Rv2397c*) gene, encoding the predicted nucleotide-binding subunit of the inorganic sulfur transporter (*subI-cysT-cysW-cysA* operon). By complementation of the *cysA* mutant with the wild-type *cysA* gene from *M. tuberculosis*, the prototrophy was fully restored and the ability to take up sulfate with the functional characteristics of an ABC transporter was returned.

Transposon mutagenesis has been used by Sassetti et al. (2003) to screen for genes important for growth by finding those genes that cannot sustain transposon insertions. Within the identified group of essential genes, 10 genes encoded for a unit of an ABC transporter and especially the four genes encoding the sulfate transporter (*subI*, *cysT*, *cysW*, and *cysA*) were found to be required for optimal growth.

Despite the possible unique sulfate transporter representation, survival of *cysA* and *subI* mutants in animals was similar to that of wild-type *M. bovis* BCG (Wooff et al., 2002). It is possible that the intracellular availability of methionine is sufficient to support the growth of these mutant strains or that sulfate is incorporated from other in vivo-induced sulfate transporters. Three other putative sulfate transporters are indeed encoded by the genome of *M. tuberculosis* and predicted to belong to the sulfate permease (SulP) family (Content et al., 2005).

Inorganic Phosphate Transport

Phosphate is another essential anion that is also transported by one or several multisubunits ABC permeases in mycobacteria. In addition to two genes (*pitA* and *pitB*) encoding putative constitutive inorganic phosphate transporters, belonging to the inorganic phosphate transporter (pit) family also found in the *M. tuberculosis* complex genomes, which are not ABC transporters (Content et al., 2005), at least three such phosphate permeases have been described and analyzed in *M. tuberculosis* on the basis of their similitude to *E. coli* orthologues. In *E. coli*, this system is composed of four proteins: the phosphate-binding protein PstS, the membrane proteins PstA and PstC, and the NBP PstB. All proteins are encoded by the corresponding genes in a single operon. The *M. tuberculosis* genome includes three genes similar to *pstS [pstS1 (Rv0934), pstS2 (Rv0932c)*, and *pstS3 (Rv0928)]*, two to *pstA [pstA1 (Rv0930)* and *pstA2 (Rv0936)]*, two to *pstC [pstC1 (Rv0936)* and *pstC2 (Rv0929)]*, and two to *pstB [pstB (Rv0933)* and *phoT (Rv0820)]*, which, except for *phoT*, are organized in three distinct operons (Braibant et al., 1994; Braibant et al., 1996a; Braibant et al., 1996b; Lefèvre et al., 1997; Peirs et al., 2005). The PstS subunits are expressed on the cell surface and are the first to recognize and capture the phosphate. Following conformation changes, they bind to the membrane-imbedded PstA and PstB subunits. These constitute a channel through which the phosphate can cross the membrane. This requires energy, which is provided by the hydrolysis of ATP by each of the two PstB subunits attached to the cytoplasmic side of PstA and PstC. The multiplication of the genes coding for the subunits of the Pst system in *M. tuberculosis* suggests an adaptation allowing the survival of the pathogen in low concentration phosphate environments such as those encountered during the infection of macrophages. However Gebhard et al. (2006) recently described the existence of at least two other high-affinity inorganic phosphate ABC transporters in *M. smegmatis* in addition to the previously discovered *Pst* system (Kriakov et al., 2003).

By using a transposon library from *M. tuberculosis*, Rengarajan et al. (2005) identified the *M. tuberculosis* genes required for intracellular survival by screening for transposon mutants that fail to grow within primary macrophages. They associated these genes into functional groups using clustering tools and found that individual members of several putative operons are each required for macrophage survival. They identified the phosphate transporter operon *pstS3-pstC2-pstA1* and the gene *phoT* as being required for macrophage survival.

The functionality of this Pst system has been investigated by biochemical approaches. Chang et al. (1994) have shown that the substrate-binding protein PstS1, expressed as recombinant protein without signal sequence and purified from *E. coli* inclusion bodies or from a soluble fraction, binds inorganic phosphate in vitro with a K_d of 0.23 μM. Sarin et al.

(2001), on the other hand, have shown that the purified recombinant PstB protein from *M. tuberculosis*, the putative NBD of the transporter, is active in vitro as a thermostable Mg^{2+}-dependent ATPase.

We have constructed two *M. tuberculosis* knockout strains by disrupting the phosphate-binding protein genes *pstS1* and *pstS2* (Peirs et al., 2005) and shown that both mutants are deficient in phosphate uptake at low phosphate concentration (0.5 to 5 μM). Recently, we observed that in a phosphate-reduced medium, the expression of *pstS3* gene from *M. tuberculosis* is more induced than the expression of *pstS1* and *pstS2* genes. The *pstS3 M. tuberculosis* knockout strain obtained by homologous recombination was found to be more sensitive to phosphate limitation in growth media than the *pstS1* and *pstS2* knockout strains (our results, not published). The *pstA1* transposon mutants isolated by Rengarajan et al. (2005) were also more sensitive to phosphate limitation than the wild type. Therefore, it became clear that the first *pst* operon (*pstS3*, *pstC2* and *pstA1*) is necessary for phosphate uptake by phosphate starvation.

Which nucleotide-binding protein, PstB or PhoT, is involved in this phosphate uptake is unclear. In fact, PhoT has a higher homology to PstB in some other prokaryotes (*M. smegmatis* and *M. avium*) than does PstB of *M. tuberculosis*. Sequencing of the *M. bovis* genome has revealed that in this member of the *M. tuberculosis* complex, the *pstB* gene is frameshifted (Garnier et al., 2003). These observations and the fact that two groups have shown that PhoT is necessary for growth at low phosphate concentrations (Collins et al., 2003; Rengarajan et al., 2005) led us to conclude that PhoT is probably the NBP that is associated with the phosphate transporter encoded by the first *pst* operon (*pstS3*, *pstC2*, and *pstA1*).

Mycolic Acid Transport

Wang et al. (2006) used a very elegant approach to identify the genes involved in mycolic acid biosynthesis and processing in *M. tuberculosis*. They used *Corynebacterium matruchotii* as a surrogate mycobacterial system, where they generate a Tn5 transposon insertion library that was screened for mutants with altered corynomycolate synthesis. One such mutant, called 319, was found presenting an interruption in a gene encoding an integral membrane protein. This protein is encoded by a gene transcriptionally coupled with three neighboring genes constituting an ABC transporter. Biochemical analysis of the mutant demonstrates that this transporter is involved in moving corynomycolic acid from the bacterial cytoplasm outside the cell and that this transport system is selective for short-chain corynomycolic acids.

Because the organization of this genetic locus is very similar in corynebacteria, nocardiae, and mycobacteria (all three belonging to the actinomycetal order), the results suggest that the orthologous *M. tuberculosis* gene locus (*Rv1459-Rv1458c-Rv1457c-Rv1456c*) could be involved in mycolic acid synthesis and transport. This remains to be demonstrated directly by targeted mutation and complementation of one of those genes. However, this work is an important step forward because this cluster was initially only annotated as a putative antibiotic transporter on the basis of in silico analysis (Braibant et al., 2000) (also discussed in chapter 4 by Marrakchi et al.).

The Amino Acid Transporter ProXVWZ

Talaat et al. (2004) examined the transcription of *M. tuberculosis* genes in three different systems (growth in exponential culture, severely combined immune-deficient [SCID] mice, and BALB/c mice). Low-dose intranasal infections (10^3 bacilli) were carried out, and lungs from 50 mice harvested at 7, 14, 21 and 28 days (100 for the early time point). A group of 122 genes were more often expressed in BALB/c mice at 21 days than in SCID mice. It is suggested that these genes are a response of the bacteria to the acquired immune response. Within this group of genes, there is one ABC transporter gene, *proZ* from the amino acid transporter *proX-proV-proW-proZ*. This transporter has also been identified as being essential by Sassetti and Rubin (2003) and is probably implicated in the survival of mycobacteria under osmotic stress due to its similarity with the glycine betaine transporters from various bacteria like *E. coli* (Gowrishankar, 1989), *Bacillus substilis* (Lin and Hansen, 1995) or *Salmonella typhimurium* (Stirling et al., 1989).

The FtsE-FtsX System

These *M. tuberculosis* genes *ftsX*(*Rv3101c*) and *ftsE*(*Rv3102c*) have been expressed in *E. coli* and used to complement efficiently an *ftsE* temperature-sensitive mutant. This suggests that the corresponding *M. tuberculosis* proteins are functional and have a function similar to their *E. coli* orthologues (Mir et al., 2006), although this function (septation, cell division, and K$^+$ transport) is not yet clearly established and seems rather related to the translocation of integral membrane proteins to the cytoplasmic membrane (Ukai et al., 1998). This is why we have recently classified the FtsE-FtsX system of *M. tuberculosis* in the type 2 secretory pathway of the TC system (Content et al., 2005).

The Lipid Exporter Rv0986-Rv0987

Pethe et al. (2004) isolated strains within a pool of *M. tuberculosis* transposon mutants that are defective in the arrest of phagosome maturation. Two of them had a mutation in an ABC transporter gene, namely *Rv1819c*, an ABC drug transporter, possibly involved in the lipid transport to the exterior of the bacterium, and *Rv0986*. The latter gene has also been identified as virulence gene by isolating *M. tuberculosis* mutants attenuated in their ability to parasitize human macrophages (Rosas-Magallanes et al., 2007).

Rosas-Magallanes et al. produced *Rv0986*::Tn and *Rv0987*::Tn *M. tuberculosis* mutants and found that these strains were affected in their ability to bind macrophages as well as nonphagocytic cells, but inside the phagocytes, the mutants were not significantly impaired in their ability to replicate. The hypothesis has been made that the Rv0986-Rv0987 transporter might secrete an adherence factor that allows bacteria's binding to macrophages and other cell types (Rosas-Magallanes et al., 2007).

The ATP-Binding Protein Mkl (Rv0665)

The *mkl* gene has been shown to be required for macrophage survival (Rengarajan et al., 2005) and also has been identified twice as being essential for survival of *M. tuberculosis* in mice (Lamichhane et al., 2005; Sassetti et al., 2003). In 2006 the group of Joshi et al. (2006) developed a method to rapidly delineate functional pathways by identifying mutations that modify each other's phenotype, i.e., "genetic interactions." Using this method, they have defined an interaction between the virulence genes *mce1* and *mce4* in this pathogen and the gene *mceG* (*mkl*) (*Rv0665*). Additionally they report that the product of this gene represents an ATPase that associates with the *mce*-encoded transmembrane proteins YrbEA and YrbEB to form multisubunit transport systems resembling ABC transporters whose putative lipid substrate is presently unknown (Joshi et al., 2006).

STRUCTURE

The structures of only two bacterial ABC transporters have been completely elucidated. The case of the *E. coli* VitB12 transporter is illustrated here (Fig. 1) (Locher and Borths, 2004).

Among the mycobacteria, the only system that has been investigated structurally so far is the ABC phosphate transporter (see "Virulence: the Case of Inorganic Phosphate Transport" and "Drug Resistance by Efflux" below). The high-affinity PstS-1 phosphate binding subunit (Chang et al., 1994) has been crystallized and its structure determined, with bound phosphate, by X-ray diffraction at a resolution of 2.16 Å (Choudhary et al., 1994; Vyas et al., 2003). Similar to its *E. coli* phosphate binding protein (PBP) orthologue, whose structure was determined previously (Luecke and Quiocho, 1990), the PstS-1 also presents some interesting features (Wang et al., 1994). Its general shape is that of a bilobar structure whose two globular domains are articulated by a hinge that determines whether the receptor is in "open" or "closed" conformation. These two domains surround a cleft where the PO$_4$ anion binds in a complex and highly specific network of 12 hydrogen bonds (instead of 13 in the *E. coli* PBP) (Luecke and Quiocho, 1990).

An additional feature of the *M. tuberculosis* PstS-1, which was predicted (Lefèvre et al., 1997) but never directly demonstrated previously in a protein structure, is its flexible 26 amino acid N-terminal extension that anchors the PstS-1 to the lipid constituents of the cell membrane and presumably allows it sufficient freedom to capture phosphate within its lobes, and to deliver it to the nearby transporter itself (Fig. 2). Another interesting characteristic of the PstS-1 structure is the existence of two aspartate residues in the ligand-binding site (instead of one in the *E. coli* PBP) that could explain the capacity acquired by *M. tuberculosis* to capture H$_2$PO$_4^-$ below pH 7. This probably contributes to its ability to survive intracellularly within the phagosome at very low phosphate concentration in an acidic environment (Vyas et al., 2003; Webb, 2003).

PstB is an ATP-binding protein (Sarin et al., 2001) which has been usually considered as the NBD subunit, the motor of the three *M. tuberculosis* inorganic phosphate permeases, although this could not be demonstrated experimentally because it has not been possible to isolate and analyze biochemically the structure and organization of these three putative bacterial permeases (see "Virulence: the Case of Inorganic Phosphate Transport" and "Drug Resistance by Efflux" below). It seems clear that this role is shared to some extent by the PhoT protein as predicted (Braibant et al., 2000) and demonstrated experimentally (Collins et al., 2003). Conversely, PstB could have an additional role in exporting some antibiotics such as the fluoroquinone ciproflaxine, at least in *M. smegmatis* (see "Concluding Remarks" below) (Banerjee et al., 2000; Banerjee et al., 1998; Bhatt et al., 2000).

The structure of PstB has not been determined, but a model has been proposed on the basis of its amino acid sequence homology to the HisP protein from *Salmonella typhimurium* (Gupta et al., 2005). A remarkable property of this ATPase, whose functional significance remains enigmatic for a mesophilic bacteria, is its thermostability (PstB retains 90% of its ATPase activity after 1 hour at 80°C) (Sarin et al., 2001; Sarin et al., 2003).

Gupta et al. (2005) have proposed that the mycobacterial PstB undergoes a conformational change on ATP binding, and on the basis of studies involving the intrinsic tryptophan fluorescence and limited proteolysis by trypsin, they suggest that this consists of the putative movement of the fourth α-helical subdomain relative to the core subdomain of the enzyme.

VIRULENCE: THE CASE OF INORGANIC PHOSPHATE TRANSPORT

Phosphate import in bacteria is accomplished through several parallel transport systems. The phosphate transport system (Pst) from *M. tuberculosis* is a tightly regulated high-affinity system encoded by three putative *pst* operons, suggesting that the bacteria are involved in subtle biochemical adaptations of *M. tuberculosis* for their survival under varying conditions during the infectious cycle. Apart from transporting phosphate and by analogy with the Pho regulon of *E. coli* and other bacteria, it has been suggested that the Pst system is involved in the regulation of genes necessary for the survival and thus contributes to the virulence of *M. tuberculosis*.

The importance of phosphate transport for the virulence of *Mycobacterium* was first shown by Collins et al. (2003). They showed that in *Mycobacterium bovis*, in which the *pstB* gene is frameshifted, the *phoT* gene is a virulence gene. An *M. bovis phoT* knockout strain is significantly less virulent than its parental strain in different animal models (Collins et al., 2003). In *M. tuberculosis*, the situation could be different because PstB is functional and because PstB and PhoT may have overlapping roles. In addition to the frameshift mutation of the *pstB* gene, *M. bovis* BCG has also one frameshift mutation in the *phoT* gene. It is thus likely that the inactivation of both *pstB* and *phoT* genes contribute to the attenuation of BCG, which may be totally devoid of high-affinity phosphate transporter (Peirs et al., unpublished).

M. tuberculosis has three different *pstS* genes encoding phosphate-binding proteins. We showed that *pstS1* and *pstS2* knockout *M. tuberculosis* strains have a reduced multiplication within the macrophages as compared with the parental strain and that they are also attenuated in their virulence in mice (Peirs et al., 2005). This reduced multiplication of the *pstS1* and *pstS2* mutants in two strains of mice suggests that these phosphate-binding proteins are functional in vivo during infection and cannot be replaced by each other, by PstS3, by the putative Pit transporter, or by any other phosphate transporter. Recently, we also observed that the *pstS3* knockout *M. tuberculosis* also has a reduced multiplication within macrophages (results not published). The phosphate transporter genes *pstS1*, *pstS2*, and *pstS3* appear to be critical for intracellular growth, suggesting that macrophages represent a phosphate-starved environment, and transport of inorganic phosphate is critical for establishing infection.

Interestingly, a similar conclusion was reached recently, using a method that screens transposon mutants that fail to grow within primary macrophages, to identify genome-wide requirements for *M. tuberculosis* intracellular survival. Among 126 genes identified, four genes (*phoT*, *pstA1*, *pstC2*, and *pstS3*), which are all responsible for inorganic phosphate transport within the macrophage and lung in vivo, in the mouse model were analyzed in more detail (Rengarajan et al., 2005). Three of them belong to the same putative operon (*pstA1*, *pstC2*, and *pstS3*). Several other genes, and in particular ABC transporter genes, have been implicated in the virulence of *M. tuberculosis* (Camacho et al., 2001; Camacho et al., 1999; Cox et al., 1999; Curry et al., 2005; Rengarajan et al., 2005; Rodriguez, 2006). However, it is striking that the inorganic phosphate importers appear to be particularly important for *M. tuberculosis* virulence in the mouse model. Further comparative assays are in progress to determine if the three putative phosphate permeases are functionally equivalent and equally important or not.

DRUG RESISTANCE BY EFFLUX

Active multidrug efflux pumps and the mycobacterial cell wall permeability barrier are the mechanisms that are thought to be potentially involved in the natural drug resistance of mycobacteria (De Rossi et al., 2006). The bacterial drug efflux pumps have been classified into two large families: the ATP-binding cassette (ABC) superfamily and the major facilitator superfamily (MFS) and three smaller families (Li et al., 2004). Here we focus on the few *M. tuberculosis* ABC transporter genes that have been characterized in laboratory assays and suggested to be potentially involved in drug resistance in *M. tuberculosis*.

M. tuberculosis contains a putative doxorubicin resistance operon, *drrABC*, which is similar to that in *Streptomyces peucetius* (Guilfoile and Hutchinson, 1991). The *drrAB* genes expressed in *M. smegmatis* confer resistance to several clinically relevant and structurally unrelated antibiotics, including ethambutol, chloramphenicol, erythromycin, norfloxacin, streptomycin, and tetracycline. The resistant phenotype is reversed by treatment with reserpine or verapamil, both of which are inhibitors of ABC transporters (Choudhuri et al., 2002). In *M. tuberculosis*, studies have suggested that the principal physiological role of the Drr proteins of *M. tuberculosis* may be the export of complex lipids like DIM to the exterior of the cell.

Bhatt et al. (2000) have revealed that the ciprofloxacin (a fluoroquinolone) resistance of a laboratory-generated *M. smegmatis* mutant results from active efflux. This mutant exhibited mRNA level overexpression (Banerjee et al., 1998), as well as chromosomal amplification of *pstB*, a gene encoding a nucleotide-binding subunit of the phosphate-specific transporter (Pst) system. This *pstB* gene from *M. smegmatis* is similar to the *pstB* gene (*Rv0933*) from *M. tuberculosis*, suggesting that this protein may be involved in phosphate import under phosphate starvation. The authors demonstrated that this mutation increases phosphate uptake and that inactivation of this gene in *M. smegmatis* resulted not only in loss of high affinity phosphate uptake but also in hypersensitivity to fluoroquinolones and to structurally unrelated compounds (Banerjee et al., 2000). These findings suggest that the same ABC transporter may be involved in both the active efflux of fluoroquinolones and phosphate transport in *M. smegmatis*.

Pasca et al. (2004) have shown that the *M. tuberculosis* Rv2686c-Rv2687c-Rv2688c operon encodes an ABC transporter responsible for fluoroquinolones efflux when overexpressed in *M. smegmatis*. This operon confers resistance to ciprofloxacin and, to a lesser extent, norfloxacin, moxifloxacin and sparfloxacin when overexpressed in *M. smegmatis*. The level of resistance decreases in the presence of known ABC transporter inhibitors (reserpine and verapamil) and carbonyl cyanide *m*-chlorophenylhydrazone (CCCP). CCCP may have an indirect effect on this ATP-driven efflux pump because CCCP-induced proton gradient uncoupling also depletes the intracellular ATP pool. A similar level of resistance has been found in *M. bovis* BCG overexpressing the *Rv2686c-Rv2687c-Rv2688c* operon. This demonstrates that these genes are truly responsible for fluoroquinolones efflux (De Rossi et al., 2006).

A word of caution is necessary because nearly all of the results presented in this section were obtained with *M. smegmatis*, whose genome is nearly twice as large as that of *M. tuberculosis*. Therefore, the *pstB*, the *phoT* and the Rv2686-8 operon genes of *M. tuberculosis* may be involved in fluoroquinolone resistance, but this hypothesis remains to be tested. It is important to stress that, so far, this mechanism of multiresistance by efflux has never been demonstrated in an antibiotic-resistant clinical strain of *M. tuberculosis* (Pai et al., 2006).

CONCLUDING REMARKS

There obviously remains much work to be done to complete our knowledge of the mycobacterial ABC transporters in the postgenomic era. First of all, it is necessary to define exactly the substrate or substrates for all the putative transporters whose real function is presently not known or not demonstrated (Table 1). Second, for all of them, it would be interesting to sort out genetically those that are essential to the bacteria and in which conditions, and to define their structural role in the architecture of the bacterial envelope and possibly in bacterial virulence (in vivo).

The remarkable amplification of certain transporters remains puzzling. For example, despite a reduced number of importers in *M. tuberculosis*, at least three of them are probably implicated in the import of inorganic phosphate, whereas in *E. coli* and *Bacillus subtilis*, only one such system is present (Lefèvre et al., 1997; Linton and Higgins, 1998; Quentin et al., 1999). This multiplication of the genes coding for the subunits of the Pst system in *M. tuberculosis* suggests an adaptation allowing the survival of the pathogen in low-concentration phosphate environments such as those encountered during the infection of macrophages. It is also possible that the multiplication of the genes is a protection of the bacterial genome against accidental loss of very important genes and allows for the emergence of new evolving genes (Meyer, 2003) or that the different paralogous components exert distinct specialized functions.

Finally, it will be interesting to focus at the most essential, bacteria specific, and vulnerable of these systems. Structural studies of the various subunits could help to screen or to define appropriate efficient inhibitors with limited toxicity, which could be developed as a new class of anti-mycobacterial agents.

REFERENCES

Banerjee, S. K., K. Bhatt, P. Misra, and P. K. Chakraborti. 2000. Involvement of a natural transport system in the process of efflux-mediated drug resistance in *Mycobacterium smegmatis*. *Mol. Gen. Genet.* **262:**949–956.

Banerjee, S. K., P. Misra, K. Bhatt, S. C. Mande, and P. K. Chakraborti. 1998. Identification of an ABC transporter gene that exhibits mRNA level overexpression in fluoroquinolone-resistant *Mycobacterium smegmatis*. *FEBS Lett.* **425:**151–156.

Bhatt, K., S. K. Banerjee, and P. K. Chakraborti. 2000. Evidence that phosphate specific transporter is amplified in a fluoroquinolone resistant *Mycobacterium smegmatis*. *Eur. J. Biochem.* **267:**4028–4032.

Borich, S. M., A. Murray, and E. Gormley. 2000. Genomic arrangement of a putative operon involved in maltose transport in the *Mycobacterium tuberculosis* complex and *Mycobacterium leprae*. *Microbios* **102:**7–15.

Braibant, M., L. De Wit, P. Peirs, M. Kalai, J. Ooms, A. Drowart, K. Huygen, and J. Content. 1994. Structure of the *Mycobacterium tuberculosis* antigen 88, a protein related to the *Escherichia coli* PstA periplasmic phosphate permease subunit. *Infect. Immun.* **62:**849–854.

Braibant, M., P. Gilot, and J. Content. 2000. The ATP binding cassette (ABC) transport systems of *Mycobacterium tuberculosis*. *FEMS Microbiol. Rev.* **24:**449–467.

Braibant, M., P. Lefevre, L. de Wit, J. Ooms, P. Peirs, K. Huygen, R. Wattiez, and J. Content. 1996a. Identification of a second *Mycobacterium tuberculosis* gene cluster encoding proteins of an ABC phosphate transporter. *FEBS Lett.* **394:**206–212.

Braibant, M., P. Lefevre, L. de Wit, P. Peirs, J. Ooms, K. Huygen, A. B. Andersen, and J. Content. 1996b. A *Mycobacterium tuberculosis* gene cluster encoding proteins of a phosphate transporter homologous to the *Escherichia coli* Pst system. *Gene* **176:**171–176.

Braun, V., and H. Killmann. 1999. Bacterial solutions to the iron-supply problem. *Trends Biochem. Sci.* **24:**104–109.

Camacho, L. R., P. Constant, C. Raynaud, M. A. Laneelle, J. A. Triccas, B. Gicquel, M. Daffe, and C. Guilhot. 2001. Analysis of the phthiocerol dimycocerosate locus of *Mycobacterium tuberculosis*. *J. Biol. Chem.* **276:**19845–19854.

Camacho, L. R., D. Ensergueix, E. Perez, B. Gicquel, and C. Guilhot. 1999. Identification of a virulence gene cluster of *Mycobacterium tuberculosis* by signature-tagged transposon mutagenesis. *Mol. Microbiol.* **34:**257–267.

Chang, Z., A. Choudhary, R. Lathigra, and F. A. Quiocho. 1994. The immunodominant 38-kDa lipoprotein of *M. tuberculosis* is a phosphate binding protein. *J. Biol. Chem.* **269:**1956–1958.

Chen, J., G. Lu, J. Lin, A. L. Davidson, and F. A. Quiocho. 2003. A tweezers-like motion of the ATP-binding cassette dimer in an ABC transport cycle. *Mol. Cell* **12:**651–661.

Choudhary, A., M. N. Vyas, N. K. Vyas, Z. Chang, and F. A. Quiocho. 1994. Crystallization and preliminary X-ray crystallographic analysis of the 38-kDa immunodominant antigen of *Mycobacterium tuberculosis*. *Protein Sci.* **3:**2450–2451.

Choudhuri, B. S., S. Bhakta, R. Barik, J. Basu, M. Kundu, and P. Chakrabarti. 2002. Overexpression and functional characterization of an ABC (ATP-binding cassette) transporter encoded by the genes drrA and drrB of *Mycobacterium tuberculosis*. *Biochem. J.* **367:**279–285.

Cole, S. T., R. Brosch, J. Parkhill, T. Garnier, C. Churcher, D. Harris, S. V. Gordon, K. Eiglmeier, S. Gas, C. E. Barry, F. Tekaia, K. Badcock, D. Basham, D. Brown, T. Chillingworth, R. Conner, R. Davies, K. Devlin, T. Feltwell, S. Gentles, N. Hamlin, S. Holroyd, T. Hornsby, K. Jagels, A. Krogh, J. Mclean, S. Moule, L. Murphy, K. Oliver, J. Osborne, M. A. Quail, M. A. Rajandream, J. Rogers, S. Rutter, K. Seeger, J. Skelton, R. Squares, S. Squares, J. E. Sulston, K. Taylor, S. Whitehead, and B. G. Barrell. 1998. Deciphering the biology of *Mycobacterium tuberculosis* from the complete genome sequence. *Nature* **396:**190–198.

Collins, D. M., R. P. Kawakami, B. M. Buddle, B. J. Wards, and G. W. de Lisle. 2003. Different susceptibility of two animal species infected with isogenic mutants of *Mycobacterium bovis* identifies phoT as having roles in tuberculosis virulence and phosphate transport. *Microbiology* **149:**3203–3212.

Content, J., M. Braibant, J. Ainsa, and N. Connell. 2005. Transport process in mycobacteria, p. 379–401. *In* S. T. Cole, K. Eisenach, D. N. McMurray, and W. R. Jacobs, Jr. (ed.), *Tuberculosis and the Tubercle Bacillus*. ASM Press, Washington, DC.

Cox, J. S., B. Chen, M. McNeil, and W. R. Jacobs. 1999. Complex lipid determine tissue specific replication of *Mycobacterium tuberculosis* in mice. *Nature* 402:79–83.

Curry, J. M., R. Whalan, D. M. Hunt, K. Gohil, M. Strom, L. Rickman, M. J. Colston, S. J. Smerdon, and R. S. Buxton. 2005. An ABC transporter containing a forkhead-associated domain interacts with a serine-threonine protein kinase and is required for growth of *Mycobacterium tuberculosis* in mice. *Infect. Immun.* 73:4471–4477.

De Rossi, E., J. A. Ainsa, and G. Riccardi. 2006. Role of mycobacterial efflux transporters in drug resistance: an unresolved question. *FEMS Microbiol. Rev.* 30:36–52.

Doeven, M. K., J. Kok, and B. Poolman. 2005. Specificity and selectivity determinants of peptide transport in *Lactococcus lactis* and other microorganisms. *Mol. Microbiol.* 57:640–649.

Fetherston, J. D., V. J. Bertolino, and R. D. Perry. 1999. YbtP and YbtQ: two ABC transporters required for iron uptake in *Yersinia pestis*. *Mol. Microbiol.* 32:289–299.

Garnier, T., K. Eiglmeier, J. C. Camus, N. Medina, H. Mansoor, M. Pryor, S. Duthoy, S. Grondin, C. Lacroix, C. Monsempe, S. Simon, B. Harris, R. Atkin, J. Doggett, R. Mayes, L. Keating, P. R. Wheeler, J. Parkhill, B. G. Barrell, S. T. Cole, S. V. Gordon, and R. G. Hewinson. 2003. The complete genome sequence of *Mycobacterium bovis*. *Proc. Natl. Acad. Sci. USA* 100:7877–7882.

Gebhard, S., S. L. Tran, and G. M. Cook. 2006. The Phn system of *Mycobacterium smegmatis*: a second high-affinity ABC-transporter for phosphate. *Microbiology* 152:3453–3465.

Gowrishankar, J. 1989. Nucleotide sequence of the osmoregulatory proU operon of Escherichia coli. *J. Bacteriol.* 171:1923–1931.

Green, R. M., A. Seth and N. D. Connell. 2000. A peptide permease mutant of *Mycobacterium bovis* BCG resistant to the toxic peptides glutathione and S-nitrosoglutathione. *Infect. Immun.* 68:429–436.

Guilfoile, P. G., and C. R. Hutchinson. 1991. A bacterial analog of the mdr gene of mammalian tumor cells is present in *Streptomyces peucetius*, the producer of daunorubicin and doxorubicin. *Proc. Natl. Acad. Sci. USA* 88:8553–8557.

Gupta, S., P. K. Chakraborti, and D. Sarkar. 2005. Nucleotide-induced conformational change in the catalytic subunit of the phosphate-specific transporter from *M. tuberculosis*: Implications for the ATPase structure. *Biochim. Biophys. Acta* 1750:112–121.

Joshi, S. M., A. K. Pandey, N. Capite, S. M. Fortune, E. J. Rubin, and C. M. Sassetti. 2006. Characterization of mycobacterial virulence genes through genetic interaction mapping. *Proc. Natl. Acad. Sci. USA* 103:11760–11765.

Koster, W. 2005. Cytoplasmic membrane iron permease systems in the bacterial cell envelope. *Front. Biosci.* 10:462–477.

Kriakov, J., S. Lee, and W. R. Jacobs, Jr. 2003. Identification of a regulated alkaline phosphatase, a cell surface-associated lipoprotein, in *Mycobacterium smegmatis*. *J. Bacteriol.* 185:4983–4991.

Lamichhane, G., S. Tyagi, and W. R. Bishai. 2005. Designer arrays for defined mutant analysis to detect genes essential for survival of *Mycobacterium tuberculosis* in mouse lungs. *Infect. Immun.* 73:2533–2540.

Lefèvre, P., M. Braibant, L. de Wit, M. Kalai, D. Roeper, J. Grotzinger, J. P. Delville, P. Peirs, J. Ooms, K. Huygen, and J. Content. 1997. Three different putative phosphate transport receptors are encoded by the *Mycobacterium tuberculosis* genome and are present at the surface of *Mycobacterium bovis* BCG. *J. Bacteriol.* 179:2900–2906.

Lefèvre, P., O. Denis, L. De Wit, A. Tanghe, P. Vandenbussche, J. Content, and K. Huygen. 2000. Cloning of the gene encoding a 22-kilodalton cell surface antigen of *Mycobacterium bovis*

BCG and analysis of its potential for DNA vaccination against tuberculosis. *Infect. Immun.* 68:1040–1047.

Li, X. Z., L. Zhang, and H. Nikaido. 2004. Efflux pump-mediated intrinsic drug resistance in *Mycobacterium smegmatis*. *Antimicrob. Agents Chemother.* 48:2415–2423.

Lin, Y., and J. N. Hansen. 1995. Characterization of a chimeric proU operon in a subtilin-producing mutant of Bacillus subtilis 168. *J. Bacteriol.* 177:6874–6880.

Linton, K. J., and C. F. Higgins. 1998. The *Escherichia coli* ATP-binding cassette (ABC) proteins. *Mol. Microbiol.* 28:5–13.

Locher, K. P., and E. Borths. 2004. ABC transporter architecture and mechanism: implications from the crystal structures of BtuCD and BtuF. *FEBS Lett.* 564:264–268.

Luecke, H., and F. A. Quiocho. 1990. High specificity of a phosphate transport protein determined by hydrogen bonds. *Nature* 347:402–406.

McAdam, R. A., T. R. Weisbrod, J. Martin, J. D. Scuderi, A. M. Brown, J. D. Cirillo, B. R. Bloom, and W. R. Jacobs, Jr. 1995. In vivo growth characteristics of leucine and methionine auxotrophic mutants of *Mycobacterium bovis* BCG generated by transposon mutagenesis. *Infect. Immun.* 63:1004–1012.

Meyer, A. 2003. Molecular evolution: Duplication, duplication. *Nature* 421:31–32.

Mir, M. A., H. S. Rajeswari, U. Veeraraghavan, and P. Ajitkumar. 2006. Molecular characterisation of ABC transporter type FtsE and FtsX proteins of *Mycobacterium tuberculosis*. *Arch. Microbiol.* 185:147–158.

Onwueme, K. C., C. J. Vos, J. Zurita, J. A. Ferreras, and L. E. N. Quadri. 2005. The dimycocerosate ester polyketide virulence factors of mycobacteria. *Prog. Lipid Res.* 44:259–302.

Pai, M. P., K. M. Momary, and K. A. Rodvold. 2006. Antibiotic drug interactions. *Med. Clin. North Am.* 90:1223–1255.

Pasca, M. R., P. Guglierame, F. Arcesi, M. Bellinzoni, E. De Rossi, and G. Riccardi. 2004. Rv2686c-Rv2687c-Rv2688c, an ABC fluoroquinolone efflux pump in *Mycobacterium tuberculosis*. *Antimicrob. Agents Chemother.* 48:3175–3178.

Peirs, P., P. Lefevre, S. Boarbi, X. M. Wang, O. Denis, M. Braibant, K. Pethe, C. Locht, K. Huygen, and J. Content. 2005. *Mycobacterium tuberculosis* with disruption in genes encoding the phosphate binding proteins PstS1 and PstS2 is deficient in phosphate uptake and demonstrates reduced in vivo virulence. *Infect. Immun.* 73:1898–1902.

Pethe, K., D. L. Swenson, S. Alonso, J. Anderson, C. Wang, and D. G. Russell. 2004. Isolation of *Mycobacterium tuberculosis* mutants defective in the arrest of phagosome maturation. *Proc. Natl. Acad. Sci. USA* 101:13642–13647.

Quentin, Y., G. Fichant, and F. Denizot. 1999. Inventory, assembly and analysis of *Bacillus subtilis* ABC transport systems. *J. Mol. Biol.* 287:467–484.

Rengarajan, J., B. R. Bloom, and E. J. Rubin. 2005. Genome-wide requirements for *Mycobacterium tuberculosis* adaptation and survival in macrophages. *Proc. Natl. Acad. Sci. USA* 102:8327–8332.

Rodriguez, G. M. 2006. Control of iron metabolism in *Mycobacterium tuberculosis*. *Trends Microbiol.* 14:320–327.

Rodriguez, G. M., and I. Smith. 2006. Identification of an ABC transporter required for iron acquisition and virulence in *Mycobacterium tuberculosis*. *J. Bacteriol.* 188:424–430.

Rosas-Magallanes, V., G. Stadthagen-Gomez, J. Rauzier, L. B. Barreiro, L. Tailleux, F. Boudou, R. Griffin, J. Nigou, M. Jackson, B. Gicquel, and O. Neyrolles. 2007. Signature-tagged transposon mutagenesis identifies novel *Mycobacterium tuberculosis* genes involved in the parasitism of human macrophages. *Infect. Immun.* 75:504–507.

Sarin, J., S. Aggarwal, R. Chaba, G. C. Varshney, and P. K. Chakraborti. 2001. B-subunit of phosphate-specific transporter from *Mycobacterium tuberculosis* is a thermostable ATPase. *J. Biol. Chem.* **276:**44590–44597.

Sarin, J., G. P. Raghava, and P. K. Chakraborti. 2003. Intrinsic contributions of polar amino acid residues toward thermal stability of an ABC-ATPase of mesophilic origin. *Protein Sci.* **12:** 2118–2120.

Sassetti, C. M., D. H. Boyd, and E. J. Rubin. 2003. Genes required for mycobacterial growth defined by high density mutagenesis. *Mol. Microbiol.* **48:**77–84.

Sassetti, C. M., and E. J. Rubin. 2003. Genetic requirements for mycobacterial survival during infection. *Proc. Natl. Acad. Sci. USA* **100:**12989–12994.

Schnappinger, D., S. Ehrt, M. I. Voskuil, Y. Liu, J. A. Mangan, I. M. Monahan, G. Dolganov, B. Efron, P. D. Butcher, C. Nathan, and G. K. Schoolnik. 2003. Transcriptional adaptation of *Mycobacterium tuberculosis* within macrophages: Insights into the phagosomal environment. *J. Exp. Med.* **198:**693–704.

Stirling, D. A., C. S. Hulton, L. Waddell, S. F. Park, G. S. Stewart, I. R. Booth, and C. F. Higgins. 1989. Molecular characterization of the proU loci of Salmonella typhimurium and Escherichia coli encoding osmoregulated glycine betaine transport systems. *Mol. Microbiol.* **3:**1025–1038.

Stratmann, J., B. Strommenger, R. Goethe, K. Dohmann, G. F. Gerlach, K. Stevenson, L. L. Li, Q. Zhang, V. Kapur, and T. J. Bull. 2004. A 38-kilobase pathogenicity island specific for *Mycobacterium avium subsp. paratuberculosis* encodes cell surface proteins expressed in the host. *Infect. Immun.* **72:**1265–1274.

Sulzenbacher, G., S. Canaan, Y. Bordat, O. Neyrolles, G. Stadthagen, V. Roig-Zamboni, J. Rauzier, D. Maurin, F. Laval, M. Daffe, C. Cambillau, B. Gicquel, Y. Bourne, and M. Jackson. 2006. LppX is a lipoprotein required for the translocation of phthiocerol dimycocerosates to the surface of *Mycobacterium tuberculosis*. *EMBO J.* **25:**1436–1444.

Sutcliffe, I. C., and D. J. Harrington. 2004. Lipoproteins of *Mycobacterium tuberculosis*: an abundant and functionally diverse class of cell envelope components. *FEMS Microbiol. Rev.* **28:** 645–659.

Talaat, A. M., R. Lyons, S. T. Howard, and S. A. Johnston. 2004. The temporal expression profile of *Mycobacterium tuberculosis* infection in mice. *Proc. Natl. Acad. Sci. USA* **101:**4602–4607.

Ukai, H., H. Matsuzawa, K. Ito, M. Yamada, and A. Nishimura. 1998. ftsE(Ts) affects translocation of K^+-pump proteins into the cytoplasmic membrane of *Escherichia coli*. *J. Bacteriol.* **180:**3663–3670.

Vyas, N. K., M. N. Vyas, and F. A. Quiocho. 2003. Crystal structure of M tuberculosis ABC phosphate transport receptor: specificity and charge compensation dominated by ion-dipole interactions. *Structure* **11:**765–774.

Waddell, S. J., G. A. Chung, K. J. Gibson, M. J. Everett, D. E. Minnikin, G. S. Besra, and P. D. Butcher. 2005. Inactivation of polyketide synthase and related genes results in the loss of complex lipids in *Mycobacterium tuberculosis* H37Rv. *Lett. Appl. Microbiol.* **40:**201–206.

Wagner, D., F. J. Sangari, A. Parker, and L. E. Bermudez. 2005. fecB, a gene potentially involved in iron transport in *Mycobacterium avium*, is not induced within macrophages. *FEMS Microbiol. Lett.* **247:**185–191.

Wang, C., B. Hayes, M. M. Vestling, and K. Takayama. 2006. Transposome mutagenesis of an integral membrane transporter in *Corynebacterium matruchotii*. *Biochem. Biophys. Res. Commun.* **340:**953–960.

Wang, Z., A. Choudhary, P. S. Ledvina, and F. A. Quiocho. 1994. Fine tuning the specificity of the periplasmic phosphate transport receptor. Site-directed mutagenesis, ligand binding, and crystallographic studies. *J. Biol. Chem.* **269:**25091–25094.

Webb, M. R. 2003. Mycobacterial ABC transport system: structure of the primary phosphate receptor. *Structure* **11:**736–738.

Wooff, E., S. L. Michell, S. V. Gordon, M. A. Chambers, S. Bardarov, W. R. Jacobs, Jr., R. G. Hewinson, and P. R. Wheeler. 2002. Functional genomics reveals the sole sulphate transporter of the *Mycobacterium tuberculosis* complex and its relevance to the acquisition of sulphur in vivo. *Mol. Microbiol.* **43:**653–663.

The Mycobacterial Cell Envelope
Edited by M. Daffé and J.-M. Reyrat
© 2008 ASM Press, Washington, DC

Chapter 12

The MmpL Protein Family

MADHULIKA JAIN, ERIC D. CHOW, AND JEFFERY S. COX

The diverse array of *Mycobacterium tuberculosis* lipids, many of which are critical for virulence and mediate interactions with host cells, exist primarily on the outermost surface of the bacterium. How do these complex lipids, which are synthesized in the cytoplasm, make their way through the cell membrane and cell wall to reach their destination on the hydrophobic surface of the mycolic acid layer? The first glimpse into the answer to this biological question has recently emerged with an understanding of the essential function of a class of transporters, known as the MmpL family, in facilitating lipid egress. The role of some of these transporters in *M. tuberculosis* virulence is becoming clear, and the lipids they secrete play key roles as modulators of the interactions between this prokaryotic pathogen and its eukaryotic host. Mechanistic studies of their function suggest that they facilitate transport differently from their drug-efflux pump cousins. Here we review this literature and highlight key unanswered questions in this emerging field.

MMPL DIVERSITY IN MYCOBACTERIA

Fourteen *mmpL* genes have been identified by analysis of the two completed *M. tuberculosis* genomic sequences (Cole et al., 1998; Tekaia et al., 1999). These genes encode for large, multitransmembrane-containing proteins and were thus given the moniker MmpL, for *m*ycobacterial *m*embrane *p*rotein, *large*. Thirteen of these genes are identical in the two sequenced strains of *M. tuberculosis*, H37Rv and CDC1551, but *mmpL14* is unique to CDC1551 (Betts et al., 2000; Fleischmann et al., 2002). Although originally identified in *M. tuberculosis*, it is clear the *mmpL* genes have expanded and radiated throughout all mycobacterial species sequenced to date. Paralogues of the *M. tuberculosis* genes are

identifiable in *Mycobacterium bovis*, *Mycobacterium leprae*, *Mycobacterium avium*, *Mycobacterium marinum*, *Mycobacterium ulcerans*, and *Mycobacterium smegmatis* (Fig. 1; Table 1). *M. marinum* has the most, with 24 MmpL homologues, whereas *M. leprae*, due to massive genome decay, contains only five intact *mmpL* genes (Cole et al., 1998). Phylogenetic analysis of the MmpL family shows an unequal distribution of the different subfamilies of MmpL transporters among these different species (Table 1). In *M. tuberculosis*, two groups emerge: the MmpL1/2/4/5/6/9 transporters are more closely related to one another, whereas the rest of the members have diverged more significantly during evolution, giving rise to generally distinct branches (Fig. 1). Interestingly, the expansion of *mmpLs* in *M. marinum* has arisen primarily from duplication of the *mmpL4/5* subfamily. It is very likely that this expansion has occurred recently because the closely related species, *M. ulcerans*, is devoid of these sequences (Table 1).

The MmpL genes of *M. bovis*, the most closely related *M. tuberculosis* species sequenced to date and a pathogen of humans and other mammals, are nearly identical to those from *M. tuberculosis*. There are a few notable exceptions. First, "MmpL13" in all sequenced *M. tuberculosis* strains has a frameshift mutation in the middle of the open reading frame, but this gene is intact in *M. bovis* (Mb1177). Thus, this transporter is likely functional in *M. bovis* but not in *M. tuberculosis*. Full-length ancestral homologues of MmpL13 are present in all of the other mycobacteria (Fig. 1, Table 1). Second, *mmpL6* lies in a variable region of the *M. tuberculosis* (TbD1) genome and is deleted in more "modern" strains but intact in ancestral strains of *M. tuberculosis* and *M. bovis* (Brosch et al., 2002). Finally, the homologues of *mmpL1* and *mmpL9* in *M. bovis* have point mutations that would lead to truncation of the protein, suggesting that these proteins may not be functional in *M. bovis*.

Madhulika Jain, Eric D. Chow, and Jeffery S. Cox • Department of Microbiology and Immunology, University of California, San Francisco, San Francisco, CA 94143-2200.

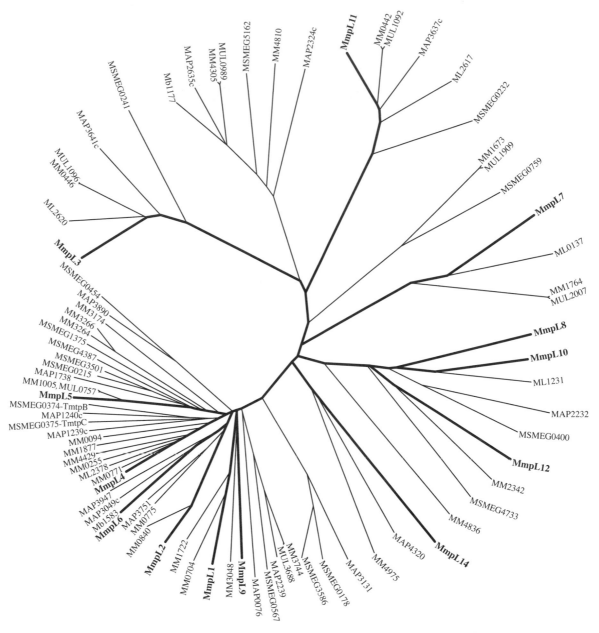

Figure 1. Phylogenetic tree of MmpL proteins from sequenced mycobacterial species. The MmpLs from *M. tuberculosis* are shown in bold. Most of the *M. bovis* orthologues of the *M. tuberculosis* MmpLs are omitted because they are identical in sequence except Mb1177, which has a frameshift mutation in *M. tuberculosis*, and Mb1583, which is a full-length version of MmpL6. Mb, *M. bovis*; ML, *M. leprae*; MAP, *M. avium*; MM, *M. marinum*; MUL, *M. ulcerans*; MSMEG, *M. smegmatis*.

The MmpL family is a subset of the larger resistance, nodulation, and division (RND) superfamily of transporters, one of the largest families of transporters with homologues present in all kingdoms of life (Tseng et al., 1999). Members of this family share similar topological features predicted from their amino acid sequences. Nearly all RND proteins contain 12 transmembrane domains with predicted extracytoplasmic loops between transmembrane domains 1 and 2, and between domains 7 and 8 (Tseng et al., 1999) (Fig. 2).

MmpL FUNCTION

Initial models of MmpL function were based on the RND transporters from gram-negative bacteria. Most of the known eubacterial RND transporters work primarily to efflux extrinsically acquired toxic molecules from cells, including heavy metals, drugs, and other noxious molecules encountered in the environment. The most well studied of these is AcrB, a major multidrug efflux pump in *Escherichia coli* that engenders resistance against a wide variety of hy-

Table 1. Distribution of MmpL transporters in mycobacteria[a]

M. tuberculosis	M. bovis	M. leprae	M. marinum	M. ulcerans	M. avium	M. smegmatis	M. vanbaalenii
MmpL1			Mm0704				
			Mm1722				
MmpL2	Mb0519		Mm0775		Ma3049c		
	Mb1583		Mm0840				
MmpL3	Mb0212c	Ml2620	Mm0446	Mu1096	Ma3641c	Ms0241	Mv0196
MmpL4	Mb0458c	Ml2378	Mm0094		Ma1239c	Ms0374	Mv3833
			Mm0255		Ma1240c	Ms0375	
			Mm0771		Ma3751	Ms0454	
			Mm1877		Ma3890		
			Mm3174		Ma3947		
			Mm4429				
MmpL5	Mb0695c		Mm1005	Mu0757	Ma1738	Ms0215	Mv1057
	Mb1787		Mm3264			Ms1375	Mv1614
			Mm3266			Ms3501	Mv1936
						Ms4387	Mv3189
MmpL7	Mb2967	Ml0137	Mm1764	Mu2007			
MmpL8	Mb3853c					Ms4733	Mv2900
MmpL9			Mm3048	Mu3688	Ma0076	Ms0178	
			Mm3744		Ma2239	Ms0567	
					Ma3131	Ms3586	
MmpL10	Mb1215	Ml1231			Ma2232	Ms0400	Mv0271
							Mv0312
MmpL11	Mb0208c	Ml2617	Mm0442	Mu1092	Ma3637c	Ms0232	Mv0188
MmpL12	Mb1549c		Mm1673	Mu1909		Ms0759	Mv0683
			Mm2342				
			Mm4836				
MmpL13a/b	Mb1177		Mm4305	Mu0989	Ma2324c	Ms5162	Mv1197
			Mm4810		Ma2635c		Mv1577
							Mv4573
MmpL14			Mm4975		Ma4320		

[a]Predicted MmpLs identified in all sequenced mycobacterial species were compared with those from M. tuberculosis and classified based on homology as determined by BLAST and ClustalW.

drophobic and hydrophilic toxic compounds and antibiotics. Numerous crystal structures of AcrB, including cocrystals with bound ligands, revealed that the protein exists as a trimer, with the transmembrane domains forming a large central pore capped by the two extracellular domains (Murakami et al., 2006; Murakami et al., 2002; Pos et al., 2004; Seeger et al., 2006; Yu et al., 2005; Yu et al., 2003). These

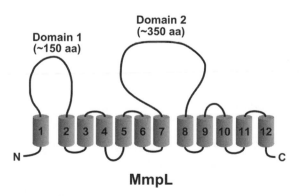

Figure 2. Predicted membrane topology of MmpLs, indicating the two non-transmembrane regions.

structural studies, coupled with biochemical studies, support a model in which hydrophobic molecules within the membrane or periplasm enter AcrB through lateral channels ("vestibules") formed between monomers. After capture, AcrB exports these molecules by handing them over to the TolC outer membrane protein through the adapter protein AcrA, completing the AcrB-AcrA-TolC efflux assembly (Higgins et al., 2004; Koronakis et al., 2000; Mikolosko et al., 2006). The M. tuberculosis MmpL proteins have more closely related homologues among other actinomycetes, such as Streptomyces coelicolor, and other more distantly related high G+C, gram-positive eubacteria. In contrast to the role of gram-negative RND family members as efflux pumps, an MmpL homologue in S. coelicolor, ActII-orf3, is involved in the synthesis and transport of an endogenously produced substrate. In particular, ActII-orf3 is required for the secretion of actinorhodin, a heterocyclic, blue pigmented antibiotic synthesized by a polyketide synthase system (Bystrykh et al., 1996). Interestingly, in the absence of actII-orf3, the bacteria accumulate a precursor of actinorhodin, raising the intriguing possibility that this subfamily of RND transporters may also be involved in substrate synthesis.

An initial clue as to the function of MmpLs was the discovery that most of the *mmpL* genes in *M. tuberculosis* are located near genes involved in polyketide synthesis or modification (Fig. 3) (Tekaia et al., 1999). Although certainly not confirmatory of function, this suggested that individual MmpL proteins may serve as transporters of the polyketides synthesized by the neighboring synthases. Furthermore, this genomic information suggests that gene duplication events took place during mycobacterial evolution, endowing *M. tuberculosis* with the ability to transport a wide variety of lipophilic molecules using a basic molecular design (Tekaia et al., 1999). Therefore it is intriguing to speculate that numerous polyketides synthesized by *M. tuberculosis* are transported to the cell surface by the MmpL family of transporters, where they may mediate host-pathogen interactions.

This idea has been borne out from experimental evidence from several studies on MmpLs, both in *M. tuberculosis* and in other mycobacterial species. Many of these *mmpL* genes have been isolated from a variety of genetic screens, and the phenotypes of these mutants demonstrate the essential function of a number of MmpLs in the synthesis or transport of endogenous polyketides to the cell surface of the bacterium (Converse et al., 2003; Cox et al., 1999; Recht et al., 2000). In *M. tuberculosis*, several of the MmpLs and their associated lipids represent critical virulence determinants.

MmpL7

MmpL7 was the first MmpL family member to be studied and was identified as required for virulence and bacterial growth in vivo in a mouse model of *M. tuberculosis* infection (Cox et al., 1999). MmpL7 is required for the subcellular localization of the key lipid virulence factor phthiocerol dimycocerosate (PDIM) to the outer layers of the cell wall. In *mmpL7* mutant cells, PDIM accumulates inside the cell, suggesting that MmpL7 acts to export PDIM across the cell membrane (Cox et al., 1999). Although PDIM and other surface-exposed lipids had previously been studied biochemically and their role in virulence was suspected (Brennan and Nikaido, 1995; Goren et al., 1974), this study provided genetic proof of the role of an *M. tuberculosis* lipid in virulence. The *mmpL7* gene is located in the genome proximally to all the biosynthetic genes required for PDIM production (Fig. 3). Mutants in many of these neighboring genes were also identified in signature-tagged transposon mutagenesis (STM) screens (Camacho et al., 1999; Cox et al., 1999) and have simi-

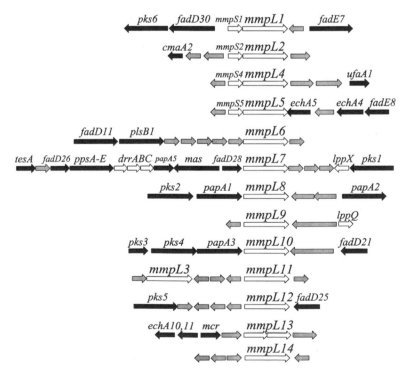

Figure 3. Gene organization of MmpLs. Genes predicted to be involved in lipid transport, including the *mmpL* and *mmpS* genes, the *lppQ* and *lppX* genes, and the *drrABC* operon are shown as white arrows. Genes predicted to be involved in lipid biosynthesis and metabolism are shown in black.

larly attenuated phenotypes as the *mmpL7* mutant. Another protein, DrrC, a member of the putative ABC transporter DrrABC has also been shown to be required for the export of PDIM to the cell wall (Camacho et al., 1999). Although the exact role of DrrC is unclear, it is possible that DrrABC provides energy for PDIM translocation across the cell membrane.

The mechanistic role of PDIM in *M. tuberculosis* pathogenesis is still not completely understood and is likely complicated. PDIM has been shown to be an important component of the cell wall for providing a physical barrier to host-induced damage, as well as for altering the immune response of the host. PDIM mutants are more sensitive to SDS and reactive nitrogen species, and have altered cytokine profiles in macrophages (Camacho et al., 2001; Reed et al., 2004; Rousseau et al., 2004). For a detailed discussion of PDIM and other related surface-exposed lipids, see chapter 17.

MmpL8

Mutants in *mmpL8* were also isolated in a signature-tagged mutagenesis screen, and MmpL8 is indispensable for growth and virulence in mice (Converse et al., 2003; Domenech et al., 2004). Initial clues to the potential function of MmpL8 came from the genomic organization of the *mmpL8* locus. The *pks2* gene, which is situated near *mmpL8* (Fig. 3), was shown previously to be required for SL-1 synthesis (Sirakova et al., 2001). Indeed, just as *mmpL7* mutants block PDIM export, *mmpL8* mutants fail to transport a well-known sulfated glycolipid, sulfolipid-1 (SL-1), to the cell surface. However, unlike *mmpL7* mutants, intact SL-1 is not found inside the cell. Instead, a precursor of SL-1 termed SL_{1278} accumulates inside *mmpL8*$^-$ cells, demonstrating that MmpL8 is also required for the complete biosynthesis of its cognate lipid (Converse et al., 2003; Domenech et al., 2004). The reason for this difference between MmpL7 and MmpL8 is not currently understood; however, we posit a model of the mechanism of MmpL function that accounts for this discrepancy later in this chapter. An important outcome of these studies was the realization that these transporters have no overlap in substrate specificity— *mmpL7* mutants transport SL-1 normally and *mmpL8* mutants secrete PDIM.

Despite the discrete requirements for PDIM and SL-1 transport, a global "lipidomic" approach identified an unexpected link between PDIM and SL-1 biosynthesis (Jain et al., 2007). The initial observation was that mutants that fail to synthesize PDIM produce higher abundance and higher mass SL-1 (Jain et al., 2007). Subsequent experiments showed that this is due to the increased metabolic flux of a shared carbon precursor, methyl malonyl CoA, through the SL-1 biosynthetic pathway, in PDIM synthesis mutants. Thus, although the MmpL-dependent transport pathways for PDIM and SL-1 are distinct, synthesis of the lipids is metabolically coupled.

Although MmpL8 is clearly critical for *M. tuberculosis* pathogenesis, the contribution of SL-1 to virulence is still somewhat murky (a complete discussion of sulfated lipids in *M. tuberculosis* can be found in chapter 18). Unlike *mmpL8* mutants, *pks2*$^-$ cells, which do not make either SL-1 or SL_{1278}, have little to no growth defects during infection of mice (Converse et al., 2003). A simple hypothesis to reconcile the disparate *mmpL8*$^-$ and *pks2*$^-$ phenotypes is that MmpL8 may function to transport an as yet unidentified substrate required for virulence. Alternatively, the attenuated virulence of the *mmpL8*$^-$ mutant could be due primarily to the buildup of SL_{1278} as SL_{1278} can be efficiently recognized by the immune system through presentation to T cells by the CD1b pathway (Gilleron et al., 2004). Finally, the metabolic coupling of SL-1 and other lipids, such as PDIM, may allow for compensatory production of other lipids in *pks2* cells but not in *mmpL8* mutants that can functionally complement the loss of SL-1 (Jain et al., 2007). Virulence studies with an *mmpL8/pks2* double knockout will help distinguish between these possibilities.

MmpL4

Although the function of MmpL4 is not known, it clearly plays a crucial role in *M. tuberculosis* virulence. *mmpL4* mutant cells show reduced bacterial numbers during growth in vivo, and infected mice have greatly increased survival (Domenech et al., 2005). In fact, *mmpL4*$^-$ is the most attenuated of all the *mmpL* mutants tested (Domenech et al., 2005). Although the nature of the molecule transported by MmpL4 is currently unknown, the regulation of *mmpL4* mRNA levels offers some clues as to the nature of its action. First, *mmpL4* transcription is induced, albeit modestly, during infection of activated macrophages (Schnappinger et al., 2003). Similarly, the gene was also induced when cells were treated with hydrogen peroxide (Schnappinger et al., 2003). Second, *mmpL4* is regulated by iron levels through direct repression by the IdeR iron-regulated repressor (Rodriguez et al., 2002). Together, these data suggest that the function of MmpL4 is manifest specifically during infection, and perhaps activated in response to the intracellular environment encountered by the bacterium. Although the details of

MmpL4's role in *M. tuberculosis* pathogenesis are currently unclear, this promises to be an active area of investigation.

MmpL3

Despite some indirect evidence that *mmpL3* mutants are viable (Sassetti et al., 2003), *mmpL3* is generally believed to be an essential gene in *M. tuberculosis* because several attempts at generating an *mmpL3* knockout have been unsuccessful (Domenech et al., 2005) (and our own unpublished observations). This is a unique feature of MmpL3 among the mycobacterial MmpLs and it is one of only four MmpLs conserved in all mycobacterial genome sequences available (Fig. 1). Although MmpL3 will be more difficult to study, understanding the function of MmpL3 in mycobacteria will likely uncover extremely important aspects of cell wall biogenesis.

TmtpB and TmtpC

Interestingly, TmtpB and TmtpC, both close homologues of MmpL4 in *M. smegmatis*, are required for the synthesis of a class of glycosylated peptidolipids (GPLs) that are absent from *M. tuberculosis* but are normally found in the outer layers of the *M. smegmatis* and *M. avium* cell wall (Recht et al., 2000; Sonden et al., 2005). *M. smegmatis* can spread on the surface of a solid growth medium by a sliding mechanism that combines both the expansive forces of a growing bacterial population as well as decreased friction between the cells and the media due to bacterial surface properties (Martinez et al., 1999). The *tmtpC* mutant was identified as a nonsliding *M. smegmatis* mutant and subsequently shown to be unable to form biofilms due to the lack of GPLs.

TmtpC is encoded in a locus containing two other genes, *tmtpA* and *tmtpB*, all of which are homologous to genes in a similar locus in *M. avium*. Although TmtpB and TmtpC are most similar to MmpL4 from *M. tuberculosis*, the genomic organization of the *tmtp* locus is much different from that of *mmpL4*. TmtpB, which is extremely similar to TmtpC, has been shown to be involved in GPL synthesis as well (Sonden et al., 2005), whereas the role of TmtpA remains untested.

In contrast to MmpL7, which is required for PDIM export and not synthesis, the requirement for TmtpB and TmtpC in lipid synthesis is similar to the role of MmpL8 in SL-1 biosynthesis (see below). Interestingly, a small integral membrane protein named Gap is specifically required for the transport of GPLs to the cell surface in *M. smegmatis* (Sonden et al., 2005). An orthologue of *gap*, *Rv3821*, is located in the MmpL8 locus and may be involved specifically in the transport of SL-1 across the cell membrane.

TRANSPORTERS OR ASSEMBLY SCAFFOLDS?

Based on the mechanism of their RND cousins such as AcrB, a straightforward model of MmpL-mediated transport is that they recognize and collect their cognate lipids from within the cell membrane and extrude them from the bilayer toward the cell wall. However, several pieces of evidence challenge this simple idea and suggest a more active role for these proteins in lipid synthesis. First, as mentioned previously, mutations in *mmpL8* in *M. tuberculosis* (as well as mutations in *tmtpB* and *tmtpC* in *M. smegmatis*, and *actII-orf3* in *S. coelicolor*) lead to lipid synthesis defects. Because MmpL8 does not contain the lipid biosynthetic domains required to complete the synthesis of SL-1 from SL_{1278}, one possibility is that MmpL8 functions to export SL_{1278} across the membrane to the periplasm, where its synthesis is completed. Although this is an attractive idea, a radically different model arises from studies of MmpL7. Although MmpL7 is not required for PDIM synthesis, it interacts with PpsE, a member of the PDIM synthetic machinery and couples PDIM synthesis and secretion (Jain and Cox, 2005). Interactions between PpsE and another member of the PDIM synthesis machinery, TesA, have also been observed (Rao and Ranganathan, 2004), leading to a model in which MmpL7 acts as a scaffold for the assembly of the entire PDIM synthesis-secretion complex (Fig. 4). Therefore, if MmpLs recruit cognate lipid biosynthetic enzymes to a common site of synthesis and export, MmpL8 may be required to recruit a polyketide biosynthetic factor that completes the synthesis of SL-1 from SL_{1278} (Fig. 4). Thus, unlike the RND drug efflux pumps, MmpLs may act as scaffolds for the assembly of polyketide synthetic machinery, thereby promoting the coupled synthesis and export of their specific lipid substrates.

A recent report showed that a collection of polyketide synthases that function together to synthesize an antibiotic in *Bacillus subtilis* assemble into a single megacomplex at the membrane (Straight et al., 2007). We proposed that this organization may be achieved through interaction with a membrane protein, consistent with our model of MmpL function. Therefore, the organization and coupling of polyketide synthases may be a common feature among prokaryotes.

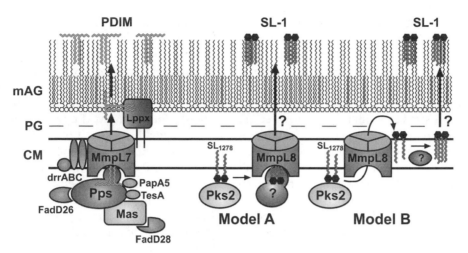

Figure 4. Model of MmpL secretion. The proposed models for PDIM and SL-1 secretion through MmpL7 and MmpL8, respectively, are shown. PpsA-E and Mas are polyketide synthases that extend straight chain fatty acids to phthiocerol and mycocerosic acid, respectively (Azad et al., 1997; Azad et al., 1996; Trivedi et al., 2005). The enzymes FadD26 and FadD28 are thought to be AMP ligases that activate straight-chain fatty acids for transfer to the Pps and Mas enzymes (Trivedi et al., 2004). The thioesterase TesA is also required for the synthesis of PDIM and interacts with PpsE (Rao and Ranganathan, 2004). PapA5 is able to catalyze the esterification of mycocerosic acids to phthiocerol to form PDIM (Onwueme et al., 2004). MmpL7 and DrrABC are required to transport PDIM across the cell membrane (Camacho et al., 1999; Cox et al., 1999). Finally, LppX is a signal sequence containing protein that is thought to transport PDIM across the periplasm to the cell wall (Sulzenbacher et al., 2006). Two models for SL-1 secretion are shown. Pks2 is required for the synthesis of SL_{1278} (Sirakova et al., 2001). In model A, MmpL8 recruits an as yet unidentified biosynthetic factor to complete the synthesis of SL-1 from SL_{1278}, before transport across the cell membrane (Jain and Cox, 2005). In model B, MmpL8 exports SL_{1278} across the cell membrane, after which it is converted to SL-1 by an as yet unidentified enzyme (Converse et al., 2003). CM, cytoplasmic membrane; PG, peptidoglycan; mAG, mycolyl-arabinogalactan. *(See the color insert for the color version of this figure.)*

SPECIFICITY OF MMPL TRANSPORT

The working model proposed above may also help explain the unique specificity of MmpL transporters. In contrast to the broad substrate specificity of RND efflux pumps such as AcrB, MmpLs appear to be involved in specific lipid pathways. Aside from PDIM export, MmpL7 does not appear to have any other lipid associated defects and secretes other surface-exposed lipids similarly to wild-type. Conversely, other MmpL mutants are able to transport PDIM like wild-type cells (Converse et al., 2003). Therefore, it is also possible that MmpLs are adapters that interface with a separate lipid export machinery. Because the interaction between MmpL7 and PpsE suggests that PDIM is coordinately synthesized and secreted (Jain and Cox, 2005), this also raises the possibility that specificity in MmpL-mediated transport is engendered by protein-protein interactions between synthase and transporter.

PERSPECTIVES AND FUTURE DIRECTIONS

Although most RND family members in gram-negative bacteria function as multidrug efflux pumps, mycobacteria use MmpLs as transporters for endogenously derived lipid substrates. Based on the role of MmpL7 in PDIM transport and the role of MmpL8 in SL-1 biosynthesis, as well as the function of MmpL homologues in other bacteria, it is tempting to speculate that all MmpLs may be involved in biosynthesis and transport of surface-exposed lipid substrates. However, this remains to be tested and based on the myriad of functions performed by members of the larger RND superfamily (Ma et al., 2002), it seems likely that some of the MmpLs may have roles in other export pathways. Likewise, although *mmpL* mutants in *M. tuberculosis* do not have dramatically increased susceptibility to various drugs tested (Domenech et al., 2005), it is also possible that MmpL family members in other mycobacteria may function as drug efflux pumps.

The mechanism of MmpL function and substrate specificity is not clearly understood. Whether MmpLs actively participate in the export of lipids across the cell membrane, or act as scaffolds to recruit lipid biosynthetic machinery prior to transport, certainly requires more analysis. The interaction between MmpL7 and PpsE raises the intriguing possibility that specificity in MmpL transport may be encoded by protein-protein interactions (Jain and Cox, 2005). Interestingly, in gram negative bacteria, the substrate specificity of RND family drug efflux pumps is encoded in the two non-transmembrane regions domains 1 and 2 (Fig. 2). Domain swaps between the drug efflux pumps AcrB and AcrD in *E. coli* resulted in the switching of their respective drug efflux specificities (Elkins and Nikaido, 2002). Similar experiments with MexB in *Pseudomonas aeruginosa* yielded similar results (Tikhonova et al., 2002). Although preliminary experiments with domain swaps between MmpL7 and MmpL8 yielded non-functional chimeras (Jain and Cox, 2005), it will be interesting to see if these domains contribute to the specificity in MmpL transport.

The energetics of MmpL-mediated transport is also uncertain. The eukaryotic RND family member, Niemann-Pick C1 (NPC1) utilizes the proton motive force to translocate cholesterol across the cell membrane (Davies et al., 2000) and it is widely thought that RND family members use proton antiport or symport as a mechanism of facilitated transport of their substrates. Indeed, AcrB has been shown to utilize proton antiport to derive energy for drug efflux (Thanassi et al., 1997; Zgurskaya and Nikaido, 1999). Surprisingly, it has also been shown that an ABC transporter LmrA in *Lactococcus lactis* can catalyze drug transport by a proton symport mechanism even when the ATP hydrolyzing domain is removed (Venter et al., 2003) leading to the hypothesis that LmrA may have evolved from an RND like transporter to an ABC transporter via acquisition of the ATP-hydrolysis domain (Kim et al., 2004). Interestingly, the proximity of MmpL7 to the ABC transporter DrrABC (Fig. 3) and the requirement of DrrC in PDIM export (Camacho et al., 1999) may mean that active transport via ATP hydrolysis is required for PDIM translocation. Thus MmpL transporters may function as passive transporters and export their substrates by facilitated diffusion or be coupled to ABC transporters that provide energy for active translocation across the membrane.

The mechanism of transport of surface-exposed lipids from the cell membrane to the bacterial surface is also poorly understood. After translocation across the cell membrane, the lipids need to be transported to the outer layers of the cell wall. This may occur simply through the process of diffusion, or perhaps there exist accessory proteins to aid in the transport of these hydrophobic lipids across the periplasm to the mycolate layer. In gram-negative bacteria, transport across the outer membrane is accomplished by outer membrane factors that associate with RND transporters. For example, AcrB interacts with the outer membrane protein TolC and the drug substrates are presumably expelled through successive export through AcrB and TolC (Tikhonova and Zgurskaya, 2004; Touze et al., 2004). Although no such outer membrane factors have been discovered in

mycobacteria, the LppX lipoprotein is an attractive candidate for a periplasmic or cell wall transporter that shuttles PDIM to the outer layers of the cell wall. The *lppX* gene is located in the PDIM locus (Fig. 3) and has recently been shown to be required for PDIM export to the cell wall (Sulzenbacher et al., 2006). Interestingly, Lppx is homologous to the lipoproteins LolA and LolB, which are found in the periplasm and the outer membrane respectively of *E. coli*. The Lol system functions to export lipoproteins across the periplasm of *E. coli* to the inner leaflet of the outer membrane (Sulzenbacher et al., 2006). LolCDE bind specifically to lipoproteins destined for the outer membrane and export them across the cell membrane, where they bind the periplasmic shuttle LolA. LolA, in turn, transfers the lipoproteins to LolB, which is located in the outer membrane. From LolB, the lipoproteins are released to the outer membrane, where they remain anchored. Interestingly, Lppx shares a similar beta half-barrel structure to LolA and LolB but has a larger substrate binding pocket that could accommodate a PDIM molecule (Sulzenbacher et al., 2006). Thus Lppx may function in a similar manner to export PDIM to the outer layer of the cell wall (Fig. 4).

Because many MmpL family members play a role in *M. tuberculosis* virulence, MmpLs may be excellent drug targets. They might be especially tractable drug targets because they reside in the membrane and are likely more accessible to drugs than key lipid synthesis and modification enzymes located inside the cell. However, the role of MmpLs during the chronic phase of infection is unclear, and if MmpLs are not required in the persistent state of mycobacteria, they may not represent a substantial improvement over existing cell wall targets.

REFERENCES

Azad, A. K., T. D. Sirakova, N. D. Fernandes, and P. E. Kolattukudy. 1997. Gene knockout reveals a novel gene cluster for the synthesis of a class of cell wall lipids unique to pathogenic mycobacteria. *J. Biol. Chem.* 272:16741–16745.

Azad, A. K., T. D. Sirakova, L. M. Rogers, and P. E. Kolattukudy. 1996. Targeted replacement of the mycocerosic acid synthase gene in *Mycobacterium bovis* BCG produces a mutant that lacks mycosides. *Proc. Natl. Acad. Sci. USA* 93:4787–4792.

Betts, J. C., P. Dodson, S. Quan, A. P. Lewis, P. J. Thomas, K. Duncan, and R. A. McAdam. 2000. Comparison of the proteome of *Mycobacterium tuberculosis* strain H37Rv with clinical isolate CDC 1551. *Microbiology* 146:3205–3216.

Brennan, P. J., and H. Nikaido. 1995. The envelope of mycobacteria. *Annu. Rev. Biochem.* 64:29–63.

Brosch, R., S. V. Gordon, M. Marmiesse, P. Brodin, C. Buchrieser, K. Eiglmeier, T. Garnier, C. Gutierrez, G. Hewinson, K. Kremer, L. M. Parsons, A. S. Pym, S. Samper, D. van Soolingen, and S. T. Cole. 2002. A new evolutionary scenario for the *Mycobacterium tuberculosis* complex. *Proc. Natl. Acad. Sci. USA* 99:3684–3689.

Bystrykh, L. V., M. A. Fernandez-Moreno, J. K. Herrema, F. Malpartida, D. A. Hopwood, and L. Dijkhuizen. 1996. Production of actinorhodin-related "blue pigments" by *Streptomyces coelicolor* A3(2). *J. Bacteriol.* **178:**2238–2244.

Camacho, L. R., P. Constant, C. Raynaud, M. A. Laneelle, J. A. Triccas, B. Gicquel, M. Daffe and C. Guilhot. 2001. Analysis of the phthiocerol dimycocerosate locus of *Mycobacterium tuberculosis*. Evidence that this lipid is involved in the cell wall permeability barrier. *J Biol. Chem.* **276:**19845–19854.

Camacho, L. R., D. Ensergueix, E. Perez, B. Gicquel, and C. Guilhot. 1999. Identification of a virulence gene cluster of *Mycobacterium tuberculosis* by signature-tagged transposon mutagenesis. *Mol. Microbiol.* **34:**257–267.

Cole, S. T., R. Brosch, J. Parkhill, T. Garnier, C. Churcher, D. Harris, S. V. Gordon, K. Eiglmeier, S. Gas, C. E. Barry, III, F. Tekaia, K. Badcock, D. Basham, D. Brown, T. Chillingworth, R. Connor, R. Davies, K. Devlin, T. Feltwell, S. Gentles, N. Hamlin, S. Holroyd, T. Hornsby, K. Jagels, B. G. A. Krogh, J. Mclean, S. Moule, L. Murphy, K. Oliver, J. Osborne, M. A. Quail, M. A. Rajandream, J. Rogers, S. Rutter, K. Seeger, J. Skelton, R. Squares, S. Squares, J. E. Sulston, K. Taylor, S. Whitehead, and B. G. Barrell. 1998. Deciphering the biology of *Mycobacterium tuberculosis* from the complete genome sequence. *Nature* **393:**537–544.

Converse, S. E., J. D. Mougous, M. D. Leavell, J. A. Leary, C. R. Bertozzi, and J. S. Cox. 2003. MmpL8 is required for sulfolipid-1 biosynthesis and *Mycobacterium tuberculosis* virulence. *Proc. Natl. Acad. Sci. USA* **100:**6121–6126.

Cox, J. S., B. Chen, M. McNeil, and W. R. Jacobs, Jr. 1999. Complex lipid determines tissue-specific replication of *Mycobacterium tuberculosis* in mice. *Nature* **402:**79–83.

Davies, J. P., F. W. Chen, and Y. A. Ioannou. 2000. Transmembrane molecular pump activity of Niemann-Pick C1 protein. *Science* **290:**2295–2298.

Domenech, P., M. B. Reed, and C. E. Barry III. 2005. Contribution of the *Mycobacterium tuberculosis* MmpL protein family to virulence and drug resistance. *Infect. Immun.* **73:**3492–3501.

Domenech, P., M. B. Reed, C. S. Dowd, C. Manca, G. Kaplan, and C. E. Barry III. 2004. The role of MmpL8 in sulfatide biogenesis and virulence of *Mycobacterium tuberculosis*. *J. Biol. Chem.* **279:**21257–21265.

Elkins, C. A., and H. Nikaido. 2002. Substrate specificity of the RND-type multidrug efflux pumps AcrB and AcrD of *Escherichia coli* is determined predominantly by two large periplasmic loops. *J. Bacteriol.* **184:**6490–6498.

Fleischmann, R. D., D. Alland, J. A. Eisen, L. Carpenter, O. White, J. Peterson, R. DeBoy, R. Dodson, M. Gwinn, D. Haft, E. Hickey, J. F. Kolonay, W. C. Nelson, L. A. Umayam, M. Ermolaeva, S. L. Salzberg, A. Delcher, T. Utterback, J. Weidman, H. Khouri, J. Gill, A. Mikula, W. Bishai, W. R. Jacobs, Jr., J. C. Venter, and C. M. Fraser. 2002. Whole-genome comparison of *Mycobacterium tuberculosis* clinical and laboratory strains. *J. Bacteriol.* **184:**5479–5490.

Gilleron, M., S. Stenger, Z. Mazorra, F. Wittke, S. Mariotti, G. Bohmer, J. Prandi, L. Mori, G. Puzo, and G. De Libero. 2004. Diacylated sulfoglycolipids are novel mycobacterial antigens stimulating CD1-restricted T cells during infection with *Mycobacterium tuberculosis*. *J. Exp. Med.* **199:**649–659.

Goren, M. B., O. Brokl, and W. B. Schaefer. 1974. Lipids of putative relevance to virulence in *Mycobacterium tuberculosis*: phthiocerol dimycocerosate and the attenuation indicator lipid. *Infect. Immun.* **9:**150–158.

Higgins, M. K., E. Bokma, E. Koronakis, C. Hughes, and V. Koronakis. 2004. Structure of the periplasmic component of a bacterial drug efflux pump. *Proc. Natl. Acad. Sci. USA* **101:** 9994–9999.

Jain, M., and J. S. Cox. 2005. Interaction between polyketide synthase and transporter suggests coupled synthesis and export of virulence lipid in *M. tuberculosis*. *PLoS Pathog.* **1:**e2.

Jain, M., C. J. Petzold, M. W. Schelle, M. D. Leavell, J. D. Mougous, C. R. Bertozzi, J. A. Leary, and J. S. Cox. 2007. Lipidomics reveals control of *Mycobacterium tuberculosis* virulence lipids via metabolic coupling. *Proc. Natl. Acad. Sci. USA* **104:**5133–5138.

Kim, S. H., A. B. Chang, and M. H. Saier, Jr. 2004. Sequence similarity between multidrug resistance efflux pumps of the ABC and RND superfamilies. *Microbiology* **150:**2493–2495.

Koronakis, V., A. Sharff, E. Koronakis, B. Luisi, and C. Hughes. 2000. Crystal structure of the bacterial membrane protein TolC central to multidrug efflux and protein export. *Nature* **405:**914–919.

Ma, Y., A. Erkner, R. Gong, S. Yao, J. Taipale, K. Basler, and P. A. Beachy. 2002. Hedgehog-mediated patterning of the mammalian embryo requires transporter-like function of dispatched. *Cell* **111:**63–75.

Martinez, A., S. Torello, and R. Kolter. 1999. Sliding motility in mycobacteria. *J. Bacteriol.* **181:**7331–7338.

Mikolosko, J., K. Bobyk, H. I. Zgurskaya, and P. Ghosh. 2006. Conformational flexibility in the multidrug efflux system protein AcrA. *Structure* **14:**577–587.

Murakami, S., R. Nakashima, E. Yamashita, T. Matsumoto, and A. Yamaguchi. 2006. Crystal structures of a multidrug transporter reveal a functionally rotating mechanism. *Nature* **443:** 173–179.

Murakami, S., R. Nakashima, E. Yamashita and A. Yamaguchi. 2002. Crystal structure of bacterial multidrug efflux transporter AcrB. *Nature* **419:**587–593.

Onwueme, K. C., J. A. Ferreras, J. Buglino, C. D. Lima, and L. E. Quadri. 2004. Mycobacterial polyketide-associated proteins are acyltransferases: proof of principle with *Mycobacterium tuberculosis* PapA5. *Proc. Natl. Acad. Sci. USA* **101:**4608–4613.

Pos, K. M., A. Schiefner, M. A. Seeger, and K. Diederichs. 2004. Crystallographic analysis of AcrB. *FEBS Lett.* **564:**333–339.

Rao, A., and A. Ranganathan. 2004. Interaction studies on proteins encoded by the phthiocerol dimycocerosate locus of *Mycobacterium tuberculosis*. *Mol. Genet. Genomics* **272:**571–579.

Recht, J., A. Martinez, S. Torello, and R. Kolter. 2000. Genetic analysis of sliding motility in *Mycobacterium smegmatis*. *J. Bacteriol.* **182:**4348–4351.

Reed, M. B., P. Domenech, C. Manca, H. Su, A. K. Barczak, B. N. Kreiswirth, G. Kaplan, and C. E. Barry, 3rd. 2004. A glycolipid of hypervirulent tuberculosis strains that inhibits the innate immune response. *Nature* **431:**84–87.

Rodriguez, G. M., M. I. Voskuil, B. Gold, G. K. Schoolnik, and I. Smith. 2002. ideR, An essential gene in *Mycobacterium tuberculosis*: role of IdeR in iron-dependent gene expression, iron metabolism, and oxidative stress response. *Infect. Immun.* **70:** 3371–3381.

Rousseau, C., N. Winter, E. Pivert, Y. Bordat, O. Neyrolles, P. Ave, M. Huerre, B. Gicquel, and M. Jackson. 2004. Production of phthiocerol dimycocerosates protects *Mycobacterium tuberculosis* from the cidal activity of reactive nitrogen intermediates produced by macrophages and modulates the early immune response to infection. *Cell. Microbiol.* **6:**277–287.

Sassetti, C. M., D. H. Boyd, and E. J. Rubin. 2003. Genes required for mycobacterial growth defined by high density mutagenesis. *Mol. Microbiol.* **48:**77–84.

Schnappinger, D., S. Ehrt, M. I. Voskuil, Y. Liu, J. A. Mangan, I. M. Monahan, G. Dolganov, B. Efron, P. D. Butcher, C. Nathan, and G. K. Schoolnik. 2003. Transcriptional adaptation of *Mycobacterium tuberculosis* within macrophages: insights into the phagosomal environment. *J. Exp. Med.* **198:**693–704.

Seeger, M. A., A. Schiefner, T. Eicher, F. Verrey, K. Diederichs, and K. M. Pos. 2006. Structural asymmetry of AcrB trimer suggests a peristaltic pump mechanism. *Science* **313:**1295–1298.

Sirakova, T. D., A. K. Thirumala, V. S. Dubey, H. Sprecher, and P. E. Kolattukudy. 2001. The *Mycobacterium tuberculosis pks2* gene encodes the synthase for the hepta- and octamethyl-branched fatty acids required for sulfolipid synthesis. *J. Biol. Chem.* **276:**16833–16839.

Sonden, B., D. Kocincova, C. Deshayes, D. Euphrasie, L. Rhayat, F. Laval, C. Frehel, M. Daffe, G. Etienne, and J. M. Reyrat. 2005. Gap, a mycobacterial specific integral membrane protein, is required for glycolipid transport to the cell surface. *Mol. Microbiol.* **58:**426–440.

Straight, P. D., M. A. Fischbach, C. T. Walsh, D. Z. Rudner, and R. Kolter. 2007. A singular enzymatic megacomplex from *Bacillus subtilis*. *Proc. Natl. Acad. Sci. USA* **104:**305–310.

Sulzenbacher, G., S. Canaan, Y. Bordat, O. Neyrolles, G. Stadthagen, V. Roig-Zamboni, J. Rauzier, D. Maurin, F. Laval, M. Daffe, C. Cambillau, B. Gicquel, Y. Bourne, and M. Jackson. 2006. LppX is a lipoprotein required for the translocation of phthiocerol dimycocerosates to the surface of *Mycobacterium tuberculosis*. *EMBO J.* **25:**1436–1444.

Tekaia, F., S. V. Gordon, T. Garnier, R. Brosch, B. G. Barrell, and S. T. Cole. 1999. Analysis of the proteome of *Mycobacterium tuberculosis* in silico. *Tuber. Lung Dis.* **79:**329–342.

Thanassi, D. G., L. W. Cheng, and H. Nikaido. 1997. Active efflux of bile salts by *Escherichia coli*. *J. Bacteriol.* **179:**2512–2518.

Tikhonova, E. B., Q. Wang, and H. I. Zgurskaya. 2002. Chimeric analysis of the multicomponent multidrug efflux transporters from gram-negative bacteria. *J. Bacteriol.* **184:**6499–6507.

Tikhonova, E. B., and H. I. Zgurskaya. 2004. AcrA, AcrB, and TolC of *Escherichia coli* form a stable intermembrane multidrug efflux complex. *J. Biol. Chem.* **279:**32116–32124.

Touze, T., J. Eswaran, E. Bokma, E. Koronakis, C. Hughes, and V. Koronakis. 2004. Interactions underlying assembly of the *Escherichia coli* AcrAB-TolC multidrug efflux system. *Mol. Microbiol.* **53:**697–706.

Trivedi, O. A., P. Arora, V. Sridharan, R. Tickoo, D. Mohanty, and R. S. Gokhale. 2004. Enzymic activation and transfer of fatty acids as acyl-adenylates in mycobacteria. *Nature* **428:**441–445.

Trivedi, O. A., P. Arora, A. Vats, M. Z. Ansari, R. Tickoo, V. Sridharan, D. Mohanty, and R. S. Gokhale. 2005. Dissecting the mechanism and assembly of a complex virulence mycobacterial lipid. *Mol. Cell* **17:**631–643.

Tseng, T. T., K. S. Gratwick, J. Kollman, D. Park, D. H. Nies, A. Goffeau, and M. H. Saier, Jr. 1999. The RND permease superfamily: an ancient, ubiquitous and diverse family that includes human disease and development proteins. *J. Mol. Microbiol. Biotechnol.* **1:**107–125.

Venter, H., R. A. Shilling, S. Velamakanni, L. Balakrishnan, and H. W. Van Veen. 2003. An ABC transporter with a secondary-active multidrug translocator domain. *Nature* **426:**866–870.

Yu, E. W., J. R. Aires, G. McDermott, and H. Nikaido. 2005. A periplasmic drug-binding site of the AcrB multidrug efflux pump: a crystallographic and site-directed mutagenesis study. *J. Bacteriol.* **187:**6804–6815.

Yu, E. W., G. McDermott, H. I. Zgurskaya, H. Nikaido, and D. E. Koshland, Jr. 2003. Structural basis of multiple drug-binding capacity of the AcrB multidrug efflux pump. *Science* **300:**976–980.

Zgurskaya, H. I., and H. Nikaido. 1999. Bypassing the periplasm: reconstitution of the AcrAB multidrug efflux pump of *Escherichia coli*. *Proc. Natl. Acad. Sci. USA* **96:**7190–7195.

The Mycobacterial Cell Envelope
Edited by M. Daffé and J.-M. Reyrat
© 2008 ASM Press, Washington, DC

Chapter 13

ESAT-6 and the Mycobacterial ESX Secretion Systems

IDA ROSENKRANDS, DARIA BOTTAI, PETER ANDERSEN, AND ROLAND BROSCH

The search for important antigens among secreted proteins from *Mycobacterium tuberculosis* present in culture filtrates led to the identification of ESAT-6 (early secretory antigenic target of 6 kDa), a small molecule that migrates in sodium dodecyl sulfate-polyacrylamide gel electrophoresis (SDS-PAGE) with an apparent molecular mass of 6 kDa (Andersen et al., 1995; Sorensen et al., 1995). When administered as a subunit vaccine in mice, the mixture of culture filtrate proteins induced protection at the same level as the live vaccine *Mycobacterium bovis* BCG (BCG) (Andersen, 1994). A strong T-cell response to ESAT-6 was observed in the first phase of infection in both mice, guinea pigs, cattle, and humans (Brandt et al., 1996; Elhay et al., 1998; Pollock and Andersen, 1997; Ravn et al., 1999), and a detailed biochemical and immunological characterization of this antigen was undertaken. ESAT-6 is released to the culture medium after short periods of growth in cultures without significant autolysis and, therefore, classified as a secreted protein, but without the conventional signal sequence (Sorensen et al., 1995). Interestingly, ESAT-6 is produced in all the virulent strains of the *M. tuberculosis* complex but not in BCG strains (Harboe et al., 1996; Sorensen et al., 1995), and subtractive genomic hybridization showed that the gene encoding ESAT-6 belong to region of difference 1 (RD1), a genomic segment deleted in all BCG strains (Mahairas et al., 1996).

It was one of the surprises during the analysis of the complete genome sequence that *M. tuberculosis* H37Rv contained several genes encoding proteins with similarities to ESAT-6 (Cole et al., 1998), and this gene family was later named ESX (Brodin et al., 2004b). Using a refined bioinformatic procedure, the ESX family was found to comprise 23 members in *M. tuberculosis* H37Rv ranging in size from 90 to 120 amino acids, with the prototype ESAT-6 being one of the less well-conserved examples. By examination of sequence similarities and gene order, it was possible to localize the *esx* genes to 11 genomic regions (Tekaia et al., 1999). As shown in Fig. 1, in most of these regions, the *esx* genes were located downstream from PE and PPE genes, encoding acidic proteins with conserved N-terminal domains bearing characteristic proline-glutamic acid (PE) or proline-proline-glutamic acid (PPE) motifs. These two novel gene families occupy approximately 8% of the coding capacity of the genome, and it has been hypothesized that the most ancestral representatives of these two families are those linked with the ESX region (Gey van Pittius et al., 2006). In the more complex organization of loci, the ESAT-6 encoding genes are part of a larger unit with apparent functional conservation (Cole et al., 1998; Tekaia et al., 1999).

It appears that some of these ESX loci (e.g., ESX-3) are present in a wide range of mycobacteria, and it seems to be essential for in vitro growth of *M. tuberculosis*, as determined by insertional mutagenesis and transposon site hybridization (TraSH) (Sassetti et al., 2003), whereas others are restricted to only a few mycobacterial species. In this chapter, we summarize what has been learned about this fascinating protein family and related proteins in the past few years since their discovery.

STRUCTURE AND LOCALIZATION OF ESAT-6

The original ESAT-6 protein was purified from *M. tuberculosis* culture filtrates by ammonium sulfate precipitation, hydrophobic interaction chromatography, followed by anion exchange chromatography. ESAT-6 is recognized by the monoclonal antibody HYB 76-8, and screening of an *M. tuberculosis*

Ida Rosenkrands and Peter Andersen • Department of Infectious Disease Immunology, Statens Serum Institut, Artillerivej 5, DK-2300 Copenhagen S, Denmark.   **Daria Bottai** • Unité de Génétique Moléculaire Bactérienne, Institut Pasteur, 25-28 rue du Dr. Roux, 75724 Paris Cedex 15, France, and Dipartimento di Patologia Sperimentale, Biotecnologie Mediche, Infettivologia ed Epidemiologia, University of Pisa, Via San Zeno 35/39, 56127 Pisa, Italy.   **Roland Brosch** • Unité de Génétique Moléculaire Bactérienne et Unité Postulante de Pathogénomique Mycobactérienne Intégrée, Institut Pasteur, 25-28 rue du Dr. Roux, 75724 Paris Cedex 15, France.

Figure 1.

	AAA-ATPase 630	1TM 540	FtsK/SpoIIIE 1330		PE 102	PPE 513	WXG100 97	WXG100 96		12TM 295	S-protease 472	S-protease 461	406	AAA-ATPase 630	
	9.3	5.6	8.1		210.1	210.1	5.8	14.1		5.24	5.7			9.3	
							Rv3905c (esxF)	Rv3904c (esxE)							
				IS*like*	Rv3622c (pe32)	Rv3621c (ppe65)	Rv3620c (esxW)	Rv3619c (esxV)							RD8
					Rv1040c (pe8)	Rv1039c (ppe15)	Rv1038c (esxJ)	Rv1037c (esxI)	IS*1560*						
					Rv1195 (pe13)	Rv1196 (ppe18)	Rv1197 (esxK)	Rv1198 (esxL)	IS*1081*						
				IS*6110*	Rv2353c (ppe39)	Rv2352c (ppe38)/ (pe29)	Rv2347c \(esxP) (plcABC)	Rv2346c (esxO) (esxR)							RD5
				IS*1081*	Rv3022c Rv3021c (ppe47)	Rv3018c (ppe46)	Rv3020c (esxS)	Rv3019c Rv3017c (esxQ)							
ESX-2		Rv3895c	Rv3894c		Rv3893c (pe36)	Rv3892c (ppe69)	Rv3891c (esxD)	Rv3890c (esxC)		Rv3889c	Rv3887c	Rv3886c Rv3883c	Rv3885c Rv3882c	Rv3884c	
ESX-5		Rv1782	Rv1783, Rv1784		(Rv1787–Rv1791) (ppe25-27, pe18-19)		Rv1792 (esxM)	Rv1793 (esxN)		Rv1794	Rv1795	Rv1796	Rv1797	Rv1798	
ESX-3	Rv0282	Rv0283	Rv0284		Rv0285 (pe5)	Rv0286 (ppe4)	Rv0287 (esxG)	Rv0288 (esxH)		Rv0289	Rv0290	Rv0291			
ESX-1	Rv3868	Rv3869	Rv3870, Rv3871		Rv3872 (pe35)	Rv3873 (ppe68)	Rv3874 (esxB)	Rv3875 (esxA)		Rv3877					RD1
ESX-4		Rv3450c	Rv3447c				Rv3445c (esxU)	Rv3444c (esxT)		Rv3448	Rv3449				

Figure 1. Genomic organization of the ESX operons in *M. tuberculosis* H37Rv, after Tekaia et al. (1999) and Gey van Pittius et al. (2001). The upper part provides information about the proteins encoded by the various genes and their corresponding partitions shown in abridged form. The figure is centered around the ESAT-6- and CFP-10-encoding genes *esxA* and *esxB* and is not to scale. Operon structures are not shown.

expression library with HYB 76-8 identified the gene encoding ESAT-6. The expected pI and mass of ESAT-6 from the deduced amino acid sequence are 4.5 and 9.8 kDa, respectively, and the determined N-terminal sequence corresponds to residues 2 to 11 in the deduced sequence (Sorensen et al., 1995). However, N-terminal truncation of ESAT-6 corresponding to removal of the first 16 residues has been noted (Weldingh and Rosenkrands, unpublished data) in agreement with observations that an N-terminal Hexahistidine tag was cleaved off from modified ESAT-6 when expressed by recombinant BCG (Brodin et al., 2005).

Several isoforms of purified ESAT-6 were observed by two-dimensional polyacrylamide gel electrophoresis, but there was no indication of glycosylation (Sorensen et al., 1995). Using a narrow-range pH 4 to 5 gradient, it is possible to resolve as many as eight ESAT-6 spots recognized by HYB 76-8 from culture filtrates. A detailed mass spectrometry analysis revealed that the four most acidic spots are acetylated at the N-terminal Thr2 residue, two ESAT-6 spots are cleaved at the C-terminus after Ala84 and two other spots have an unknown modification on peptide Thr75-Ala95 (Okkels et al., 2004). Although it was previously believed that N-acetylation of proteins was rather rare in prokaryotes, in a recent paper, it was shown that N-acetylation was present in three of the 13 tested proteins (23%) (Rison et al.,

2007), suggesting that in the mycobacteria this modification may be common, and an important regulator of protein function. As described below, N-acetylation of ESAT-6 might influence the binding capacities of ESAT-6 with protein partners, but the exact biological function of this modification in ESAT-6 remains for the moment unknown. *M. tuberculosis* H37Rv mutants in which Thr2 is replaced with His in the ESAT-6 protein do not affect the secretion of ESAT-6 and the growth rate in severely combined immune-deficient (SCID) mice, indicating that the acetylation of the Thr2 residue is not required for its normal function (Brodin et al., 2005).

CFP-10 is another important T-cell antigen present in culture filtrates and identified as a member of the ESX protein family. The genes encoding CFP-10 and ESAT-6 are adjacent to each other and were found to be cotranscribed (Berthet et al., 1998). Copurification of coexpressed recombinant ESAT-6 and CFP-10 suggested an interaction between the two proteins (Dillon et al., 2000), and Renshaw et al. (2002) showed that ESAT-6 and CFP-10 expressed in *Escherichia coli* form a tight ($K_d \leq 1.1 \times 10^{-8}$M), 1:1 heterodimeric complex (Fig. 2A). Brodin et al. (2005) confirmed the presence of this complex in culture filtrates and cytosol extracts from recombinant BCG expressing ESAT-6 and CFP-10, indicating that the complex is present shortly after synthesis as well as after secretion. Interestingly, only the nonacety-

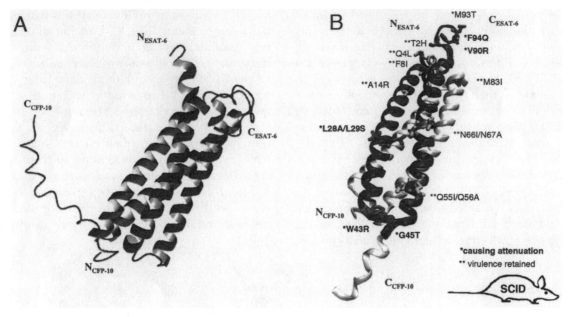

Figure 2. A ribbon representation of the backbone topology of the CFP-10·ESAT-6 complex based on the solution structure (Renshaw et al., 2005) (A) and the model of the ESAT-6 CFP-10 complex (Brodin et al., 2005) (B), showing experimentally introduced mutations into ESAT-6 mutations and their effect in the SCID mouse model.

lated forms of ESAT-6 are able to interact with recombinant CFP-10 in 2-DE blot overlay assays, and the acetylation of ESAT-6 could be a way to dissociate the formed complex in vivo (Okkels et al., 2004). Rv0287 and Rv0288 (TB10.4) are two other members of the ESAT-6 family that are cotranscribed, and the interaction of Rv0287 and Rv0288 was demonstrated by protein blot overlay assays (Okkels and Andersen, 2004). This interaction was also observed by yeast two hybrid analysis, and the interaction of Rv3019c and Rv3020c was shown by this methodology as well (Lightbody et al., 2004).

Using *M. smegmatis* as the host, a screening system termed mycobacterial protein fragment complementation (M-PFC) was developed to study protein-protein interactions in mycobacteria (Singh et al., 2006). Screening of an *M. tuberculosis* library for proteins that interact with CFP-10 identified several proteins including secretory components from the SecA/SecYEG pathways, ESAT-6 and Pks13 (Rv3800c). Two of these secretory components and Pks13 also associate with ESAT-6. Pks13 is a polyketide synthase, and it was hypothesized that the acetyltransferase domain of Pks13 could be responsible for acetylation of ESAT-6 and possibly also CFP-10 (Singh et al., 2006).

The solution structure of the CFP-10-ESAT-6 complex determined by NMR spectroscopy was recently reported by Renshaw and colleagues (2005). The complex shows a four-helix bundle structure, in which the individual proteins lie antiparallel to each other and display a hydrophobic contact surface. In both proteins, the N- and C-termini are disordered and form flexible arms at the ends of the core structure. The overall structure is in agreement with a model of the complex constructed by bioinformatics on the basis of mutations in ESAT-6 that interfered with induction of specific T-cell responses and virulence (Fig. 2) (Brodin et al., 2005).

The Secretion Machinery ESX-1

As described earlier, ESAT-6 and its protein partner CFP-10 are absent from the BCG vaccine due to deletion of the RD1 region (Mahairas et al., 1996). Interestingly, other attenuated members of the *M. tuberculosis* complex like *Mycobacterium microti* (Brodin et al., 2002) and the Dassie bacillus (Mostowy et al., 2004) also lack ESAT-6 and CFP-10, suggesting that this genomic region carries genes that are important for the virulence of *M. tuberculosis*. However, the first experiments intended to complement BCG with the 10 kb RD1 region from *M. bovis* resulted in ESAT-6 expression, but the recombinant BCG strain failed to show increased virulence in mice (Mahairas et al., 1996). It was only when BCG or *M. microti* was complemented with the integrating cosmid pRD1-2F9, which carries a 32-kb fragment of *M. tuberculosis* spanning RD1 together with large portions of its flanking regions, that an increased virulence phenotype was observed in SCID mice (Pym et al., 2002). The same recombinant BCG strain was found to persist to a greater degree in the organs of immunocompetent mice, but it induced considerably less pathology than *M. tuberculosis* H37Rv. However, after aerosol infection, pRD1-2F9 complemented BCG (BCG::RD1) multiplied more extensively in lungs and spleens than parental BCG and induced noticeable lesions in immunocompetent mice (Majlessi et al., 2005). By complementary approaches, deletion of RD1 from *M. tuberculosis* resulted in an attenuated phenotype in mice, for which a decreased bacterial load and fewer lesions were observed (Hsu et al., 2003; Lewis et al., 2003).

These data indicated that the proteins encoded in the extended RD1-region were involved in the pathogenesis of *M. tuberculosis* and related tubercle

bacilli. Experiments involving strains that only partially complemented the RD1 region showed that it was absolutely essential that ESAT-6 and CFP-10 were secreted to obtain an RD1 associated, more virulent phenotype, and these data led to the description of ESX-1 as a novel secretion system required for the export of ESAT-6 and CFP-10 (Guinn et al., 2004; Pym et al., 2003; Stanley et al., 2003) (Fig. 3). In contrast to other bacterial secretion systems implicated in pathogenesis such as bacterial type III and IV secretion systems, the ESX-1 system of *M. tuberculosis* is also functioning during in vitro growth of the bacterium and does not require host cell interaction for being expressed (Brodin et al., 2006).

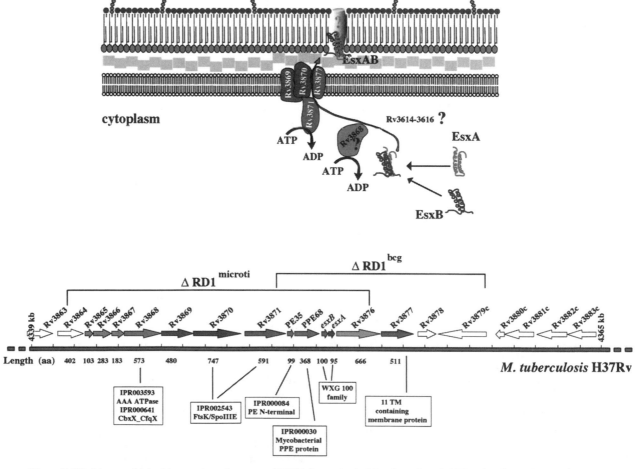

Figure 3. Working model for biogenesis and export of ESAT-6 proteins in *M. tuberculosis* H37Rv plus the positions of various genes and deletions (Brodin et al., 2004b). The upper part presents a possible functional model indicating predicted subcellular localization and known or potential protein-protein interactions. Rosetta stone analysis indicates direct interaction between proteins Rv3870 and Rv3871; Rv3868 is an AAA-ATPase that is likely to act as a chaperone. ESAT-6 and CFP-10 are predicted to be exported through a transmembrane channel, consisting of at least Rv3870, Rv3871, and Rv3877 and possibly Rv3869 in a process catalyzed by ATP hydrolysis. The lower part shows key genes, the various proteins from the RD1 region, their sizes (number of amino acid residues), and the protein families. *(See the color insert for the color version of this figure.)*

There has been major progress in identifying genes required for ESAT-6 and CFP-10 secretion. Inactivation or deletion of Rv3868-Rv3871 and Rv3877 abolished ESAT-6 and CFP-10 export, which indicates that they are directly involved in secretion (Brodin et al., 2006; Guinn et al., 2004; Pym et al., 2003; Stanley et al., 2003). Rv3868 and Rv3871 contain ATP binding sites; the former was predicted to function as an ATPase associated with various cellular activities (AAA). Rv3869, Rv3870, and Rv3877 are three putative membrane proteins with 1, 3, and 11 predicted transmembrane domains, respectively. Together with Rv3871, a cytosolic component of the ESX-1 system, they are thought to form a membrane-bound ESX-1 secretory complex, with protein export driven by ATP hydrolysis (Brodin et al., 2004b). In a recent study, a C-terminal signal sequence was identified that allowed the unstructured C terminus of the CFP-10 substrate being recognized by Rv3871, that itself interacts with the membrane protein Rv3870. Point mutations in the signal abolished binding of CFP-10 to Rv3871 and prevented secretion of the CFP-10/ESAT-6 complex (Champion et al., 2006). Because Rv3870 and Rv3871 both show similarity to proteins of the SpoIIIE/FtsK ATPase family (Tekaia et al., 1999), these proteins might perform an essential part of the work necessary to secrete ESX-1 substrates. This mechanism resembles type IV secretion systems in gram-negative bacteria, in which a membrane-bound SpoIIIE/FtsK-like ATPase recognizes an unstructured C-terminal signal sequence and directs the secreted substrate to the cytoplasmic membrane (Christie et al., 2005).

Apart from the RD1 region, a locus that is elsewhere encoded in the genome of *M. tuberculosis* has also been described to be implicated in the secretion of ESAT-6. Indeed, the proteins Rv3614c-Rv3616c, which show between 36 to 50% amino acid sequence similarity with proteins from the RD1 region Rv3867-Rv3864 have been shown to be linked to the secretion of ESAT-6 (Fortune et al., 2005; MacGurn et al., 2005). Mutant strains that had any of these three genes inactivated due to transposon insertions did not secrete ESAT-6 and CFP-10 in spite of a complete RD1 locus. Similarly, in a *M. tuberculosis* ESAT-6 knockout strain, no Rv3616 was detectable in the supernatant, arguing that the two proteins might be both exported by the ESX-1 system (Fortune et al., 2005).

Several effects have been described to be associated with the expression of ESX-1 proteins, in particular the two early secreted proteins ESAT-6 and CFP-10. As such, it was observed that tubercle bacilli expressing ESX-1 were engulfed much more efficiently by murine macrophages than ESX-1 lacking

bacilli (Brodin et al., 2006). Stanley and coworkers reported that murine macrophages infected with *M. tuberculosis* ESX-1 knockout mutants elicited substantially more IL-12 p40 than macrophages infected with wild-type bacilli, suggesting that ESX-1 proteins were involved in IL-12 suppression (Stanley et al., 2003). It was also reported that ESX-1 proteins were involved in cytotoxic or pore-forming activity (Derrick and Morris, 2007; Hsu et al., 2003) and type I interferon response (Stanley et al., 2007). As far as the individual contribution of different ESX-1 region is concerned, it is clear that ESAT-6 is one of the key players involved in the pathogenicity of *M. tuberculosis*. It was shown that the exchange of a single residue in ESAT-6 can result in attenuation of *M. tuberculosis* (Brodin et al., 2005). As regards genes PE35 and PPE68, both localized in the RD1 region of *M. tuberculosis* H37Rv, some indications on their function are now available. PE35, which is encoded directly upstream from PPE68, *esxB* (CFP-10), and *esxA* (ESAT-6), was shown to be involved in the expression of these genes. Indeed, two independently derived *M. tuberculosis* strains with transposon insertions in the PE35 gene did not express PPE68, ESAT-6, or CFP-10, and were both attenuated. In contrast, transposon mutagenesis or partial deletion of PPE68 did not have an attenuating effect in a mouse infection model. It was observed that for PPE68 knockout mutants more ESAT-6 was found in the culture supernatant and the strains were rather more pathogenic than the wild-type strain (Brodin et al., 2006; Demangel et al., 2004). As one of the possible explanations of this phenomenon, it was hypothesized that PPE68 eventually could act as a gating protein for the secretion apparatus. Absence or truncation of the protein could then result in an increased export of ESAT-6 and CFP-10. It should be mentioned that the PPE68 protein was never present in the culture supernatant, instead it was found to be associated to the membrane/cell wall fraction but not secreted (Brodin et al., 2006; Okkels et al., 2003). In contrast, PE35 was well identified in the culture filtrate of *M. tuberculosis* cultures (Fortune et al., 2005).

However, although there is good agreement in the literature that ESX-1 has a dramatic effect on pathogenicity of tubercle bacilli, at the time of writing, the exact biological role of ESX-1 proteins in the infection process remains largely uncertain.

ESX-1 in Other Mycobacterial Species

In the original paper on the identification of ESAT-6, Sorensen et al. (2005) have shown that apart from *M. tuberculosis*, several other mycobacterial species like *Mycobacterium kansasii*, *Mycobacterium szulgai*, and *Mycobacterium marinum* also contained proteins that reacted with the monoclonal ESAT-6 antibody HYB 76-8. A region syntenous with the RD1 region of *M. tuberculosis* also was later identified in the nonpathogenic *M. smegmatis* (Gey Van Pittius et al., 2001; Marmiesse et al., 2004), whose essential parts have been characterized by gene knockout strategies (Converse and Cox, 2005). In addition, the *M. smegmatis* RD1 proteins have been described as being part of a DNA-conjugation system (Flint et al., 2004), which suggest that pathogenic mycobacteria might have adapted an ancestral conjugation system for protein secretion involved in pathogenicity.

We know from the genome sequence of *Mycobacterium leprae* that the causing agent of leprosy harbors a highly similar region of the ESX-1 system (Cole et al., 2001). The fact that this bacterium, which has undergone such a tremendous reductive evolution, has retained a functional copy of ESX-1 greatly emphasizes the importance of this system for the bacterium, and it suggests that also this system is involved in host-pathogen interaction and needed for full virulence in *M. leprae* as well.

The ESX system has also been extensively studied in *M. marinum*, a slow-growing mycobacterium that is well known as a fish pathogen and that is sporadically isolated from water-borne skin infections in humans. *M. marinum* causes disease in zebra fish, and it is through this infection model that biological processes involved in pathogenicity have been best studied. Indeed, the bacterium is able to cause RD1-dependent macrophage aggregation as part of granuloma formation, and this ability has been tightly linked to intercellular bacterial dissemination (Volkman et al., 2004). Increased bacterial spreading of wild-type versus ESX-1 knockout mutants of *M. marinum* was earlier found by the group of Eric Brown (Gao et al., 2004), and this ability was linked to the extraordinary capacity of *M. marinum* to escape from the phagosomal compartment into the cytosol of an infected macrophage. However, preliminary inspection of the yet unpublished genome sequence shows that *M. marinum* contains two highly similar gene couples that encode ESAT-6 and CFP-10 proteins (http://genolist.pasteur.fr/MarinoList). One pair is localized inside the region orthologous to the RD1 region, whereas the second copy is found outside this region. It remains to be determined whether the second copy is functional. However, the higher copy number may have some impact on gene regulation and gene expression, eventually also explaining some of the observed ESX-1 related differences between *M. marinum* and the tubercle bacilli (Brodin et al., 2006).

M. marinum was also the host in which a close link between ESX-5, a system which in *M. tuberculosis* spans genes Rv1782-Rv1798 (Gey van Pittius et al., 2001), and PPE genes (PPE41; Rv2430c) was first demonstrated (Abdallah et al., 2006). PPE41, a hydrophilic secreted protein that does not carry a signal sequence, has been shown to induce a strong B-cell response in humans (Choudhary et al., 2003). PPE41 and PE25 (Rv2431c) are expressed from an operon (Tundup et al., 2006), and they form a strong heterodimeric protein complex, whose structure has recently been determined (Strong et al., 2006). *M. marinum*, which does not contain a orthologue of PPE41, produced and secreted PPE41 on introduction of a vector encoding PE25/PPE41 under control of an Hsp60 promoter, whereas complementation of *M. smegmatis* with the same construct resulted in expression but not in secretion of the PPE41 protein. The genes identified by the use of high-density Himar1 transposon mutagenesis as being involved in the secretion of PPE41 protein were located in region ESX-5. Further complementation experiments of the *M. smegmatis* PE25/PP41 containing construct with the ESX-5 genes resulted finally in secretion of the PPE41 protein, confirming the data from the *M. marinum* experiments. These experiments have paved the way for a more in depth analysis of the link between ESX and PE/PPE protein families. It may well be possible that ESX-5 and other ESX systems are also implicated in the export or secretion of other PE/PPE proteins.

Interaction of ESX Proteins with the Host

Binding to host proteins

The active secretion of ESAT-6 and CFP-10 and the surface features of the CFP-10-ESAT-6 complex suggested a function based on specific binding to host proteins rather than a pore-forming role of the complex to mediate cell lysis activity (Renshaw et al., 2005). Specific binding of the CFP-10-ESAT-6 complex to the surface of macrophage and monocyte cells but not to fibroblast cell lines was seen by fluorescence microscopy, suggesting that the host target protein is found on the surface of the cells. However, because it is secreted by an intracellular pathogen, it is not completely clear how the complex is translocated to the surface of the host cell. Truncated CFP-10 (residues 1-84) and full-length ESAT-6 form a complex, but the flexible C-terminal arm of CFP-10 is missing. In U937 cells, binding of the truncated CFP-10-ESAT-6 complex is reduced compared with the intact complex, indicating a role for the C-terminus of CFP-10 in binding to the host target protein (Renshaw et al., 2005). However, in a recent study, the C-terminal tail of CFP-10 was described as bearing a specific C-terminal signal sequence that is needed for interaction with membrane protein Rv3871 of the ESX-1 secretion machinery (Champion et al., 2006) and might therefore play a more specific role for the export of the CFP-10-ESAT-6 complex than for the interaction with host proteins. Furthermore, attachment of the signal to yeast ubiquitin was sufficient for secretion from M. tuberculosis cells, demonstrating that this ESX-1 signal is portable.

In another observation, by yeast two-hybrid analysis ESAT-6 was found to interact with rat syntenin-1, and the interaction was confirmed in vitro by protein overlay and by surface plasmon resonance analysis. These data suggest that syntenin could play a role as a possible cellular interaction partner of ESAT-6 (Schumann et al., 2006).

Immunological recognition

The initial encouraging results in mice, which identified ESAT-6 as a major T-cell antigen, led to a more detailed investigation of the immunological potential of ESAT-6 in different animal models and in humans. In aerosol challenge experiments in mice with M. tuberculosis ESAT-6 induced a protective immune response both as a DNA vaccine and a subunit vaccine administered in the adjuvant dimethyl dioctadecylammonium bromide and monophosphoryl lipid A (Brandt et al., 2000; Kamath et al., 1999). Protection induced by ESAT-6 has also been observed with other delivery systems, such as an attenuated Salmonella typhimurium strain, which secretes ESAT-6 and intranasal immunization with an attenuated influenza virus expressing ESAT-6 (Mollenkopf et al., 2001; Stukova et al., 2006). Because it is introduced through the aerosol route, ESAT-6 expressing BCG::RD1 has the capacity to induce antimycobacterial innate and adaptive immunity comparable to that of virulent M. tuberculosis H37Rv. Indeed, BCG::RD1, but not parental BCG, triggered substantial influx of activated/effector T cells and dendritic cells into lungs and initiated a proinflammatory program in lung dendritic cells (Majlessi et al., 2005). In patients with tuberculosis, the T-cell recognition of ESAT-6 has been studied intensively, and both in high and low endemic regions, a very high frequency of donors were reported to recognize this antigen (Arend et al., 2000; Lalvani et al., 2001; Mori et al., 2004; Mustafa et al., 1998; Ravn et al., 1999; van Pinxteren et al., 2000). The frequent and strong recognition is strongly indicative of high levels of this molecule being available during infection. In support of this, aerosol infection of mice induce a higher number of specific CD4 T-cells against ESAT-6 in the lung compared with Ag85B, another secreted T-cell

antigen, and this was associated with a higher transcription level of esat-6 (Rogerson et al., 2006). Furthermore, ESAT-6 is blessed by the presence of numerous human T-cell epitopes that can be presented to T cells in association with multiple HLA-DR molecules (Mustafa et al., 2000; Mustafa et al., 2003; Ravn et al., 1999). This characteristically strong antigenicity makes ESAT-6 a very important antigen for tuberculosis vaccines but has also been used as the basis for IFN-γ based tuberculosis diagnosis (Pai, 2005). ESAT-6 was also identified as a target for B-cell responses in a variety of species, such as humans, elephants, nonhuman primates, and cattle (Brusasca et al., 2003; Lyashchenko et al., 1998a; Lyashchenko et al., 1998b; Lyashchenko et al., 2006). CFP-10, encoded by the gene adjacent to ESAT-6, appeared to possess the same immunodominant properties as ESAT-6 and was found as a strong B- and T-cell target in mice, guinea pigs, cattle and humans (Arend et al., 2000; Colangeli et al., 2000; Dillon et al., 2000; Skjøt et al., 2000; van Pinxteren et al., 2000).

Therefore, the antigenic qualities of ESAT-6 seem in fact to be a phenomenon shared by a large number of the other ESAT-6 family members as discussed elsewhere (Brodin et al., 2004a). Rv0288 is strongly recognized by peripheral blood mononuclear cells from human patients with tuberculosis as well as BCG-vaccinated donors (Skjøt et al., 2000). Rv3017c and Rv3019c are two ESAT-6 proteins with high amino acid sequence similarity to Rv0288 (74 and 84%, respectively), which were also identified as human T-cell antigens, but despite the sequence similarity, unique epitopes were identified in all three proteins (Skjøt et al., 2002). In cattle, subunit vaccination with Rv3019c was found to induce significant protection against M. bovis challenge (Hogarth et al., 2005). The five members of the Mtb9.9 subfamily show an even higher degree of amino acid sequence similarity (≥93%), and were all identified as T-cell antigens that induce IFN-γ in peripheral blood mononuclear cells from PPD+ human donors. Interestingly, heterogeneity in the T-cell responses to each of the antigens were observed (Alderson et al., 2000), and it could be a result of antigen variation for the several duplicated genes in the ESAT-6 family if the small variations in highly immunogenic molecules lead to different epitope patterns. To be an advantage for the pathogen and allow immune evasion, an expression switching mechanism is then a prerequisite. However, comparative studies of the expression of the individual genes in vivo to investigate this hypothesis have not yet been reported. Until now, 10 of the 23 members of the ESAT-6 family proteins have been identified as T-cell antigens, but whether the remaining members of the family are also B- or T-cell antigens also remains to be explored.

Apart from ESAT-6 (Rv3875) and CFP-10 (Rv3874), several other T-cell antigens have been identified in the RD1 region: Rv3871, Rv3872, Rv3873, Rv3878 and Rv3879c (Agger et al., 2003; Cockle et al., 2002; Demangel et al., 2004; Mustafa et al., 2002; Okkels et al., 2003). The frequent occurrence of T-cell antigens in these regions is striking and has led to the hypothesis of antigenic gene clusters or immunogenicity islands on the *M. tuberculosis* genome (Brodin et al., 2004b; Okkels et al., 2003), possibly a result of simultaneous presentation of these proteins to the immune system caused by coordinated expression and protein interactions.

Three other immunogenicity islands related to ESX family proteins were also proposed (Okkels et al., 2003): Rv0286 and Rv0288 from ESX-3, Rv3017c, Rv3018c, Rv3019c and 3021c from the highly homologous region suggested to be duplicated from ESX-3 (Gey van Pittius et al., 2006), and Rv1196 and Rv1198 from one of the regions suggested to be duplicated from ESX-5 (Gey van Pittius et al., 2006).

The many recently published studies on ESAT-6 structure and secretion have also provided new insights on the immunological aspects of ESAT-6. However, expression of ESAT-6 by the mycobacterium is not enough to induce a T-cell response, secretion of the protein is needed to obtain immune recognition in mice (Brodin et al., 2005; Brodin et al., 2006). Notably, investigation of a BCG mutant expressing truncated ESAT-6 lacking 12 C-terminal amino acid residues has shown that the mutant strain was able to secrete the truncated ESAT-6, but unable to induce a T-cell response to the ESAT-6:1-20 peptide, suggesting that host cell interaction with the C-terminus of ESAT-6 may be required to obtain a specific immune response (Brodin et al., 2006).

Because ESAT-6 and CFP-10 exist as 1:1 complex in mycobacteria after secretion of the proteins, the immunological recognition of the complex would be relevant to study. In vitro digestion with proteolytic enzymes demonstrated that the complex was more resistant to processing than ESAT-6 and CFP-10 alone, and in patients with tuberculosis, the T-cell responses after restimulation with the complex were lower compared with the ESAT-6-stimulated cells (Marei et al., 2005). This suggests that in vivo, the complex may be less susceptible for antigen processing by the host immune system or, alternatively, that the complex may eventually dissociate at some stage (de Jonge et al., 2007).

Most of the studies described earlier deal with CD4 T-cell recognition. However, CD8 epitopes have also been observed. In Rv0288 and Rv3019c the GYAGTLQSL epitope was identified by cytotoxic T-cells from mycobacterium-infected mice (Majlessi et al., 2003), and after aerosol infection of C3H mice with *M. tuberculosis,* the CD8 epitope VESTAGSL was identified in CFP-10 (Kamath et al., 2004). CD8 epitopes were also identified in ESAT-6, both in human TB patients (Pathan et al., 2000; Smith et al., 2000) and in mice after immunization with an adenovirus vector expressing a fusion protein of Ag85B and ESAT-6. However, the induced CD8 response was not protective against a primary tuberculosis infection in mice (Bennekov et al., 2006).

The strong immunological recognition of ESAT-6 has stimulated research and development into a range of applications of ESAT-6 for diagnostic and vaccine use. Tests detecting the release of IFN-γ or the presence of IFN-γ-producing cells on contact with ESAT-6 and CFP-10 have been developed and commercialized (QuantiFERON-TB Gold, manufactured by Cellestis Limited, Carnegie, Victoria, Australia, or T-SPOT-TB, manufactured by Oxford Immunotec, Oxford, United Kingdom). The absence of ESAT-6 and CFP-10 in BCG allows discrimination between patients with tuberculosis and BCG-vaccinated individuals. Furthermore, a correlation between ESAT-6 specific IFN-γ levels and the risk of developing clinical TB postexposure has been observed in several studies (reviewed by Andersen et al., in press), suggesting ESAT-6 as a prognostic marker for development of active tuberculosis. Finally, ESAT-6 has demonstrated a potential as skin test reagent (Pollock et al., 2003; van Pinxteren et al., 2000) and as a serodiagnostic antigen in human patients with tuberculosis (Brusasca et al., 2001; Lyashchenko et al., 1998a).

The ESAT-6 family proteins are also found to be important in the development of vaccines against tuberculosis. One approach is the recombinant BCG and *M. microti* vaccines that express ESX-1 proteins and that leads to induction of specific immune responses against ESAT-6 and enhanced protection against tuberculosis challenge compared with vaccination with BCG (Brodin et al., 2004a; Pym et al., 2003). A subunit vaccine approach based on a fusion protein between Ag85B and ESAT-6 was found to induce significant protection against aerosol challenge with *M. tuberculosis* in mice, guinea pigs, and nonhuman primates (Langermans et al., 2005; Olsen et al., 2000; Olsen et al., 2004) and is currently being evaluated in a clinical trial. Another efficient subunit vaccine was obtained after replacing ESAT-6 with TB10.4 (Rv0288), and thereby ESAT-6 could be used for diagnostic purposes only and continue discrimination between infected and vaccinated individuals in the future (Dietrich et al., 2005).

CONCLUDING REMARKS

M. tuberculosis is certainly one of the most successful and best-adapted bacterial pathogens of humans. The infectious cycle of *M. tuberculosis* ranges from lifelong asymptomatic parasitism to the rapid development of tuberculous cavities in the lungs of patients that have become susceptible to develop the disease. This cycle allows the efficient spread of high numbers of bacteria through aerosol to new hosts. From the perspective of the bacterium, this cycle is absolutely essential to maintain its so successful life cycle, and it seems likely that this adaptation occurred during a long-lasting coevolution between the bacterium and its host. In agreement with this hypothesis, it was recently shown that *Mycobacterium canettii* and other tubercle bacilli with smooth colony morphology show extensive nucleotide polymorphisms and may represent extant members of the progenitor species of the *M. tuberculosis* complex (Gutierrez et al., 2005). Because proteins from the ESX family are strongly involved in host-pathogen interaction, it will be very interesting to investigate the repertoire of ESX proteins in these smooth tubercle bacilli. Such a comparative analysis has the potential to gain important insights into evolution of ESX systems in relation to host adaptation. Together with continued efforts to functionally characterize ESX systems in *M. tuberculosis*, this information should allow us to better understand the role of ESX systems in mycobacteria and design new approaches to fight the disease they cause.

Acknowledgments. We are grateful to S. T. Cole, M. I. de Jonge, C. Demangel, and T. Garnier for advice and fruitful discussions.

This work received support of the Institut Pasteur, the European Union (LHSP-CT-2005-018923), and the Association Française Raoul Follereau.

REFERENCES

Abdallah, A. M., T. Verboom, F. Hannes, M. Safi, M. Strong, D. Eisenberg, R. J. Musters, C. M. Vandenbroucke-Grauls, B. J. Appelmelk, J. Luirink, and W. Bitter. 2006. A specific secretion system mediates PPE41 transport in pathogenic mycobacteria. *Mol. Microbiol.* **62**:667–679.

Agger, E. M., I. Brock, L. M. Okkels, S. M. Arend, C. S. Aagaard, K. N. Weldingh, and P. Andersen. 2003. Human T-cell responses to the RD1-encoded protein TB27.4 (Rv3878) from *Mycobacterium tuberculosis*. *Immunology* **110**:507–512.

Alderson, M. R., T. Bement, C. H. Day, L. Zhu, D. Molesh, Y. A. Skeiky, R. Coler, D. M. Lewinsohn, S. G. Reed, and D. C. Dillon. 2000. Expression cloning of an immunodominant family of *Mycobacterium tuberculosis* antigens using human CD4(+) T cells. *J. Exp. Med.* **191**:551–560.

Andersen, P. 1994. The T cell response to secreted antigens of *Mycobacterium tuberculosis*. *Immunobiology* **191**:537–547.

Andersen, P., A. B. Andersen, A. L. Sorensen, and S. Nagai. 1995. Recall of long-lived immunity to *Mycobacterium tuberculosis* infection in mice. *J. Immunol.* **154**:3359–3372.

Andersen, P., T. M. Doherty, M. Pai, and K. Weldingh. 2007. The prognosis of latent tuberculosis: can disease be predicted? *Trends Mol. Med.* **13**:175–182.

Arend, S. M., P. Andersen, K. E. van Meijgaarden, R. L. Skjøt, Y. W. Subronto, J. T. van Dissel, and T. H. Ottenhoff. 2000. Detection of active tuberculosis infection by T cell responses to early-secreted antigenic target 6-kDa protein and culture filtrate protein 10. *J. Infect. Dis.* **181**:1850–1854.

Bennekov, T., J. Dietrich, I. Rosenkrands, A. Stryhn, T. M. Doherty, and P. Andersen. 2006. Alteration of epitope recognition pattern in Ag85B and ESAT-6 has a profound influence on vaccine-induced protection against *Mycobacterium tuberculosis*. *Eur. J. Immunol.* **36**:3346–3355.

Berthet, F. X., P. B. Rasmussen, I. Rosenkrands, P. Andersen, and B. Gicquel. 1998. A *Mycobacterium tuberculosis* operon encoding ESAT-6 and a novel low-molecular-mass culture filtrate protein (CFP-10). *Microbiology* **144**:3195–3203.

Brandt, L., M. Elhay, I. Rosenkrands, E. B. Lindblad, and P. Andersen. 2000. ESAT-6 subunit vaccination against *Mycobacterium tuberculosis*. *Infect. Immun.* **68**:791–795.

Brandt, L., T. Oettinger, A. Holm, A. B. Andersen, and P. Andersen. 1996. Key epitopes on the ESAT-6 antigen recognized in mice during the recall of protective immunity to *Mycobacterium tuberculosis*. *J. Immunol.* **157**:3527–3533.

Brodin, P., M. I. de Jonge, L. Majlessi, C. Leclerc, M. Nilges, S. T. Cole, and R. Brosch. 2005. Functional analysis of early secreted antigenic target-6, the dominant T-cell antigen of *Mycobacterium tuberculosis*, reveals key residues involved in secretion, complex formation, virulence, and immunogenicity. *J. Biol. Chem.* **280**:33953–33959.

Brodin, P., K. Eiglmeier, M. Marmiesse, A. Billault, T. Garnier, S. Niemann, S. T. Cole, and R. Brosch. 2002. Bacterial artificial chromosome-based comparative genomic analysis identifies *Mycobacterium microti* as a natural ESAT-6 deletion mutant. *Infect. Immun.* **70**:5568–5578.

Brodin, P., L. Majlessi, R. Brosch, D. Smith, G. Bancroft, S. Clark, A. Williams, C. Leclerc, and S. T. Cole. 2004a. Enhanced protection against tuberculosis by vaccination with recombinant *Mycobacterium microti* vaccine that induces T cell immunity against region of difference 1 antigens. *J. Infect. Dis.* **190**:115–122.

Brodin, P., L. Majlessi, L. Marsollier, M. I. de Jonge, D. Bottai, C. Demangel, J. Hinds, O. Neyrolles, P. D. Butcher, C. Leclerc, S. T. Cole, and R. Brosch. 2006. Dissection of ESAT-6 system 1 of *Mycobacterium tuberculosis* and impact on immunogenicity and virulence. *Infect. Immun.* **74**:88–98.

Brodin, P., I. Rosenkrands, P. Andersen, S. T. Cole, and R. Brosch. 2004b. ESAT-6 proteins: protective antigens and virulence factors? *Trends Microbiol.* **12**:500–508.

Brusasca, P. N., R. Colangeli, K. P. Lyashchenko, X. Zhao, M. Vogelstein, J. S. Spencer, D. N. McMurray, and M. L. Gennaro. 2001. Immunological characterization of antigens encoded by the RD1 region of the *Mycobacterium tuberculosis* genome. *Scand. J. Immunol.* **54**:448–452.

Brusasca, P. N., R. L. Peters, S. L. Motzel, H. J. Klein, and M. L. Gennaro. 2003. Antigen recognition by serum antibodies in non-human primates experimentally infected with *Mycobacterium tuberculosis*. *Comp. Med.* **53**:165–172.

Champion, P. A., S. A. Stanley, M. M. Champion, E. J. Brown, and J. S. Cox. 2006. C-terminal signal sequence promotes virulence factor secretion in *Mycobacterium tuberculosis*. *Science* **313**:1632–1636.

Choudhary, R. K., S. Mukhopadhyay, P. Chakhaiyar, N. Sharma, K. J. Murthy, V. M. Katoch, and S. E. Hasnain. 2003. PPE antigen Rv2430c of *Mycobacterium tuberculosis* induces a strong B-cell response. *Infect. Immun.* **71**:6338–6343.

Christie, P. J., K. Atmakuri, V. Krishnamoorthy, S. Jakubowski, and E. Cascales. 2005. Biogenesis, architecture, and function of bacterial type IV secretion systems. *Annu. Rev. Microbiol.* **59:** 451–485.

Cockle, P. J., S. V. Gordon, A. Lalvani, B. M. Buddle, R. G. Hewinson, and H. M. Vordermeier. 2002. Identification of novel *Mycobacterium tuberculosis* antigens with potential as diagnostic reagents or subunit vaccine candidates by comparative genomics. *Infect. Immun.* **70:** 6996–7003.

Colangeli, R., J. S. Spencer, P. Bifani, A. Williams, K. Lyashchenko, M. A. Keen, P. J. Hill, J. Belisle, and M. L. Gennaro. 2000. MTSA-10, the product of the Rv3874 gene of *Mycobacterium tuberculosis*, elicits tuberculosis-specific, delayed-type hypersensitivity in guinea pigs. *Infect. Immun.* **68:** 990–993.

Cole, S. T., R. Brosch, J. Parkhill, T. Garnier, C. Churcher, D. Harris, S. V. Gordon, K. Eiglmeier, S. Gas, C. E. Barry, 3rd, F. Tekaia, K. Badcock, D. Basham, D. Brown, T. Chillingworth, R. Connor, R. Davies, K. Devlin, T. Feltwell, S. Gentles, N. Hamlin, S. Holroyd, T. Hornsby, K. Jagels, A. Krogh, J. McLean, S. Moule, L. Murphy, K. Oliver, J. Osborne, M. A. Quail, M. A. Rajandream, J. Rogers, S. Rutter, K. Seeger, J. Skelton, R. Squares, S. Squares, J. E. Sulston, K. Taylor, S. Whitehead, and B. G. Barrell. 1998. Deciphering the biology of *Mycobacterium tuberculosis* from the complete genome sequence. *Nature* **393:** 537–544.

Cole, S. T., K. Eiglmeier, J. Parkhill, K. D. James, N. R. Thomson, P. R. Wheeler, N. Honore, T. Garnier, C. Churcher, D. Harris, K. Mungall, D. Basham, D. Brown, T. Chillingworth, R. Connor, R. M. Davies, K. Devlin, S. Duthoy, T. Feltwell, A. Fraser, N. Hamlin, S. Holroyd, T. Hornsby, K. Jagels, C. Lacroix, J. Maclean, S. Moule, L. Murphy, K. Oliver, M. A. Quail, M. A. Rajandream, K. M. Rutherford, S. Rutter, K. Seeger, S. Simon, M. Simmonds, J. Skelton, R. Squares, S. Squares, K. Stevens, K. Taylor, S. Whitehead, J. R. Woodward, and B. G. Barrell. 2001. Massive gene decay in the leprosy bacillus. *Nature* **409:** 1007–1011.

Converse, S. E., and J. S. Cox. 2005. A protein secretion pathway critical for *Mycobacterium tuberculosis* virulence is conserved and functional in *Mycobacterium smegmatis. J. Bacteriol.* **187:** 1238–1245.

de Jonge, M. I., G. Pehau-Arnaudet, M. M. Fretz, F. Romain, D. Bottai, P. Brodin, N. Honoré, G. Marchal, W. Jiskoot, P. England, S. T. Cole, and R. Brosch. 2007. ESAT-6 from *Mycobacterium tuberculosis* dissociates from its putative chaperone CFP-10 under acidic conditions and exhibits membrane-lysing activity. *J. Bacteriol.* **189:** 6028–6034.

Demangel, C., P. Brodin, P. J. Cockle, R. Brosch, L. Majlessi, C. Leclerc, and S. T. Cole. 2004. Cell envelope protein PPE68 contributes to *Mycobacterium tuberculosis* RD1 immunogenicity independently of a 10-kilodalton culture filtrate protein and ESAT-6. *Infect. Immun.* **72:** 2170–2176.

Derrick, S. C., and S. L. Morris. 2007. The ESAT6 protein of *Mycobacterium tuberculosis* induces apoptosis of macrophages by activating caspase expression. *Cell. Microbiol.* **9:** 1547–1555.

Dietrich, J., C. Aagaard, R. Leah, A. W. Olsen, A. Stryhn, T. M. Doherty, and P. Andersen. 2005. Exchanging ESAT6 with TB10.4 in an Ag85B fusion molecule-based tuberculosis subunit vaccine: efficient protection and ESAT6-based sensitive monitoring of vaccine efficacy. *J. Immunol.* **174:** 6332–6339.

Dillon, D. C., M. R. Alderson, C. H. Day, T. Bement, A. Campos-Neto, Y. A. Skeiky, T. Vedvick, R. Badaro, S. G. Reed, and R. Houghton. 2000. Molecular and immunological characterization of *Mycobacterium tuberculosis* CFP-10, an immunodiagnostic antigen missing in *Mycobacterium bovis* BCG. *J. Clin. Microbiol.* **38:** 3285–3290.

Elhay, M. J., T. Oettinger, and P. Andersen. 1998. Delayed-type hypersensitivity responses to ESAT-6 and MPT64 from *Mycobacterium tuberculosis* in the guinea pig. *Infect. Immun.* **66:** 3454–3456.

Flint, J. L., J. C. Kowalski, P. K. Karnati, and K. M. Derbyshire. 2004. The RD1 virulence locus of *Mycobacterium tuberculosis* regulates DNA transfer in *Mycobacterium smegmatis. Proc. Natl. Acad. Sci. USA* **101:** 12598–12603.

Fortune, S. M., A. Jaeger, D. A. Sarracino, M. R. Chase, C. M. Sassetti, D. R. Sherman, B. R. Bloom, and E. J. Rubin. 2005. Mutually dependent secretion of proteins required for mycobacterial virulence. *Proc. Natl. Acad. Sci. USA* **102:** 10676–10681.

Gao, L. Y., S. Guo, B. McLaughlin, H. Morisaki, J. N. Engel, and E. J. Brown. 2004. A mycobacterial virulence gene cluster extending RD1 is required for cytolysis, bacterial spreading and ESAT-6 secretion. *Mol. Microbiol.* **53:** 1677–1693.

Gey Van Pittius, N. C., J. Gamieldien, W. Hide, G. D. Brown, R. J. Siezen, and A. D. Beyers. 2001. The ESAT-6 gene cluster of *Mycobacterium tuberculosis* and other high G+C Gram-positive bacteria. *Genome Biol.* **2:** RESEARCH0044.

Gey van Pittius, N. C., S. L. Sampson, H. Lee, Y. Kim, P. D. van Helden, and R. M. Warren. 2006. Evolution and expansion of the *Mycobacterium tuberculosis* PE and PPE multigene families and their association with the duplication of the ESAT-6 (esx) gene cluster regions. *BMC Evol. Biol.* **6:** 95.

Guinn, K. I., M. J. Hickey, S. K. Mathur, K. L. Zakel, J. E. Grotzke, D. M. Lewinsohn, S. Smith, and D. R. Sherman. 2004. Individual RD1-region genes are required for export of ESAT-6/CFP-10 and for virulence of *Mycobacterium tuberculosis. Mol. Microbiol.* **51:** 359–370.

Gutierrez, M. C., S. Brisse, R. Brosch, M. Fabre, B. Omais, M. Marmiesse, P. Supply, and V. Vincent. 2005. Ancient origin and gene mosaicism of the progenitor of *Mycobacterium tuberculosis. PLoS Pathog.* **1:** e5.

Harboe, M., T. Oettinger, H. G. Wiker, I. Rosenkrands, and P. Andersen. 1996. Evidence for occurrence of the ESAT-6 protein in *Mycobacterium tuberculosis* and virulent *Mycobacterium bovis* and for its absence in *Mycobacterium bovis* BCG. *Infect. Immun.* **64:** 16–22.

Hogarth, P. J., K. E. Logan, H. M. Vordermeier, M. Singh, R. G. Hewinson, and M. A. Chambers. 2005. Protective immunity against *Mycobacterium bovis* induced by vaccination with Rv3109c—a member of the esat-6 gene family. *Vaccine* **23:** 2557–2564.

Hsu, T., S. M. Hingley-Wilson, B. Chen, M. Chen, A. Z. Dai, P. M. Morin, C. B. Marks, J. Padiyar, C. Goulding, M. Gingery, D. Eisenberg, R. G. Russell, S. C. Derrick, F. M. Collins, S. L. Morris, C. H. King, and W. R. Jacobs, Jr. 2003. The primary mechanism of attenuation of bacillus Calmette-Guerin is a loss of secreted lytic function required for invasion of lung interstitial tissue. *Proc. Natl. Acad. Sci. USA* **100:** 12420–12425.

Kamath, A. B., J. Woodworth, X. Xiong, C. Taylor, Y. Weng, and S. M. Behar. 2004. Cytolytic CD8+ T cells recognizing CFP10 are recruited to the lung after *Mycobacterium tuberculosis* infection. *J. Exp. Med.* **200:** 1479–1489.

Kamath, A. T., C. G. Feng, M. Macdonald, H. Briscoe, and W. J. Britton. 1999. Differential protective efficacy of DNA vaccines expressing secreted proteins of *Mycobacterium tuberculosis. Infect. Immun.* **67:** 1702–1707.

Lalvani, A., A. A. Pathan, H. McShane, R. J. Wilkinson, M. Latif, C. P. Conlon, G. Pasvol, and A. V. Hill. 2001. Rapid detection of *Mycobacterium tuberculosis* infection by enumeration of antigen-specific T cells. *Am. J. Respir. Crit. Care Med.* **163:** 824–828.

Langermans, J. A., T. M. Doherty, R. A. Vervenne, T. van der Laan, K. Lyashchenko, R. Greenwald, E. M. Agger, C. Aagaard,

H. Weiler, D. van Soolingen, W. Dalemans, A. W. Thomas, and P. Andersen. 2005. Protection of macaques against *Mycobacterium tuberculosis* infection by a subunit vaccine based on a fusion protein of antigen 85B and ESAT-6. *Vaccine* 23:2740–2750.

Lewis, K. N., R. Liao, K. M. Guinn, M. J. Hickey, S. Smith, M. A. Behr, and D. R. Sherman. 2003. Deletion of RD1 from *Mycobacterium tuberculosis* mimics bacille Calmette-Guerin attenuation. *J. Infect. Dis.* 187:117–123.

Lightbody, K. L., P. S. Renshaw, M. L. Collins, R. L. Wright, D. M. Hunt, S. V. Gordon, R. G. Hewinson, R. S. Buxton, R. A. Williamson, and M. D. Carr. 2004. Characterisation of complex formation between members of the *Mycobacterium tuberculosis* complex CFP-10/ESAT-6 protein family. Towards an understanding of the rules governing complex formation and thereby functional flexibility. *FEMS Microbiol. Lett.* 238:255–262.

Lyashchenko, K., R. Colangeli, M. Houde, H. Al Jahdali, D. Menzies, and M. L. Gennaro. 1998a. Heterogeneous antibody responses in tuberculosis. *Infect. Immun.* 66:3936–3940.

Lyashchenko, K. P., R. Greenwald, J. Esfandiari, J. H. Olsen, R. Ball, G. Dumonceaux, F. Dunker, C. Buckley, M. Richard, S. Murray, J. B. Payeur, P. Andersen, J. M. Pollock, S. Mikota, M. Miller, D. Sofranko, and W. R. Waters. 2006. Tuberculosis in elephants: antibody responses to defined antigens of *Mycobacterium tuberculosis*, potential for early diagnosis, and monitoring of treatment. *Clin. Vaccine Immunol.* 13:722–732.

Lyashchenko, K. P., J. M. Pollock, R. Colangeli, and M. L. Gennaro. 1998b. Diversity of antigen recognition by serum antibodies in experimental bovine tuberculosis. *Infect. Immun.* 66:5344–5349.

MacGurn, J. A., S. Raghavan, S. A. Stanley, and J. S. Cox. 2005. A non-RD1 gene cluster is required for Snm secretion in *Mycobacterium tuberculosis*. *Mol. Microbiol.* 57:1653–1663.

Mahairas, G. G., P. J. Sabo, M. J. Hickey, D. C. Singh, and C. K. Stover. 1996. Molecular analysis of genetic differences between *Mycobacterium bovis* BCG and virulent *M. bovis*. *J. Bacteriol.* 178:1274–1282.

Majlessi, L., P. Brodin, R. Brosch, M. J. Rojas, H. Khun, M. Huerre, S. T. Cole, and C. Leclerc. 2005. Influence of ESAT-6 Secretion System 1 (RD1) of *Mycobacterium tuberculosis* on the interaction between mycobacteria and the host immune system. *J. Immunol.* 174:3570–3579.

Majlessi, L., M. J. Rojas, P. Brodin, and C. Leclerc. 2003. CD8+-T-cell responses of *Mycobacterium*-infected mice to a newly identified major histocompatibility complex class I-restricted epitope shared by proteins of the ESAT-6 family. *Infect. Immun.* 71:7173–7177.

Marei, A., A. Ghaemmaghami, P. Renshaw, M. Wiselka, M. Barer, M. Carr, and L. Ziegler-Heitbrock. 2005. Superior T cell activation by ESAT-6 as compared with the ESAT-6-CFP-10 complex. *Int. Immunol.* 17:1439–1446.

Marmiesse, M., P. Brodin, C. Buchrieser, C. Gutierrez, N. Simoes, V. Vincent, P. Glaser, S. T. Cole, and R. Brosch. 2004. Macroarray and bioinformatic analyses reveal mycobacterial 'ore'-genes, variation in the ESAT-6 gene family and new phylogenetic markers for the *Mycobacterium tuberculosis* complex. *Microbiology* 150:483–496.

Mollenkopf, H. J., D. Groine-Triebkorn, P. Andersen, J. Hess, and S. H. Kaufmann. 2001. Protective efficacy against tuberculosis of ESAT-6 secreted by a live *Salmonella typhimurium* vaccine carrier strain and expressed by naked DNA. *Vaccine* 19:4028–4035.

Mori, T., M. Sakatani, F. Yamagishi, T. Takashima, Y. Kawabe, K. Nagao, E. Shigeto, N. Harada, S. Mitarai, M. Okada, K. Suzuki, Y. Inoue, K. Tsuyuguchi, Y. Sasaki, G. H. Mazurek, and I. Tsuyuguchi. 2004. Specific detection of tuberculosis infection with an interferon-gamma based assay using new antigens. *Am. J. Respir. Crit. Care Med.* 70:59–64.

Mostowy, S., D. Cousins, and M. A. Behr. 2004. Genomic interrogation of the dassie bacillus reveals it as a unique RD1 mutant within the *Mycobacterium tuberculosis* complex. *J. Bacteriol.* 186:104–109.

Mustafa, A. S., H. A. Amoudy, H. G. Wiker, A. T. Abal, P. Ravn, F. Oftung, and P. Andersen. 1998. Comparison of antigen-specific T-cell responses of tuberculosis patients using complex or single antigens of *Mycobacterium tuberculosis*. *Scand. J. Immunol.* 48:535–543.

Mustafa, A. S., P. J. Cockle, F. Shaban, R. G. Hewinson, and H. M. Vordermeier. 2002. Immunogenicity of *Mycobacterium tuberculosis* RD1 region gene products in infected cattle. *Clin. Exp. Immunol.* 130:37–42.

Mustafa, A. S., F. Oftung, H. A. Amoudy, N. M. Madi, A. T. Abal, F. Shaban, I. Rosenkrands, and P. Andersen. 2000. Multiple epitopes from the *Mycobacterium tuberculosis* ESAT-6 antigen are recognized by antigen-specific human T cell lines. *Clin. Infect. Dis.* 30:S201–S205.

Mustafa, A. S., F. A. Shaban, R. Al-Attiyah, A. T. Abal, A. M. El-Shamy, P. Andersen, and F. Oftung. 2003. Human Th1 cell lines recognize the *Mycobacterium tuberculosis* ESAT-6 antigen and its peptides in association with frequently expressed HLA class II molecules. *Scand. J. Immunol.* 57:125–134.

Okkels, L. M., and P. Andersen. 2004. Protein-protein interactions of proteins from the ESAT-6 family of *Mycobacterium tuberculosis*. *J. Bacteriol.* 186:2487–2491.

Okkels, L. M., I. Brock, F. Follmann, E. M. Agger, S. M. Arend, T. H. Ottenhoff, F. Oftung, I. Rosenkrands, and P. Andersen. 2003. PPE protein (Rv3873) from DNA segment RD1 of *Mycobacterium tuberculosis*: strong recognition of both specific T-cell epitopes and epitopes conserved within the PPE family. *Infect. Immun.* 71:6116–6123.

Okkels, L. M., E. C. Muller, M. Schmid, I. Rosenkrands, S. H. Kaufmann, P. Andersen, and P. R. Jungblut. 2004. CFP10 discriminates between nonacetylated and acetylated ESAT-6 of *Mycobacterium tuberculosis* by differential interaction. *Proteomics* 4:2954–2960.

Olsen, A. W., P. R. Hansen, A. Holm, and P. Andersen. 2000. Efficient protection against *Mycobacterium tuberculosis* by vaccination with a single subdominant epitope from the ESAT-6 antigen. *Eur. J. Immunol.* 30:1724–1732.

Olsen, A. W., A. Williams, L. M. Okkels, G. Hatch, and P. Andersen. 2004. Protective effect of a tuberculosis subunit vaccine based on a fusion of antigen 85B and ESAT-6 in the aerosol guinea pig model. *Infect. Immun.* 72:6148–6150.

Pai, M. 2005. Alternatives to the tuberculin skin test: interferon-gamma assays in the diagnosis of *Mycobacterium tuberculosis* infection. *Indian J. Med. Microbiol.* 23:151–158.

Pathan, A. A., K. A. Wilkinson, R. J. Wilkinson, M. Latif, H. McShane, G. Pasvol, A. V. Hill, and A. Lalvani. 2000. High frequencies of circulating IFN-gamma-secreting CD8 cytotoxic T cells specific for a novel MHC class I-restricted *Mycobacterium tuberculosis* epitope in *M. tuberculosis*-infected subjects without disease. *Eur. J. Immunol.* 30:2713–2721.

Pollock, J. M., and P. Andersen. 1997. Predominant recognition of the ESAT-6 protein in the first phase of interferon with *Mycobacterium bovis* in cattle. *Infect. Immun.* 65:2587–2592.

Pollock, J. M., J. McNair, H. Bassett, J. P. Cassidy, E. Costello, H. Aggerbeck, I. Rosenkrands, and P. Andersen. 2003. Specific delayed-type hypersensitivity responses to ESAT-6 identify tuberculosis-infected cattle. *J. Clin. Microbiol.* 41:1856–1860.

Pym, A. S., P. Brodin, R. Brosch, M. Huerre, and S. T. Cole. 2002. Loss of RD1 contributed to the attenuation of the live tuberculosis vaccines *Mycobacterium bovis* BCG and *Mycobacterium microti. Mol. Microbiol.* **46:**709–717.

Pym, A. S., P. Brodin, L. Majlessi, R. Brosch, C. Demangel, A. Williams, K. E. Griffiths, G. Marchal, C. Leclerc, and S. T. Cole. 2003. Recombinant BCG exporting ESAT-6 confers enhanced protection against tuberculosis. *Nat. Med.* **9:**533–539.

Ravn, P., A. Demissie, T. Eguale, H. Wondwosson, D. Lein, H. A. Amoudy, A. S. Mustafa, A. K. Jensen, A. Holm, I. Rosenkrands, F. Oftung, J. Olobo, F. von Reyn, and P. Andersen. 1999. Human T cell responses to the ESAT-6 antigen from *Mycobacterium tuberculosis. J. Infect. Dis.* **179:**637–645.

Renshaw, P. S., K. L. Lightbody, V. Veverka, F. W. Muskett, G. Kelly, T. A. Frenkiel, S. V. Gordon, R. G. Hewinson, B. Burke, J. Norman, R. A. Williamson, and M. D. Carr. 2005. Structure and function of the complex formed by the tuberculosis virulence factors CFP-10 and ESAT-6. *EMBO J.* **24:**2491–2498.

Renshaw, P. S., P. Panagiotidou, A. Whelan, S. V. Gordon, R. G. Hewinson, R. A. Williamson, and M. D. Carr. 2002. Conclusive evidence that the major T-cell antigens of the *Mycobacterium tuberculosis* complex ESAT-6 and CFP-10 form a tight, 1:1 complex and characterization of the structural properties of ESAT-6, CFP-10, and the ESAT-6*CFP-10 complex. Implications for pathogenesis and virulence. *J. Biol. Chem.* **277:**21598–21603.

Rison, S. C., J. Mattow, P. R. Jungblut, and N. G. Stoker. 2007. Experimental determination of translational starts using peptide mass mapping and tandem mass spectrometry within the proteome of *Mycobacterium tuberculosis. Microbiology* **153:**521–528.

Rogerson, B. J., Y. J. Jung, R. LaCourse, L. Ryan, N. Enright, and R. J. North. 2006. Expression levels of *Mycobacterium tuberculosis* antigen-encoding genes versus production levels of antigen-specific T cells during stationary level lung infection in mice. *Immunology* **118:**195–201.

Sassetti, C. M., D. H. Boyd, and E. J. Rubin. 2003. Genes required for mycobacterial growth defined by high density mutagenesis. *Mol. Microbiol.* **48:**77–84.

Schumann, G., S. Schleier, I. Rosenkrands, N. Nehmann, S. Hälbich, P. F. Zipfel, M. I. de Jonge, S. T. Cole, T. Munder, and U. Möllmann. 2006. *Mycobacterium tuberculosis* secreted protein ESAT-6 interacts with the human protein syntenin-1. *Centr. Eur. J. Biol.* **1:**183–202.

Singh, A., D. Mai, A. Kumar, and A. J. Steyn. 2006. Dissecting virulence pathways of *Mycobacterium tuberculosis* through protein-protein association. *Proc. Natl. Acad. Sci. USA* **103:**11346–11351.

Skjøt, R. L. V., T. Oettinger, I. Rosenkrands, P. Ravn, I. Brock, S. Jacobsen, and P. Andersen. 2000. Comparative evaluation of low-molecular-mass proteins from *Mycobacterium tuberculosis*

identifies members of the ESAT-6 family as immunodominant T-cell antigens. *Infect. Immun.* **68:**214–220.

Skjøt, R. L., I. Brock, S. M. Arend, M. E. Munk, M. Theisen, T. H. Ottenhoff, and P. Andersen. 2002. Epitope mapping of the immunodominant antigen TB10.4 and the two homologous proteins TB10.3 and TB12.9, which constitute a subfamily of the esat-6 gene family. *Infect. Immun.* **70:**5446–5453.

Smith, S. M., M. R. Klein, A. S. Malin, J. Sillah, K. Huygen, P. Andersen, K. P. McAdam, and H. M. Dockrell. 2000. Human CD8(+) T cells specific for *Mycobacterium tuberculosis* secreted antigens in tuberculosis patients and healthy BCG-vaccinated controls in The Gambia. *Infect. Immun.* **68:**7144–7148.

Sorensen, A. L., S. Nagai, G. Houen, P. Andersen, and A. B. Andersen. 1995. Purification and characterization of a low-molecular-mass T-cell antigen secreted by *Mycobacterium tuberculosis. Infect. Immun.* **63:**1710–1717.

Stanley, S. A., J. E. Johndrow, P. Manzanillo, and J. S. Cox. 2007. The type I IFN response to infection with *Mycobacterium tuberculosis* requires ESX-1-mediated secretion and contributes to pathogenesis. *J. Immunol.* **178:**3143–3152.

Stanley, S. A., S. Raghavan, W. W. Hwang, and J. S. Cox. 2003. Acute infection and macrophage subversion by *Mycobacterium tuberculosis* require a specialized secretion system. *Proc. Natl. Acad. Sci. USA* **100:**13001–13006.

Strong, M., M. R. Sawaya, S. Wang, M. Phillips, D. Cascio, and D. Eisenberg. 2006. Toward the structural genomics of complexes: crystal structure of a PE/PPE protein complex from *Mycobacterium tuberculosis. Proc. Natl. Acad. Sci. USA* **103:**8060–8065.

Stukova, M. A., S. Sereinig, N. V. Zabolotnyh, B. Ferko, C. Kittel, J. Romanova, T. I. Vinogradova, H. Katinger, O. I. Kiselev, and A. Egorov. 2006. Vaccine potential of influenza vectors expressing *Mycobacterium tuberculosis* ESAT-6 protein. *Tuberculosis* (Edinburgh) **86:**236–246.

Tekaia, F., S. V. Gordon, T. Garnier, R. Brosch, B. G. Barrell, and S. T. Cole. 1999. Analysis of the proteome of *Mycobacterium tuberculosis in silico. Tuber. Lung Dis.* **79:**329–342.

Tundup, S., Y. Akhter, D. Thiagarajan, and S. E. Hasnain. 2006. Clusters of PE and PPE genes of *Mycobacterium tuberculosis* are organized in operons: evidence that PE Rv2431c is co-transcribed with PPE Rv2430c and their gene products interact with each other. *FEBS Lett.* **580:**1285–1293.

van Pinxteren, L. A., P. Ravn, E. M. Agger, J. Pollock, and P. Andersen. 2000. Diagnosis of tuberculosis based on the two specific antigens ESAT-6 and CFP10. *Clin. Diagn. Lab. Immunol.* **7:**155–160.

Volkman, H. E., H. Clay, D. Beery, J. C. Chang, D. R. Sherman, and L. Ramakrishnan. 2004. Tuberculous granuloma formation is enhanced by a mycobacterium virulence determinant. *PLoS Biol.* **2:**e367.

Chapter 14

Mycobacterial Sigma Factors and Surface Biology

SAHADEVAN RAMAN, ALESSANDRO CASCIOFERRO, ROBERT N. HUSSON, AND RICCARDO MANGANELLI

Sigma factors are interchangeable subunits of bacterial RNA polymerase that are required for promoter selectivity and transcription initiation. All eubacterial genomes encode at least one primary sigma factor, which is an essential protein and is responsible for the transcription of housekeeping genes. In addition to the primary sigma factor, bacterial genomes encode a variable number of dispensable alternative sigma factors responsible for transcription of specific regulons in response to different environmental conditions (Gruber and Gross, 2003). Recently, sigma factors have been divided into four different groups depending on their structure and function. Group 1 includes the primary, essential sigma factors. Group 2 includes the primary-like sigma factors, which are structurally similar to those belonging to group 1. They are usually involved in the transcription of general stress response and stationary phase survival genes (Helmann, 2002), with the exception of those encoded by *Bacteroidetes*, which function as primary, essential, sigma factors (Vingadassalom et al., 2005). Group 3 includes sigma factors involved either in the general stress response, flagellum biosynthesis, or bacterial differentiation (e.g., sporulation). Group 4 is the most numerous and heterogeneous. Because many of the sigma factors of this family control functions associated with some aspect of the cell surface or transport, they were named extracytoplasmic function (ECF) sigma factors (Lonetto et al., 1994).

Regulation of alternative sigma factor activity is very fine tuned, and in some cases, complex regulatory networks involving several sigma factors have been described (Hilbert and Piggot, 2004; Rodrigue et al., 2006). Sigma factor activity can be regulated at transcriptional, posttranscriptional, and posttranslational levels. Transcriptional regulation often involves positive feedback due to the recognition by the sigma factor of the promoter of its own structural gene, but there are examples of sigma factor genes whose transcription is regulated by a two-component system or by another sigma factor. The best characterized example of posttranscriptional regulation among sigma factors is represented by *Escherichia coli rpoS*, regulated at least by three small RNAs (Repoila et al., 2003).

Posttranslational regulation is usually due to proteins, anti-sigma factors that specifically bind to a sigma factor and thereby prevent its interaction with RNA polymerase until a specific signal allows its release. This mechanism may be modulated by other proteins, anti-anti-sigma factors, that can interact with anti-sigma factors and block their activity. Posttranslational modification of the anti- or anti-anti-sigma factors allows another layer of regulation by altering the ability of these proteins to bind to each other or the sigma factor (Hughes and Mathee, 1998).

A well-characterized example of posttranslational sigma factor regulation is that regulating *E. coli* σ^E activity in response to periplasmic stresses such as heat-shock or ethanol that could damage cell envelope, preventing the correct folding of outer membrane porins (Ades, 2004). The signal transduction pathway linking periplasmic stress with σ^E activity is based on the widely conserved signaling pathway called regulated intramembrane proteolysis (RIP). In this pathway, two proteases, known as site 1 and site 2 protease (S1P and S2P, respectively), act sequentially, processing the transmembrane anti-sigma factor RseA, which results in the release of the active form of the sigma factor. The S1P senses signals to initiate proteolysis, whereas the S2P acts on the partially degraded protein (Brown et al., 2000). These cleavage events release the cytoplasmic domain of RseA from the membrane, but this inhibitory domain still binds and exerts its action on σ^E, because this partially degraded anti-sigma factor still binds σ^E with a higher affinity than core RNA polymerase

Sahadevan Raman and Robert N. Husson • Children's Hospital Boston and Division of Infectious Diseases, Harvard Medical School, 300 Longwood Avenue, Boston, MA 02115. **Alessandro Cascioferro and Riccardo Manganelli** • Department of Histology, Microbiology and Medical Biotechnologies, University of Padova Medical School, Via Gabelli 63, 35121 Padua, Italy.

(Campbell et al., 2003). ClpXP, Lon, and other ATP-dependent cytoplasmic proteases are involved in the final degradation of RseA and σ^E release (Fig. 1) (Chaba et al., 2007).

SIGMA FACTORS AND THE CELL SURFACE

Sigma factors regulate processes involved in cell surface biology in many bacterial species. Well-characterized examples include *E. coli* σ^E (also named RpoE), and FecI in gram-negative bacteria and by *Bacillus subtilis* σ^X, σ^W, and σ^M in gram-positive bacteria.

E. coli σ^E is primarily involved in responding to periplasmic stress. Most genes in the σ^E regulon encode proteins located in the inner and outer membranes as well as in the periplasm. These proteins can be divided in two groups: one group is involved in folding or degradation of misfolded polypeptides in the bacterial envelope, whereas the other group is involved in biosynthesis and transport of lipopolysaccharide, one of the main components of the gram-negative outer membrane (Dartigalongue et al., 2001). Interestingly, it has been recently demonstrated that alteration of lipopolysaccharide structure can induce σ^E activity (Tam and Missiakas, 2005).

FecI is involved in the induction of a ferric citrate transport system. The inducer, ferric citrate, elicits a signal that is transmitted across the outer membrane, the periplasm, and the cytoplasmic membrane into the cytoplasm by binding to an outer membrane transport protein. The signal is then transferred across the three subcellular compartments by an outer membrane transport protein that interacts in the periplasm with a cytoplasmic transmembrane protein, which is required for activation of FecI (Braun and Mahren, 2005).

In *B. subtilis*, σ^X, σ^W, and σ^M react to a different but overlapping spectrum of stimuli, which act at different locations in the cell envelope. Their regulons also partially overlap as the result of the similar sequence recognition properties of the corresponding holoenzymes (Helmann, 2002).

B. subtilis σ^X controls modifications of the cell envelope that regulate the net charge of both cell wall and cytoplasmic membrane. This sigma factor regulates the expression of genes encoding proteins involved in modification of teichoic acids by esterification with D-alanine and incorporation of phosphatidylethanolamine into the cell membrane. The consequence is that in both cases, positively charged amino groups are introduced into the cell envelope. The final result is a reduction in the net negative charge of the cell envelope, which has been previously implicated as a resistance mechanism specific for cationic antimicrobial peptide (Cao and Helmann, 2004).

σ^W controls an "antibiosis" regulon (about 60 genes) including both defensive (detoxification) and offensive (bacteriocin expression) components (Cao et al., 2002a). Interestingly, most of the detoxifying enzymes encoded by the σ^W regulon are specific for antibiotics that target cell wall biosynthesis. Accordingly, the σ^W regulon is induced by cell wall-active antibiotics such as vancomycin, cephalosporin, the mammalian cationic peptides LL37 and PG-1, and poly-L-lysine (Cao et al., 2002b), and provide resistance to bacterial compounds produced by various *Bacillus* strains (Butcher and Helmann, 2006).

Finally, σ^M switches on its regulon following exposure of bacteria to high osmolarity, low pH, paraquat, ethanol, bacitracin, vancomycin, phosphomycin, and heat, most of which interfere with structure and biosynthesis of the cell membrane, envelope, and wall (Thackray and Moir, 2003). Among the genes under its control are *bcrC*, encoding a bacitracin resistance determinant (Mascher et al., 2003); *yiqL*, encoding a putative hydrolase conferring paraquat resistance; and *divIC*, involved in cell-division initiation (Minnig et al., 2003).

The characterization of stress response systems used by bacteria to respond to surface stress is extremely important because it can contribute to a bet-

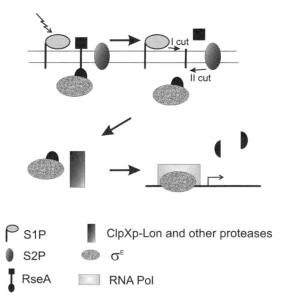

S1P

S2P

RseA

ClpXp-Lon and other proteases

σ^E

RNA Pol

Figure 1. Regulated intramembrane proteolysis (RIP) of RseA in *E. coli*. Two proteases, S1P and S2P, act sequentially in processing the transmembrane anti-sigma factor RseA. These cleavage events release the cytoplasmic domain of RseA, which is still able to bind and exert its action on σ^E. ClpXP, Lon and other ATP-dependent cytoplasmic proteases are involved in the final degradation of RseA and σ^E release.

ter understanding of both bacterial sensing and signaling through different cell compartments, and cell envelope physiology and biogenesis. The characterization of mycobacterial surface response systems is of particular interest owing to the unusual structure and physiological importance of the cell envelope of the members of this genus.

MYCOBACTERIAL SIGMA FACTORS

All sequenced mycobacterial genomes encode a group 1 primary sigma factor (σ^A), and a group 2 primary-like sigma factor (σ^B). Group 3 sigma factors are represented by two proteins in *Mycobacterium avium* and *Mycobacterium paratuberculosis* but are not represented in *Mycobacterium leprae*. All other mycobacterial species for which the complete genome sequence is available encode a single group 3 sigma factor: σ^F. The number of group 4 sigma factors varies from two in *M. leprae* (σ^E, and σ^C) to 23 in *Mycobacterium smegmatis*. *Mycobacterium tuberculosis* encodes 10 sigma factors of this group (Manganelli et al., 2004b; Rodrigue et al., 2006; Waagmeester et al., 2005).

REGULATION OF MYCOBACTERIAL SIGMA FACTORS BY ANTI-SIGMA FACTORS

To date, five sigma factor/anti-sigma factor couples have been identified in *M. tuberculosis*: σ^F-UsfX (Beaucher et al., 2002); σ^K-RskA (Said-Salim et al., 2006); σ^H-RshA (Song et al., 2003), σ^L-RslA (Dainese et al., 2006; Hahn et al., 2005), and σ^E-RseA (Rodrigue et al., 2006). Interestingly, only RskA and RslA appear to be transmembrane proteins, (Dainese et al., 2006; Hahn et al., 2005; Said-Salim et al., 2006).

Usually, anti-sigma factors are encoded by genes lying downstream from the gene encoding their cognate sigma factor and are often cotranscribed with the sigma factor gene. An in silico analysis of *M. tuberculosis* genome revealed putative anti-sigma factors genes 3′ of *sigD*, *sigM*, and *sigG* (Rv3413c, Rv3912, and Rv0181c, respectively) (A. Cascioferro and R. Manganelli, unpublished data). Moreover, an analysis performed with the Search Pattern function available on Tuberculist (http://genolist.pasteur.fr/TubercuList/) enabled us to identify a gene (Rv0093c) encoding a protein containing a typical anti-sigma factor signature previously described in *Streptomyces* (Paget et al., 2001; Cascioferro and Manganelli, unpublished). Interestingly, this protein was recently shown to bind σ^C, suggesting that it could represent

its cognate anti-sigma factor (Thakur et al., 2007). Analysing the predicted protein products of these four genes using the TMHMM and TMpred topology prediction algorithms (www.expasy.org), we found that Rv3413c and Rv0093c are predicted as transmembrane proteins by both software programs, whereas Rv3912, and Rv0181c were predicted as transmembrane proteins only from TMpred. Thus, nearly all of the group 3 and group 4 sigma factors of *M. tuberculosis* have been shown or predicted to be regulated by anti-sigma factors. In addition to σ^L and σ^K, σ^D, σ^M, σ^G and σ^C are predicted to be posttranslationally regulated by transmembrane anti-sigma factors (Table 1). The mechanism of regulation of anti-sigma factor activity has been shown for RshA, which functions as a redox switch and is inactivated by oxidation, and for UsfX, whose activity is regulated by interaction with an anti-anti-sigma factor that is subject to reversible phosphorylation (Beaucher et al., 2002; Song et al., 2003). It is anticipated that a range of mechanisms, potentially including receptor-ligand interactions, RIP, and posttranslational modifications will modulate sigma-anti-sigma interactions and, thereby, control specific sigma factor-regulated transcription.

In order to predict if RIP could be involved in the regulation of anti-sigma factor activity in *M. tuberculosis*, we searched its genome for S1P and S2P orthologues: we found five genes encoding putative S1P and 3 genes encoding putative S2P. Among the five putative S1P (Rv1043c, Rv3671, PepA, PepD, and HtrA), only HtrA had all the expected characteristics (transmembrane localization and a single PDZ domain). Moreover, studying its topology using translational fusions to the cellular localization markers LacZ and PhoA (Green and Cutting, 2000), we could show that the catalytic and the PDZ domains of this protein are indeed localized on the extracytoplasmic side of the cell, as expected for S1P (Cascioferro and Manganelli, unpublished).

Among the S2P orthologues (Rv2869c, Rv2625c, and Rv0359c), Rv2869c is a transmembrane protease containing all the typical features of S2P, and it was recently shown to be involved in the regulation by RIP of *M. tuberculosis* cell envelope composition, as well as growth and persistence in vivo (Makinoshima and Glickman, 2005).

Taken together, these observations suggest that *M. tuberculosis* genome encodes orthologues of both S1P and S2P (HtrA and Rv2869c, respectively), strongly supporting the possibility that RIP could be involved in the surface stress signal transduction and in posttranslational regulation of sigma factors also in this important pathogen.

Table 1. *M. tuberculosis* sigma factors

Sigma factor	Anti-sigma factor	Anti-sigma factor putative cellular localization	Anti-anti-sigma factor	Inducing conditions[a]
σ^A				
σ^B				Heat shock, oxidative stress, surface stress
σ^C	Rv0093c[b]	Membrane		
σ^D	Rv3413c[b]	Membrane		
σ^E	RseA (Rv1222)	Cytoplasm		Surface stress, oxidative stress, heat shock, growth in macrophages
σ^F	UsfX (RsbW, Rv3287c)	Cytoplasm	Rv1635c, Rv3687c	
σ^G	Rv0181c[b]	Membrane		Growth in macrophages
σ^H	RshA (Rv3221A)	Cytoplasm		Heat shock, oxidative stress, growth in macrophages
σ^I				
σ^J				Stationary phase, growth in macrophages
σ^K	RskA (Rv0444c)	Membrane		
σ^L	RslA (Rv0736)	Membrane		
σ^M	Rv3912[b]	Membrane		

[a]Reviewed by Rodrigue et al. (2006).
[b]Putative anti-sigma factors.

M. TUBERCULOSIS SIGMA FACTORS INVOLVED IN CELL SURFACE BIOLOGY

σ^F

σ^F is the only group 3 sigma factor encoded in the *M. tuberculosis* genome. A strain in which its structural gene was deleted was able to grow to a threefold higher density in stationary phase than the wild-type strain and did not exhibit any lag phase after dilution from a dense culture into fresh medium (Chen et al., 2000). When tested for resistance to stress, the *sigF* mutant showed the same sensitivity to heat shock, cold shock, hypoxia and long-term stationary phase as the parental strain, but when tested in a mouse model of infection, it was clearly attenuated in the late-stage of the disease (Geiman et al., 2004). The transcriptional profiles of the *sigF* mutant and that of the wild-type parental strain were compared using DNA microarrays. Only 38 genes were down-regulated in the *sigF* mutant in exponential phase, whereas in the early and late stationary phases, their number increased to 187 and 277, respectively, suggesting a major role of σ^F in the adaptation to stationary phase (Geiman et al., 2004). Interestingly, several genes whose expression in stationary phase required σ^F were involved in the biosynthesis and structure of the cell envelope, or in the biosynthesis and degradation of surface polysaccharides and lipopolysaccharides. Moreover, the *sigF* mutant showed reduced expression of several transcriptional regulators, some of which (TetR, GntR, and MarR) control the expression of efflux pumps in other bacteria. Accordingly, the *sigF* mutant showed several phenotypes, suggesting a defect in surface composition and/or permeability, such as the inability to retain the neutral red stain, which correlates with a reduction in the expression of envelope-associated mycobacterial sulfolipids (see chapter 18) (Geiman et al., 2004), hypersusceptibility to rifampin and rifapentine, and diminished uptake of cheno-deoxycholate (Chen et al., 2000).

σ^F was recently shown to regulate the biosynthesis of a surface-associated lipidic pigment in *M. smegmatis* (R. Provvedi, D. Kocincova, G. Etienne, V. Donà, D. Euphrasie, M. Daffe, R. Manganelli, and J.-M. Reyrat, unpublished data). Interestingly, the production of a novel cell wall-associated pigment after long-term anaerobic culture mimicking dormancy, was shown in tubercle bacilli (Cunningham et al., 2004). It is tempting to speculate that σ^F could be involved in the production of such pigment.

In summary, it is possible to conclude that σ^F governs functions that are active in stationary phase and in late-stage disease, regulating the expression of genes involved in the structure and function of the mycobacterial cell wall and its complex network of lipids and polysaccharides, including pigments and virulence-related sulfolipids.

σ^D

Like many other *M. tuberculosis* group 4 sigma factors, σ^D appears to regulate a limited number of specific genes directly but has a broad range of indirect effects on transcription. The nature of this regulon provides insight into potential physiological roles in the bacterium. The genes that appear to be directly regulated by σ^D include Rv1815, Rv1816, Rv1884c (*rpfC*), Rv1883c, Rv3414c (*sigD*), and Rv3413c (*rsdA*) (Raman et al., 2004). Although the structure and function of the products of these genes have not been experimentally demonstrated, bioinformatic analysis of these proteins, and functional analysis of related proteins, provides insight into their probable function.

σ^D (Rv3414c) regulates its own expression and that of the immediate downstream gene *rsdA* (Rv3413c). RsdA is a putative membrane protein with predicted anti-sigma function, and thus may control the activity of σ^D in response to extracellular or membrane localized signals. Further evidence for the involvement of σ^D in cell envelope processes is provided by the regulation by σ^D of a major resuscitation promoting factor (*rpf*) gene *rpfC* (Rv1884c). Rpf proteins were first characterized in *Micrococcus luteus*, where they were shown to be required for resumption of growth from stationary phase or from cultures diluted to low density (Mukamolova et al., 1998). Recent studies of Rpf proteins provide evidence for their biochemical function as peptidoglycan hydrolases and transglycosylases that can degrade the cell wall (Mukamolova et al., 2006). Although there appears to be functional overlap among the five *rpf* genes of *M. tuberculosis*, *rpfC*, the only *rpf* gene regulated by σ^D, is expressed at a much higher level than any of the other *rpf* genes (Raman et al., 2004; Tufariello et al., 2004). The regulation by σ^D of *rpfC* in the context of decreased *sigD* expression in response to hypoxia suggests a role for σ^D in regulating cell wall structure during the entry into or emergence from hypoxia-induced dormancy.

The protein encoded by the σ^D-regulated gene Rv1815 has a predicted protease function. Rv1816, encoded in the same operon, has a DNA-binding domain encompassing a helix-turn-helix motif at its N-terminal region. Sequence and motif analysis suggests that it has a transcriptional regulatory role similar to TetR family proteins (http://www.ebi.ac.uk/InterProScan/index.html). This protein is also predicted to contain a single transmembrane domain. The coregulation by SigD of a membrane-based transcriptional regulator (Rv1816) and a protease (Rv1815) suggests that the protease activity of Rv1815 may contribute to the transduction of cell surface or membrane based signalling events mediated by Rv1816.

In addition to the above-described genes that are completely σ^D-dependent, evidence also indicates a partial role for σ^D in regulation of genes encoding other cell envelope-associated proteins, including

iniB, iniA and *iniC*. The three-gene *iniBAC* operon (Rv0341, Rv0342, and Rv0343) was shown to be induced in response to inhibitory concentrations of cell envelope-disrupting antibiotics, including isoniazid, which inhibits mycolic acid biosynthesis in actively growing mycobacteria (Alland et al., 2000). σ^D regulation of this operon thus suggests a role for σ^D or regulatory proteins encoded by σ^D-dependent genes in activating mechanisms that respond to cell surface injury.

Cell integrity stress, like other forms of stress, may lead to increased expression of chaperone/heat shock genes, and the activity of these heat shock proteins (Hsps) is likely to be essential for maintaining cell envelope homeostasis. In *M. tuberculosis*, direct regulation of chaperone protein genes by σ^E and σ^H occurs in response to oxidative, envelope, and heat stress (Manganelli et al., 2001; Manganelli et al., 2002; Raman et al., 2001). Notable among the upregulated genes in wild-type relative to a *sigD* mutant strain of *M. tuberculosis* were the *dnaK, grpE, groEL1, groES,* and *groEL2* genes encoding major *M. tuberculosis* Hsps. In the setting of the cell envelope functions of SigD-regulated genes and the transmembrane anti-sigma factor RsdA, these data suggest that increased expression of these Hsp genes when SigD is expressed may be required for adaptive responses to cell envelope stress. Alterations in the expression of the genes regulated directly and indirectly by SigD resulted in a moderately attenuated virulence phenotype of an *M. tuberculosis sigD* mutant in the intravenous mouse model of infection (Raman et al., 2004; Tufariello et al., 2004).

σ^E

The sigma factor σ^E was first described in *M. smegmatis*, where it was shown to be involved in survival under conditions of high temperature, acidic pH, exposure to detergents, and oxidative stress (Wu et al., 1997). Disruption of *sigE* in *M. tuberculosis* leads to an increased sensitivity to heat shock, sodium dodecyl sulfate (SDS), oxidative stress (Manganelli et al., 2001), and vancomycin (R. Provvedi, F. Boldrin, I. Smith, and R. Manganelli, unpublished data). Moreover, the *sigE* mutant is unable to grow in both resting THP-1-derived macrophages and human dendritic cells, is more sensitive to the bactericidal activity of activated mice macrophages, and is attenuated in mice (Manganelli et al., 2004a). In agreement with these data, *sigE* transcription is induced after exposure to heat shock, alkaline pH, detergents, oxidative stress (diamide), vancomycin, or during growth in human macrophages (Schnappinger et al., 2003). Although the induction following heat shock and exposure to diamide is mediated by σ^H, the induction following exposure to detergents and alkaline pH is mediated by a two-component system (MprA/MprB), suggesting that, in this case, the signal recognized by the cell is extracytoplasmic (He et al., 2006). In *M. tuberculosis*, *sigE* is followed by a gene encoding its anti-sigma factor (RseA) (Rodrigue et al., 2006), which is predicted to be a cytoplasmic protein. Interestingly, the gene encoding RseA (Rv1222) is followed by two genes: the first encodes the putative S1P HtrA, whereas the second encodes an integral membrane protein belonging to the twin-arginine-translocation system. The presence of two arginine residues in the RseA N-terminus suggested the hypothesis that this protein could be targeted to the bacterial membrane through the twin-arginine-translocation system to allow its degradation by HtrA. However, several attempts to show RseA-TatB-HtrA interaction failed (Dainese and Manganelli, unpublished).

The transcription profile of the *sigE* mutant was analyzed using DNA microarrays in the absence of specific stimuli inducing σ^E activity: in this condition, the *sigE* mutant was affected in the expression of 38 genes, four of which were predicted to be involved in mycolic acid biosynthesis. However, following exposure of *M. tuberculosis* to SDS, 23 genes belonging to 13 putative transcriptional units were induced in the wild-type parental strain but not in the *sigE* mutant, suggesting direct or indirect regulation by σ^E (Manganelli et al., 2001). The presence in this group of at least four genes involved in fatty acid degradation (*aceA*, *fadB2*, *fadE23*, and *fadE24*) suggests the possibility that these genes are not only involved in the metabolic exploitation of the fatty acids but also may function as components of a shunt pathway that degrades/detoxifies the fatty acids that accumulate as a result of inhibition of biosynthesis of cell surface lipids or damage to the fatty acid-rich envelope of *M. tuberculosis*. Consistent with this hypothesis, *fadE23* and *fadE24* were are also induced by isoniazid, an inhibitor of mycolic acid biosynthesis (Wilson et al., 1999) (also see chapter 4).

Another protein involved in surface biology under σ^E transcriptional control is HtpX, an intramembrane metalloprotease of the M48 family. Its function in *M. tuberculosis* is unknown, but its orthologue in *E. coli* participates in the proteolytic quality control of membrane proteins in conjunction with FtsH (Sakoh et al., 2005).

Finally, a surface-related operon under σ^E control is that including Rv2743c, Rv2744c, and Rv2745c. The product of Rv2744c is highly homologous to phage shock protein A (PspA) part of a cytoplasmic membrane protection system in gram-negative bacteria (Darwin, 2005). In *E. coli*, PspA is a cytoplasmic protein and binds a transcriptional regulator. In the presence of a specific surface stress signal (probably a decrease of the proton-motive force due to cytoplasmic membrane permeabilization), a transmembrane protein activates PspA, which releases the transcriptional factor and associates to the internal surface of the cytoplasmic membrane, causing its stabilization. The first and the last genes of this *M. tuberculosis* σ^E-regulated operon encode a transmembrane protein and a DNA-binding protein, respectively. Although they do not share any similarity to the transcriptional regulator and to the transmembrane protein involved in the PspA system of gram-negative bacteria, it is tempting to speculate that they represent functional analogs of these *E. coli* proteins.

In summary, σ^E responds to several stresses, including surface damage, such as those induced by detergents, alkaline pH, or vancomycin. Among the proteins encoded by σ^E-regulated genes, those probably involved in surface biology are responsible for fatty acid degradation, membrane protein quality control, and membrane stabilization, suggesting that σ^E is not only responsible for controlling surface stability following the exposure to damaging environmental conditions but can also control its composition. In support of this hypothesis, the finding that phagocytosis of live or heat killed *sigE* mutants by dendritic cells stimulates production of high levels of IL-10 suggests that these bacteria could present modifications in cell envelope structure, favoring the interaction with cellular molecules known to promote robust IL-10 production, such as DC-Sign, Toll-like receptor 2, and mannose receptor (Giacomini et al., 2006). An accurate study of cell surface chemical composition of the *M. tuberculosis sigE* mutant would be of great value to confirm this hypothesis.

σ^K

Two recent papers from the same group recently described different polymorphisms related to the σ^K molecule. The first is a start codon mutation present in some *Mycobacterium bovis* BCG strains (Charlet et al., 2005); the others are different single nucleotide polymorphisms in the gene encoding the σ^K-specific anti-sigma factor (RskA), which compromise its functionality (Said-Salim et al., 2006). Studying the transcription profile of the mutants expressing a nonfunctional RskA by DNA microarrays, the authors described two small regions under σ^K transcriptional control. One region included *sigK* and *rskA*, together with genes encoding a putative amine oxidase, a cyclopropane-fatty-acyl-phospholipid synthase, and four other proteins of unknown function. The other region included genes encoding the surface-associated lipoprotein Mpt83; the integral membrane proteins DipZ, Rv2876, and Rv2877c; and finally the secreted proteins Mpt70, and Mpt53. Mpt83, and Mpt70 C-terminal domains are highly homologous. Their structure revealed a complex and novel bacterial fold structurally homologous to the two C-terminal FAS1 domains of the cell adhesion protein fasciclin I. It was proposed that they could bind to cell surface host proteins to induce changes in host cell behavior advantageous to the pathogen, perhaps through the modulation of signaling pathways (Carr et al., 2003).

Mpt53 is a secreted DsbE-like protein containing a thioredoxin-active site. This protein was recently characterized and hypothesized to be a functional homologue of DsbA, able to catalyze the formation of disulfide bonds in unfolded secreted proteins (Goulding et al., 2004). However, Rv2877c is an integral membrane protein with a conserved CcdA domain. Proteins containing this domain are usually involved in the transfer of a reducing potential from the cytoplasm to secreted disulfide bond isomerases or to the cytochrome c thioreduction pathway (Appia-Ayme and Berks, 2002; Le Brun et al., 2000). Interestingly, DipZ is also an integral membrane protein with a CcdA domain. It was recently shown that transcription of Mpt53 and Rv2877c structural genes is also controlled by σ^L (Dainese et al., 2006; Hahn et al., 2005). In summary, σ^K regulates two small genetic loci, encoding transmembrane or secreted proteins that could be important for the interactions with host cells and for the control of exported proteins redox state.

σ^L

The σ^L protein is of particular interest in considering the role of sigma factor-regulated transcription in modulating the cell envelope for two reasons. First, σ^L has been shown to directly regulate transcription of a major cluster of polyketide synthase genes, and second, a *sigL* mutant is highly attenuated in the mouse model of infection (Dainese et al., 2006; Hahn et al., 2005). Despite identification of its regulon and demonstration that σ^L is regulated by a transmembrane anti-sigma factor, the conditions under which *sigL* is induced in vivo remain unknown.

Through microarray and other transcriptional analyses, a σ^L regulon has been identified that can be grouped into a number of functional clusters, including: (i) *sigL* and its cotranscribed anti-sigma factor gene *rslA*, (ii) the *pks10-pks7-pks8-pks17-pks9-pks11* operon and the Rv1138c-Rv1139c operon, and (iii) the Rv2877c-*mpt53* operon. These σ^L-regulated genes encode membrane or secreted proteins and enzymes that synthesize surface lipids, indicating

a critical role for σ^L-regulated gene expression in modulating the *M. tuberculosis* cell envelope. In addition to these operons, for which σ^L-dependent in vivo promoters have been demonstrated, a number of other loci, based on microarray data, appear to be regulated by σ^L. Like the operons listed earlier, these genes encode a range of secreted and membrane proteins.

σ^L and its cognate anti-sigma factor RslA have been shown to physically interact, and RslA has been shown to inhibit σ^L-dependent transcription in vitro (Dainese et al., 2006; Hahn et al., 2005). RslA has sequence features characteristic of other anti-sigma factors, including RshA the anti-sigma factor that regulates σ^H activity (Song et al., 2003). In particular, RslA contains the HLXXCXXC zinc-binding motif that is critical for RshA function as a redox switch, regulating σ^H in response to oxidative and heat stress. Despite the presence of this motif and preliminary evidence that RslA also binds zinc (Hahn and Husson, unpublished), the SigL-RslA interaction does not appear to be redox regulated and σ^L does not play a major role in transcription regulation in response to oxidative stress (Hahn, 2005). In contrast to RshA, which is cytoplasmic, RslA has a single transmembrane domain, an intracellular domain that binds σ^L, and an extracellular domain of unknown function. This topology suggests that the extracellular domain is required for sensing extracellular environmental signals that are transduced to alter the σ^L-RslA interaction and thus activate transcription. The signals sensed and the mechanism of signal transduction, however, remain unknown.

The *pks10-pks7-pks8-pks17-pks9-pk11* genes appear to be transcribed as a σ^L-dependent operon based on microarray data and the identification of a σ^L-dependent in vivo promoter 5′ of *pks10*. *pks10*, and *pks7* have been directly shown to be transcribed as a single transcriptional unit and no additional promoter could be identified 5′ of *pks7* (Hahn et al., 2005). Microarray data indicate that the four other genes are also expressed from the σ^L-dependent *pks10* promoter (Dainese et al., 2006). The products of this large cluster of polyketide synthases are not known. Based on data from inactivating mutations of *pks7* and *pks10*, these two genes have been suggested to play a role in synthesis of phthiocerol dimycocerosic acid (PDIM), with *pks10* specifically implicated in phthiocerol synthesis (Rousseau et al., 2003; Sirakova et al., 2003). In contrast to these results, a strain of *M. tuberculosis* in which *sigL* is inactivated is not deficient in PDIM synthesis, and recent in vitro characterization of the *fadD26-ppsA-E* suggests that this locus is sufficient for PDIM biosynthesis (Trivedi et al., 2005; T. Y. Chen, M.-Y. Hahn, R. N. Husson and D. B. Moody, unpublished). Based on a strain in which these two genes were inactivated, the *pks8* and *pks17* gene products have been suggested to catalyze the synthesis of monomethyl branched unsaturated fatty acids (Dubey et al., 2003). The exact nature of the lipids produced by the *pks* gene cluster and whether downstream genes within this cluster may be transcribed both as part of a single operon and independently of σ^L remain to be determined.

The σ^L-regulated Rv1139c-Rv1138c operon may also be linked to the function of this *pks* cluster. In other mycobacteria, in *Streptomyces coelicolor*, and in other gram-positive organisms, *pks10* homologues are linked to homologues of Rv1139c, and in mycobacteria and *Streptomyces*, to homologues of Rv1138c. Although the function of these genes is not known, Rv1139c has at least two predicted membrane spanning segments and incorporates an isoprenyl-cysteine carboxymethyl transferase domain (Mulder et al., 2005). This domain functions in eukaryotes to posttranslationally modify proteins by methylating carboxy-teminal cysteines of prenylated proteins. Rv1138c encodes a protein that is predicted to be an aromatic ring monooxygenase. The manner in which these enzymes could participate in synthesis or modification of products of the σ^L-regulated *pks* locus is not known.

Another σ^L-dependent locus, the *mpt53*-Rv2877c operon, encodes the secreted antigen Mpt53, which is a predicted disulfide interchange protein, and the Rv2877c product, which incorporates seven transmembrane segments and is a predicted cytochrome c biogenesis family protein. This family of proteins also includes the disulfide exchange protein DsbD of *E. coli*. A structural and biochemical analysis suggested that Mpt53 may be a functional orthologue of *E. coli* DsbA, a periplasmic disulfide interchange protein (Goulding et al., 2004). If so, Rv2877c may function in this pathway, rather than in cytochrome c biogenesis. These analyses suggest that Mpt53 and Rv2877c function to maintain proper disulfide bond formation in proteins in the extracellular environment of *M. tuberculosis*, an environment that has been likened to a periplasmic space (Niederweis, 2003).

Among the other loci that appear to be regulated directly or indirectly by σ^L, the Rv1145-Rv1147 operon may also have a major role in modulating the *M. tuberculosis* cell envelope. The first two genes of this operon encode predicted lipid transporters (*mmpL13a* and *mmpL13b*) (see chapter 12), whereas Rv1147 is predicted to encode a *S*-adenosyl-L-methionine methyltransferase. Thus, the products of these genes are predicted to function in transport and possibly modification of cell envelope lipids of *M. tuberculosis*.

Although characterization of the function of the σ^L regulon will require substantial additional experimental data, results to date and the sequence-based bioinformatic analyses described earlier suggest a major role for σ^L in modulating the *M. tuberculosis* cell surface, particularly the cell envelope lipid content. Based on regulation of σ^L activity by a membrane-based anti-sigma factor, this σ^L-dependent regulation is likely to occur in response to specific environmental signals encountered by the bacterium during infection.

σ^M

Like many other ECF sigma factor genes in *M. tuberculosis*, *sigM* is autoregulated from its own promoter (Raman et al., 2006). Unlike most other ECF sigma factors, however, a putative anti-sigma factor, RsmA, encoded by Rv3912 immediately downstream from *sigM* is not coexpressed as part of an operon with *sigM*. Bioinformatic topology analysis indicates a single transmembrane domain in this protein, suggesting that it may function to transduce extracellular or surface signals to alter transcription by interaction of its intracellular domain with σ^M.

Based on microarray and primer extension analysis comparing *sigM* overexpression with mutant strains, genes that are likely to be under direct regulation of σ^M include few obvious membrane proteins (Agarwal et al., 2007; Raman et al., 2006). Instead, the σ^M-dependent regulons encode many secreted proteins. The genes that have been shown to have in vivo σ^M-dependent promoters are Rv0096, Rv2515c, Rv3235, Rv3440c, Rv3445c, Rv3906, and Rv3911 (Raman et al., 2006). With the exception of *sigM* itself, all of these σ^M-regulated genes have one or more downstream genes that are coexpressed in a σ^M-dependent manner.

Rv0096 (PPE1) is the first gene in a five-gene operon (Rv0096-Rv0101) that includes a nonribosomal peptide synthase (NRPS) cluster. In *Streptomyces* spp., in which they have been extensively studied, NRPSs synthesize a wide range of bioactive molecules including antibiotics, most of which are secreted. In *M. tuberculosis*, a distinct NRPS locus synthesizes mycobactin, a secreted iron-chelating siderophore that is required for mycobacterial replication in macrophages (De Voss et al., 2000; Quadri et al., 1998). Sequence- and structure-based analysis of the σ^M-regulated *nrp* locus showed three sets of the domains required for nonribosomal peptide synthesis. The product of this NRPS is not known. In addition to being positively regulated by σ^M, this NRPS-encoding operon is negatively regulated by a two-component regulatory system of *M. tuberculosis*, SenX3-RegX3 (Parish et al., 2003).

The only σ^M-regulated gene that encodes protein with a clear transmembrane domain is Rv2515c. This unknown hypothetical protein harbors an N-terminal helix-turn-helix motif and a C-terminal H-E-X-X-H motif indicative of a Zn metalloprotease active site (Mulder et al., 2005). These predicted features suggest that this DNA-binding protein belongs to the xenobiotic-response-element family of transcriptional regulators. A downstream gene Rv2514c, which is cotranscribed with Rv2515c, encodes a cytoplasmic protein of unknown function.

The rest of the genes that are likely to be directly regulated by σ^M encode small secreted proteins of the ESAT-6-like (Esx) family and adjacent genes. The *M. tuberculosis* genome contains 11 pairs of *esx* genes plus one unpaired *esx* gene (Cole et al., 1998). The function of the Esx proteins is not known; however, the best studied of these proteins, EsxA (ESAT-6) and EsxB (Cfp10), are important virulence factors and major antigens of *M. tuberculosis*, and are secreted by a novel secretion system (Brodin et al., 2004; Converse and Cox, 2005; Hsu et al., 2003). σ^M regulates genes encoding two sets of Esx proteins, Rv3445c-Rv3444c (EsxU-EsxT) and Rv3905c-Rv3904c (EsxF-EsxE), and three other genes, *Rv3906c, Rv3440,* and *Rv3235*, of unknown function. *Rv3906* is the first gene in the *esxF-esxE* operon, and *Rv3440* lies in close proximity to the *esxT-esxU* operon. The specific function of these secreted proteins is not known (see chapter 13).

An interesting finding of the microarray experiments comparing global gene expression in *sigM* mutant versus wild-type strains of *M. tuberculosis* H37Rv was the moderately increased expression of several genes encoding enzymes for the biosynthesis of surface lipids (Raman et al., 2006). Among these were several genes that are involved in PDIM biosynthesis. Consistent with the transcriptional data, the *sigM* mutant strain produced substantially more PDIM than the wild type or the *sigM* overexpresser.

These data suggest that σ^M regulates the expression of genes that produce several secreted or cell surface molecules and that may be important in host-bacillus interactions during specific stages of infection. The low basal expression of *sigM*, its increased expression in the late stationary phase, and the finding that it is not attenuated in the animal models of infection (Agarwal et al., 2007; Karls et al., 2006) suggest the possibility that the σ^M regulon may play a role in latency or other late stages of *M. tuberculosis* infection.

CONCLUDING REMARKS

Understanding the complex biology of the *M. tuberculosis* surface is one of the most interesting challenges of the mycobacterial field. The peculiar structure and composition of the *M. tuberculosis* cell envelope is involved in the resistance to environmen-

tal stresses such as desiccation or exposure to detergents and disinfectants, as well as in tolerance to several antibacterial drugs. Moreover, some of its components have been shown to be directly involved in the modulation of the immune response (Briken et al., 2004), and in the interference with macrophage trafficking events (Russell, 2007).

Because infection is a dynamic process, it is possible to hypothesize that *M. tuberculosis* adaptation involves modulation of its surface structure and composition in response to the changing environment. The finding that the exposure of *M. tuberculosis* to low oxygen tension stimulates the production of a cell wall-associated pigment supports this hypothesis (Cunningham et al., 2004).

Sigma factors are often involved in regulation of surface structure and composition in response to external signals. As shown in this chapter, at least six of the 13 sigma factors encoded in the *M. tuberculosis* genome are involved in surface biology. Even though at the current state of knowledge it is difficult to assign many of them specific functions, we can hypothesize that some of them, such as σ^F, σ^D and σ^E, regulate genes responsible for the maintenance of surface homeostasis following damage that can occur during stationary phase or following exposure to surface-damaging host effector molecules. Others such as σ^L and σ^M appear to be involved in the regulation of the composition of the mycobacterial surface in response to still unknown stimuli and can play an important role in modulating *M. tuberculosis* interactions with the host. We predict that the continuously increasing amount of data available on mycobacterial sigma factors will be of extreme value for understanding the biology of *M. tuberculosis* infection that may lead to novel approaches to fight tuberculosis.

REFERENCES

Ades, S. E. 2004. Control of the alternative sigma factor SigE in *Escherichia coli. Curr. Opin. Microbiol.* 7:157–162.

Agarwal, N., S. C. Woolwine, S. Tyagi, and W. R. Bishai. 2007. Characterization of the *Mycobacterium tuberculosis* sigma factor SigM by assessment of virulence and identification of SigM-dependent genes. *Infect. Immun.* 75:452–461.

Alland, D., A. J. Steyn, T. Weisbrod, K. Aldrich, and W. R. Jacobs, Jr. 2000. Characterization of the *Mycobacterium tuberculosis iniBAC* promoter, a promoter that responds to cell wall biosynthesis inhibition. *J. Bacteriol.* 182:1802–1811.

Appia-Ayme, C., and B. C. Berks. 2002. SoxV, an orthologue of the CcdA disulfide transporter, is involved in thiosulfate oxidation in *Rhodovulum sulfidophilum* and reduces the periplasmic thioredoxin SoxW. *Biochem. Biophys. Res. Commun.* 296:737–741.

Beaucher, J., S. Rodrigue, P. E. Jacques, I. Smith, R. Brzezinski, and L. Gaudreau. 2002. Novel *Mycobacterium tuberculosis* anti-sigma factor antagonists control SigF activity by distinct mechanisms. *Mol. Microbiol.* 45:1527–1540.

Braun, V., and S. Mahren. 2005. Transmembrane transcriptional control (surface signalling) of the *Escherichia coli* Fec type. *FEMS Microbiol. Rev.* 29:673–684.

Briken, V., S. A. Porcelli, G. S. Besra, and L. Kremer. 2004. Mycobacterial lipoarabinomannan and related lipoglycans: from biogenesis to modulation of the immune response. *Mol. Microbiol.* 53:391–403.

Brodin, P., I. Rosenkrands, P. Andersen, S. T. Cole, and R. Brosch. 2004. ESAT-6 proteins: protective antigens and virulence factors? *Trends Microbiol.* 12:500–508.

Brown, M. S., J. Ye, R. B. Rawson, and J. L. Goldstein. 2000. Regulated intramembrane proteolysis: a control mechanism conserved from bacteria to humans. *Cell* 100:391–398.

Butcher, B. G., and J. D. Helmann. 2006. Identification of *Bacillus subtilis* sigma-dependent genes that provide intrinsic resistance to antimicrobial compounds produced by Bacilli. *Mol. Microbiol.* 60:765–782.

Campbell, E. A., J. L. Tupy, T. M. Gruber, S. Wang, M. M. Sharp, C. A. Gross, and S. A. Darst. 2003. Crystal structure of *Escherichia coli* SigE with the cytoplasmic domain of its anti-sigma RseA. *Mol. Cell* 11:1067–1078.

Cao, M., P. A. Kobel, M. M. Morshedi, M. F. Wu, C. Paddon, and J. D. Helmann. 2002a. Defining the *Bacillus subtilis* SigW regulon: a comparative analysis of promoter consensus search, run-off transcription/macroarray analysis (ROMA), and transcriptional profiling approaches. *J. Mol. Biol.* 316:443–457.

Cao, M., T. Wang, R. Ye, and J. D. Helmann. 2002b. Antibiotics that inhibit cell wall biosynthesis induce expression of the *Bacillus subtilis* SigW and SigM regulons. *Mol. Microbiol.* 45:1267–1276.

Cao, M., and J. D. Helmann. 2004. The *Bacillus subtilis* extracytoplasmic-function SigX factor regulates modification of the cell envelope and resistance to cationic antimicrobial peptides. *J. Bacteriol.* 186:1136–1146.

Carr, M. D., M. J. Bloemink, E. Dentten, A. O. Whelan, S. V. Gordon, G. Kelly, T. A. Frenkiel, R. G. Hewinson, and R. A. Williamson. 2003. Solution structure of the *Mycobacterium tuberculosis* complex protein MPB70: from tuberculosis pathogenesis to inherited human corneal desease. *J. Biol. Chem.* 278:43736–43743.

Chaba, R., I. L. Grigorova, J. M. Flynn, T. A. Baker, and C. A. Gross. 2007. Design principles of the proteolytic cascade governing the SigE-mediated envelope stress response in *Escherichia coli*: keys to graded, buffered, and rapid signal transduction. *Genes Dev.* 21:124–136.

Charlet, D., S. Mostowy, D. Alexander, L. Sit, H. G. Wiker, and M. A. Behr. 2005. Reduced expression of antigenic proteins MPB70 and MPB83 in *Mycobacterium bovis* BCG strains due to a start codon mutation in *sigK*. *Mol. Microbiol.* 56:1302–1313.

Chen, P., R. E. Ruiz, Q. Li, R. F. Silver, and W. R. Bishai. 2000. Construction and characterization of a *Mycobacterium tuberculosis* mutant lacking the alternate sigma factor gene, *sigF*. *Infect. Immun.* 68:5575–5580.

Cole, S. T., R. Brosch, J. Parkhill, T. Garnier, C. Churcher, D. Harris, S. V. Gordon, K. Eiglmeier, S. Gas, C. E. Barry, III, F. Tekaia, K. Badcock, D. Basham, D. Brown, T. Chillingworth, R. Connor, R. Davies, K. Devlin, T. Feltwell, S. Gentles, N. Hamlin, S. Holroyd, T. Hornsby, K. Jagels, A. Krogh, J. McLean, S. Moule, L. Murphy, K. Oliver, J. Osborne, M. A. Quail, M. A. Rajandream, J. Rogers, S. Rutter, K. Seeger, J. Skelton, R. Squares, S. Squares, J. E. Sulston, K. Taylor, S. Whitehead, and B. G. Barrell. 1998. Deciphering the biology of *Mycobacterium tuberculosis* from the complete genome sequence. *Nature* 393:537–544.

Converse, S. E., and J. S. Cox. 2005. A protein secretion pathway critical for *Mycobacterium tuberculosis* virulence is conserved and functional in *Mycobacterium smegmatis*. *J. Bacteriol.* 187:1238–1245.

Cunningham, A. F., P. R. Ashton, C. L. Spreadbury, D. A. Lammas, R. Craddock, C. W. Wharton, and P. R. Wheeler. 2004. Tubercle bacilli generate a novel cell wall-associated pigment after long-term anaerobic culture. *FEMS Microbiol. Lett.* 235:191–198.

Dainese, E., S. Rodrigue, G. Delogu, R. Provvedi, L. Laflamme, R. Brzezinski, G. Fadda, I. Smith, L. Gaudreau, G. Palú, and R. Manganelli. 2006. Posttranslational regulation of *Mycobacterium tuberculosis* extracytoplasmic-function sigma factor SigL and roles in virulence and in global regulation of gene expression. *Infect. Immun.* 74:2457–2461.

Dartigalongue, C., D. Missiakas, and S. Raina. 2001. Characterization of the *Escherichia coli* SigE regulon. *J. Biol. Chem.* 276:20866–20875.

Darwin, A. J. 2005. The phage-shock-protein response. *Mol. Microbiol.* 57:621–628.

De Voss, J. J., K. Rutter, B. G. Schroeder, H. Su, Y. Zhu, and C. E. Barry III. 2000. The salicylate-derived mycobactin siderophores of *Mycobacterium tuberculosis* are essential for growth in macrophages. *Proc. Natl. Acad. Sci. USA* 97:1252–1257.

Dubey, V. S., T. D. Sirakova, M. H. Cynamon, and P. E. Kolattukudy. 2003. Biochemical function of *msl5* (*pks8* plus *pks17*) in *Mycobacterium tuberculosis* H37Rv: biosynthesis of monomethyl branched unsaturated fatty acids. *J. Bacteriol.* 185:4620–4625.

Geiman, D. E., D. Kaushal, C. Ko, S. Tyagi, Y. C. Manabe, B. G. Schroeder, R. D. Fleischmann, N. E. Morrison, P. J. Converse, P. Chen, and W. R. Bishai. 2004. Attenuation of late-stage disease in mice infected by the *Mycobacterium tuberculosis* mutant lacking the SigF alternate sigma factor and identification of SigF-dependent genes by microarray analysis. *Infect. Immun.* 72:1733–1745.

Giacomini, E., A. Sotolongo, E. Iona, M. Severa, M. E. Remoli, V. Gafa, R. Lande, L. Fattorini, I. Smith, R. Manganelli, and E. M. Coccia. 2006. Infection of human dendritic cells with a *Mycobacterium tuberculosis sigE* mutant stimulates production of high levels of interleukin-10 but low levels of CXCL10: impact on the T-cell response. *Infect. Immun.* 74:3296–3304.

Goulding, C. W., M. I. Apostol, S. Gleiter, A. Parseghian, J. Bardwell, M. Gennaro, and D. Eisenberg. 2004. Gram-positive DsbE proteins function differently from Gram-negative DsbE homologs. A structure to function analysis of DsbE from *Mycobacterium tuberculosis*. *J. Biol. Chem.* 279:3516–3524.

Green, D. H., and S. M. Cutting. 2000. Membrane topology of the *Bacillus subtilis* pro-SigK processing complex. *J. Bacteriol.* 182:278–285.

Gruber, T. M., and C. A. Gross. 2003. Multiple sigma subunits and the partitioning of bacterial transcription space. *Annu. Rev. Microbiol.* 57:441–466.

Hahn, M. Y., S. Raman, M. Anaya, and R. N. Husson. 2005. The *Mycobacterium tuberculosis* extracytoplasmic-function sigma factor SigL regulates polyketide synthases and secreted or membrane proteins and is required for virulence. *J. Bacteriol.* 187:7062–7071.

He, H., R. Hovey, J. Kane, V. Singh, and T. C. Zahrt. 2006. MprAB is a stress-responsive two-component system that directly regulates expression of sigma factors SigB and SigE in *Mycobacterium tuberculosis*. *J. Bacteriol.* 188:2134–2143.

Helmann, J. D. 2002. The extracytoplasmic function (ECF) sigma factors. *Adv. Microb. Physiol.* 46:47–110.

Hilbert, D. W., and P. J. Piggot. 2004. Compartmentalization of gene expression during *Bacillus subtilis* spore formation. *Microbiol. Mol. Biol. Rev.* 68:234–262.

Hsu, T., S. M. Hingley-Wilson, B. Chen, M. Chen, A. Z. Dai, P. M. Morin, C. B. Marks, J. Padiyar, C. Goulding, M. Gingery,

D. Eisenberg, R. G. Russell, S. C. Derrick, F. M. Collins, S. L. Morris, C. H. King, and W. R. Jacobs, Jr. 2003. The primary mechanism of attenuation of bacillus Calmette-Guerin is a loss of secreted lytic function required for invasion of lung interstitial tissue. *Proc. Natl. Acad. Sci. USA* 100:12420–12425.

Hughes, K. T., and K. Mathee. 1998. The anti-sigma factors. *Annu. Rev. Microbiol.* 52:231–286.

Karls, R. K., J. Guarner, D. N. McMurray, K. A. Birkness, and F. D. Quinn. 2006. Examination of *Mycobacterium tuberculosis* sigma factor mutants using low-dose aerosol infection of guinea pigs suggests a role for SigC in pathogenesis. *Microbiology* 152:1591–1600.

Le Brun, N. E., J. Bengtsson, and L. Hederstedt. 2000. Genes required for cytochrome c synthesis in *Bacillus subtilis*. *Mol. Microbiol.* 36:638–650.

Lonetto, M. A., K. L. Brown, K. E. Rudd, and M. J. Buttner. 1994. Analysis of the *Streptomyces coelicolor sigE* gene reveals the existence of a subfamily of eubacterial RNA polymerase sigma factors involved in the regulation of extracytoplasmic functions. *Proc. Natl. Acad. Sci. USA* 91:7573–7577.

Makinoshima, H., and M. S. Glickman. 2005. Regulation of *Mycobacterium tuberculosis* cell envelope composition and virulence by intramembrane proteolysis. *Nature* 436:406–409.

Manganelli, R., M. I. Voskuil, G. K. Schoolnik, and I. Smith. 2001. The *Mycobacterium tuberculosis* ECF sigma factor SigE: role in global gene expression and survival in macrophages. *Mol. Microbiol.* 41:423–437.

Manganelli, R., M. I. Voskuil, G. K. Schoolnik, E. Dubnau, M. Gomez, and I. Smith. 2002. Role of the extracytoplasmic-function sigma factor SigH in *Mycobacterium tuberculosis* global gene expression. *Mol. Microbiol.* 45:365–374.

Manganelli, R., L. Fattorini, D. Tan, E. Iona, G. Orefici, G. Altavilla, P. Cusatelli, and I. Smith. 2004a. The extra cytoplasmic function sigma factor SigE is essential for *Mycobacterium tuberculosis* virulence in mice. *Infect. Immun.* 72:3038–3041.

Manganelli, R., R. Provvedi, S. Rodrigue, J. Beaucher, L. Gaudreau, and I. Smith. 2004b. Sigma factors and global gene regulation in *Mycobacterium tuberculosis*. *J. Bacteriol.* 186:895–902.

Mascher, T., N. G. Margulis, T. Wang, R. W. Ye, and J. D. Helmann. 2003. Cell wall stress responses in *Bacillus subtilis*: the regulatory network of the bacitracin stimulon. *Mol. Microbiol.* 50:1591–1604.

Minnig, K., J. L. Barblan, S. Kehl, S. B. Moller, and C. Mauel. 2003. In *Bacillus subtilis* W23, the duet SigX-SigM, two sigma factors of the extracytoplasmic function subfamily, are required for septum and wall synthesis under batch culture conditions. *Mol. Microbiol.* 49:1435–1447.

Mukamolova, G. V., A. S. Kaprelyants, D. I. Young, M. Young, and D. B. Kell. 1998. A bacterial cytokine. *Proc. Natl. Acad. Sci. USA* 95:8916–8921.

Mukamolova, G. V., A. G. Murzin, E. G. Salina, G. R. Demina, D. B. Kell, A. S. Kaprelyants, and M. Young. 2006. Muralytic activity of *Micrococcus luteus* Rpf and its relationship to physiological activity in promoting bacterial growth and resuscitation. *Mol. Microbiol.* 59:84–98.

Mulder, N. J., R. Apweiler, T. K. Attwood, A. Bairoch, A. Bateman, D. Binns, P. Bradley, P. Bork, P. Bucher, L. Cerutti, R. Copley, E. Courcelle, U. Das, R. Durbin, W. Fleischmann, J. Gough, D. Haft, N. Harte, N. Hulo, D. Kahn, A. Kanapin, M. Krestyaninova, D. Lonsdale, R. Lopez, I. Letunic, M. Madera, J. Maslen, J. McDowall, A. Mitchell, A. N. Nikolskaya, S. Orchard, M. Pagni, C. P. Ponting, E. Quevillon, J. Selengut, C. J. Sigrist, V. Silventoinen, D. J. Studholme, R. Vaughan, and C. H. Wu. 2005. InterPro, progress and status in 2005. *Nucleic Acids Res.* 33:D201–D205.

Niederweis, M. 2003. Mycobacterial porins: new channel proteins in unique outer membranes. *Mol. Microbiol.* 49:1167–1177.

Paget, M. S., J. B. Bae, M. Y. Hahn, W. Li, C. Kleanthous, J. H. Roe, and M. J. Buttner. 2001. Mutational analysis of RsrA, a zinc-binding anti-sigma factor with a thiol-disulphide redox switch. *Mol. Microbiol.* 39:1036–1047.

Parish, T., D. A. Smith, G. Roberts, J. Betts, and N. G. Stoker. 2003. The *senX3-regX3* two-component regulatory system of *Mycobacterium tuberculosis* is required for virulence. *Microbiology* 149:1423–1435.

Quadri, L. E., J. Sello, T. A. Keating, P. H. Weinreb, and C. T. Walsh. 1998. Identification of a *Mycobacterium tuberculosis* gene cluster encoding the biosynthetic enzymes for assembly of the virulence-conferring siderophore mycobactin. *Chem. Biol.* 5:631–645.

Raman, S., T. Song, X. Puyang, S. Bardarov, W. R. Jacobs, Jr., and R. N. Husson. 2001. The alternative sigma factor SigH regulates major components of oxidative and heat stress responses in *Mycobacterium tuberculosis*. *J. Bacteriol.* 183:6119–6125.

Raman, S., R. Hazra, C. C. Dascher, and R. N. Husson. 2004. Transcription regulation by the *Mycobacterium tuberculosis* alternative sigma factor SigD and its role in virulence. *J. Bacteriol.* 186:6605–6616.

Raman, S., X. Puyang, T. Y. Cheng, D. C. Young, D. B. Moody, and R. N. Husson. 2006. *Mycobacterium tuberculosis* SigM positively regulates Esx secreted protein and nonribosomal peptide synthetase genes and down regulates virulence-associated surface lipid synthesis. *J. Bacteriol.* 188:8460–8468.

Repoila, F., N. Majdalani, and S. Gottesman. 2003. Small noncoding RNAs, co-ordinators of adaptation processes in *Escherichia coli*: the RpoS paradigm. *Mol. Microbiol.* 48:855–861.

Rodrigue, S., R. Provvedi, P. E. Jacques, L. Gaudreau, and R. Manganelli. 2006. The sigma factors of *Mycobacterium tuberculosis*. *FEMS Microbiol. Rev.* 30:926–941.

Rousseau, C., T. D. Sirakova, V. S. Dubey, Y. Bordat, P. E. Kolattukudy, B. Gicquel, and M. Jackson. 2003. Virulence attenuation of two Mas-like polyketide synthase mutants of *Mycobacterium tuberculosis*. *Microbiology* 149:1837–1847.

Russell, D. G. 2007. Who puts the tubercle in tuberculosis? *Nat. Rev. Microbiol.* 5:39–47.

Said-Salim, B., S. Mostowy, A. S. Kristof, and M. A. Behr. 2006. Mutations in *Mycobacterium tuberculosis* Rv0444c, the gene encoding anti-SigK, explain high level expression of MPB70 and MPB83 in *Mycobacterium bovis*. *Mol. Microbiol.* 62:1251–1263.

Sakoh, M., K. Ito, and Y. Akiyama. 2005. Proteolytic activity of HtpX, a membrane-bound and stress-controlled protease from *Escherichia coli*. *J. Biol. Chem.* 280:33305–33310.

Schnappinger, D., S. Ehrt, M. I. Voskuil, Y. Liu, J. A. Mangan, I. M. Monahan, G. Dolganov, B. Efron, P. D. Butcher, C. Nathan, and G. K. Schoolnik. 2003. Transcriptional adaptation of *Mycobacterium tuberculosis* within macrophages: insights into the phagosomal environment. *J. Exp. Med.* 198:693–704.

Sirakova, T. D., V. S. Dubey, M. H. Cynamon, and P. E. Kolattukudy. 2003. Attenuation of *Mycobacterium tuberculosis* by disruption of a *mas*-like gene or a chalcone synthase-like gene, which causes deficiency in dimycocerosyl phthiocerol synthesis. *J. Bacteriol.* 185:2999–3008.

Song, T., S. L. Dove, K. H. Lee, and R. N. Husson. 2003. RshA, an anti-sigma factor that regulates the activity of the mycobacterial stress response sigma factor SigH. *Mol. Microbiol.* 50:949–959.

Tam, C., and D. Missiakas. 2005. Changes in lipopolysaccharide structure induce the SigE-dependent response of *Escherichia coli*. *Mol. Microbiol.* 55:1403–1412.

Thackray, P. D., and A. Moir. 2003. SigM, an extracytoplasmic function sigma factor of *Bacillus subtilis*, is activated in response to cell wall antibiotics, ethanol, heat, acid, and superoxide stress. *J. Bacteriol.* **185:**3491–3498.

Thakur, K. G., A. M. Joshi, and B. Gopal. 2007. Structural and biophysical studies on two promoter recognition domains of the extra-cytoplasmic function sigma factor SigC from *Mycobacterium tuberculosis*. *J. Biol. Chem.* **282:**4711–4718.

Trivedi, O. A., P. Arora, A. Vats, M. Z. Ansari, R. Tickoo, V. Sridharan, D. Mohanty, and R. S. Gokhale. 2005. Dissecting the mechanism and assembly of a complex virulence mycobacterial lipid. *Mol. Cell.* **17:**631–643.

Tufariello, J. M., W. R. Jacobs, Jr., and J. Chan. 2004. Individual *Mycobacterium tuberculosis* resuscitation-promoting factor homologues are dispensable for growth in vitro and in vivo. *Infect. Immun.* **72:**515–526.

Vingadassalom, D., A. Kolb, C. Mayer, T. Rybkine, E. Collatz, and I. Podglajen. 2005. An unusual primary sigma factor in the *Bacteroidetes* phylum. *Mol. Microbiol.* **56:**888–902.

Waagmeester, A., J. Thompson, and J. M. Reyrat. 2005. Identifying sigma factors in *Mycobacterium smegmatis* by comparative genomic analysis. *Trends Microbiol.* **13:**505–509.

Wilson, M., J. DeRisi, H. H. Kristensen, P. Imboden, S. Rane, P. O. Brown, and G. K. Schoolnik. 1999. Exploring drug-induced alterations in gene expression in *Mycobacterium tuberculosis* by microarray hybridization. *Proc. Natl. Acad. Sci. USA* **96:**12833–12838.

Wu, Q. L., D. Kong, K. Lam, and R. N. Husson. 1997. A mycobacterial extracytoplasmic function sigma factor involved in survival following stress. *J. Bacteriol.* **179:**2922–2929.

The Mycobacterial Cell Envelope
Edited by M. Daffé and J.-M. Reyrat
© 2008 ASM Press, Washington, DC

Chapter 15

Biosynthesis of Mycobacterial Lipids by Multifunctional Polyketide Synthases

VIVEK T. NATARAJAN, DEBASISA MOHANTY, AND RAJESH S. GOKHALE

The cell envelope of mycobacterium consists of a dense network of unusual lipids and sugars. This impermeable barrier imparts resistance against hostile environments and to the commonly used antimicrobial agents. The complex cell envelope is proposed to be a major determinant of virulence and plays an active role in modulating the host immune response (Asselineau and Laneelle, 1998; Brennan and Nikaido, 1995; Daffe and Draper, 1998; Gokhale et al., 2007). The plasma membrane is made up of the typical triacyl glycerols and forms the base of the cell envelope. The lipomannan and lipoarabinomannan lipopolysaccharides are anchored to this bilayer through phosphatidyl inositol mannosides (PIMs) (Besra and Brennan, 1997; Takayama et al., 2005). Long-chain α-alkyl β-hydroxy mycolic acids are the core lipid constitutents of mycobacterial cell envelope (Takayama et al., 2005). These are esterified to the arabinogalactan to form the mycolic arabinogalactan (MAG), which link to the peptidoglycan through a phosphodiester bond. In addition, mycolic acids also exist as free glycolipids in the form of trehalose monomycolate (TMM) and trehalose dimycolate (TDM) (Rao et al., 2006). Other mycobacterial cell envelope lipids include sulfolipids (SL), polyacyl trehalose (PAT), mannosyl-β-1-phosphomycoketide (MPM), and diacyl trehalose (DAT), all of which require polyketide enzymatic machinery for their biosynthesis (Gokhale et al., 2007; Goren, 1972; Jackson et al., 2006; Matsunaga et al., 2004; Minnikin et al., 2002; Sirakova et al., 2001). Other polyketide-derived lipids include two structurally related members, phenolphthiocerol glycolipid (PGL) and phthiocerol dimycoserosates (PDIMs) (Onwueme et al., 2005), mycobacterial siderophores that are involved in iron sequestration (Ratledge, 2004) and glycopeptidolipids (GPLs) (Etienne et al., 2002) isolated from opportunistic environmental mycobacteria. Interestingly, the final assembly of mycolic acid biosynthesis also uses polyketide thiotemplate enzymology (Portevin et al., 2004). Fig. 1 depicts chemical structures of representative polyketide-derived mycobacterial lipids.

POLYKETIDE SYNTHASES IN MYCOBACTERIA

In 1998, the *Mycobacterium tuberculosis* genome sequencing project revealed a large number of proteins homologous to polyketide synthases (PKSs) (Cole et al., 1998), which typically produce secondary metabolites in *Streptomyces* and fungi (Hopwood, 1999; Walsh, 2004). For decades, polyketide natural products have been known for their importance in pharmaceuticals, as antibiotics, or as immunosuppressive agents or in veterinary feed. Because mycobacteria are not known to produce secondary metabolites, the repertoire of PKSs and nonribosomal peptide synthetases (NRPSs) in mycobacteria came as a surprise. Mycobacteria on the other hand were known for their unusually large collection of lipids that constitute the unusual cell envelope. A combination of chromatographic, mass spectrometric and nuclear magnetic resonance (NMR) studies have revealed the identity of many mycobacterial lipids, whose variety puzzled the researchers for over 80 years. In recent years, novel biosynthetic mechanisms involved in their production have been elucidated. It is now clear that mycobacteria use PKS machinery in alliance with the fatty acid synthases (FASs) to produce esoteric cell envelope lipids (Gokhale et al., 2007).

Before genome sequence information, Kolattukudy and coworkers (1997) had suggested involvement of pks loci in the production of DIM lipids.

Vivek T. Natarajan, Debasisa Mohanty, and Rajesh S. Gokhale • National Institute of Immunology, Aruna Asaf Ali Marg, New Delhi 110 067, India.

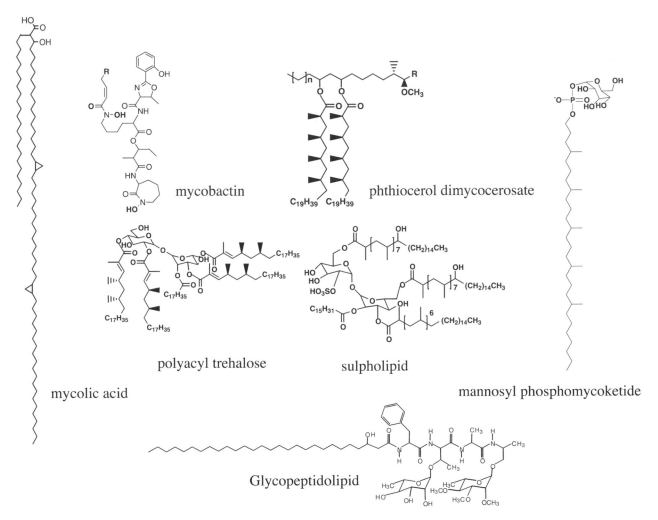

Figure 1. Chemical structures of polyketide derived lipids from mycobacteria.

Earliest indications emerged from the knockout of the *pps* cluster in *Mycobacterium bovis*, which implicated their role in PDIM synthesis. Postgenome, genetic disruption of *pks2* from *M. tuberculosis* resulted in the abrogation of sulfolipid production (Converse et al., 2003; Domenech et al., 2004; Sirakova et al., 2001). Similarly, disruption of pks13 homologue of *M. tuberculosis* in *Corynebacterium glutamicum,* which is a related organism, revealed its function in the formation of mycolates (Portevin et al., 2004). Although gene disruption studies of PKS-like genes from genetically tractable species across mycobacteria have discovered their importance in the synthesis of specific lipids, these were not informative in divulging the exact functioning of the PKS gene products. Polar effects during gene disruption and spontaneous variants arising in laboratory strains complicated the interpretations in some of these studies. Clarity in their functioning had to await biochemical characterization of PKS enzymes.

THE MULTICOMPONENT POLYKETIDE SYNTHASE MACHINERY

Polyketide synthases function as multifunctional enzymatic assembly lines (Bentley and Bennett, 1999; Cane and Walsh, 1999; Gokhale, 2001; Khosla et al., 1999; Walsh, 2004). They use a limited number of simple substrates like acyl-coenzyme A thioesters and synthesize a vast array of metabolites broadly termed polyketides. Polyketide biosynthesis resembles fatty acid synthesis and is carried out by a collection of enzymatic domains. The carbon backbone is synthesized by a stepwise decarboxylative condensation of small carboxylic acid thioesters using the chemistry of Claisen-type condensation (Fig. 2). In the case of fatty acid synthases, the two-carbon extender unit is provided by malonyl coenzyme A (MCoA), thereby extending the length of the acyl chain by a C_2 unit. The first step in the reaction involves the coupling of acetyl CoA (starter unit) with malonyl CoA (extender

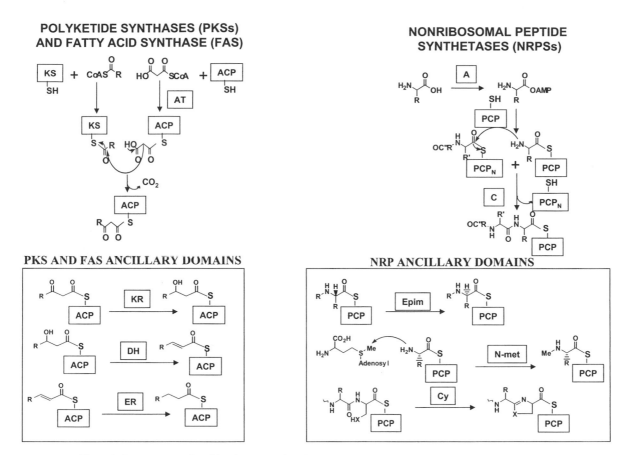

Figure 2. Reactions catalyzed by the essential and ancillary domains of PKS, FAS, and NRPS enzymes.

unit) to form an enzyme bound acetoacetyl thioester at the phosphopantetheine (P-pant) arm attached to the acyl carrier protein (ACP). Various catalytic domains of PKS and FAS are defined based on their function. The Acyl transferase (AT) domain selects the extender unit and transfers it to the growing polyketide chain, whereas the ketoacyl synthase (KS) domain catalyzes the decarboxylative condensation of the growing chain. KS, AT, and ACP form the core domains that are essential for -CH_2-CO- addition. The P-pant group serves as a flexible tether for both monomeric extender units as well as the growing acyl chain. This facilitates the movement of the growing polyketide to the different active sites while still being tethered covalently to the enzyme, a hallmark of enzymes that follow thioester-based assembly line synthesis.

NRPSs use similar template-based assembly line enzymology for synthesis of peptides (Challis and Naismith, 2004; Walsh, 2004). The core domains for NRPSs include an adenylation (A) domain for the selection and activation of amino acid monomers, a condensation (C) domain required for the formation of the peptide bond, and a thiolation or peptidyl carrier protein (T or PCP) domain with the swinging P-

pant arm to transfer the growing chain to various active sites. These correspond to the AT, KS and ACP domains, respectively, of PKS enzymes. The peptide bond formation in NRPS modules involves nucleophillic attack of the amino group of an amino acyl-S-PCP donor on the acyl group of the upstream electrophillic acyl- or peptidyl acyl-S-PCP chain, by the C domain (Fig. 2).

Ancillary domains in PKSs carry out ketoreductase (KR), dehydratase (DH), or enoyl reductase (ER) reactions. The presence of these additional domains dictates the degree of reduction observed in the polyketides. In FAS, all of these domains are functional, which results in a completely reduced hydrocarbon chain. Certain PKSs in addition possess methyl transferase domain and acyl CoA ligase domains. Ancillary domains present in NRPS modules include epimerization, N-methylation, or heterocyclization activities (Fig. 2).

Based on the protein architecture, PKSs can be classified into three main categories (Gokhale, 2001; Khosla et al., 1999). Type I PKSs are assemblies of multifunctional polypeptides; type II PKSs contain these domains on individual polypeptides that form a noncovalent complex. Type III PKSs are recently discovered in bacteria and now encompasses the plant chalcone synthases (Austin and Noel, 2003). These are structurally and mechanistically distinct from type I and II members and do not possess acyl carrier domains. It is interesting to note that mycobacteria contain all three types of proteins. Type I PKSs are further divided into modular and iterative. Modular proteins consist of minimal domains (KS, AT, and ACP) with various ancillary domains, and these are used once during the entire catalytic cycle. Thereby the final polyketide production requires as many modules as the number of ketide units it possesses. This is exemplified by the Pps A-E proteins from the Pps cluster, which is involved in PDIM synthesis (Fig. 3) (Cox et al., 1999; Trivedi et al., 2005). Iterative enzymes use the same set of active sites repeatedly (similar to eukaryote type FAS) to form the final product. Mycocerosic acid synthase (MAS) from the same Pps cluster is an excellent example of a type I iterative enzyme (Mathur and Kolattukudy, 1992; Trivedi et al., 2005). In contrast, type III PKSs function iteratively, using a single active site that carries out decarboxylation, condensation and cyclization reactions. The acyl CoA starter is covalently acylated to the active site Cys, and the extender unit is bound noncovalently at the active site pocket. PKS18 protein (Rv1372) is an example of type III PKS (Sankaranarayanan et al., 2004; Saxena et al., 2003).

A KNOWLEDGE-BASED APPROACH FOR UNDERSTANDING PKS AND NRPS ENZYMES

In the last two decades, several gene clusters of polyketide and nonribosomally derived peptide natural products have been sequenced. The correspondence of enzymatic domains to chemical structures provided an opportunity to computationally rationalize biosynthesis of these natural products. Development of such a database resource also provides a prospect to predict functions of newly identified PKS and NRPS gene clusters. We have developed a web-based PKS-NRPS prediction server (http://www.nii.res.in/nrps-pks.html) (Ansari et al., 2004; Yadav et al., 2003a, 2003b). It is a knowledge-based software that apart from correctly identifying catalytic domains in PKSs and NRPSs, also facilitates prediction of substrate specificity for the starter and extender moieties. It predicts the presence of ACP domains that are difficult to assign based on Conserved Domain Database (CDD) searches. NRPS-PKS can identify different reductive domains in PKS proteins. CDD search often fails to distinguish between such reductive domains like DH, ER and KR. Another major challenge in deciphering the chemical structure

of the polyketide product of an unknown PKS gene cluster is to predict the substrate specificity of the AT domain. The program does this by analyzing the chemical structures of known PKS products and the structure of the E. coli FAS AT domain. Using a homology modeling approach, the putative active site residues of AT domains of known specificity are identified and substrate specificity of a query AT domain is predicted based on a match of active site residues. The knowledge base used in various predictions made by NRPS-PKS is based on a comprehensive analysis of various families of PKSs and NRPSs having known metabolic products. Information on chemical structures of these complex metabolites, the sequence stretches corresponding to various catalytic domains and their active site residues as well as substrate specificities are organized as a set of inter-linked databases. Highlights and utility of each of the databases is summarized briefly as follows.

PKSDB

This database contains the sequences of various type I modular PKS clusters having known polyketide products. It gives intuitive pictorial depiction of domain organization along with chemical structures of the polyketide products for easy identification of chemical moieties added by various modules. NRPS-PKS server provides novel clues about putative polyketide products of new modular PKS clusters based on comparison with sequence and structural features of PKS domains in PKSDB.

ITERDB

Combines information from 21 type I iterative PKSs. Apart from depiction of catalytic domains, ITERDB provides information on number of iterative condensation steps catalyzed by each of the iterative PKSs cataloged in ITERDB.

CHSDB

By virtue of being a single monofunctional protein, it is relatively easy to identify type III PKS proteins in a genome. However, prediction of their polyketide products remain a challenging task in view of the diversity of potential starter and extender units and various different types of cyclization reactions catalyzed by this enzyme family. The wealth of information from chalcone synthase (CHS)-like proteins from plant kingdom and their homology-based structural analysis have resulted in the identification of 32 residues that determine CoA binding, selection of starter/extender substrates, and various types of

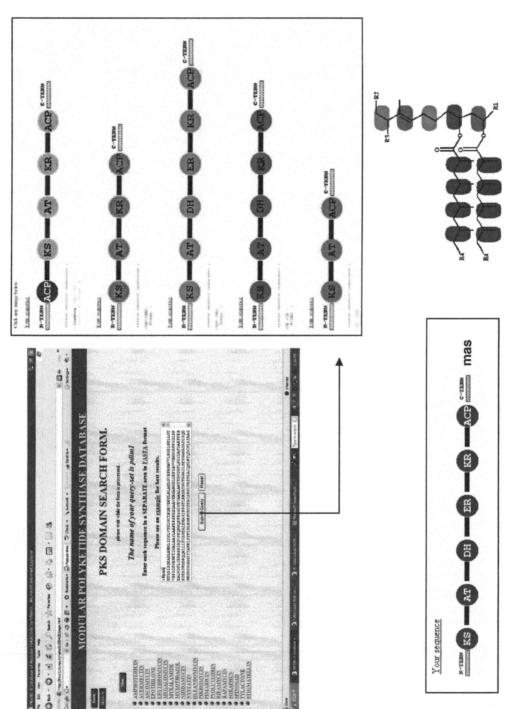

Figure 3. Pictorial depiction of domains predicted in *PpsA-E* genes and Mas, by PKSDB and ITERDB. The ketide units ir the final PDIM structure are indicated by colors similar to the synthesizing modules. (*See the color insert for the color version of this figure.*)

cyclization of the polyketide chains. CHSDB stores this information for various experimentally characterized plant and bacterial CHS proteins.

NRPSDB

The sequence similarity of the C, A, T, cyclization (Cy), epimerization (E), methyl transferase (MT) and thioesterase (TE) domains is poor among their respective families; hence, this program uses the results from fold recognition by Genthreader to assign identities and boundaries for various catalytic domains in NRPS proteins. Similar to the AT domains of PKSs, NRPSDB also catalogs information about substrate specificities of various adenylation domains based on the comparison of their putative active site residues with the substrate bound crystal structure of grsA adenylation domain. Thus, domain organization and substrate specificities of NRPS proteins can be predicted based on comparison with NRPS clusters cataloged in this database.

With the availability sequence information of mycobacterial species, such a knowledge-based approach for function prediction has facilitated research on novel PKS and NRPS genes, and has been a tremendous resource for characterizing novel enzymatic functions. Table 1 lists the type I multifunctional PKS proteins that have been identified in 11 different mycobacterial genomes. Fig. 3 demonstrates a combined PKSDB and ITERDB output illustrated for mycobacterial *pps* cluster. Cell-free reconstitution of these modular proteins substantiated the predictions of these domains (Trivedi et al., 2005). The software also suggested that the ER domain was missing from the PpsD protein (Yadav et al., 2003a). This was strange because DIM products consist of a saturated carbon chain at this position (shown in red color in Fig. 3 in the color insert). Thus, this raised the possibility that an independent protein was converting a double bond into a saturated chain. Indeed, a recent study has confirmed that another open reading frame (ORF) from this Pps cluster is involved this reduction (C. Chalut, personal communication). Similarly, analysis through CHSDB for Rv1372 had predicted that this protein would not be able to use typical plant type III starter units. Experimental studies indeed confirmed this and showed that this PKS18 protein used long-chain acyl CoAs to produce unusually long tri- and tetraketide pyrones (Saxena et al., 2003).

The PKS-NRPS is also a powerful tool for engineering so-called designer natural products. The extender unit selection is mediated by the AT domain, and insights into the substrate-specifying residues indicated that the discrimination between MCoA and methyl malonyl CoA (MMCoA) is determined by a single residue (Yadav et al., 2003a). Based on this, mutation of Ser 726 to a bulkier Phe in mycocerosic acid synthase (Mas) resulted in the production of linear fatty acid chain instead of methyl branched chain (Trivedi et al., 2005). Changing the specificity of substrate, in addition to giving insights into the functioning of the enzyme, paves the way for making novel me-tabolites. This in turn could alter cell functions and/or its interaction with the host.

FUNCTIONAL INTERPLAY BETWEEN FAAL AND PKS IN THE GENERATION OF POLYKETIDE LIPIDS

PKS and NRPS enzymes do not function in isolation and require a battery of other enzymatic activities for making polyketide lipids. One such enzymatic cross-talk between homologues of FadD proteins with PKS became clear only recently. The *M. tuberculosis* genome contains a large number of genes that are homologous to *Escherichia coli* FadD proteins, which convert fatty acids to fatty acyl CoAs, the first step toward their degradation by β-oxidation. Sequence analysis revealed only a weak similarity to adenylation domains of NRPS and the acyl-activating enzymes like firefly luciferase. Although the sequence similarity among the proteins is low, the 34 proteins could be clustered into two distinct groups. Interestingly, most of the proteins belonging to the smaller cluster are found to be proximal to *pks* or *nrps* loci in the mycobacterial genome (Trivedi et al., 2004). Biochemical assays with several proteins revealed that although proteins from both clusters form acyl-adenylate, many of the proteins from the larger cluster converted these acyl-adenylate intermediates to acyl CoA upon addition of CoASH. In contrast, the proteins from the other cluster failed to convert them to their CoA derivates. The former cluster represents classical fatty acyl-CoA ligases (FACLs) that are omnipresent. The proteins from the smaller cluster constituted a new family of fatty acyl-adenylate ligase (FAAL) analogous to the adenylation domains of NRPSs, which form amino acyl-adenylates. The PKS proteins present adjacent to FAALs revealed an interesting pattern of the presence of an extra N-terminal ACP domain. Similar to aminoacyl-adenylates that are transferred to the thiolation domain for biosynthesis, we proposed a similar function for FAAL proteins.

FAAL proteins were shown to transfer the long-chain fatty acyl chain to their cognate PKS proteins in an ATP-dependent manner (Fig. 4A). A functional

Table 1. Type I polyketide synthases across various species and strains of mycobacteria[a]

Strain and synthase	Functional domains
M. avium 104	
MAV_0218	ACP-KS-AT-TE
MAV_1321	KS-AT-DH-ER-KR-ACP
MAV_1763	KS-AT-KR-ACP-KS-AT-DH-ER-KR-ACP
MAV_2011	KS
MAV_2012	AT-KR-ACP
MAV_2370	KS-AT-DH-ER-KR-ACP
MAV_2450	KS-AT-DH-ER-KR-ACP-KS-AT-DH-ER-KR-ACP
MAV_3107	KS-AT-ACP
MAV_3108	KS-AT-DH-ER-KR-ACP
MAV_3109	KS-AT-DH-ER-KR-ACP
MAV_4343	KS
M. bovis BCG strain Pasteur 1173P2	
BCG_0443	ACP-KS
BCG_0444	AT-ACP-TE
BCG_1243	KS-AT-DH-ER-KR-ACP
BCG_1579c	KS-AT-DH-ER-KR-ACP
BCG_1700	KS-AT-DH-ER-KR-ACP
BCG_1701	KS-AT-DH-ER
BCG_1702	ER-KR-ACP
BCG_1703	KS-AT-ACP
BCG_2067c	KS-AT-DH-ER-KR-ACP-KS-AT-DH-ER-KR-ACP
BCG_2395c	AT-KR-ACP
BCG_2396c	KS
BCG_2953	ACP-KS-AT-KR-ACP
BCG_2954	KS-AT-KR-ACP
BCG_2955	KS-AT-DH-ER-KR-ACP
BCG_2956	KS-AT-DH-KR-ACP
BCG_2957	KS-AT-ACP-C
BCG_2962c	KS-AT-DH-ER-KR-ACP
BCG_2968c	KS-AT-DH-ER-KR-ACP
BCG_3862c	ACP-KS-AT-TE
BCG_3888c	KS-AT-DH-ER-KR-ACP
M. leprae TN	
ML0101	ACP-KS-AT-TE
ML0135	KS-AT-DH-ER-KR-ACP
ML0139	KS-AT-DH-ER-KR-ACP
ML1229	KS-AT-DH-ER-KR-ACP
ML2353	KS-AT-ACP-C
ML2354	KS-AT-KR-ACP
ML2355	KS-AT-DH-ER-KR-ACP
ML2356	KS-AT-KR-ACP
ML2357	ACP-KS-AT-KR-ACP
Mycobacterium sp. strain MCS	
Mmcs_0244	KS-AT-KR-ACP-KS-AT-DH-ER-KR-ACP
Mmcs_0783	KS
Mmcs_2832	KS-AT-ACP-C
Mmcs_2833	KS-AT-KR-ACP
Mmcs_2834	KS-AT-ACP
Mmcs_2835	ACP-KS-AT-KR-ACP
Mmcs_3119	KS-AT-DH-ER-KR-ACP
Mmcs_3467	AT-KR-ACP
Mmcs_3468	KS
Mmcs_3472	KS-AT-KR-ACP
Mmcs_5010	ACP-KS-AT-TE
M. bovis AF2122/97	
Mb0412	ACP-KS
Mb0413	AT-ACP-TE

Continued on following page

Table 1. *Continued*

Strain and synthase	Functional domains
Mb1213	KS-AT-DH-ER-KR-ACP
Mb1554c	KS-AT-DH-ER-KR-ACP
Mb1689	KS-AT-DH-ER-KR-ACP
Mb1690	KS-AT-DH-ER
Mb1691	ER-KR-ACP
Mb1692	KS-AT-ACP
Mb2074c	KS-AT-DH-ER-KR-ACP-KS-AT-DH-ER-KR-ACP
Mb2402c	AT-KR-ACP
Mb2403c	KS
Mb2956	ACP-KS-AT-KR-ACP
Mb2957	KS-AT-KR-ACP
Mb2958	KS-AT-DH-ER-KR-ACP
Mb2959	KS-AT-DH-KR-ACP
Mb2960	KS-AT-ACP-C
Mb2965c	KS-AT-DH-ER-KR-ACP
Mb2971c	KS-AT-DH-ER-KR-ACP
Mb3830c	ACP-KS-AT-TE
Mb3855c	KS-AT-DH-ER-KR-ACP
M. smegmatis strain MC^2 155	
MSMEG_0408	KS-AT-KR-ACP-KS-AT-DH-ER-KR-ACP
MSMEG_1204	KS
MSMEG_4512	AT-KR-ACP
MSMEG_4513	KS
MSMEG_4727	KS-AT-DH-ER-KR-ACP
MSMEG_6392	ACP-KS-AT-TE
MSMEG_6629	AT
MSMEG_6767	KS-AT-KR-ACP
M. avium subsp. *paratuberculosis* K-10	
MAP0220	ACP-KS-AT-TE
MAP1370	KS-AT-DH-ER-KR-ACP
MAP1371	KS-AT-ACP
MAP1796c	KS-AT-DH-ER-KR-ACP-KS-AT-DH-ER-KR-ACP
MAP1867c	KS-AT-DH-ER-KR-ACP
MAP2174c	AT-KR-ACP
MAP2175c	KS
MAP2230c	KS-AT-KR-ACP-KS-AT-DH-ER-KR-ACP
MAP2603	DH-ER-KR-ACP
MAP2604c	KS-AT
MAP3485	KS
MAP3764c	KS-AT-DH-ER-KR-ACP
M. ulcerans Agy99	
MUL_1655	KS-AT-ACP
MUL_2005	KS-AT-DH-ER-KR-ACP
MUL_2010	KS-AT-DH-ER-KR-ACP
MUL_2015	KS-AT-ACP-C
MUL_2016	KS-AT-DH-KR-ACP
MUL_2017	KS-AT-DH-ER-KR-ACP
MUL_2018	KS-AT-KR-ACP
MUL_2019	ACP-KS-AT-KR-ACP
MUL_2266	KS-AT-DH-ER-KR-ACP-KS-AT-DH-ER-KR-ACP
MUL_3637	KS
MUL_3638	AT-KR-ACP
MUL_4983	ACP-KS-AT-TE
M. vanbaalenii PYR-1	
Mvan_0269	KS-AT-KR-ACP-KS-AT-DH-ER-KR-ACP
Mvan_1000	KS-AT-ACP-C-A-ACP-C
Mvan_1003	KS-AT-KR-ACP-KS-AT-DH-ER-KR-ACP
Mvan_3123	KS-AT-ACP-C

Table 1. *Continued*

Strain and synthase	Functional domains
Mvan_3124	KS-AT-KR-ACP
Mvan_3125	KS-AT-KR-ACP
Mvan_3126	ACP-KS-AT-KR-ACP
Mvan_3128	ACP-KS-AT-KR-ACP
Mvan_3754	AT
Mvan_3846	AT-KR-ACP
Mvan_3847	KS
Mvan_3852	KS-AT-KR-ACP
Mvan_4950	KS
Mvan_5640	ACP-KS-AT-TE
M. tuberculosis H37Rv	
Rv0405	ACP-KS-AT-ACP-TE
Rv1180	KS
Rv1181	AT-DH-ER-KR-ACP
Rv1527c	KS-AT-DH-ER-KR-ACP
Rv1661	KS-AT-DH-ER-KR-ACP
Rv1662	KS-AT-DH-ER
Rv1663	ER-KR-ACP
Rv1664	KS-AT-ACP
Rv2048c	KS-AT-DH-ER-KR-ACP-KS-AT-DH-ER-KR-ACP
Rv2381c	AT-KR-ACP
Rv2382c	KS
Rv2931	ACP-KS-AT-KR-ACP
Rv2932	KS-AT-KR-ACP
Rv2933	KS-AT-DH-ER-KR-ACP
Rv2934	KS-AT-DH-KR-ACP
Rv2935	KS-AT-ACP-C
Rv2940c	KS-AT-DH-ER-KR-ACP
Rv2946c	AT-DH-ER-KR-ACP
Rv2947c	KS
Rv3800c	ACP-KS-AT-TE
Rv3825c	KS-AT-DH-ER-KR-ACP
M. tuberculosis CDC1551	
MT0418	ACP-KS-AT-ACP-TE
MT1218	KS-AT-DH-ER-KR-ACP
MT1701	KS-AT-DH-ER-KR-ACP
MT1702	KS-AT-DH-ER
MT1703	ER-KR-ACP
MT1704	KS-AT-ACP
MT2108	KS-AT-DH-ER-KR-ACP-KS-AT-DH-ER-KR-ACP
MT2449	AT-KR-ACP
MT2450	KS
MT3000	ACP-KS-AT-KR-ACP
MT3002	KS-AT-KR-ACP
MT3003	KS-AT-DH-ER-KR-ACP
MT3004	KS-AT-DH-KR-ACP
MT3005	KS-AT-ACP-C
MT3010	KS-AT-DH-ER-KR-ACP
MT3018	AT-DH-ER-KR-ACP
MT3021.1	KS
MT3907	ACP-KS-AT-TE
MT3933	KS-AT-DH-ER-KR-ACP

*a*Different domains are annotated based on the PKSDB prediction server.

interplay between FadD26 with PpsA, FadD32 with PKS13, and FadD30 with PKS6 was thus established (Trivedi et al., 2004). Perhaps this justifies the non-redundant role of multiple FadD proteins in the genome of mycobacteria.

PDIM SYNTHESIS BY PPSA-E AND MAS

PDIM synthesis machinery highlights the biochemical logic of a modular polyketide lipid biosynthesis. This carbon backbone synthesis can be deconvoluted

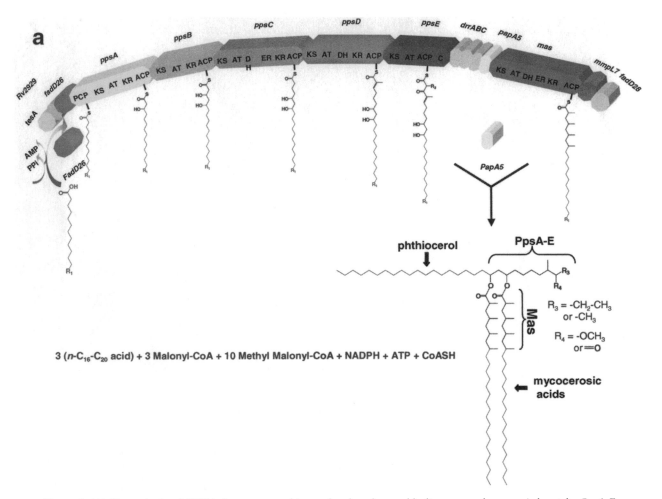

Figure 4. (A) Biosynthesis of PDIM demonstrates thiotemplate-based assembly line enzymology carried out by PpsA-E, FadD26, Mas, and PapA5 proteins. (B) Mycobactin synthesis by mbt-1 and -2 clusters. mbt-1 cluster proteins synthesize the mycobactin core, and mbt-2 cluster proteins along with MbtG modify the core structure of mycobactin. *(See the color insert for the color version of this figure.)*

into four key steps: priming long-chain fatty acid and synthesis of the diol-component of phthiocerol, synthesis of phthiocerol, synthesis of mycocerosic acid, and finally, the condensation of mycocerosic acid to the diol of phthiocerol (Fig. 4a). FadD26 initiates the priming of PpsA by transferring acyl group (C_{12}-C_{16}) to the PCP domain. PpsA converts it to acyl diketide. The next PpsB protein extends the length by another ketide unit. Formation of the diol results from the absence of DH and ER domains in PpsA and PpsB. The AT domains of PpsC and PpsD have MCoA and MMCoA specificities, respectively. The final step to synthesize the diol-pentaketide product is performed by PpsE, which has flexibility to incorporate either MCoA or MMCoA to form a β-keto acid of the diol.

Synthesis of the mycocerosic acid component by *M. tuberculosis* MAS demonstrated that the enzyme can use fatty acids C_6-C_{16} as a starter unit and MMCoA as the extender. Four cycles of iteration would result in the formation of tetramethyl-

branched mycocerosic acid. In vitro studies suggest that these enzymes could have tremendous flexibility, which, in turn, could provide diversity in vivo. MAS enzyme could use starters of varying lengths and also carry out varying cycles of iterations, thereby producing a spectrum of products differing in -CH_2- units and/or the number of methyl branched fortifications in these lipids. Biosynthesis of the final PDIM molecule involves transferring mycocerosic acid to the diol. An open reading frame in the neighborhood (predicted to be a polyketide associated protein) PapA5 was able to transesterify the enzyme-bound MAS products. All of these reactions were recently demonstrated in a cell-free system, by cloning and expression of these enzymes and demonstration of the metabolites by mass spectrometry (Trivedi et al., 2004; Trivedi et al., 2005). Other open reading frames adjacent to the *pps* cluster are involved in PGL synthesis, and they are discussed in subsequent chapters.

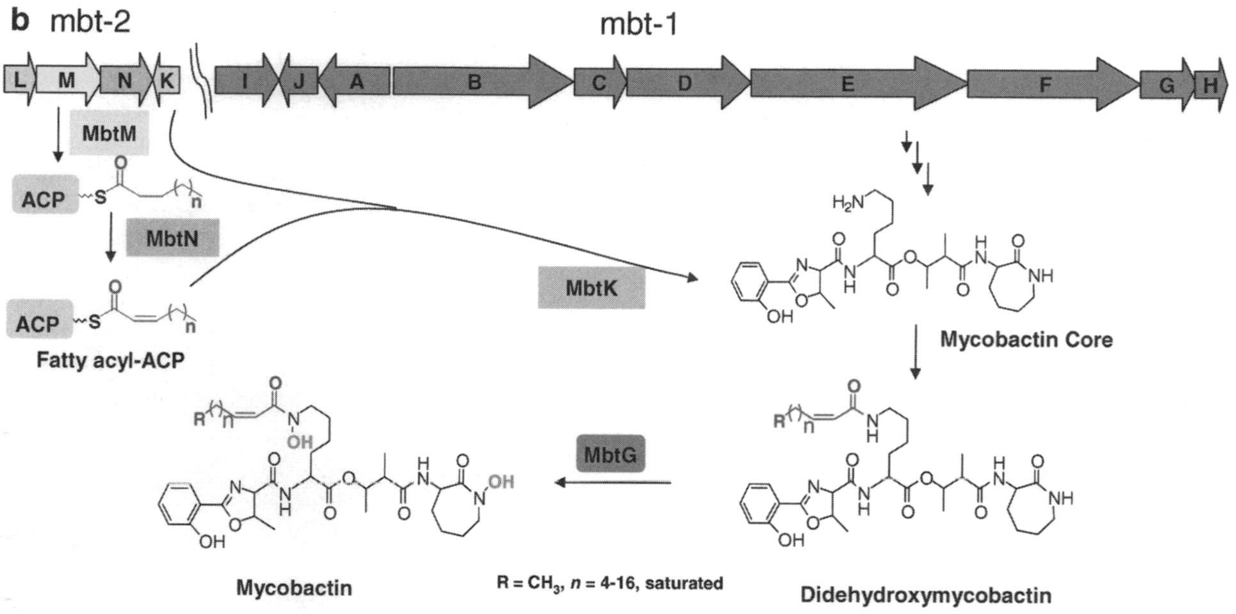

Figure 4. *Continued*

MYCOBACTIN SYNTHESIS BY MBT1 AND MBT2 CLUSTER

Mycobactin is a lipopeptide containing six amino acids including a lysine that gets modified by tailoring enzymes. The major biosynthetic locus Mbt1 consists of a hybrid NRPS-PKS cluster that also encodes salicylate-AMP ligase, isochorismate synthetase, and a hydroxylase (De Voss et al., 2000; LaMarca et al., 2004; Quadri et al., 1998). These enzymes synthesize the core of the mycobactin, which is further modified by another cluster Mbt2 positioned at a different locus (Fig. 4b) (Krithika et al., 2006). Coexpression of the two clusters is believed to be regulated by an iron-dependent repressor (IdeR) (Rodriguez and Smith, 2003). Unconventional tailoring enzyme activities from this Mbt2 cluster follow the thioester template-based catalysis, conceivably an efficient way of generating specificity by channelizing the metabolites for biosynthesis. FadD33 (MbtM) activates lauroyl AMP and loads it to the ACP protein. The N-acylation of the lysine (Nε) is carried out by a novel protein Rv1347c (MbtK). A FadE homologue, FadE14 (MbtN), brings about unsaturation of the fatty acyl chain at the Cα-Cβ bond using the reducing equivalents of FADH$_2$. FadE14 protein prefers ACP-bound substrates and thus differs from atypical acyl-CoA dehydrogenases that catalyze the formation of a trans-double bond during β-oxidation of fatty acids. However, the stereochemistry of the double bond produced by FadE14 protein has not been elucidated. It would be interesting to understand how a *cis*-double bond is formed in the final

mycobactin structure. Formation of the iron-chelating hydroxamate group is generated by *mbtG*, a protein predicted to be a lysine N6-hydroxylase (Krithika et al., 2006). This enzyme belongs to a class of flavoprotein monooxygenase and uses molecular oxygen for the hydroxylation. These exquisite tailoring enzymes modify the mycobactin core and render an amphiphillic iron-chelating property together with dynamic diffusibility to the pks-nrps metabolite.

Sulfolipid Synthesis by PKS2

Sulfolipids consist of phthioceronic acid and hydroxyphthioceronic acids acylated along with fatty acids to a sulfated trehalose core to form sulfolipids. Knockout studies indicated the role of the *pks2* gene in the formation of these sulfolipids. PKS2 is an iterative enzyme, and the NRPS-PKS database predicts AT domain selectivity toward MMCoA. We have recently reconstituted PKS2 enzyme and have demonstrated biosynthesis of methyl-branched phthioceronic acids. Hydroxylation could be brought about by an external P$_{450}$ hydroxylase or more elegantly by skipping the DH and ER domains during the first cycle of iteration. We also show that fatty acid starters could be loaded by a FAAL homologue, FadD23 (T. Chopra and R. S. Gokhale, unpublished results). Sulfation of trehalose moiety involves a sulfotransferase, Stf0 (Mougous et al., 2002; Mougous et al., 2006), and the genetic inactivation of PapA1 and PapA2 suggests their requirement for the transfer of phthioceronic acids (Bhatt et al., 2007).

MYCOKETIDE SYNTHESIS BY PKS12

Rv2048c (PKS12) is a bimodular enzyme containing two complete sets of catalytic sites in a single large open reading frame (Table 1). Genetic studies have indicated its role in mannosyl-β-1-mycoketide synthesis (Matsunaga et al., 2004). The mycoketide chain contains an acyl chain with branching at every fifth carbon. It can be envisaged that the two modules alternately add a methyl malonyl and a malonyl group to result in such a branching pattern. Based on analysis from NRPS-PKS software, the AT domain of the first module shows specificity for MMCoA and that of the second module to MCoA (Chopra and Gokhale, unpublished results). However, Pks12 poses an interesting paradigm for iteration because it contains two modules instead of one module; it would be interesting to understand the mechanism of chain transfer in this enzyme system.

GLYCOPEPTIDOLIPID SYNTHESIS

Glycopeptidolipids constitute a major class of outer-layer lipids in various species of mycobacteria, including *M. avium* and *M. smegmatis* (Chatterjee and Khoo, 2001). GPL consists of a peptide core, a lipidic side chain, and sugar units. The hydroxyl groups of D-allo Thr and L-alaninol are modified by glycosylation with deoxytalose and rhamnose sugars. The locus identified by gene disruption studies in *M. smegmatis* suggests the involvement of glycosyl transferase (Deshayes et al., 2005) and O-methyl transferase (Jeevarajah et al., 2004) in GPL production. The locus also contains genes involved in lipid transport, NRPS proteins, a FadE homologue, and PKS; these could be involved in synthesizing the core of GPL molecule. Biochemical characterization is required to fully elucidate the role of different enzymes in the biosynthesis of GPL.

PROSPECTS

Recent years have provided significant insights into the mechanism by which mycobacteria synthesize an array of complex lipids. The clever ways in which mycobacteria use polyketide biosynthetic machinery to produce these metabolites is quite remarkable. Before this, PKS proteins were not known to participate in the biosynthesis of lipids. Mycobacterial PKS proteins have thus provided a paradigm shift in terms of our understanding of these assembly-line megasynthases. Interestingly, a type III PKS from *Azotobacter* has also been recently demonstrated to produce phenolic lipids, which are important for the formation of a dormant cyst (Funa et al., 2006). A mycobacterial cell envelope, along with complex lipids and polysaccharides, is also associated with numerous proteins that influence cellular structure and its interactions with host cells. Despite considerable progress in characterization of the major cellular components of *M. tuberculosis*, there is little information on the nature of proteins associated with the cell envelope. Because the cell envelope constitutes the key interface between pathogen and host, the cell wall-associated proteins and lipids are presumably key determinants of pathogenesis and immunogenicity. There is also mounting evidence that despite the lack of significant genetic heterogeneity between strains of *M. tuberculosis,* the variable expression leading to phenotypic heterogeneity contributes toward virulence and persistence of this major killer of humankind (Rao et al., 2006; Reed et al., 2004). With tremendous repertoire of lipid-modifying enzymes, many of which show promiscuity in their substrate specificity (Arora et al., 2005), it is tempting to speculate that this pathogen remodels its surface by using alternative substrates to suit a changing host environment. Dynamic remodeling also brings forth the possibility of these lipids to act as a source of nutrition. The highly reduced carbon backbone of these lipids and their water-insoluble property render them as excellent storage materials and can be used at times of starvation, probably through the β-oxidation pathway.

In the present genomic era, many promising targets for developing antitubercular drugs are being characterized. Polyketide enzymes are excellent targets for drug discovery for a multitude of reasons. However, presently specific inhibitors of PKSs are not known. FAAL proteins that feed long-chain fatty acids into PKS enzymes for the production of complex lipids also could prove to be important drug targets. We have designed new inhibitors of FadD proteins that simultaneously target multiple enzymes and thus show potent inhibition of mycobacterial growth (P. Arora and R. S. Gokhale, unpublished results). The so-called one-disease-one-drug-one-target paradigm that has dominated thinking in the pharmaceutical industry for the past few decades is now being increasingly challenged by compounds designed to bind to more than one target (Morphy and Rankovic, 2007). It is now widely acknowledged that high specificity for a single target might not always deliver the required efficacy versus side effect profile. The philosophy of developing a single entity such that it could act as multiple ligands by inhibiting many enzymes is in tune with the emerging drug discovery paradigm.

Acknowledgment. Rajesh S. Gokhale is an honorary member of the faculty at Jawaharlal Nehru Centre for Advanced Scientific Research, Bangalore, India.

REFERENCES

Ansari, M. Z., G. Yadav, R. S. Gokhale, and D. Mohanty. 2004. NRPS-PKS: a knowledge-based resource for analysis of NRPS/PKS megasynthases. *Nucleic Acids Res.* **32:**W405–W413.

Arora, P., A. Vats, P. Saxena, D. Mohanty, and R. S. Gokhale. 2005. Promiscuous fatty acyl CoA ligases produce acyl-CoA and acyl-SNAC precursors for polyketide biosynthesis. *J. Am. Chem. Soc.* **127:**9388–9389.

Asselineau, J., and G. Laneelle. 1998. Mycobacterial lipids: a historical perspective. *Front. Biosci.* **3:**e164–e174.

Austin, M. B., and J. P. Noel. 2003. The chalcone synthase superfamily of type III polyketide synthases. *Nat. Prod. Rep.* **20:**79–110.

Bentley, R., and J. W. Bennett. 1999. Constructing polyketides: from collie to combinatorial biosynthesis. *Annu. Rev. Microbiol.* **53:**411–446.

Besra, G. S., and P. J. Brennan. 1997. The mycobacterial cell wall: biosynthesis of arabinogalactan and lipoarabinomannan. *Biochem. Soc. Trans.* **25:**845–850.

Bhatt, K., S. S. Gurcha, A. Bhatt, G. S. Besra, and W. R. Jacobs, Jr. 2007. Two polyketide-synthase-associated acyltransferases are required for sulfolipid biosynthesis in Mycobacterium tuberculosis. *Microbiology* **153:**513–520.

Brennan, P. J., and H. Nikaido. 1995. The envelope of mycobacteria. *Annu. Rev. Biochem.* **64:**29–63.

Cane, D. E., and C. T. Walsh. 1999. The parallel and convergent universes of polyketide synthases and nonribosomal peptide synthetases. *Chem. Biol.* **6:**R319–R325.

Challis, G. L., and J. H. Naismith. 2004. Structural aspects of nonribosomal peptide biosynthesis. *Curr. Opin. Struct. Biol.* **14:**748–756.

Chatterjee, D., and K. H. Khoo. 2001. The surface glycopeptidolipids of mycobacteria: structures and biological properties. *Cell Mol. Life Sci.* **58:**2018–2042.

Cole, S. T., R. Brosch, J. Parkhill, T. Garnier, C. Churcher, D. Harris, S. V. Gordon, K. Eiglmeier, S. Gas, C. E. Barry III, F. Tekaia, K. Badcock, D. Basham, D. Brown, T. Chillingworth, R. Connor, R. Davies, K. Devlin, T. Feltwell, S. Gentles, N. Hamlin, S. Holroyd, T. Hornsby, K. Jagels, A. Krogh, J. McLean, S. Moule, L. Murphy, K. Oliver, J. Osborne, M. A. Quail, M. A. Rajandream, J. Rogers, S. Rutter, K. Seeger, J. Skelton, R. Squares, S. Squares, J. E. Sulston, K. Taylor, S. Whitehead, and B. G. Barrell. 1998. Deciphering the biology of Mycobacterium tuberculosis from the complete genome sequence. *Nature* **393:**537–544.

Converse, S. E., J. D. Mougous, M. D. Leavell, J. A. Leary, C. R. Bertozzi, and J. S. Cox. 2003. MmpL8 is required for sulfolipid-1 biosynthesis and Mycobacterium tuberculosis virulence. *Proc. Natl. Acad. Sci. USA* **100:**6121–6126.

Cox, J. S., B. Chen, M. McNeil, and W. R. Jacobs, Jr. 1999. Complex lipid determines tissue-specific replication of Mycobacterium tuberculosis in mice. *Nature* **402:**79–83.

Daffe, M., and P. Draper. 1998. The envelope layers of mycobacteria with reference to their pathogenicity. *Adv. Microb. Physiol.* **39:**131–203.

Deshayes, C., F. Laval, H. Montrozier, M. Daffe, G. Etienne, and J. M. Reyrat. 2005. A glycosyltransferase involved in biosynthesis of triglycosylated glycopeptidolipids in Mycobacterium smegmatis: impact on surface properties. *J. Bacteriol.* **187:**7283–7291.

De Voss, J. J., K. Rutter, B. G. Schroeder, H. Su, Y. Zhu, and C. E. Barry III. 2000. The salicylate-derived mycobactin siderophores of Mycobacterium tuberculosis are essential for growth in macrophages. *Proc. Natl. Acad. Sci. USA* **97:**1252–1257.

Domenech, P., M. B. Reed, C. S. Dowd, C. Manca, G. Kaplan, and C. E. Barry III. 2004. The role of MmpL8 in sulfatide biogenesis and virulence of Mycobacterium tuberculosis. *J. Biol. Chem.* **279:**21257–21265.

Etienne, G., C. Villeneuve, H. Billman-Jacobe, C. Astarie-Dequeker, M. A. Dupont, and M. Daffe. 2002. The impact of the absence of glycopeptidolipids on the ultrastructure, cell surface and cell wall properties, and phagocytosis of Mycobacterium smegmatis. *Microbiology* **148:**3089–3100.

Funa, N., H. Ozawa, A. Hirata, and S. Horinouchi. 2006. Phenolic lipid synthesis by type III polyketide synthases is essential for cyst formation in Azotobacter vinelandii. *Proc. Natl. Acad. Sci. USA* **103:**6356–6361.

Gokhale, R. S., and Tuteja, D. 2001. Biochemistry of polyketide synthases, p. 341–372. *In* H.-J. Rehm and G. Reed (ed.), *Biotechnology*, 2nd ed., vol. 10. Wiley-VCH, Weinheim, Germany.

Gokhale, R. S., P. Saxena, T. Chopra, and D. Mohanty. 2007. Versatile polyketide enzymatic machinery for the biosynthesis of complex mycobacterial lipids. *Nat. Prod. Rep.* **24:**267–277.

Goren, M. B. 1972. Mycobacterial lipids: selected topics. *Bacteriol. Rev.* **36:**33–64.

Hopwood, D. A. 1999. Forty years of genetics with Streptomyces: from in vivo through in vitro to in silico. *Microbiology* **145:** 2183–2202.

Jackson, M., G. Stadthagen, and B. Gicquel. 2006. Long-chain multiple methyl-branched fatty acid-containing lipids of Mycobacterium tuberculosis: Biosynthesis, transport, regulation and biological activities. *Tuberculosis* (Edinburgh) **87:**78–86.

Jeevarajah, D., J. H. Patterson, E. Taig, T. Sargeant, M. J. McConville, and H. Billman-Jacobe. 2004. Methylation of GPLs in Mycobacterium smegmatis and Mycobacterium avium. *J. Bacteriol.* **186:**6792–6799.

Khosla, C., R. S. Gokhale, J. R. Jacobsen, and D. E. Cane. 1999. Tolerance and specificity of polyketide synthases. *Annu. Rev. Biochem.* **68:**219–253.

Kolattukudy, P. E., N. D. Fernandes, A. K. Azad, A. M. Fitzmaurice, and T. D. Sirakova. 1997. Biochemistry and molecular genetics of cell-wall lipid biosynthesis in mycobacteria. *Mol. Microbiol.* **24:**263–270.

Krithika, R., U. Marathe, P. Saxena, M. Z. Ansari, D. Mohanty, and R. S. Gokhale. 2006. A genetic locus required for iron acquisition in Mycobacterium tuberculosis. *Proc. Natl. Acad. Sci. USA* **103:**2069–2074.

LaMarca, B. B., W. Zhu, J. E. Arceneaux, B. R. Byers, and M. D. Lundrigan. 2004. Participation of fad and mbt genes in synthesis of mycobactin in Mycobacterium smegmatis. *J. Bacteriol.* **186:**374–382.

Mathur, M., and P. E. Kolattukudy. 1992. Molecular cloning and sequencing of the gene for mycocerosic acid synthase, a novel fatty acid elongating multifunctional enzyme, from Mycobacterium tuberculosis var. bovis Bacillus Calmette-Guerin. *J. Biol. Chem.* **267:**19388–19395.

Matsunaga, I., A. Bhatt, D. C. Young, T. Y. Cheng, S. J. Eyles, G. S. Besra, V. Briken, S. A. Porcelli, C. E. Costello, W. R. Jacobs, Jr., and D. B. Moody. 2004. Mycobacterium tuberculosis pks12 produces a novel polyketide presented by CD1c to T cells. *J. Exp. Med.* **200:**1559–1569.

Minnikin, D. E., L. Kremer, L. G. Dover, and G. S. Besra. 2002. The methyl-branched fortifications of Mycobacterium tuberculosis. *Chem. Biol.* **9:**545–553.

Morphy, R., and Z. Rankovic. 2007. Fragments, network biology and designing multiple ligands. *Drug Discov. Today* **12:**156–160.

Mougous, J. D., R. E. Green, S. J. Williams, S. E. Brenner, and C. R. Bertozzi. 2002. Sulfotransferases and sulfatases in mycobacteria. *Chem. Biol.* **9:**767–776.

Mougous, J. D., R. H. Senaratne, C. J. Petzold, M. Jain, D. H. Lee, M. W. Schelle, M. D. Leavell, J. S. Cox, J. A. Leary, L. W. Riley, and C. R. Bertozzi. 2006. A sulfated metabolite produced by stf3 negatively regulates the virulence of Mycobacterium tuberculosis. *Proc. Natl. Acad. Sci. USA* **103**:4258–4263.

Onwueme, K. C., C. J. Vos, J. Zurita, J. A. Ferreras, and L. E. Quadri. 2005. The dimycocerosate ester polyketide virulence factors of mycobacteria. *Prog. Lipid Res.* **44**:259–302.

Portevin, D., C. De Sousa-D'Auria, C. Houssin, C. Grimaldi, M. Chami, M. Daffe, and C. Guilhot. 2004. A polyketide synthase catalyzes the last condensation step of mycolic acid biosynthesis in mycobacteria and related organisms. *Proc. Natl. Acad. Sci. USA* **101**:314–319.

Quadri, L. E., J. Sello, T. A. Keating, P. H. Weinreb, and C. T. Walsh. 1998. Identification of a Mycobacterium tuberculosis gene cluster encoding the biosynthetic enzymes for assembly of the virulence-conferring siderophore mycobactin. *Chem. Biol.* **5**:631–645.

Rao, V., F. Gao, B. Chen, W. R. Jacobs, Jr., and M. S. Glickman. 2006. Trans-cyclopropanation of mycolic acids on trehalose dimycolate suppresses Mycobacterium tuberculosis-induced inflammation and virulence. *J. Clin. Invest.* **116**:1660–1667.

Ratledge, C. 2004. Iron, mycobacteria and tuberculosis. *Tuberculosis* (Edinburgh) **84**:110–130.

Reed, M. B., P. Domenech, C. Manca, H. Su, A. K. Barczak, B. N. Kreiswirth, G. Kaplan, and C. E. Barry III. 2004. A glycolipid of hypervirulent tuberculosis strains that inhibits the innate immune response. *Nature* **431**:84–87.

Rodriguez, G. M., and I. Smith. 2003. Mechanisms of iron regulation in mycobacteria: role in physiology and virulence. *Mol. Microbiol.* **47**:1485–1494.

Sankaranarayanan, R., P. Saxena, U. B. Marathe, R. S. Gokhale, V. M. Shanmugam, and R. Rukmini. 2004. A novel tunnel in

mycobacterial type III polyketide synthase reveals the structural basis for generating diverse metabolites. *Nat. Struct. Mol. Biol.* **11**:894–900.

Saxena, P., G. Yadav, D. Mohanty, and R. S. Gokhale. 2003. A new family of type III polyketide synthases in Mycobacterium tuberculosis. *J. Biol. Chem.* **278**:44780–44790.

Sirakova, T. D., A. K. Thirumala, V. S. Dubey, H. Sprecher, and P. E. Kolattukudy. 2001. The Mycobacterium tuberculosis pks2 gene encodes the synthase for the hepta- and octamethyl-branched fatty acids required for sulfolipid synthesis. *J. Biol. Chem.* **276**:16833–16839.

Takayama, K., C. Wang, and G. S. Besra. 2005. Pathway to synthesis and processing of mycolic acids in Mycobacterium tuberculosis. *Clin. Microbiol. Rev.* **18**:81–101.

Trivedi, O. A., P. Arora, V. Sridharan, R. Tickoo, D. Mohanty, and R. S. Gokhale. 2004. Enzymic activation and transfer of fatty acids as acyl-adenylates in mycobacteria. *Nature* **428**:441–445.

Trivedi, O. A., P. Arora, A. Vats, M. Z. Ansari, R. Tickoo, V. Sridharan, D. Mohanty, and R. S. Gokhale. 2005. Dissecting the mechanism and assembly of a complex virulence mycobacterial lipid. *Mol. Cell* **17**:631–643.

Walsh, C. T. 2004. Polyketide and nonribosomal peptide antibiotics: modularity and versatility. *Science* **303**:1805–1810.

Yadav, G., R. S. Gokhale, and D. Mohanty. 2003a. Computational approach for prediction of domain organization and substrate specificity of modular polyketide synthases. *J. Mol. Biol.* **328**:335–363.

Yadav, G., R. S. Gokhale, and D. Mohanty. 2003b. SEARCHPKS: a program for detection and analysis of polyketide synthase domains. *Nucleic Acids Res.* **31**:3654–3658.

Chapter 16

The Constituents of the Cell Envelope and Their Impact on the Host Immune System

Warwick J. Britton and James A. Triccas

Mycobacterium tuberculosis is a remarkably successful human pathogen. The organism has established latent infection in nearly one third of humans, where it survives within pulmonary granulomas in the face of a strong host cellular immune response. This survival is dependent on the mycobacterium modulating the host response, and this effect is largely mediated by components of the cell envelope. *M. tuberculosis* is transmitted by the respiratory route, and the bacilli lodge in the most distal airways, the alveoli. The mycobacteria are phagocytosed by alveolar macrophages and then cross the respiratory epithelium and are taken up by interstitial macrophages and dendritic cells (DCs), the major antigen-presenting cells (APCs) for the activation of T lymphocytes (Flynn and Chan, 2001). On encountering the mycobacteria, DCs are stimulated to upregulate major histocompatibility complex (MHC) class II and costimulatory molecules, along with the chemokine receptor CCR7, so that they migrate through lymphatics to the draining lymph nodes (Demangel and Britton, 2000). In addition, the activated DCs secrete proinflammatory and immunoregulatory cytokines, including interleukin-12 (IL-12), IL-18, and IL-23. Within the lymph node, naïve CD4$^+$ T cells recognize peptide/MHC class II complexes on the surface of *M. tuberculosis*-infected DCs and, under the influence of IL-12 and other cytokines, differentiate into interferon-γ (IFN-γ) secreting Th1-like T cells (Flynn and Chan, 2001) (Fig. 1). In addition, *M. tuberculosis*-infected DCs activate MHC class I-restricted CD8$^+$ T cells, nonclassically restricted CD8$^+$ T cells and CD1-restricted T cells. The activated T cells change their pattern of adhesion molecule expression so that they no longer circulate through lymph nodes but are recruited to the site of the original *M. tuberculosis* infection in the lung. There, the effector T cells recognize *M. tuberculosis*-infected macrophages and secrete cytokines, notably IFN-γ, lymphotoxin-α (LT-α), and tumor necrosis factor (TNF), which activate the bacterial killing mechanisms in infected macrophages (Bean et al., 1999; Cooper et al., 1993; Roach et al., 2001) (Fig. 1). These mechanisms include the activation of phagocyte oxidase and inducible nitric oxide synthase (iNOS) to produce reactive nitrogen intermediates (RNI) and reactive oxygen intermediates (ROI) respectively, as well as other IFN-γ response genes including the GTPase, LRG47 (MacMicking et al., 2003). The mycobacteria, chiefly through cell wall constituents, resist macrophage-killing mechanisms, and although the bacterial replication is reduced, some mycobacteria enter a dormant state associated with transcription to selected "dormancy" genes and survive within the host cell (Schnappinger et al., 2003). This results in persistent antigenic stimulation and the further recruitment of monocytes and lymphocytes. Under the influence of TNF and chemokines, the activated macrophages and lymphocytes aggregate to form granulomas, which are characterized in humans by central necrosis or caseation, giant cells, and fibrosis (Saunders and Britton, 2007). This complex process is regulated at each stage by a balance of stimulatory and inhibitory cytokines (reviewed in O'Garra and Britton, 2007).

Mycobacterial cell wall components influence this process, both through activation signals, which initiate macrophage and DC responses to infection, and inhibitory signals, which enhance survival of the organisms (Karakousis et al., 2004). Paradoxically, the primary host cell for these facultative intracellular bacteria is the macrophage, which also has the potential to kill the pathogen. Therefore, virulent mycobacteria must manipulate the normal macrophage responses to intracellular pathogens, and cell

Warwick J. Britton • Centenary Institute of Cancer Medicine and Cell Biology, Locked Bag No. 6, Newtown, NSW 2042, and Department of Medicine, University of Sydney (D06), Sydney, NSW 2006, Australia. James A. Triccas • Department of Infectious Diseases and Immunology, University of Sydney (D06), Sydney, NSW 2006, Australia.

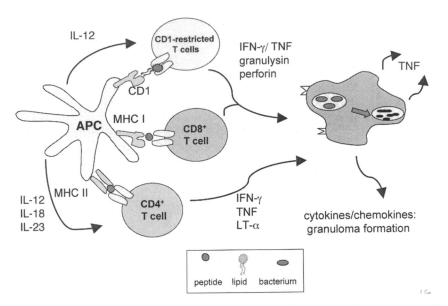

Figure 1. T-cell immunity to mycobacteria. Professional antigen-presenting cells (APCs), such as dendritic cells, can present peptide fragments to classical CD8 and CD4 T cells through major histocompatibility (MHC) class I and II molecules, respectively. Mycobacterial lipids can also be presented through CD1 molecules to T-cell subsets including γδ T cells, natural killer (NK) T cells and other non classic T cells lacking expression of CD4 and CD8 surface molecules. Activated APCs release soluble meditators, such as the cytokines IL-12, IL-18, and IL-23, which act to enhance the effector function of mycobacteria-specific T cells. Release of IFN-γ and TNF serve to activate macrophages infected with mycobacteria to promote bacterial killing. Perforin is cytolytic for infected cells, and granulysin is bactericidal for mycobacteria. Activated macrophages release cytokines and chemokines, which are required for granuloma formation in order to contain the infection effectively.

wall components interfere with these responses at multiple stages, including the initial process of phagocytosis, the pattern of macrophage cytokine responses to infection, maturation of the mycobacterial phagosome, antigen presentation by DCs and macrophages, and the response of macrophages to activation. In addition, components of the cell wall become targets for both T-cell and humoral responses, and also contribute to the process of granuloma formation. This review focuses on recent developments in our understanding of how different components of the cell envelope from virulent mycobacteria, in particular *M. tuberculosis*, interact with each stage of these innate and adaptive immune responses.

INITIAL ENCOUNTER OF MYCOBACTERIA BY MACROPHAGES

Uptake by Macrophages

The major receptors for phagocytosis of mycobacteria by macrophages are the mannose receptor (MR) and complement receptor 3 (CR3) (Schlesinger, 1993), and characteristically they mediate phagocytosis without activating a proinflammatory response as occurs through Fcγ receptor-mediated phagocytosis. The MR is a C-type lectin, which binds mannose (Taylor et al., 2005). The mannose-containing glycol-

ipids in the mycobacterial cell wall include mannose-capped lipoarabinomannan (manLAM), lipomannan, and phosphatidyl-*myo*-inositol mannosides (PIMs), and all of these molecules contribute to the binding of *M. tuberculosis* to MR. The terminal mannose caps on the arabinose side chains of manLAM bind to the MR, and mediate phagocytosis of *M. tuberculosis*, and manLAMs from virulent and attenuated strains of *M. tuberculosis* show varying degrees of binding (Schlesinger et al., 1994; Schlesinger et al., 1996). By contrast, the phosphatidyl-*myo*-inositol caps on the arabinose side chains of LAM (PILAM) from avirulent mycobacteria, such as *M. smegmatis*, do not bind MR (Schlesinger et al., 1994). The PIMs, which are a significant component of the cell envelope, vary in their number of mannose molecules and fatty acids. Higher order PIMs with five or six mannose units bind to MR, and this binding is increased in more highly acylated PIMs (Torrelles et al., 2006). These higher order PIMs are more common in virulent than in avirulent strains of *M. tuberculosis*, and contribute to their uptake by macrophages. The engagement of MR on human macrophages by LAM influences the maturation of the phagosome following phagocytosis, leading to a reduction in phagolysosomal fusion (Kang et al., 2005).

Macrophages express multiple complement receptors (CR), and it is established that CR1, CR3, and CR4 bind C3-opsonized mycobacteria and facilitate their phagocytosis (Fenton et al., 2005; Schlesinger et al., 1990; Schlesinger, 1993). In addition, *M. tuberculosis* binds directly to CR3 independent of complement deposition, and there is considerable variability in this CR3-mediated phagocytosis between clinical strains (Cywes et al., 1997; Stokes et al., 1993). This nonopsonic binding of *M. tuberculosis* and other mycobacteria (Cywes et al., 1997; Le Cabec et al., 2000) to CR3 occurs through the binding of capsular polysaccharides, glyco-peptidolipids, and PIMs (Cywes et al., 1997; Villeneuve et al., 2005) to the lectin-binding domain of CR3, which is separate from the complement site. There is considerable redundancy in the macrophage binding receptors for mycobacteria, and although CR3 is an important phagocytic receptor for *M. tuberculosis* in in vitro studies, CR3 appears dispensable for the control of mycobacterial infections. A reduction in the uptake of *M. tuberculosis* by CR3$^{-/-}$ macrophages in vitro was observed (Rooyakkers and Stokes, 2005). However, CR3 deficiency had no effect on the control of *M. tuberculosis* infection either in macrophages in vitro or in mice (Hu et al., 2000). Indeed, blockade of CR1, CR2, CR3, and MR by antibodies had no effect on the survival of *M. tuberculosis* in human macrophages (Zimmerli et al., 1996), indicating that the critical steps controlled by *M. tuberculosis* to enhance its survival are downstream of phagocytosis.

Human dendritic cells express an additional C-type lectin, dendritic cell-specific ICAM-3-grabbing nonintegrin (DC-SIGN), which is not present on macrophages but serves as a major receptor for *M. tuberculosis* on DCs (Tailleux et al., 2003). The terminal mannose caps on ManLAM of *M. tuberculosis* bind DC-SIGN, thereby facilitating uptake, whereas the phosphatidyl-*myo*-inositol caps on PILAM from *M. smegmatis* and other nonpathogenic mycobacteria do not associate with DC-SIGN (Maeda et al., 2003). Lower order PIMs containing fewer mannose molecules and LM also bind DC-SIGN providing multiple interactions between the cell envelope and this DC-specific receptor (Torrelles, 2006).

The C-type lectin used by mycobacteria for phagocytosis influences the outcome of infection. Phagocytosis of virulent *M. tuberculosis* or Man-LAM-coated beads through MR leads to uptake into phagosomes, which interact with early endosomes. These fail to mature and recruit markers of late endosomes or fuse with lysosomes (Kang et al., 2005). PIMs from *M. tuberculosis* containing 5 or 6 mannose units also bind MR and limit fusion of the mycobacterial phagosome with lysosomes (Torrelles et al., 2006). By contrast, if ManLAM from *M. tuber-*

culosis binds to DC-SIGN, the mycobacteria are targeted to lysosomes in DCs (Geijtenbeek et al., 2003; Tailleux et al., 2003), and this would facilitate antigen processing for the activation of T cells. This suggests that macrophages, which express MR but not DC-SIGN, may provide a more protected site for mycobacterial survival than DCs, which express high levels of DC-SIGN. However, binding of ManLAM by DC-SIGN has additional effects, which prevent the maturation of the DC in response to activation by mycobacteria or other Toll-like receptor (TLR) signals. This ManLAM-mediated suppression results in reduced IL-12 and increased IL-10 production by human DCs (Geijtenbeek et al., 2003). Therefore, DC-SIGN serves both as a receptor for uptake of virulent mycobacteria and a target for modulation of the innate response of DCs by ManLAM from *M. tuberculosis*. ManLAM can also inhibit IL-12 production and stimulate the anti-inflammatory cytokine IL-10 in human DC by providing a negative signal through the MR (Chieppa et al., 2003; Gagliardi et al., 2005; Nigou et al., 2001).

Other receptors that contribute to the binding and uptake of mycobacteria by macrophages include class A scavenger receptors (Zimmerli et al., 1996), CD14 (Khanna et al., 1996), and the transmembrane glycoproteins CD43 and CD36 (Philips et al., 2005; Randhawa et al., 2005), but the components of the cell wall engaging these receptors are as yet not determined.

Activation of Macrophages and Dendritic Cells by Cell Envelope Components

In addition to phagocytic receptors, DCs and macrophages express activation receptors, which on engagement by pathogen ligands activate signaling pathways leading to phenotypic and functional changes of the cells without promoting phagocytosis. Cell wall components of mycobacteria bind to these pathogen recognition receptors on DCs and macrophages, and initiate the host response to infection. The best characterized of these are the TLRs (reviewed in Takeda et al., 2003). Mammalian TLRs were identified as homologues of the *Drosophila* Toll protein, which contribute to insect immunity against fungal infections, as well as being critical for dorsal-ventral patterning during fly development (Lemaitre et al., 1996). The cytoplasmic domain of Toll is homologous with this domain of the mammalian IL-1 and IL-18 receptors, and signals through a shared pathway using the MyD88 adaptor protein and IL-1 receptor-associated kinases to activate NFκB, AP-1, and mitogen-activated protein kinases (Takeda et al., 2003). The first characterized mammalian TLR was

TLR4, which is the critical component of the cellular receptor for LPS. However, there are now 10 defined human and 11 murine TLRs recognizing a wide range of microbial and host ligands. These use MyD88, as well as other adaptor proteins, for signaling increased gene transcription in response to infection (Takeda et al., 2003). There have been two approaches to determining the significance of TLRs in the host response to mycobacterial infection. First, whole mycobacteria and purified cell wall components have been tested as ligands for individual TLRs in vitro, and second, the requirement of selected TLRs for host control of mycobacterial infections has been examined using gene-deficient mice. The heterodimers with TLR2/TLR1 and TLR2/TLR6, TLR4 and TLR9 have been demonstrated to contribute to the recognition of mycobacterial infection (Krutzik and Modlin, 2004).

Multiple cell wall constituents have been identified as ligands for TLR2, including certain forms of LAM (Means et al., 1999), PIMs (Jones et al., 2001), lipomannan (Quesniaux et al., 2004), and lipoproteins. These molecules are readily available on the surface of infecting mycobacteria to activate innate immune responses. LAM was first identified as a ligand for TLR2 (Means et al., 1999), and it was recognized that nonmannose-capped PILAM (previously termed Ara-LAM) from rapidly growing mycobacteria, such as *M. smegmatis*, activated TLR2 but not TLR4, whereas ManLAM from the slow-growing *M. tuberculosis* or *M. bovis* BCG failed to stimulate TLR2, even though extracts from these virulent mycobacteria did so (Means et al., 1999). Viable *M. tuberculosis*, but not *M. avium*, activated TLR4 as well as TLR2, but the exact component in the protease-sensitive cell wall extract of *M. tuberculosis* that activated TLR4 was not identified (Means et al., 1999). Subsequently, it was found that PIMs, which are abundant in the cell wall of virulent as well as avirulent mycobacteria, are potent activators of TLR2 (Jones et al., 2001). The capacity of mycobacterial glycoconjugates to have diverse effects is illustrated by the effects of LM, which activated CD86 expression and cytokine and iNOS secretion by macrophages through TLR2/MyD88 but inhibited LPS-activation of macrophages through a TLR-independent pathway (Quesniaux et al., 2004).

A second group of TLR2 agonists are mycobacterial lipoproteins, which are secreted and associated with the cell wall. The 19-kDa LpqH (Rv3763) was the first identified lipoprotein of *M. tuberculosis*, which activated human macrophages to secrete IL-12 and generate nitric oxide in a TLR2-dependent fashion (Brightbill et al., 1999). It is now clear that lipoproteins of both *M. leprae* and *M. tuberculosis*

can mediate the activation of both macrophages and dendritic cells in a process dependent on TLR1 association with TLR2 (Hertz et al., 2001; Krutzik et al., 2003; Takeuchi et al., 2002). Other mycobacterial lipoproteins that signal through TLR2 include LprG (Rv1411c; Gehring et al., 2004) and LprA (Rv1270c; Pecora et al., 2006), whereas PstS1 (Rv0934; Jung et al., 2006) can activate macrophages through both TLR2 and TLR4.

Activation of macrophages or DCs by mycobacterial glycolipids through TLR2 has a wide range of biological effects, including the secretion of IL-12 and proinflammatory cytokines, such as TNF, IL-1, and IL-6 (Brightbill et al., 1999; Jung et al., 2006; Means et al., 1999); the increased maturation of DCs with upregulation of the costimulatory molecules CD80 and CD86 (Hertz et al., 2001; Pecora et al., 2006); the induction of iNOS and direct antibacterial activity in macrophages (Brightbill et al., 1999; Quesniaux et al., 2004; Thoma-Uszynski et al., 2001); and the induction of apoptosis in infected cells. In addition, signaling through TLR2 by lipoproteins has inhibitory effects on macrophages by reducing their responses to IFN-γ stimulation (Gehring et al., 2003; Gehring et al., 2004), and this counterregulatory effect is discussed later in the context of antigen presentation and bacterial killing.

The biological relevance of TLR activation by cell wall components has been examined by mycobacterial infection of gene-deficient mice. Deletion of the adaptor protein MyD88 profoundly increases the susceptibility to *M. tuberculosis* and *M. avium* infection (Feng et al., 2003; Fremond et al., 2004; Scanga et al., 2004); however, this interferes with both TLR and cytokine signaling by IL-1 and IL-18. There have been variable results in infection studies with mice lacking a single TLR, but in general, TLR2$^{-/-}$ mice are more susceptible to low-dose aerosol *M. tuberculosis* infection with a mild reduction in the activation of IFN-γ-secreting CD4 T cells (Bafica et al., 2005; Salgame et al., 2005). By contrast, mice deficient in TLR4 signaling have shown increased susceptibility to *M. tuberculosis* infection in the chronic phase in one study (Abel et al., 2002). TLR9 recognizes CpG DNA motifs present in bacterial DNA and signals through MyD88 from intracellular vacuoles, including phagosomes (Takeda et al., 2003). TLR9$^{-/-}$ mice show minimal defect in the control of *M. tuberculosis* infection. However, TLR2$^{-/-}$ TLR9$^{-/-}$ double-mutant mice were profoundly susceptible, with marked reduction in specific CD4 T cell responses (Bafica et al., 2005). This indicates that there is redundancy in mechanisms of DC activation by *M. tuberculosis*. However, TLR2 activation by cell wall PIMs and lipoproteins contributes to the innate host response regulating the development of protective T-cell immunity.

BCG-derived peptidoglycan and related acylated muramyl dipeptides also activate TLR2 and TLR4 (Uehori et al., 2005). However, blocking TLR2 and TLR4 fails to account fully for the innate cytokine response of macrophages to BCG peptidoglycan (Uehori et al., 2003), suggesting that other pathogen recognition receptors may recognize components of mycobacteria cell wall. Recently, a family of intracellular proteins, variously termed nucleotide-binding oligomerization domain (NOD) or CARD, transcription enhancer, R (purine)-binding, pyrin, lots of leucine repeats (CATERPILLER), have been identified as pathogen recognition receptors, which recognize muramyl dipeptide from gram-positive bacteria and other components of peptidoglycan (Ting and Williams, 2005). These are encoded by genes linked to resistance genes in plants and encode a combined nucleotide-binding domain/leucine-rich region (NBD/LRR) motif. These are crucial for the control of cytokine and inflammatory responses, and have been linked to genetic susceptibility to Crohn's disease and other inflammatory syndromes. Expression of monarch-1, a member of this family, acts as a negative regulator of the NFκB and cytokine responses to TLR2/4 activation and *M. tuberculosis* infection by interference with the MyD88 signaling pathway (Williams et al., 2005). *M. tuberculosis* infection downregulates monarch-1 expression, thus permitting a proinflammatory response to the infection. In a separate study, *M. tuberculosis* was found to activate NOD2 transfected cells, and human macrophages with a loss of function mutation in NOD2 showed an 80% reduction in cytokine responses to stimulation with *M. tuberculosis* (Ferwerda et al., 2005). Moreover, the TLR2 ligand, mycobacterial 19-kDa lipoprotein, and NOD2 ligand, muramyl dipeptide, synergized for the maximal induction of cytokine responses, suggesting that the TLR and NOD pathways are nonredundant mechanisms for the recognition of *M. tuberculosis* infection and induction of innate immunity.

Activation of Cytokine Responses by Cell Envelope Components

The net effect of stimulation of TLR and NOD pathways by mycobacterial infection is the secretion of proinflammatory cytokines, including TNF, IL-1β, IL-6, and IL-18, by macrophages and DCs. In addition, macrophages infected with *M. tuberculosis* in the absence of T cell stimulation produce IL-10 without IL-12, whereas infected DCs express IL-12p40 and produce IL-12 and IL-23 with minimal IL-10

(Salgame, 2005). There are two important caveats to this general pattern. First, there is a fine balance between the effects of different cell wall components and their receptors, so that in different cytokines milieu infection may have variable effects. For example, engagement of DC-SIGN on DCs by ManLAM induces IL-10, which balances the stimulation of IL-12 production by the effects of PIMs and lipoproteins on TLR2. Inhibition of endogenous IL-10 production by BCG-infected DCs markedly enhances the production of IL-12 and activation of CD4 T cells (Demangel et al., 2001) and increases DC migration to draining lymph nodes (Demangel et al., 2002). Conversely, prior activation of macrophages by IFN-γ switches the response to stimulation with live *M. tuberculosis* from IL-10 to IL-12, and this may occur in the context of chronic pulmonary *M. tuberculosis* infection.

The second caveat is that in vitro studies have generally focused on the macrophage responses to laboratory strains of *M. tuberculosis* and their purified cell wall components, but recently, it has been recognized that clinical isolates of *M. tuberculosis* vary widely in their capacity to stimulate macrophage secretion of proinflammatory (TNF and IL-1) or anti-inflammatory (IL-10) cytokines (Chacon-Salinas et al., 2005; Hoal-van Helden et al., 2001; Lopez et al., 2003; Theus et al., 2005). Strains associated with low TNF and high IL-10 production showed more rapid growth rates in human macrophages compared with strains stimulating high TNF levels (Theus et al., 2005). Infection of mice with *M. tuberculosis* strains representing the four major genotypes identified by IS6110 fingerprinting resulted in variable expression of TNF and iNOS in the lungs and markedly different immunopathologies, with the Beijing strain producing a weak TNF response and extensive, progressive pneumonia (Lopez et al., 2003). Moreover, prior BCG immunization was less protective against the Beijing strain than the laboratory strain, *M. tuberculosis* H37Rv. Although multiple cytokine and cellular interactions are likely to be responsible for this outcome, it is probable that variations in the components of the mycobacterial cell envelope are responsible for variable cytokine response to different *M. tuberculosis* strains.

This principle has been dramatically illustrated in the host responses to selected strains of *M. tuberculosis*, HN878, and the related W/Beijing isolates, which are hypervirulent in both mouse and rabbit models of infection (Reed et al., 2004; Tsenova et al., 2005). These isolates induced less IL-12 production and Th1-like T cells than the CDC1551 isolate and the H37Rv laboratory strain, and preferentially stimulated IL-4, IL-13, and type I IFN production by human monocytes (Manca et al., 2001; Manca et al., 2004). This hypervirulence and switch in the pattern of innate immunity were associated with the production of a surface phenolic glycolipid (PGL), absent from laboratory strains, which was derived from a polyketide synthase. Disruption in the polyketide synthase resulted in loss of the hypervirulent phenotype in mice and the increased stimulation of IL-12, TNF, and IL-6 release from infected cells (Reed et al., 2004). Therefore, the expression of this surface PGL had a profound effect on the pattern of innate immunity and control of infection.

MODULATION OF THE MYCOBACTERIAL PHAGOSOME BY THE CELL ENVELOPE

It has been recognized since the seminal experiments of D'Arcy Hart (Armstrong et al., 1971) that an important property of virulent mycobacteria is their capacity to arrest the normal maturation of the phagocytic vacuole so that it does not fuse with lysosomes and undergo acidification, thus protecting the mycobacteria from effects of active lysosomal hydro-

lases and other killing mechanisms. The molecular regulation of phagocytosis is increasingly understood (Fig. 2A), and mycobacterial cell wall components and proteins that interfere with this process have been defined (Nguyen and Pieters, 2005; Vergne et al., 2004a). Briefly, extracellular bacteria or opsonized latex particles bind to phagocytic receptors and are taken into phagosomes by rearrangement of the cytoskeleton in a process controlled by a complex series of regulatory molecules (Aderem et al., 2003). Following phagocytosis, several proteins are recruited to the phagosome membrane including the GTP-binding protein Rab5 and subunits of the phosphatidylinositol (PI) 3-kinase, hVPS34. This leads to the local generation of PI3-phosphate (PI3P), and the early endosome antigen 1 (EEA1) recognizes the PI3P through its FYVE binding domain (Lawe et al., 2000) and localizes to the phagosome. The maturing phagosome acquires the GTPases Rab 7, typical of late endosomes, and proteins of the SNARE (soluble N-ethylmaleimide-sensitive factor attachment protein receptor) complex, including syntaxin 6. EEA1 is essential for fusion with lysosomes, whereas syntaxin 6 is required for the transport from the *trans*-Golgi of

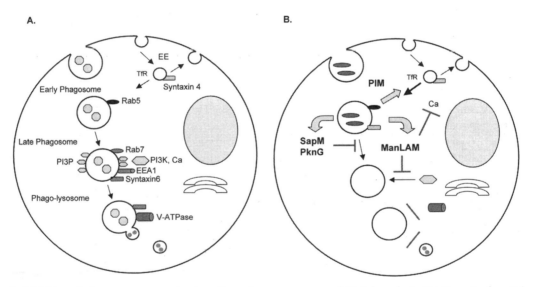

Figure 2. Inhibition of phagosome maturation by cell envelope components of *M. tuberculosis*. (A) Opsonized particles and extracellular bacteria are taken into phagosomes, which mature and fuse with lysosomes containing acid hydrolases necessary to digest the bacteria. The early phagosomes acquire the GTPase Rab5 and have access to transferring receptors (TfR) and a supply of iron from the early endosome (EE) compartment. Subunits of PI3 kinase (PI3K) are recruited and generate PI3P on the maturing phagosome, which acquires Rab7 and the early endosome antigen 1 (EEA1) which is essential for fusion with lysosomes. The late phagosome also recruits syntaxin 6, which is required for acquisition of the vacuolar proton ATPase (V-ATPase), necessary for acidification of the phagosome. (B) Phagosomes containing virulent mycobacteria, such as *M. tuberculosis*, do not mature beyond the early phagosome stage characterized by Rab5. They fail to acquire markers of late phagosomes, fuse with lysosomes or undergo acidification. ManLAM inhibits PI3-kinase and prevents the generation of PI3P on the surface of the phagosome and the acquisition of the tethering proteins and the V-ATPase. ManLAM also blocks the Ca²⁺ flux required for recruitment of PI3-kinase to the phagosome. *M. tuberculosis* also secretes the acid phosphatase SapM, which hydrolyzes PI3P and prevents PI3P accumulating on the surface of phagosomes containing live, virulent mycobacteria. The production of the serine-threonine kinase G (PknG) by virulent mycobacteria also inhibits phagolysosomal fusion. By contrast, lower order PIMs increase the fusion of early phagosomes with early endosomes, so they acquire syntaxin 4 and access to iron for mycobacterial replication.

the vacuolar proton ATPase necessary for acidification and lysosomal hydrolases to the phagosome (Fig. 2A). Subunits of phagocyte oxidase assemble on the phagolysosomal membrane and utilize cytoplasmic NADPH to generate intravacuolar ROI, which are further catalysed by lysosomal myeloperoxidase to produce potent metabolites. This leads to the degradation of the phagocytosed bacteria. The mycobacterial phagosome is arrested at an early stage identified by Rab5 and fails to acquire the later markers of Rab7 (Via et al., 1997) and syntaxin 6 (Fratti et al., 2003) or to accumulate PI3P necessary for docking with lysosomes (Deretic et al., 2006; Vergne et al., 2004a).

Components of the cell wall of virulent mycobacteria are partially responsible for this inhibition of phagosome maturation (Fig. 2B). ManLAM, which is abundant in the phagosome and intercalates into the host membranes (Beatty et al., 2000), inhibits phagolysomal fusion and the recruitment of late endosomal proteins to the mycobacterial phagosome (Fratti et al., 2001; Fratti et al., 2003). For example, phagosomes containing ManLAM-coated latex beads fail to acquire EEA1 and fuse with lysosomes. ManLAM inhibits the PI3-kinase and prevents the generation of PI3P on the phagosome surface, and the subsequent recruitment of the tethering molecules, EEA1 (Fratti et al., 2001) and the multivesicular body endosome-sorting regulator Hrs (Vieira et al., 2004), and syntaxin 6 (Fratti et al., 2003). This inhibits phagosome fusion with lysosomes and their acidification. In addition, ManLAM from virulent, but not avirulent, mycobacteria modulates Ca^{2+} intracellular signaling within mycobacteria infected macrophages (Vergne et al., 2003). This Ca^{2+} flux and calmodulin are required for the recruitment of hVPS34 and deposition of PI3P on the phagosome. Live, but not dead, M. tuberculosis inhibits this Ca^{2+} signaling and prevents phagosome maturation and fusion (Malik et al., 2000). Another mechanism by which virulent mycobacteria manipulate phagosome maturation is the secretion of SapM (Rv3310), a mycobacterial acid phosphatase (Vergne et al., 2005). This hydrolyzes PI3P and prevents its accumulation on phagosomes containing live, but not dead, mycobacteria, and so inhibits the recruitment of effector molecules leading to phagolysosomal fusion. Therefore, virulent M. tuberculosis uses a variety of strategies to target PI3P to prevent the maturation of the phagosome (Fig. 2B). In contrast to the effects of LAM, mycobacterial lower order PIMs stimulate fusion of the mycobacterial phagosome with early endosomes (Vergne et al., 2004b). PIM-coated latex beads acquire the markers of early endosomes, syntaxin 4, and transferrin receptors, and this early en-

dosomal fusion supplies the mycobacterial compartment with iron and other nutrients that enhance mycobacterial survival.

In addition to cell wall components, mycobacterial proteins can interfere with phagosomal maturation and confer a survival advantage on virulent mycobacteria. For example, the eukaryotic-like serine/threonine protein kinase G (PknG or Rv0410c), which is the only soluble kinase shared by M. tuberculosis and M. leprae, is secreted within macrophage phagosomes and inhibits phagosome-lysosome fusion, and mediates intracellular survival of mycobacteria (Walburger et al., 2004). Inactivation of PknG by gene disruption or chemical inhibition resulted in delivery of pathogenic mycobacteria to lysosomes and mycobacterial killing, and conversely, expression of PknG in nonpathogenic prevented phagosomal maturation. Other genetic screens have identified additional mycobacterial proteins that appear to inhibit the maturation and acidification of mycobacterial phagosomes (Pethe et al., 2004; Stewart et al., 2005). This reinforces the importance of manipulating the phagosome environment for the survival of virulent mycobacteria.

MODULATION OF ANTIGEN PRESENTATION BY CELL WALL-ASSOCIATED LIPOPROTEINS

The induction of adaptive immunity to mycobacteria and activation of infected macrophages by IFN-γ can overcome the maturation arrest of mycobacteria-containing phagosomes (Schaible et al., 1998). However, virulent mycobacteria can also manipulate adaptive immunity by interfering with antigen processing and presentation. The most important mechanism in macrophages is to prevent the delivery of mycobacteria to lysosomes for degradation, and the generation of peptide fragments and lipid components for antigen presentation. In addition, M. tuberculosis can inhibit the processing and presentation of peptides by MHC class II molecules to CD4 T cells through mycobacterial lipoproteins, which have the paradoxical effects of both activating DCs through TLR2/TLR1 (see earlier) and inhibiting the induction of cellular immunity.

Lipoproteins are an important family of cell wall-associated molecules that encode various functions, which include modulators of mycobacterial virulence and host immunity (Sutcliffe and Harrington, 2004). The most studied member of the mycobacterial lipoprotein family is the 19-kDa LpqH of M. tuberculosis. Early studies centered on the ability of the protein to interfere with the induction of antimycobacterial protective immunity. Expression of the M. tuberculosis LpqH in the nonpathogenic

M. smegmatis or *Mycobacterium vaccae* induced an immune response against the recombinant protein in mice. However, the presence of the antigen abrogated the ability of the recombinant strains to protect mice against *M. tuberculosis* infection (Abou-Zeid et al., 1997; Yeremeev et al., 2000). This inhibition was associated with the reduced ability of the vaccine strains to induce delayed-type hypersensitivity responses. Similarly, *M. tuberculosis* LprG suppressed immune responses and reduced vaccine-induced protective immunity (Hovav et al., 2003; Hovav et al., 2006).

One mechanism of this immune suppression is reduced capacity for antigen presentation to T cells. Infection with *M. tuberculosis* or exposure to *M. tuberculosis* lipoproteins inhibits the IFN-γ-induced expression of and antigen presentation by MHC class II molecules by macrophages, and thus promotes evasion of CD4$^+$ T-cell-mediated immunity. LpqH was the first lipoprotein found to inhibit expression of MHC class II and the class II transactivator CIITA, as well as antigen processing by murine and human macrophages through TLR2 (Gehring et al., 2003; Noss et al., 2001; Pai et al., 2003). This occurs through TLR2-induced signaling through the MAPKs, p38, and ERK, leading to inhibition of chromatin remodeling and histone acetylation at the gene encoding CIITA (Pennini et al., 2006). LpqH also inhibited antigen processing for alternate MHC class I presentation by macrophages without reducing class I expression (Tobian et al., 2003). Similarly, the *M. tuberculosis* LprG (Gehring et al., 2004) and LprA (Pecora et al., 2006) lipoproteins are also TLR2 agonists that induce early cytokine expression and DC maturation but inhibit IFN-γ-induced MHC class II antigen processing and presentation by macrophages. This may have a beneficial effect for the host by limiting macrophage activation of effector CD4$^+$ T cell at the site of infection and so preventing inflammatory damage to the lung.

Therefore, mycobacterial lipoproteins can either stimulate innate immune responses, particularly in DCs, or suppress antimycobacterial immunity, in both cases through their interactions with TLR2. Mycobacterial strains that lack individual lipoproteins or are unable to process lipoproteins have been developed to define more precisely the role of these proteins during infection. Deletion of *lpqH* from *M. tuberculosis* resulted in a slight decrease in the replication of the mutant strain in human macrophages and reduced induction of cytokine IL-1β, whereas overexpression of LpqH stimulated increased cytokine release by infected cells (Stewart et al., 2005). *M. tuberculosis* strains deficient in the LpqH protein displayed a reduced ability to induce apoptosis of human macrophages compared with the wild-type strain (Ciaramella et al., 2004). This effect may have been due to the lack of LpqH signaling or reduced bacterial loads caused by attenuation of the *lpqH*-deficient strains (Lathigra et al., 1996; Stewart et al., 2005). Overexpression or deletion of *lpqH* did not affect the protective efficacy of the BCG vaccine (Yeremeev et al., 2000), indicating that the generation of vaccine-induced immunity, or indeed the suppression of such responses, was independent of LpqH. The transfer of mature lipoproteins to the cell wall is dependent on *lspA* (Rv1539), which encodes the lipoprotein signal peptidase (Sutcliffe and Harrington, 2004). *M. tuberculosis* mutants deficient in *lspA* were unable to process lipoproteins required for TLR2 signaling. However, this did not affect the capacity of the *M. tuberculosis* mutants to inhibit the macrophage response to IFN-γ (Banaiee et al., 2006). Therefore, although there is evidence that mycobacterial lipoproteins can modulate antimycobacterial immunity, some of the functions ascribed to lipoproteins can be performed by other mycobacterial components.

EFFECTS OF THE MYCOBACTERIAL CELL ENVELOPE ON MACROPHAGE ACTIVATION

Although virulent mycobacteria may modulate antigen presentation, infection with *M. tuberculosis* (Flynn and Chan, 2001) or *M. leprae* (Britton and Lockwood, 2004) stimulates a strong Th1-like T-cell response dominated by IFN-γ and TNF, both of which are essential for protection against mycobacteria. Nevertheless, virulent *M. tuberculosis* can survive in both mice and humans in the face of these potent IFN-γ-mediated T-cell responses (Flynn and Chan, 2001; Jung et al., 2002), resulting in progressive infection. There are a number of host factors that may contribute to this, including the timing and magnitude of IFN-γ response; the effects of inhibitory cytokines, such as IL-10 and TGF-β; and possibly regulatory T cells (see later). However, mycobacterial mechanisms employed to evade host responses and the capacity to switch gene expression in the face of a hostile environment are fundamental to the survival of virulent mycobacteria. Important among these is the ability of *M. tuberculosis* to inhibit the multiple effects of IFN-γ-mediated activation on macrophages and thereby reduce their capacity to kill virulent *M. tuberculosis* (Fortune et al., 2004). These evasion mechanisms can operate in three broad ways; inhibition of the transcriptional response to IFN-γ, interference with intracellular signaling within activated macrophages, and protective barriers against the effects of bactericidal ROI and RNI (Fig. 3).

First, *M. tuberculosis* infection or exposure to LpgH inhibited 42% and 36% of the 347 macrophage genes induced by IFN-γ, including those involved in inflammatory and killing responses as well as genes committed to antigen presentation (Pai et al., 2004). Live *M. tuberculosis* blocks the transcriptional response to IFN-γ in infected macrophages by disrupting the interaction of STAT1 with the transcriptional coactivators CREB binding protein and p300, while leaving signaling through STAT1 intact (Ting et al., 1999) (Fig. 3). This inhibitory effect occurs in part through MyD88-dependent signaling through TLR-2 activation by LprH and in part through MyD88-independent mechanisms activated by the cell wall mycolyl-arabinogalactan peptidoglycan complex (Fortune et al., 2004). Furthermore, *M. tuberculosis* Δ*lspA* mutants inhibited IFN-γ activation of macrophages as effectively as wild-type bacteria, indicating that other cell wall components contribute to the inhibition. These include PIMs, which provide TLR2-dependent inhibitory signals to macrophage activation (Banaiee et al., 2006).

Second, components of the mycobacterial cell wall can interfere with signaling within the activated macrophages (Fig. 3). The best characterized of these

is LAM, which has been recognized to block macrophage enzymes in signaling cascades, such as protein kinase C (Briken et al., 2004; Chan et al., 1991). The mechanism of this inhibition has been elucidated (Knutson et al., 1998). LAM activates the Src homology 2-containing tyrosine phosphatase (SHP-1), which then translocates to the macrophage membrane and inactivates phosphorylated signalling kinases, including MAP kinase. This results in reduced TNF and IL-12 production by activated macrophages, as well as decreased MHC class II expression.

Third, components of the cell wall may act as scavengers for ROI, as in the case of LAM (Chan et al., 1991) or RNI (Fig. 3). For example, mutants of *M. tuberculosis* lacking phthiocerol dimycocerosates (DIMs) from their cell wall are attenuated in their growth in activated macrophages and in mice. This is not caused by differences in the transcriptional response of macrophages to the DIM-deficient mutant and virulent *M. tuberculosis* (Scandurra et al., 2007) but is related to increased sensitivity of the mutant strain to the bactericidal effects of RNI released by activated macrophages (Rousseau et al., 2004). This effect did not reduce the activation of Th1 T cell responses by the mutant, which proved a more effective and durable vaccine against subsequent *M. tuberculosis* infection than BCG (Pinto et al., 2004).

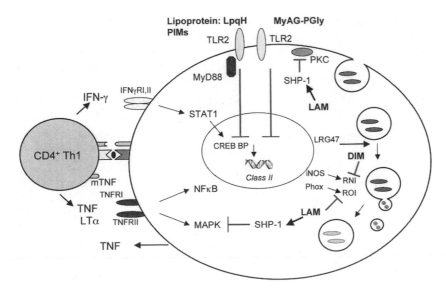

Figure 3. Inhibition of macrophage activation by cell envelope components of *M. tuberculosis*. Infection with virulent *M. tuberculosis* inhibits the transcriptional response of macrophages to the activation by IFN-γ secreted by mycobacterium-specific CD4 T cells by disrupting the interaction of STAT1 with the transcriptional coactivators CREB-binding protein (CREB BP) and p300, as demonstrated by reduced MHC class II expression. This inhibition is dependent on activation of TLR2/MyD88 by lipoproteins, such as LpgH, and PIMS, and also on stimulation of TLR2 by the mycolyl-arabinogalactan peptidoglycan complex through an MyD88-independent pathway. LAM interferes with intracellular signaling by activating the Src homology containing tyrosine phosphatase SHP-1, which inactivates phosphorylated signaling kinases, such as protein kinase C (PKC) and MAP kinase. In addition, cell wall components, such as LAM and DIM, can act as scavengers for ROI and RNI, respectively, produced by phagocyte oxidase (Phox) and inducible nitric oxide synthase (iNOS).

PRESENTATION OF MYCOBACTERIAL CELL ENVELOPE MOLECULES TO THE IMMUNE SYSTEM

The classic pathway of antigen recognition involves the presentation of protein fragments to T cells by MHC class I and class II antigen presentation molecules. Both CD4 and CD8 T cells recognize mycobacterial antigens, which include secreted and intracellular proteins (Sable et al., 2007). However, other T-cell populations, including $\gamma\delta$ TcR T cells, natural killer (NK) T cells and "double-negative" T cells, which lack expression of CD4 and CD8, also participate in the immune response to mycobacteria (Boom et al., 2003). In addition, mycobacterial lipids and other cell wall-associated molecules are presented to the immune system by nonclassic antigen presenting molecules, such as the CD1 antigens (Dascher and Brenner, 2005).

Presentation of Mycobacterial Antigens to T Cells by Members of the CD1 Family

CD1 proteins, which are antigen presenting molecules encoded by genes located outside of the MHC, recognize nonpeptide lipid or glycolipid structures, including components of mycobacteria (reviewed in Ulrichs and Porcelli, 2000). The CD1 isoforms are composed of two groups; group 1, which contains CD1a, CD1b, and CD1c, and group 2 which contains only the CD1d isoform (Calabi et al., 1989). The first report of an antigen recognized by CD1 molecules was provided by (Porcelli et al., 1992), who demonstrated that a CD4$^-$ CD8$^-$ $\alpha\beta^+$ T-cell line recognized a mycobacterial antigen in a CD1b-restricted fashion (Porcelli et al., 1992). The antigen recognized was the major cell wall component mycolic acid (Beckman et al., 1994), demonstrating that mycobacterial lipids are target of the immune response. Two $\alpha\beta^+$ CD4$^-$ CD8$^-$ CD1b-restricted T cells, one derived from a leprosy skin lesion, were found to recognize LAM, and recognition required internalization and endosomal acidification of the antigen (Sieling et al., 1995). T cells activated by LAM secrete proinflammatory cytokines and are cytolytic, suggesting that LAM-reactive T cells may play some role in the establishing protective immunity (Sieling et al., 1995). Recognition of LAM by CD1b is dependent in part on the MRs present on host cells (Prigozy et al., 1997). MRs were required for the internalization of LAM by APCs, and blocking these receptors with anti-MRs antibodies inhibited antigen presentation to LAM-reactive T cells. MRs colocalize with CD1b molecules, suggesting the receptors are able to transport LAM to the sites of antigen loading. The multifunctional glycoprotein saposin C (SAP-C) also plays a role in the pres-

entation of M. tuberculosis lipids through CD1b, because SAP-C-deficient fibroblasts transfected with CD1b are unable to activate lipid-specific T cells (Winau et al., 2004). The lipid antigen and SAP-C also colocalize in lysosomal compartments, and SAP-C was found to interact directly with CD1b, thereby permitting direct loading of CD1b with the lipid antigen. CD1b is able to recognize other mycobacterial cell wall components. For example, human T-cell clones specific for the diacylated M. tuberculosis sulfolipids, 2-palmitoyl or 2-stearoyl-3-hydroxyphthioceranoyl-2′-sulfate-α-α′-D-trehalose (Ac2SGL), predominantly recognized the antigen in the context of CD1b molecules (Gilleron et al., 2004). Ac2SGL-specific T cells were capable of cytokine release and recognition of M. tuberculosis-infected cells.

Other CD1 molecules appear to play a role in the generation of antimycobacterial immunity. Human CD1c-restricted T-cell lines recognize mycobacterial lipids antigens (Beckman et al., 1996), and the antigens commonly recognized by CD1c-restricted T-cell lines are members of the isoprenoid glycolipid family (Moody et al., 2000). CD1d, which commonly participates in antigen presentation to NK T cells, is also involved in presentation of mycobacterial lipids. Cell wall-derived PIM was able to activate both human and murine NKT cells through CD1d, and these cells were cytotoxic and released immune mediators, such as IFN-γ (Fischer et al., 2004). Mice deficient in CD1d showed no increased susceptibility to M. tuberculosis infection (Behar et al., 1999; Sousa et al., 2000). However, because CD1d is the only CD isoform present in mice, a role for the CD1 proteins in protective immunity in humans infected with mycobacteria cannot be excluded. CD1a presents members of the lipopeptide class (didehydroxymycobactins), from M. tuberculosis to $\alpha\beta^+$ T cells (Moody et al., 2004), suggesting that CD1 proteins can also present structurally diverse, pathogen-derived lipopeptides to T cells.

The final member of the CD1 family, CD1e, has recently been found to participate in the processing of hexamannosylated PIMs (PIM6) for recognition by CD1b-restricted T cells (de la Salle et al., 2005). CD1e appears to enhance digestion of the oligomannose moiety by lysosomal alpha-mannosidase, and this function serves to increase the repertoire of glycolipid T-cell antigens which can be presented to the immune system. Mycobacteria can augment the capacity of CD1-bearing cells to present lipid components, because exposure of CD1-myeloid precursors to mycobacterial cell wall products converted the cells to efficient APCs, which expressed CD1 proteins (Roura-Mir et al., 2005). This process was dependent on TLR2 activation, because the induction of CD1 expression was inhibited by neutralizing anti-

TLR2 antibodies. These results demonstrate that the CD1 family comprises an inducible antigen processing system, whose function is regulated by microbial products through the TLR2 pattern recognition receptor.

The identification of mycobacterial cell wall components as novel antigens has resulted in further definition of the T-cell response to pathogenic bacteria. CD8 αβ T cells generally recognize peptide in the context of MHC class I molecules. However, nonclassic CD8 αβ T cell lines specific for *M. tuberculosis* lipid antigens have been identified (Rosat et al., 1999). Antigen presentation to these CD8 T cells is through CD1a or CD1c molecules, and these T cells are cytotoxic and display a Th1-like T cell cytokine response. *M. tuberculosis* infected individuals contain nonclassically restricted CD8 T cells, and in some individuals, these represent the majority of CD8 T cells detected during latent infection (Lewinsohn et al., 2000). These CD8 T cells recognize a cell-associated antigen of *M. tuberculosis* in the context of the HLA-E molecule (Heinzel et al., 2002). The precise nature of the antigen has not been determined. However, it is most likely a hydrophobic peptide, possibly modified by lipidation or glycosylation (Kawashima et al., 2003).

Role of the Mycobacterial Cell Envelope in the Suppression of T-Cell Immunity

One of the notable aspects of mycobacteria is their capacity to suppress the host immune response at multiple levels, including the effector T-cell function. This is most evident in leprosy, in which T-cell responses to *M. leprae* antigens are selectively suppressed in the lepromatous form of the disease (Britton and Lockwood, 2004). Early reports suggested that phenolic glycolipid and LAM from *M. leprae* could suppress T-cell responses (Bloom and Mehra, 1984). This immunosuppressive action was considered to be mediated by suppressor CD8 T cells, which were capable of inhibiting the proliferation of helper T cells (Kaplan et al., 1987; Modlin et al., 1986). LAM from *M. tuberculosis* was reported to have similar effects (Kaplan et al., 1987). Research into the nature of these suppressor T cells waned, owing in part to the observation that contaminating lipopolysaccharide in mycobacterial lipid preparations was responsible for this suppressive effect in some cases (Molloy et al., 1990). More recently, research has refocused on regulatory T cells, which function to inhibit the activation and function of helper T-cell responses, thereby maintaining immune homeostasis (O'Garra and Viera, 2004). These cells, which are typically characterized by the expression of the cell surface markers CD4 and CD25 and the transcription factor FoxP3, play a role in suppressing antimalarial (Walther et al., 2005) and antileishmainal (Belkaid et al., 2002) immunity, particularly through their secretion of the effector molecules IL-10 and TGF-B. CD25$^+$CD4$^+$ FoxP3$^+$ T cells have been detected in the peripheral blood of patients with tuberculosis, and these cells were capable of suppressing IFN-γ responses (Chen et al., 2007; Guyot-Revol et al., 2006). This suggests a potential role for regulatory T cells in modulating host immunity during human tuberculosis. However, it is to be determined whether these cells contribute to active tuberculosis disease.

LAM is the most studied of components of the mycobacterial cell wall with immunosuppressive properties (Briken et al., 2004). ManLAM can inhibit the phagosome maturation and so create a niche for the survival of virulent mycobacteria, and also suppress macrophage function (see earlier). In addition, LAM can act directly on T cells to downregulate transcription of mRNA for the cytokines IL-2, IL-3, and granulocyte-macrophage colony-stimulating factor (GM-CSF) and the IL-2 receptor alpha chain (Chujor et al., 1992). LAM may act on T cells due to its ability to be inserted directly into the plasma membrane of target cells through the acyl chains of PI anchors (Ilangumaran, 1995). The capacity of LAM to insert into T-helper cell rafts resulted in decreased release of type 1 cytokines (IL-2 and IFN-γ) and an increase in production of type 2 cytokines (IL-4 and IL-5) (Shabaana et al., 2005). The suppressive effects of *M. tuberculosis* LAM may be due in part to the ability of the molecule to induce IL-4 secretion in the bone marrow, which may have a suppressive effect on the generation of protective T-helper type 1 responses (Collins et al., 1998).

The ability of mycobacterial cell wall components to inhibit T-cell responses is not restricted to LAM. The glycolipid 2,3-di-O-acyl-trehalose (DAT) has been shown to reduce the proliferation of murine T cells (Saavedra et al., 2001). DAT is also able to reduce antigen-induced proliferation of human CD4 and CD8 T cell subsets, and this effect is associated with decreased expression of the activation markers CD25 and CD69 and reduced cytokine release (Saavedra et al., 2006).

Antibody Responses Induced by Components of the Mycobacterial Cell Envelope

The intracellular nature of pathogenic mycobacteria dictates that a cell-mediated immune response is required to activate host cells and kill resident bacteria. Although the role of CD4 and CD8 effector

T cells in the protective immune response to mycobacteria is well established, the role of antibodies is less clear (reviewed in Glatman-Freedman, 2006). Aggregations of B cells are evident in the lungs of humans and mice during active tuberculosis infection (Tsai et al., 2006). Early reports indicated that mice deficient in B cells displayed either a small increase in bacterial loads following *M. tuberculosis* infection (Vordermeier et al., 1996) or exhibited no difference in susceptibility (Johnson et al., 1997). Infection of B-cell-deficient mice with the virulent CDC1551 strain of *M. tuberculosis* did not alter the bacterial growth. However, less severe pulmonary granuloma formation with reduced lymphocyte infiltration and delayed dissemination of bacteria from the lungs to the peripheral organs developed in the absence of B cells (Bosio et al., 2000). Reduced immune responses in the lungs of immunoglobulin A (IgA)-deficient mice, however, increased the susceptibility of these mice to infection with *Mycobacterium bovis* BCG (Rodriguez et al., 2005).

Despite the controversy surrounding the role of antibodies in mycobacterial protection, it is clear that cell wall components are major targets of humoral immunity during mycobacterial infection. A predominant target of the antimycobacterial humoral immune response is LAM. Humans and animals infected with *M. tuberculosis* or vaccinated with BCG generate anti-LAM antibodies (Brown et al., 2003; Da Costa et al., 1993; de Valliere et al., 2005; Watanabe et al., 2006), and disseminated tuberculosis in children was associated with reduced anti-LAM immunoglobulin G (IgG) antibodies (Costello et al., 1992). Anti-LAM antibodies may activate the classic complement pathway, and deposition of C3b on the mycobacteria enhances their phagocytosis through CRs (Hetland et al., 1998). Indeed, anti-LAM antibodies increased the phagocytosis of *M. bovis* BCG. Intriguingly, in one study, human anti-LAM antibodies enhanced the growth inhibitory effects of macrophages on mycobacteria and also increased the effector function of antimycobacterial T cells (de Valliere et al., 2005). Monoclonal antibodies (MAbs) recognizing LAM, when delivered to mice either before infection or together with *M. tuberculosis*, reduced bacterial growth and prolonged survival compared with treatment with control antibodies (Hamasur et al., 2004). This effect may be related to the ability of the antibody to limit the pathological effects of LAM, because treatment of mice with an anti-LAM MAb followed by injection of LAM enhanced the clearance of LAM from the circulation (Glatman-Freedman et al., 2000).

Other cell wall components of mycobacteria are targets of the antibody response. Phenolic glycolipid 1 (PGL-1) is an abundant component of the *M. leprae* cell wall and is involved in the invasion of Schwann cells of the peripheral nerve by *M. leprae* (Ng et al., 2000). PGL-1 is a major target of the immune response in *M. leprae*-infected individuals (Cho et al., 1983). Immunoglobulin M (IgM) anti-PGL-1 antibodies are commonly used for the serodiagnosis of leprosy, and a fall in IgM anti-PGL-1 antibodies or serum *M. leprae* PGL-1 antigen levels can be used as markers of successful antimicrobial therapy (Roche et al., 1991; Roche et al., 1993; Cho et al., 2001). The production of PGL by *M. tuberculosis* appears to be important in determining the virulence of *M. tuberculosis* isolates, because hypervirulence of certain *M. tuberculosis* strains is dependent on a polyketide synthase-derived PGL (Reed et al., 2004). Major PGLs of *M. tuberculosis* are targets of the humoral immune response. Antibodies to the PGLTB1 antigen, which resembles PGL-1 from *M. leprae*, are present in tuberculosis patients (Chanteau et al., 1992; Simonney et al., 1995; Watanabe et al., 1995). PGLtb1 and PGL-tbK, which were described as being specific to *M. tuberculosis*, are widely distributed among *M. tuberculosis* strains. However, only a small proportion of patients with tuberculosis develop antibodies to these specific glycolipids (Watanabe et al., 1995). A large proportion of healthy controls contained anti-PGL-tb Abs, suggesting exposure to *M. tuberculosis* or environmental mycobacteria containing structurally related glycolipids.

Other cell wall lipids participate in the humoral recognition of mycobacteria, and a summary of the classes of lipids recognized by the immune response to mycobacteria are outlined in Table 1. Most studies have attempted to use the humoral response to cell surface components for the specific detection of *M. tuberculosis* infection. Mice infected with *M. tuberculosis* developed high IgM antibody responses to sulfolipid I (SL-I), which were greater than those to the immunodominant 30-kDa antigen (Cardona et al., 2002). Therefore, the presence of antisulfolipid antibodies was tested for specific immunodiagnosis of tuberculosis. In individuals infected with *M. tuberculosis*, sulfolipids induce IgM, IgG, and IgA subclasses (Table 1). Some reports demonstrated satisfactory specificity for assays detecting anti-SL-I and sulfolipid V antibodies (Cruaud et al., 1989; Julian et al., 2002), and the sensitivity of the assay was improved by combining it with serological assays using additional proteins and glycolipids (Julian et al., 2004). However, the fact that leprosy patients recognize *M. tuberculosis* SL-I may limit the usefulness of assays based on sulfolipids for the specific detection of tuberculosis (Luna-Herrera et al., 1996).

Table 1. Classes of mycobacterial cell wall molecules recognized by the humoral immune response during mycobacterial infection[a]

Molecule	Species	Antibody isotype(s)	Reference(s)
LAM	*M. tuberculosis*	IgA, IgG, IgM	Brown et al. (2003)
	M. leprae	IgG	Meeker et al. (1990), Roche et al. (1993)
Sulfolipids			
Sulfolipid I[b]	*M. tuberculosis*	IgA, IgG, IgM	Julian et al. (2002), Julian et al. (2004)
Sulfolipid IV	*M. tuberculosis*	IgM	Cruaud et al. (1989)
Trehalose-containing glycolipids			
DAT, TAT	*M. tuberculosis*	IgA, IgG, IgM	Julian et al. (2002), Julian et al. (2004), Simonney et al. (1995)
TDM, TMM	*M. tuberculosis*	IgA, IgG, IgM	Fujita et al. (2005), Julian et al. (2002), Julian et al. (2004)
Phenolic glycolipids (PGLs)			
PGL-1	*M. leprae*	IgM	Cho et al. (1983), Roche et al. (1993), Cho et al. (2001)
PGLtb1	*M. tuberculosis*	IgG, IgM	Chanteau et al. (1992), Simonney et al. (1995)
PGLtbK	*M. tuberculosis*	IgM	Watanabe et al. (1995)
PIMs	*M. avium*	IgG	Fujita et al. (2005)
Lipoproteins			
19 kDa[c]	*M. avium* subsp. *paratuberculosis*	Not defined	Huntley et al. (2005)
38 kDa	*M. tuberculosis*	IgA, IgG, IgM	Julian et al. (2004), Silva et al. (2003), Uma Devi et al. (2001)

[a]The table is representative of the molecules recognized by the human immune response and is not an exhaustive list of all studies performed. LAM, lipoarabinomannan; DAT, 2,3-diacyltrehalose; TAT, 2,3,6-triacyl trehalose; TDM, trehalose 6,6′ dimycolate; TMM, trehalose 6-monomycolate; PIMs, phosphatidylinositol dimannosides.
[b]Sulfolipids derived from *M. tuberculosis* are also recognized by individuals infected with *M. leprae* (Cruaud et al., 1990).
[c]Reactivity was defined in infected cattle.

CELL MIGRATION INDUCED BY MYCOBACTERIAL COMPONENTS OF THE CELL ENVELOPE

The generation of effective immunity to pathogenic mycobacteria requires the coordinated migration of cellular components in an attempt to contain the infection. This coordinated process involves the accumulation of T cells, macrophages, neutrophils, and other inflammatory cells types at the sites of infection and is most apparent in the process of granuloma formation (discussed later). Mycobacterial cell wall components have been shown to invoke the chemotaxis of specific cell types directly. AraLAM from *M. tuberculosis* induced the migration of human blood monocytes and monocyte-derived macrophages by a process dependent on the CD14 signaling receptor (Mazurek et al., 2001). ManLAM did not induce migration of either cell type, highlighting that mannose capping may be employed by virulent mycobacterial strains to reduce inflammatory cell migration. A separate analysis of the effect of ManLAM and AraLAM, however, revealed no difference in monocyte/macrophage migration induced by either form of LAM (Fietta et al., 2000). The ability of LAM to induce migration of phagocytic cells may relate to its role in the induction of chemoattractant chemokines, such as MCP3 and IL-8 (Zhang et al., 1995; Vouret-Craviari et al., 1997). Trehalose dimycolate (TDM, or cord factor) also contributes to the recruitment of monocytes (Lima et al., 2001), whereas injection of mice with TDM (Lima et al., 2001) or mycolic acid (Korf et al., 2005) induces the recruitment of neutrophils to the site of inoculation. Mycolic acid was also able to induce the formation of foamy macrophages, which have been predicted to play a role in the containment of infection with virulent mycobacteria (Korf et al., 2005).

Cell wall components may also contribute to the stimulation of T-cell responses through their ability to influence the migration of leukocyte subsets. In vivo studies in mice demonstrated that lipids derived from BCG could influence the migration of leukocytes into the pleural cavity of immunized mice, a property that was not shared by lipids derived from *M. leprae* (Moura et al., 1999). LAM has been demonstrated to participate in the chemotaxis of T cells from human peripheral blood. Culture supernatants from human alveolar macrophages infected with *M. tuberculosis* induced migration of T cells, and a MAb directed against LAM blocked a large proportion of this chemotactic activity (Berman et al., 1996). Intriguingly, AraLAM and ManLAM from *M. tuberculosis*, but not LAM from *M. bovis* BCG, induced T-cell chemotaxis in vitro, further highlighting that differences in LAM structure have a profound influence of the interaction of the molecule with the immune system.

The Mycobacterial Cell Envelope and Granulomas

The containment of mycobacterial infection requires the formation of granulomas (or tubercles), which are nodular aggregations of lymphocytes, macrophages, and epithelioid cells. Granulomas are characteristic of the chronic inflammatory responses to a variety of intracellular pathogens and noninfectious agents, and the cellular and cytokine requirements for granuloma formation have been recently reviewed (Saunders and Britton, 2007). There are differences in the structure of granulomas in mice and humans during *M. tuberculosis* infection. In humans, the infected macrophages, epithelioid cells, and multinucleate giant cells are admixed and surrounded by CD4 T cells with a cuff of CD8 T cells, and the central region may undergo caseous necrosis, leading to cavity formation. In contrast, in mice, the granulomas develop dense aggregations of lymphocytes surrounding macrophages without giant cells or central caseation, or fail to form cavities. Nevertheless, inbred strains of mice provide an excellent model for analyzing both host and mycobacterial factors, which are involved in granuloma formation (Russell, 2007; Saunders and Britton, 2007).

TDM (or cord factor) is the best characterized cell wall component that stimulates granuloma formation. When delivered to mice intravenously after immersion in oil (Bekierkunst, 1968) or coupled with particles (Actor et al., 2001; Geisel et al., 2005; Lima et al., 2001), TDM rapidly induces granulomatous inflammation, a property shared with trehalose monomycolate (Geisel et al., 2005). TDM-induced granuloma formation is associated with the induction of cytokines and chemokines, which are required for inflammatory cell recruitment and the maintenance of granuloma formation (Geisel et al., 2005; Lima et al., 2001; Perez et al., 2000; Takimoto et al., 2006; Yamagami et al., 2001). Accordingly, TDM-coated beads induced a cascade of inflammatory cells responsible for granuloma formation, including neutrophils, macrophages, dendritic cells, T cells, NK cells, and B cells (Rhoades et al., 2005). TNF and the related LT-α are essential for the formation of granulomas in tuberculosis infection (Bean et al., 1999; Roach et al., 2001), and anti-TNF antibody neutral-

ization studies confirmed that TNF was required for TDM-induced granuloma formation (Takimoto et al., 2006). Interestingly, granuloma formation was still induced by TDM in mice lacking IFN-γ or unable to signal through the IL-12 receptor (Takimoto et al., 2006). Moreover, mice lacking TLR2 and TLR4 also exhibited no defect in TDM-induced granuloma formation (Geisel et al., 2005). However, mice deficient in STAT4, which is required for the development of IL-12-dependent T-cell responses, and MyD88, which is necessary for TLR signaling, showed defective granulomatous responses (Geisel et al., 2005; Oiso et al., 2005). CD1d-restricted T cells may also contribute to granuloma formation, because CD1d$^{-/-}$ mice injected with TDM displayed less organized granulomas despite the more rapid induction of mRNA for proinflammatory cytokines (Actor et al., 2001; Guidry et al., 2004). Therefore, although the requirement for TNF in the inflammatory responses to mycobacterial lipids is well established, the role of other host factors remains to be elucidated. Fine structural variations in the mycolic acid component of TDM have a marked effect on its proinflammatory activity (Rao et al., 2006). Mutation of the *M. tuberculosis* cyclopropane-mycolic acid synthase (ΔcmaA2) prevented the transcyclopronation of the mycolic acid in TDM. This ΔcmaA2 strain was hypervirulent in mice owing to an excessive granulomatous response caused by a greatly increased macrophage inflammatory response to the modified TDM. Therefore, there is a fine balance between the stereochemistry of the cell wall lipids and the host inflammatory response.

Although the trehalose mycoserates are key determinants for stimulating the granuloma formation during mycobacterial infections, other cell wall components contribute to this response. Sepharose beads coated with *M. tuberculosis* PPD, a crude mixture of mycobacterial antigens containing cell wall components, induced formation of granulomatous structures when injected into mice (Boring et al., 1997) or incubated with human peripheral blood mononuclear cells (PBMCs) in a cell culture model (Puissegur et al., 2004). Both LAM and PIM from *M. tuberculosis* induced the formation of pulmonary granulomas after injection into mice, and this phenomenon was associated with TNF production in the lungs but was independent of IFN-γ (Takimoto et al., 2006). A direct comparison of LAM and PIM with TDM in this study revealed that TDM was the most potent stimulator of pulmonary TNF release and the dominant molecule promoting granuloma formation (Takimoto et al., 2006). Cellular recruitment in the lung after administration of LAM was dependent on IL-1, a key cytokine involved in granuloma formation

(Juffermans et al., 2000). PIM may contribute to granuloma formation by binding directly to T cells and promoting binding of the cells to fibronectin of the extracellular matrix (Sanchez et al., 2006). Both PIM2 and PIM6 can induce granuloma formation, and the presence of at least one fatty acid chain on the molecule was required to mediate this effect (Gilleron et al., 2001). In addition, muramyl dipeptide of *M. tuberculosis* can induce granuloma formation in mice when delivered conjugated to branched fatty acids (Emori et al., 1985). Although the administration of SL-I from *M. tuberculosis* did not stimulate granulomatous responses in mice, codelivery of SL-I with TDM inhibited TDM-induced granuloma formation and TNF release (Okamoto et al., 2006). It is not established if sulfolipid suppresses granuloma function in vivo during infection with virulent *M. tuberculosis*, and this may not be the case because sulfolipid deficiency did not influence the virulence of *M. tuberculosis* in mice and guinea pigs (Rousseau et al., 2003).

CONCLUSION

In summary, multiple components of the mycobacterial cell envelope can influence the host innate and adaptive immune responses to virulent mycobacteria. A major challenge is to understand the relative importance of the different cell components, which have immunomodulatory properties in vitro, in determining the behavior of pathogenic mycobacteria within the host. These effects combine to promote the survival of mycobacteria within an ecological niche in macrophages during latent infection and may inhibit the activation of infected macrophages by T-cell signals, resulting in replication of mycobacteria and active disease in some individuals. In human *M. tuberculosis* infection, the propensity of TDM and other cell wall components to promote granuloma formation in association with a chronic T-cell response leads to caseating granulomas, which may erode into airways. The resulting cavities provide the route for aerosol transmission of the tubercle bacilli to other susceptible hosts and so ensure the life cycle of the mycobacteria. Therefore, the capacity of mycobacterial cell wall components to modulate host immune responses is fundamental to the maintenance of pathogenic mycobacteria in human populations.

Acknowledgments. We acknowledge the financial support of the NHMRC of Australia, the Australian Respiratory Council and the NSW Department of Health through its infrastructure grant to the Centenary Institute for studies in our laboratories.

REFERENCES

Abel, B., N. Thieblemont, V. J. Quesniaux, N. Brown, J. Mpagi, K. Miyake, F. Bihl, and B. Ryffel. 2002. Toll-like receptor 4 expression is required to control chronic *Mycobacterium tuberculosis* infection in mice. *J. Immunol.* **169:**3155–3162.

Abou-Zeid, C., M. P. Gares, J. Inwald, R. Janssen, Y. Zhang, D. B. Young, C. Hetzel, J. R. Lamb, S. L. Baldwin, I. M. Orme, V. Yeremeev, B. V. Nikonenko, and A. S. Apt. 1997. Induction of a type 1 immune response to a recombinant antigen from *Mycobacterium tuberculosis* expressed in *Mycobacterium vaccae*. *Infect. Immun.* **65:**1856–1862.

Actor, J. K., M. Olsen, R. L. Hunter, Jr., and Y. J. Geng. 2001. Dysregulated response to mycobacterial cord factor trehalose-6,6'-dimycolate in CD1D$^{-/-}$ mice. *J. Interferon Cytokine Res.* **21:**1089–1096.

Aderem, A., and D. M. Underhill. 1999. Mechanisms of phagocytosis in macrophages. *Annu. Rev. Immunol.* **17:**593–623.

Armstrong, J. A., and P. D. Hart. 1971. Response of cultured macrophages to *Mycobacterium tuberculosis*, with observations on fusion of lysosomes with phagosomes. *J. Exp. Med.* **134:**713–740.

Bafica, A., C. A. Scanga, C. G. Feng, C. Leifer, A. Cheever, and A. Sher. 2005. TLR9 regulates Th1 responses and cooperates with TLR2 in mediating optimal resistance to *Mycobacterium tuberculosis*. *J. Exp. Med.* **202:**1715–1724.

Banaiee, N., E. Z. Kincaid, U. Buchwald, W. R. Jacobs, Jr., and J. D. Ernst. 2006. Potent inhibition of macrophage responses to IFN-gamma by live virulent *Mycobacterium tuberculosis* is independent of mature mycobacterial lipoproteins but dependent on TLR2. *J. Immunol.* **176:**3019–3027.

Bean, A. G., D. R. Roach, H. Briscoe, M. P. France, H. Korner, J. D. Sedgwick, and W. J. Britton. 1999. Structural deficiencies in granuloma formation in TNF gene-targeted mice underlie the heightened susceptibility to aerosol *Mycobacterium tuberculosis* infection, which is not compensated for by lymphotoxin. *J. Immunol.* **162:**3504–3511.

Beatty, W. L., E. R. Rhoades, H. J. Ullrich, D. Chatterjee, J. E. Heuser, and D. G. Russell. 2000. Trafficking and release of mycobacterial lipids from infected macrophages. *Traffic* **1:**235–247.

Beckman, E. M., S. A. Porcelli, C. T. Morita, S. M. Behar, S. T. Furlong, and M. B. Brenner. 1994. Recognition of a lipid antigen by CD1-restricted alpha beta+ T cells. *Nature* **372:**691–694.

Beckman, E. M., A. Melian, S. M. Behar, P. A. Sieling, D. Chatterjee, S. T. Furlong, R. Matsumoto, J. P. Rosat, R. L. Modlin, and S. A. Porcelli. 1996. CD1c restricts responses of mycobacteria-specific T cells. Evidence for antigen presentation by a second member of the human CD1 family. *J. Immunol.* **157:**2795–2803.

Behar, S. M., C. C. Dascher, M. J. Grusby, C.-R. Wang, and M. B. Brenner. 1999. Susceptibility of mice deficient in CD1D or Tap1 to infection with *Mycobacterium tuberculosis*. *J. Exp. Med.* **189:**1973–1980.

Bekierkunst, A. 1968. Acute granulomatous response produced in mice by trehalose-6,6-dimycolate. *J. Bacteriol.* **96:**958–961.

Belkaid, Y., C. A. Piccirillo, S. Mendez, E. M. Shevach, and D. L. Sacks. 2002. CD4+CD25+ regulatory T cells control Leishmania major persistence and immunity. *Nature* **420:**502–507.

Berman, J. S., R. L. Blumenthal, H. Kornfeld, J. A. Cook, W. W. Cruikshank, M. W. Vermeulen, D. Chatterjee, J. T. Belisle, and M. J. Fenton. 1996. Chemotactic activity of mycobacterial lipoarabinomannans for human blood T lymphocytes in vitro. *J. Immunol.* **156:**3828–3835.

Bloom, B. R., and V. Mehra. 1984. Immunological unresponsiveness in leprosy. *Immunol. Rev.* **80:**5–28.

Boom, W. H., D. H. Canaday, S. A. Fulton, A. J. Gehring, R. E. Rojas, and M. Torres. 2003. Human immunity to M. tuberculosis: T cell subsets and antigen processing. *Tuberculosis* (Edinburgh) **83:**98–106.

Boring, L., J. Gosling, S. W. Chensue, S. L. Kunkel, R. V. Farese, Jr., H. E. Broxmeyer, and I. F. Charo. 1997. Impaired monocyte migration and reduced type 1 (Th1) cytokine responses in C-C chemokine receptor 2 knockout mice. *J. Clin. Investig.* **100:** 2552–2561.

Bosio, C. M., D. Gardner, and K. L. Elkins. 2000. Infection of B cell-deficient mice with CDC 1551, a clinical isolate of *Mycobacterium tuberculosis*: delay in dissemination and development of lung pathology. *J. Immunol.* **164:**6417–6425.

Brightbill, H. D., D. H. Libraty, S. R. Krutzik, R. B. Yang, J. T. Belisle, J. R. Bleharski, M. Maitland, M. V. Norgard, S. E. Plevy, S. T. Smale, P. J. Brennan, B. R. Bloom, P. J. Godowski, and R. L. Modlin. 1999. Host defense mechanisms triggered by microbial lipoproteins through toll-like receptors. *Science* **285:** 732–736.

Briken, V., S. A. Porcelli, G. S. Besra, and L. Kremer. 2004. Mycobacterial lipoarabinomannan and related lipoglycans: from biogenesis to modulation of the immune response. *Mol. Microbiol.* **53:**391–403.

Britton, W. J., and D. N. Lockwood. 2004. Leprosy. *Lancet* **363:**1209–1219.

Brown, R. M., O. Cruz, M. Brennan, M. L. Gennaro, L. Schlesinger, Y. A. Skeiky, and D. F. Hoft. 2003. Lipoarabinomannan-reactive human secretory immunoglobulin A responses induced by mucosal bacille Calmette-Guerin vaccination. *J. Infect. Dis.* **187:**513–517.

Calabi, F., J. M. Jarvis, L. Martin, and C. Milstein. 1989. Two classes of CD1 genes. *Eur. J. Immunol.* **19:**285–292.

Cardona, P. J., E. Julian, X. Valles, S. Gordillo, M. Munoz, M. Luquin, and V. Ausina. 2002. Production of antibodies against glycolipids from the *Mycobacterium tuberculosis* cell wall in aerosol murine models of tuberculosis. *Scand. J. Immunol.* **55:**639–645.

Chacon-Salinas, R., J. Serafin-Lopez, R. Ramos-Payan, P. Mendez-Aragon, R. Hernandez-Pando, D. Van Soolingen, L. Flores-Romo, S. Estrada-Parra, and I. Estrada-Garcia. 2005. Differential pattern of cytokine expression by macrophages infected in vitro with different *Mycobacterium tuberculosis* genotypes. *Clin. Exp. Immunol.* **140:**443–449.

Chan, J., X. D. Fan, S. W. Hunter, P. J. Brennan, and B. R. Bloom. 1991. Lipoarabinomannan, a possible virulence factor involved in persistence of *Mycobacterium tuberculosis* within macrophages. *Infect. Immun.* **59:**1755–1761.

Chanteau, S., P. Glaziou, and R. Chansin. 1992. Assessment of the diagnostic value of the native PGLTB1, its synthetic neoglyco-conjugate PGLTB0 and the sulfolipid IV antigens for the serodiagnosis of tuberculosis. *Int. J. Lepr. Other Mycobact. Dis.* **60:** 1–7.

Chen, X., B. Zhou, M. Li, Q. Deng, X. Wu, X. Le, C. Wu, N. Larmonier, W. Zhang, H. Zhang, H. Wang, and E. Katsanis. 2007. CD4(+)CD25(+)FoxP3(+) regulatory T cells suppress *Mycobacterium tuberculosis* immunity in patients with active disease. *Clin. Immunol.* **123:**50–59.

Chieppa, M., G. Bianchi, A. Doni, A. Del Prete, M. Sironi, G. Laskarin, P. Monti, L. Piemonti, A. Biondi, A. Mantovani, M. Introna, and P. Allavena. 2003. Cross-linking of the mannose receptor on monocyte-derived dendritic cells activates an anti-inflammatory immunosuppressive program. *J. Immunol.* **171:**4552–4560.

Cho, S. N., D. L. Yanagihara, S. W. Hunter, P. H. Gelber, and P. J. Brennan. 1983. Serological specificity of phenolic glycolipid I from *Mycobacterium leprae* and use in serodiagnosis of leprosy. *Infect. Immun.* **41:**1077–1083.

Cho, S. N., R. V. Cellona, L. G. Villahermosa, T. T. Fajardo, Jr., M. V. Balagon, R. M. Abalos, E. V. Tan, G. P. Walsh, J. D. Kim, and P. J. Brennan. 2001. Detection of phenolic glycolipid I of

Mycobacterium leprae in sera from leprosy patients before and after start of multidrug therapy. *Clin. Diagn. Lab. Immunol.* **8:**138–142.

Chujor, C. S., B. Kuhn, B. Schwerer, H. Bernheimer, W. R. Levis, and D. Bevec. 1992. Specific inhibition of mRNA accumulation for lymphokines in human T cell line Jurkat by mycobacterial lipoarabinomannan antigen. *Clin. Exp. Immunol.* **87:**398–403.

Ciaramella, A., A. Cavone, M. B. Santucci, S. K. Garg, N. Sanarico, M. Bocchino, D. Galati, A. Martino, G. Auricchio, M. D'Orazio, G. R. Stewart, O. Neyrolles, D. B. Young, V. Colizzi, and M. Fraziano. 2004. Induction of apoptosis and release of interleukin-1 beta by cell wall-associated 19-kDa lipoprotein during the course of mycobacterial infection. *J. Infect. Dis.* **190:**1167–1176.

Collins, H. L., U. E. Schaible, and S. H. Kaufmann. 1998. Early IL-4 induction in bone marrow lymphoid precursor cells by mycobacterial lipoarabinomannan. *J. Immunol.* **161:**5546–5554.

Cooper, A. M., D. K. Dalton, T. A. Stewart, J. P. Griffin, D. G. Russell, and I. M. Orme. 1993. Disseminated tuberculosis in interferon gamma gene-disrupted mice. *J. Exp. Med.* **178:**2243–2247.

Costello, A. M., A. Kumar, V. Narayan, M. S. Akbar, S. Ahmed, C. Abou-Zeid, G. A. Rook, J. Stanford, and C. Moreno. 1992. Does antibody to mycobacterial antigens, including lipoarabinomannan, limit dissemination in childhood tuberculosis? *Trans. R. Soc. Trop. Med. Hyg.* **86:**686–692.

Cruaud, P., C. Berlie, J. Torgal Garcia, F. Papa, and H. L. David. 1989. Human IgG antibodies immunoreacting with specific sulfolipids from *Mycobacterium tuberculosis*. *Zentbl. Bakteriol.* **271:**481–485.

Cruaud, P., M. C. Potar, H. L. David, F. Papa, J. Torgal-Garcia, F. Maroja, and A. T. Orsi-Souza. 1990. IgG and IgM antibodies immunoreacting with a 2,3-diacyl trehalose-2′-sulphate in sera from leprosy patients. *Zentbl. Bakteriol.* **273:**209–215.

Cywes, C., H. C. Hoppe, M. Daffe, and M. R. Ehlers. 1997. Nonopsonic binding of *Mycobacterium tuberculosis* to complement receptor type 3 is mediated by capsular polysaccharides and is strain dependent. *Infect. Immun.* **65:**4258–4266.

Da Costa, C. T., S. Khanolkar-Young, A. M. Elliott, K. M. Wasunna, and K. P. McAdam. 1993. Immunoglobulin G subclass responses to mycobacterial lipoarabinomannan in HIV-infected and non-infected patients with tuberculosis. *Clin. Exp. Immunol.* **91:**25–29.

Dascher, C. C., and M. B. Brenner. 2005. CD1 and tuberculosis, p. 475–487. *In* S. T. Cole, K. D. Eisenbach, D. N. McMurray, and W. R. Jacobs, Jr. (ed.), *Tuberculosis and the Tubercle Bacillus.* ASM Press, Washington, DC.

de la Salle, H., S. Mariotti, C. Angenieux, M. Gilleron, L. F. Garcia-Alles, D. Malm, T. Berg, S. Paoletti, B. Maitre, L. Mourey, J. Salamero, J. P. Cazenave, D. Hanau, L. Mori, G. Puzo, and G. De Libero. 2005. Assistance of microbial glycolipid antigen processing by CD1e. *Science* **310:**1321–1324.

Demangel, C., and W. J. Britton. 2000. Interaction of dendritic cells with mycobacteria: where the action starts. *Immunol. Cell Biol.* **78:**318–324.

Demangel, C., U. Palendira, C. G. Feng, A. W. Heath, A. G. Bean, and W. J. Britton. 2001. Stimulation of dendritic cells via CD40 enhances immune responses to *Mycobacterium tuberculosis* infection. *Infect. Immun.* **69:**2456–2461.

Demangel, C., P. Bertolino, and W. J. Britton. 2002. Autocrine IL-10 impairs dendritic cell (DC)-derived immune responses to mycobacterial infection by suppressing DC trafficking to draining lymph nodes and local IL-12 production. *Eur. J. Immunol.* **32:** 994–1002.

Deretic, V., S. Singh, S. Master, J. Harris, E. Roberts, G. Kyei, A. Davis, S. de Haro, J. Naylor, H. H. Lee, and I. Vergne. 2006.

Mycobacterium tuberculosis inhibition of phagolysosome biogenesis and autophagy as a host defence mechanism. *Cell Microbiol.* 8:719–727.

de Valliere, S., G. Abate, A. Blazevic, R. M. Heuertz, and D. F. Hoft. 2005. Enhancement of innate and cell-mediated immunity by antimycobacterial antibodies. *Infect. Immun.* 73:6711–6720.

Emori, K., S. Nagao, N. Shigematsu, S. Kotani, M. Tsujimoto, T. Shiba, S. Kusumoto, and A. Tanaka. 1985. Granuloma formation by muramyl dipeptide associated with branched fatty acids, a structure probably essential for tubercle formation by *Mycobacterium tuberculosis*. *Infect. Immun.* 49:244–249.

Feng, C. G., C. A. Scanga, C. M. Collazo-Custodio, A. W. Cheever, S. Hieny, P. Caspar, and A. Sher. 2003. Mice lacking myeloid differentiation factor 88 display profound defects in host resistance and immune responses to *Mycobacterium avium* infection not exhibited by Toll-like receptor 2 (TLR2)- and TLR4-deficient animals. *J. Immunol.* 171:4758–4764.

Fenton, M. J., L. W. Riley, and L. S. Schesinger. 2005. Receptor-mediated recognition of *Mycobacterium tuberculosis* by host cells, p 427–436. *In* S. T. Cole, K. D. Eisenbach, D. N. McMurray, and W. R. Jacobs, Jr. (ed.), *Tuberculosis and the Tubercle Bacillus*, ASM Press, Washington, DC.

Ferwerda, G., S. E. Girardin, B. J. Kullberg, L. Le Bourhis, D. J. de Jong, D. M. Langenberg, R. van Crevel, G. J. Adema, T. H. Ottenhoff, J. W. Van der Meer, and M. G. Netea. 2005. NOD2 and toll-like receptors are nonredundant recognition systems of *Mycobacterium tuberculosis*. *PLoS Pathog.* 1:279–285.

Fietta, A., C. Francioli, and G. Gialdroni Grassi. 2000. Mycobacterial lipoarabinomannan affects human polymorphonuclear and mononuclear phagocyte functions differently. *Haematologica* 85:11–18.

Fischer, K., E. Scotet, M. Niemeyer, H. Koebernick, J. Zerrahn, S. Maillet, R. Hurwitz, M. Kursar, M. Bonneville, S. H. Kaufmann, and U. E. Schaible. 2004. Mycobacterial phosphatidylinositol mannoside is a natural antigen for CD1d-restricted T cells. *Proc. Natl. Acad. Sci. USA* 101:10685–10690.

Flynn, J. L., and J. Chan. 2001. Immunology of tuberculosis. *Annu. Rev. Immunol.* 19:93–129.

Fortune, S. M., A. Solache, A. Jaeger, P. J. Hill, J. T. Belisle, B. R. Bloom, E. J. Rubin, and J. D. Ernst. 2004. *Mycobacterium tuberculosis* inhibits macrophage responses to IFN-gamma through myeloid differentiation factor 88-dependent and -independent mechanisms. *J. Immunol.* 172:6272–6280.

Fratti, R. A., J. M. Backer, J. Gruenberg, S. Corvera, and V. Deretic. 2001. Role of phosphatidylinositol 3-kinase and Rab5 effectors in phagosomal biogenesis and mycobacterial phagosome maturation arrest. *J. Cell Biol.* 154:631–644.

Fratti, R. A., J. Chua, I. Vergne, and V. Deretic. 2003. *Mycobacterium tuberculosis* glycosylated phosphatidylinositol causes phagosome maturation arrest. *Proc. Natl. Acad. Sci. USA* 100:5437–5442.

Fremond, C. M., V. Yeremeev, D. M. Nicolle, M. Jacobs, V. F. Quesniaux, and B. Ryffel. 2004. Fatal *Mycobacterium tuberculosis* infection despite adaptive immune response in the absence of MyD88. *J. Clin. Investig.* 114:1790–1799.

Fujita, Y., T. Doi, K. Sato, and I. Yano. 2005. Diverse humoral immune responses and changes in IgG antibody levels against mycobacterial lipid antigens in active tuberculosis. *Microbiology* 151:2065–2074.

Gagliardi, M. C., R. Teloni, F. Giannoni, M. Pardini, V. Sargentini, L. Brunori, L. Fattorini, and R. Nisini. 2005. *Mycobacterium bovis* Bacillus Calmette-Guerin infects DC-SIGN-dendritic cell and causes the inhibition of IL-12 and the enhancement of IL-10 production. *J. Leukoc. Biol.* 78:106–113.

Gehring, A. J., R. E. Rojas, D. H. Canaday, D. L. Lakey, C. V. Harding, and W. H. Boom. 2003. The *Mycobacterium tuberculosis* 19-kilodalton lipoprotein inhibits gamma interferon-regulated HLA-DR and Fc gamma R1 on human macrophages through Toll-like receptor 2. *Infect. Immun.* 71:4487–4497.

Gehring, A. J., K. M. Dobos, J. T. Belisle, C. V. Harding, and W. H. Boom. 2004. *Mycobacterium tuberculosis* LprG (Rv1411c): a novel TLR-2 ligand that inhibits human macrophage class II MHC antigen processing. *J. Immunol.* 173:2660–2668.

Geijtenbeek, T. B., S. J. Van Vliet, E. A. Koppel, M. Sanchez-Hernandez, C. M. Vandenbroucke-Grauls, B. Appelmelk, and Y. Van Kooyk. 2003. Mycobacteria target DC-SIGN to suppress dendritic cell function. *J. Exp. Med.* 197:7–17.

Geisel, R. E., K. Sakamoto, D. G. Russell, and E. R. Rhoades. 2005. In vivo activity of released cell wall lipids of *Mycobacterium bovis* bacillus Calmette-Guerin is due principally to trehalose mycolates. *J. Immunol.* 174:5007–5015.

Gilleron, M., C. Ronet, M. Mempel, B. Monsarrat, G. Gachelin, and G. Puzo. 2001. Acylation state of the phosphatidylinositol mannosides from *Mycobacterium bovis* bacillus Calmette Guerin and ability to induce granuloma and recruit natural killer T cells. *J. Biol. Chem.* 276:34896–34904.

Gilleron, M., S. Stenger, Z. Mazorra, F. Wittke, S. Mariotti, G. Bohmer, J. Prandi, L. Mori, G. Puzo, and G. De Libero. 2004. Diacylated sulfoglycolipids are novel mycobacterial antigens stimulating CD1-restricted T cells during infection with *Mycobacterium tuberculosis*. *J. Exp. Med.* 199:649–659.

Glatman-Freedman, A., A. J. Mednick, N. Lendvai, and A. Casadevall. 2000. Clearance and organ distribution of *Mycobacterium tuberculosis* lipoarabinomannan (LAM) in the presence and absence of LAM-binding immunoglobulin M. *Infect. Immun.* 68:335–341.

Glatman-Freedman, A. 2006. The role of antibody-mediated immunity in defense against *Mycobacterium tuberculosis*: advances toward a novel vaccine strategy. *Tuberculosis* (Edinburgh) 86:191–197.

Guidry, T. V., M. Olsen, K. S. Kil, R. L. Hunter, Jr., Y. J. Geng, and J. K. Actor. 2004. Failure of CD1D-/- mice to elicit hypersensitive granulomas to mycobacterial cord factor trehalose 6,6'-dimycolate. *J. Interferon Cytokine Res.* 24:362–371.

Guyot-Revol, V., J. A. Innes, S. Hackforth, T. Hinks, and A. Lalvani. 2006. Regulatory T cells are expanded in blood and disease sites in patients with tuberculosis. *Am. J. Respir. Crit. Care Med.* 173:803–810.

Hamasur, B., M. Haile, A. Pawlowski, U. Schroder, G. Kallenius, and S. B. Svenson. 2004. A mycobacterial lipoarabinomannan specific monoclonal antibody and its F(ab') fragment prolong survival of mice infected with *Mycobacterium tuberculosis*. *Clin. Exp. Immunol.* 138:30–38.

Heinzel, A. S., J. E. Grotzke, R. A. Lines, D. A. Lewinsohn, A. L. McNabb, D. N. Streblow, V. M. Braud, H. J. Grieser, J. T. Belisle, and D. M. Lewinsohn. 2002. HLA-E-dependent presentation of Mtb-derived antigen to human CD8+ T cells. *J. Exp. Med.* 196:1473–1481.

Hertz, C. J., S. M. Kiertscher, P. J. Godowski, D. A. Bouis, M. V. Norgard, M. D. Roth, and R. L. Modlin. 2001. Microbial lipopeptides stimulate dendritic cell maturation via Toll-like receptor 2. *J. Immunol.* 166:2444–2450.

Hetland, G., H. G. Wiker, K. Hogasen, B. Hamasur, S. B. Svenson, and M. Harboe. 1998. Involvement of antilipoarabinomannan antibodies in classical complement activation in tuberculosis. *Clin. Diagn. Lab. Immunol.* 5:211–218.

Hoal-van Helden, E. G., L. A. Stanton, R. Warren, M. Richardson, and P. D. van Helden. 2001. Diversity of in vitro cytokine responses by human macrophages to infection by *Mycobacterium tuberculosis* strains. *Cell Biol. Int.* 25:83–90.

Hovav, A. H., J. Mullerad, L. Davidovitch, Y. Fishman, F. Bigi, A. Cataldi, and H. Bercovier. 2003. The *Mycobacterium tuberculosis* recombinant 27-kilodalton lipoprotein induces a strong

Th1-type immune response deleterious to protection. *Infect. Immun.* 71:3146–3154.

Hovav, A. H., J. Mullerad, A. Maly, L. Davidovitch, Y. Fishman, and H. Bercovier. 2006. Aggravated infection in mice co-administered with *Mycobacterium tuberculosis* and the 27-kDa lipoprotein. *Microbes Infect.* 8:1750–1757.

Hu, C., T. Mayadas-Norton, K. Tanaka, J. Chan, and P. Salgame. 2000. *Mycobacterium tuberculosis* infection in complement receptor 3-deficient mice. *J. Immunol.* 165:2596–2602.

Huntley, J. F., J. R. Stabel, and J. P. Bannantine. 2005. Immunoreactivity of the *Mycobacterium avium* subsp. paratuberculosis 19-kDa lipoprotein. *BMC Microbiol.* 5:3.

Ilangumaran, S., S. Arni, M. Poincelet, J. M. Theler, P. J. Brennan, D. Nasir ud, and D. C. Hoessli. 1995. Integration of mycobacterial lipoarabinomannans into glycosylphosphatidylinositol-rich domains of lymphomonocytic cell plasma membranes. *J. Immunol.* 155:1334–1342.

Johnson, C. M., A. M. Cooper, A. A. Frank, C. B. Bonorino, L. J. Wysoki, and I. M. Orme. 1997. *Mycobacterium tuberculosis* aerogenic rechallenge infections in B cell-deficient mice. *Tuber. Lung Dis.* 78:257–261.

Jones, B. W., T. K. Means, K. A. Heldwein, M. A. Keen, P. J. Hill, J. T. Belisle, and M. J. Fenton. 2001. Different Toll-like receptor agonists induce distinct macrophage responses. *J Leukoc. Biol.* 69:1036–1044.

Juffermans, N. P., A. Verbon, J. T. Belisle, P. J. Hill, P. Speelman, S. J. van Deventer, and T. van der Poll. 2000. Mycobacterial lipoarabinomannan induces an inflammatory response in the mouse lung. A role for interleukin-1. *Am. J. Respir. Crit. Care Med.* 162:486–489.

Julian, E., L. Matas, A. Perez, J. Alcaide, M. A. Laneelle, and M. Luquin. 2002. Serodiagnosis of tuberculosis: comparison of immunoglobulin A (IgA) response to sulfolipid I with IgG and IgM responses to 2,3-diacyltrehalose, 2,3,6-triacyltrehalose, and cord factor antigens. *J. Clin. Microbiol.* 40:3782–3788.

Julian, E., L. Matas, J. Alcaide, and M. Luquin. 2004. Comparison of antibody responses to a potential combination of specific glycolipids and proteins for test sensitivity improvement in tuberculosis serodiagnosis. *Clin. Diagn. Lab. Immunol.* 11:70–76.

Jung, S. B., C. S. Yang, J. S. Lee, A. R. Shin, S. S. Jung, J. W. Son, C. V. Harding, H. J. Kim, J. K. Park, T. H. Paik, C. H. Song, and E. K. Jo. 2006. The mycobacterial 38-kilodalton glycolipoprotein antigen activates the mitogen-activated protein kinase pathway and release of proinflammatory cytokines through Toll-like receptors 2 and 4 in human monocytes. *Infect. Immun.* 74:2686–2696.

Jung, Y. J., R. LaCourse, L. Ryan, and R. J. North. 2002. Virulent but not avirulent *Mycobacterium tuberculosis* can evade the growth inhibitory action of a T helper 1-dependent, nitric oxide Synthase 2-independent defense in mice. *J. Exp. Med.* 196:991–998.

Kang, P. B., A. K. Azad, J. B. Torrelles, T. M. Kaufman, A. Beharka, E. Tibesar, L. E. DesJardin, and L. S. Schlesinger. 2005. The human macrophage mannose receptor directs *Mycobacterium tuberculosis* lipoarabinomannan-mediated phagosome biogenesis. *J. Exp. Med.* 202:987–999.

Kaplan, G., R. R. Gandhi, D. E. Weinstein, W. R. Levis, M. E. Patarroyo, P. J. Brennan, and Z. A. Cohn. 1987. *Mycobacterium leprae* antigen-induced suppression of T cell proliferation in vitro. *J. Immunol.* 138:3028–3034.

Karakousis, P. C., W. R. Bishai, and S. E. Dorman. 2004. *Mycobacterium tuberculosis* cell envelope lipids and the host immune response. *Cell Microbiol.* 6:105–116.

Kawashima, T., Y. Norose, Y. Watanabe, Y. Enomoto, H. Narazaki, E. Watari, S. Tanaka, H. Takahashi, I. Yano, M. B. Brenner, and M. Sugita. 2003. Cutting edge: major CD8 T cell response to live bacillus Calmette-Guerin is mediated by CD1 molecules. *J. Immunol.* 170:5345–5348.

Khanna, K. V., C. S. Choi, G. Gekker, P. K. Peterson, and T. W. Molitor. 1996. Differential infection of porcine alveolar macrophage subpopulations by nonopsonized *Mycobacterium bovis* involves CD14 receptors. *J. Leukoc. Biol.* 60:214–220.

Knutson, K. L., Z. Hmama, P. Herrera-Velit, R. Rochford, and N. E. Reiner. 1998. Lipoarabinomannan of *Mycobacterium tuberculosis* promotes protein tyrosine dephosphorylation and inhibition of mitogen-activated protein kinase in human mononuclear phagocytes. Role of the Src homology 2 containing tyrosine phosphatase 1. *J. Biol. Chem.* 273:645–652.

Korf, J., A. Stoltz, J. Verschoor, P. De Baetselier, and J. Grooten. 2005. The *Mycobacterium tuberculosis* cell wall component mycolic acid elicits pathogen-associated host innate immune responses. *Eur. J. Immunol.* 35:890–900.

Krutzik, S. R., M. T. Ochoa, P. A. Sieling, S. Uematsu, Y. W. Ng, A. Legaspi, P. T. Liu, S. T. Cole, P. J. Godowski, Y. Maeda, E. N. Sarno, M. V. Norgard, P. J. Brennan, S. Akira, T. H. Rea, and R. L. Modlin. 2003. Activation and regulation of Toll-like receptors 2 and 1 in human leprosy. *Nat. Med.* 9:525–532.

Krutzik, S. R., and R. L. Modlin. 2004. The role of Toll-like receptors in combating mycobacteria. *Semin. Immunol.* 16:35–41.

Lathigra, R., Y. Zhang, M. Hill, M. J. Garcia, P. S. Jackett, and J. Ivanyi. 1996. Lack of production of the 19-kDa glycolipoprotein in certain strains of *Mycobacterium tuberculosis*. *Res. Microbiol.* 147:237–249.

Lawe, D. C., V. Patki, R. Heller-Harrison, D. Lambright, and S. Corvera. 2000. The FYVE domain of early endosome antigen 1 is required for both phosphatidylinositol 3-phosphate and Rab5 binding. Critical role of this dual interaction for endosomal localization. *J. Biol. Chem.* 275:3699–3705.

Le Cabec, V., C. Cols, and I. Maridonneau-Parini. 2000. Nonopsonic phagocytosis of zymosan and Mycobacterium kansasii by CR3 (CD11b/CD18) involves distinct molecular determinants and is or is not coupled with NADPH oxidase activation. *Infect. Immun.* 68:4736–4745.

Lemaitre, B., E. Nicolas, L. Michaut, J. M. Reichhart, and J. A. Hoffmann. 1996. The dorsoventral regulatory gene cassette spatzle/Toll/cactus controls the potent antifungal response in Drosophila adults. *Cell* 86:973–983.

Lewinsohn, D. M., A. L. Briden, S. G. Reed, K. H. Grabstein, and M. R. Alderson. 2000. *Mycobacterium tuberculosis*-reactive CD8+ T lymphocytes: the relative contribution of classical versus nonclassical HLA restriction. *J. Immunol.* 165:925–930.

Lima, V. M., V. L. Bonato, K. M. Lima, S. A. Dos Santos, R. R. Dos Santos, E. D. Goncalves, L. H. Faccioli, I. T. Brandao, J. M. Rodrigues-Junior, and C. L. Silva. 2001. Role of trehalose dimycolate in recruitment of cells and modulation of production of cytokines and NO in tuberculosis. *Infect. Immun.* 69:5305–5312.

Lopez, B., D. Aguilar, H. Orozco, M. Burger, C. Espitia, V. Ritacco, L. Barrera, K. Kremer, R. Hernandez-Pando, K. Huygen, and D. van Soolingen. 2003. A marked difference in pathogenesis and immune response induced by different *Mycobacterium tuberculosis* genotypes. *Clin. Exp. Immunol.* 133:30–37.

Luna-Herrera, J., O. Rojas-Espinosa, and S. Estrada-Parra. 1996. Recognition of lipid antigens by sera of mice infected with *Mycobacterium lepraemurium*. *Int. J. Lepr. Other Mycobact. Dis.* 64:299–305.

MacMicking, J. D., G. A. Taylor, and J. D. McKinney. 2003. Immune control of tuberculosis by IFN-gamma-inducible LRG-47. *Science* 302:654–659.

Maeda, N., J. Nigou, J.-L. Herrmann, M. Jackson, A. Amara, H. Lagrange, G. Puzo, B. Gicquel, and O. Neyrolles. 2003. The cell surface receptor DC-sign discriminates between *Mycobacte-*

rium species through selective recognition of the mannose caps on lipoarabinomannan. *J. Biol. Chem.* **278:**5513–5516.

Malik, Z. A., G. M. Denning, and D. J. Kusner. 2000. Inhibition of Ca(2+) signaling by *Mycobacterium tuberculosis* is associated with reduced phagosome-lysosome fusion and increased survival within human macrophages. *J. Exp. Med.* **191:**287–302.

Manca, C., L. Tsenova, A. Bergtold, S. Freeman, M. Tovey, J. M. Musser, C. E. Barry, III, V. H. Freedman, and G. Kaplan. 2001. Virulence of a *Mycobacterium tuberculosis* clinical isolate in mice is determined by failure to induce Th1 type immunity and is associated with induction of IFN-alpha/beta. *Proc. Natl. Acad. Sci. USA* **98:**5752–5757.

Manca, C., M. B. Reed, S. Freeman, B. Mathema, B. Kreiswirth, C. E. Barry, 3rd, and G. Kaplan. 2004. Differential monocyte activation underlies strain-specific *Mycobacterium tuberculosis* pathogenesis. *Infect. Immun.* **72:**5511–5514.

Mazurek, G. H., P. A. LoBue, C. L. Daley, J. Bernardo, A. A. Lardizabal, W. R. Bishai, M. F. Iademarco, and J. S. Rothel. 2001. Comparison of a whole-blood interferon gamma assay with tuberculin skin testing for detecting latent *Mycobacterium tuberculosis* infection. *JAMA* **286:**1740–1747.

Means, T. K., E. Lien, A. Yoshimura, S. Wang, D. T. Golenbock, and M. J. Fenton. 1999. The CD14 ligands lipoarabinomannan and lipopolysaccharide differ in their requirement for Toll-like receptors. *J. Immunol.* **163:**6748–6755.

Means, T. K., S. Wang, E. Lien, A. Yoshimura, D. T. Golenbock, and M. J. Fenton. 1999. Human toll-like receptors mediate cellular activation by *Mycobacterium tuberculosis*. *J. Immunol.* **163:**3920–3927.

Meeker, H. C., G. Schuller-Levis, F. Fusco, M. A. Giardina-Becket, E. Sersen, and W. R. Levis. 1990. Sequential monitoring of leprosy patients with serum antibody levels to phenolic glycolipid-I, a synthetic analog of phenolic glycolipid-I, and mycobacterial lipoarabinomannan. *Int. J. Lepr. Other Mycobact. Dis.* **58:**503–511.

Modlin, R. L., H. Kato, V. Mehra, E. E. Nelson, F. Xue-dong, T. H. Rea, P. K. Pattengale, and B. R. Bloom. 1986. Genetically restricted suppressor T-cell clones derived from lepromatous leprosy lesions. *Nature* **322:**459–461.

Molloy, A., G. Gaudernack, W. R. Levis, Z. A. Cohn, and G. Kaplan. 1990. Suppression of T-cell proliferation by *Mycobacterium leprae*: the role of lipopolysaccharide. *Proc. Natl. Acad. Sci.* **87:**973–977.

Moody, D. B., T. Ulrichs, W. Muhlecker, D. C. Young, S. S. Gurcha, E. Grant, J. P. Rosat, M. B. Brenner, C. E. Costello, G. S. Besra, and S. A. Porcelli. 2000. CD1c-mediated T-cell recognition of isoprenoid glycolipids in *Mycobacterium tuberculosis* infection. *Nature* **404:**884–888.

Moody, D. B., D. C. Young, T. Y. Cheng, J. P. Rosat, C. Roura-Mir, P. B. O'Connor, D. M. Zajonc, A. Walz, M. J. Miller, S. B. Levery, I. A. Wilson, C. E. Costello, and M. B. Brenner. 2004. T cell activation by lipopeptide antigens. *Science* **303:**527–531.

Moura, A. C., P. S. Leonardo, M. G. Henriques, and R. S. Cordeiro. 1999. Opposite effects of M. leprae or M. bovis BCG delipidation on cellular accumulation into mouse pleural cavity. Distinct accomplishment of mycobacterial lipids in vivo. *Inflamm. Res.* **48:**308–313.

Ng, V., G. Zanazzi, R. Timpl, J. F. Talts, J. L. Salzer, P. J. Brennan, and A. Rambukkana. 2000. Role of the cell wall phenolic glycolipid-1 in the peripheral nerve predilection of *Mycobacterium leprae*. *Cell* **103:**511–524.

Nguyen, L., and J. Pieters. 2005. The Trojan horse: survival tactics of pathogenic mycobacteria in macrophages. *Trends Cell Biol.* **15:**269–276.

Nigou, J., C. Zelle-Rieser, M. Gilleron, M. Thurnher, and G. Puzo. 2001. Mannosylated lipoarabinomannans inhibit IL-12 produc-

tion by human dendritic cells: evidence for a negative signal delivered through the mannose receptor. *J. Immunol.* **166:**7477–7485.

Noss, E. H., R. K. Pai, T. J. Sellati, J. D. Radolf, J. Belisle, D. T. Golenbock, W. H. Boom, and C. V. Harding. 2001. Toll-like receptor 2-dependent inhibition of macrophage class II MHC expression and antigen processing by 19-kDa lipoprotein of *Mycobacterium tuberculosis*. *J. Immunol.* **167:**910–918.

O'Garra, A., and P. Vieira. 2004. Regulatory T cells and mechanisms of immune system control. *Nat. Med.* **10:**801–805.

O'Garra, A., and W. J. Britton. 2007. Cytokines in tuberculosis, *In* S. H. E. Kaufmann, E. Rubin, W. J. Britton, and P. van Helden (ed.), *Handbook of Tuberculosis*, vol. 2. Wiley, Mannheim, Germany.

Oiso, R., N. Fujiwara, H. Yamagami, S. Maeda, S. Matsumoto, S. Nakamura, N. Oshitani, T. Matsumoto, T. Arakawa, and K. Kobayashi. 2005. Mycobacterial trehalose 6,6'-dimycolate preferentially induces type 1 helper T cell responses through signal transducer and activator of transcription 4 protein. *Microb. Pathog.* **39:**35–43.

Okamoto, Y., Y. Fujita, T. Naka, M. Hirai, I. Tomiyasu, and I. Yano. 2006. Mycobacterial sulfolipid shows a virulence by inhibiting cord factor induced granuloma formation and TNF-alpha release. *Microb. Pathog.* **40:**245–253.

Pai, R. K., M. Convery, T. A. Hamilton, W. H. Boom, and C. V. Harding. 2003. Inhibition of IFN-gamma-induced class II transactivator expression by a 19-kDa lipoprotein from *Mycobacterium tuberculosis*: a potential mechanism for immune evasion. *J. Immunol.* **171:**175–184.

Pai, R. K., M. E. Pennini, A. A. Tobian, D. H. Canaday, W. H. Boom, and C. V. Harding. 2004. Prolonged toll-like receptor signaling by *Mycobacterium tuberculosis* and its 19-kilodalton lipoprotein inhibits gamma interferon-induced regulation of selected genes in macrophages. *Infect. Immun.* **72:**6603–6614.

Pecora, N. D., A. J. Gehring, D. H. Canaday, W. H. Boom, and C. V. Harding. 2006. *Mycobacterium tuberculosis* LprA is a lipoprotein agonist of TLR2 that regulates innate immunity and APC function. *J. Immunol.* **177:**422–429.

Pennini, M. E., R. K. Pai, D. C. Schultz, W. H. Boom, and C. V. Harding. 2006. *Mycobacterium tuberculosis* 19-kDa lipoprotein inhibits IFN-gamma-induced chromatin remodeling of MHC2TA by TLR2 and MAPK signaling. *J. Immunol.* **176:**4323–4330.

Perez, R. L., J. Roman, S. Roser, C. Little, M. Olsen, J. Indrigo, R. L. Hunter, and J. K. Actor. 2000. Cytokine message and protein expression during lung granuloma formation and resolution induced by the mycobacterial cord factor trehalose-6,6'-dimycolate. *J. Interferon Cytokine Res.* **20:**795–804.

Pethe, K., D. L. Swenson, S. Alonso, J. Anderson, C. Wang, and D. G. Russell. 2004. Isolation of *Mycobacterium tuberculosis* mutants defective in the arrest of phagosome maturation. *Proc. Natl. Acad. Sci. USA* **101:**13642–13647.

Philips, J. A., E. J. Rubin, and N. Perrimon. 2005. Drosophila RNAi screen reveals CD36 family member required for mycobacterial infection. *Science* **309:**1251–1253.

Pinto, R., B. M. Saunders, L. R. Camacho, W. J. Britton, B. Gicquel, and J. A. Triccas. 2004. *Mycobacterium tuberculosis* defective in phthiocerol dimycocerosate translocation provides greater protective immunity against tuberculosis than the existing bacille Calmette-Guerin vaccine. *J. Infect. Dis.* **189:**105–112.

Porcelli, S., C. Morita, and M. B. Brenner. 1992. CD1b restricts the response of human CD4-8- T lymphocytes to a microbial antigen. *Nature* **360:**593–597.

Prigozy, T. I., P. A. Sieling, D. Clemens, P. L. Stewart, S. M. Behar, S. A. Porcelli, M. B. Brenner, R. L. Modlin, and M. Kronenberg.

1997. The mannose receptor delivers lipoglycan antigens to endosomes for presentation to T cells by CD1b molecules. *Immunity* **6:**187–197.

Puissegur, M. P., C. Botanch, J. L. Duteyrat, G. Delsol, C. Caratero, and F. Altare. 2004. An in vitro dual model of mycobacterial granulomas to investigate the molecular interactions between mycobacteria and human host cells. *Cell. Microbiol.* **6:**423–433.

Quesniaux, V., C. Fremond, M. Jacobs, S. Parida, D. Nicolle, V. Yeremeev, F. Bihl, F. Erard, T. Botha, M. Drennan, M. N. Soler, M. Le Bert, B. Schnyder, and B. Ryffel. 2004. Toll-like receptor pathways in the immune responses to mycobacteria. *Microbes Infect.* **6:**946–959.

Randhawa, A. K., H. J. Ziltener, J. S. Merzaban, and R. W. Stokes. 2005. CD43 is required for optimal growth inhibition of *Mycobacterium tuberculosis* in macrophages and in mice. *J. Immunol.* **175:**1805–1812.

Rao, V., F. Gao, B. Chen, W. R. Jacobs, Jr., and M. S. Glickman. 2006. Trans-cyclopropanation of mycolic acids on trehalose dimycolate suppresses *Mycobacterium tuberculosis*-induced inflammation and virulence. *J. Clin. Investig.* **116:**1660–1667.

Reed, M. B., P. Domenech, C. Manca, H. Su, A. K. Barczak, B. N. Kreiswirth, G. Kaplan, and C. E. Barry III. 2004. A glycolipid of hypervirulent tuberculosis strains that inhibits the innate immune response. *Nature* **431:**84–87.

Rhoades, E. R., R. E. Geisel, B. A. Butcher, S. McDonough, and D. G. Russell. 2005. Cell wall lipids from *Mycobacterium bovis* BCG are inflammatory when inoculated within a gel matrix: characterization of a new model of the granulomatous response to mycobacterial components. *Tuberculosis* (Edinburgh) **85:**159–176.

Roach, D. R., H. Briscoe, B. Saunders, M. P. France, S. Riminton, and W. J. Britton. 2001. Secreted lymphotoxin-alpha is essential for the control of an intracellular bacterial infection. *J. Exp. Med.* **193:**239–246.

Roche, P. W., W. J. Britton, S. S. Failbus, W. J. Theuvenet, M. Lavender, and R. B. Adiga. 1991. Serological responses in primary neuritic leprosy. *Trans. R. Soc. Trop. Med. Hyg.* **85:**299–302.

Roche, P. W., W. J. Britton, S. S. Failbus, K. D. Neupane, and W. J. Theuvenet. 1993. Serological monitoring of the response to chemotherapy in leprosy patients. *Int. J. Lepr. Other Mycobact. Dis.* **61:**35–43.

Rodriguez, A., A. Tjarnlund, J. Ivanji, M. Singh, I. Garcia, A. Williams, P. D. Marsh, M. Troye-Blomberg, and C. Fernandez. 2005. Role of IgA in the defense against respiratory infections IgA deficient mice exhibited increased susceptibility to intranasal infection with *Mycobacterium bovis* BCG. *Vaccine* **23:**2565–2572.

Rooyakkers, A. W., and R. W. Stokes. 2005. Absence of complement receptor 3 results in reduced binding and ingestion of *Mycobacterium tuberculosis* but has no significant effect on the induction of reactive oxygen and nitrogen intermediates or on the survival of the bacteria in resident and interferon-gamma activated macrophages. *Microb. Pathog.* **39:**57–67.

Rosat, J. P., E. P. Grant, E. M. Beckman, C. C. Dascher, P. A. Sieling, D. Frederique, R. L. Modlin, S. A. Porcelli, S. T. Furlong, and M. B. Brenner. 1999. CD1-restricted microbial lipid antigen-specific recognition found in the CD8+ alpha beta T cell pool. *J. Immunol.* **162:**366–371.

Roura-Mir, C., L. Wang, T. Y. Cheng, I. Matsunaga, C. C. Dascher, S. L. Peng, M. J. Fenton, C. Kirschning, and D. B. Moody. 2005. *Mycobacterium tuberculosis* regulates CD1 antigen presentation pathways through TLR-2. *J. Immunol.* **175:**1758–1766.

Rousseau, C., O. C. Turner, E. Rush, Y. Bordat, T. D. Sirakova, P. E. Kolattukudy, S. Ritter, I. M. Orme, B. Gicquel, and M. Jackson. 2003. Sulfolipid deficiency does not affect the viru-

lence of *Mycobacterium tuberculosis* H37Rv in mice and guinea pigs. *Infect. Immun.* **71:**4684–4690.

Rousseau, C., N. Winter, E. Pivert, Y. Bordat, O. Neyrolles, P. Ave, M. Huerre, B. Gicquel, and M. Jackson. 2004. Production of phthiocerol dimycocerosates protects *Mycobacterium tuberculosis* from the cidal activity of reactive nitrogen intermediates produced by macrophages and modulates the early immune response to infection. *Cell. Microbiol.* **6:**277–287.

Russell, D. G. 2007. Who puts the tubercle in tuberculosis? *Nat. Rev. Microbiol.* **5:**39–47.

Saavedra, R., E. Segura, R. Leyva, L. A. Esparza, and L. M. Lopez-Marin. 2001. Mycobacterial di-O-acyl-trehalose inhibits mitogen- and antigen-induced proliferation of murine T cells in vitro. *Clin. Diagn. Lab. Immunol.* **8:**1081–1088.

Saavedra, R., E. Segura, E. P. Tenorio, and L. M. Lopez-Marin. 2006. Mycobacterial trehalose-containing glycolipid with immunomodulatory activity on human CD4+ and CD8+ T-cells. *Microbes Infect.* **8:**533–540.

Sable, S. B., B. B. Plikaytis, and T. M. Shinnick. 2007. Tuberculosis subunit vaccine development: Impact of physicochemical properties of mycobacterial test antigens. *Vaccine* **25:**1553–1566.

Salgame, P. 2005. Host innate and Th1 responses and the bacterial factors that control *Mycobacterium tuberculosis* infection. *Curr. Opin. Immunol.* **17:**374–380.

Sanchez, M. D., Y. Garcia, C. Montes, S. C. Paris, M. Rojas, L. F. Barrera, M. A. Arias, and L. F. Garcia. 2006. Functional and phenotypic changes in monocytes from patients with tuberculosis are reversed with treatment. *Microbes Infect.* **8:**2492–2500.

Saunders, B. M., and W. J. Britton. 2007. Life and death in the granuloma: immunopathology of tuberculosis. *Immunol. Cell Biol.* **85:**103–111.

Scandurra, G. M., R. B. Williams, J. A. Triccas, R. Pinto, B. Gicquel, B. Slobedman, A. Cunningham, and W. J. Britton. 2007. Effect of phthiocerol dimycocerosate deficiency on the transcriptional response of human macrophages to *Mycobacterium tuberculosis*. *Microbes Infect.* **9:**87–95.

Scanga, C. A., A. Bafica, C. G. Feng, A. W. Cheever, S. Hieny, and A. Sher. 2004. MyD88-deficient mice display a profound loss in resistance to *Mycobacterium tuberculosis* associated with partially impaired Th1 cytokine and nitric oxide synthase 2 expression. *Infect. Immun.* **72:**2400–2404.

Schaible, U. E., S. Sturgill-Koszycki, P. H. Schlesinger, and D. G. Russell. 1998. Cytokine activation leads to acidification and increases maturation of *Mycobacterium avium*-containing phagosomes in murine macrophages. *J. Immunol.* **160:**1290–1296.

Schlesinger, L., C. G. Bellinger-Kawahara, N. R. Payne, and M. A. Horwitz. 1990. Phagocytosis of *Mycobacterium tuberculosis* is mediated by human monocyte complement receptors and complement component C3. *J. Immunol.* **144:**2771–2780.

Schlesinger, L. S. 1993. Macrophage phagocytosis of virulent but not attenuated strains of *Mycobacterium tuberculosis* is mediated by mannose receptors in addition to complement receptors. *J. Immunol.* **150:**2920–2930.

Schlesinger, L. S., S. R. Hull, and T. M. Kaufman. 1994. Binding of the terminal mannosyl units of lipoarabinomannan from a virulent strain of *Mycobacterium tuberculosis* to human macrophages. *J. Immunol.* **152:**4070–4079.

Schlesinger, L. S., T. M. Kaufman, S. Iyer, S. R. Hull, and L. K. Marchiando. 1996. Differences in mannose receptor-mediated uptake of lipoarabinomannan from virulent and attenuated strains of *Mycobacterium tuberculosis* by human macrophages. *J. Immunol.* **157:**4568–4575.

Schnappinger, D., S. Ehrt, M. I. Voskuil, Y. Liu, J. A. Mangan, I. M. Monahan, G. Dolganov, B. Efron, P. D. Butcher, C. Nathan, and G. K. Schoolnik. 2003. Transcriptional adaptation of *My-*

cobacterium tuberculosis within macrophages: insights into the phagosomal environment. *J. Exp. Med.* **198:**693–704.

Shabaana, A. K., K. Kulangara, I. Semac, Y. Parel, S. Ilangumaran, K. Dharmalingam, C. Chizzolini, and D. C. Hoessli. 2005. Mycobacterial lipoarabinomannans modulate cytokine production in human T helper cells by interfering with raft/microdomain signalling. *Cell. Mol. Life Sci.* **62:**179–187.

Sieling, P. A., D. Chatterjee, S. A. Porcelli, T. I. Prigozy, R. J. Mazzaccaro, T. Soriano, B. R. Bloom, M. B. Brenner, M. Kronenberg, P. J. Brennan, and R. L. Modlin. 1995. CD1-restricted T cell recognition of microbial lipoglycan antigens. *Science* **269:**227–230.

Silva, V. M., G. Kanaujia, M. L. Gennaro, and D. Menzies. 2003. Factors associated with humoral response to ESAT-6, 38 kDa and 14 kDa in patients with a spectrum of tuberculosis. *Int. J. Tuberc. Lung Dis.* **7:**478–484.

Simonney, N., J. M. Molina, M. Molimard, E. Oksenhendler, C. Perronne, and P. H. Lagrange. 1995. Analysis of the immunological humoral response to *Mycobacterium tuberculosis* glycolipid antigens (DAT, PGLTb1) for diagnosis of tuberculosis in HIV-seropositive and -seronegative patients. *Eur. J. Clin. Microbiol. Infect. Dis.* **14:**883–891.

Sousa, A. O., R. J. Mazzaccaro, R. G. Russell, F. K. Lee, O. C. Turner, S. Hong, L. Van Kaer, and B. R. Bloom. 2000. Relative contributions of distinct MHC class I-dependent cell populations in protection to tuberculosis infection in mice. *Proc. Natl. Acad. Sci. USA* **97:**4204–4208.

Stewart, G. R., J. Patel, B. D. Robertson, A. Rae, and D. B. Young. 2005. Mycobacterial mutants with defective control of phagosomal acidification. *PLoS Pathog.* **1:**269–278.

Stewart, G. R., K. A. Wilkinson, S. M. Newton, S. M. Sullivan, O. Neyrolles, J. R. Wain, J. Patel, K. L. Pool, D. B. Young, and R. J. Wilkinson. 2005. Effect of deletion or overexpression of the 19-kilodalton lipoprotein Rv3763 on the innate response to *Mycobacterium tuberculosis*. *Infect. Immun.* **73:**6831–6837.

Stokes, R. W., I. D. Haidl, W. A. Jefferies, and D. P. Speert. 1993. Mycobacteria-macrophage interactions. Macrophage phenotype determines the nonopsonic binding of *Mycobacterium tuberculosis* to murine macrophages. *J. Immunol.* **151:**7067–7076.

Sutcliffe, I. C., and D. J. Harrington. 2004. Lipoproteins of *Mycobacterium tuberculosis*: an abundant and functionally diverse class of cell envelope components. *FEMS Microbiol. Rev.* **28:**645–659.

Tailleux, L., O. Schwartz, J. L. Herrmann, E. Pivert, M. Jackson, A. Amara, L. Legres, D. Dreher, L. P. Nicod, J. C. Gluckman, P. H. Lagrange, B. Gicquel, and O. Neyrolles. 2003. DC-SIGN is the major *Mycobacterium tuberculosis* receptor on human dendritic cells. *J. Exp. Med.* **197:**121–127.

Takeda, K., T. Kaisho, and S. Akira. 2003. Toll-like receptors. *Annu. Rev. Immunol.* **21:**335–376.

Takeuchi, O., S. Sato, T. Horiuchi, K. Hoshino, K. Takeda, Z. Dong, R. L. Modlin, and S. Akira. 2002. Cutting edge: role of Toll-like receptor 1 in mediating immune response to microbial lipoproteins. *J. Immunol.* **169:**10–14.

Takimoto, H., H. Maruyama, K. I. Shimada, Y. Yakabe, I. Yano, and Y. Kumazawa. 2006. Interferon-gamma independent formation of pulmonary granuloma in mice by injections with trehalose dimycolate (cord factor), lipoarabinomannan and phosphatidylinositol mannosides isolated from *Mycobacterium tuberculosis*. *Clin. Exp. Immunol.* **144:**134–141.

Taylor, P. R., S. Gordon, and L. Martinez-Pomares. 2005. The mannose receptor: linking homeostasis and immunity through sugar recognition. *Trends Immunol.* **26:**104–110.

Theus, S. A., M. D. Cave, and K. D. Eisenach. 2005. Intracellular macrophage growth rates and cytokine profiles of *Mycobacte-*

rium tuberculosis strains with different transmission dynamics. *J. Infect. Dis.* **191:**453–460.

Thoma-Uszynski, S., S. Stenger, O. Takeuchi, M. T. Ochoa, M. Engele, P. A. Sieling, P. F. Barnes, M. Rollinghoff, P. L. Bolcskei, M. Wagner, S. Akira, M. V. Norgard, J. T. Belisle, P. J. Godowski, B. R. Bloom, and R. L. Modlin. 2001. Induction of direct antimicrobial activity through mammalian toll-like receptors. *Science* **291:**1544–1547.

Ting, J. P., and K. L. Williams. 2005. The CATERPILLER family: an ancient family of immune/apoptotic proteins. *Clin. Immunol.* **115:**33–37.

Ting, L. M., A. C. Kim, A. Cattamanchi, and J. D. Ernst. 1999. *Mycobacterium tuberculosis* inhibits IFN-gamma transcriptional responses without inhibiting activation of STAT1. *J. Immunol.* **163:**3898–3906.

Tobian, A. A., N. S. Potter, L. Ramachandra, R. K. Pai, M. Convery, W. H. Boom, and C. V. Harding. 2003. Alternate class I MHC antigen processing is inhibited by Toll-like receptor signaling pathogen-associated molecular patterns: *Mycobacterium tuberculosis* 19-kDa lipoprotein, CpG DNA, and lipopolysaccharide. *J. Immunol.* **171:**1413–1422.

Torrelles, J. B., A. K. Azad, and L. S. Schlesinger. 2006. Fine discrimination in the recognition of individual species of phosphatidyl-myo-inositol mannosides from *Mycobacterium tuberculosis* by C-type lectin pattern recognition receptors. *J. Immunol.* **177:**1805–1816.

Tsai, M. C., S. Chakravarty, G. Zhu, J. Xu, K. Tanaka, C. Koch, J. Tufariello, J. Flynn, and J. Chan. 2006. Characterization of the tuberculous granuloma in murine and human lungs: cellular composition and relative tissue oxygen tension. *Cell. Microbiol.* **8:**218–232.

Tsenova, L., E. Ellison, R. Harbacheuski, A. L. Moreira, N. Kurepina, M. B. Reed, B. Mathema, C. E. Barry, 3rd, and G. Kaplan. 2005. Virulence of selected *Mycobacterium tuberculosis* clinical isolates in the rabbit model of meningitis is dependent on phenolic glycolipid produced by the bacilli. *J. Infect. Dis.* **192:**98–106.

Uehori, J., M. Matsumoto, S. Tsuji, T. Akazawa, O. Takeuchi, S. Akira, T. Kawata, I. Azuma, K. Toyoshima, and T. Seya. 2003. Simultaneous blocking of human Toll-like receptors 2 and 4 suppresses myeloid dendritic cell activation induced by *Mycobacterium bovis* bacillus Calmette-Guerin peptidoglycan. *Infect. Immun.* **71:**4238–4249.

Uehori, J., K. Fukase, T. Akazawa, S. Uematsu, S. Akira, K. Funami, M. Shingai, M. Matsumoto, I. Azuma, K. Toyoshima, S. Kusumoto, and T. Seya. 2005. Dendritic cell maturation induced by muramyl dipeptide (MDP) derivatives: monoacylated MDP confers TLR2/TLR4 activation. *J. Immunol.* **174:**7096–7103.

Ulrichs, T., and S. A. Porcelli. 2000. CD1 proteins: targets of T cell recognition in innate and adaptive immunity. *Rev. Immunogenet.* **2:**416–432.

Uma Devi, K. R., B. Ramalingam, P. J. Brennan, P. R. Narayanan, and A. Raja. 2001. Specific and early detection of IgG, IgA and IgM antibodies to *Mycobacterium tuberculosis* 38kDa antigen in pulmonary tuberculosis. *Tuberculosis* (Edinburgh) **81:**249–253.

Vergne, I., J. Chua, and V. Deretic. 2003. Tuberculosis toxin blocking phagosome maturation inhibits a novel Ca2+/calmodulin-PI3K hVPS34 cascade. *J. Exp. Med.* **198:**653–659.

Vergne, I., J. Chua, S. B. Singh, and V. Deretic. 2004a. Cell biology of *Mycobacterium tuberculosis* phagosome. *Annu. Rev. Cell Dev. Biol.* **20:**367–394.

Vergne, I., R. A. Fratti, P. J. Hill, J. Chua, J. Belisle, and V. Deretic. 2004b. *Mycobacterium tuberculosis* phagosome maturation arrest: mycobacterial phosphatidylinositol analog phosphatidylinositol mannoside stimulates early endosomal fusion. *Mol. Biol. Cell* **15:**751–760.

Vergne, I., J. Chua, H. H. Lee, M. Lucas, J. Belisle, and V. Deretic. 2005. Mechanism of phagolysosome biogenesis block by viable *Mycobacterium tuberculosis*. *Proc. Natl. Acad. Sci. USA* **102:** 4033–4038.

Via, L. E., D. Deretic, R. J. Ulmer, N. S. Hibler, L. A. Huber, and V. Deretic. 1997. Arrest of mycobacterial phagosome maturation is caused by a block in vesicle fusion between stages controlled by rab5 and rab7. *J. Biol. Chem.* **272:**13326–13331.

Vieira, O. V., R. E. Harrison, C. C. Scott, H. Stenmark, D. Alexander, J. Liu, J. Gruenberg, A. D. Schreiber, and S. Grinstein. 2004. Acquisition of Hrs, an essential component of phagosomal maturation, is impaired by mycobacteria. *Mol. Cell Biol.* **24:**4593–4604.

Villeneuve, C., M. Gilleron, I. Maridonneau-Parini, M. Daffe, C. Astarie-Dequeker, and G. Etienne. 2005. Mycobacteria use their surface-exposed glycolipids to infect human macrophages through a receptor-dependent process. *J. Lipid Res.* **46:**475–483.

Vordermeier, H. M., N. Venkataprasad, D. P. Harris, and J. Ivanyi. 1996. Increase of tuberculous infection in the organs of B cell-deficient mice. *Clin. Exp. Immunol.* **106:**312–316.

Vouret-Craviari, V., S. Cenzuales, G. Poli, and A. Mantovani. 1997. Expression of monocyte chemotactic protein-3 in human monocytes exposed to the mycobacterial cell wall component lipoarabinomannan. *Cytokine* **9:**992–998.

Walburger, A., A. Koul, G. Ferrari, L. Nguyen, C. Prescianotto-Baschong, K. Huygen, B. Klebl, C. Thompson, G. Bacher, and J. Pieters. 2004. Protein kinase G from pathogenic mycobacteria promotes survival within macrophages. *Science* **304:**1800–1804.

Walther, M., J. E. Tongren, L. Andrews, D. Korbel, E. King, H. Fletcher, R. F. Andersen, P. Bejon, F. Thompson, S. J. Dunachie, F. Edele, J. B. de Souza, R. E. Sinden, S. C. Gilbert, E. M. Riley, and A. V. Hill. 2005. Upregulation of TGF-beta, FOXP3, and CD4+CD25+ regulatory T cells correlates with more rapid parasite growth in human malaria infection. *Immunity* **23:**287–296.

Watanabe, M., I. Honda, K. Kawajiri, S. Niinuma, S. Kudoh, and D. E. Minnikin. 1995. Distribution of antibody titres against phenolic glycolipids from *Mycobacterium tuberculosis* in the sera from tuberculosis patients and healthy controls. *Res. Microbiol.* **146:**791–797.

Watanabe, Y., E. Watari, I. Matsunaga, K. Hiromatsu, C. C. Dascher, T. Kawashima, Y. Norose, K. Shimizu, H. Takahashi, I. Yano, and M. Sugita. 2006. BCG vaccine elicits both T-cell mediated and humoral immune responses directed against mycobacterial lipid components. *Vaccine* **24:**5700–5707.

Williams, K. L., J. D. Lich, J. A. Duncan, W. Reed, P. Rallabhandi, C. Moore, S. Kurtz, V. M. Coffield, M. A. Accavitti-Loper, L. Su, S. N. Vogel, M. Braunstein, and J. P. Ting. 2005. The CATERPILLER protein monarch-1 is an antagonist of toll-like receptor-, tumor necrosis factor alpha-, and *Mycobacterium tuberculosis*-induced pro-inflammatory signals. *J. Biol. Chem.* **280:**39914–39924.

Winau, F., V. Schwierzeck, R. Hurwitz, N. Remmel, P. A. Sieling, R. L. Modlin, S. A. Porcelli, V. Brinkmann, M. Sugita, K. Sandhoff, S. H. Kaufmann, and U. E. Schaible. 2004. Saposin C is required for lipid presentation by human CD1b. *Nat. Immunol.* **5:**169–174.

Yamagami, H., T. Matsumoto, N. Fujiwara, T. Arakawa, K. Kaneda, I. Yano, and K. Kobayashi. 2001. Trehalose 6,6'-dimycolate (cord factor) of *Mycobacterium tuberculosis* induces foreign-body- and hypersensitivity-type granulomas in mice. *Infect. Immun.* **69:**810–815.

Yeremeev, V. V., I. V. Lyadova, B. V. Nikonenko, A. S. Apt, C. Abou-Zeid, J. Inwald, and D. B. Young. 2000. The 19-kD antigen and protective immunity in a murine model of tuberculosis. *Clin. Exp. Immunol.* **120:**274–279.

Zhang, Y., M. Broser, H. Cohen, M. Bodkin, K. Law, J. Reibman, and W. N. Rom. 1995. Enhanced interleukin-8 release and gene expression in macrophages after exposure to *Mycobacterium tuberculosis* and its components. *J. Clin. Investig.* **95:**586–592.

Zimmerli, S., S. Edwards, and J. D. Ernst. 1996. Selective receptor blockade during phagocytosis does not alter the survival and growth of *Mycobacterium tuberculosis* in human macrophages. *Am. J. Respir. Cell Mol. Biol.* **15:**760–770.

II. SPECIFIC FEATURES

The Mycobacterial Cell Envelope
Edited by M. Daffé and J.-M. Reyrat
© 2008 ASM Press, Washington, DC

Chapter 17

Biosynthesis and Roles of Phenolic Glycolipids and Related Molecules in *Mycobacterium tuberculosis*

CHRISTOPHE GUILHOT, CHRISTIAN CHALUT, AND MAMADOU DAFFÉ

During their efforts to associate immunizing properties of fractions of tubercle bacilli with their gross structural features revealed by infrared, Smith and colleagues (1960) discovered and classified a number of mycobacterial lipids found only in particular species; these lipids were named mycosides. For instance, mycoside A referred to particular substances characteristic of photochromogenic mycobacterial strains (mainly *Mycobacterium kansasii*), whereas mycoside B typified bovine strains. Further structural studies, mostly by Gastambide-Odier and colleagues (1967), showed that mycoside A (from *M. kansasii*), mycoside B (from *Mycobacterium bovis*), and mycoside G (from *Mycobacterium marinum*) are closely related; they were all subsequently classified into the family of phenolic or phenol glycolipids (PGLs). PGLs are now known to be produced in significant amounts by *M. leprae* (phenol glycolipid 1 [PGL-1]), *M. kansasii* (mycoside A), a few strains of *Mycobacterium tuberculosis* (PGL-tb), many strains of *M. bovis* (mycoside B) and a few other slow-growing mycobacteria (*Mycobacterium ulcerans, M. marinum, Mycobacterium gastri, Mycobacterium microti,* and *Mycobacterium haemophilum*). A family of lipids structurally related to PGLs has been isolated from waxes of the human tubercle bacillus, and have been found in all PGL-producing mycobacterial strains examined. They are nonglycosylated nonphenolic substances that have the same lipid core as PGL, a phthiocerol diester (Daffé, 1988a). More recently, a third group of compounds related to PGL was isolated from various *M. tuberculosis* and *M. bovis* strains and characterized by Constant et al. (2002). These molecules, called *p*-hydroxybenzoic acid derivatives (*p*-HBADs), are released into the culture medium and contain the same glycosylated phenolic moiety as PGL.

STRUCTURE OF THE VARIOUS COMPOUNDS AND LOCALIZATION

PGLs consist of a conserved lipid core (Daffé, 1988a; Minnikin, 1982) and a variable carbohydrate moiety (Brennan, 1988; Daffé and Lemassu, 2000); the lipid core is composed of a family of long-chain β-diols (C_{33}-C_{41}), phenolphthiocerol, and related molecules (Fig. 1). The aliphatic and nonglycosylated derivatives of the diols are called phthiocerols (Fig. 1). The major β-diols are usually accompanied by structural variants of these alcohols containing either a keto group in place of the methoxy group (phenolphthiodiolone and phthiodiolone) or a methyl group rather than an ethyl group at the terminus of the molecules and near the methoxyl group (phenolphthiocerol B and phthiocerol B) (Fig. 1). The various β-diols are esterified by polymethyl-branched (C_{27}-C_{34}) fatty acids. When the configuration of the asymmetric centers bearing the methyl branches are of the D series, the fatty acids are called mycocerosic acids whereas those of the L series are known as phthioceranic acids (Brennan, 1988). The corresponding phthiocerol diesters are known as DIMs (or PDIMs) and DIPs. DIPs have only been isolated from *M. marinum* and *M. ulcerans,* whereas the other PGL-containing mycobacterial species produce DIMs (Daffé, 1988a). A methoxylated phenolphthiocerol, the so-called attenuation indicator lipid, was isolated from some attenuated tubercle bacilli (Goren et al., 1974) but, despite its name, the link between its presence and reduced virulence is unclear. The sugar moiety of PGL consists of one to four sugar residues, according to the species, and most are *O*-methylated deoxysugars (Brennan, 1988; Daffé and Lemassu, 2000). The same PGL structures may be found in several phylogenetically related mycobacterial species,

Christophe Guilhot, Christian Chalut, and Mamadou Daffé • Department of Molecular Mechanisms of Mycobacterial Infections, Institut de Pharmacologie et de Biologie Structurale du Centre National de la Recherche Scientifique (CNRS), and Université Paul Sabatier, 205 route de Narbonne, 31077 Toulouse Cedex 4, France.

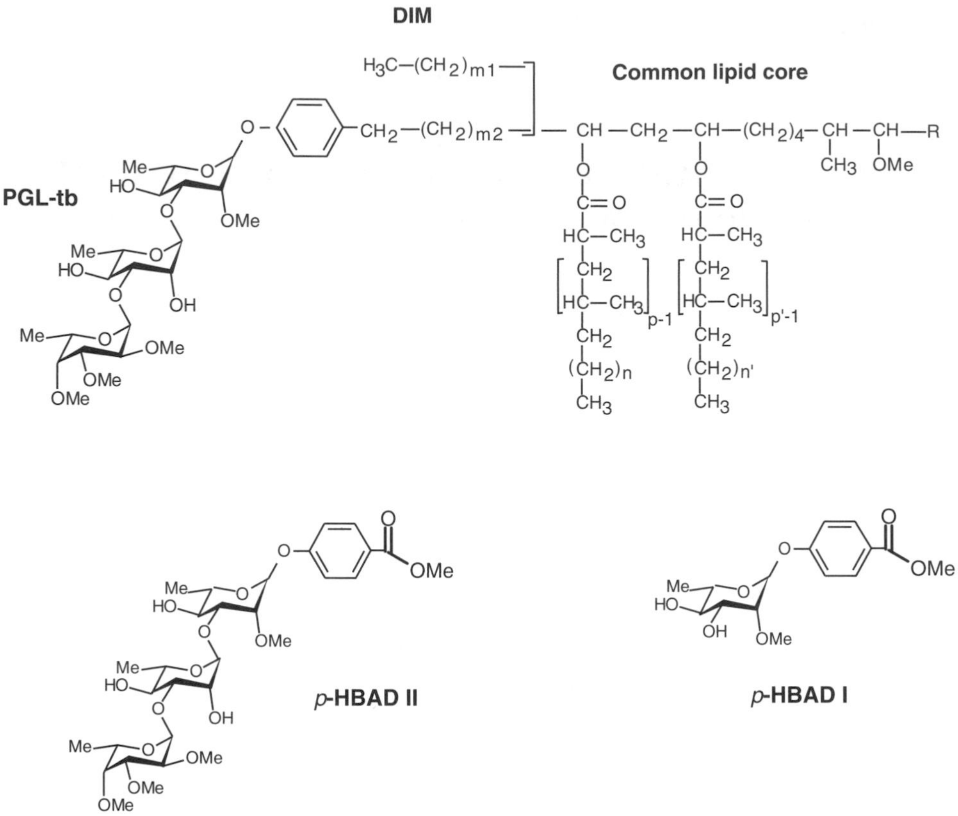

Figure 1. Structures of PGL-tb, *p*-HBAD I, *p*-HBAD II and DIM from *M. tuberculosis*. In *M. tuberculosis*, p and p′ = 3 to 5; n and n′ = 16 to 18; m2 = 15 to 17; m1 = 20 to 22; R = CH$_2$-CH$_3$ or CH$_3$.

for example in the pairs *M. bovis* and *M. microti*, *M. kansasii* and *M. gastri*, and *M. marinum* and *M. ulcerans*. A PGL that is abundant in one species may also be present in small amounts in a related species; this is the case of the PGL of *M. bovis*, which is one of the minor PGL found in strains of *M. tuberculosis* producing PGL (Daffé, 1988b). As indicated earlier, a third group of products, the *p*-HBADs, are synthesized by *M. tuberculosis* and *M. bovis* BCG (Constant et al., 2002). Their structure was established by Constant et al. (2002) as corresponding to glycosylated *p*-hydroxybenzoic acid methyl esters, structurally related to the type-specific PGL (Fig. 1).

Using various mechanical and detergent treatments, Daffé and colleagues investigated the subcellular localization of various lipids of *M. tuberculosis* (Ortalo-Magné et al., 1996). They developed a method based on sequential treatments with glass beads and detergents to release the materials more or less buried in the mycobacterial cell envelope. Their study demonstrated that PGL-tb and DIM are located in the capsule of *M. tuberculosis*: part of them exposed to the cell surface and part of them contributing to the outer layer of the pseudo-outer membrane of the *M. tuberculosis* cell envelope (Ortalo-

Magné et al., 1996). In contrast, to these compounds that remained attached to the bacterial cell, *p*-HBADs are released in the culture medium (Constant et al., 2002).

BIOSYNTHESIS AND TRANSLOCATION

The structure of the backbone of PGLs and the related DIMs was established more than 30 years ago, but their synthesis remained elusive until the last decade. The common lipid core found in PGL and DIM (Fig. 1) suggested a common part in their biosynthesis pathway in which either n-C$_{20}$ fatty-acyl chains or *p*-hydroxyphenylalkanoates would be elongated to form the long-chain β-diols phthiocerol or phenolphthiocerol, respectively. However the nature of the enzymes involved, the sequence of reactions, and the source of substrates were all unknown. The same was true for the fatty acids esterifying these β-diols, the mycocerosic acids, which have a similar structure in DIMs and PGLs (Fig. 1). Renewal of interest in the synthesis of DIM and PGL originated from the pioneering work of Kolattukudy and colleagues (1997), who suggested that the biosynthesis pathways of these compounds are analogous to that of erythronolide formation. These authors were the first to implicate a family of enzymes called polyketide synthases in the biosynthesis of PGLs and DIMs. They also cloned the first DIM and PGL biosynthetic genes (Azad et al., 1996; Mathur and Kolattukudy, 1992). The completion of the *M. tuberculosis* genome sequence (Cole et al., 1998) and the development of powerful genetic tools (Pelicic et al., 1998) together opened new possibilities for the study of the biology of *M. tuberculosis*. Using these new tools, several groups have revisited the biosynthesis pathways of DIM and PGL. The picture which has emerged from their work is that the genes encoding the various enzymes of the DIM and PGL pathways are clustered on a 73-kbp fragment of the *M. tuberculosis* chromosome, the DIM+PGL locus (Fig. 2). The genes and organization of this locus are highly conserved in all sequenced mycobacteria producing PGL and related molecules, with the exception of *M. leprae*, in which the DIM+PGL locus is split into two loci. In the last decade, the combination of genetic manipulation and biochemical characterization of recombinant strains have provided an almost complete view of the biosynthesis pathways of DIM and PGL in *M. tuberculosis* and related mycobacteria.

Formation of *p*-Hydroxyphenylalkanoate

The aromatic nucleus in PGL was first suspected to come from tyrosine because metabolic labeling with [3-^{14}C]tyrosine led to radioactive PGL, although the incorporation of radioactivity into PGL was low (Gastambide-Odier et al., 1967). An alternative pathway was proposed by Minnikin (Minnikin, 1982), who suggested that this aromatic nucleus may come from chorismic acid through the formation of *p*-hydroxybenzoic acid. Indeed, in various microorganisms, including *Escherichia coli*, *p*-hydroxybenzoic acid is formed after the removal of the pyruvyl moiety from chorismate, a reaction catalyzed by a chorismate pyruvate lyase, product of the *ubiC* gene in *Escherichia coli* (Siebert et al., 1992; Siebert et al., 1994). An orthologue of this gene, *Rv2949c*, was recently identified in the DIM+PGL locus. Insertion of a transposon into this gene led to a *M. tuberculosis* strain unable to synthesize *p*-HBAD (Stadthagen et al., 2005) or PGL-tb (our unpublished data). The enzyme encoded by *Rv2949c* catalyzes the formation of *p*-hydroxybenzoic acid from chorismate in vitro and is the only enzymatic source of this product in *M. tuberculosis*. This enzyme was purified after production in surrogate hosts. It is active over a wide pH range, with an optimum pH at 7.5, and is inhibited by *p*-hydroxybenzoic acid, the product of the reaction. This finding definitively established the source of *p*-hydroxybenzoic acid as being chorismate rather than tyrosine. This conclusion was confirmed by a direct and massive incorporation of labeled *p*-hydroxybenzoic acid into the PGL of *M. tuberculosis* (Reed et al., 2004) or *M. microti* and its aglycosyl precursors, that is, phenolphthiocerol and phenolphthiodione dimycocerosates (P. Draper, unpublished data).

Constant et al. (2002) proposed that *p*-hydroxybenzoic acid is then elongated by a type 1 polyketide

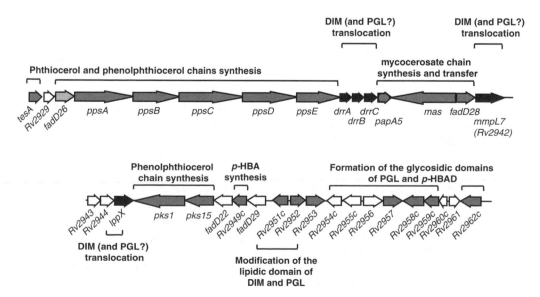

Figure 2. Genetic organization of the DIM+PGL locus. The genes are represented by horizontal arrows. Genes encoding proteins involved in DIM, PGL, or *p*-HBAD biosynthesis are in gray. Genes encoding proteins involved in the translocation of these molecules are in black. The roles of the various genes are indicated.

synthase, product of the gene *pks15/1*, to form *p*-hydroxyphenylalkanoate; the reaction may involve eight or nine elongation cycles using malonyl-CoA as the extender unit. This proposal was based on the findings that (i) Pks15/1 contains all the catalytic domains required for catalyzing such elongation; that is, a β-ketoacyl synthase domain (KS), an acyl-transferase domain (AT), a β-ketoacyl reductase domain (KR), a dehydratase domain (DH), an enoyl-reductase domain (ER) and an acyl carrier protein domain (ACP); (ii) the AT domain, which determines the substrate specificity for the elongator molecules, is predicted to be selective for malonyl-CoA; and (iii) disruption of the *pks15/1* gene in *M. bovis* BCG abolishes the production of PGL without affecting the synthesis of DIM and *p*-HBAD (Constant et al., 2002). These authors also demonstrated that the only reason for the inability of most *M. tuberculosis* isolates, including the reference strain H37Rv, to produce the type-specific PGL of *M. tuberculosis* (PGL-tb) is a frameshift mutation in the *pks15/1* gene. Recently, a fatty acid, possibly the product of Pks15/1, was purified from *M. microti*, a species of the *M. tuberculosis* complex producing PGL similar to mycoside B from *M. bovis* (Thurman et al., 1993). Indeed, a previously unknown compound was isolated following metabolic labelling of *M. microti* with [^14^C]-*p*-hydroxybenzoate (Draper, unpublished). This putative lipid precursor is a *p*-hydroxyphenyl fatty acyl unit (with an alkyl chain length of 11 to 23 carbon atoms) that esterifies glycerol together with two C_{16}-C_{18} fatty acids to form an unusual triglyceride (Draper, unpublished).

These studies identified two enzymes catalyzing two early steps of the biosynthesis pathway of PGL and *p*-HBADs. These two proteins are specific for the formation of PGL and *p*-HBADs because disruption of *Rv2949c* or *pks15/1* does not affect the formation of DIM. Several questions remain to be solved concerning these early steps of the PGL pathway. First, how is *p*-hydroxybenzoic acid loaded onto Pks15/1? Trivedi et al. (2004) demonstrated that a new class of long-chain fatty acyl-AMP ligases (FAALs) from *M. tuberculosis* activates long chain fatty acids as acyl-adenylates and loads them onto Pks. Usually, the gene *fadD* encoding the FAAL maps adjacent to the *pks* gene. Interestingly, there are two *fadD* genes upstream from *pks15/1* (Fig. 2). One of these genes, *fadD29*, encodes a protein similar to the FAALs identified by Trivedi et al. (2004). The other, *fadD22*, encodes a polypeptide more distantly related to FAALs. Therefore, it is plausible that one of these two enzymes is involved in the activation and loading of *p*-hydroxybenzoate onto Pks15/1. Another intriguing question came from the finding by Draper et al.

(P. Draper et al., unpublished data), who purified the *p*-hydroxyphenyl fatty acid intermediates as part of a triglyceride. The authors suggested that this compound may serve as either a storage substance or a carrier molecule, but its function has not been demonstrated.

Formation of Phthiocerol and Phenolphthiocerol

This part of the biosynthesis pathway corresponds to the formation of the common lipid core of PGLs and DIMs (Fig. 1). Early work by Gastambide-Odier and colleagues (1963, 1967) to decipher the phthiocerol and phenolphthiocerol biosynthesis pathway were based on metabolic labeling either with [1- or 3-^14^C]propionate or with [^14^C]methionine, followed by chemical degradation and tracing the radioactivity. These elegant studies showed that the O-methyl group of phthiocerol and phenolphthiocerol is provided by *S*-adenosylmethionine (SAM) and that both the four terminal carbons and the methyl branches were from propionate (Gastambide-Odier et al., 1963). The next major advance came from the work of Azad et al. (1997), who postulated that the formation of phthiocerol and phenolphthiocerol involves a series of modular polyketide synthases. Using DNA probes corresponding to either the KS or AT domains from the fatty acid synthase Fas or the newly discovered Mas enzyme responsible for mycocerosic acid formation (Mathur and Kolattukudy, 1992), they screened a cosmid library of *M. bovis* BCG and identified a genomic locus that might contain the series of *pks* genes sought. This locus contains five *pks* genes and is conserved in *M. tuberculosis* and *M. leprae* (Azad et al., 1997). Insertion/deletion in two of these *pks* genes, *ppsB* and *ppsC*, by allelic exchange in *M. bovis* BCG produced a mutant strain unable to synthesize both DIMs and PGLs: thus, these *pks* are involved in the DIM and PGL biosynthetic pathway (Azad et al., 1997). Similarly, a mutant containing an insertion into the *ppsE* gene of *M. tuberculosis* was recently shown to contain no DIMs, confirming these initial results (Simeone et al., 2007a). The modular Pks system first identified by Kolattukudy and coworkers (Azad et al., 1997) contains five *pks* genes encoding type-1 modular polyketide synthases, named *ppsA-E* in *M. tuberculosis* (Cole et al., 1998). They are part of a predicted large operon containing the *ppsA-E* genes and also *fadD26*, which belongs to the FAAL family of proteins; three genes, *drrABC*, encoding a putative ABC transporter; and *papA5*, a gene of unknown function when the operon was described (Cole et al., 1998). Camacho et al. (1999, 2001) obtained a transposon insertion within the *fadD26* gene, and this mutation abolished the formation of

DIM in *M. tuberculosis*. A similar result was described by Cox et al. (1999) for a transposon insertion just upstream from *fadD26*. Complementation of the *fadD26* mutant with a functional *fadD26* gene partially restored the production of DIM demonstrating the direct involvement of FadD26 (Camacho et al., 2001). The authors suggested that this enzyme is involved in the activation of n-C_{20} fatty acid, the PpsA-E substrate for the formation of phthiocerol. This initial proposition was confirmed by Trivedi et al. (2004) who demonstrated that FadD26 is a fatty acyl AMP ligase, involved in the activation of long-chain fatty acids as acyl-adenylates and in the loading of these substances onto PpsA. It is still unclear whether FadD26 is also responsible for the activation of the other PpsA substrates, that is the *p*-hydroxyphenyl fatty acids. The work of Camacho et al. (2001) was performed using a *M. tuberculosis* strain unable to synthesize PGL, so conclusions about the role of FadD26 in the biosynthesis of PGL could not be drawn. Similarly, Trivedi and colleagues (2004) did not investigate the role of FadD26 in the activation of *p*-hydroxyphenyl fatty acids or in its transfer onto PpsA.

Bioinformatic analysis of the PpsA-E protein sequences revealed that these enzymes harbor the catalytic domains required for the formation of a 10-carbon chain containing a keto function on C3, a methyl group on C4, and a hydroxyl group on each C9 and C11 (Minnikin et al., 2002; Trivedi et al., 2005). Trivedi et al. (2005) provided experimental evidence supporting the predicted activities of PpsA, PpsB, and PpsE. First, they purified these proteins from a heterologous host, made competent for post-translational activation of Pks. They then demonstrated that PpsA in vitro can elongate activated lauric acid, provided by FadD26, with malonyl-CoA to form 3-hydroxytetradecanoic acid. Their findings suggest a selectivity of PpsA or FadD26 for medium- to long-chain fatty acids, consistent with the proposed in vivo activities of these enzymes. In the same study, the authors demonstrated that PpsB can use the product of PpsA and elongate it with malonyl-CoA to form a mixture of 3, 5-dihydroxyhexadecanoic acid and 3-keto, 5-hydroxyhexadecanoic acid. This suggests partial reduction of the keto group at position 3 by the KR domain of PpsB in this in vitro assay. Finally, they used an artificial substrate derived from *N*-acetyl cysteamine (NAC), the 9-hydroxy decanoyl-NAC, as a substrate for PpsE and showed that this enzyme can use either malonyl-CoA or methylmalonyl-CoA to elongate this substrate and produce either 11-hydroxy 3-keto dodecanoic acid or 11-hydroxy 2-methyl 3-keto dodecanoic acid. This flexibility of PpsE in the choice of the extender unit is consistent with the heterogeneity at the ter-

mini of phthiocerol and phenolphthiocerol molecules (group R2, Fig. 3B). These in vitro experiments very elegantly confirmed the predicted function of three of the five PpsA-E proteins. However, they provide no information about PpsC and PpsD.

Bioinformatics analyses of the amino acid sequences of PpsC and PpsD by various groups indicate that PpsC contains all of the catalytic domains (KS-AT-DH-ER-KR-ACP) required for elongation with malonyl-CoA and PpsD for that with methyl-malonyl-CoA, followed by the full reduction of the keto group (Minnikin et al., 2002; Trivedi et al., 2005). However, we and others (Onwueme et al., 2005a; Yadav et al., 2003) have been unable to identify the enoyl reductase domain in PpsD, although this domain can be easily found in PpsC. This raised the possibility that the reduction of the double bond between C4 and C5 is catalyzed by an enoyl reductase not part of PpsD. We obtained data indicating that *Rv2953* might be involved in this reaction (Simeone et al., 2007b). Indeed, we found that disruption of the *Rv2953* gene in *M. tuberculosis* H37Rv led to the production of DIM-like compounds with an unsaturation between C4 and C5. This structural modification of DIM in the mutant strain depended on the deletion of the *Rv2953* gene because complementation of the mutation by reintroduction of a functional *Rv2953* gene rescued DIM production. Transfer of a functional *pks15/1* gene into this mutant strain resulted in accumulation of phenolglycolipids with molecular masses corresponding to PGL-tb with an additional double bond. These findings are consistent with PpsD having no functional ER domain, and strongly suggest that *Rv2953* provides the ER activity in trans to PpsD during phthiocerol and phenolphthiocerol biosynthesis in *M. tuberculosis* (Simeone et al., 2007b).

Another intriguing and unsolved question concerns the release of the phthiocerol chain formed by the PpsE synthase. PpsE contains an additional domain at its carboxy-terminal end. This domain shares limited similarities with the condensation domain of various non-ribosomal peptide synthases (NRPS). Trivedi et al. (2005) proposed that this catalytic domain may be involved in polyketide release from the synthase as observed for other NRPS. However, the β-keto acid produced by PpsE remained attached to the synthase in their *in vitro* enzymatic assays, raising doubts about the involvement of the carboxy-terminal domain of PpsE in the release of phthiocerol or phenolphthiocerol chain. Another candidate protein, encoded by the gene *tesA*, may fulfill this function. Indeed, TesA exhibits sequence similarities with type-II thioesterase, and Waddell et al. (2005) showed that the insertion of a transposon into *tesA* in H37Rv led

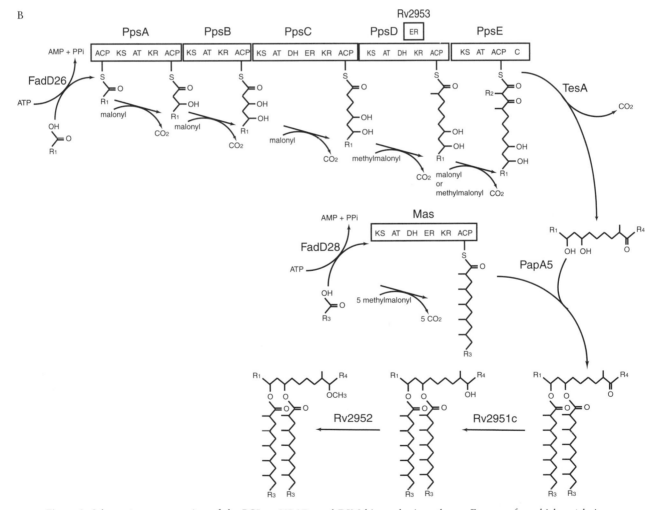

Figure 3. Schematic representation of the PGL, *p*-HBAD, and DIM biosynthesis pathway. Enzymes for which a role is suspected but has not been demonstrated experimentally are indicated by a question mark. (A) Formation of *p*-HBA and *p*-hydroxyphenylalkanoate. (B) Formation of the lipid core common to DIM and PGL-tb. R_1, CH_3-$(CH_2)_{20-22}$; R_2, H or CH_3; R_3, CH_3-$(CH_2)_{16-18}$; R_4, CH_2-CH_3 or CH_3. (C) Formation of the terminal species-specific carbohydrate domain common to PGL-tb and *p*-HBAD. R_5, $COOCH_3$ or phthiocerol dimycocerosate.

C

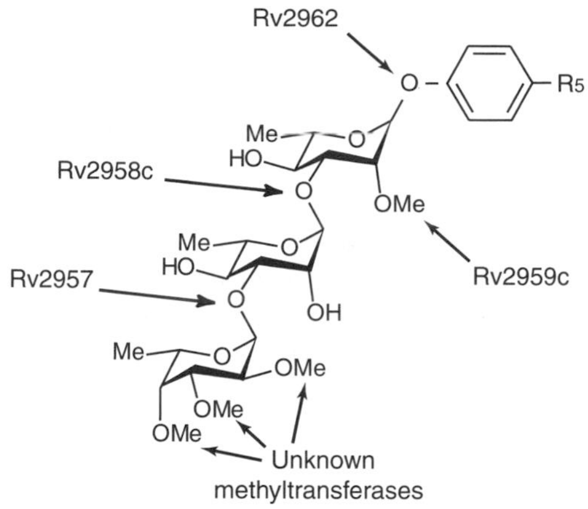

Figure 3. *Continued*

to a DIM-less strain. Although complementation experiments are required to prove the role of *tesA* in DIM synthesis, and by extrapolation in PGL formation, this initial finding suggests that this enzyme may provide the thioesterase activity lacking in PpsE. Consistent with this, Rao and Ranganathan (2004) using two independent systems demonstrated that TesA interacts with the carboxyl-terminal end of PpsE. However, it is not known whether TesA alone is sufficient for the release of the phthiocerol and phenolphthiocerol chains or whether association with the additional pseudo-condensation domain of PpsE is required.

As indicated earlier, PpsE catalyzes the last condensation step of phthiocerol and phenolphthiocerol synthesis and leaves an unreduced keto group on C3. These products, called phthiodiolone and phenolphthiodiolone, are found as diesters of mycocerosates in the cell envelope of *M. tuberculosis* complex strains (Daffé, 1988a). However, in the major products DIM and PGL-tb, this carbon C3 bears a methoxy group, suggested to be derived from phthiodiolone and phenolphthiodiolone after a two-step process in which the keto group is first reduced and then methylated by a SAM-dependent methyltransferase (Gastambide-Odier et al., 1963). This model was validated by the identification of the two enzymes catalyzing these reactions. First, Perez et al. (2004a) demonstrated that mutation of gene *Rv2952* in *M. tuberculosis* H37Rv led to a strain accumulating phthiotriol dimycocerosates and, after transfer of a functional *pks15/1* gene, glycosylated phenolphthiotriol dimycocerosates (Perez et al., 2004a), the two expected intermediates in the biosynthesis of DIM A and PGL-tb. This enzyme encoded by *Rv2952* belongs to a group of methyltransferases involved in the methylation of hydroxyl groups on fatty acids (Jeevarajah et al., 2002). Thus, it was con-

cluded that *Rv2952* encoded the methyltransferase catalyzing the SAM-dependent methylation of phthiotriol and phenolphthiotriol to give phthiocerol and phenolphthiocerol. Second, the ketoreductase enzyme proposed to act before Rv2952 was identified independently by two groups in *M. tuberculosis* and in *M. ulcerans* (Simeone et al., 2007a; Onwueme et al., 2005b). Starting from the observation that *M. ulcerans* and some strains of *M. kansasii* are unable to produce DIM A but produce phthiodiolone dimycocerosates, Onwueme et al. (2005b) looked at gene sequences at the DIM+PGL locus in *M. ulcerans*: they identified a mutation introducing a premature stop codon into the ortholog of *Rv2951c*. Complementation of *M. ulcerans* with the functional *Rv2951c* ortholog from *M. marinum* endows the recombinant strain with the ability to synthesize phthiotriol, and this is strong evidence that *Rv2951c* encodes the phthiodiolone ketoreductase (Onwueme et al., 2005b). The same conclusion was reached by Simeone et al. (2007a), who produced a *M. tuberculosis* mutant in which the *Rv2951c* gene is disrupted. This mutant strain is unable to synthesize DIM A but accumulates DIM B. When a functional *pks15/1* was transferred into this mutant strain, glycosylated phenolphthiodiolone was produced. This phenotype was fully reversed on complementation with a functional *Rv2951c* gene, thereby excluding polar effect on expression of neighboring genes. This group concluded that *Rv2951c* encodes the ketoreductase catalyzing the first step of the modification of phthiodiolone and phenolphthiodiolone in *M. tuberculosis*.

Production and Transfer of Mycocerosates

Another branch of the biosynthesis of PGL and related molecules concerns the formation of mycocerosates. Here again, the first insight into the biosynthesis of mycocerosates was the pioneering work of Gastambide-Odier and coworkers (1963). Using [14C]propionate labeling, the authors demonstrated that the methyl branches on mycocerosates are derived from elongation of C_{16} and C_{20} fatty acids with 3 or 4 propionate units probably first converted into methylmalonyl-CoA. This finding was confirmed 20 years later by Rainwater and Kolattukudy (1983), who showed that cell-free extracts of *M. bovis* BCG incorporated [14C]methylmalonyl-CoA into mycocerosic acids. In a series of illuminating papers, the group of Kolattukudy purified the enzyme responsible for this elongation reaction and identified it as an iterative type 1 polyketide synthase containing six catalytic domains (KS-AT-DH-ER-KR-ACP) (Mathur and Kolattukudy, 1992; Rainwater and

Kolattukudy, 1983, 1985). This enzyme, named Mas, exhibited preference in vitro for long-chain acyl-CoA primers and specificity for methylmalonyl-CoA, as expected from its proposed function in vivo. In addition, this enzyme requires NADPH as a cofactor for the ketoreductase and enoylreductase activities (Rainwater and Kolattukudy, 1985). It was suggested that the specificity for methylmalonyl-CoA was determined by the AT and KS domain of Mas (Fernandes and Kolattukudy, 1997). Indeed, overexpression of the AT and KS domains of Mas from *M. bovis* BCG in *E. coli* increased the incorporation of methylmalonyl-CoA into fatty acids by crude recombinant *E. coli* extract (Fernandes, 1997). Later, Yadav et al. (2003) suggested that a phenylalanine residue in the substrate-binding pocket of the AT domain contributes to the specificity for malonyl-CoA rather than methylmalonyl-CoA. This role of the AT domain was experimentally confirmed by Trivedi et al. (2005) who demonstrated that mutation of the serine S726 residue of the AT domain to a phenylalanine switched the specificity of Mas from methylmalonyl-CoA to malonyl-CoA.

The role of Mas in the formation of PGL and DIM was directly established by the group of P. Kolattukudy (Azad et al., 1996) through the construction of a *M. bovis* BCG *mas::hyg* mutant. This mutant was unable to synthesize mycocerosic acids and, therefore, PGL. Surprisingly this group reported that the *M. bovis* BCG *mas::hyg* mutant still produced DIM but with shorter fatty acids. This raised the question of whether other fatty acids can be transferred onto the phthiocerol chain in the absence of mycocerosates. However, this observation has not been confirmed and other authors have raised doubts about the method used by Azad et al. (1996) to purify the DIM molecules (Onwueme et al., 2005b). In *M. tuberculosis*, the substrates of Mas are C_{16} and C_{20} fatty acids. The initial work by Rainwater and Kolatukkudy (1983) suggested that the preferred substrates were acyl-CoA. The obvious candidate protein for catalyzing the activation of C_{16} and C_{20} fatty acids and transfer of acyl-CoA onto Mas was FadD28, a protein encoded by a gene adjacent to *mas* and shown in three independent studies to be required for both PGL (Fitzmaurice et al., 1998) and DIM biosynthesis (Camacho et al., 2001; Cox et al., 1999). Fitzmaurice and Kolattukudy (1997, 1998) indicated that purified FadD28 exhibited an acyl-CoA synthase activity using fatty acids of various chain lengths (C_{10} to C_{18}). However, Trivedi et al. (2004) demonstrated more recently in vitro that FadD28 is indeed involved in the activation of fatty acids but as acyl-AMP rather than acyl-CoA. Indeed,

incubation of purified FadD28 with fatty acids, ATP, and coenzyme A led to formation of acyl-AMP but not acyl-CoA. These findings are in apparent conflict with the previous observation by Fitzmaurice and Kolattukudy (1997), but the structure of the FadD28 products was not established in this initial report, so their exact nature is unknown. These findings and the close chromosomal localization of FadD28 and Mas strongly suggest that FadD28 is indeed involved in the activation of the Mas substrates. However, there is still no direct evidence of an interaction between FadD28 and Mas and the loading onto the synthase.

Analysis of the Mas sequence failed to identify a thioesterase domain that could be involved in the release of mycocerosates from the Pks. Indeed, they were subsequently shown to be directly transferred from Mas onto their final acceptors, the phthiocerol or phenolphthiocerol chains. The enzyme catalyzing this transfer was identified as PapA5 (Onwueme et al., 2004). A first series of in vitro assays demonstrated that purified PapA5 transfers various chain length acyl-CoA onto various nucleophile acceptors with a strong preference for saturated medium-chain-length alcohols (Onwueme et al., 2004). Insertional inactivation of the *papA5* gene in *M. tuberculosis* yielded a strain unable to synthesize DIMs. This mutation was complemented by reintroducing a functional copy of *papA5*, demonstrating the involvement of PapA5 in the synthesis of DIM in vivo (Onwueme et al., 2004). This mutation study was performed in a *M. tuberculosis* strain unable to synthesize PGL because of a mutation in the *pks15/1* gene; therefore, it was not possible to confirm the role of PapA5 in PGL synthesis. However, it is likely that PapA5 is also involved in the transfer of mycocerosates onto phenolphthiocerol during PGL biosynthesis. The crystal structure of PapA5 has been established at 2.75 Å resolution (Buglino et al., 2004), and revealed that it shares structural similarities with various acetyltransferase including chloramphenicol acetyltransferase and carnitine acetyltransferase. The PapA5 active site contains conserved histidine and aspartic residues, and mutation of these residues into alanine abolished the enzymatic activities (Onwueme et al., 2004). The three-dimensional structure revealed two hydrophobic channels linking the PapA5 surface to the active sites, a feature consistent with its enzymatic activity (Buglino et al., 2004). These findings established the role of PapA5 in the transfer of mycocerosates onto their alcohol acceptors. However, a question remaining was whether PapA5 uses mycocerosates previously released from the synthase or transfers the Mas-bound acyl directly onto the acceptor. Consistent with the second model, Trivedi et al. (2005) established in vitro that PapA5 exhibited more than 100-fold higher affinity for acylated-Mas than for acyl-CoA. They also showed that the transfer reaction is inhibited by unacylated-Mas, consistent with an interaction between Mas and PapA5. The same authors modeled the interaction of the ACP domain of Mas with PapA5 and identified two amino acids of PapA5 as being important for this interaction: Arg234 and Arg312. Site-specific mutation of these two residues, R312E and R234E, increased the K_M of mutant PapA5 for Mas-ACP by 40-fold and fourfold respectively, although the K_{cat} values were similar to those of the wild-type PapA5. These results established that these two residues were indeed important for the recognition of Mas. Therefore, it is likely that mycocerosates are not released from Mas by a conventional thioesterase but rather transferred directly by PapA5 onto their phthiocerol or phenolphthiocerol acceptors.

Synthesis of the Saccharide Appendage of PGL-tb and *p*-HBAD

DIM and PGL share a common lipid core synthesized by the same enzymatic machinery, but unlike PGL, DIM has no saccharide domain. In contrast, the carbohydrate part of PGL is found in the *p*-HBAD released into the culture supernatants of *M. tuberculosis* and related species (Constant et al., 2002). Thurman et al. (1993) showed that phenolphthiocerol dimycocerosates, recovered from lipid extracts of *M. microti*, could be glycosylated in vitro by acellular extracts of the bacterium. Therefore, the enzymes responsible for the formation of the saccharide appendage can use phenolphthiocerol dimycocerosates as substrates. However, the conservation of structure in this saccharide domain between PGL and *p*-HBADs produced by the same mycobacterial species suggested the involvement of the same biosynthetic enzymes. This notion was confirmed by Perez et al. (2004a, 2004b), who identified four enzymes, three glycosyltransferases and one methyltransferase, involved in the formation of the glycosyl moiety of both PGL and *p*-HBADs. They produced and biochemically analyzed four independent *M. tuberculosis* mutants in which genes *Rv2962c*, *Rv2957*, *Rv2958c*, and *Rv2959c* were inactivated. A ΔRv2962c::km mutant of *M. tuberculosis* H37Rv (made competent for PGL-tb biosynthesis through the transfer of a functional *pks15/1* gene) no longer produced *p*-HBADs or PGL-tb; however, it accumulated two aglycosyl derivatives of PGL, *p*-hydroxyphenolphthiocerol dimycocerosates and *p*-methoxylated hydroxyphenolphthio-

cerol dimycocerosates. This is consistent with the glycosyltransferase encoded by *Rv2962* being involved in transfer of the first rhamnosyl residue onto *p*-hydroxyphenolphthiocerol dimycocerosates and *p*-hydroxyphenolmethylester. In contrast, *Rv2957* and *Rv2958c* mutants synthesized monoglycosylated *p*-HBAD and PGL, that is, *p*-HBAD I and mycoside B (Fig. 1), suggesting that the glycosyltransferases encoded by these two genes catalyze the transfer of the terminal two sugar residues. However, the reaction sequence—formation of a disaccharide followed by the transfer or sequential transfer of two monosaccharides—and the exact role of each enzyme remained to be established. *M. bovis* BCG was used as a surrogate host to investigate *Rv2958* function (Perez et al., 2004b). *M. bovis* strains synthesize mostly mycoside B and *p*-HBAD I, and Perez et al. (2004b) noticed that the orthologue of *Rv2958* is mutated in both sequenced *M. bovis* strains (the vaccine BCG Pasteur strain and the AF2122/97 isolate). In these strains, a single base insertion at nucleotide position 867 leads to a frameshift and a 186-bp shorter open reading frame for the *Rv2958c* orthologue. These authors transferred the functional *Rv2958c* gene from *M. tuberculosis* into *M. bovis* BCG and found that the recombinant strain synthesized mono-*O*-methyl diglycosyl-phenolphthiocerol, independent of the presence or absence of a functional *Rv2957* gene (Perez, 2004b). This establishes that the Rv2958c protein can transfer an additional sugar onto mycoside B and therefore is probably the glycosyltransferase involved in the transfer of the second rhamnosyl residue onto mycoside B or *p*-HBAD I. However, the exact roles of the Rv2957 and Rv2958 proteins remain to be completely established.

Perez et al. (2004a) also identified the gene *Rv2959c* as encoding the methyltransferase involved in the methylation of the position 2 of the first rhamnosyl residue of PGL-tb and *p*-HBAD. They constructed an unmarked mutation within this gene in *M. tuberculosis* H37Rv in which the *pks15/1* mutation was complemented by a functional *pks15/1* gene. The resulting mutant produced only unmethylated rhamnosyl-phenolphthiocerol dimycocerosates and tri-*O*-methylfucosyl-di-rhamnosyl-phenophthiocerol dimycocerosates. Thus, methyltransferases other than Rv2959c are responsible for the *O*-methylation of the terminal fucosyl residue. Close examination of the DIM+PGL locus in *M. tuberculosis* revealed that three genes, *Rv2954*, *Rv2955c*, and *Rv2956*, putatively encode polypeptides sharing similarities with methyltransferases from other bacterial species (our unpublished data). In view of their location within this chromosomal locus dedicated to DIM, PGL and p-HBAD biosynthesis, it is plausible that these three genes are responsible for the methylation of the hydroxyl groups at positions 2, 3 and 4 of the terminal fucosyl residue in PGL-tb and p-HBAD II. Furthermore, *M. marinum* and *M. leprae*, two mycobacterial species that synthesized PGL with no terminal methylated fucosyl residue do not contain orthologues of these three genes (Onwueme, 2005b). However, the exact roles of these three putative proteins have yet to be demonstrated.

Translocation of PGL and Related Molecules

PGL-tb and DIM are found in the outermost layer of the *M. tuberculosis* envelope (Ortalo-Magné et al., 1996) and p-HBADs are secreted into the culture medium (Constant et al., 2002). However, several lines of evidence suggest that these compounds are either entirely or partially formed in the cytoplasm of *M. tuberculosis*. First, some of the enzymes required for the biosynthesis of DIM and PGL, including Pks and FadD enzymes, required cofactors NADPH and ATP respectively, which are only found in the cytosol. Second, various proteins are required for the translocation of preformed DIM in *M. tuberculosis* Mt103 and Erdman (two strains that do not synthesize PGL-tb). Camacho et al. (1999) obtained two strains derived from Mt103 by insertion of a transposon into the *mmpL7* or *drrC* genes, both in the DIM+PGL locus (Fig. 2). Both the *drrC* and *mmpL7* mutant strains synthesized DIM structurally identical to that in the wild-type strain, but they failed to translocate these compounds to the cell surface (Camacho et al., 2001). Subcellular fractionation revealed that the DIM were retained mostly in a fraction corresponding to plasma membrane, cytosol and putative periplasm (PMCPP) in the *drrC* and *mmpL7* mutants (Camacho et al., 2001). Cox et al. (1999) reported similar findings for another mutant derived from the Erdman strain by transposon insertion into the *mmpL7* gene. These results established that the enzymatic machinery required for the formation of DIM is located in the cytoplasm of *M. tuberculosis*.

The gene *drrC* encodes an integral membrane protein predicted to be part of an ABC transporter involving two other subunits encoded by genes *drrA* and *drrB* mapping upstream from *drrC* (Camacho et al., 1999). Thorough bioinformatic analysis of the DrrA, DrrB and DrrC proteins revealed that DrrA contains an ABC transporter signature and the Walker A and B motifs involved in nucleotide binding (Braibant et al., 2000). In contrast, DrrB and C encode integral membrane pro-

teins with five and six predicted transmembrane segments, respectively. Both of these proteins contain the ABC-2 motif found in subclasses of ABC transporters involved in drug or carbohydrate export (Braibant et al., 2000). Therefore, all of the sequence features typical of an ABC transporter involved in metabolite or drug export are present in the DrrABC proteins.

The MmpL7 protein belongs to another class of transporters, known as RND (Resistance, nodulation and cell division) permeases (Tseng et al., 1999). Like other members of this family, MmpL7 is predicted to contain 12 transmembrane domains and two large loops. Using a yeast two-hybrid system, Jain and Cox (2005) showed that the loop between the seventh and eighth transmembrane domains interacts with the polyketide synthase, PpsE, involved in DIM and PGL synthesis. Unexpectedly, overproduction of the MmpL7 interacting loop in *M. tuberculosis* inhibited the synthesis of DIM in an MmpL7-dependent fashion. This was surprising because inactivation of MmpL7 prevented DIM translocation but not synthesis (Camacho et al., 2001; Cox et al., 1999; Jain and Cox, 2005). To explain these unexpected results, Jain and Cox (2005) proposed a model in which synthesis and transport are coupled; MmpL7 initially exerts an inhibitory effect on DIM synthesis until the entire synthesis-transport machinery is assembled and then relieves this inhibition to allow efficient formation and export of these compounds. However, this model has yet to be validated by experimental data.

Another gene, *lppX*, encoding a lipoprotein has been found to be required for correct localization of DIM in the cell envelope (Sulzenbacher et al., 2006). Insertional inactivation of this gene prevents the exportation of DIM to the cell surface and leads to accumulation of DIM in deeper layers of the cell envelope. The structure of the LppX protein was established: it displays a U-shaped β-half barrel in which a large hydrophobic cavity is suitable for accommodating a DIM molecule. The subcellular distribution of DIM in a Δ*lppX* mutant strain has been investigated. DIM accumulated in the cell wall (44%) and PMCPP fraction (56%), a distribution differing slightly from those in the Δ*drrC* and Δ*mmpL7* mutants, in which 78% and 92%, respectively, of DIM were found in the PMCPP fraction (cytosol, plus plasma membrane and putative periplasm) (Camacho et al., 2001). Although these two fractionation experiments were performed independently, making a direct comparison hazardous, it is possible that LppX acts downstream from the MmpL7 and DrrABC transporters in the translocation of DIM. MmpL7 and DrrABC might be involved in transport across the plasma membrane and LppX in transfer to the outer layer of the cell envelope. This agrees with the localization of these proteins, MmpL7 and DrrABC being predicted to be within the plasma membrane and LppX having been found mostly in the cell wall (Lefèvre et al., 2000). However, the precise role of each of these proteins remains to be determined.

All of this work on the translocation process involved DIM because the *M. tuberculosis* strains used were unable to synthesize PGL-tb. The presence or absence of *p*-HBADs in the culture media of the various Δ*drrC*, Δ*mmpL7*, and Δ*lppX* has not been reported. Therefore, no data are available concerning the transport of PGL and *p*-HBAD. Because the enzymes involved in DIM biosynthesis are also required for formation of the lipid core of PGL, it is likely that at least part of this compound is also synthesized in the cytoplasm. However, whether the saccharide domain is added before or after translocation of phenolphthiocerol dimycocerosates and the site of *p*-HBAD synthesis remain to be addressed.

Other Genes Involved in DIM and PGL Biosynthesis

Twenty-three genes have already been demonstrated to contribute to the biosynthesis and translocation of DIM, PGL-tb, and *p*-HBADs (this includes genes involved only in this pathway and not in similar pathways, such as the *fas* gene, for instance). Other genes, including *pks11* (Waddell et al., 2005), *pks12* (Sirakova et al., 2003b), *pks10* (Sirakova et al., 2003a), *mb0100* (Hotter et al., 2005) and *pks7* (Rousseau et al., 2003), have also been associated with the biosynthesis of DIM; however, this was on the basis of insertional inactivation of genes and lipid analysis but complementation experiments allowing more definitive conclusions are lacking. In addition, no biosynthetic role has been suggested for these genes. Therefore, the set of genes unambiguously demonstrated to be involved in DIM, PGL-tb, and *p*-HBAD biosynthesis is composed of the 23 genes discussed earlier. Remarkably, all of these genes are clustered on the genome of *M. tuberculosis* (Fig. 2). This DIM+PGL locus covers 73 kbp of the *M. tuberculosis* chromosome, meaning that 1.7% of the *M. tuberculosis* genome is specifically devoted to the biosynthesis and translocation of these compounds. Not only are the genes required for DIM, PGL-tb, and *p*-HBAD biosynthesis clustered but, even within the DIM+PGL locus, the genes are grouped according to their roles in the biosynthetic pathway (Fig. 2). This organization into functional

groups is consistent with the proposed functions of several still uncharacterized genes in the locus: *fadD22* and *fadD29* in the formation of *p*-hydroxyphenylalkanoate, and *Rv2954c*, *Rv2955c*, and *Rv2956* in the synthesis of the saccharide domains of PGL-tb and *p*-HBAD.

ROLE OF PHENOLIC GLYCOLIPIDS AND RELATED COMPOUNDS IN THE BIOLOGY OF *M. TUBERCULOSIS*

Role of PGL-tb and DIM in the Organization of the Cell Envelope

PGL-tb and DIM are located in the outermost layers of the *M. tuberculosis* envelope and are thought to be a component of the capsule of the tubercle bacillus (Ortalo-Magné et al., 1996). As such, several PGL (perhaps all) are serologically active, but this seems unlikely to be important for pathogenesis. They are also thought to be involved in the outer permeability layer that typifies mycobacteria and related species (Daffé and Draper, 1998; Liu et al., 1996; Minnikin, 1982). However, most clinical isolates of *M. tuberculosis* do not produce PGL-tb, and therefore, it is unlikely that this compound plays a major role in the structural organization of the *M. tuberculosis* cell envelope. In contrast, DIM have been found in all clinical isolates tested (Daffé, 1988a; Goren et al., 1974). Their nonamphipathic character and abundance in the cell envelope have suggested that they are structural, providing a hydrophobic barrier around the *M. tuberculosis* cell and possibly a platform for anchoring other components of the cell envelope (Minnikin, 1982). Although this anchoring role remains to be demonstrated, one study has provided evidence that DIM is involved in the permeability barrier. Camacho et al. (Camacho, 2001) demonstrated that mutation of the *fadD26* gene, causing a DIM-less phenotype, increased the permeability of *M. tuberculosis* to chenodeoxycholate, a negatively charged hydrophobic compound that diffuses passively through lipid domains (Liu et al., 1996). This modification of the cell envelope also resulted in greater sensitivity to detergent (Camacho et al., 2001), a feature associated with an increase in permeability in gram-negative bacteria. These findings support the notion that DIM contributes to the permeability barrier formed by the cell envelope. However, the loss of DIM does not affect susceptibility to various antibiotics or reactive nitrogen derivatives, indicating that the overall protection conferred by the cell envelope was not greatly affected (Camacho et al., 2001).

Role of PGL-tb and Related Compounds in *M. tuberculosis* Virulence

The subcellular location of PGL-tb and DIM, partly exposed to the bacterial surface (Ortalo-Magné et al., 1996), makes them ideally positioned to interact with host cells and contribute to *M. tuberculosis* virulence. The same is true for *p*-HBAD which is released into the culture medium of *M. tuberculosis* (Constant et al., 2002). Several studies provide evidence that these three substances are virulence factors of *M. tuberculosis*, playing different roles during the infection of the host.

DIM and early multiplication within the host

The first work associating DIM with virulence was by Goren et al. (1974), who isolated a spontaneous mutant of H37Rv deficient in DIM synthesis after serial passage in vitro. This mutant strain was strongly attenuated in guinea pigs, with fewer lung lesions and lower bacilli counts in the spleen than a virulent *M. tuberculosis* Trudeau strain. However, the parental strain was lost during this study making direct comparison with the DIM-less mutant impossible and therefore any conclusion about the role of DIM hazardous. Nevertheless, these findings were supported 2 years later by the observation that coating avirulent *M. tuberculosis* H37Ra with a mixture of DIM and cholesteryl oleate resulted in the bacilli persisting at higher counts in mouse organs (Kondo and Kanai, 1976).

The direct demonstration of the importance of DIM for full virulence of *M. tuberculosis* came from the simultaneous work of two separate groups. Using a signature-transposon mutagenesis strategy, they looked for mutants attenuated for their multiplication in the mouse lung (Camacho et al., 1999; Cox et al., 1999). Among the strains with abnormal replication in vivo were mutants with insertions into genes involved in DIM biosynthesis and translocation. Intravenous infection of mice with a DIM-less mutant or a strain unable to translocate these lipids properly showed that both types of mutant were defective for growth in the lung during the initial phase of the infection (first 3 weeks) and that this defect was restricted to the lung (Cox et al., 1999). Indeed, wild-type and mutant strains multiplied similarly in both the spleen and liver (Cox et al., 1999). However, Rousseau et al. (2004) subsequently raised doubts about this role in tissue-specific replication. They observed that, following intranasal inoculation, DIM-less mutants exhibited a growth defect both in the lung and spleen of mice. The in vivo growth curves indicated that DIM was required for optimal multi-

plication during the first three weeks of infection in both organs but also for persistence in the lung. The reason for the discrepancy between the two studies is unclear but might be a consequence of the different protocols used. Indeed, Cox et al. (1999) used a high infectious dose (10^6 CFU) and multiplication in the spleen and liver was only a factor 10 between day 1 and day 21. In contrast, Rousseau et al. (2004) used a low infectious dose (10^4 CFU) and the wild-type CFU count increased by 100- and 1,000-fold in the lung and spleen during the first three weeks. This higher multiplication may have uncovered a difference masked in the first study. Alternatively, the different infection routes or genetic backgrounds of the strains used might explain the different behavior observed.

Rousseau et al. (2004) tried to identify the step of the *M. tuberculosis* infectious cycle in which DIM is required. They showed that a DIM-less mutant replicated like the wild-type parental strain in resting mouse macrophages but exhibited a very slight defect in macrophages previously activated by interferon-γ (IFN-γ) and tumor necrosis factor alpha (TNF-α) cytokines. This defect seemed to be associated with an increase in sensitivity to RNI: pretreatment of the activated macrophages with an inhibitor of RNI production abolished the difference detected in untreated cells. However, this result conflicts with previous findings that, in vitro, the same DIM-less mutant and the WT strain were equally sensitive to RNI (Camacho et al., 2001). Further effort is required to resolve this issue. In the same study, Rousseau et al. (2004) showed that infection of mouse macrophages or dendritic cells with DIM-less mutants induced a greater secretion of the proinflammatory cytokines TNF-α and interleukin-6 (IL-6) than that induced by WT *M. tuberculosis*. This implicates these molecules in the modulation of the early immune response. Pethe et al. (2004) identified two mutants with transposon insertions into the *fadD26* and *fadD28* genes by screening for strain unable to block the acidification of their phagosome (unlike the parental CDC1551 strain). A careful analysis of the pH in the vacuole containing the *fadD26* mutant showed that 3 hours after phagocytosis, the bacillus environment was pH 5.8, whereas for the wild-type strain, the phagosome pH was 6.5. This was not associated with a growth defect: the *fadD26* mutant and wild-type strains exhibited similar growth curves in mouse macrophages, consistent with previous findings by Rousseau et al. (2004). This defect in acidification arrest suggests that DIM may contribute to the blockage of phagosome maturation. However, in this study, as in others investigating the role of DIM during infection of the host, there were no experiments involving complementation of the mutation and analysis of the phenotype of the complemented strains. Therefore, the possibility of secondary mutations responsible for one or another of these phenotypes cannot be excluded.

The conclusion from these various studies is that DIM are clearly required for optimal early multiplication within the host, possibly by modulating trafficking of the phagosome and the cross-talk between the infected phagocyte and other cells of the immune system. However, whether this effect is direct through the interaction of the surface-exposed DIM with host cellular partners or indirect through the structuration of the mycobacterial cell envelope remains to be established. In either case, the molecular mechanisms of DIM action are still completely unknown.

PGL-tb and *p*-HBADs and the modulation of the host immune response

Unlike DIM, which have been found in all clinical isolates of *M. tuberculosis* tested, PGL-tb are produced by only a small proportion of strains (Constant et al., 2002; Daffé, 1988a). This suggests that this glycolipid is not required for pathogenesis. However, Goren et al. (1974) reported that synthesis of a truncated form of PGL-tb, a methoxy-phenolphthiocerol dimycocerosate, was associated with an attenuated phenotype in a guinea pig. Therefore, the production of compounds related to PGL-tb might interfere with the interaction between *M. tuberculosis* and its host. Consistent with this idea, several reports demonstrate that PGL obtained from various nontuberculous strains exhibited bioactivities in vitro that might contribute to virulence. For instance, the PGL-1 from *M. leprae* can neutralize hydroxyl- and superoxide radicals in vitro, a property shared by deacylated PGL-1 and to some extent by the carbohydrate moiety of the substance (Chan et al., 1989). The radicals involved are believed to be responsible in part for the bactericidal activities of phagocytic cells. Neill and Klebanoff (1988) showed that this property of PGL-1 could operate in vivo: the lipid prevented killing of staphylococci by human macrophages, apparently because it was able to neutralize hydroxyl radicals. *M. leprae* produces PGL-1 in abundance, possibly explaining its resistance to intracellular killing. Whether the related lipids of *M. kansasii* and *M. bovis* (and of some strains of *M. tuberculosis*) have similar capabilities is unknown. Moreover, PGL-1 is able to suppress the oxidative response, measured as the release of superoxide by human macrophages (Vachula et al., 1990), possibly explaining why this response is abnormally low in leprosy patients. This activity seems to be specific to

PGL-1 of *M. leprae*: PGL from *M. microti* and *M. kansasii*, which have different carbohydrate moieties, have no such activity, nor does the related nonglycosylated molecule (Vachula et al., 1989). PGL-1 has also been reported to be active in an indirect test of specific immunosuppression in leprosy: it inhibits the concanavalin A stimulation of lymphocytes from patients with lepromatous leprosy (Mehra et al., 1984). PGLs from *M. bovis* and *M. kansasii* were not active in this test. Specific suppression of cell-mediated immunity to the pathogen is a feature of lepromatous leprosy, but the relevance of the model used to test PGL-1 to human disease is not clear. In addition, PGLs from several species of mycobacteria, including PGL-1, nonspecifically inhibit lymphoproliferative responses to various stimuli, including several antigens and mitogens (Fournié et al., 1989). However, PGL-tb was not investigated in any of these studies.

A clear demonstration of the role of PGL-tb in *M. tuberculosis* virulence came from the work of Reed et al. (2004), who established that the production of PGL in *M. tuberculosis* was associated with a hypervirulent phenotype in the mouse model. Disruption of the *pks15/1* gene in a highly virulent clinical isolate (HN878) of the Beijing family synthesizing PGL-tb, resulted in a mutant strain attenuated for its ability to kill mice following aerosol infection. Interestingly, this phenotype was not associated with a defect in multiplication or persistence within the lung or spleen. Reed et al. (2004) showed that the synthesis of PGL-tb inhibited the production of the proinflammatory cytokines TNF-α, IL-12, IL-6, and MCP-1 by infected mouse macrophages. The authors were able to reproduce this effect in an in vitro assay in which mouse bone-marrow macrophages were incubated with apolar lipid extracts containing various amounts of PGL-tb. They observed a dose-dependent inhibition of proinflammatory cytokine release. Remarkably, this effect was dependent on the saccharide domain of PGL-tb, and the related PGL from BCG and DIM had no such effect. Consistent with these findings, Manca et al. (2004) showed that human monocytes stimulated with lipid extract from HN878, which contained PGL-tb, preferentially induced IL-4 and IL-13, two cytokines that deactivate phagocytes. In contrast, lipid extract from CDC 1551, unable to synthesize PGL-tb because of a mutation within *pks15/1*, induced more IL-12 and other molecules important for phagocyte activation (Manca et al., 2004). Although the role of PGL-tb was not directly tested in this study, these results, together with those of Reed et al. (2004), clearly support the idea that PGL-tb modulates the host immune response, delaying or impairing protective immunity and thereby leading to increased virulence. Extending these results, Tsenova et al. (2005) tested the role of PGL-tb in infection of the central nervous system (CNS). They demonstrated that the longer persistence of HN878 than CDC1551 or H37Rv in the CNS and its ability to induce more severe tuberculous meningitis following intrathecal infection was dependent on PGL-tb production. A HN878 mutant harboring an insertion into the *pks15/1* gene was attenuated in the rabbit model of tuberculous meningitis. In addition, this mutant strain also appeared to be impaired for dissemination from the brain to the lung. These results provide the first hint that PGL-tb might also be important for dissemination across the blood-brain barrier and multiplication within the CNS.

The saccharide moiety of PGL-tb seems to be required for the immunomodulation activities, and indeed, DIM and mycoside B do not exhibit the same bioactivities in vitro. This saccharide portion is found in *p*-HBAD II, therefore, it was interesting to question the role of this compound in virulence and immunomodulation. Stadthagen et al. (2006) addressed this issue by comparing the behavior of three mutants of *M. tuberculosis* strain Mt103 with transposon insertions within *Rv2959c*, and *Rv2949c* and upstream from *Rv2958c*. The *Rv2949c* mutant did not produce any form of *p*-HBAD but the *Rv2958c* mutant synthesized *p*-HBAD I and unmethylated *p*-HBAD I (UM-*p*-HBAD I) and the *Rv2959c* mutant only UM-*p*-HBAD I. The three mutants were not affected in their growth in macrophages or in mice but induced differential lung lesions at late time points (42 days postinfection). The most severe lesions were observed for the *Rv2958c* mutant, whereas the *Rv2959c* mutant behaved like the wild-type strain and the *Rv2949c* mutant exhibited an intermediate phenotype (Stadthagen et al., 2006). These phenotypes in mice were associated with the ability of the *Rv2958c* and *Rv2949c* mutants to induce higher proinflammatory cytokine production by infected bone marrow macrophages than the third mutant or wild-type strains (Stadthagen et al., 2006). These results showed that some forms of *p*-HBADs may, like PGL-tb, inhibit the proinflammatory response of infected macrophages and may thus contribute to the pathogenesis of *M. tuberculosis*. Like PGL also, the structure of the carbohydrate moieties seems to be crucial for the immunomodulatory activity of *p*-HBADs.

Although these studies provide insights into the structural determinants of PGL-tb and *p*-HBADs required for their functions, the overall mechanisms of action of these molecules remain to be characterized. The requirement for the saccharide moiety and the importance of its structure for the functions of PGL-tb and *p*-HBADs suggest that these compounds may interact with a phagocyte receptor. However, the cellular partner remains to be identified.

CONCLUSIONS

The availability of the *M. tuberculosis* genome sequence and the concomitant development of powerful genetic tools for mycobacteria were enormous leaps forward for studying the biology of *M. tuberculosis*, including the biosynthesis and role of PGL-tb and related compounds. The last 2 decades have been extremely fruitful, and it seems clear that almost all the genes and enzymes involved in the biosynthesis and translocation of PGL-tb and related compounds have now been identified. However, there is still a long way to go to decipher how these extremely complex enzymatic pathways work at the molecular level. For instance, the supramolecular organization of the proteins involved is unknown. More than 20 enzymes are required for the formation of PGL-tb, but we do not know how these enzymes work together. Several studies presented in this review indicate that some of them interact allowing the hand-to-hand transfer of the growing lipid chain. Whether these interactions are strong enough to allow the formation of large protein complexes remains to be established. If there are complexes, understanding their architecture, subcellular location, interaction with the transport machinery and changes in their composition according to the physiological state of the bacterium are important issues to address.

Other intriguing questions concern the interspecies diversity of the structure of PGL-tb and related compounds. For instance, the lipid core of DIM and PGL is conserved among the various mycobacterial species producing these compounds, but there are differences concerning the stereochemistry of several of the chiral centers (Daffé, 1988a; Daffé et al., 1984). This suggests that the enzymatic machinery catalyzing the formation of this lipid core exhibits subtle variations in its mechanism of action, variations that remain to be elucidated. The same is true for the saccharide domain of PGLs, which are either species specific or specific to a few species. So what are the enzymes involved in the formation of these domains? And what is the significance of this diversity in the lipid core structure and on the saccharide domain? Several studies have provided the first hints that these differences may have a profound impact on the role of PGL-tb and related compounds during the infectious process. However, the molecular mechanisms explaining these apparently different roles have not been elucidated.

Solving these questions may uncover the link between the structure of compounds exposed at the mycobacterial cell surface and the specific features of the various diseases provoked by the mycobacterial species producing PGL.

REFERENCES

Azad, A. K., T. D. Sirakova, N. D. Fernandes, and P. E. Kolattukudy. 1997. Gene knockout reveals a novel gene cluster for the synthesis of a class of cell wall lipids unique to pathogenic mycobacteria. *J. Biol. Chem.* 272:16741–16745.

Azad, A. K., T. D. Sirakova, L. M. Rogers, and P. E. Kolattukudy. 1996. Targeted replacement of the mycocerosic acid synthase gene in *Mycobacterium bovis* BCG produces a mutant that lacks mycosides. *Proc. Natl. Acad. Sci. USA* 93:4787–4792.

Braibant, M., P. Gilot, and J. Content. 2000. The ATP binding cassette (ABC) transport systems of *Mycobacterium tuberculosis*. *FEMS Microbiol. Rev.* 24:449–467.

Brennan, P. J. 1988. Mycobacterium and other actinomycetes, p. 203–298. *In* C. Ratledge and S. G. Wilkinson (ed.), *Microbial Lipids*. Academic Press, London, United Kingdom.

Buglino, J., K. C. Onwueme, J. A. Ferraras, L. E. N. Quadri, and C. D. Lima. 2004. Crystal structure of PapA5, a phthiocerol dimycocerosyl transferase from *Mycobacterium tuberculosis*. *J. Biol. Chem.* 279:30634–30642.

Camacho, L. R., P. Constant, C. Raynaud, M.-A. Lanéelle, J.-A. Triccas, B. Gicquel, M. Daffé, and C. Guilhot. 2001. Analysis of phthiocerol dimycocerosate locus of *Mycobacterium tuberculosis*: evidence that this lipid is involved in the cell wall permeability barrier. *J. Biol. Chem.* 276:19845–19854.

Camacho, L. R., D. Ensergueix, E. Perez, B. Gicquel, and C. Guilhot. 1999. Identification of a virulence gene cluster of *Mycobacterium tuberculosis* by signature-tagged transposon mutagenesis. *Mol. Microbiol.* 34:257–267.

Chan, J., T. Fujiwara, P. Brennan, M. McNiel, S. J. Turco, and J.-C. Sibille. 1989. Microbial glycolipids: possible virulence factors that scavenge oxygen radicals. *Proc. Natl. Acad. Sci. USA* 86:2453–2457.

Cole, S. T., R. Brosch, J. Parkhill, T. Garnier, C. Churcher, D. Harris, S. V. Gordon, K. Eiglmeier, S. Gas, C. E. Barry III, F. Tekaia, K. Badcock, D. Basham, D. Brown, T. Chillingworth, R. Connor, R. Davies, K. Devlin, T. Feltwell, S. Gentles, N. Hamlin, S. Holroyd, T. Hornsby, K. Jagels, A. Krogh, J. McLean, S. Moule, L. Murphy, K. Oliver, J. Osborne, M. A. Quail, M.-A. Rajandream, J. Rogers, S. Rutter, K. Seeger, J. Skelton, R. Squares, S. Squares, J. E. Sulston, K. Taylor, S. Whitehead, and B. G. Barrell. 1998. Deciphering the biology of *Mycobacterium tuberculosis* from the complete genome sequence. *Nature* 393:537–544.

Constant, P., E. Perez, W. Malaga, M.-A. Lanéelle, O. Saurel, M. Daffé, and C. Guilhot. 2002. Role of the *pks15/1* gene in the biosynthesis of phenolglycolipids in the *M. tuberculosis* complex: evidence that all strains synthesize glycosylated p-hydroxybenzoic methyl esters and that strains devoid of phenolglycolipids harbour a frameshift mutation in the *pks15/1* gene. *J. Biol. Chem.* 277:38148–38158.

Cox, J. S., B. Chen, M. McNeil, and W. R. Jacobs, Jr. 1999. Complex lipid determines tissue-specific replication of *M. tuberculosis* in mice. *Nature* 402:79–83.

Daffé, M., and P. Draper. 1998. The envelope layers of mycobacteria with reference to their pathogenicity. *Adv. Microb. Physiol.* 39:131–203.

Daffé, M., and M. A. Lanéelle. 1988a. Distribution of phthiocerol diester, phenolic mycosides and related compounds in Mycobacteria. *J. Gen. Microbiol.* 134:2049–2055.

Daffé, M., M.-A. Lanéelle, C. Lacave, and G. Lanéelle. 1988b. Monoglycosyl diacyl phenol-phthiocerol of *Mycobacterium tuberculosis* and *Mycobacterium bovis*. *Biochem. Biophys. Acta* 958:443–449.

Daffé, M., M.-A. Laneelle, J. Roussel, and C. Asselineau. 1984. Specific lipids from *Mycobacterium ulcerans*. *Ann. Microbiol.* 135A:191–201.

Daffé, M., and A. Lemassu. 2000. Glycomicrobiology of the mycobacterial cell surface: structure and biological activities of the cell envelope glycoconjugates, p. 225–273. *In* R. J. Doyle (ed.), *Glycomicrobiology*. Plenum Press, New York, NY.

Fernandes, N. D., and P. E. Kolattukudy. 1997. Methylmalonyl coenzyme A selectivity of cloned and expressed acyltransferase and β-ketoacyl synthase domains of mycocerosic acid synthase from *Mycobacterium bovis* BCG. *J. Bacteriol.* 179:7538–7543.

Fitzmaurice, A. M., and P. E. Kolattukudy. 1997. Open reading frame 3, which is adjacent to the mycoserosic acid synthase gene, is expressed as an acyl coenzyme A synthase in *Mycobacterium bovis* BCG. *J. Bacteriol.* 179:2608–2615.

Fitzmaurice, A. M., and P. E. Kolattukudy. 1998. An acyl-CoA synthase (*acoas*) gene adjacent to the mycoserosic acid synthase (*mas*) locus is necessary for mycoserosyl lipid synthesis in *Mycobacterium tuberculosis* var. *bovis* BCG. *J. Biol. Chem.* 273: 8033–8039.

Fournié, J.-J., E. Adams, R. J. Mullins, and A. Basten. 1989. Inhibition of human lymphoproliferative responses by mycobacterial phenolic glycolipids. *Infect. Immun.* 57:3653–3659.

Gastambide-Odier, M., J. M. Delaumény, and E. Lederer. 1963. Biosynthesis of phthiocerol: incorporation of methionine and propionic acid. *Chem. Ind.* 69:1285–1286.

Gastambide-Odier, M., P. Sarda, and E. Lederer. 1967. Biosynthèse des aglycones des mycosides A et B. *Bull. Soc. Chim. Biol.* 49:849–864.

Goren, M. B., O. Brokl, and W. B. Schaefer. 1974. Lipids of putative relevance to virulence in *Mycobacterium tuberculosis*: phthiocerol dimycoserosate and the attenuation indicator lipid. *Infect. Immun.* 9:150–158.

Hotter, G. S., B. J. Wards, P. Mouat, G. S. Besra, J. Gomes, M. Singh, S. Bassett, P. Kawakami, P. R. Wheeler, G. W. de Lisle, and D. M. Collins. 2005. Transposon mutagenesis of Mb0100 at the *ppe-nrp* locus in *Mycobacterium bovis* disrupts phthiocerol dimycocerosate (PDIM) and glycosylphenol-PDIM biosynthesis, producing an avirulent strain with vaccine properties at least equal to those of *M. bovis* BCG. *J. Bacteriol.* 187:2267–2277.

Jain, M., and J. S. Cox. 2005. Interaction between polyketide synthase and transporter suggests coupled synthesis and export of virulence lipid in *M. tuberculosis*. *PLoS Pathog.* 1:12–19.

Jeevarajah, D., J. H. Patterson, M. J. McConville, and H. Billman-Jacobe. 2002. Modification of glycopeptidolipids by an O-methyltransferase of *Mycobacterium smegmatis*. *Microbiology* 148:3079–3087.

Kolattukudy, P. E., N. D. Fernandes, A. K. Azad, A. M. Fitzmaurice, and T. D. Sirakova. 1997. Biochemistry and molecular genetics of cell-wall lipid biosynthesis in mycobacteria. *Mol. Microbiol.* 24:263–270.

Kondo, E., and K. Kanai. 1976. A suggested role of a host-parasite lipid complex in mycobacterial infection. *Jpn. J. Med. Sci. Biol.* 29:199–210.

Lefèvre, P., O. Denis, L. De Wit, A. Tanghe, P. Vandenbussche, J. Content, and K. Huygen. 2000. Cloning of the gene encoding a 22-kilodalton cell surface antigen of *Mycobacterium bovis* BCG and analysis of its potential for DNA vaccination against tuberculosis. *Infect. Immun.* 68:3674–3679.

Liu, J., C. E. Barry III, G. S. Besra, and H. Nikaido. 1996. Mycolic acid structure determines the fluidity of the mycobacterial cell wall. *J. Biol. Chem.* 271:29545–29551.

Manca, C., M. B. Reed, S. Freeman, B. Mathema, B. Kreiswirth, C. E. Barry III, and G. Kaplan. 2004. Differential monocyte activation underlies strain-specific *Mycobacterium tuberculosis* pathogenesis. *Infect. Immun.* **72:**5511–5514.

Mathur, M., and P. E. Kolattukudy. 1992. Molecular cloning and sequencing of the gene for mycocerosic acid synthase, a novel fatty acid elongating multifunctional enzyme, from *Mycobacterium tuberculosis* var. *bovis* Bacillus Calmette-Guerin. *J. Biol. Chem.* **267:**19388–19395.

Mehra, V., P. J. Brennan, E. Rada, J. Convit, and B. R. Bloom. 1984. Lymphocyte suppression in leprosy induced by unique *M. leprae* glycolipid. *Nature* **308:**194–196.

Minnikin, D. E. 1982. Lipids: complex lipids, their chemistry, biosynthesis and roles, p. 95–184. *In* C. Ratledge and J. Stanford (ed.), *The Biology of Mycobacteria.* Academic Press, London, United Kingdom.

Minnikin, D. E., L. Kremer, L. G. Dover, and G. S. Besra. 2002. The methyl-branched fortifications of *Mycobacterium tuberculosis*. *Chem. Biol.* **9:**545–553.

Neill, M. A., and S. J. Klebanoff. 1988. The effect of phenolic glycolipid-1 from *Mycobacterium leprae* on the antimicrobial activity of human macrophages. *J. Exp. Med.* **167:**30–42.

Onwueme, K. C., J. A. Ferraras, J. Buglino, C. D. Lima, and L. E. N. Quadri. 2004. Mycobacterial polyketide-associated proteins are acyltransferases: proof of principle with *Mycobacterium tuberculosis* PapA5. *Proc. Natl. Acad. Sci. USA* **101:**4608–4613.

Onwueme, K. C., C. J. Vos, J. Zurita, J. A. Ferraras, and L. E. N. Quadri. 2005a. The dimycocerosate ester polyketide virulence factors of mycobacteria. *Prog. Lipid. Res.* **44:**259–302.

Onwueme, K. C., C. J. Vos, J. Zurita, C. E. Soll, and L. E. N. Quadri. 2005b. Identification of phthiodiolone ketoreductase, an enzyme required for production of mycobacterial diacyl phthiocerol virulence factors. *J. Bacteriol.* **187:**4760–4766.

Ortalo-Magné, A., A. Lemassu, M.-A. Lanéelle, F. Bardou, G. Silve, P. Gounon, G. Marchal, and M. Daffé. 1996. Identification of the surface-exposed lipids on the cell envelope of *Mycobacterium tuberculosis* and other mycobacterial species. *J. Bacteriol.* **178:**456–461.

Pelicic, V., J.-M. Reyrat, and B. Gicquel. 1998. Genetic advances for studying *Mycobacterium tuberculosis* pathogenicity. *Mol. Microbiol.* **28:**413–420.

Perez, E., P. Constant, F. Laval, A. Lemassu, M. A. Lanéelle, M. Daffé, and C. Guilhot. 2004a. Molecular dissection of the role of two methyltransferases in the biosynthesis of phenolglycolipids and phthiocerol dimycoserosate in the *Mycobacterium tuberculosis* complex. *J. Biol. Chem.* **279:**42584–42592.

Perez, E., P. Constant, A. Lemassu, F. Laval, M. Daffé, and C. Guilhot. 2004b. Characterization of three glycosyltransferases involved in the biosynthesis of the phenolic glycolipid antigens from the *Mycobacterium tuberculosis* complex. *J. Biol. Chem.* **279:**42574–42583.

Pethe, K., D. L. Swenson, S. Alonso, J. Anderson, C. Wang, and D. G. Russell. 2004. Isolation of *Mycobacterium tuberculosis* mutants defective in the arrest of phagosome maturation. *Proc. Natl. Acad. Sci. USA* **103:**13642–13647.

Rainwater, D. L., and P. E. Kolattukudy. 1983. Synthesis of mycocerosic acids from methylmalonyl Coenzyme A by cell-free extracts of *Mycobacterium tuberculosis* var. *bovis* BCG. *J. Biol. Chem.* **258:**2979–2985.

Rainwater, D. L., and P. E. Kolattukudy. 1985. Fatty acid biosynthesis in *Mycobacterium tuberculosis* var. *bovis* BCG: purification and characterization of a novel fatty acid synthase, mycocerosic acid synthase, which elongates n-fatty acyl-CoA with methylmalonyl-CoA. *J. Biol. Chem.* **260:**616–623.

Rao, A., and A. Ranganathan. 2004. Interaction studies on proteins encoded by the phthiocerol dimycocerosate locus of *Mycobacterium tuberculosis*. *Mol. Gen. Genomics* **272:**571–579.

Reed, M. B., P. Domenech, C. Manca, H. Su, A. K. Barczak, B. N. Kreiswirth, G. Kaplan, and C. E. Barry III. 2004. A glycolipid of hypervirulent tuberculosis strains that inhibits the innate immune response. *Nature* **431:**84–87.

Rousseau, C., T. D. Sirakova, V. S. Dubey, P. Bordat, P. E. Kolattukudy, B. Gicquel, and M. Jackson. 2003. Virulence attenuation of two Mas-like polyketide synthase mutants of *Mycobacterium tuberculosis*. *Microbiology* **149:**1837–1847.

Rousseau, C., N. Winter, N. Pivert, Y. Bordat, O. Neyrolles, P. Avé, M. Huerre, B. Gicquel, and M. Jackson. 2004. Production of phthiocerol dimycocerosates protects *Mycobacterium tuberculosis* from the cidal activity of reactive nitrogen intermediates produced by macrophages and modulates the early immune response to infection. *Cell. Microbiol.* **6:**277–287.

Siebert, M., A. Bechthold, M. Melzer, U. May, U. Berger, G. Schröder, J. Schröder, K. Severin, and L. Heide. 1992. Ubiquinone biosynthesis. Cloning of the genes coding for chorismate pyruvate-lyase and 4-hydroxybenzoate octaprenyl transferase from *Escherichia coli*. *FEBS Lett.* **307:**347–350.

Siebert, M., K. Severin, and L. Heide. 1994. Formation of 4-hydroxybenzoate in *Escherichia coli*: characterization of the *ubiC* gene and its encoded enzyme chorismate pyruvate-lyase. *Microbiology* **140:**897–904.

Simeone, R., P. Constant, W. Malaga, C. Guilhot, M. Daffé, and C. Chalut. 2007a. Molecular dissection of the biosynthetic relationship between phthiocerol and phthiodiolone dimycocerosates and their critical role in the virulence and permeability of *Mycobacterium tuberculosis*. *FEBS J.* **274:**1957–1969.

Simeone, R., P. Constant, C. Guilhot, M. Daffé, and C. Chalut. 2007b. Identification of the missing trans-acting enoyl reductase required for phthiocerol dimycocerosate and phenolglycolipid biosynthesis in *Mycobacterium tuberculosis*. *J. Bacteriol.* **189:**4597–4602.

Sirakova, T. D., V. S. Dubey, M. H. Cynamon, and P. E. Kolattukudy. 2003a. Attenuation of *Mycobacterium tuberculosis* by disruption of a *mas*-like gene or a chalcone synthase-like gene, which causes deficiency in dimycocerosyl phthiocerol synthesis. *J. Bacteriol.* **185:**2999–3008.

Sirakova, T. D., V. S. Dubey, H. J. Kim, M. H. Cynamon, and P. E. Kolattukudy. 2003b. The largest open reading frame (*pks12*) in the *Mycobacterium tuberculosis* genome is involved in pathogenesis and dimycocerosyl phthiocerol synthesis. *Infect. Immun.* **71:**3794–3801.

Smith, D. W., H. M. Randall, A. P. MacLennan, and E. Lederer. 1960. Mycosides: a new class of type-specific glycolipids of mycobacteria. *Nature* **186:**887–888.

Stadthagen, G., M. Jackson, P. Charles, F. Boudou, N. Barilone, M. Huerre, P. Constant, A. Liav, I. Bottova, J. Nigou, T. Brando, G. Puzo, M. Daffé, P. Benjamin, S. Coade, R. S. Buxton, R. E. Tascon, A. Rae, B. D. Robertson, D. B. Lowrie, D. B. Young, B. Gicquel, and R. Griffin. 2006. Comparative investigation of the pathogenicity of three *Mycobacterium tuberculosis* mutants defective in the synthesis of *p*-hydroxybenzoic acid derivatives. *Microbes Infect.* **8:**2245–2253.

Stadthagen, G., J. Kordulakova, R. Griffin, P. Constant, I. Bottova, N. Barilone, B. Gicquel, M. Daffé, and M. Jackson. 2005. *p*-Hydroxybenzoic acid synthesis in *Mycobacterium tuberculosis*. *J. Biol. Chem.* **49:**40699–40706.

Sulzenbacher, G., S. Canaan, Y. Bordat, O. Neyrolles, G. Stadthagen, V. Roig-Zamboni, J. Rauzier, D. Maurin, F. Laval, M. Daffe, C. Cambillau, B. Gicquel, Y. Bourne, and M. Jackson. 2006. LppX is a lipoprotein required for the translocation of phthiocerol dimycocerosates to the surface of *Mycobacterium tuberculosis*. *EMBO J.* **25:**1436–1444.

Thurman, P. F., W. Chai, J. R. Rosankiewicz, H. J. Rogers, A. M. Lawson, and P. Draper. 1993. Possible intermediates in the biosynthesis of mycoside B by *Mycobacterium microti. Eur. J. Biochem.* **212:**705–711.

Trivedi, O. A., P. Arora, V. Sridharan, R. Tickoo, D. Mohanty, and R. S. Gokhale. 2004. Enzymatic activation and transfer of fatty acids as acyl-adenylates in mycobacteria. *Nature* **428:**441–445.

Trivedi, O. A., P. Arora, A. Vats, M. Z. Ansari, R. Tickoo, V. Sridharan, D. Mohanty, and R. S. Gokhale. 2005. Dissecting the mechanism and assembly of a complex virulence mycobacterial lipid. *Mol. Cell* **17:**631–643.

Tseng, T.-T., K. S. Gratwick, J. Kollman, D. Park, D. H. Nies, A. Goffeau, and M. H. Saier, Jr. 1999. The RND permease superfamily: an ancient, ubiquitous and diverse family that includes human disease and development proteins. *J. Mol. Microbiol. Biotechnol.* **1:**107–125.

Tsenova, L., E. Ellison, R. Harbacheuski, A. L. Moreira, N. Kurepina, M. B. Reed, B. Mathema, C. E. Barry III, and G. Kaplan. 2005. Virulence of selected mycobacterium tuberculosis clinical isolates in the rabbit model of meningitis is dependent on phenolic glycolipid produced by the bacilli. *J. Infect. Dis.* **192:**98–106.

Vachula, M., T. J. Holzer, and B. R. Andersen. 1989. Suppression of monocyte oxidative response by phenolic glycolipid I of *Mycobacterium leprae. J. Immunol.* **142:**1696–1701.

Vachula, M., T. J. Holzer, L. Kizlaitis, and B. R. Andersen. 1990. Effect of *Mycobacterium leprae*'s phenolic glycolipid-I on interferon-gamma augmentation of monocyte oxidative responses. *Int. J. Lepr. Other Mycobact. Dis.* **58:**342–346.

Waddell, S. J., G. A. Chung, K. J. C. Gibson, M. J. Everett, D. E. Minnikin, G. S. Besra, and P. D. Butcher. 2005. Inactivation of polyketide synthase and related genes results in the loss of complex lipids in *Mycobacterium tuberculosis* H37Rv. *Lett. Appl. Microbiol.* **40:**201–206.

Yadav, G., R. S. Gokhale, and D. Mohanty. 2003. Computational approach for prediction of domain organization and substrate specificity of modular polyketide synthases. *J. Mol. Biol.* **328:**335–363.

The Mycobacterial Cell Envelope
Edited by M. Daffé and J.-M. Reyrat
© 2008 ASM Press, Washington, DC

Chapter 18

Sulfated Metabolites from *Mycobacterium tuberculosis*: Sulfolipid-1 and Beyond

Carolyn R. Bertozzi and Michael W. Schelle

Sulfated metabolites are abundant in higher eukaryotes, where they play roles in cell-to-cell communication. By contrast, sulfated metabolites are rare in prokaryotes, with only a handful identified to date. Mycobacteria are one of only three genera in which sulfated molecules have been found, the other two being the plant pathogen *Xanthomonas oryzae* and the plant symbiont *Sinorhizobium meliloti* (Fisher and Long, 1992; Schelle and Bertozzi, 2006; Shen et al., 2002). In *Mycobacterium tuberculosis*, the abundant cell wall component sulfolipid-1 (SL-1) has received the most attention. This curious molecule was associated with the virulence of *M. tuberculosis* shortly after its identification. However, as discussed in this chapter, a convincing link has yet to be established. Unlike many of the other molecules discussed in this book, such as phthiocerol dimycocerosate (PDIM) and trehalose-6,6′-dimycolate (TDM), a function for SL-1 has not yet been proposed and the phenotypes of SL-1-deficient mutants in model organisms have elicited contradictory interpretations. Many of the genes involved in SL-1 production have now been identified, providing a platform for defining its functions with respect to pathogenesis. This chapter highlights the recent advances in the understanding of SL-1 biosynthesis and discusses the potential biological significance of SL-1 and other sulfated metabolites in the context of both historical observations and modern experiments.

DISCOVERY AND CHARACTERIZATION OF SULFOLIPID-1

M. tuberculosis isolates that exhibit high virulence were shown early on to fix the cationic dye neutral red (Dubos and Middlebrook, 1948). Interestingly, the bacteria were able to absorb the dye from neutral or even alkali solutions, indicating the presence of a highly acidic surface component that could form a salt with the positively charged dye. The correlation of neutral red fixation and virulence led Gardner Middlebrook of the National Jewish Hospital in Denver, Colorado, to fractionate virulent *M. tuberculosis* and identify the substances responsible for coordinating the dye (Middlebrook et al., 1959). He found that the neutral red dye would selectively partition to the organic layer of an aqueous/organic cell wall extract from virulent H37Rv bacteria. Initially, the anionic material was assumed to be a phospholipid. However, elemental analysis of the fraction exhibiting the greatest dye fixing ability indicated the presence of sulfur, not phosphorus. The incorporation of radiolabeled sulfate into the same neutral red-fixing fraction and the appearance of a characteristic sulfuric acid absorption band upon analysis by infrared spectroscopy supplied further evidence of a sulfate-containing lipid. Thus, the neutral red-fixing metabolite of *M. tuberculosis* was determined to be a sulfolipid, also termed sulfatide (not to be confused with the human sulfolipid of the same name), and proposed to be vital to the pathogen's virulence.

Association of Sulfolipids with Virulence

Shortly after this initial discovery, Gangadharam and colleagues (1963) explored the relationship between sulfatide content and strain virulence. Clinical isolates from Britain and India were analyzed for sulfolipid production and screened for virulence in guinea pigs. In most of the strains tested, the index of infectivity paralleled the sulfatide content, that is, strains producing greater amounts of sulfolipid also

Carolyn R. Bertozzi • Departments of Chemistry and Molecular and Cell Biology, Howard Hughes Medical Institute, University of California, Berkeley, Berkeley, CA 94720-1460.   Michael W. Schelle • Department of Chemistry, University of California, Berkeley, Berkeley, CA 94720-1460.

gave rise to larger bacterial burdens in the lungs of the guinea pigs. A follow-up study by Goren and coworkers (1974) revealed a similar trend in his assessment of a much larger pool of isolates. This study examined *M. tuberculosis* phage-type A and I strains from Britain and India, respectively. The phage-type A strains showed increased virulence and enhanced sulfatide production when compared with the phage-type I strains. These two initial studies sparked curiosity regarding the function of the sulfatides and provoked Goren and colleagues to elucidate the chemical structures of the sulfolipids.

Despite these initial reports suggesting a link between sulfatide content and virulence, the story became more complicated when the correlation was again probed by Goren years after his initial study (Goren et al., 1982). He noted that in the *M. tuberculosis* phage-type B strains, virulence and sulfolipid content were not linked. Indeed, the place of origin of the strain more strongly correlated to virulence than the amount of sulfolipid production, casting doubt on the validity of previous claims associating pathogenicity and sulfatide level. Nonetheless, early speculation that sulfatides were involved in virulence catalyzed numerous studies aimed at probing their functions.

Elucidation of the SL-1 Structure

Ten years after the initial identification of a sulfolipid from *M. tuberculosis*, Goren extended the work of Middlebrook by further purifying and structurally characterizing the sulfated constituents of *M.*

tuberculosis (Goren, 1970a, 1970b; Goren et al., 1971; Goren et al., 1976a). The sulfatide previously isolated was found to be, in fact, a collection of several molecules closely related in structure and composition. Five sulfatide families were evident from thin-layer chromatography (TLC) analysis and named sulfolipids I to V according to their mobility on the TLC plate. The most abundant and lipophilic sulfatide, SL-1 (Fig. 1), was found to constitute 0.7% of the dry weight of the bacteria. SL-1 is easily extracted from cells with hexane and is thought to reside as a free lipid associated with the mycolic acid layer of the cell envelope. Goren achieved enhanced purification of SL-1 by separation of crude hexane-decylamine extracts on DEAE anion exchange resin, rather than fractionation by silica chromatography as employed by Middlebrook. The improved separation allowed for the isolation of pure SL-1 in sufficient quantities to undertake structural studies.

To determine the structure of SL-1, Goren relied heavily on infrared spectroscopy, nuclear magnetic resonance spectroscopy (NMR), gas chromatography-mass spectrometry (GC-MS), and chemical degradation studies. He determined that SL-1 consisted of a nonreducing core disaccharide acylated with four lipid moieties and elaborated with a characteristic sulfate ester. The disaccharide is the common metabolite trehalose, which consists of two glucose monomers joined through an 1,1-α,α linkage (Fig. 1). Trehalose is predominantly found in bacteria and plants, where it serves as an osmoprotectant and thermoprotectant (Wolf et al., 2003; Woodruff et al., 2004). Mycobacteria incorporate trehalose into a

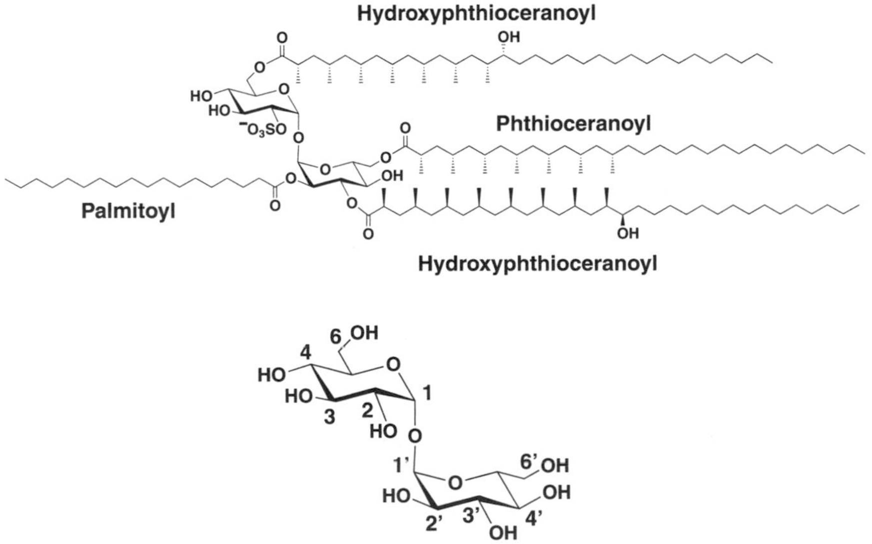

Figure 1. The chemical structure of sulfolipid-1 (SL-1) and trehalose. The hydroxy-bearing carbons in the trehalose ring are numbered.

variety of important metabolites, including TDM which, when cyclopropanated, is responsible for cording of mycobacterial cultures (Glickman et al., 2000; Noll et al., 1956). The position of the sulfate ester was deduced by saponification and permethylation of SL-1. Spontaneous desulfation followed by hydrolysis of the glycosidic bond gave rise to 2,3,4,6-tetra-O-methylglucose and 3,4,6-tri-O-methylglucose, thus fixing the sulfate ester at the 2-position of trehalose.

SL-1 is actually a family of lipoforms based on the structure illustrated in Fig. 1. The heterogeneity in SL-1 stems from variations in the lengths and elaborations of the acyl substituents, typically varying by the addition of two or three methylene units (Goren et al., 1971). The lipid moieties on SL-1 consist of a straight chain acyl group, typically a C_{16} palmitoyl or a C_{18} stearoyl group, and three unique methyl-branched fatty acids. Termed phthioceranic acids, these exotic lipids are isotactic polypropylene oligomers consisting of four to 10 isopropylene repeats, capped with a C_{16-18} straight-chain alkyl tail. A hydroxy group often modifies the first carbon of the straight chain, giving rise to a more polar acid dubbed hydroxyphthioceranic acid. The methyl-branched series of the (hydroxy)phthioceranates are dextrorotatory, indicating the L stereochemistry seen in the homologous *M. tuberculosis*-derived phthienoic acids (Fray and Polgar, 1956; Stallberg-Stenhagen, 1954). The D configuration for the hydroxy group of the hydroxyphthioceranates was suggested based on the positive shift in optical rotation compared with unsubstituted phthioceranate (Asselineau and Asselineau, 1978). However, the absolute stereochemistry of the hydroxy and methyl groups has not been definitively assigned.

The placement of the acyl substituents on trehalose was determined by an elaborate study in which SL-1 was partially degraded by methanolysis and the resulting products analyzed by chemical derivatization and mass spectrometry (Goren et al., 1976a). Interestingly, the sulfatides SL-III, SL-IV, and SL-V were found to be produced by the selective methanolysis of SL-1 and identified as tri-, di-, and mono-acyl trehalose-2-sulfates, respectively. Further analysis of the desulfated degradation products identified the position of each acyl chain on the trehalose-2-sulfate (T2S) core. The straight-chain acyl group of SL-1 resides at the 2′-position of trehalose with a phthioceranic acid group at the 3′-position and two hydroxyphthioceranic acids occupying the 6- and 6′-positions. SL-II was revealed to be a close relative of SL-1 with a trehalose-2-sulfate-2′-palmitate core but containing only hydroxyphthioceranic acids at the 3′-, 6′- and 6-positions. The definition of SL-1 has

since grown to encompass SL-II, which is now considered an SL-1 lipoform.

BIOSYNTHESIS OF SL-1

Despite the elucidation of the structure of SL-1 in 1976, the biosynthesis of the metabolite remained a mystery for the next 25 years. The sequencing of the *M. tuberculosis* genome in 1998 provided the tools necessary to draw links from gene to protein and metabolite (Cole et al., 1998). Figures 2 and 3 represent our current understanding of the SL-1 biosynthetic pathway. To provide a historical picture, each step is described in the order in which they were identified.

Pks2, the SL-1 Polyketide Synthase

The first gene identified in the SL-1 biosynthetic pathway encodes the polyketide synthase Pks2 (see chapter 6) (Sirakova et al., 2001). *pks2* (*Rv3825c*) was discovered by Mathur and Kolattukudy (1992) based on its homology to the previously described polyketide synthase Mas, which is responsible for forming the methyl-branched mycocerosic acids found on PDIM. Based on this homology, Pks2 was assumed to be responsible for the synthesis of the methyl-branched acyl groups of SL-1, the (hydroxy)phthioceranic acids. Indeed, disruption of *pks2* in *M. tuberculosis* provided a mutant lacking SL-1, as determined by metabolic labeling with radiolabeled propionate or sulfate. Furthermore, GC-MS analysis of saponified total lipid extracts from the wild-type H37Rv and Δ*pks2* strains revealed the absence of (hydroxy)phthioceranic acids from the mutant.

The Lipid Transporter MmpL8 and the SL$_{1278}$ Intermediate

Genomic screens have identified many critical pathways and metabolites used by *M. tuberculosis* to establish a successful infection in the host. A signature-tag mutagenesis (STM) screen identified *mmpL7* as a gene essential for *M. tuberculosis* growth in a mouse model of infection (Cox et al., 1999). A member of a class of large transmembrane proteins (the MmpL family, see chapter 11), MmpL7 was found to be essential for transporting PDIM to the cell wall. In an extension of the previous STM screen, a second *mmpL* gene was determined to be essential for virulence in mice, *mmpL8* (*Rv3823c*) (Converse et al., 2003). The location of *mmpL8* in the *M. tuberculosis* genome immediately suggested a function for the

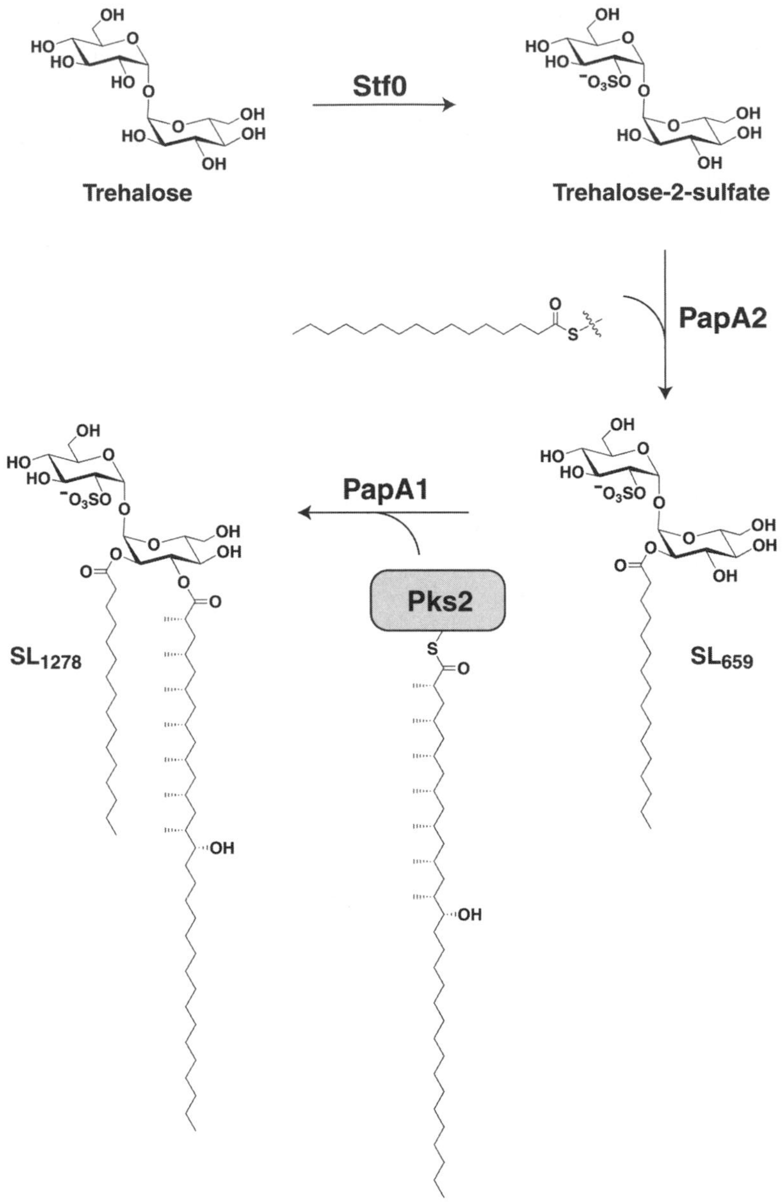

Figure 2. The biosynthesis of SL_{1278}. Trehalose is sulfated by Stf0 to form trehalose-2-sulfate (T2S), which then is acylated with a palmitoyl group by PapA2 at the 2′ position to form SL_{659}. This product is then acylated at the 3′ position by PapA1 with a hydroxyphthioceranoyl group synthesized by Pks2 to yield SL_{1278}.

protein. *mmpL8* lies downstream of the *pks2* gene, apparently in the same operon, suggesting a role for MmpL8 in SL-1 biosynthesis. Indeed, an *M. tuberculosis* strain with a mutation in *mmpL8* was not able to synthesize SL-1. Unlike the Δ*pks2* mutant, mutants lacking MmpL8 accumulated an unidentified lipid that was significantly more polar than SL-1. This novel lipid incorporated both radiolabeled propionate and sulfate, indicating a methyl-branched sulfolipid. Analysis by high-resolution Fourier transform-ion cyclotron resonance mass spectrometry (FT-ICR MS) identified a lipid envelope ranging from an m/z of 1278 to 1432. Thus, the lipid was termed

SL_{1278} based on the exact mass of the lowest molecular weight species. This mass corresponds to a sulfated trehalose acylated with a palmitoyl and a hydroxyphthioceranoyl group. A later study by Barry and coworkers (Domenech et al., 2004) defined the structure of SL_{1278} through purification and analysis by GC-MS and NMR spectroscopy. Corroborating Goren's SL-1 structure, SL_{1278} was shown to contain a T2S core acylated at the 2′- and 3′-positions with a palmitoyl and either a hydroxyphthioceranoyl or phthioceranoyl group. Unfortunately, the authors were unable to precisely define the positions of these lipids and, therefore, deferred to the structure by Goren

A. Intracellular acylation

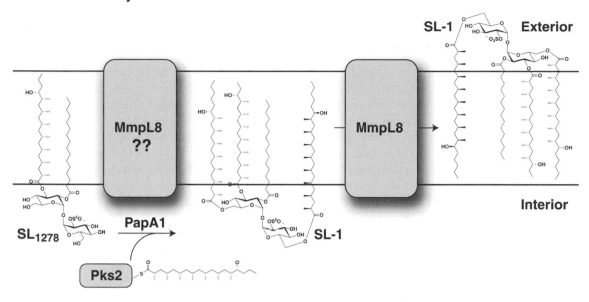

B. Extracellular acylation

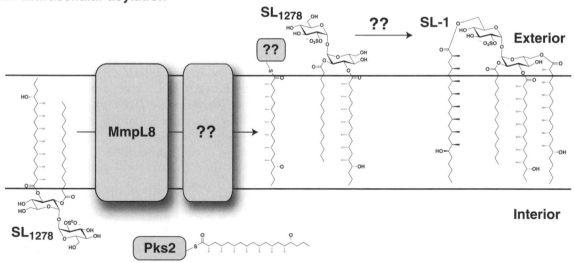

Figure 3. Two possible pathways for completion of SL-1 biosynthesis. (A) Intracellular acylation of SL_{1278} is accomplished in the presence of MmpL8 by PapA1 and Pks2. SL-1 is then transported by MmpL8. (B) Extracellular acylation is accomplished by transport of SL_{1278} by MmpL8. The hydroxyphthioceranoyl moiety is also transported by an unknown protein and then coupled to SL_{1278} by an extracellular acyltransferase.

placing the palmitoyl group at the 2′-position with the (hydroxy)phthioceranoyl group at the 3′-position.

SL_{1278} was shown to be an intermediate along the pathway to SL-1 by pulsing wild-type cells with radiolabeled sulfate (Converse et al., 2003). TLC analysis showed the metabolic incorporation of sulfate into SL_{1278}, which, in turn, was transformed into SL-1. Further demonstration that SL_{1278} is a natural intermediate on the SL-1 biosynthesis pathway

comes from the identification of the SL_{1278} envelope in an earlier FT-ICR MS-based screen for sulfated metabolites in *M. tuberculosis* (Mougous et al., 2002b). The analysis of the $\Delta mmpL8$ mutant indicated that, similar to MmpL7, MmpL8 was likely to be responsible for lipid transport. Surface lipid extracts from $\Delta mmpL8$ mutants demonstrated that SL_{1278} is found only in the interior of the cell. Thus, it was proposed that MmpL8 transports SL_{1278} to the surface of the cell, where SL-1 synthesis is completed (Fig. 3B).

StfO and the Stf Family of Sulfotransferases

The sulfate ester functionality distinguishes SL-1 from other well-characterized mycobacterial lipids. The sulfate group has stimulated interest as to the molecular function of SL-1, as well as aided in the detection of the molecule due to specific labeling with heavy (^{34}S) or radioactive (^{35}S) sulfate. A mass spectrometry-based screen for uncharacterized sulfated metabolites in mycobacteria used heavy sulfate to identify SL_{1278} as well as T2S (Mougous et al., 2002b; Mougous et al., 2004). The presence of T2S in *M. tuberculosis* extracts suggests that sulfation of trehalose is the initial step in SL-1 biosynthesis. The structure of SL_{1278} added validity to this hypothesis (Domenech et al., 2004). However, confirmation that T2S is an intermediate in SL-1 biosynthesis awaited the identification of the associated sulfotransferase.

Mycobacteria encode a large number of putative sulfotransferases, collectively termed the Stf family (Mougous et al., 2002a). The *M. tuberculosis* genome encodes four *stf* genes (*stf0* to *stf3*; *Rv0295c*, *Rv3529c*, *Rv1373*, and *Rv2267c*, respectively). As a comparison, the fast-growing avirulent *M. smegmatis* encodes only two *stf* genes (*stf0,1*) whereas the opportunistic human pathogen *M. avium* encodes nine *stf* genes (*stf0*, *stf1*, and *stf4* through *stf10*). Three of the *M. tuberculosis* Stf family members (*stf1* to *stf3*) were discovered based on their homology to human GlcNAc-6-sulfotransferase (Mougous et al., 2002a). A fourth sulfotransferase, *stf0*, was identified based on homology to LpsS, a sulfotransferase from *Sinorhizobium meliloti* that sulfates lipopolysaccharide (Cronan and Keating, 2004; Mougous et al., 2004). These enzymes contain characteristic sulfotransferase motifs, including the 5′-phosphosulfate binding loop (5′PSB) and the 3′-phosphate binding domain (3′PB), which are essential for recognition of the universal sulfate donor 3′-phosphoadenosine-5′-phosphosulfate (PAPS) (Chapman et al., 2004). Although this chapter focuses on SL-1-associated processes, the enzymes involved in PAPS production are also essential to sulfated metabolite production and a wealth of literature is emerging regarding the structure, function, and regulation of the sulfate assimilation machinery (Carroll et al., 2005a; Carroll et al., 2005b; Mougous et al., 2006a; Pinto et al., 2004; Schelle and Bertozzi, 2006; Sun et al., 2005; Sun and Leyh, 2006; Williams et al., 2002).

To identify the sulfotransferase responsible for SL-1 production, an *M. tuberculosis* mutant in *stf0*

was generated and Stf0 was expressed, purified, and crystallized (Mougous et al., 2004). The Δ*stf0* mutant failed to produce both SL-1 and T2S. Biochemical characterization of StfO confirmed its sulfotransferase activity with trehalose as the preferred substrate. This selectivity was further demonstrated by analysis of the crystal structure of StfO bound to trehalose. The protein exists as a dimer both in solution and in the crystal structure. The 2′- and 3′-hydroxys of trehalose make extensive contacts with amino acid side chains from both protomers of the StfO dimer, indicating that acylation of trehalose would preclude binding to the enzyme. Collectively, these data indicate that sulfation of free trehalose by StfO is the first step of SL-1 biosynthesis.

The discovery of StfO raises some intriguing questions about the role of sulfation and the function of T2S in nontuberculous bacteria. StfO and T2S can be found in both *M. smegmatis* and *M. avium* (Mougous et al., 2002a; Mougous et al., 2002b). Indeed, a gene showing high homology to *stf0* has been found in every sequenced mycobacterial genome except in the decaying genome of *M. leprae*, which contains a pseudogene. The prevalence of *stf0* indicates a role for T2S in mycobacteria other than simply as a metabolic precursor of SL-1. T2S is known to be an osmoprotectant in haloalkaliphilic archaea and may serve a similar function in mycobacteria, which have the ability to survive desiccation and high osmolite concentrations (Desmarais et al., 1997). Alternatively, T2S may be acylated to form SL-1-like structures, although it is important to note that neither *M. avium* or *M. smegmatis* encode orthologues of *pks2* or *mmpL8*.

Acylation of the T2S Core

After the identification of StfO as the enzyme responsible for the first step of SL-1 biosynthesis, research refocused on the order of addition of lipids onto the T2S core. Based on the structure of SL_{1278}, the 2′-palmitoyl and 3′-(hydroxy)phthioceranoyl groups must be added before the acylation of the 6- and 6′-positions (Converse et al., 2003; Domenech et al., 2004). The order of addition of the acyl chains was solved by examining the region of the genome already attributed to SL-1 biosynthesis. *pks2* and *mmpL8* bookend a third gene in the *M. tuberculosis* genome, *papA1* (*Rv3824c*) (Fig. 4) (Cole et al., 1998). The *M. tuberculosis* genome encodes five genes annotated as polyketide-associated proteins (termed *papA1-5*). As their name suggests, each *papA* gene resides in close proximity to a *pks* gene. The PapA-associated Pks proteins lack the thioesterase domain

Figure 4. Genetic arrangement of the SL-1 biosynthetic cluster.

necessary for cleavage of the polyketide from the protein and the resulting polyketides are not found as free acids (Kolattukudy et al., 1997; Minnikin et al., 2002). In a proof-of-principle experiment, the mycocerosic acid synthase (Mas)-associated PapA5 was shown to be an acyltransferase involved in PDIM biosynthesis (Buglino et al., 2004; Onwueme et al., 2004; Trivedi et al., 2005). PapA5 selectively acylated the phthiocerol backbone of PDIM with the mycocerosic acids synthesized by Mas.

By analogy, PapA1 was assumed to transfer the (hydroxy)phthioceranic acid produced by Pks2 to T2S (Kumar et al., 2007). However, purified PapA1 showed no activity toward T2S or trehalose. Further examination of the SL-1 operon revealed a second *papA* gene downstream of the *pks2-papA1-mmpL8* gene cluster annotated as *papA2* (*Rv3820c*). Purification and biochemical characterization identified PapA2 as capable of adding a straight-chain fatty acid to the 2′-position of T2S. Importantly, PapA2 was not able to acylate unsulfated trehalose. Analysis of the *M. tuberculosis* Δ*papA2* mutant revealed a novel sulfated metabolite on the pathway to SL-1. As shown in Fig. 2, addition of the straight chain lipid to T2S produces an intermediate termed SL_{659} based on its mass. This intermediate was observed at low levels in wild-type *M. tuberculosis* extracts but was absent from the Δ*papA2* mutant. In addition, the mutant was unable to produce SL_{1278} or SL-1, but retained the ability to synthesize T2S.

With the second step in SL-1 biosynthesis established, it was suspected that the esterification of the 3′-(hydroxy)phthioceranic acid would follow and that this addition would be accomplished by PapA1. Indeed, incubation of synthetic SL_{659} with purified PapA1 and an acyl donor led to the production of an SL_{1278} analogue with acyl groups at the 2′- and 3′-positions (Kumar et al., 2007). Further evidence came from examination of the Δ*papA1* mutant. The mutant strain was unable to produce SL_{1278} and SL-1 but retained the ability to synthesize SL_{659} and T2S as expected based on the scheme in Fig. 2. These observations establish the first steps of the biosynthesis of SL-1, beginning with the sulfation of the 2-position of trehalose by Stf0 to form T2S, followed by the installation of the 2′-palmitate onto T2S by PapA2 to form SL_{659}. The synthesis of SL_{1278} is completed by the PapA1-mediated addition of the (hydroxy)phthioceranoyl group produced by Pks2 to SL_{659}.

The Missing Link between SL_{1278} and SL-1

Despite having defined a clear biosynthetic pathway that converts trehalose to SL_{1278}, subsequent steps elaborating this molecule to form SL-1, as well as transport to the cell surface, remain to be elucidated. Recently, Gap, a small integral membrane protein from *M. smegmatis* was shown to be required for transport of glycopeptidolipids to the cell surface (Sonden et al., 2005). A *gap* homologue is present in the SL-1 biosynthetic gene cluster (*Rv3821*), although it remains to be seen whether this gene is associated with SL-1 transport or synthesis. Indeed, there are a number of proteins that may be responsible for mediating the final steps in SL-1 biosynthesis. Two hypotheses are currently favored: intracellular lipid addition (Fig. 3A) or extracellular lipid addition (Fig. 3B). In the first case, SL_{1278} would be elaborated to SL-1 on the interior of the cell, and then fully formed SL-1 would be transported to the exterior by MmpL8. However, the phenotype of the Δ*mmpL8* mutant indicates that the addition of the 6- and 6′-(hydroxy)phthioceranoyl groups is dependent on MmpL8 (Converse et al., 2003; Domenech et al., 2004). In order to account for this observation, MmpL8 may facilitate acylation by recruiting the necessary acyltransferases or by directly catalyzing acyl transfer itself. Indeed, MmpL7 has been shown to associate with the polyketide synthase PpsE and is thought to act as both a lipid transporter and a molecular scaffold for PDIM biosynthesis (Jain and Cox, 2005). As the lipids occupying the 6- and 6′-positions of SL-1 are synthesized by Pks2, a logical acyltransferase would be PapA1 (Kumar et al., 2007; Sirakova et al., 2001). However, biochemical assays have shown PapA1 to modify the 3′-position of SL_{659}, not the 6- and 6′-positions. Thus, to invoke PapA1 as the final SL-1 acyltransferase would require that its association with MmpL8 alters its regiospecificity.

The second possibility is that an extracellular acyltransferase converts SL_{1278} to SL-1 after transport of SL_{1278} to the cell's exterior by MmpL8 (Fig. 3B).

Extracellular acyltransferases are known in *M. tuberculosis*, most notably the antigen 85 complex (Belisle et al., 1997). This enzyme catalyzes the esterification of mycolic acids to trehalose to form trehalose-6,6'-dimycolate. In fact, the antigen 85 complex has been suggested as a possible SL-1 acyltransferase (Converse et al., 2003). Several problems exist with this hypothesis. First, antigen 85 would have to display tremendous substrate flexibility to accommodate the heavily elaborated SL_{1278} in place of free trehalose. Because the (hydroxy)phthioceranoyl groups added in the last step of SL-1 biosynthesis are likely produced by Pks2 within the cell's interior, these lipids would require transport to the outer membrane in order to serve as substrates of an extracellular acyltransferase. No candidates for this transporter have been identified to date. Although there is meager evidence supporting either of these hypotheses, neither possibility can be ruled out. Current efforts to identify the missing acyltransferase(s) are ongoing.

DEFINING THE MOLECULAR FUNCTION OF SL-1

A review of the literature in search of a defined molecular function for SL-1 reveals few firm conclusions. In one of the earliest biological studies, Goren et al. (1976b) reported that administration of 200 to 400 μg of SL-1/ml to mouse peritoneal macrophages results in enlarged and foamy cells. The SL-1-treated macrophages also acquired the ability to fix neutral red dye, indicating that some proportion of SL-1 was adsorbed on the macrophage cell surface. However, the most interesting claim from this study was that SL-1-treated macrophages were unable to fuse yeast-containing phagosomes with lysosomes. This observation suggests that SL-1 is a critical virulence factor, because inhibition of phago-lysosomal fusion is a hallmark of *M. tuberculosis* infection (Armstrong and Hart, 1971; Goren, 1977). Yet, later these researchers retracted this claim, stating that the observed effects in macrophages were artifacts (Goren et al., 1987a, 1987b). Also, the high concentration of SL-1 used in the study casts doubt on the physiological relevance of any observations.

Conflicting data regarding the biological effects of SL-1 have been reported in other studies of immune cells as well. Several studies have tried to address the role of SL-1 in leukocyte priming and superoxide (O_2^-) release. A collection of papers by Pabst et al. (Brozna et al., 1991; Pabst et al., 1988) indicate that purified SL-1 inhibits superoxide release by primed human monocytes upon secondary stimulation. SL-1 was able to reverse monocyte activation

by a variety of agents including lipopolysaccharide (LPS), interferon-γ, tumor necrosis factor alpha, and interleukin-1β. However, data from Andersen et al. (Zhang et al., 1988; Zhang et al., 1991; Zhang et al., 1994) suggest the opposite response of monocytes to SL-1. In a series of studies, the authors describe an increase of superoxide release from human neutrophils primed with either LPS or a substimulatory dose of SL-1. Both groups used subtly different leukocyte differentiation and isolation procedures, as well as different assay conditions, which were cited as potential reasons for the disparity in their results.

Many studies of SL-1 have investigated its synergism with the structurally similar lipid TDM (cord factor) as a toxic immunostimulant. An early report by Kato and Goren (1974) suggested that toxicity of TDM is enhanced when coadministered to mice with SL-1. In the same report, SL-1 was found to decrease respiration and oxidative phosphorylation in purified murine liver mitochondria. This result was not entirely expected because SL-1 alone does not display toxicity in mice. The addition of albumin to the mitochondria reaction medium inhibited the toxicity of SL-1, and therefore, it was speculated that in vivo, SL-1 is nontoxic owing to nonspecific adsorption by albumin or related proteins.

Contrary reports suggest that SL-1 decreases TDM toxicity (Yarkoni et al., 1979). As in the previous discrepancies with SL-1 function, the differences in the experimental design, specifically the use of an oil:water emulsion and the breed of mice, may have contributed to the disparate results. Recently, oil:water emulsions of SL-1 and TDM were again studied for granulomatogenicity in mice (Okamoto et al., 2006). Consistent with the work of Yarkoni and coworkers, SL-1 decreased granuloma formation induced by TDM, leading the authors to speculate that the immunostimulatory role of TDM could be offset by the immunosuppressive activity of SL-1 (Yarkoni et al., 1979). Although this is an attractive hypothesis for the role of SL-1, the relevance of these toxicity studies to an actual *M. tuberculosis* infection is not clear.

THE PHENOTYPES OF SL-1 DEFICIENT MUTANTS

Mutation of *pks*2 Shows No Obvious Change in Phenotype

The recent availability of various mutants in SL-1 biosynthesis has provided a new platform from which to study the function of SL-1 in vivo. Mutants from each known step in the pathway have been tested for virulence in animal models of *M. tubercu-*

losis infection. The initial correlation between the abundance of SL-1 and the virulence of the strain suggested an important role for SL-1 (Gangadharam et al., 1963; Goren et al., 1974). However, when the Δ*pks2* mutant was analyzed by aerosol infection of BALB/c mice, the phenotype matched that of the wild-type bacteria (Rousseau et al., 2003). The mutant strain was able to replicate in the lung and persist through 105 days of the infection with no significant difference from wild-type H37Rv. A second study assessing the virulence of the Δ*pks2* strain involved injection of the bacteria in the tail vein of the more resistant C57BL/6 mouse strain (Converse et al., 2003). Again, the mutant displayed the same phenotype as wild-type bacteria, with no replication or persistence defect in the lung, liver, or spleen. Recently, the virulence of the Δ*pks2* strain was again examined, this time assessing time to death. This parameter is often the most sensitive measure of the virulence of a strain, and more precisely measures the ability of the bacteria to persist to the termination of infection. Again, mice infected with the Δ*pks2* mutant survived the infection as well as mice infected with wild-type bacteria (Schelle and Bertozzi, personal communication).

Although the mouse model of *M. tuberculosis* infection can provide an accurate measure of the virulence of certain strains, other animal models more closely mimic the human disease. The guinea pig offers more human-like granuloma structures than the mouse, including a more complete repertoire of CD1 molecules, and is a common model for evaluating mutants for virulence as well as testing vaccines. Given the lack of distinct phenotype for the Δ*pks2* mutant in mouse models of infection, the mutant was further analyzed in a guinea pig model of infection (Rousseau et al., 2003). However, the mutant again showed no difference in virulence from wild-type bacteria in any organ analyzed. Furthermore, the granuloma structures after 100 days of infection mimicked granulomas formed by wild-type bacteria. These extensive studies concluded that SL-1 has no role in either the mouse or guinea pig models of *M. tuberculosis* infection. However, a few points must be considered before casting judgment on SL-1. First, the recently elucidated SL-1 biosynthetic pathway indicates that the Δ*pks2* mutant still produces the sulfolipid SL$_{659}$. Furthermore, the acyltransferase PapA1 is still functional in the Δ*pks2* mutant, leaving the possibility that SL$_{659}$ may be further elaborated into an SL-1 analog.

Mutants Lacking MmpL8 Show Attenuated Virulence

In vivo analysis of the Δ*pks2* mutants left the SL-1 community questioning the importance of the molecule that had previously garnered so much attention. However, a twist occurred in the SL-1 story with the subsequent analysis of the Δ*mmpL8* mutant. The *mmpL8* gene was identified from a screen for mutants unable to survive in the context of an infection (Converse et al., 2003). When the Δ*mmpL8* mutant was examined in a tail vein infection of C57BL/6 mice, it showed decreased persistence in the mouse as compared with the wild-type Erdman strain. This virulence defect was specific for the persistent stage of the infection, because bacterial replication early in the infection was not effected by the mutation. A second study of the Δ*mmpL8* mutant in the H37Rv background through aerosol infection in C57BL/6 or B6D2/F1 mice again showed attenuated persistence, as manifested by a significantly extended time to death (Domenech et al., 2004). Unlike the previous study, in which the Δ*mmpL8* mutant showed attenuation at day 40 in the lungs of the mice, the second study saw no attenuation up to 182 days postinfection. The authors indicated that this difference is likely due to differences in the infections, with tail vein injection leading to much higher initial bacterial loads.

The attenuation of the Δ*mmpL8* mutant in the persistent stage of *M. tuberculosis* infection contrasts with the reported phenotype of the Δ*pks2* mutants. However, as noted earlier, the Δ*pks2* mutant has not been assessed to the terminal stage of infection and, therefore, may also prove to be attenuated in persistence. Yet, another potential reason for the observed differences is the dramatic accumulation of SL$_{1278}$ in the Δ*mmpL8* mutant strain, an intermediate that may interact with the immune system (Converse et al., 2003; Gilleron et al., 2004). The CD1 family of antigen-presentation molecules resemble major histocompatibility complex (MHC) class I molecules and activate both CD8$^+$ and CD4$^+$ T cells. In humans, five CD1 variants have been found, CD1a through -e, each showing specificity for certain lipids (De Libero and Mori, 2005). SL$_{1278}$ was found to be presented by CD1b on the surface of *M. tuberculosis*-infected antigen-presentation cells. Also, the presentation of SL$_{1278}$ by CD1b stimulated bactericidal CD8$^+$ T cells from patients infected with *M. tuberculosis* (Gilleron et al., 2004).

Given the potential activity of SL$_{1278}$ as a T cell-restricted antigen and the resulting bactericidal effects of activated T cells, it stands to reason that the Δ*mmpL8* mutant would be attenuated in infection models. However, mice lack the CD1b variant and only possess the related CD1d lipid presentation molecule, which has not been shown to bind SL$_{1278}$ (De Libero and Mori, 2005). The CD1d variant differs from CD1b predominantly in the size of the lipid binding groove, with CD1b accommodating a larger lipid than CD1d (Moody et al., 2005). However, mycobacterial-derived phosphatidylinositol mannoside, known to be presented by human CD1b, was recently shown to also activate NK T cells through both human and murine CD1d presentation (Fischer et al., 2004; Sieling et al., 1995). Analysis of the $\Delta mmpL8$ mutant in the CD1d$^{-/-}$ mouse will help resolve this potential conflict. These studies are currently under way.

Sulfolipid Deficiency Does Not Alter Virulence

Interpretation of the phenotypes of both the $\Delta pks2$ and $\Delta mmpL8$ mutants is difficult owing to the presence of SL-1 intermediates in the bacteria. In the case of the $\Delta mmpL8$ mutant, the overabundance of SL$_{1278}$, rather than the absence of SL-1, may be responsible for the phenotypic differences. Likewise, in the $\Delta pks2$ mutant, further analysis of its metabolome has revealed the presence of SL$_{659}$, which may have a biological activity similar to SL$_{1278}$ or could be further elaborated by PapA1 to form SL-1 analogs. For these reasons, virulence analysis of a true sulfolipid-deficient mutant was necessary. Considering the pathway in Fig. 2, two enzymes are logical targets for the creation of sulfolipid-free bacteria, Stf0 and PapA2. The $\Delta stf0$ mutant lacks the ability to synthesize T2S and, therefore, blocks any potential modifications resulting in a functional SL-1 analogue. However, given the prevalence of $stf0$ in mycobacteria other than $M.$ $tuberculosis$, one could argue that any $\Delta stf0$ phenotype is a result of the lack of T2S rather than SL-1. Therefore, the $\Delta papA2$ mutant is a logical control because that strain retains the ability to synthesize T2S but lacks any elaborated sulfolipids.

Data from our lab show that neither $\Delta stf0$ or $\Delta papA2$ contribute to virulence in a mouse model of infection (Kumar et al., 2007; Schelle and Bertozzi, personal communication). BALB/c mice infected through aerosol with wild-type Erdman, $\Delta stf0$, or $\Delta papA2$ strains showed no significant difference in the ability of the bacteria to grow or persist through the time to death of the animals. These results suggest that SL-1 is either of no importance to $M.$ $tuberculosis$ or that it only contributes to tuberculosis disease in a human host. Indeed, sulfated host-specific metabolites have been identified in other organisms, including the pathogenic rice pathogen $Xanthomonas$ $oryzae$ (Shen et al., 2002). A summary of all the SL-1 mutants reported to date, and their metabolite profiles and phenotypes, is provided in Table 1.

ALTERNATE SULFATED METABOLITES IN MYCOBACTERIA

Mycobacteria are known to produce additional sulfated structures. As stated previously, most mycobacterial species possess multiple sulfotransferases (Mougous et al., 2002a). However, matching a sulfotransferase to a specific sulfated metabolite is difficult. This is mainly due to the relative dearth of data on sulfated metabolites in mycobacteria. Radiolabeling with ^{35}S-sulfate and analysis by TLC has been the most popular means of identifying sulfated metabolites, ever since the initial work of Middlebrook in the late 1950s. Not until recently has a more sensitive and sophisticated approach been adopted. Mass spectrometry has been used by a number of groups to help identify sulfolipids from $M.$ $tuberculosis$ (Converse et al., 2003; Domenech et al., 2004; Kumar et al., 2006; Mougous et al., 2002b; Mougous et al., 2004; Mougous et al., 2006b). The most sensitive and accurate of these techniques is the labeling of two identical cultures with either ^{32}S-sulfate or heavy ^{34}S-sulfate, followed by analysis of the extracts with FT-ICR MS (Mougous et al., 2002b). Peaks corresponding to metabolites that contain a sulfur atom will shift by exactly 2.00 mass units in the extracts from ^{34}S-sulfate-labeled cultures. This technique was used to identify SL-1, T2S, SL$_{1278}$, SL$_{659}$, and an uncharacterized sulfated molecule with a mass of 881, termed S881, from crude cell extracts.

S881 is the first sulfated molecule identified from $M.$ $tuberculosis$ extracts that is not related to SL-1. The presence of additional sulfated metabolites was expected due to the three remaining Stf sulfo-

Table 1. Metabolite profiles and phenotypes of SL-1 mutants

Strain	Trehalose	T2S	SL$_{687}$	SL$_{1278}$	SL-1	Surface sulfolipid	Effect on phenotype[a]
Wild-type	+	+	+	+	+	+	−
$\Delta stf0$	+	−	−	−	−	−	None
$\Delta papA2$	+	+	−	−	−	−	None
$\Delta pks2$	+	+	+	−	−	?	None
$\Delta papA1$	+	+	+	−	−	−	None
$\Delta mmpL8$	+	+	+	+++	−	−	↑TTD

[a]Relative to wild-type phenotype.

transferases. Disruption of the *stf3* gene eliminated synthesis of S881, identifying Stf3 as the S881 sulfotransferase (Mougous et al., 2006b). Mutants in *stf3* were evaluated for virulence by both tail vein infection of BALB/c mice and aerosol infection of C57BL/6 mice. Surprisingly, the Δ*stf3* mutants showed accelerated bacterial growth in the initial stages of the infection as well as a shortened time to death of the animals, indicating that the deletion of *stf3* causes hypervirulence. These unexpected results have hastened efforts to elucidate the structure of S881 because it may be involved in modulating the host response to *M. tuberculosis* infection.

While the precise roles of Stf1 and Stf2 have not yet been elucidated, Stf2 is suggested to be a glycolipid sulfotransferase, as partially purified Stf2 sulfates glucosylceramide and galactosylceramide (Rivera-Marrero et al., 2002). Stf2, like Stf3, is an *M. tuberculosis*-specific sulfotransferase. However, no *M. tuberculosis*-specific lipids are known to be sulfated by Stf2 and mutants in Δ*stf2* show no aberrant sulfation (Mougous et al., 2002a). The ability of Stf2 to sulfate a mammalian glycolipid suggests that the sulfotransferase may act on host molecules in order to alter the host glycolipid repertoire. Unfortunately, there are no reports yet that address this interesting hypothesis.

Sulfated Metabolites of Nontuberculous Mycobacteria

Although *M. tuberculosis* is the most extensively studied species of the mycobacterial genus, several other mycobacteria have medical importance owing to their synergism with human immunodeficiency virus. One such species is the opportunistic environmental mycobacterium *M. avium*, which preferentially infects individuals with compromised immunity. This pathogen also encodes nine putative sulfotransferases, the largest number of sulfotransferases of any sequenced mycobacterial species (Mougous et al., 2002a). Furthermore, a clinical isolate of *M. avium* synthesizes a novel sulfated glycopeptidolipid (GPL), dubbed S-GPL (Fig. 5) (Khoo et al., 1999). Interestingly, S-GPL was overproduced in the same strain after the infected individual underwent ethambutol treatment. The structure of the metabolite is based on the common GPL core structure discussed extensively in chapter 19. This core structure is elaborated with an N-terminal β-hydroxy-C$_{33}$-acyl chain, a methylated rhamnose group appended to the alaninol residue, and a 6-deoxytalose moiety attached to the *allo*-threonine residue. The sulfate addition was found at the 4-position of the deoxytalose residue. Unfortunately, owing to the genomic plasticity of the

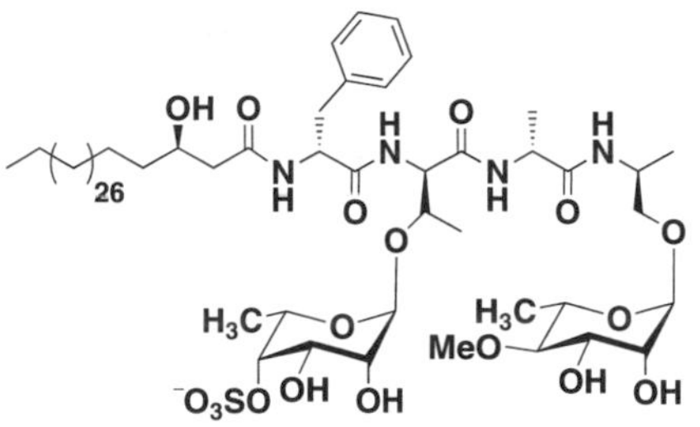

Figure 5. The chemical structure of S-GPL.

M. avium genome, reliable gene replacement technologies have only recently been described (Otero et al., 2003; Wu et al., 2006). Therefore, the Stf responsible for the sulfation of S-GPL has not yet been identified. Also, the function of a sulfated GPL structure can only be hypothesized because little data exist for this class of sulfated compounds. Possible roles involve modulating known effects of GPLs, including sliding motility and biofilm formation (Carter et al., 2003; Yamazaki et al., 2006). Apart from *M. avium*, a distinct sulfated GPL was identified from a clinical isolate of *M. fortuitum*, strengthening the hypothesis that sulfated metabolites enable mycobacteria to cause disease in humans (Lopez Marin et al., 1992).

CONCLUDING REMARKS

Recently, genetic tools have sped the discovery of the biomolecules essential for SL-1 biosynthesis. Unfortunately, this has not yet increased our understanding of the role of SL-1 in *M. tuberculosis* infection. Current animal models do not reflect many important aspects of tuberculosis disease, including the ability of the bacteria to spread between individuals, persist for long periods of time within a human granuloma, and reactivate after persistence. As our knowledge of human immunity grows, better cellular and in vivo models for *M. tuberculosis* virulence will be created. It is hoped that these models will generate new lines of inquiry and help reveal the roles of SL-1 and the yet unclassified sulfated metabolites in *M. tuberculosis* and other medically relevant mycobacteria.

REFERENCES

Armstrong, J. A., and P. D. Hart. 1971. Response of cultured macrophages to Mycobacterium tuberculosis, with observations on fusion of lysosomes with phagosomes. *J. Exp. Med.* **134:** 713–740.

Asselineau, C., and J. Asselineau. 1978. Trehalose-containing glycolipids. *Prog. Chem. Fats Other Lipids* **16:**59–99.

Belisle, J. T., V. D. Vissa, T. Sievert, K. Takayama, P. J. Brennan, and G. S. Besra. 1997. Role of the major antigen of Mycobacterium tuberculosis in cell wall biogenesis. *Science* **276:**1420–1422.

Brozna, J. P., M. Horan, J. M. Rademacher, K. M. Pabst, and M. J. Pabst. 1991. Monocyte responses to sulfatide from Mycobacterium tuberculosis: inhibition of priming for enhanced release of superoxide, associated with increased secretion of interleukin-1 and tumor necrosis factor alpha, and altered protein phosphorylation. *Infect. Immun.* **59:**2542–2548.

Buglino, J., K. C. Onwueme, J. A. Ferreras, L. E. Quadri, and C. D. Lima. 2004. Crystal structure of PapA5, a phthiocerol dimycocerosyl transferase from Mycobacterium tuberculosis. *J. Biol. Chem.* **279:**30634–30642.

Carroll, K. S., H. Gao, H. Chen, J. A. Leary, and C. R. Bertozzi. 2005a. Investigation of the iron-sulfur cluster in Mycobacterium tuberculosis APS reductase: implications for substrate binding and catalysis. *Biochemistry* **44:**14647–14657.

Carroll, K. S., H. Gao, H. Chen, C. D. Stout, J. A. Leary, and C. R. Bertozzi. 2005b. A conserved mechanism for sulfonucleotide reduction. *PLoS Biol.* **3:**e250.

Carter, G., M. Wu, D. C. Drummond, and L. E. Bermudez. 2003. Characterization of biofilm formation by clinical isolates of Mycobacterium avium. *J. Med. Microbiol.* **52:**747–752.

Chapman, E., M. D. Best, S. R. Hanson, and C. H. Wong. 2004. Sulfotransferases: structure, mechanism, biological activity, inhibition, and synthetic utility. *Angew. Chem. Int. Ed. Engl.* **43:**3526–3548.

Cole, S. T., R. Brosch, J. Parkhill, T. Garnier, C. Churcher, D. Harris, S. V. Gordon, K. Eiglmeier, S. Gas, C. E. Barry, 3rd, F. Tekaia, K. Badcock, D. Basham, D. Brown, T. Chillingworth, R. Connor, R. Davies, K. Devlin, T. Feltwell, S. Gentles, N. Hamlin, S. Holroyd, T. Hornsby, K. Jagels, A. Krogh, J. McLean, S. Moule, L. Murphy, K. Oliver, J. Osborne, M. A. Quail, M. A. Rajandream, J. Rogers, S. Rutter, K. Seeger, J. Skelton, R. Squares, S. Squares, J. E. Sulston, K. Taylor, S. Whitehead, and B. G. Barrell. 1998. Deciphering the biology of Mycobacterium tuberculosis from the complete genome sequence. *Nature* **393:**537–544.

Converse, S. E., J. D. Mougous, M. D. Leavell, J. A. Leary, C. R. Bertozzi, and J. S. Cox. 2003. MmpL8 is required for sulfolipid-1 biosynthesis and Mycobacterium tuberculosis virulence. *Proc. Natl. Acad. Sci. USA* **100:**6121–6126.

Cox, J. S., B. Chen, M. McNeil, and W. R. Jacobs, Jr. 1999. Complex lipid determines tissue-specific replication of Mycobacterium tuberculosis in mice. *Nature* **402:**79–83.

Cronan, G. E., and D. H. Keating. 2004. Sinorhizobium meliloti sulfotransferase that modifies lipopolysaccharide. *J. Bacteriol.* **186:**4168–4176.

De Libero, G., and L. Mori. 2005. Recognition of lipid antigens by T cells. *Nat. Rev. Immunol.* **5:**485–496.

Desmarais, D., P. E. Jablonski, N. S. Fedarko, and M. F. Roberts. 1997. 2-Sulfotrehalose, a novel osmolyte in haloalkaliphilic archaea. *J. Bacteriol.* **179:**3146–3153.

Domenech, P., M. B. Reed, C. S. Dowd, C. Manca, G. Kaplan, and C. E. Barry. 2004. The role of MmpL8 in sulfatide biogenesis and virulence of Mycobacterium tuberculosis. *J. Biol. Chem.* **279:**21257–21265.

Dubos, R. J., and G. Middlebrook. 1948. Cytochemical reaction of virulent tubercle bacilli. *Am. Rev. Tuberc.* **58:**698–699.

Fischer, K., E. Scotet, M. Niemeyer, H. Koebernick, J. Zerrahn, S. Maillet, R. Hurwitz, M. Kursar, M. Bonneville, S. H. Kaufmann, and U. E. Schaible. 2004. Mycobacterial phosphatidylinositol mannoside is a natural antigen for CD1d-restricted T cells. *Proc. Natl. Acad. Sci. USA* **101:**10685–10690.

Fisher, R. F., and S. R. Long. 1992. Rhizobium-plant signal exchange. *Nature* **357:**655–660.

Fray, G. I., and N. Polgar. 1956. Synthesis of (+)-2(L)-4(L)-dimethyldocosanoic acid, an oxidation product of mycolipenic acid. *Chem. Ind.* 22–23.

Gangadharam, P. R., M. L. Cohn, and G. Middlebrook. 1963. Infectivity, pathogenicity and sulpholipid fraction of some Indian and British strains of tubercle bacilli. *Tubercle* **44:**452–455.

Gilleron, M., S. Stenger, Z. Mazorra, F. Wittke, S. Mariotti, G. Bohmer, J. Prandi, L. Mori, G. Puzo, and G. De Libero. 2004. Diacylated sulfoglycolipids are novel mycobacterial antigens stimulating CD1-restricted T cells during infection with Mycobacterium tuberculosis. *J. Exp. Med.* **199:**649–659.

Glickman, M. S., J. S. Cox, and W. R. Jacobs. 2000. A novel mycolic acid cyclopropane synthetase is required for cording, persistence, and virulence of Mycobacterium tuberculosis. *Mol. Cell* **5:**717–727.

Goren, M. B. 1970a. Sulfolipid-I of Mycobacterium tuberculosis, strain H37Rv.1. Purification and properties. *Biochim. Biophys. Acta* **210:**116–126.

Goren, M. B. 1970b. Sulfolipid-I of Mycobacterium tuberculosis, strain H37Rv.2. Structural studies. *Biochim. Biophys. Acta* **210:**127–138.

Goren, M. B. 1977. Phagocyte lysosomes: interactions with infectious agents, phagosomes, and experimental perturbations in function. *Annu. Rev. Microbiol.* **31:**507–533.

Goren, M. B., O. Brokl, and B. C. Das. 1976a. Sulfatides of Mycobacterium tuberculosis: the structure of the principal sulfatide (SL-I). *Biochemistry* **15:**2728–2735.

Goren, M. B., O. Brokl, B. C. Das, and E. Lederer. 1971. Sulfolipid-I of Mycobacterium tuberculosis, Strain-H37Rv—nature of acyl substituents. *Biochemistry* **10:**72–81.

Goren, M. B., O. Brokl, and W. B. Schaeffe. 1974. Lipids of putative relevance to virulence in *Mycobacterium tuberculosis*: correlation of virulence with elaboration of sulfatides and strongly acidic lipids. *Infect. Immun.* **9:**142–149.

Goren, M. B., P. D'Arcy Hart, M. R. Young, and J. A. Armstrong. 1976b. Prevention of phagosome-lysosome fusion in cultured macrophages by sulfatides of Mycobacterium tuberculosis. *Proc. Natl. Acad. Sci. USA* **73:**2510–2514.

Goren, M. B., J. M. Grange, V. R. Aber, B. W. Allen, and D. A. Mitchison. 1982. Role of lipid content and hydrogen peroxide susceptibility in determining the guinea-pig virulence of Mycobacterium tuberculosis. *Br. J. Exp. Pathol.* **63:**693–700.

Goren, M. B., A. E. Vatter, and J. Fiscus. 1987a. Polyanionic agents as inhibitors of phagosome-lysosome fusion in cultured macrophages: evolution of an alternative interpretation. *J. Leukoc. Biol.* **41:**111–121.

Goren, M. B., A. E. Vatter, and J. Fiscus. 1987b. Polyanionic agents do not inhibit phagosome-lysosome fusion in cultured macrophages. *J. Leukoc. Biol.* **41:**122–129.

Jain, M., and J. S. Cox. 2005. Interaction between polyketide synthase and transporter suggests coupled synthesis and export of virulence lipid in M. tuberculosis. *PLoS Pathog.* **1:**e2.

Kato, M., and M. B. Goren. 1974. Enhancement of cord factor toxicity by sulfolipid. *Jpn. J. Med. Sci. Biol.* **27:**120–124.

Khoo, K. H., E. Jarboe, A. Barker, J. Torrelles, C. W. Kuo, and D. Chatterjee. 1999. Altered expression profile of the surface glycopeptidolipids in drug-resistant clinical isolates of Mycobacterium avium complex. *J. Biol. Chem.* **274:**9778–9785.

Kolattukudy, P. E., N. D. Fernandes, A. K. Azad, A. M. Fitzmaurice, and T. D. Sirakova. 1997. Biochemistry and molecular genetics of cell-wall lipid biosynthesis in mycobacteria. *Mol. Microbiol.* **24:**263–270.

Kumar, P., M. W. Schelle, M. Jain, F. L. Lin, C. J. Petzold, M. D. Leavell, J. A. Leary, J. S. Cox, and C. R. Bertozzi. 2007. PapA1 and PapA2 are acyltransferases essential for the biosynthesis of the Mycobacterium tuberculosis virulence factor sulfolipid-1. *Proc. Natl. Acad. Sci. USA* **104:**11221–11226.

Lopez Marin, L. M., M. A. Laneelle, D. Prome, G. Laneelle, J. C. Prome, and M. Daffe. 1992. Structure of a novel sulfate-containing mycobacterial glycolipid. *Biochemistry* 31:11106–11111.

Mathur, M., and P. E. Kolattukudy. 1992. Molecular cloning and sequencing of the gene for mycocerosic acid synthase, a novel fatty acid elongating multifunctional enzyme, from Mycobacterium tuberculosis var. bovis Bacillus Calmette-Guerin. *J. Biol. Chem.* 267:19388–19395.

Middlebrook, G., C. M. Coleman, and W. B. Schaefer. 1959. Sulfolipid from virulent tubercle bacilli. *Proc. Natl. Acad. Sci. USA* 45:1801–1804.

Minnikin, D. E., L. Kremer, L. G. Dover, and G. S. Besra. 2002. The methyl-branched fortifications of Mycobacterium tuberculosis. *Chem. Biol.* 9:545–553.

Moody, D. B., D. M. Zajonc, and I. A. Wilson. 2005. Anatomy of CD1-lipid antigen complexes. *Nat. Rev. Immunol.* 5:387–399.

Mougous, J. D., R. E. Green, S. J. Williams, S. E. Brenner, and C. R. Bertozzi. 2002a. Sulfotransferases and sulfatases in mycobacteria. *Chem. Biol.* 9:767–776.

Mougous, J. D., M. D. Leavell, R. H. Senaratne, C. D. Leigh, S. J. Williams, L. W. Riley, J. A. Leary, and C. R. Bertozzi. 2002b. Discovery of sulfated metabolites in mycobacteria with a genetic and mass spectrometric approach. *Proc. Natl. Acad. Sci. USA* 99:17037–17042.

Mougous, J. D., D. H. Lee, S. C. Hubbard, M. W. Schelle, D. J. Vocadlo, J. M. Berger, and C. R. Bertozzi. 2006a. Molecular basis for G protein control of the prokaryotic ATP sulfurylase. *Mol. Cell* 21:109–122.

Mougous, J. D., C. J. Petzold, R. H. Senaratne, D. H. Lee, D. L. Akey, F. L. Lin, S. E. Munchel, M. R. Pratt, L. W. Riley, J. A. Leary, J. M. Berger, and C. R. Bertozzi. 2004. Identification, function and structure of the mycobacterial sulfotransferase that initiates sulfolipid-1 biosynthesis. *Nat. Struct. Mol. Biol.* 11:721–729.

Mougous, J. D., R. H. Senaratne, C. J. Petzold, M. Jain, D. H. Lee, M. W. Schelle, M. D. Leavell, J. S. Cox, J. A. Leary, L. W. Riley, and C. R. Bertozzi. 2006b. A sulfated metabolite produced by stf3 negatively regulates the virulence of Mycobacterium tuberculosis. *Proc. Natl. Acad. Sci. USA* 103:4258–4263.

Noll, H., H. Bloch, J. Asselineau, and E. Lederer. 1956. The chemical structure of the cord factor of Mycobacterium tuberculosis. *Biochim. Biophys. Acta* 20:299–309.

Okamoto, Y., Y. Fujita, T. Naka, M. Hirai, I. Tomiyasu, and I. Yano. 2006. Mycobacterial sulfolipid shows a virulence by inhibiting cord factor induced granuloma formation and TNF-alpha release. *Microb. Pathog.* 40:245–253.

Onwueme, K. C., J. A. Ferreras, J. Buglino, C. D. Lima, and L. E. Quadri. 2004. Mycobacterial polyketide-associated proteins are acyltransferases: proof of principle with Mycobacterium tuberculosis PapA5. *Proc. Natl. Acad. Sci. USA* 101:4608–4613.

Otero, J., W. R. Jacobs, Jr., and M. S. Glickman. 2003. Efficient allelic exchange and transposon mutagenesis in Mycobacterium avium by specialized transduction. *Appl. Environ. Microbiol.* 69:5039–5044.

Pabst, M. J., J. M. Gross, J. P. Brozna, and M. B. Goren. 1988. Inhibition of macrophage priming by sulfatide from Mycobacterium tuberculosis. *J. Immunol.* 140:634–640.

Pinto, R., Q. X. Tang, W. J. Britton, T. S. Leyh, and J. A. Triccas. 2004. The Mycobacterium tuberculosis cysD and cysNC genes form a stress-induced operon that encodes a tri-functional sulfate-activating complex. *Microbiology* 150:1681–1686.

Rivera-Marrero, C. A., J. D. Ritzenthaler, S. A. Newburn, J. Roman, and R. D. Cummings. 2002. Molecular cloning and expression of a novel glycolipid sulfotransferase in Mycobacterium tuberculosis. *Microbiology* 148:783–792.

Rousseau, C., O. C. Turner, E. Rush, Y. Bordat, T. D. Sirakova, P. E. Kolattukudy, S. Ritter, I. M. Orme, B. Gicquel, and M. Jackson. 2003. Sulfolipid deficiency does not affect the virulence of Mycobacterium tuberculosis H37Rv in mice and guinea pigs. *Infect. Immun.* 71:4684–4690.

Schelle, M. W., and C. R. Bertozzi. 2006. Sulfate metabolism in mycobacteria. *Chembiochem* 7:1516–1524.

Shen, Y., P. Sharma, F. G. da Silva, and P. Ronald. 2002. The Xanthomonas oryzae pv. lozengeoryzae raxP and raxQ genes encode an ATP sulphurylase and adenosine-5′-phosphosulphate kinase that are required for AvrXa21 avirulence activity. *Mol. Microbiol.* 44:37–48.

Sieling, P. A., D. Chatterjee, S. A. Porcelli, T. I. Prigozy, R. J. Mazzaccaro, T. Soriano, B. R. Bloom, M. B. Brenner, M. Kronenberg, P. J. Brennan, et al. 1995. CD1-restricted T cell recognition of microbial lipoglycan antigens. *Science* 269:227–230.

Sirakova, T. D., A. K. Thirumala, V. S. Dubey, H. Sprecher, and P. E. Kolattukudy. 2001. The Mycobacterium tuberculosis pks2 gene encodes the synthase for the hepta- and octamethyl-branched fatty acids required for sulfolipid synthesis. *J. Biol. Chem.* 276:16833–16839.

Sonden, B., D. Kocincova, C. Deshayes, D. Euphrasie, L. Rhayat, F. Laval, C. Frehel, M. Daffe, G. Etienne, and J. M. Reyrat. 2005. Gap, a mycobacterial specific integral membrane protein, is required for glycolipid transport to the cell surface. *Mol. Microbiol.* 58:426–440.

Stallberg-Stenhagen, S. 1954. Optically active higher aliphatic compounds. 12. Trans-2,5D-dimethyl-delta-2-3-heneicosenoic acid and trans-2,4d-dimethyl-delta-2-3-heneicosenoic acid. *Ark. Kemi* 6:537–559.

Sun, M., J. L. Andreassi II, S. Liu, R. Pinto, J. A. Triccas, and T. S. Leyh. 2005. The trifunctional sulfate-activating complex (SAC) of Mycobacterium tuberculosis. *J. Biol. Chem.* 280:7861–7866.

Sun, M., and T. S. Leyh. 2006. Channeling in sulfate activating complexes. *Biochemistry* 45:11304–11311.

Trivedi, O. A., P. Arora, A. Vats, M. Z. Ansari, R. Tickoo, V. Sridharan, D. Mohanty, and R. S. Gokhale. 2005. Dissecting the mechanism and assembly of a complex virulence mycobacterial lipid. *Mol. Cell* 17:631–643.

Williams, S. J., R. H. Senaratne, J. D. Mougous, L. W. Riley, and C. R. Bertozzi. 2002. 5′-Adenosinephosphosulfate lies at a metabolic branch point in mycobacteria. *J. Biol. Chem.* 277:32606–32615.

Wolf, A., R. Kramer, and S. Morbach. 2003. Three pathways for trehalose metabolism in Corynebacterium glutamicum ATCC13032 and their significance in response to osmotic stress. *Mol. Microbiol.* 49:1119–1134.

Woodruff, P. J., B. L. Carlson, B. Siridechadilok, M. R. Pratt, R. H. Senaratne, J. D. Mougous, L. W. Riley, S. J. Williams, and C. R. Bertozzi. 2004. Trehalose is required for growth of Mycobacterium smegmatis. *J. Biol. Chem.* 279:28835–28843.

Wu, C. W., J. Glasner, M. Collins, S. Naser, and A. M. Talaat. 2006. Whole-genome plasticity among Mycobacterium avium subspecies: insights from comparative genomic hybridizations. *J. Bacteriol.* 188:711–723.

Yamazaki, Y., L. Danelishvili, M. Wu, E. Hidaka, T. Katsuyama, B. Stang, M. Petrofsky, R. Bildfell, and L. E. Bermudez. 2006. The ability to form biofilm influences Mycobacterium avium invasion and translocation of bronchial epithelial cells. *Cell. Microbiol.* 8:806–814.

Yarkoni, E., M. B. Goren, and H. J. Rapp. 1979. Effect of sulfolipid I on trehalose-6,6′-dimycolate (cord factor) toxicity and antitumor activity. *Infect. Immun.* 24:586–588.

Zhang, L., D. English, and B. R. Andersen. 1991. Activation of human neutrophils by Mycobacterium tuberculosis-derived sulfolipid-1. *J. Immunol.* **146:**2730–2736.

Zhang, L., J. C. Gay, D. English, and B. R. Andersen. 1994. Neutrophil priming mechanisms of sulfolipid-I and N-formyl-methionyl-leucyl-phenylalanine. *J. Biomed. Sci.* **1:**253–262.

Zhang, L., M. B. Goren, T. J. Holzer, and B. R. Andersen. 1988. Effect of Mycobacterium tuberculosis-derived sulfolipid I on human phagocytic cells. *Infect. Immun.* **56:**2876–2883.

The Mycobacterial Cell Envelope
Edited by M. Daffé and J.-M. Reyrat
© 2008 ASM Press, Washington, DC

Chapter 19

The Mycobacterial Heparin-Binding Hemagglutinin: a Virulence Factor and Antigen Useful for Diagnostics and Vaccine Development

CAMILLE LOCHT, DOMINIQUE RAZE, CARINE ROUANET, CHRISTOPHE GENISSET, JÉRÔME SEGERS, AND FRANÇOISE MASCART

We dedicate this chapter to the memory of Franco D. Menozzi, the codiscoverer of HBHA, who passed away suddenly 2 years ago at the age of 43.

Most key steps of mycobacterial pathogenesis rely heavily on the interactions of the pathogens with more or less specific target cells within the infected host. The components of the microbial cell wall obviously play major roles in these interactions. Although mostly composed of lipids and carbohydrates, both exerting important functions in the host-pathogen interactions, the mycobacterial cell wall also contains a number of specific proteins that are likely to participate in the virulence mechanisms. Several of these proteins, including porins, the Erp, PE, PPE, and MmpL protein families, as well as proteins that are components of specialized transport systems, are described in detail in other chapters of this volume. However, the precise functions and role in virulence are known only for a limited set of these proteins.

One of the most recently characterized mycobacterial cell wall proteins is the heparin-binding hemagglutinin (HBHA). Initially found on the surface of *Mycobacterium tuberculosis* and *Mycobacterium bovis* bacillus Calmette-Guérin (BCG), HBHA is also present on the surface of other mycobacteria, including other pathogenic and nonpathogenic species (see later). Roughly 10 years ago, it was identified in *M. tuberculosis* as one of the major adhesins involved in binding of the mycobacteria to epithelial cells and fibroblasts, but not to macrophage-like cell lines (Menozzi et al., 1996).

The adherence of *M. tuberculosis*, as well as of other members of the *M. tuberculosis* complex, to fibroblasts and epithelial cells was found to be inhibitable in vitro by sulfated carbohydrates, such as heparin, fucoidan, chondroitin sulfate, and dextran sulfate but not by their nonsulfated counterparts (Menozzi et al., 1996). The inhibition by sulfated sugars was dose dependent and was maximal with as little as 10 µg/ml of heparin when the binding of the mycobacteria to Chinese hamster ovary cells was assayed. These sulfated carbohydrates did not inhibit adherence of the mycobacteria to macrophage-like cells, even at high concentrations. These observations suggest that *M. tuberculosis* has evolved adherence mechanisms to nonprofessional phagocytes that are distinct from the binding mechanisms to macrophages. They also suggest that the mycobacteria contain at their surface adhesin molecules able to interact with sulfated glycoconjugates on the surface of these cells.

Using heparin-Sepharose chromatography, a surface protein with an apparent molecular mass of 28,000 kDa was isolated from mycobacterial cell wall fractions. Like many adhesins from other pathogenic microorganisms, the purified 28-kDa heparin-binding protein from *M. tuberculosis* exhibits hemagglutination activity toward rabbit erythrocytes. Therefore, the protein was named heparin-binding hemagglutinin.

The cell-surface location of HBHA on the mycobacteria was confirmed by immunoelectron microscopy using anti-HBHA monoclonal antibodies, and later by atomic force microscopy (Dupres et al., 2005). Hemagglutination of rabbit erythrocytes is

Camille Locht, Dominique Raze, Carine Rouanet, Christophe Genisset, and Jérôme Segers • INSERM U629, Lille, Institut Pasteur de Lille, 1, rue du Prof. Calmette, F-59019 Lille Cedex, France. Françoise Mascart • Laboratory of Vaccinology and Mucosal Immunity, Erasme Hospital, Université Libre de Bruxelles, 808, route de Lennik, B-1070 Brussels, Belgium.

totally inhibited in the presence of these monoclonal antibodies or in the presence of heparin or dextran sulfate. Also the cell binding of the mycobacteria can be inhibited significantly by anti-HBHA antibodies.

The direct role of HBHA in bacterial adherence to epithelial cells was confirmed by the use of isogenic *M. tuberculosis* mutant strains (Pethe et al., 2001). A *M. tuberculosis* or a BCG strain in which the gene encoding HBHA (*hbhA*) is disrupted by the insertion of a kanamycin-resistance gene, expresses reduced adherence to the human type II pneumocyte cell line A549 compared to the respective isogenic parent strain, whereas the binding to macrophage-like cell lines is not affected by the mutation. When the mutation is complemented by the natural *hbhA* gene under the control of its own promoter, binding to the A549 cells is restored to wild-type levels. These observations confirm that HBHA is one of the key adhesins for cells other than professional phagocytes, suggesting that interactions of *M. tuberculosis* with cells other than macrophages may play an important role in the pathogenesis of tuberculosis.

STRUCTURE-FUNCTION RELATIONSHIP OF HBHA

Although the electrophoretic mobility of the *M. tuberculosis* HBHA (MT-HBHA) suggested an apparent molecular mass of 28,000 Da, the calculated molecular weight based on the amino acid sequence is 21,402, which is substantially lower than the apparent molecular weight (Menozzi et al., 1998). This discrepancy is due to both an aberrant electrophoretic mobility of the protein, and a complex post-translational modification, consisting of several mono- and dimethylations of lysine residues (Pethe et al., 2002). The protein consists of 198 amino acids. It contains no methionine, no histidine, no cysteine, and no tryptophan, and the first amino acid of the mature protein determined by N-terminal amino acid sequencing of purified HBHA corresponds to the codon immediately following the ATG initiation codon of the HBHA open reading frame.

According to the amino acid sequence, HBHA can be subdivided into two distinct domains (Fig. 1). The 159-residue-long N-terminal domain contains a large predicted coiled-coil region that extends over 81 amino acids (Delogu and Brennan, 1999), as well as a putative leucine-rich transmembrane segment near the N-terminus of HBHA (residues 13 to 34). The putative transmembrane segment may perhaps serve to anchor HBHA in the outer lipid layer of the mycobacterial surface, although it lacks the typical LPXTGX consensus sequence known to anchor pro-

teins at the surface of gram-positive bacteria (Schneewind et al., 1992). The predicted coiled-coil region is composed of heptad repeats with a number of predicted α helical turns. This suggests that the protein has an extended rather than a globular conformation, perhaps accounting for its aberrant electrophoretic mobility. This hypothesis is consistent with the fact that even a truncated form of HBHA, containing only the N-terminal domain and lacking residues 160 to 198, maintains aberrant electrophoretic mobility during sodium dodecyl sulfate (SDS)-polyacrylamide gel electrophoresis (Pethe et al., 2000).

The predicted coiled-coil region of HBHA may be involved in homotypic protein-protein interactions to form oligomeric structures. Evidence for this has been provided by the fact that a histidine-tagged recombinant N-terminal domain of HBHA bound to a nickel-chelating resin is able to bind and pull out native HBHA from an *M. tuberculosis* extract (Delogu and Brennan, 1999). By the use of atomic force microscopy applied to the surface of the mycobacteria, the HBHA-heparin adhesion forces have been shown to display a bimodal distribution and to increase with contact time, indicating that the heparin-binding site of HBHA is surface exposed and suggesting that the interaction involves multiple intermolecular bridges (Dupres et al., 2005). This is consistent with the presence of a dimeric or a multimeric form of HBHA on the surface of the mycobacteria. Atomic force microscopy also showed that the distribution of HBHA on the mycobacterial surface is not homogeneous but is concentrated in nanodomains.

In addition to serving as an adhesin of *M. tuberculosis* to epithelial cells, HBHA is capable of mediating mycobacterial autoaggregation, and anti-HBHA monoclonal antibodies are able to inhibit the autoagglutination of *M. tuberculosis*. This autoagglutination activity has been shown to be expressed by the N-terminal domain containing the coiled-coil region, suggesting that the bacteria are interconnected by these coiled-coil structures. As shown for other pathogenic bacteria (Tyewska et al., 1988), the ability to form intermycobacterial aggregates may participate in virulence by promoting the formation of microcolonies at the initial site of infection.

The approximately 40-amino-acid-residue-long C-terminal domain is essentially composed of alanines, prolines, and lysines, organized in two types of lysine-rich repeats, named R1 and R2, respectively (Pethe et al., 2000). R1 (KKAAPA) is directly repeated three times between residues 160 and 177, and R2 (KKAAAKK) is present twice between residues 178 and 194. These repeats constitute the high-affinity heparin-binding site of HBHA and are

A

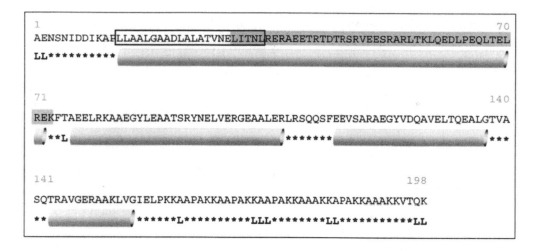

B

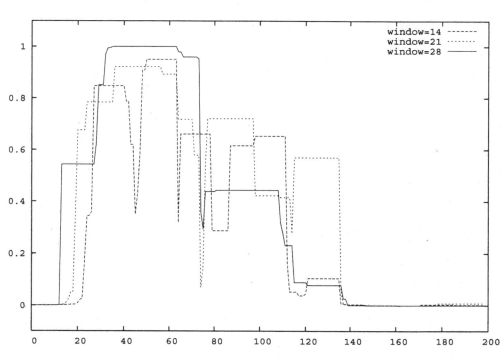

Figure 1. Predicted secondary structure of HBHA. (A) The secondary structure of HBHA was predicted by using PROF (University of Wales, Aberystwyth Computational Biology Group). Predicted alpha helices are depicted as cylinders below the amino acid sequence, L represents predicted loop structures, and * represents unpredicted structures due to low reliability of structure prediction. The open box represents the hydrophobic region, and the gray box represents the predicted coiled-coil region. (B) Coiled-coil propensity. The probability that the protein will adopt a coiled-coil conformation was calculated by the Coils server (ch.EMBnet.org) and was obtained by using default options in scanning windows of 14, 21, and 28 residues, as indicated.

responsible for the binding of HBHA to heparan sulfate glycosaminoglycans, including those of the proteoglycan decorin (Delogu and Brennan, 1999), a protein commonly found in interstitial tissues. The deletion of these repeats from a recombinant form of HBHA totally abolished the heparin-binding ability of HBHA under conditions in which the full-length molecule strongly binds to heparin (Pethe et al., 2000). A single R1 motif is not sufficient for heparin binding. An HBHA molecule with two R1 motifs binds weakly to heparin, and the addition of a third R1 motif strongly strengthens the interaction of HBHA with heparin, but the addition of the R2 motifs is required to recover full binding capacity. When the entire C-terminal region of HBHA was grafted onto an irrelevant protein, such as the maltose-binding protein MalE, the hybrid protein acquired heparin-binding activity that was quantitatively similar to that of full-length HBHA, as evaluated by surface plasmon resonance measurements. These results indicate that all the molecular determinants responsible for heparin binding are located within the C-terminal lysine-rich repeat region of the protein. The affinity constants of both full-length HBHA and grafted MalE were estimated to be in the range of 20 to 25 nM (Pethe et al., 2000), similar to those found for other heparin-binding proteins. Furthermore, grafted MalE bound well to the surface of epithelial cells, as did full-length HBHA, whereas neither truncated HBHA nor ungrafted MalE bound to epithelial cell surfaces to a significant extent, indicating that the heparin-binding site of HBHA is responsible for epithelial cell binding of the adhesin. Furthermore, treatment of the epithelial cells with heparinase III led to a strong reduction of HBHA binding, indicating that proteoglycans containing heparan sulfate chains serve as HBHA receptors on the surface of the epithelial cells.

From atomic force microscopy studies (Dupres et al., 2005) and immunogold-labeling using monoclonal antibodies that recognize the C-terminal domain of HBHA (Menozzi et al., 1996), it appears that this region is exposed on the surface of *M. tuberculosis*, consistent with its important role in adherence function.

Taken together, these observations make HBHA one of the members of the growing family of microbial heparin-binding adhesins, including among other adhesins in *Chlamydia trachomatis* (Su et al., 1996), *Neisseria gonorrhoeae* (Grant et al., 1999), *Bordetella pertussis* (Menozzi et al., 1994), *Borrelia burgdorferi* (Leong et al., 1995), group B streptococci (Baron et al., 2004), *Plasmodium falciparum* (Gysin et al., 1997), human immunodeficiency virus (Mondor et al., 1998), hepatitis B virus (Cooper et al., 2005) and many others (Menozzi et al., 2002). However, despite the widespread presence of heparin-binding adhesins, both in prokaryotes and in eukaryotes, no clear universal consensus sequence of heparin-binding sites has emerged. Cardin and Weintraub (1989) have proposed two heparin-binding consensus sequences for various eukaryotic proteins, consisting of the motifs XBBXBX and XBBBXXBX, where B represents a basic amino acid and X any other amino acid. However, the heparin-binding site of HBHA fits neither of these motifs, although the high density of lysine residues in the HBHA heparin-binding domain provides some resemblance to the Cardin and Weintraub consensus and is consistent with their role as basic residues interacting with the negatively charged groups present on the sulfated proteoglycans.

In addition to binding to sulfated proteoglycans at the surface of epithelial cells, endothelial cells and fibroblasts, HBHA is also able to adhere to complement component C3 in human serum (Mueller-Oritz et al., 2001). Because C3 can also bind to complement receptors at the surface of mononuclear phagocytes, the adherence activity of HBHA to C3 may indirectly lead to binding of the mycobacteria to mononuclear phagocytes. C3 has indeed been identified as a major serum component in mediating phagocytosis of *M. tuberculosis* by human monocytes (Schlesinger et al., 1990), and using a C3 ligand affinity blot protocol, HBHA was identified as the major C3-binding protein in *M. tuberculosis* cell wall fractions (Mueller-Oritz et al., 2001). Binding of HBHA to C3 occurred only after complement activation and appeared to depend on the presence of the internal thioester of C3. Similar to the heparin-binding activity of HBHA, binding to C3 also depends on the C-terminal lysine-rich repeats. Truncated HBHA, lacking these repeats did not bind to C3, whereas MalE onto which the repeats were grafted acquired C3 binding. Interestingly, Mueller-Oritz et al. (2001) found that beads coated with recombinant HBHA are able to bind to the J774-A1 macrophage-like cell line, even in the absence of complement. However, this binding was enhanced significantly in the presence of human serum, and the enhancement could be blocked by anti-C3 antibodies, raising the question about the specificity of HBHA-coated beads binding to macrophage-like cell lines in the absence of C3.

Purified HBHA also binds to concanavalin A, which was initially proposed to indicate that HBHA may be a glycoprotein (Menozzi et al., 1998; Mueller-Oritz et al., 2001). However, it has subsequently been established that the posttranslational modification of HBHA is a complex methylation pattern (Pethe et al., 2002), totally accounting for the observed electrophoretic mobility differences between the recombinant and the native forms of the protein, which raises the possibility that the observed concanavalin A binding may also be nonspecific.

The three-dimensional structure of HBHA has not yet been determined. However, structural predictions by using methods such as the secondary structure prediction system PROF (University of Wales, Aberystwyth Computational Biology Group), suggest that the N-terminal domain of the protein is rich in alpha helices, whereas the C-terminal lysine-rich region is relatively unstructured. Using PROF several α helices are predicted, separated by small loops (Fig. 1A). Using the Coils program (Lupas et al., 1991), a coiled-coil region is strongly predicted from residue 30 to 73 (Fig. 1B). This region is located immediately downstream of the highly hydrophobic region (Fig. 1A). After residue 74, the coiled-coil prediction score decreases gradually to reach 0 at position 140 (Fig. 1B).

THE BIOGENESIS OF HBHA

The structural gene of HBHA in the *M. tuberculosis* chromosome (*rv0475*) is followed by two open reading frames of unknown function and preceded by an open reading frame coding for a potential transcriptional regulator (Fig. 2). The distance between the stop codon of the *hbhA* gene and the initiation codon of the downstream gene is 111 bp, and the distance between the stop codon of the upstream gene and the initiation codon of *hbhA* is 353 bp, a region large enough to contain promoter sequences. When inserted upstream of a promoterless green fluorescence protein gene used as a reporter gene, the *hbhA* upstream region is able to drive expression of the reporter gene in mycobacteria (C. Rouanet, unpublished observations), suggesting that this region indeed contains a functional promoter. Reverse transcriptase PCR (RT-PCR) used to amplify the intergenic region between the *hbhA* gene and the downstream open reading frame resulted in the generation of an amplicon of the expected size (Mueller-Oritz et al., 2002), indicating that *hbhA* is cotranscribed with the downstream gene *rv0476*. The distances between *rv0476* and *rv0477*, and between *rv0477* and *rv0478* are 7 and 2 bp, respectively, suggesting that all four genes are transcribed into a polycistronic mRNA. The latter open reading frame (*deoC*) codes for a putative deoxyboribose-phosphate aldolase. The *deoC* downstream gene is oriented in the opposite direction (Fig. 2).

Evidence that the activity of the *hbhA* promoter may be regulated by environmental signals comes from in vivo expression analysis of the *M. tuberculosis hbhA* gene during growth in axenic cultures and during infection of mice or various cell lines (Delogu et al., 2006). RT-PCR have been performed on the *hbhA* transcripts in comparison to mycobacterial control transcripts using *M. tuberculosis* RNA extracted from either the exponential or the stationary growth phase to show that *hbhA* expression was upregulated approximately 100-fold during stationary phase. When expression was analyzed after *M. tuberculosis* infection of bone marrow-derived macrophages or A549 type II pneumocytes, cell-specific upregulation of *hbhA* expression was found. Strong upregulation was already seen 1 day, but more strongly 3 days, after infection of the A549 cells, whereas no significant upregulation was seen after infection of the macrophages, indicating that *M. tuberculosis* upregulates *hbhA* expression upon infection of epithelial cells but not of macrophages.

When expression of *hbhA* was examined after low-dose aerosol infection of mice (approximately 200 CFU per C57BL/6 mouse), an approximately 40-fold increase in *hbhA* expression was measured in the lungs 14 days after infection. This expression decreased at day 28 after infection to reach a stable level of approximately 10-fold enhanced expression over the control. In contrast, no significant increase in *hbhA* expression was detected in the spleen of the same mice, resulting in an overall 10- to 40-fold higher *hbhA* expression in the lungs compared to the spleen.

This differential regulation is rather substantial compared with the regulation of other *M. tuberculosis* genes that have been studied so far, suggesting that strongly active regulatory proteins are involved. These proteins remain to be identified. One potential candidate may be Rv0474, encoded by the most proximal gene. Rv0474 is a putative transcriptional regulator of the Xre family (Fig. 2). Current studies are being carried out in our laboratory to address this possibility.

After transcription, the HBHA open reading frame is directly translated into the mature protein. No proteolytic processing occurs, except that the initiation methionine appears to be removed after translation, because the first amino acid residue identified by HBHA protein sequencing corresponds to the second codon of the open reading frame (Menozzi et al., 1996; Menozzi et al., 1998). The fact that HBHA is surface exposed despite the lack of a cleavable N-terminal signal peptide suggests that its cellular localization is Sec independent. The localization of HBHA at the mycobacterial membrane may be mediated by the putative hydrophobic N-proximal transmembrane domain (see earlier). However, this would not explain the surface accessibility of the C-terminal domain, implying that the major portion of the protein crosses the mycobacterial membrane. It is conceivable that the subcellular localization of HBHA requires the help of one or several accessory proteins that remain to be identified.

M. tuberculosis is known to secrete a number of different proteins in a Sec-independent manner. One well-studied example is the 6-kDa early secreted antigenic target (ESAT-6) and other members of this family, whose secretion requires a specific secretion

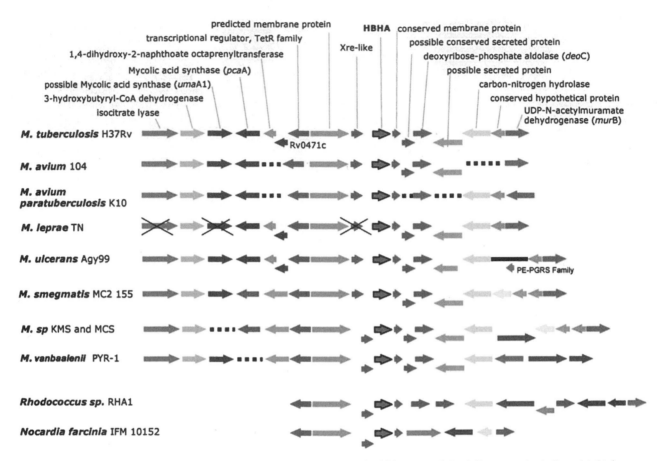

Figure 2. Structure of the *hbhA* locus in different bacterial species. The *hbhA* gene of the different species indicated is highlighted as the arrow in bold. Conserved open reading frames between the different species are shown by the same color. The corresponding putative proteins encoded by the genes are indicated on the top. Chromosomal deletions or absence of genes are indicated by the dotted lines, and pseudogenes in *M. leprae* are depicted by the crosses. (*See the color insert for the color version of this figure.*)

system encoded by genes that are located in the vicinity of the ESAT-6 structural gene (Guinn et al., 2004).

Comparative genomics have revealed that *M. tuberculosis* produces two different SecA proteins. One of them, SecA2, has been shown to be involved in the secretion of the superoxide dismutase SodA, a protein also lacking a cleavable signal peptide (Braunstein et al., 2003). Moreover, a twin-arginine translocation (TAT) system has also been identified in *M. tuberculosis* (Saint-Joanis et al., 2006). TAT systems are used by many bacteria to secrete fully folded proteins containing a twin arginine signal peptide. However, HBHA contains no readily identifiable twin arginine signal sequence. Whether any of these, or any yet to be discovered, Sec-independent mechanisms are involved in the cellular localization of HBHA remains an open question.

An important feature of HBHA is its posttranslational modification, consisting of a complex methylation pattern of the C-terminal lysine-rich repeats (Pethe et al., 2002). The methylation is catalyzed by one or several mycobacterial S-adenosyl-L-methionine-dependent HBHA:methyltransferases (Pethe et al., 2002, Host et al., 2007). HBHA-specific methyltransferase activity can be detected in mycobacterial cell extracts, and in vitro methylation of recombinant HBHA reproduces a complex methylation pattern at the C-terminal domain of the protein, as is found in the native protein (Host et al., 2007). Anti-HBHA monoclonal antibodies that decorate the surface of *M. tuberculosis* recognize the C-terminal, methylated region of the protein, and the epitopes specific for these antibodies can be generated by enzymatic in vitro methylation of a recombinant form of the protein.

Where exactly the methylation of HBHA occurs within the mycobacterial cell is not definitively known. However, enzymatic activity has been found both in the soluble cell extracts and in the cell wall fractions (Pethe et al., 2002), suggesting that at least some of the methylations occur at the cell wall. Studies on truncated recombinant forms of HBHA produced in mycobacteria, in particular forms that lack the putative transmembrane domain, have indicated that the presence of this domain is required for the methylation to occur within the mycobacteria (Delogu et al., 2004). This raised the possibility that the N-terminal domain of HBHA is required either for the recognition by the methyltransferases, or for targeting the protein to the proper cellular compartment, where methylation can occur. Support in favor of the second hypothesis came from recently described in vitro methylation experiments (Host et al., 2007). When MalE, onto which the C-terminal lysine-rich repeats of HBHA were grafted, was subjected to in vitro methylation by mycobacterial cell extracts, the protein could readily be methylated, and mass spectrometry analysis indicated that the methylation of the grafted MalE resulted also in a complex pattern, similar to that seen after methylation of full-length HBHA. Because the grafted MalE protein did not contain the putative transmembrane domain of HBHA, these observations indicate that this domain is not required for recognition by the methyltransferases but is likely to be involved in targeting the protein to the cell wall, where it then can be properly methylated. Taken together, these studies suggest that after translation, HBHA is targeted to the mycobacterial membrane, where it is methylated. Whether the methylation of HBHA contributes to its final destination awaits further investigation. The identification and mutational inactivation of the methyltransferases should shed light on this issue.

ROLE OF HBHA IN THE PATHOGENESIS OF TUBERCULOSIS

As shown earlier, HBHA binds to several nonprofessional phagocytic target cells, as well as extracellular structures, which raises the issue of its relevance for the pathogenesis of tuberculosis. Although professional phagocytes of the monocytic lineage, including alveolar macrophages, have long been recognized as important players in the pathogenesis of tuberculosis and the physiology of latent M. tuberculosis infection (Russell, 2001), other cell types, including epithelial cells, endothelial cells, fibroblasts, and, as most recently shown, adipocytes (Neyrolles et al., 2006), are also now being considered as important M. tuberculosis targets during infection. Half a century ago, Shepard (1957) reported that pathogenic M. tuberculosis can invade HeLa cells and multiply within these cells. He also found a direct correlation between the apparent pathogenicity of the tubercle bacillus strains for humans and their growth in the HeLa cells.

More recently, several groups have shown that M. tuberculosis can invade and replicate within lung epithelial cells, such as the A549 cell line, in vitro (McDonough and Kress, 1995; Bermudez and Goodman, 1996; Metha et al., 1996). Only virulent M. tuberculosis H37Rv and not attenuated M. bovis BCG was cytotoxic for these cells, although both were able to enter the A549 cells (McDonough and Kress, 1995). Cytotoxicity appeared to be independent of host factors and was thus proposed to be driven by bacterial virulence factors (Dobos et al., 2000). By using a two-layer transwell system it was found that bacteria that had invaded and destroyed A549 cells could subsequently be taken up by endothelial cells. This led to mycobacterial crossing of a bicellular layer (Bermudez et al., 2002). Therefore, this process can ultimately result in extrapulmonary dissemination by direct interactions of the mycobacteria with epithelial and endothelial cells.

In addition to playing a potentially important role in extrapulmonary dissemination, the interactions of mycobacteria with epithelial and endothelial cells may also be involved in latency of M. tuberculosis infection. Latency is by far the most frequent outcome of M. tuberculosis infections, because more than 90% of the infected individuals will not develop any sign of disease during their lifetime. The precise location of the mycobacteria during latency has been a matter of debate since Robert Koch's time. An attractive candidate for hosting the mycobacteria is the characteristic tuberculous granuloma, in which the bacteria are shielded from the sterilizing immune response of the host and which protects the host from uncontrolled bacterial outgrowth.

However, in situ PCR experiments on lung tissues from latently infected subjects who had died from causes other than tuberculosis have recently indicated the presence of M. tuberculosis DNA within type II pneumocytes, endothelial cells, and fibroblasts (Hernandez-Pando et al., 2000), providing further evidence for a role of these nonprofessional phagocytes in M. tuberculosis infection.

HBHA-mediated binding of M. tuberculosis to these nonprofessional phagocytes therefore suggests a relevant role of this protein in the pathogenesis of tuberculosis. Although initially identified as an adhesin for epithelial cells and fibroblasts, it is also a likely candidate to play an important role in the interactions with adipocytes, the most recently identified cell type hosting M. tuberculosis during latency (Neyrolles et al., 2006). Indeed, adherence of M. tuberculosis to adipocytes can be inhibited by sulfated carbohydrates that had previously been shown to inhibit M. tuberculosis binding to epithelial cells via HBHA (Menozzi et al., 1996). However, the direct role of HBHA in *M. tuberculosis*-adipocyte interactions has not been experimentally investigated yet.

Identification of the role of HBHA during *M. tuberculosis* infection in vivo became possible through the use of isogenic mutants that lack the *hbhA* gene. Isogenic mutants have been constructed on several different genetic backgrounds: *M. bovis* BCG (Pethe et al., 2001), *M. tuberculosis* H37Rv (Mueller-Oritz et al., 2002), and the clinical *M. tuberculosis* isolate MT103 (Pethe et al., 2001). The mutations did not affect the ability of the mycobacteria to grow in liquid broth or on solid growth medium in vitro. When BALB/c mice were infected intranasally with parental, HBHA-deficient mutant or complemented BCG (10^6 CFU) or *M. tuberculosis* MT103 (10^3 CFU), and the CFUs were monitored over time in the lungs, the colonization profiles of the mutant strains were found similar to those of the corresponding parental and complemented strains (Pethe et al., 2001), although a slight delay in lung colonization was observed for the MT103 mutant strain, compared with the corresponding HBHA-producing strains. A slightly more pronounced effect on lung colonization was seen 21 days after aerosol infection of C57BL/6 mice with 50 to 1,000 CFU of a HBHA-deficient *M. tuberculosis* H37Rv strain compared with its parental counterpart (Mueller-Oritz et al., 2002).

The most striking effect of the mutations was observed when the colonization of deeper organs, such as the spleen and the liver, were monitored after intranasal or aerosol infection. Three weeks after infection, approximately 100-fold less mutant *M. tuberculosis* was recovered in the spleen (Pethe et al., 2001; Mueller-Oritz et al., 2002) and in the liver (Mueller-Oritz et al., 2002), compared with the parental or complemented strains. However, when the BALB/c mice were infected intravenously with 10^6 CFU, the mutant strain colonized the spleen as efficiently as the parental and the complemented strains (Pethe et al., 2001). These observations suggest that HBHA plays a role in dissemination from the lungs to deeper tissues. However, the mutations do not totally prevent extrapulmonary dissemination, and spleen colonization by mutant *M. tuberculosis* gradually increases over time after intranasal infection, suggesting that other factors may also contribute to extrapulmonary dissemination.

In addition to the use of isogenic mutant strains, the in vivo role of HBHA was also studied by the use of anti-HBHA monoclonal antibodies (Pethe et al., 2001). When C57BL/6 mice were infected intranasally with BCG precoated with the anti-HBHA monoclonal antibodies, lung colonization was not significantly affected by the antibodies. However, precoating with the antibodies led to a substantial reduction in spleen colonization, when analyzed 7 days after infection and compared to precoating with an irrelevant antibody. These results confirm the conclusions drawn from the observations made with the isogenic mutant strains and suggest, in addition, that the HBHA-mediated extrapulmonary dissemination step occurs rather early during infection. Bacterial multiplication either extracellularly or intracellularly within alveolar macrophages before extrapulmonary dissemination would have resulted in uncoating the mycobacteria, and subsequently, no effect of the antibodies on dissemination would have been seen.

That extrapulmonary dissemination of *M. tuberculosis* in mice occurs within days after infection has been seen by investigating the kinetics of organ colonization after aerosol infection. Depending on the mouse strain used, bacteria appear in the pulmonary lymph nodes 9 to 11 days and in the spleen and liver 10 to 14 days after aerosol infection with roughly 300 to 400 CFU (Chackerian et al., 2002). These observations suggest that *M. tuberculosis* spreads to the local lymph nodes through the lymphatics and then disseminates hematogenously to deeper organs. By using T- and B-cell-deficient mice, the extrapulmonary dissemination has been shown to be independent of lymphocyte-induced inflammatory lesions. Nevertheless, the earlier dissemination occurs, the earlier pulmonary lesions can be detected in the infected lungs, but the lesions do not occur before dissemination. Again, the appearance of the lesions is independent of B and T cells. Interestingly, when only one lung was infected initially, the contralateral lung became infected only after the mycobacteria had spread to the spleen and liver (Mischenki et al., 2004), suggesting that the infection of initially intact lung tissues is secondary to extrapulmonary dissemination. The initiation of T-cell immune responses also occurs only after dissemination of the mycobacteria to the pulmonary lymph nodes, indicating that dissemination is an essential event for both the pathogenesis of tuberculosis and the induction of T-cell-mediated immunity.

Interruption or deletion of the *hbhA* gene did not appear to affect direct adherence of the mycobacteria to macrophages, as evidenced by studies on the J774.A1 murine macrophage-like cell line (Pethe et al., 2001, Mueller-Oritz et al., 2002) and, surprisingly, also did not significantly affect mycobacterial binding to human complement component C3 (Mueller-Oritz et al., 2002). In contrast, the mutation resulted in a significant reduction of *M. tuberculosis* adherence to epithelial cells, such as the A549 pneumocytes (Pethe et al., 2001). Therefore, the mycobacterial interactions with pulmonary epithelial cells may constitute an important mechanism of extrapulmonary dissemination, although it is most likely not an exclusive mechanism. Other cells, such as macrophages and dendritic cells may also be involved, perhaps in an HBHA-independent fashion. However, it has been proposed that the role of these cells may come into play at later stages of infection, when the numbers of infected alveolar macrophages have increased (Sato et al., 2003). Thus, epithelial cells, including M cells have been suggested as possible early ports of entry into the host, and entry through M cells occurs very rapidly (Teitelbaum et al., 1999). From there, the bacilli can subsequently be conveyed to the draining lymph nodes, which would then rapidly lead to both dissemination and the induction of local immune responses.

The mechanism of how HBHA-mediated *M. tuberculosis* adherence to epithelial cells can lead to extrapulmonary spreading has been investigated by the use of a transwell model of cellular barriers and HBHA-coated colloidal gold particles (Menozzi et al., 2006). HBHA can induce the reorganization of the actin filament network in confluent cell layers, but it does not appear to affect the tight junctions of epithelial and endothelial cell layers. HBHA-coated gold particles attach rapidly to the membrane of epithelial cells, both nonpolarized HEp-2 and polarized A549 cells, and are then internalized in membrane-bound vesicles, which migrate across the polarized cells to reach the basal side. Both the adherence of the particles to the epithelial cell layer and the subsequent transcytosis of the particles were abolished when the cells were pretreated with heparinase III, or when the particles were coated with HBHA derivatives that lack the C-terminal lysine-rich repeats. Thus, these observations indicate that HBHA induces receptor-mediated endocytosis through the interaction of its C-terminal domain with heparan sulfate-containing glycoconjugates at the surface of the cells and suggest therefore that *M. tuberculosis* can transcytose epithelial cell layers through the HBHA-binding mechanisms. Although *M. tuberculosis* can invade and cross A549 monolayers, it is not efficient in invading endothelial cell monolayers (Bermudez et al., 2002). However, after the bacteria have invaded A549 cells, they are subsequently able to be taken up by endothelial cells much more efficiently. This may thus be the path to macrophage-independent extrapulmonary dissemination.

The fact that the *hbhA* gene expression is upregulated several fold when *M. tuberculosis* has invaded pneumocytes suggests that the protein is not only important for adherence to epithelial cells but may play an additional role once the mycobacteria are within the epithelial cells. This role remains to be analyzed, but it is tempting to speculate that increased production of HBHA within the epithelial cells may be involved in the mycobacterial escape from these cells into the endothelial cell layer, which then results in spreading to the lymphatics or the general bloodstream.

After crossing the epithelial and endothelial cell layers, disseminated bacteria can be kept in check by a protective, albeit not sterilizing, immune response, and the infection may become latent, as is observed for most *M. tuberculosis*-infected human individuals. If the protective immune responses are not strong enough, the mycobacteria can escape immunity and colonize vulnerable zones, preferentially the apical regions of the lungs (Balasubramanian et al., 1994). Thus, extrapulmonary dissemination represents a major step in the pathogenesis of tuberculosis and for the induction of a protective immune response.

HBHA-LIKE MOLECULES IN OTHER ACTINOMYCETES

The sequence of HBHA is strictly conserved among clinical *M. tuberculosis* isolates, as evidenced by sequence analysis of 16 different isolates representing the breadth of genomic diversity in the species and containing representative samples of the three principal genetic groups of *M. tuberculosis* (Musser et al., 2000).

Although mostly studied in *M. tuberculosis*, HBHA-like molecules are also produced by other mycobacteria. Cell extracts of several other mycobacterial species, including *Mycobacterium avium*, *Mycobacterium intracellulare*, and *Mycobacterium smegmatis*, have been analyzed by heparin-Sepharose chromatography for the presence of HBHA-like molecules (Delogu et al., 1999). In each case, a protein binding to heparin-Sepharose could be detected. The heparin-binding proteins from the *M. avium*, *M. intracellulare*, and *M. smegmatis* appeared to migrate slightly slower during SDS-polyacrylamide gel electrophoresis. The heparin-binding proteins from all three nontuberculous mycobacterial species were recognized by the anti-HBHA monoclonal antibodies. In this context, it is interesting to note that these anti-HBHA monoclonal antibodies were in fact generated against *M. avium* antigens (Rouse et al., 1991). Both *M. avium* and *M. smegmatis* lysates also contain complement component C3-binding proteins with apparent M_r upon SDS-polyacrylamide gel electrophoresis similar to those of the anti-HBHA cross-reactive proteins (Mueller-Oritz et al., 2001). Similar to the MT-HBHA, the *M. avium* C3-binding protein was found in both the soluble and particulate mycobacterial subcellular fractions.

The *M. avium* HBHA had previously been identified as one of the two major *M. avium* adhesins involved in binding to the human respiratory epithelial cell line HEp-2 (Reddy and Kumar, 2000). The *hbhA* gene from *M. avium* subsp. *paratuberculosis* has been cloned and sequenced and was found to be closely related to the *M. tuberculosis hbhA* gene (Sechi et al., 2006). If the initiation codon of the *M. avium* subsp. *paratuberculosis hbhA* gene is the same as that of the *M. tuberculosis hbhA* gene, and if, like for MT-HBHA, the initiation methionine is removed from the mature protein, the *M. avium* subsp. *paratuberculosis* HBHA (MAMP-HBHA) is 189 amino acid residues long, whereas the *M. avium* subsp. *avium* HBHA (MAMA-HBHA) is 204 residues long. The additional six residues of the MAMA-HBHA over the MT-HBHA may account for the difference in electrophoretic mobility observed between the two proteins. Intriguingly, the homogeneity of the HBHA molecules among the different species within the *M. tuberculosis* complex contrasts somewhat with the relatively substantial difference in sizes between the HBHA proteins of the two subspecies of the *M. avium* complex. However, despite this difference in size, sequence alignments indicate that the HBHA sequences of both *M. avium* subspecies are nearly identical in their N-terminal domains, with a single Glu to Gly substitution at amino acid position 50. The size difference between the two proteins is due to the presence of two additional KKAPA repeats and one KKAAA motif in the MAMA-HBHA as compared with MAMP-HBHA.

When produced in recombinant *M. smegmatis*, the MAMP-HBHA was found to cross-react with the anti-MT-HBHA monoclonal antibodies and migrated slower during SDS-polyacrylamide gel electrophoresis than predicted by its calculated molecular weight (Sechi et al., 2006), suggesting that the MAMP-HBHA is also methylated in its C-terminal domain. The recombinant *M. smegmatis* producing the MAMP-HBHA adhered more efficiently to the Caco2 intestinal epithelial cell line than nonrecombinant *M. smegmatis*, or *M. smegmatis* that contained the *M. paratuberculosis hbhA* gene inserted in the opposite orientation with respect to the promoter. These observations suggest that the MAMP-HBHA is surface exposed in recombinant *M. smegmatis* and that it serves as an adhesin for intestinal epithelial cells. Intestinal epithelial cells are relevant target cells for *M. avium* subsp. *paratuberculosis*, because this mycobacterial species is responsible for chronic intestinal granulomatous infections in ruminants (Chacon et al., 2004).

Attempts to purify the *M. smegmatis* HBHA (MS-HBHA) have initially been unsuccessful (Pethe et al., 2001), and compared with *M. tuberculosis*, *M. smegmatis* binds less well to epithelial cells (Bermudez et al., 1995). Nevertheless, a surface-associated HBHA cross-reactive protein was identified in *M. smegmatis* and found to correspond to the laminin-binding protein (Pethe et al., 2001). This laminin-binding protein was isolated by heparin-Sepharose chromatography and found to react with anti-HBHA antibodies directed against the C-terminal, lysine-rich region of HBHA. It reacted poorly with polyclonal anti-HBHA antibodies that predominantly recognize epitopes located in the N-terminal moiety of the molecule. Sequence analyses indicated that the laminin-binding protein indeed contains a lysine-rich region, similar to that of HBHA, and, like those of HBHA, these lysines are either mono- or dimethylated.

However, recently, comparative genome analyses have revealed that the *M. smegmatis* genome also contains a bona fide *hbhA* homologue (Biet et al., 2007). A 699-bp open reading frame in the *M. smegmatis* genome shows 75% sequence identity to the *M. tuberculosis hbhA* open reading frame. The MS-HBHA is 231 residues long. Compared with MT-HBHA, MS-HBHA contains a nine amino acid insertion, from position 84 to 92 of the mature protein, corresponding to the almost exact repetition of the previous nine residues. The sequence similarities among the HBHA proteins are particularly high up to residue 162, with reference to the *M. tuberculosis* sequence. Polyclonal anti-HBHA antibodies raised against a truncated form the *M. tuberculosis* protein that lacks the C-terminal domain, recognize the MS-HBHA, indicating strong cross-reactivity of the N-terminal domain of the HBHA proteins from the two species. The sequences are more divergent in the C-terminal regions of the proteins. The MS-HBHA contains a large insertion immediately upstream of the lysine-rich repeats. This insertion contains several acidic amino acids.

Moreover, the C-terminal end also shows several differences, in particular with respect to some of the lysine pairs. Despite these differences, however, MS-HBHA is recognized by the two anti-TB-HBHA monoclonal antibodies, specific for the C-terminal domain. Similar to MT-HBHA, the recombinant form of MS-HBHA produced in *Escherichia coli* migrates faster during SDS-polyacrylamide electrophoresis than the native form. These two observations suggest that MS-HBHA is also methylated in its C-terminal domain, which was confirmed by mass spectrometry analysis of this region. This analysis provided evidence that MS-HBHA contains both mono- and di-methyl lysines in its C-terminal region.

The sequence differences in the C-terminal domains between MT-HBHA and MS-HBHA may account for differences seen in their affinity for heparin. Whereas MT-HBHA elutes from heparin-Sepharose at roughly 300 mM, MS-HBHA elutes already at approximately 75 mM NaCl, indicating that the *M. tuberculosis* protein has higher affinity for heparin than its *M. smegmatis* homologue.

Despite the fact that, like MT-HBHA, MS-HBHA is also cell wall associated, it does not appear to mediate adherence of mycobacteria to epithelial cells. Binding of *M. smegmatis* to A549 cells could not be inhibited by the presence of heparin. In contrast, recombinant *M. smegmatis* producing the MT-HBHA bound more strongly to these cells than nonrecombinant *M. smegmatis*, and this increased binding could be inhibited by the presence of heparin. Therefore, it has been proposed that HBHA molecules produced by pathogenic mycobacteria may have evolved as adhesins from HBHA homologues in nonpathogenic saprophytes, where these molecules do not exert the same functions.

Comparative genomics allows us now to analyze *hbhA* genes in other mycobacteria. Although less well characterized than in *M. tuberculosis*, *M. avium* and *M. smegmatis*, *hbhA* genes are present in all currently sequenced mycobacterial genomes (Fig. 3A). The HBHA sequences can be aligned throughout the *Mycobacterium* genus and beyond, and the *hbhA* flanking genes can be compared among the different species. Genes homologous to *hbhA* are also present in *Nocardia farcinia*, *Rhodococcus*, and *Alcanivorax borkumensis* (Fig. 3B).

The genes located in the immediate neighborhood of *hbhA* are also conserved within the *Mycobacterium* genus (Fig. 2). The immediate downstream gene, coding for a putative conserved membrane protein, is invariably present in all sequenced mycobacterial genomes, as well as in the genomes of *Rhodococcus* and *Nocardia farcinia*. The two following genes of the same operon are also present in all the other mycobacteria, except for the first of these two genes, coding for a putative conserved secreted protein that appears to be absent in the genome of *M. avium* subsp. *paratuberculosis*. In this species, the start of the *deoC* gene is therefore located 48 base pairs downstream of the stop codon of the second gene of the operon. The gene is also absent in the chromosomes of *Rhodococcus* and *Nocardia farcinia*. The *deoC* gene in these two species is separated from the *hbhA* gene by 8 and 3 open reading frames, respectively.

Upstream of the *hbhA* gene, a conserved gene encoding the putative Xre-like regulator is present in all the mycobacterial species, as well as in *Nocardia farcinia* and *Rhodococcus*. However, this gene is a pseudogene in *Mycobacterium leprae*. In *Mycobacterium* sp. strains KMS and MCS and in *Mycobacte-rium vanbaalenii*, as well as in *Rhodococcus* and *Nocardia farcinia*, the stop codon of this gene overlaps the initiation codon of the *hbhA* gene, strongly suggesting therefore that in these species the two genes are part of the same polycistronic operon.

Upstream of this putative regulatory gene, a strictly conserved pair of genes is found, transcribed in opposite orientations from each other. The first of these genes encodes a putative regulator of the TetR family, whereas the second gene encodes a putative membrane protein. Further upstream, the genes are less well conserved, at least between the mycobacteria on the one side, and *Rhodococcus* and *Nocardia farcinia* on the other side.

IMMUNE RESPONSES TO HBHA: POTENTIAL FOR A VACCINE AND DIAGNOSTICS

B-Cell Responses

The first evidence that HBHA is recognized by the human immune system came from immunoblot analyses of sera from seven patients with active tuberculosis. All seven sera contained antibodies that reacted with purified HBHA, whereas none of the five tested negative control sera contained detectable antibodies to HBHA (Menozzi et al., 1996). These initial observations were confirmed in a later study with sera from 14 tuberculosis patients and from 14 *M. tuberculosis*-infected healthy individuals. Anti-HBHA immunonoglobulin G (IgG) were detected in 82% of the sera from tuberculosis patients and in 36% of the sera from infected healthy subjects (Masungi et al., 2002). Again, uninfected controls or BCG vaccinees did not produce detectable anti-HBHA antibodies.

More recently, a total number of 179 human sera were analyzed by enzyme-linked immunosorbent assays (ELISA) for the presence of anti-HBHA antibodies (Zanetti et al., 2005). These sera were obtained from 52 uninfected controls, from 38 healthy subjects that reacted to purified protein derivative (PPD) and from 111 patients with pulmonary tuberculosis. The sera were probed with two forms of HBHA, a non-methylated recombinant form produced in *E. coli*, and a methylated form produced in recombinant *M. smegmatis*, in comparison to the other mycobacterial antigens PPE44, PE_PGRS33 and antigen 85B. Whereas low or undetectable levels of anti-PPE44, anti-PE_PGRS33 and anti-antigen 85B antibodies were found in the sera from the tuberculosis patients, the sera of 44 out of the 111 tuberculosis patients (39%) contained anti-HBHA antibody levels that yielded high optical density readings when the methylated form of the protein was

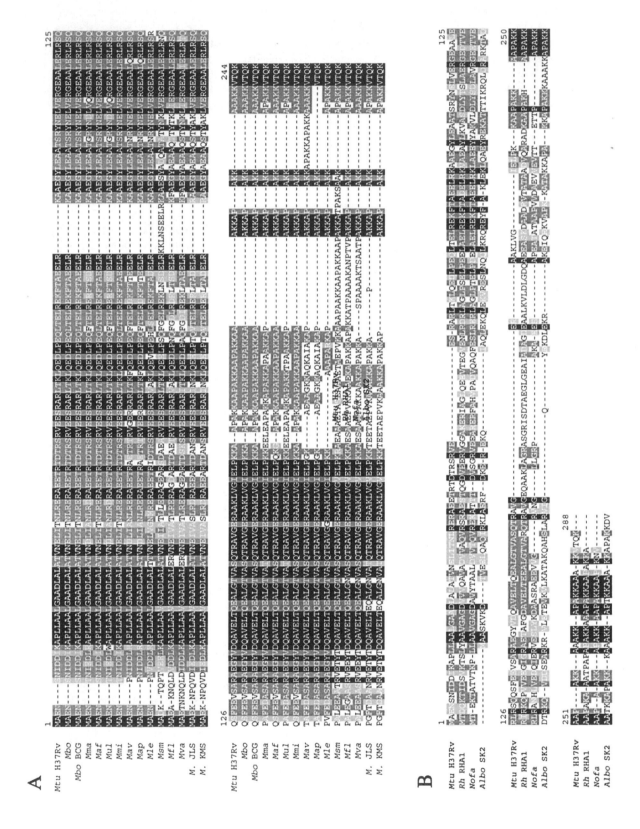

used. Interestingly, when nonmethylated HBHA was used, only 24 out of the 111 were positive. Two of the tuberculosis patients with high anti-HBHA antibody levels were negative in the tuberculin skin test, suggesting that HBHA may have a value for the development of serodiagnostic tests in conjunction with the tuberculin skin test to detected patients suffering from active tuberculosis.

These studies not only show that HBHA is immunogenic in tuberculosis patients, but they also revealed important differences in the immunogenicity of methylated compared with nonmethylated HBHA. These differences had already been seen by the differential reactivities of anti-HBHA monoclonal antibodies that react strongly with methylated HBHA but very weakly with the nonmethylated form of the protein (Pethe et al., 2002). Methylation-dependent B-cell epitopes could be generated by in vitro methylation of recombinant HBHA using mycobacterial extracts that contain HBHA:methyltransferases in the presence of S-adenosyl-methionine (Host et al., 2007) or by chemical methylation using $NaBH_4$ and formalin.

Relatively limited studies have addressed the isotype profiles of the anti-HBHA antibodies. However, recently HBHA was identified as the major Triton X-114-soluble M. tuberculosis antigen strongly reacting with immunoglobulin M (IgM) in sera from tuberculosis patients (Shin et al., 2006). The IgM reactivity was much stronger than the IgG reactivity in these sera. Again, the IgM from the tuberculosis patients recognized HBHA only in its methylated form. By comparing the anti-HBHA IgM and IgG reactivities with those to antigen 85B in chronic, early tuberculosis patients and healthy controls, the IgG and IgM responses to both antigens were higher in chronic than in early tuberculosis patients. Very low responses were seen in the sera from the healthy controls. Whereas the IgG responses to antigen 85B in the chronic patients were substantially higher than the IgG responses to HBHA, the levels of the IgM to these two antigens showed the opposite picture. Some of the anti-HBHA IgM-negative subjects were positive for anti-HBHA IgG. When the IgM and IgG levels were combined, the sensitivity for the detection of chronic and early tuberculosis were 52% and 24%, respectively.

Antibodies to HBHA from other mycobacterial species have also been considered for serodiagnostic use. The MAMP-HBHA has been tested by ELISA for reactivity of sera from cattle with evidence of M. paratuberculosis infection (Sechi et al., 2006). The sera from five out of seven infected cattle contained antibodies to recombinant MAMP-HBHA produced in M. smegmatis (thus presumably methylated), whereas only four out of 46 control cattle were positive for anti-MAMP-HBHA antibodies. Because the infected animals were not clinically ill, the detection of anti-MAMP-HBHA antibodies have been suggested to be useful for the early diagnosis of Johne's disease in cattle, as well as possibly for Crohn's disease in humans.

The role of antibodies in antimycobacterial immunity is a subject of intense discussion (Casadevall et al., 2004). Evidence that anti-HBHA antibodies can influence the course of M. tuberculosis infection was provided by the fact that coating the bacilli with anti-HBHA monoclonal antibodies before nasal infection substantially reduces the extrapulmonary dissemination of the bacteria (Pethe et al., 2001). The same anti-HBHA monoclonal antibodies also are able to reduce binding of M. tuberculosis to epithelial cells. Human studies have shown that sera from tuberculosis patients that contain anti-HBHA antibodies are able to inhibit in a dose-dependent manner the invasion of M. tuberculosis into A549 cells (Shin et al., 2006). Sera from patients with high levels of anti-HBHA IgM inhibited A549 cell invasion significantly better than sera from patients with lower anti-HBHA IgM levels. This observation suggests that anti-HBHA antibodies in tuberculosis patients can inhibit M. tuberculosis interactions with epithelial cells and thereby may contribute to a reduction of extrapulmonary dissemination if present in the respiratory tract at the time of infection.

Observations like these and others have led to the suggestion that there may be a role for antibodies in protection against mycobacterial infections, and that antibody therapies against diseases such as tuberculosis may be considered (Casadevall et al., 2004). However, intravenous administration of the anti-HBHA monoclonal antibodies to mice before aerosol infection with M. tuberculosis did not result in any significant reduction in bacterial counts in lungs and spleens, compared to antibody-naïve mice (Parra et al., 2004). Although these observations speak against a protective role of anti-HBHA

Figure 3. Alignments of the HBHA sequences form different mycobacterial (A) and nonmycobacterial (B) species. Strictly conserved residues are boxed in black, similar residues are boxed in gray, and dashes indicate gaps. Mtb H37Rv, M. tuberculosis H37Rv; Mbo, M. bovis; Mbo BCG, M. bovis BCG; Mma, M. marinum; Maf, M. africanum; Mul, M. ulcerans; Mmi, M. microti; Mav, M. avium; Map, M. avium subsp. paratuberculosis; Mle, M. leprae; Msm, M. smegmatis; Mfl, M. flavescens; Mva, M. vanbaalenii; M. JLS, Mycobacterium strain JLS; M. KMS, Mycobacterium strain KMS; Rh RHA1, Rhodococcus strain RHA1; Nofa 10152, Nocardia farcinia; Albo SK2, Alcanivorax borkumensis.

antibodies, the anatomical location or the isotype of anti-HBHA antibodies may be important for their anti-infectious activity. Especially, the IgA isotype may be relevant in that regard, if it can be used at mucosal sites, and passive IgA immunoprophylaxis has been proposed particularly for immunocompromized hosts, when vaccination is not possible (Williams et al., 2004). However, immunoglobulin A (IgA) has a relatively short half-life, and effective use of these antibodies may depend on finding ways to stabilize them sufficiently to ensure a minimal threshold level able to neutralize HBHA activity in the respiratory mucosa at the time of arrival of the pathogen.

T-Cell Responses

The most relevant protective immune responses to *M. tuberculosis* infections are without doubt T-cell responses. Both CD4$^+$ T cells and CD8$^+$ T cells have been shown to contribute to protection against tuberculosis (Flynn and Chan, 2001; Woodworth and Behar, 2006). Both T cell types can produce interferon-γ (IFN-γ), which is known to be essential for protection against *M. tuberculosis*. CD8$^+$ T cells can, in addition, express cytotoxic activities toward *M. tuberculosis*-infected cells, as well as bactericidal activities.

T cell responses to HBHA have been studied in *M. tuberculosis*-infected individuals. The majority of *M. tuberculosis*-infected healthy individuals have been reported to mount a strong T-cell response to HBHA, as evidenced by high levels of IFN-γ secreted by their peripheral lymphocytes upon stimulation with native HBHA (Masungi et al., 2002), whereas control subjects who were not infected did not. Curiously, similar to the B-cell responses, the T-cell responses in latently infected human subjects were much stronger to the native, methylated form than to the recombinant, nonmethylated form, but the response to the recombinant form could be enhanced when it was chemically methylated (Temmerman et al., 2004).

In contrast to infected healthy subjects, only very low HBHA-specific IFN-γ responses were detected in patients with active tuberculosis. The HBHA T-cell response thus provides an interesting immunological difference between infected healthy subjects and patients with active tuberculosis, suggesting that the HBHA T-cell response may contribute to protection against disease. These observations suggest also that HBHA may have a potential as a diagnostic antigen for the detection of latent *M. tuberculosis* infection (Hougardy et al., 2007a). Interestingly, when the tuberculosis patients were successfully treated, most of them acquired the abil-

ity to produce high levels of HBHA-specific IFN-γ, indicating that they had no intrinsic restriction to mount a HBHA-specific T-cell response. Therefore, the inability of their lymphocytes to produce IFN-γ upon HBHA stimulation is most likely due to reversible antigen-specific immunosuppression in the tuberculosis patients.

It has recently been shown that this immunosuppression of the HBHA T-cell response is due to the generation of regulatory T cells with the CD4$^+$CD25^{high}FoxP3$^+$ phenotype (Hougardy et al., 2007b). Depletion of these cells from the peripheral blood lymphocytes resulted in an increase of the HBHA-induced IFN-γ secretion to the levels found for the infected healthy subjects. That these regulatory T cells may participate in the pathogenesis of the disease is supported by the fact that they can be induced in vitro from peripheral blood mononuclear cells of infected healthy subjects but not of noninfected individuals, by incubating them with BCG in the presence of transforming growth factor β (TGF-β) (Hougardy et al., 2007c). Thus, the induction of these cells may be related to progression from latent infection to active disease by downregulating T-cell immunity to *M. tuberculosis*, which may tip the balance in favor of the mycobacteria. This may represent a mechanism of immune-evasion used by *M. tuberculosis*.

Phenotypic analyses of the HBHA-specific T cells from infected healthy subjects have shown that the IFN-γ secreted upon in vitro stimulation with this antigen comes from both the CD4$^+$ T and the CD8$^+$ T cells, as well as from a small proportion of CD4$^-$CD8$^-$ T cells (Masungi et al., 2002). Consistent with this observation, incubation of the lymphocytes with blocking antibodies to major histocompatibility (MHC) molecules indicated that both class I and class II MHC molecules are involved in HBHA presentation.

Phenotypical analyses of the IFN-γ-producing CD8$^+$ T cells from infected healthy subjects indicated that most of them expressed CD45RO, whereas the expression of CD57 was generally faint or nondetectable, indicating that these cells were activated in an antigen-specific manner and had expanded from memory T cells (Temmerman et al., 2005). The HBHA-specific CD8$^+$ T cells from latently infected individuals also were shown to express cytolytic activity on HBHA-loaded autologous monocyte-derived macrophages. This cytolytic activity depended on the release of granules containing granzyme and perforin. Consistent with these results, flow cytometry analyses indicated that HBHA-stimulated CD8$^+$ T cells produced elevated levels of perforin and granzyme. Interestingly, double staining for both IFN-γ and perforin indicated that most of the HBHA-stimulated CD8$^+$ T cells producing perforin were distinct from those that produced IFN-γ. Furthermore, no correlation was found between the percentages of HBHA-induced IFN-γ- and perforin-producing CD8$^+$ T cells, suggesting that *M. tuberculosis* infection induces at least two distinct HBHA-specific CD8$^+$ T cell subsets. Finally, HBHA-expanded CD8$^+$ T cells from peripheral blood of latently infected individuals were also able to inhibit growth of the mycobacteria within infected autologous macrophages.

Altogether these observations strongly argue in favor of a protective role of HBHA-specific T cells during latent infection with *M. tuberculosis*, suggesting that HBHA may not only be a potent antigen for the immunodetection of latently infected individuals, but also a potential candidate antigen for the development of new antituberculosis vaccines (Locht et al., 2006). CD8$^+$ T lymphocytes, in particular, have been reported to be involved in the control of latent *M. tuberculosis* infection by limiting its reactivation in mouse models (Van Pinxteren et al., 2000). The protective potential of HBHA has been demonstrated against *M. tuberculosis* challenge in several murine models (Locht et al., 2006).

However, initial attempts to induce protection with HBHA in mice have failed. These experiments made use of the genetic immunization with HBHA-encoding plasmids (Li et al., 1999). Administration of these constructs to C57BL/6 mice induced strong antibody and IL-4 responses to HBHA. However, there was a low IFN-γ response, and no significant protection was obtained, although similar constructs with other mycobacterial antigens reduced the bacterial counts up to 10-fold compared to the negative control animals.

However, when native, methylated HBHA was used as an immunogen in the presence of strong Th1 adjuvants, significant protection was induced in BABL/c mice against infection with *M. tuberculosis* (Temmerman et al., 2004). Protection levels seen with three injections of 5 µg of HBHA each were similar to those induced by vaccination with BCG. In contrast, when nonmethylated HBHA was used as an immunogen, no significant protection was seen under the same conditions. Similar results were obtained in the aerosol challenge model with C57BL/6 mice (Parra et al., 2004). These data indicate a crucial importance of the methylation of HBHA for its immunoprotective properties.

In contrast to DNA vaccination, immunization with the purified native or recombinant proteins induced a robust HBHA-specific IFN-γ response, in addition to a strong antibody response. In fact, mice immunized with the nonmethylated form produced comparable antibody and IFN-γ levels to those immunized with the methylated form (Temmerman et al., 2004). In addition, the methylation did not appear to influence the isotype profiles of the anti-HBHA antibodies. The mechanisms explaining the differences in the ability of the two forms of HBHA to induce protective immunity are not yet fully understood, but only T cells from mice immunized with the native form were able to induce IFN-γ production upon incubation with macrophages that were infected with *M. tuberculosis*, and only T cells from mice vaccinated with the native form were able to lyse BCG-infected macrophages and kill the intracellular mycobacteria (Temmerman et al., 2004). These observations indicate that important effector functions of the HBHA-specific T cells require the protein to be methylated.

Taken together, all of these observations suggest that native, methylated HBHA may be a good antigen for the immunodiagnosis of *M. tuberculosis* infections, that the peripheral blood HBHA-specific T-cell responses are indicative of a latent infection, and that HBHA is a powerful protective antigen and therefore may be proposed as one of the components of future, acellular vaccines against tuberculosis. The observations also indicate that measuring mere antibody or IFN-γ responses to protective proteins, such as HBHA, may not be sufficient to be used as correlates of protection but that other effector functions of T cells should also be considered.

CONCLUDING REMARKS AND FUTURE DIRECTIONS

Described for the first time approximately 10 years ago, HBHA has evolved into one of the few characterized *M. tuberculosis* virulence factors and antigens. Obviously, many questions remain unanswered and certainly constitute exciting topics for future work. The entire HBHA regulon needs to be worked out, its methyltransferases remain to be identified, and its mechanism of secretion is still to be uncovered. The precise role of HBHA in pathogenesis of tuberculosis, but also of other mycobacterial infections, is still an open issue, including the most relevant cell targets and receptors of HBHA. The functional role of HBHA-like molecules in genera other than *Mycobacterium*, and in nonpathogenic bacteria also remains unknown.

Although promising as an immunodiagnostic tool, especially for the detection of latently infected individuals, prospective clinical studies need to be carried out in a number of different target populations to evaluate the true diagnostic potential of this antigen, in particular in comparison, or in addition to other diagnostic tools. As protective immunogenicity appears to depend on the methylation of HBHA, the precise methylation pattern required for protection still awaits identification, as well as the mechanistic role of the methylation in T and B cell immunogenicity.

Answers to these questions may undoubtedly not only advance our basic understanding of the molecular biology of *M. tuberculosis* but also potentially provide us with new tools for diagnosis and immunoprophylaxis, which are essential for the control of one of the most devastating infectious diseases worldwide.

REFERENCES

Balasubramanian, V., E. H. Wiegeshaus, T. Taylor, and D. W. Smith. 1994. Pathogenesis of tuberculosis: pathway to apical localization. *Tuber. Lung Dis.* 75:168–178.

Baron, M. J., G. R. Bolduc, M. B. Goldberg, T. C. Aupérin, and L. C. Madoff. 2004. Alpha C protein of group B *Streptococcus* binds host cell surface glycosaminoglycan and enters cells by an actin-dependent mechanism. *J. Biol. Chem.* 279:24714–24723.

Bermudez, L. E., and J. Goodman. 1996. *Mycobacterium tuberculosis* invades and replicates within type II alveolar cells. *Infect. Immun.* 64:1400–1406.

Bermudez, L. E., F. J. Sangari, P. Kolonoski, M. Petrofsky, and J. Goodman. 2002. The efficiency of the translocation of *Mycobacterium tuberculosis* across a bilayer of epithelial and endothelial cells as a model for the alveolar wall is a consequence of transport within mononuclear phagocytes and invasion of alveolar epithelial cells. *Infect. Immun.* 70:140–146.

Bermudez, L. E., K. Shelton, and L. S. Young. 1995. Comparison of the ability of *Mycobacterium avium*, *M. smegmatis* and *M. tuberculosis* to invade and replicate within HEp-2 epithelial cells. *Tuber. Lung Dis.* 76:240–247.

Biet, F., M. A. de Melo Marques, M. Grayon, E. K. X. da Silveira, P. J. Brennan, H. Drobecq, D. Raze, M. C. Vidal Pessolani, C. Locht, and F. D. Menozzi. 2007. *Mycobacterium smegmatis* produces a HBHA homologue which is not involved in epithelial adherence. *Microbes Infect.* 9:175–182.

Braunstein, M., B. J. Espinosa, J. Chan, J. T. Belisle, and W. R. Jacobs, Jr. 2003. SecA2 functions in the secretion of superoxide dismutase A and in the virulence of *Mycobacterium tuberculosis*. *Mol. Microbiol.* 48:453–464.

Cardin, A. D., and H. J. R. Weintraub. 1989. Molecular modeling of protein-glycosaminoglycan interactions. *Arteriosclerosis* 9:21–32.

Casadevall, A., E. Dadachova, and L.-A. Pirofski. 2004. Passive antibody therapy for infectious diseases. *Nat. Rev. Microbiol.* 2:695–703.

Chackerian, A. A., J. M. Alt, T. V. Perera, C. C. Dascher, and S. M. Behar. 2002. Dissemination of *Mycobacterium tuberculosis* is influenced by host factors and precedes the initiation of T-cell immunity. *Infect. Immun.* 70:4501–4509.

Chacon, O., L. E. Bermudez, and R. G. Barletta. 2004. Johne's disease, inflammatory bowel disease, and *Mycobacterium paratuberculosis*. *Ann. Rev. Microbiol.* 58:329–363.

Cooper, A., G. Tal, O. Lider, and Y. Shaul. 2005. Cytokine induction by the hepatitis B virus capsid in macrophages is facilitated by membrane heparan sulfate and involves TLR2. *J. Immunol.* 175:3165–3176.

Delogu, G., and M. J. Brennan. 1999. Functional domains present in the mycobacterial hemagglutinin HBHA. *J. Bacteriol.* 181:7464–7469.

Delogu, G., A. Bua, C. Pusceddu, M. Parra, G. Fadda, M. J. Brennan, and S. Zanetti. 2004. Expression and purification of recombinant methylated HBHA in *Mycobacterium smegmatis*. *FEMS Microbiol. Lett.* 239:33–39.

Delogu, G., M. Sanguinetti, B. Posteraro, S. Rocca, S. Zanetti, and G. Fadda. 2006. The *hbhA* gene of *Mycobacterium tuberculosis* is specifically upregulated in the lungs but not in the spleens of aerogenically infected mice. *Infect. Immun.* 74:3006–3011.

Dobos, K. M., E. A. Spotts, F. D. Quinn, and C. H. King. 2000. Necrosis of lung epithelial cells during infection with *Mycobacterium tuberculosis* is preceded by cell permeation. *Infect. Immun.* 68:6300–6310.

Dupres, V., F. D. Menozzi, C. Locht, B. H. Clare, N. L. Abbott, S. Guenot, C. Bompard, D. Raze, and Y. F. Dufrêne. 2005. Nanoscale mapping and functional analysis of individual adhesions on living bacteria. *Nat. Methods* 2:515–520.

Flynn, J. L., and J. Chan. 2001. Immunology of tuberculosis. *Annu. Rev. Immunol.* 19:93–129.

Grant, C. C. R., M. P. Bos, and R. J. Belland. 1999. Proteoglycan receptor binding by *Neisseria gonorrhoeae* MS11 is determined by the HV-1 region of OpaA. *Mol. Microbiol.* 32:233–242.

Guinn, K. M., M. J. Hickey, S. K. Mathur, K. L. Zakel, J. E. Grotzke, D. M. Lewinsohn, S. Smith, and D. R. Sherman. 2004. Individual RD1-region genes are required for export of ESAT-6/CFP-10 and for virulence of *Mycobacterium tuberculosis*. *Mol. Microbiol.* 51:359–370.

Gysin, J., B. Pouvelle, M. Le Tonqueze, L. Edelman, and M. C. Boffa. 1997. Condroitin sulfate of thrombomodulin is an adhesin receptor for *Plasmodium falciparum*-infected erythrocytes. *Mol. Biochem. Parasitol.* 88:267–271.

Hernandez-Pando, R., M. Jeyanathan, G. Mengistu, D. Aguilar, H. Orozco, M. Harboe, G. A. W. Rook, and G. Bjune. 2000. Persistence of DNA from *Mycobacterium tuberculosis* in superficially normal lung tissue during latent infection. *Lancet* 356:2133–2137.

Host, H., H. Drobecq, C. Locht, and F. D. Menozzi. 2007. Enzymatic methylation of the *Mycobacterium tuberculosis* heparin-binding haemagglutinin. *FEMS Microbiol. Lett.* 268:144–150.

Hougardy, J. M., K. Schepers, S. Place, A. Drowart, V. Lechevin, V. Verscheure, A.-S. Debire, T. M. Doherty, J. P. Van Vooren, C. Locht, and F. Mascart. 2007a. Heparin-binding hemagglutinin-induced IFN-γ release as a diagnostic tool for latent tuberculosis. *PLoS One* 2:e926.

Hougardy, J. M., S. Place, M. Hildebrand, A. Drowart, A.-S. Debrie, C. Locht, and F. Mascart. 2007b. Regulatory T cells depress immune responses to protective antigens in active tuberculosis. *Am. J. Respir. Crit. Care Med.* 176:409–416.

Hougardy, J. M., V. Verscheure, C. Locht, and F. Mascart. 2007c. In vivo expansion of CD4+CD25^high^FOXP3+CD127^low/−^ regulatory T cells from peripheral blood lymphocytes of healthy *Mycobacterium tuberculosis*-infected humans. *Microbes Infect.* 9:1325–1332.

Leong, J. M., P. E. Morrisscy, E. Ortega-Barria, M. E. A. Pereira, and J. Coburn. 1995. Hemagglutination and proteoglycan binding of the Lyme disease spirochete, *Borrelia burgdorferi*. *Infect. Immun.* 63:874–883.

Li, Z., A. Howard, C. Kelley, G. Delogu, F. Collins, and S. Morris. 1999. Immunogenicity of DNA vaccines expressing tuberculosis proteins fused to tissue plasminogen activator signal sequences. *Infect. Immun.* 67:4780–4786.

Locht, C., J.-M. Hougardy, C. Rouanet, S. Place, and F. Mascart. 2006. Heparin-binding hemagglutinin, from an extrapulmonary dissemination factor to a powerful diagnostic and protective antigen against tuberculosis. *Tuberculosis* 86:303–309

Lupas, A., M. Van Dyke, and J. Stock. 1991. Predicting coiled coils from protein sequences. *Science* 252:1162–1164.

Masungi, C., S. Temmerman, J. P. Van Vooren, A. Drowart, K. Pethe, F. D. Menozzi, C. Locht, and F. Mascart. 2002. Differential T and B cell responses against *Mycobacterium tuberculosis* heparin-binding hemagglutinin adhesin in infected healthy individuals and patients with tuberculosis. *J. Infect. Dis.* 185:513–520.

McDonough, K. A., and Y. Kress. 1995. Cytotoxicity of lung epithelial cells is a virulence-associated phenotype of *Mycobacterium tuberculosis*. *Infect. Immun.* 63:4802–4811.

Menozzi, F. D., R. Bischoff, E. Fort, M. J. Brennan, and C. Locht. 1998. Molecular characterization of the mycobacterial heparin-binding hemagglutinin, a mycobacterial adhesin. *Proc. Natl. Acad. Sci. USA* 95:12625–12630.

Menozzi, F. D., R. Mutomo, G. Renauld, C. Gantiez, J. H. Hannah, E. Leininger, M. J. Brennan, and C. Locht. 1994. Heparin-inhibitable lectin activity of the filamentous hemagglutinin adhesin of *Bordetella pertussis*. *Infect. Immun.* 62:769–778.

Menozzi, F. D., K. Pethe, P. Bifani, F. Soncin, M. J. Brennan, and C. Locht. 2002. Enhanced bacterial virulence through exploitation of host glycosaminoglycans. *Mol. Microbiol.* 43:1379–1396.

Menozzi, F. D., V. M. Reddy, D. Cayet, D. Raze, A.-S. Debrie, M.-P. Dehouck, R. Cecchelli, and C. Locht. 2006. *Mycobacterium tuberculosis* heparin-binding hemagglutinin adhesin (HBHA) triggers receptor-mediated transcytosis without altering the integrity of tight junctions. *Microbes Infect.* 8:1–9.

Menozzi, F. D., J. H. Rouse, M. Alavi, M. Laude-Sharp, J. Muller, R. Bischoff, M. J. Brennan, and C. Locht. 1996. Identification of a heparin-binding hemagglutinin present in mycobacteria. *J. Exp. Med.* 184:993–1001.

Metha, P. K., C. H. King, E. H. White, J. J. Murtagh, Jr., and F. D. Quinn. 1996. Comparison of in vitro models for the study of *Mycobacterium tuberculosis* invasion and intracellular replication. *Infect. Immun.* 64:2673–2679.

Mischenki, V. V., M. A. Kapina, E. B. Eruslanov, E. V. Kondratieva, I. V. Lyadova, D. B. Young, and A. S. Apt. 2004. Mycobacterial dissemination and cellular responses after 1-lobe restricted tuberculosis infection of genetically susceptible and resistant mice. *J. Infect. Dis.* 190:2137–2145.

Mondor, I., S. Ugolini, and Q. J. Sattentau. 1998. Human immunodeficiency virus type I attachment to HeLa CD4 cells is CD4 independent and gp120 dependent and requires cell surface heparans. *J. Virol.* 72:3623–3634.

Mueller-Oritz, S. L., E. Sepulveda, M. R. Olsen, C. Jagannath, A. R. Wanger, and S. J. Norris. 2002. Decreased infectivity despite unaltered C3 binding by a Δhbha mutant of *Mycobacterium tuberculosis*. *Infect. Immun.* 70:6751–6760.

Mueller-Oritz, S. L., A. R. Wanger, and S. J. Norris. 2001. Mycobacterial protein HBHA binds human complement component C3. *Infect. Immun.* 69:7501–7511.

Musser, J. M., A. Amin, and S. Ramaswamy. 2000. Negligible genetic diversity of *Mycobacterium tuberculosis* host immune system protein targets: evidence of limited selective pressure. *Genetics* 155:7–16.

Neyrolles, O., R. Hernandez-Pando, F. Pietri-Rouxel, P. Fomès, L. Tailleux, J. A. B. Payan, E. Pivert, Y. Bordat, D. Aguilar, M.-C. Prévost, C. Petit, and B. Gicquel. 2006. Is adipose tissue a place for *Mycobacterium tuberculosis* persistence? *PLoS ONE* 1:e43.

Parra, M., T. Pickett, G. Delogu, V. Dheenadhayalan, A.-S. Debrie, C. Locht, and M. J. Brennan. 2004. The mycobacterial heparin-binding hemagglutinin is a protective antigen in the

mouse aerosol challenge model of tuberculosis. *Infect. Immun.* 72:6799–6805.

Pethe, K., S. Alonso, F. Biet, G. Delogu, M. J. Brennan, C. Locht, and F. D. Menozzi. 2001. The heparin-binding haemagglutinin of *M. tuberculosis* is required for extrapulmonary dissemination. *Nature* 412:190–194.

Pethe, K., M. Aumercier, E. Fort, C. Gatot, C. Locht, and F. D. Menozzi. 2000. Characterization of the heparin-binding site of the mycobacterial heparin-binding hemagglutinin adhesin. *J. Biol. Chem.* 275:14273–14280.

Pethe, K., P. Bifani, H. Drobecq, C. Sergheraert, A.-S. Debrie, C. Locht, and F. D. Menozzi. 2002. Mycobacterial heparin-binding hemagglutinin and laminin-binding protein share antigenic methyllysines that confer resistance to proteolysis. *Proc. Natl. Acad. Sci. USA* 99:10759–10764.

Pethe, K., V. Puech, M. Daffé, C. Josenhans, H. Drobecq, C. Locht, and F. D. Menozzi. 2001. *Mycobacterium smegmatis* laminin-binding glycoprotein shares epitopes with *Mycobacterium tuberculosis* heparin-binding haemagglutinin. *Mol. Microbiol.* 39:89–99.

Reddy, V. M., and B. Kumar. 2000. Interaction of *Mycobacterium avium* complex with human respiratory epithelial cells. *J. Infect. Dis.* 181:1189–1193.

Rouse, D. A., S. L. Morris, A. B. Karpas, J. C. Mackall, P. G. Probst, and S. D. Chaparas. 1991. Immunological characterization of recombinant antigens isolated from a *Mycobacterium avium* lambda gt11 expression library by using monoclonal antibody probes. *Infect. Immun.* 59:2595–2600.

Russell, D. G. 2001. *Mycobacterium tuberculosis*: here today, and here tomorrow. *Nat. Rev. Mol. Cell Biol.* 2:569–577.

Saint-Joanis, B., C. Demangel, M. Jackson, P. Brodin, L. Marsollier, H. Boshoff, and S. T. Cole. 2006. Inactivation of Rv2525c, a substrate for the twin arginine translocation (Tat) system of *Mycobacterium tuberculosis*, increases beta-lactam susceptibility and virulence. *J. Bacteriol.* 188:6669–6679.

Sato, J., J. Schorey, V. A. Ploplis, E. Haalboom, L. Krahule, and F. J. Castellino. 2003. The fibrinolytic system in dissemination and matrix protein deposition during a *Mycobacterium* infection. *Am. J. Pathol.* 163:517–531.

Schlesinger, L. S., C. G. Bellinger-Kawahara, N. R. Payne, and M. A. Horwitz. 1990. Phagocytosis of *Mycobacterium tuberculosis* is mediated by human monocyte complement receptors and complement component C3. *J. Immunol.* 144:2771–2780.

Schneewind, O., P. Model, and V. A. Fischetti. 1992. Sorting of protein A to the staphylococcal cell wall. *Cell* 70:267–281.

Sechi, L. A., N. Ahmed, G. E. Felis, I. Duprè, S. Cannas, G. Fadda, A. Bua, and S. Zanetti. 2006. Immunogenicity and cytoadherence of recombinant heparin binding haemagglutinin (HBHA) of *Mycobacterium avium* subsp. *paratuberculosis*: functional promiscuity or a role in virulence? *Vaccine* 24:236–243.

Shepard, C. C. 1957. Growth characteristics of tubercule bacilli and certain other mycobacteria in HeLa cells. *J. Exp. Med.* 105:39–48.

Shin, A. R., K. S. Lee, J. S. Lee, S. Y. Kim, C. H. Song, S. B. Jung, C. S. Yang, E. K. Jo, J. K. Park, T. H. Paik, and H. J. Kim. 2006. *Mycobacterium tuberculosis* HBHA protein reacts strongly with the serum immunoglobulin M of tuberculosis patients. *Clin. Vaccine Immunol.* 13:869–875.

Su, H., L. Raymond, D. D. Rockey, E. Fischer, T. Hackstadt, and H. D. Caldwell. 1996. A recombinant *Chlamydia trachomatis* major outer membrane protein binds to heparan sulfate receptors on epithelial cells. *Proc. Natl. Acad. Sci. USA* 93:11143–11148.

Teitelbaum, R., W. Schubert, L. Gunther, Y. Kress, F. Macaluso, J. W. Pollard, D. N. McMurray, and B. R. Bloom. 1999. The

M cells as a portal of entry to the lung for the bacterial pathogen *Mycobacterium tuberculosis*. *Immunity* **10**:641–650.

Temmerman, S., K. Pethe, M. Parra, S. Alonso, C. Rouanet, T. Picket, A. Drowart, A.-S. Debrie, G. Delogu, F. D. Menozzi, C. Sergheraert, M. J. Brennan, F. Mascart, and C. Locht. 2004. Methylation-dependent T cell immunity to Mycobacterium tuberculosis heparin-binding hemagglutinin. *Nat. Med.* **10**:935–941.

Temmerman, S. T., S. Place, A.-S. Debrie, C. Locht, and F. Mascart. 2005. Effector functions of heparin-binding hemagglutinin-specific CD8+ T lymphocytes in latent human tuberculosis. *J. Infect. Dis.* **192**:226–232.

Tyewska, S. K., V. A. Fischetti, and R. J. Gibbons. 1988. Binding selectivity of *Streptococcus pyogenes* and M-proteins to epithelial cells differs from that of lipoteichoic acid. *Curr. Microbiol.* **16**:209–216.

Van Pinxteren, L. A., J. P. Cassidy, B. H. Smedegaard, E. M. Agger, and P. Andersen. 2000. Control of latent *Mycobacterium tuberculosis* infection is dependent on CD8+ T cells. *Eur. J. Immunol.* **30**:3689–3698.

Williams, A., R. Reljic, I. Naylor, S. O. Clark, G. Falero-Diaz, M. Singh, S. Challacombe, P. D. Marsh, and J. Ivanyi. 2004. Passive protection with immunoglobulin A antibodies against tuberculosis early infection in lungs. *Immunology* **111**:328–333.

Woodworth, J. S., and S. M. Behar. 2006. *Mycobacterium tuberculosis*-specific CD8+ T cells and their role in immunity. *Crit. Rev. Immunol.* **26**:317–352.

Zanetti, S., A. Bua, G. Delogu, C. Pusceddu, M. Mura, F. Saba, P. Pirina, C. Garzelli, C. Vertuccio, L. A. Sechi, and G. Fadda. 2005. Patients with pulmonary tuberculosis develop a strong humoral response against methylated heparin-binding hemagglutinin. *Clin. Diagn. Lab. Immunol.* **12**:1135–1138.

Chapter 20

The Role of Mycobacterial Kinases and Phosphatases in Growth, Pathogenesis, and Cell Wall Metabolism

Anil K. Tyagi, Ramandeep Singh, and Vibha Gupta

A bacterial pathogen is a microorganism that is able to evade the various normal defenses of a particular host to cause infection. To survive and prosper are the fundamental objectives for both bacteria and host. Two opposing forces—the host immunity and the virulence factors of the pathogen—decide the outcome of bacterial infections (Bliska et al., 1993; Medzhitov and Janeway, 2000; Hornef et al., 2002; Rosenberger and Finlay, 2003; Coombes et al., 2004; Hilleman, 2004). *Mycobacterium tuberculosis*, a gram-positive bacterium and the causative agent of tuberculosis (TB), is a successful pathogen that overcomes numerous challenges presented by the immune system of the host (Houben et al., 2006; Cosma et al., 2003; Pieters and Gatfield, 2002). Within macrophages and other phagocytic cells, *M. tuberculosis* encounters a hostile environment, which is a defense mechanism of the host to limit the growth and spread of bacterial pathogens to neighboring cells. Infection with *M. tuberculosis* does not always result in the onset of disease. The host arrests the infection by developing a cell-mediated immune response, resulting in the formation of lesions characteristic of mycobacterial infections (Dannenberg, 1991).

To be well adapted to the numerous environmental conditions that *M. tuberculosis* encounters in the host, it must be able to induce or repress a number of genes for a quick adjustment to new conditions. This series of events occurs coordinately by means of numerous mechanisms of signal transduction and transcriptional regulation. The entry of *M. tuberculosis* into the host involves cross-talk of signals between the host and the mycobacterium, resulting in reprogramming of host signaling network. The mycobacterium mediates the transduction of several intracellular and extracellular cues to appropriate cellular responses by way of protein phosphoryla-

tion, which enables it to regulate important cellular functions and mount an adaptive response to cope with diverse environmental stresses in humans.

Protein phosphorylation/dephosphorylation represents the central mechanism for distribution of signals to various parts of the cell for regulation of cell growth, differentiation, mobility, and survival (Stock et al., 1989). The complex and diverse bacterial signaling proteins include histidine kinases, Ser/Thr/Tyr protein kinases (STYK), Ser/Thr/Tyr phosphatases, methyl-accepting chemotaxis receptors, adenylate and diguanylate cyclases, and c-di-GMP phosphodiesterases (Stock et al., 2000; Galyov et al., 1993; Hunter, 1995; Stone and Dixon, 1994; Cozzone, 1993, 2005; Römling et al., 2005). Cyclic nucleotides and other diverse molecules such as (p)ppGpp, Ca^{2+}, inositol triphosphate, and diacylglycerol have also been recognized to function as second messengers in signal transduction processes (Galperin et al., 2001).

The cell wall of *M. tuberculosis* plays a crucial role in its defense against the host (see chapter 14). *M. tuberculosis* has approximately 250 genes involved in its lipid metabolism, and the lipid contents of the pathogen contribute to 60% of the cell dry weight (Daffe and Draper, 1998; Cole et al., 1998) (see chapter 1). A number of genes involved in the synthesis of essential components of cell envelope for maintaining appropriate cell wall architecture of *M. tuberculosis* have been identified, and their requirement for the virulence of *M. tuberculosis* has been established (Bhakta et al., 2004; Camacho et al., 1999; Dubnau et al., 2000; Glickman et al., 2000). Changes in cell wall composition in response to various environmental stimuli are central to the adaptation of *M. tuberculosis* during infection. Many of these stimuli are transduced in the bacteria by sensor

Anil K. Tyagi and Vibha Gupta • Department of Biochemistry, University of Delhi South Campus, Benito Juarez Road, New Delhi 110 021, India. **Ramandeep Singh** • Laboratory of Immunogenetics, Tuberculosis Research Section, NIAID, National Institutes of Health, Rockville, MD 20852.

kinases on the mycobacterial membrane, enabling the pathogen to modify itself for survival in the hostile environment. This chapter summarizes studies on the role of mycobacterial kinases and phosphatases (i) in the growth and pathogenesis of mycobacterium and (ii) in the cell wall metabolism of the pathogen.

ROLE OF MYCOBACTERIAL KINASES AND PHOSPHATASES IN THE GROWTH AND PATHOGENESIS OF MYCOBACTERIA

Protein Kinases

Protein kinases can be classified into two super families based on their sequence similarities and enzymatic specifications: (i) the histidine kinase superfamily, which autophosphorylates a conserved histidine residue, the best known examples being histidine kinases belonging to a two-component family found in bacteria, fungi, and plants (Stock et al., 2000; Alex and Simon, 1994), and (ii) the superfamily of serine, threonine, and tyrosine kinases that phosphorylate on serine, threonine, and tyrosine residues, respectively (Hunter, 1995; Hanks et al., 1988; Stock et al., 1989). No sequence similarities exist between the proteins from these two superfamilies, although some sequence conservation does exist within the members of the same superfamily.

Earlier, the two-component systems (TCSs) involving a histidine kinase (HK) and a response regulator (RR) were considered to play a key role in phosphotransfer mechanism for signal transduction in bacteria, whereas serine/threonine protein kinases (STPKs) and their associated phosphatases were more relevant to signal transduction pathways in eukaryotes. However, with the inflow of bacterial genome sequences, it is now known that these eukaryotic-like protein kinases and phosphatases are present in prokaryotes also and play an important role in bacterial metabolism and pathogenesis. The presence of these eukaryotic-like signaling systems, in addition to TCSs, raises curiosity about their physiological roles and biochemical mechanisms. However, based on the chemical stability of various phosphorylated amino acids, it is now thought that whereas two-component signals involving histidine are often employed for the generation of transient/short-term signals, Ser/Thr/Tyr phosphorylations usually produce long-term signals.

Table 1 describes the presence of Ser/Thr/Tyr kinases and phosphatases and TCSs in various species of mycobacteria. It is clear that several of these enzymes such as PknA, PknB, PknG, PknL, PstP, RegX3, MtrA, MtrB, MprA, MprB, PrrA, PrrB, SenX3, PdtaR, and PdtaS are present uniformly in all

mycobacteria, whose genome sequence is available. Interestingly, the highest identity is represented in the case of various orthologues from *M. bovis*, emphasizing its close relationship with *M. tuberculosis*. *Mycobacterium leprae* has the smallest number of orthologues present, reflecting on its dependence on the host due to reductive evolution by shedding off the genes (Table 1).

Two-component systems and their role in mycobacterial growth

This signaling mechanism is widespread in prokaryotes and is also found in some eukaryotes. The basic TCS is composed of two proteins: (i) sensor HK that senses external signals, and (ii) an RR to which the signal is transmitted (Fig. 1). The stimulus causes the HK to autophosphorylate a conserved histidine residue, forming a highly reactive phosphoramidate bond. The cognate RR catalyzes the phosphoryl transfer to a conserved aspartate within its own receiver domain. The phosphorylated response regulator interacts with transcription factors, which in turn up or down regulate number of genes (Stock et al., 1995). The resulting changes in global gene expression direct the organism to respond to the initial signal sensed by the histidine kinase. The TCSs are normally cotranscribed in pairs. Most histidine kinases and response regulators are highly conserved; nevertheless, there is little sequence similarity in the sensing domain of different histidine kinases. This variety of extracellular, intracellular, and transmembrane sensor domains accounts for the repertoire of molecules that can initiate the signal.

A number of studies have been carried out to understand the role of TCSs in the physiology of mycobacteria (Zahrt and Deretic, 2001; Haydel and Clark-Curtiss, 2004; Tyagi and Sharma, 2004). *M. tuberculosis* has only 12 paired TCS homologues compared with more than 30 present in *Escherichia coli* or *Bacillus subtilis*, which perhaps is a reflection on the presence of a relatively much higher number of eukaryotic serine/threonine kinases in *M. tuberculosis*. In addition to 12 TCSs, *M. tuberculosis* genome also has six orphan unlinked RR genes and 2 HK genes, pointing to the complexity of coordinated regulation in this organism (Cole et al., 1998).

Based on the global expression analysis during the growth of *M. tuberculosis* in human macrophages, the TCS genes can be classified into three groups: (i) constitutively expressed (*rv1626, rv2027c* and *mtrA*); (ii) differentially expressed (*regX3, phoP, prrA, mprA, kdpE, trcR, devR,* and *tcrX* and orphan HK *Rv3220c*); and (iii) unexpressed (not showing any detectable expression) during the growth of

Table 1. Presence of Ser/Thr/Tyr kinases and phosphatases and two-component systems in mycobacteria

S. no.	M. tuberculosis[a] H37Rv	Presence or absence (identity) of orthologue						
		M. ulcerans Agy99	M. avium 104	M. avium subsp. paratuberculosis K-10	M. leprae TN	M. smegmatis mc²155	M. vanbaalenii PYR-1	M. bovis AF2122/97
1	PknA Rv0015c	+ (77%) MUL_0019	+ (92%) MAV_0019	+ (95%) MAP0018c	+ (88%) ML0017	+ (91%) MSMEG_0030	+ (83%) Mvan_0024	+ (100%) Mb0015c
2	PknB Rv0014c	+ (89%) MUL_0018	+ (86%) MAV_0017	+ (86%) MAP0016c	+ (86%) ML0016	+ (69%) MSMEG_0028	+ (75%) Mvan_0023	+ (100%) Mb0014c
3	PknD Rv0931c	P MUL_4438	+ (67%) MAV_4238	+ (66%) MAP3387c	P ML0743	+ (53%) MSMEG_1200	+/−	+ (100%) Mb0955c
4	PknE Rv1743	+ (72%) MUL_3180	+ (70%)[b] MAV_1417	+ (57%) MAP1049c	P ML0992	−	+/−	+ (99%) Mb1772
5	PknF Rv1746	+ (70%) MUL_4471	+ (64%) MAV_3145	+ (65%) MAP1332	P ML1386	+ (68%) MSMEG_3677	+ (68%) Mvan_5635	+ (99%) Mb1775
6	PknG Rv0410c	+ (87%) MUL_2810	+ (85%) MAV_4751	+ (85%) MAP3893c	+ (83%) ML0304	+ (79%) MSMEG_0786	+ (80%) Mvan_0703	+ (100%) Mb0418c
7	PknH Rv1266c	+ (70%) MUL_3375	+ (70%)[b] MAV_1417	+ (88%) MAP2504	−	+/−	+/−	+ (100%) Mb1297c
8	PknI Rv2914c	+ (51%) MUL_2042	−	+/−	P ML1620	+/−	+/−	+ (100%) Mb2938c
9	PknJ Rv2088	+/−	+/−	+/−	−	−	+/−	+ (100%) Mb2115
10	PknK Rv3080c	+/−	−	−	−	+ (52%) MSMEG_0529	+ (58%) Mvan_1076	+ (99%) Mb3107c
11	PknL Rv2176	+ (78%) MUL_3519	+ (72%) MAV_2318	+ (73%) MAP1914	+ (75%) ML0897	+ (63%) MSMEG_4243	+ (65%) Mvan_3539	+ (99%) Mb2198
12	PPP/PstP Rv0018c	+ (89%) MUL_0022	+ (82%) MAV_0022	+ (82%) MAP0021c	+ (81%) ML0020	+ (77%) MSMEG_0033	+ (78%) Mvan_0027	+ (99%) Mb0018c
13	MptpA Rv2234	+ (80%) MUL_1319	+ (80%) MAV_2206	+ (81%) MAP1985	P ML1643	+ (71%) MSMEG_4309	+ (72%) Mvan_3622	+ (100%) Mb2258
14	MptpB Rv0153c	+ (73%) MUL_4739	+ (74%) MAV_5145	+ (77%)[c] MAP3568c-MAP3569c	−	+ (54%) MSMEG_0100	+ (53%) Mvan_0115	+ (99%) Mb0158c
15	PhoP Rv0757	+ (95%) MUL_0463	+ (94%) MAV_0701	+ (93%) MAP0591	P ML2239	+ (93%) MSMEG_5872	+ (90%) Mvan_5175	+ (100%) Mb0780
16	PhoR Rv0758	+ (74%) MUL_0464	+ (76%) MAV_0703	+ (76%) MAP0592	P ML2238	+ (71%) MSMEG_5870	+ (71%) Mvan_5174	+ (99%) Mb0781
17	TcrX Rv3765c	+ (92%) MUL_4391	+ (94%) MAV_0304	+ (94%) MAP0259	P ML2361	+ (85%) MSMEG_4990	+ (85%) Mvan_4429	+ (100%) Mb3791c
18	TcrY Rv3764c	P MUL_4390	+ (82%) MAV_0305	+ (81%) MAP0260	P ML2362	+ (58%) MSMEG_4989	+ (58%) Mvan_4428	+ (99%) Mb3790c
19	SenX3 Rv0490	+ (83%) MUL_4559	+ (84%) MAV_4661	+ (83%) MAP3982	+ (84%) ML2440	+ (72%) MSMEG_0936	+ (76%) Mvan_0830	+ (100%) Mb0500
20	RegX3 Rv0491	+ (98%) MUL_4560	+ (99%) MAV_4660	+ (99%) MAP3983	+ (95%) ML2439	+ (92%) MSMEG_0937	+ (94%) Mvan_0831	+ (100%) Mb0501

Continued on following page

Table 1. *Continued*

S. no.	*M. tuberculosis*[a] H37Rv	*M. ulcerans* Agy99	*M. avium* 104	*M. avium* subsp. *paratuberculosis* K-10	*M. leprae* TN	*M. smegmatis* mc²155	*M. vanbaalenii* PYR-1	*M. bovis* AF2122/97
				Presence or absence (identity) of orthologue				
21	MtrA Rv3246c	+ (99%) MUL_2582	+ (99%) MAV_4209	+ (99%) MAP3360c	+ (98%) ML0773	+ (96%) MSMEG_1874	+ (96%) Mvan_1752	+ (100%) Mb3274c
22	MtrB Rv3245c	+ (90%) MUL_2581	+ (87%) MAV_4208	+ (87%) MAP3359c	+ (88%) ML0774	+ (82%) MSMEG_1875	+ (84%) Mvan_1753	+ (99%) Mb3273c
23	MprA Rv0981	+ (94%) MUL_4701	+ (97%) MAV_1094	+ (97%) MAP0916	+ (93%) ML0174	+ (90%) MSMEG_5488	+ (92%) Mvan_4844	+ (99%) Mb1007
24	MprB Rv0982	+ (86%) MUL_4700	+ (83%) MAV_1095	+ (84%) MAP0917	+ (81%) ML0175	+ (78%) MSMEG_5487	+ (75%) Mvan_4843	+ (99%) Mb1008
25	PrrA Rv0903c	+ (96%) MUL_0248	+ (96%) MAV_1022	+ (96%) MAP0834c	+ (95%) ML2123	+ (93%) MSMEG_5662	+ (94%) Mvan_5009	+ (100%) Mb0927c
26	PrrB Rv0902c	+ (94%) MUL_0249	+ (82%) MAV_1021	+ (80%) MAP0833c	+ (92%) ML2124	+ (81%) MSMEG_5663	+ (79%) Mvan_5010	+ (100%) Mb0926c
27	KdpD Rv1028c	–	+ (80%) MAV_1170	+ (80%) MAP0996c	–	+ (77%) MSMEG_5395	+ (79%) Mvan_4762	+ (99%) Mb1056c
28	KdpE Rv1027c	+/–	+ (91%) MAV_1169	+ (91%) MAP0995c	–	+ (85%) MSMEG_5396	+ (83%) Mvan_4763	+ (100%) Mb1055c
29	TrcR Rv1033c	+ (90%) MUL_4624	+ (85%) MAV_1176	+ (85%) MAP1002c	P ML0261	+ (71%) MSMEG_2916	+ (75%) Mvan_2296	+ (100%) Mb1062c
30	TrcS Rv1032c	+ (79%) MUL_4625	+ (73%) MAV_1175	+ (73%) MAP1001c	P ML0260	+ (52%) MSMEG_2915	+ (54%) Mvan_2295	+ (100%) Mb1061c
31	DevR Rv3133c	+ (90%) MUL_2423	+ (85%) MAV_4109	+ (85%) MAP3271c	–	+ (85%) MSMEG_5244	+ (86%) Mvan_2548	+ (100%) Mb3157c
32	DevS Rv3132c	+ (78%) MUL_2422	+/–	+/–	–	+ (66%) MSMEG_5241	+/–	+ (99%) Mb3156c
33	DosT Rv2027c	–	–	–	–	–	+ (67%) Mvan_1395	+ (99%) Mb2052c
34	PdtaR Rv1626	+ (91%) MUL_1608	+ (93%) MAV_3158	+ (93%) MAP1319	+ (90%) ML1286	+ (87%) MSMEG_3246	+ (88%) Mvan_2840	+ (100%) Mb1652
35	PdtaS Rv3220c	+ (85%) MUL_2542	+ (84%) MAV_4168	+ (84%) MAP3321c	+ (81%) ML0803	+ (75%) MSMEG_1918	+ (77%) Mvan_1781	+ (99%) Mb3246c
36	NarL Rv0844c	+ (92%) MUL_0351	+ (87%) MAV_0872	+ (87%) MAP0689c	–	+ (78%) MSMEG_0105	+/–	+ (99%) Mb0867c
37	NarS Rv0845	+ (74%) MUL_0350	+ (71%) MAV_0874	+ (71%) MAP0690	–	+/–	+/–	+ (99%) Mb0868
38	TcrA RV0602c	–	+ (50%) MAV_3406	+ (50%) MAP1102c	–	–	–	+ (99%) Mb0618c
39	Rv0600c	–	–	–	–	–	+/–	+ (100%) Mb0616c

		MUL	MAV	MAP	ML	MSMEG	Mvan	Mb
40	Rv0601c	−	+/−	−	+/−	−	−	+ (100%) Mb0617c
41	Rv0818	+ (89%) MUL_0435	+ (80%) MAV_0760	+ (80%) MAP0649	P ML2194	+ (81%) MSMEG_5784	+ (83%) Mvan_5097	+ (99%) Mb0841
42	Rv2884	+ (64%) MUL_2073	+ (64%) MAV_3735	+ (65%) MAP2948	P ML1592	−	+/−	+ (100%) Mb2908
43	Rv0260c	+ (82%) MUL_1183	+ (78%) MAV_4900	+ (78%) MAP3706c	−	+ (74%) MSMEG_0432	+ (72%) Mvan_0301	+ (100%) Mb0266c
44	Rv3143	+ (87%) MUL_2453	+ (83%) MAV_4032	+ (83%) MAP3200	−	+ (73%) MSMEG_2064	+ (76%) Mvan_1886	+ (100%) Mb3167

^{a} Refers to annotations along with genomic locus tags for various Ser/Thr/Tyr kinases and phosphatases and two-component systems of *M. tuberculosis* H$_{37}$Rv. Other columns in the table describe the presence (+, with >50% identity) or absence (−, with <15% identity) of various orthologues in different mycobacterial species based on homology search performed by BLAST (not necessarily validated experimentally). In cases where more than one orthologue was found, only the one with the highest identity is listed in the table. +/− indicates that identity ranges between 50 and 15%. P indicates a pseudogene of an *M. tuberculosis* orthologue.

^{b} If two different proteins of *M. tuberculosis* exhibited homology with a single protein from another mycobacterial species, the latter has been shown as an orthologue of *M. tuberculosis* protein with which it exhibited higher degree of identity. However, MAV_1417 of *M. avium* exhibited exactly 70% of identity with both PknE and PknH of *M. tuberculosis*, and hence MAV_1417 has been depicted as an orthologue of both proteins of *M. tuberculosis*.

^{c} In this case, MAP3669c represented almost half the size of MptpB and exhibited high degree of identity (76%) with the N-terminal half of MptpB from *M. tuberculosis*. Interestingly, the C-terminal half of MptpB showed the similar extent of identity (78%) with the adjacent gene MAP3568c.

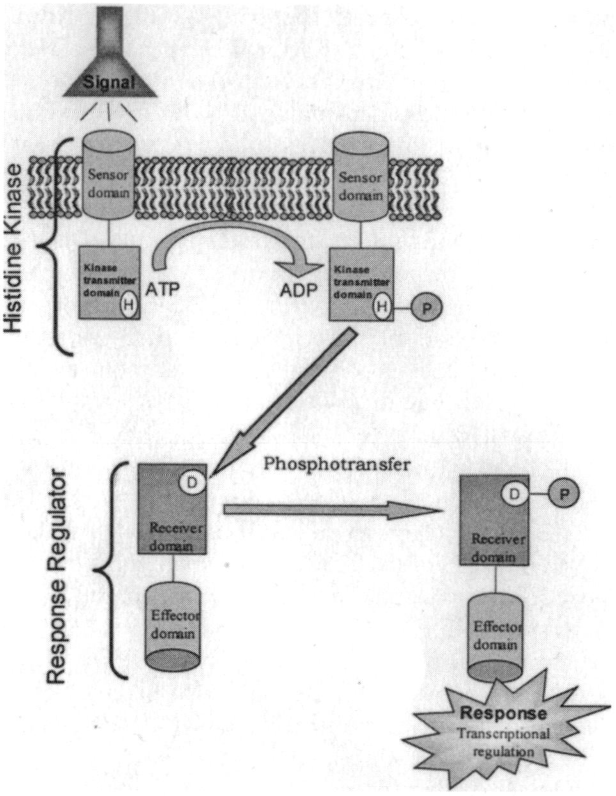

Figure 1. Two-component signal transduction. Two-component signal transduction systems are a mechanism that bacteria use to sense and respond to their environment. These modular and conserved systems are typically composed of a histidine kinase (HK) generally anchored in the cell membrane and a cytoplasmic response regulator (RR). Both proteins harbor two functional important domains (HK comprises sensor and kinase transmitter domains, whereas RR comprises receiver and effector domains). The HK detects a specific environmental stimulus through its sensor domain leading to ATP-dependent autophosphorylation of a histidine residue in the cytoplasmic kinase transmitter domain. The phosphoryl group from the activated transmitter domain is then transferred to an aspartic acid residue in the receiver domain of its cognate RR, resulting in the activation of the effector domain that mediates the cellular response (changes in gene expression or protein function).

pathogen in human macrophages (*tcrA*, *narL*, *rv0195*, *rv0260c*, *rv0818*, *rv2884,* and *rv3143*). Constitutive expression in macrophages indicates that these genes are probably involved in survival of the pathogen within macrophages (Haydel and Clark-Curtiss, 2004). Genetic and biochemical analysis have established the roles of DevR (DosR) as a regulator of hypoxia-responsive genes and PhoP as a modulator of acylated mannosylated lipoarabinomannans (Park et al., 2003; Ludwiczak et al., 2002). The studies involving high-density mutagenesis of *M. tuberculosis* genes (Sassetti and Rubin, 2003; Sassetti et al., 2001, 2003) support the role of MtrA response regulator and SenX3 and KdpD histidine kinase genes in the survival of the pathogen in mice. These studies also suggested the role of PhoP, KdpE, Rv1626, MtrA RRs, and MprB, DevS (DosS), MtrB HKs in optimal in vitro growth of the bacilli (Sassetti et al., 2003). Targeted mutagenesis of several TCSs also reveals the importance of these regulators in virulence. It has been shown that *mtrA* component of the mtrA-mtrB TCS of *M. tuberculosis* is essential for viability of *M. tuberculosis* (Zahrt and Deretic, 2000). This response regulator is differentially expressed in virulent and avirulent strains. It is located in an operon, along with its kinase MtrB, although the kinase does not appear to be required for growth and viability (Via et al., 1996). Global analysis of gene expression in the nutrition starvation model of persistence revealed that among the signal transduction systems, only the *kdpE* gene was induced (Betts et al., 2002). Although KdpD-KdpE (Rv1027c-Rv1028c) in *M. tuberculosis* has not been functionally characterized, the role of the analogous system in *E. coli* is established in controlling expression of the potassium transport (Polarek et al., 1992) and hence this TCS is likely to be involved in regulating expression of the *kdpF-kdpA-kdpB-kdpC* locus, encoding the putative K$^+$ transporter, to which it is divergently placed in the genome. Interestingly, although a *kdpE* mutant of *M. tuberculosis* showed normal growth in human macrophages, another *kdpE* mutant exhibited a hypervirulent phenotype in severely combined immune-deficient (SCID) mice (Fontan et al., 2004; Parish et al., 2003). Similar is the case in which mice infected with *devR* mutant, *tcrXY* mutant and *trcS* mutant die more rapidly than mice infected with the parental *M. tuberculosis* strain, thereby suggesting that *M. tuberculosis* actively maintains a balance between the survival of its host and itself (Parish et al., 2003a).

The two-component regulatory system *devR-devS* (*rv3133c-rv3132c*) was found to be overexpressed in the virulent strain *M. tuberculosis* H37Rv in comparison to its avirulent counterpart, H37Ra. Subsequent disruption of *devR* in *M. tuberculosis* H37Rv resulted in attenuated virulence in the guinea pig model (Dasgupta et al., 2000; Malhotra et al., 2004) but hypervirulence in mice as compared to the parental strain (Parish et al., 2003a). Another orphan sensor DosT (Rv2027c) is also able to activate DevR (Roberts et al., 2004). The DevRST system (three-component system involving DevR, DevS, and DosT) plays an important role in the response of *M. tuberculosis* to hypoxia and nitric oxide in which DevS is a histidine kinase, which, in response to an environmental cue, transduces the signal to its cognate response regulator DevR (Saini et al., 2004). Both PYP-like sensor domains (PAS) and GAF domains are known to be present in different HKs. The GAF acronym comes from the names of the first three different classes of proteins identified to contain them: cGMP-specific and -regulated cyclic nucleotide phosphodiesterase, adenylyl cyclase, and *E. coli* transcription factor FhlA. Usually, it is perceived that the GAF domain binds to cyclic nucleotides and the PAS domain frequently binds heme. The sensor module of DevS contains two GAF domains. Sardiwal and coworkers (2005) have now shown that the proximal GAF domain of the DevS sensor binds heme. These observations for the first time show that in addition to their known structural similarity, GAF and PAS domains can also function in an analogous manner. It is proposed that binding of heme to either oxygen or nitric oxide plays an important role in the regulation of the DevS sensor. Molecular oxygen and nitric oxide play negative and positive regulators, respectively, for the sensor; hence, the presence of the former inactivates the sensor, whereas the presence of the latter induces it.

The *prrA-prrB* pair (*rv0903c-rv0902c*) is one of the five TCS gene pairs conserved in all mycobacterial species and can consequently be assumed to play a fundamental role in signal transduction in mycobacteria (Tyagi and Sharma, 2004). The *prrA* mutant strain of *M. tuberculosis*, identified from a transposon mutant library, was impaired in its ability to survive in early stages of infection in macrophages (Ewann et al., 2002). These genes are transcribed in response to host-cell interaction, making these TCS proteins prime targets for the development of new antitubercular agents (Graham and Clark-Curtiss, 1999).

Studies with the *senX3-regX3* (*Rv0490-Rv0491*) TCS mutant have shown that this TCS may play an important role in the virulence because the mutation in the associated genes renders the pathogen defective in its survival in mice (Rickman et al., 2004; Parish et al., 2003b). Using prediction and structural modeling techniques, an atypical PAS domain in the N-terminal region of *M. tuberculosis* SenX3 sensor was identified that lacked two helices, leading to a more open structure (Rickman et al., 2004). This PAS-like domain is found in a variety of prokaryotic and eukaryotic sensors of oxygen and redox potential. Examples include the Aer protein, an *E. coli* signal transducer that responds to changes in the concentration of oxygen, redox carriers, and carbon sources (Repik et al., 2000), and FixL from *Rhizobium meliloti*, which has been shown to be an oxygen-sensing protein (Lois et al., 1993). This led to the postulation that SenX3-RegX3 may play a role in oxygen sensing and therefore may be required for the protective response to oxidative stress in *M. tuberculosis*.

The function of the PhoP-PhoQ TCS is to detect the concentration of Mg^{2+} in the bacterial cell and to upregulate transcription of high affinity Mg^{2+} transport systems in a low Mg^{2+} environment, allowing the bacteria to overcome Mg^{2+} starvation (Vescovi et al., 1997; Groisman, 2001). Disruption of the *phoP* component of *phoP-phoR* (the TCS that controls transcription of virulence genes in a number of intracellular bacterial pathogens such as *Salmonella*, *Shigella*, and *Yersinia*) in *M. tuberculosis*, resulted in impaired multiplication of the mutant in the host. This mutant was found to be attenuated in a mouse model as compared to the parental strain, suggesting that PhoP is required for intracellular growth of *M. tuberculosis* (Perez et al., 2001). Recent studies provide evidence that the PhoP-PhoR (Rv0757-Rv0758) regulon controls the biosynthesis of polyketide-derived lipids in *M. tuberculosis* (Walters et al., 2006; Gonzalo Asensio et al., 2006). Determination and identification of specific environmental cues and signals associated with PhoR are likely to help us understand the regulation involved in the biosynthesis of these important lipids. None of the structures of the proteins from this TCS has been solved. PhoP belongs to the OmpR subfamily of response regulators, and the receiver domains of this subfamily (PhoB from *E. coli*, DrrD from *Thermotoga maritima*, and PhoP from *B. subtilis*) have been structurally characterized (Itou and Tanaka, 2001). Although all three proteins have a similar fold, there exists extreme variability in conformation of helix α4 and loops associated with it. In fact, in the case of PhoP from *B. subtilis*, this region plays a critical functional role in dimerization (Birck et al., 2003). It may be expected that this region in the PhoP from the *M. tuberculosis* counterpart might hold the key to the mechanism of regulation by this protein.

The MprA-MprB (Rv0981-Rv0982) TCS system has been found to be essential for growth of *M. tuberculosis* during persistence (Zahrt and Deretic, 2001). Microarray and quantitative reverse transcriptase PCR (RT-PCR) studies demonstrate that MprA regulates in vivo expression of *sigB* and *sigE* by binding to their promoter regions (He et al., 2006). In the same study, it was reported that the sensor kinase MprB recognizes the stress conditions and induces the expression of SigE by phosphorylating MprA. The extracytoplasmic loop domain of the cognate sensor kinase MprB was required to sense the stress and phosphorylate MprA. Structural elucidation will provide further insights into the mechanism of regulation by this TCS, and it is hoped that further study will offer new targets for designing of inhibitors specifically efficient against persistent mycobacteria.

The TCS TrcS-TrcR (Rv1032c-Rv1033c) of *M. tuberculosis* is composed of the TrcS histidine kinase and the TrcR response regulator, which is homologous to the OmpR class of DNA binding response regulators. Both mutants have been tested for survival in vivo, and neither were found to be attenuated in macrophages nor in immunocompetent mice (Fontan et al., 2004; Ewann et al., 2002). Interestingly, infection of SCID mice with a *trcS* mutant of *M. tuberculosis* resulted in hypervirulence (Parish et al., 2003a). The logical explanation for this observation may lie in the ability of this TCS to repress genes necessary for bacterial pathogenicity. A recent study has demonstrated that the TrcR response regulator represses expression of *rv1057* (Haydel and Clark-Curtiss, 2006). This gene is transcribed by a σ^E promoter, which is already known to be involved in environmental stress response and virulence. In addition to several genes involved in cellular and lipid metabolism and fatty acid degradation, *rv1057* is induced when *M. tuberculosis* is grown in the presence of free fatty acid as the sole carbon source, suggesting a possible role of this gene in intraphagosomal survival and lipid metabolism (Schnappinger et al., 2003). Based on bioinformatic analysis, Rv1057 is a seven-bladed β-propeller protein. This family of proteins carries out diverse functions including enzyme catalysis, signal transduction, ligand binding, transport, mediation of protein-protein interactions, control of cell division, and modulation of gene expression (Fulop and Jones, 1999).

The crystal structure of one of the orphan response regulators, Rv1626, has been determined (Morth et al., 2004). This gene, along with two other orphan HK genes, namely *rv2027c* and *mtrA*, has been observed to be constitutively expressed throughout growth in human macrophages (Haydel and Clark-Curtiss, 2004), indicating that these genes are likely to be involved in adaptation of *M. tuberculosis* within macrophages. The structure of the Rv1626 gene product is highly similar in both domains to antitermination factor AmiR from *Pseudomonas aeruginosa*. This protein has been subsequently identified as the first member of a new class of proteins termed as phosphorylation-dependent transcriptional antitermination regulators (PdtaR). Further work identified another orphan HK, Rv3220c, as a cognate HK of Rv1626 and demonstrated the existence of a specific phosphotransfer between PdtaR and Rv3220c, designated PdtaS (a phosphorylation-dependent transcriptional antitermination sensor) (Morth et al., 2004).

Serine/Threonine Protein Kinases

In humans, there are more than 500 putative Ser/Thr or Ser/Tyr protein kinases (Manning et al., 2002) and more than 130 putative protein phosphatases (Alonso et al., 2004). Eukaryotic STPKs regulate a variety of substrate proteins, and about one third of eukaryotic proteins are phosphorylated (Meggio and Pinna, 2003). In 1979, 25 years after the discovery of protein phosphorylation in eukaryotes, isocitrate dehydrogenase was the first bacterial enzyme reported to be phosphorylated on a serine residue (Garnak and Reeves, 1979). However, since then, a number of bacterial physiological processes have been known to be regulated by phosphorylation and dephosphorylation of serine, threonine, and tyrosine residues. YpkA, a serine/threonine kinase of *Yersinia pseudotuberculosis*, is an essential virulence determinant (Galyov et al., 1993; Cozzone, 1993). YpkA has been shown to be translocated to the inner surface of the host plasma membrane and helps in establishing the disease by disrupting the eukaryotic cytoskeleton or by reprogramming the host signaling pathways (Hakansson et al., 1996). Actin is a cellular activator of YpkA and translocation of this kinase into epithelial cells causes disruption of the actin skeleton, thereby inhibiting macrophage functions and phagocytosis (Juris et al., 2000). *Yersinia pestis* phospho-tyrosine phosphatase (YopH) is secreted into macrophages to weaken the immune response of the host by dephosphorylating some of its P-tyrosine proteins (Yuan et al., 2005). Cozzone (2005) has reviewed the implications of Ser/Thr/Tyr protein phosphorylation in bacterial virulence.

Serine and threonine kinases from the eukaryotic organisms fall into a single superfamily called Hanks-type protein kinases. These are characterized by a conserved lysine residue in the active site motif (DXKPXN, X is any amino acid) and 11 conserved domains (Hanks and Quinn, 1991). Genome analyses have revealed that numerous bacteria also contain eukaryotic-like Ser/Thr protein kinases (Kennelly, 2002). The first such enzyme in prokaryotes was discovered in 1991 in *Myxococcus xanthus*, which is required for optimal spore formation (Munoz-Dorado et al., 1991). Based on the genome sequences, *M. xanthus* is believed to contain the highest number of potential eukaryotic protein kinases. Some cyanobacteria also contain a large number of eukaryotic protein kinases, which seem to be related to their lifestyle (Zhang and Shi, 2005). These Ser/Thr kinases and phosphatases have been shown to regulate a number of cellular processes such as morphological differentiation and secondary metabolism (Umeyama et al., 2002), oxidative stress response (Neu et al., 2002), purine biosynthesis (Rajagopal et al., 2005), sugar transport, glycogen consumption and glucose metabolism (Nariya and Inouye, 2002), cell growth and septum formation (Treuner-Lange et al., 2001; Deol et al., 2005), cell density and viability (Gaidenko et al., 2002), and carbon catabolite re-

pression (Poncet et al., 2004). Similarly protein tyrosine phosphorylation is implicated in the control of heat shock response (Klein et al., 2003), adaptation to cold (Ray et al., 1994), flagellin export (South et al., 1994), and cell division and differentiation (Wu et al., 1999).

The eleven STPKs of *M. tuberculosis*, namely PknA to PknL, belong to the PKN2 family of prokaryotic protein kinases (Cole et al., 1998; Leonard et al., 1998). Most of the mycobacterial STPKs (PknA, PknB, PknD, PknE, PknF, PknG, PknH, and PknI) have been biochemically well characterized, and for some, their physiological roles has also been investigated (Av-Gay et al., 1999; Av-Gay and Everett, 2000; Koul et al., 2001; Peirs et al., 1997; Molle et al., 2003; Chaba et al., 2002; Gopalaswamy et al. 2004). During its evolution, *M. leprae* has lost a number of genes (it has 2,400 fewer genes than *M. tuberculosis*) including several kinases (Eiglmeier et al., 2001). However, genes encoding PknA, PknB, PknG, and PknL are still present in its genome, implying that these kinases may be essential for its growth in vivo. Transposon-insertion mutagenesis experiments carried out to identify the genes required for optimal in vitro mycobacterial growth resulted in the identification of only three out of the 11 mycobacterial kinases, namely PknA, PknB, and PknG (Sassetti et al., 2003). The other STPKs may possibly be involved in adaptation of the pathogen to the hostile intracellular environment.

It has been observed that the genes encoding fork head-associated (FHA) proteins cluster along with Ser/Thr protein kinases and phosphatases in the *M. tuberculosis* genome. The FHA domain binds to phosphothreonine residues on target proteins and mediates protein-protein interaction (Pallen et al., 2002; Durocher and Jackson, 2002) (Fig. 2). Hence, the interaction of the FHA domains with their cognate proteins is likely to be influenced by respective kinases/phosphatases involved in the phosphorylation cascade. Recent studies on Rv1747 (an ABC transporter containing two predicted FHA domains) phosphorylation by PknF indicate that autophosphorylation of STPK, in addition to activating the kinase domains, creates binding sites for substrate proteins with FHA domains leading to relatively simple, linear signaling pathways (Molle et al., 2003; Molle et al., 2004). Several Ser/Thr kinases such as PknB, PknD, PknE, and PknF have been shown to phosphorylate the FHA domain of GarA, a regulator of glycogen degradation during cell growth (Villarino et al., 2005; Belanger and Hatfull, 1999). All of these findings suggest that mycobacterial kinases upon their activation by phosphorylation, in turn, phosphorylate bacterial proteins with FHA domains. These

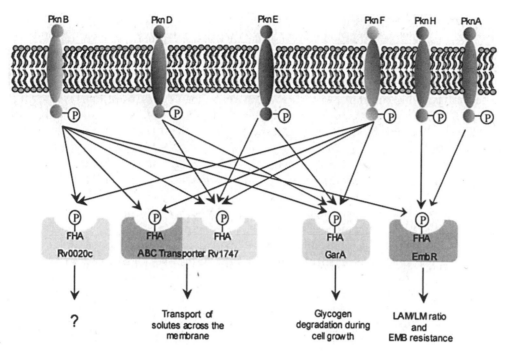

Figure 2. Phosphorylation of FHA-domain proteins by serine/threonine protein kinases (STPKs) in vitro. FHA domains are ubiquitous phosphothreonine peptide recognition motifs that play diverse roles in STPK signal transduction. PknF senses extracellular signals and regulates transport of solutes across the cellular membrane through phosphorylation of Rv1747 (ABC transporter). Rv1747 has two FHA domains, one of which is phosphorylated by PknB, PknD, PknE, and PknF, whereas the other domain is more restrictively phosphorylated. The above-mentioned STPKs also phosphorylate GarA, a regulator of glycogen degradation during cell growth. PknB and PknF have also been shown to phosphorylate Rv0020c in vitro. PknH, PknB, and PknA phosphorylates the FHA-containing protein EmbR, which, in turn, induces transcription from the *embCAB* operon, leading to a higher LAM/LM ratio. LAM is known to be an important determinant of virulence and modulator of host immune responses.

phosphorylated proteins then regulate a variety of cellular processes such as bacterial growth, cell division, transport of nutrients across the membrane, and the glycogen degradation pathway.

Most mycobacterial kinases are predicted to be transmembrane proteins with intracellular kinase domains; however, PknG lacks such a transmembrane segment. Phagosomes containing *pknG* mutant of *M. bovis* BCG fused with the lysosomes and the mutant mycobacteria could not survive in the macrophages (Walburger et al., 2004). By subcellular localization, it has also been shown that PknG is present in the phagosomal lumen as well as in the cytosol of the infected macrophages (Walburger et al., 2004). These results suggested that PknG is actively secreted into the cytosol of host cells and might interfere with phagosomal-lysosomal fusion. In another study, *pknG* mutant of *M. tuberculosis* showed reduced virulence in immunocompetent mice as compared with the wild-type strain (Cowley et al., 2004). PknG has also been implicated in sensing and regulating glutamate/glutamine levels and shown to mediate mycobacterial survival in host cells (Cowley et al., 2004). All of these findings suggest that STPKs PknA, PknB,

PknF, PknG, and PknH appear to regulate multiple aspects of bacterial metabolism and enable the bacilli to adapt to various stresses (Av-Gay and Everett, 2000).

The structure-function studies of protein phosphorylation by this family of kinases/phosphatases have been reviewed by Greenstein et al. (2005). The topology of various membrane-associated S/T/Y kinases clearly indicates that these enzymes are generally composed of an extracellular/periplasmic domain and a cytosolic domain, the latter being responsible for its catalytic activity. Recently determined structures of Ser/Thr/Tyr phosphor-signaling domains from mycobacteria provide mechanistic understanding of STPK and phosphatase modules. The kinase domain from *M. tuberculosis* PknB, adopts a classic STPK structure that includes the P-loop (which binds the phosphate groups of the nucleotide), the C helix (which orients key residues in the active site), and the activation loop, which is mostly disordered in the PknB crystal structures (Young et al., 2003; Ortiz-Lombardia, 2003). The most obvious structural difference from the classic kinases involves the glycine-rich loop present in the mycobacterial enzyme, which owing to the absence of ATP, moves further toward the C-terminal lobe. In silico screening for PknB inhibitors, identified a compound, earlier used in cancer treatment, that also prevents mycobacterial cell growth. This compound was later revealed to be a PknB inhibitor by virtue of competition with ATP. In the crystal structure of PknB and mitoxantrone complex, PknB crystallizes as a "back-to-back" homodimer, as seen in case of its crystal structure when complexed with ATP analogues (Wehenkel et al., 2006). Such homodimers have the kinase domain in an overall closed conformation, and the activation loop is disordered. This structural organization of PknB, which resembles that of the RNA-dependent protein kinase PKR, allows to hypothesize a possible role of such homodimerization in the regulation of PknB activity (Dar et al., 2005; Dey et al., 2005). It is proposed that the monomeric state of PknB may render it inactive, possibly due to a misplaced helix, as proposed in the cases of other eukaryotic protein kinases. Ligand binding possibly induces "back-to-back" dimerization of the catalytic domain, as observed in the crystal structure. Such a dimerization has been demonstrated for a PknB analogue from *B. subtilis*, namely protein kinase PrkC (Madec et al., 2002).

Protein Phosphatases

Owing to the limited information available on protein phosphatases, studies on microbial phosphatases represent an evolving area of scientific investigations. Phosphatases represent a diverse family of biologically active proteins involved in dephosphorylation reactions occurring in signal transduction as well as in several metabolic pathways (Stock et al., 1995; Rossolini et al., 1998). Some of these enzymes are secreted outside the cell and are usually able to dephosphorylate a broad group of structurally unrelated substrates at acidic-to-neutral pH range. They are referred to as nonspecific acid phosphatases (NSAPs).

Studies on several intracellular pathogens have demonstrated the important role of these acid phosphatases in pathogenicity (Remaley et al., 1985; Baca et al., 1993; Reilly et al., 1996). The culture filtrate of *M. tuberculosis* is known to have an acid phosphatase activity now designated as secreted acid phosphatase of *M. tuberculosis* (SapM) (Raynaud et al., 1998; Saleh and Belisle, 2000). SapM represents a new class of bacterial NSAPs because it does not possess the signature motifs used to define any of the known classes of bacterial NSAPs. Actually, it shows greater sequence homology with fungal acid phosphatases than with bacterial acid phosphatases

(Saleh and Belisle, 2000). A recent study has revealed that the difference between live and dead intracellular mycobacteria lies in that the latter cannot remove PI3P from their phagosomal membrane (Vergne et al., 2005). PI3P is a membrane-tagging signal leading to phagosomal maturation (Fratti et al., 2001; Vieira et al., 2001). Live mycobacteria secrete a phosphatase that dephosphorylates PI3P and inhibits phagosome-late endosome fusion. SapM is now shown to be responsible for PI3Pase activity of *M. tuberculosis* and provides a new target for drug development (Vergne et al., 2005).

Another phosphatase family, called the protein tyrosine phosphatase (PTP) superfamily, in general is divided into the following three families based on their molecular masses and substrate specificities: (i) the high-molecular-mass PTPase family, whose members have a conserved 30-kDa catalytic domain; (ii) the dual-specificity PTPase, whose members are able to dephosphorylate both serine-threonine and tyrosine residues; and (iii) the low-molecular-weight PTPase, whose members have a single 18-kDa catalytic domain (Denu and Dixon, 1995; Yuvaniyama et al., 1996; Zhang et al., 1995).

Enzymological and structural studies have led to the conclusion that reactions catalyzed by the PTPases share a common chemical mechanism. Most of the PTPs contain conserved structural elements such as phosphate-binding loop encompassing the PTPase signature motif $(H/V)C(X)_5R(S/T)$ and an essential acid/base Asp residue on a surface loop. Site-directed mutagenesis has revealed that the cysteine residue in the PTP loop and Asp residue in the surface loop is required for phosphatase activity, for example, Cys 403 in YopH, Cys 215 in PTP1, Cys 124 in dual specificity phosphatase VHR, and Cys12 in bovine low-molecular-weight phosphatase (Guan and Dixon, 1990; Gautier et al., 1991; Zhou et al., 1994). The cysteine is phosphorylated through a thiophosphate linkage $(-S-PO_3^{2-})$ during catalytic turnover. This phosphoenzyme complex is then hydrolyzed by water. Mutations in the arginine residues in the PTPase signature motif also result in the loss of enzymatic activity for two receptor-like PTPases, LAR and CD45 (Streuli et al., 1990; Johnson et al., 1992), and low-molecular-weight phosphatase from bovine liver (Cirri et al., 1993).

It has been shown that several bacterial pathogens have eukaryotic phosphatases. These include the conventional IphP from *Nostoc commune* and the conventional PTPs YopH from *Yersinia pestis* and SptP from *Salmonella typhimurium* (Guan and Dixon, 1990; Kaniga et al., 1996; Potts et al., 1993; Howell et al., 1996). The pathogenic microorganisms *Yersinia* and *Salmonella* secrete YopH and SptP, respectively, into the host, allowing these phosphatases to interfere with the host signal transduction pathways. In addition, vaccinia virus encodes a dual-specificity phosphatase VH1, which is essential for viral transcription and infectivity (Liu et al., 1995). Myxoma virus also encodes dual-specificity phosphatases essential for virus viability (Mossman et al., 1995). PTPases have also been detected in *Leishmania donovani* and *Trypanosoma cruzi* (Bakalara et al., 1995).

The *M. tuberculosis* genome has a protein Ser/Thr phosphatase, PstP belonging to the diverse Mg^{2+}- or Mn^{2+}-dependent protein phosphatase (PPM) family (Bork et al., 1996). In prokaryotes, these enzymes are involved in controlling different bacterial processes such as spore formation, the development of fruiting bodies, vegetative growth, or cell segregation. PstP is a membrane-bound, lone Ser/Thr phosphatase that dephosphorylates targets of the 11 STPKs in *M. tuberculosis* (Chopra et al., 2003; Boitel et al., 2003).

The *M. tuberculosis* genome has two genes, *mptpA* (Rv2234) and *mptpB* (Rv0153c), encoding low-molecular-weight and conventional/dual-specific tyrosine phosphatases, respectively (Cole et al., 1998). Both MptpA and MptpB of *M. tuberculosis* are secretory proteins (Koul et al., 2000). Transcriptional analysis has revealed that expression of MptpA in *M. bovis* BCG is upregulated in the stationary phase and human monocytes (Cowley et al., 2002). A recent study has implicated the role of MptpA in phagocytosis and actin polymerization (Castandet et al., 2005). Disruption of *mptpA* impaired the ability of *M. tuberculosis* to survive in activated macrophages and guinea pigs. In activated macrophages at 2 days postinfection, an approximately twofold reduction in the survival of the intracellular *mptpA* mutant was observed in comparison to the intracellular parental strain. This difference in survival increased to approximately 14-fold at 6 days postinfection (Singh et al, unpublished results). In guinea pig model system, at 3 weeks postinfection, an approximately 10-fold difference was observed in bacillary loads in the spleens and lungs of guinea pigs infected with *mptpA* mutant strain when compared with the bacillary load in spleens and lungs of animals infected with the parental strain. However, at 6 weeks postinfection, this difference increased from 10- to 90-fold. This 90-fold difference was statistically significant (Singh et al, unpublished results). Singh and coworkers reported that disruption of *mptpB* had no significant effect on the morphology and growth of *M. tuberculosis* in defined liquid cultures, suggesting that MptpB is not required for the growth of *M. tuberculosis* under in vitro conditions

(Singh et al., 2003a). It was also shown that the parental strain and *mptpB* mutant were comparable in their ability to infect and survive in the resting macrophages. However, *mptpB* mutant was more sensitive to killing in activated macrophages as compared with the parental strain. An approximately fivefold reduction in the survival of intracellular *mptpB* mutant was observed in comparison to the survival of the parental strain at 4 days postinfection. At 6 days postinfection, an approximately sevenfold reduction was observed in the survival of the intracellular *mptpB* mutant strain in comparison to the internalized parental strain (Singh et al., 2003a). Infection of guinea pigs with the mutant strain resulted in a 70-fold reduction in bacillary load of spleens in infected animals as compared with the bacillary load in animals infected with the parental strain at 6 weeks postinfection. On reintroduction of the *mptpB* gene in the mutant strain, the complemented strain was able to establish infection and survive in guinea pigs (Singh et al., 2003a). Tyrosine phosphorylation is an important constituent of signal transduction pathways. Because mycobacteria have no tyrosine kinase, and tyrosine phosphatases from this organism are shown to be essential for the survival of the pathogen in the host, in all likelihood they may target a component of the host signal transduction system to exert their influence.

ROLE OF MYCOBACTERIAL KINASES AND PHOSPHATASES IN CELL WALL METABOLISM

The mycobacterial cell wall plays a crucial role in mediating interactions between mycobacteria and their environment (Daffe and Draper, 1998) (see chapter 14). The mycobacterial cell envelope is mainly composed of mycolic acids, and they represent key virulence factors required for intracellular survival and pathophysiology of tuberculosis (Dubnau et al., 2000; Gao et al., 2003). They consist of very long chains of α-branched β-hydroxy fatty acids (C_{60}-C_{90}), whose biosynthesis is controlled by two elongation systems (see chapter 1). The type I fatty acid system (FAS-I) is typified by the existence of a multifunctional enzyme and is present in mammals. The type II fatty acid system (FAS-II) is found in most bacteria and plants, and contains distinct enzymes encoded by unique genes (Kremer et al., 2000). The main locus of mycobacterial FAS-II is an operon comprising of five genes. The third and fourth open reading frames (ORFs) *kasA* and *kasB* encode the β-ketoacyl-ACP synthases that elongate the growing meromycolate precursor. In a recent study, it has been shown that PknA, PknB, PknE, PknF, and PknH phosphorylate KasA, KasB, and *mt*FabD on three different sites. In the same study, it has been shown that these phosphorylated FAS-II components could be dephosphorylated by *M. tuberculosis* Ser/Thr phosphatase PstP. Differential phosphorylation by the mycobacterial kinases/serine threonine phosphatase in response to stress conditions is also shown to directly affect the phosphorylation profile of KasA and KasB, and as a result modulate mycolic acid biosynthesis, thereby enabling mycobacteria to adapt and survive in various environmental conditions (Molle et al., 2006).

We have identified and reported the *mymA* operon (*rv3083-rv3089*) of *M. tuberculosis*, which is regulated by an AraC/XylS transcriptional regulator *virS* (*rv3082c*) placed in a divergent manner to this operon (Gupta et al., 1999; Singh et al., 2003b). We have shown that on exposure to acidic pH, the promoters of the *mymA* operon as well as *virS* are upregulated in macrophages. The induction of the *mymA* operon by VirS under acidic conditions is important for maintaining the appropriate cell envelope structure of *M. tuberculosis,* as substantiated by a significantly reduced ability of Mtb∆*virS* (a *virS* mutant of *M. tuberculosis*) and Mtb*mym:hyg* (a *mymA* mutant of *M. tuberculosis*) to survive in the activated macrophages and in caseating granuloma as compared with the parental strain. Moreover, a drastic reduction (800-fold) was observed in the ability of the mutant strains to specifically survive in guinea pig spleen as compared with the parental strain (Singh et al., 2005). Understanding the mechanisms involved in the regulation of *mymA* operon by VirS might help in the identification of new targets against persistent mycobacteria. The presence of several Ser/Thr phosphorylation sites in VirS and the placement of *pknK* (*rv3080c*) upstream to *virS* makes it tempting to hypothesize the involvement of PknK-mediated phosphorylation of VirS as a mechanism of *mymA* operon regulation (Fig. 3).

Ser/Thr protein kinases PknA and PknB are overexpressed in exponential phase of growth, and the overexpression of these kinases causes cell growth and morphological changes, possibly due to defects in cell wall synthesis and cell division. Both *pknA* and *pknB* were reported to be downregulated in response to starvation, suggesting their involvement in the regulation of replication (Betts et al., 2002). By proteomic analysis, it was found that PknA and PknB phosphorylated a conserved hypothetical protein, Rv1422, which is part of an operon that also encodes a homologue of the UvrC excision repair endonuclease and a homologue of *Streptomyces coelicolor* transcription factor WhiA. Wag31 (Rv2145c) is also phosphorylated by PknA alone, and PknA and PknB together. Wag31 is a homologue of the cell shape/cell division protein DivIVA and is essential for the growth of *M. tuberculosis* (Sassetti et al., 2003). These results suggest a novel function of PknA and PknB in the regulation of cell shape in mycobacteria (Kang et al., 2005). Phosphorylated PknA and PknB are shown to be dephosphorylated by PstP, a serine/threonine phosphatase (Chopra et al., 2003). The genes encoding *pknA*, *pknB* and *pstP* are present in an operon that also includes *rodA* and *pbpA*, two genes encoding proteins involved in peptidoglycan synthesis during cell growth (Cole et al., 1998; Matsuhashi et al., 1990). The operon also encodes two proteins harboring FHA domains (Rv0019c and Rv0020c). Cross-linked peptidoglycan, a major component of the bacterial cell wall, is synthesized by the penicillin-binding proteins (PBPs) during both cell elongation and cell division. PBPs are involved in cell wall expansion, cell shape maintenance, septum formation, and cell division. A mycobacterial PBP homologue (PBPA, or Rv0016c) was found to be phosphorylated by PknB and dephosphorylated by PstP (Dasgupta et al., 2006). Thus, PknA, PknB and PstP provide tight regulation of cell elongation by phosphorylation or dephosphorylation of specific bacterial proteins.

By using yeast two-hybrid systems, it has been shown that Rv1747, an ABC transporter (the only mycobacterial protein with two FHA domains), interacts with protein kinase PknF (Fig. 2). This interaction was found to be phosphorylation dependent and abrogated by site directed mutagenesis in the activation loop of PknF and first FHA domain of Rv1747 (Curry et al., 2005). In the same study, it was shown that Rv1747 mutant (belonging to the class of *giv* mutants) exhibited reduced bacillary loads in the mouse lungs and spleen in comparison to the parental strain. It has been postulated that PknF senses environmental signals and regulates the transport of solutes across the cellular membrane through phosphorylation of Rv1747 (Curry et al., 2005). Moreover, PknF is shown to play a direct/indirect role in the regulation of glucose transport, cell growth, and septum formation in *M. tuberculosis* (Deol et al., 2005) (Fig. 3). Interestingly, this kinase is absent from *M. smegmatis,* a saprophytic nonpathogenic species of mycobacterium (Deol et al., 2005).

In addition, PknH, another Ser/Thr kinase, has also been shown to phosphorylate the FHA-containing protein EmbR (Fig. 2), a putative transcriptional regulator of arabinosyl transferases (EmbC, EmbA, and EmbB) (Molle et al., 2003). PknH was found to be upregulated in *M. tuberculosis* when it infected macrophages (Sharma et al., 2006a). Activation of EmbR on phosphorylation by PknH induces transcription from the *embCAB* operon, leading to a higher LAM/LM ratio (Fig. 3). This LAM/LM ratio

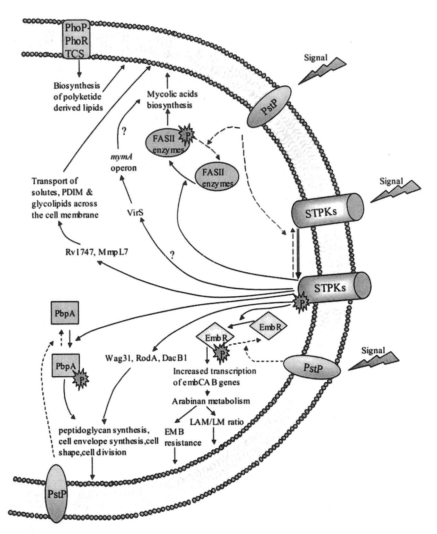

Figure 3. Role of kinases and phosphatases in cell wall metabolism of mycobacteria. In response to various environmental cues, STPKs become autophosphorylated (double arrow) and, in turn, phosphorylate their target proteins, enabling them to carry out their respective functions in the cell. Of the many target proteins of STPKs, only those known to be involved in cell wall metabolism are shown in the figure. PstP (a Ser/Thr phosphatase) in response to environmental cues dephosphorylates (dashed arrows) phosphorylated STPKs and some of the phosphorylated target proteins such as EmbR and PbpA, thus rendering them inactive. The PhoP-PhoR two-component system is involved in the biosynthesis of polyketide derived lipids, which are important constituents of the cell wall. ? indicates that the pathway is proposed but not experimentally validated.

regulates the synthesis of arabinan, an important component of arabinogalactan, which is essential for the structural integrity of mycobacterial cell wall (see chapters 3 and 7). LAM is an important determinant of virulence and modulator of host immune response; thus, the PknH/EmbR pair, by upregulating LAM/LM ratio, plays an important role in mycobacterial pathogenesis (Chatterjee and Khoo, 1998, Nigou et al., 2003). However, the *pknH* mutant of *M. tuberculosis* is more hypervirulent in mouse organs than its parental and complemented strain and showed higher resistance to acidified nitrite stress than the parental strain, suggesting that PknH could control the in vivo growth of mycobacterium in response to nitric oxide (Papavinasasundaram et al., 2005). EmbR is now shown to be an additional sub-

strate for multiple mycobacterial STPKs, namely PknA and PknB, as well as for the Ser/Thr phosphatase PstP, indicating that such multiple-signal pathways may be involved in generating a global response by integrating diverse signals (Sharma et al., 2006b). Interestingly, EmbR is also involved in the development of resistance against ethambutol (Sreevatsan et al., 1997). Furthermore, a recent study by Zheng and coworkers (2007) shows that pknH phosphorylates more substrates, namely Rv0681, a TetR-class transcription factor, and Rv3330-DacB1, a probable PBP. DacB1 in *B. subtilis* is a sporulation-specific protein involved in cell envelope biosynthesis. It has been postulated that phosphorylation of DacB1 by PknH regulates the synthesis of peptidoglycan and the cell envelope (Zheng et al., 2007).

Bioinformatic analysis has revealed the presence of 40 putative substrates for PknH belonging to diverse classes and thus suggesting the involvement of PknH kinase in regulating multiple cellular pathways by phosphorylating various substrates. Some of these putative PknH substrates such as Dacb1, Esx1 locus, FecB, lppI, mmpL8, Rv1333, Rv1517, Rv1824, Rv2091c, and Rv2333c are involved in cell wall processes (Zheng et al., 2007).

Protein kinase PknD, which has an extracellular β-propeller motif in its structure and is postulated to act as an anchoring sensor domain, phosphorylates MmpL7 protein (Perez et al., 2006). MmpL7 and other MmpLs (mycobacterial membrane protein large) belong to the RND (resistance, nodulation and cell division) family of transporters (Domenech et al., 2005) (see chapter 11). MmpL7 has been determined to be essential for virulence (Camacho et al., 2001) presumably owing to its involvement in the transport of phthiocerol dimycocerosate (PDIM) and a related but distinct phenolic glycolipid to the cell wall (Cox et al., 1999; Reed et al., 2004). By mass spectroscopy, it was identified that four out of the five MmpL7 peptides were phosphorylated. Besides MmpL7, other MmpL transporters contain serine or threonine phosphorylation sites, suggesting that various kinases could possibly play an important role in the regulation of such important cell wall constituents (Perez et al., 2006).

Another role of Ser/Thr kinases in the control of exopolysaccharide production arises from the findings that Stk1 phosphorylates phosphoglucosamine mutase GlmM. This enzyme catalyzes the first step in the synthesis of UDP-N-acetylglucosamine, an essential common precursor to cell envelope components (Novakova et al., 2005). Stk1 has been shown to be implicated in the virulence of *Streptococcus pneumoniae* (Echenique et al., 2004). By bioinformatic analysis, PknI of *M. tuberculosis* has been shown to be homologous to Stk1; hence, it is tempting to propose that PknI possibly has a role to play in the virulence of *M. tuberculosis* (Gopalaswamy et al., 2004). However, although the kinases associated with virulence are generally upregulated during macrophage infection, real-time PCR studies show that the expression of *pknI* is actually downregulated in macrophages (Singh et al., 2006).

The information about the role of kinases and phosphatases in the metabolism of mycobacterial cell wall is just beginning to emerge, and several examples of phophorylation of a component by multiple kinases and phosphorylation of multiple substrates by a single kinase are already cited above. In *Streptomyces,* another gram-positive organism, around 34 Ser/Thr kinases appear to regulate the synthesis of polyketide secondary metabolites (Umeyama et al., 2002). Recent studies indicate that *M. tuberculosis* PhoP-PhoR TCS positively regulates the enzymes involved in the biosynthesis of complex lipids of pathogenic mycobacterial cell wall (Walters et al., 2006; Gonzalo Asensio et al., 2006) (Fig. 3). In view of the unique cell wall structure of mycobacteria and the presence of a large repertoire of polyketides and complex lipids (see chapters 1, 4, and 6), kinases and phosphatases are bound to play an important role in the regulation of the cell wall metabolism of this pathogen.

MYCOBACTERIAL KINASES AND PHOSPHATASES AS DRUG TARGETS

Structure-based designing of inhibitors and consequent development of drugs constitute a very popular strategy that has resulted in successful examples of drug development (for example, the neuraminidase inhibitors zanamivir [Relenza] and oseltamivir [Tamiflu] [von Itzstein et al., 1993; Kim et al., 1997] and the human immunodeficiency virus [HIV]-protease inhibitors nelfinavir [Viracept], amprenavir [Agenerase], and lopinavir [Aluviran] [Kaldor et al., 1997; Kim et al., 1995; Sham et al., 1998]).

Signal transduction pathways regulate innumerable biological activities in a cell and involve protein kinases and protein phosphatases. Many of these kinases and phosphatases have been identified as promising therapeutic targets for the development of novel drugs against several diseases (Bridges, 2001; Shawver et al., 2002; Grosios and Traxler, 2003; Noble et al., 2004). The recognition of the role of TCS, STPKs, and PTPs in mycobacterial virulence and persistence has led TB research to focus on discovering/identifying molecules that would inhibit these classes of mycobacterial proteins.

The protein kinase inhibitor 1-(5-isoquinolinesulfonyl)-2-methylpiperazine (H7) inhibited the growth of two different mycobacterial strains, the slow-growing *M. bovis* and the fast-growing saprophyte *M. smegmatis* mc^2 155. Micromolar concentration of the piperazine compound was also found to inhibit the enzymatic activity of mycobacterial serine threonine kinase PknB (Drews et al., 2001). Moreover, it has been shown that inhibitors of protein kinases can prevent uptake of *M. leprae* by peritoneal macrophages in mice (Prabhakaran et al., 2000). This suggests that conventional protein kinase inhibitors can provide a starting framework for developing specific antimycobacterial inhibitors.

In a recent study, a number of representative indolizine derivatives against mycobacterial phosphatases in an in vitro assay system have been discovered (Fig. 4). In this study, 5′ substituted 3-benzoyl indolizine-1-carbonitrile was found to inhibit MptpB but not MptpA, whereas 5-(phenoxymethyl)indolizine was found to be 10-fold more selective for MptpB. The IC_{50} values of 5-(phenoxymethyl)indolizine for MptpB and MptpA was 7.5 μM and 74.9 μM, respectively (Weide et al., 2006). In another study, a roseophilin derivative was found to be a potent inhibitor of MptpA with an IC_{50} value of 9.4 μM (Manger et al., 2005) (Fig. 4). Further investigation of nonylprodigiosin analogs identified a compound that inhibited the enzymatic activity of MptpA with a MIC value of 28.7 μM (Manger et al., 2005, Fig. 4). In an alternative approach, MptpA inhibitors were identified from the rationally assorted fragment library of 20,000 compounds. Out of these compounds 2-dimethyl pyrrol-1-yl-benzoic acid derivatives were found to inhibit MptpA with IC_{50} values of 1.9 and 1.6 μM, respectively (Manger et al., 2005) (Fig. 4).

Although activities of this new class of compounds have to be evaluated in vivo as antitubercular drugs, the scaffolds provided by these molecules in conjunction with available three-dimensional struc-tural information and future medicinal chemical methodologies should allow for accelerated development of more specific and potent inhibitors of mycobacterial kinases and phosphatases.

CONCLUSIONS

M. tuberculosis is an intracellular pathogen well equipped to adapt and reside within the phagosomes of human macrophages and to persist within the host for many years. To understand its pathogenesis, it is essential to understand the regulatory processes that control these adaptive responses. It is clear that mycobacterium establishes a successful infection by disrupting the host-signaling machinery and is competently aided in this strategy by its various kinases and phosphatases.

Studies involving gene knockouts and transposon mutagenesis have proven beyond a doubt that these kinases and phosphatases are essential elements for growth and virulence of *M. tuberculosis*. Although these studies have provided us with rich information about mycobacterial kinases and phosphatases, there remains a tremendous challenge in understanding the

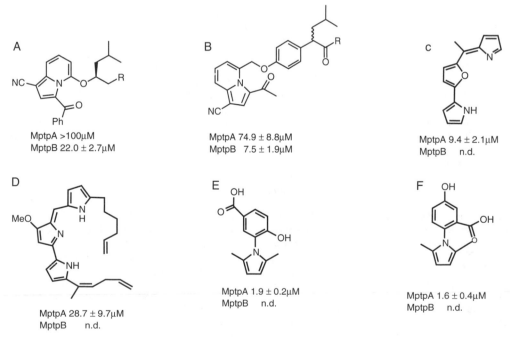

Figure 4. Structures and IC_{50} values (μM) of various compounds that have been identified as inhibitors of mycobacterial tyrosine phosphatases. (A and B) 3′ substituted indolizine-1-carbonitrile derivatives have been identified as inhibitors of mycobacterial tyrosine phosphatase B (MptpB) having IC_{50} values of 22.0 and 7.5 μM, respectively. Roseophilin and prodigiosin represent a new class of natural products and have been shown to inhibit tyrosine phosphatases. An analogue of roseophilin (C) has been shown to inhibit mycobacterial tyrosine phosphatase A (MptpA) with an IC_{50} value of 9.4 μM. Compound D, a prodigiosin derivative, inhibited the enzymatic activity of MptpA with an IC_{50} value of 28.7 μM. In an alternative approach 2-dimethylpyrrol-1-yl benzoic acid derivatives (E and F) have been identified as MptpA inhibitors from the rationally assorted fragment based FMP library. Compounds E and F had IC_{50} values of 1.9 and 1.6 μM, respectively.

biology and signaling functions that these proteins carry out. Also, recent information has provided a great impetus to the role of kinases and phosphatases in the genesis and maintenance of the unique cell wall structure of this pathogen. However, these studies provide a clear indication that these enzymes play a far more important role in the metabolism of mycobacterial cell wall than what is already known. Future work will expose the mechanistic details and proteins used by this pathogen to downregulate the host signaling pathways. Structural analysis of complexes of these signaling proteins may provide the key to designing molecules for selective disruption of signal transduction. Research in this area is likely to accelerate as more information on mycobacterial kinases/phosphatases becomes available, which can be examined for understanding the regulatory switches employed by *M. tuberculosis* for in vivo growth, survival, long-term persistence in the host, and cell wall metabolism.

Acknowledgments. We thank Garima Khare for helping with the preparation of figures and proofreading of the manuscript and David Kastrinsky for helping with the ChemDraw figures. Poonam Snotra is acknowledged for her help with the bibliography. Rajiv Chawla is acknowledged for secretarial help.

REFERENCES

Alex, L. A., and M. I. Simon. 1994. Protein histidine kinases and signal transduction in prokaryotes and eukaryotes. *Trends Genet.* **10:**133–138.

Alonso, A., J. Sasin, N. Bottini, I. Friedberg, I. Friedberg, A. Osterman, A. Godzik, T. Hunter, J. Dixon, and T. Mustelin. 2004. Protein tyrosine phosphatases in the human genome. *Cell* **117:**699–711.

Av-Gay, Y., and M. Everett. 2000. The eukaryotic like Ser/Thr protein kinases of *Mycobacterium tuberculosis*. *Trends Microbiol.* **8:**238–244.

Av-Gay, Y., S. Jamil, and S. J. Drews. 1999. Expression and characterization of the *Mycobacterium tuberculosis* serine/threonine protein kinase PknB. *Infect. Immun.* **67:**5676–5682.

Baca, O. G., M. J. Roman, R. H. Glew, R. F. Christner, J. E. Buhler, and A. S. Aragon. 1993. Acid phosphatase activity in *Coxiella burnetii*: a possible virulence factor. *Infect. Immun.* **61:**4232–4239.

Bakalara, N., A. Seyfang, C. Davis, and T. Baltz. 1995. Characterization of a life-cycle-stage-regulated membrane protein tyrosine phosphatase in *Trypanosoma brucei*. *Eur. J. Biochem.* **234:**871–877.

Belanger, A. E., and G. F. Hatfull. 1999. Exponential-phase glycogen recycling is essential for growth of *Mycobacterium smegmatis*. *J. Bacteriol.* **181:**6670–6678.

Betts, J. C., P. T. Lukey, L. C. Robb, R. A. McAdam, and K. Duncan. 2002. Evaluation of a nutrient starvation model of *Mycobacterium tuberculosis* persistence by gene and protein expression profiling. *Mol. Microbiol.* **43:**717–731.

Bhakta, S., G. S. Besra, A. M. Upton, T. Parish, C. Sholto-Douglas-Vernon, K. J. C. Gibson, S. Knutton, S. Gordon, R. P. daSilva, M. C. Anderton, and E. Sim. 2004. Arylamine N-acetyltransferase is required for synthesis of mycolic acids and complex lipids in *Mycobacterium bovis* BCG and represents a novel drug target. *J. Exp. Med.* **199:**1191–1199.

Birck, C., Y. Chen, F. M. Hulett, and J. P. Samama. 2003. The crystal structure of the phosphorylation domain in PhoP reveals a functional tandem association mediated by an asymmetric interface. *J. Bacteriol.* **185:**254–261.

Bliska, J. B., J. E. Galan, and S. Falkow. 1993. Signal transduction in the mammalian cell during bacterial attachment and entry. *Cell* **73:**903–920.

Boitel, B., M. Ortiz-Lombardia, R. Duran, F. Pompeo, S. T. Cole, C. Cervenansky, and P. M. Alzari. 2003. PknB kinase activity is regulated by phosphorylation in two Thr residues and dephosphorylation by PstP, the cognate phospho-Ser/Thr phosphatase, in *Mycobacterium tuberculosis*. *Mol. Microbiol.* **49:**1493–1508.

Bork, P., N. P. Brown, H. Hegyi, and J. Schultz. 1996. The protein phosphatase 2C (PP2C) superfamily: detection of bacterial homologues. *Protein Sci.* **5:**1421–1425.

Bridges, A. 2001. Chemical inhibitors of protein kinases. *Chem. Rev.* **101:**2541–2572.

Camacho, L. R., P. Constant, C. Raynaud, M. A. Laneelle, J. A. Triccas, B. Gicquel, M. Daffe, and C. Guilhot. 2001. Analysis of the phthiocerol dimycocerosate locus of *Mycobacterium tuberculosis*. Evidence that this lipid is involved in the cell wall permeability barrier. *J. Biol. Chem.* **276:**19845–19854.

Camacho, L. R., D. Ensergueix, E. Perez, B. Gicquel, and C. Guilhot. 1999. Identification of a virulence gene cluster of *Mycobacterium tuberculosis* by signature-tagged transposon mutagenesis. *Mol. Microbiol.* **34:**257–267.

Castandet, J., J. F. Prost, P. Peyron, C. Astarie-Dequeker, E. Anes, A. J. Cozzone, G. Griffiths, and I. Maridonneau-Parini. 2005. Tyrosine phosphatase MptpA of *Mycobacterium tuberculosis* inhibits phagocytosis and increases actin polymerization in macrophages. *Res. Microbiol.* **156:**1005–1013.

Chaba, R., M. Raje, and P. K. Chakraborti. 2002. Evidence that a eukaryotic-type serine/threonine protein kinase from *Mycobacterium tuberculosis* regulates morphological changes associated with cell division. *Eur. J. Biochem.* **269:**1078–1085.

Chatterjee, D., and K. H. Khoo. 1998. Mycobacterial lipoarabinomannan: an extraordinary lipoheteroglycan with profound physiological effects. *Glycobiology* **8:**113–120.

Chopra, P., B. Singh, R. Singh, R. Vohra, A. Koul, L. S. Meena, H. Koduri, M. Ghildiyal, P. Deol, T. K. Das, A. K. Tyagi, and Y. Singh. 2003. Phosphoprotein phosphatase of *Mycobacterium tuberculosis* dephosphorylates serine-threonine kinases PknA and PknB. *Biochem. Biophys. Res. Commun.* **311:**112–120.

Cirri, P., P. Chiarugi, G. Camici, G. Manao, L. Pazzagli, A. Caselli, I. Barghini, G. Cappugi, G. Raugei, and G. Ramponi. 1993. The role of Cys-17 in the pyridoxal 5′-phosphate inhibition of the bovine liver low M(r) phosphotyrosine protein phosphatase. *Biochim. Biophys. Acta* **1161:**216–222.

Cole, S. T., R. Brosch, J. Parkhill, T. Garnier, C. Churcher, D. Harris, S. V. Gordon, K. Eiglmeier, S. Gas, C. E. Barry, F. Tekaia, K. Badcock, D. Basham, D. Brown, T. Chillingworth, R. Connor, R. Davies, K. Devlin, T. Feltwell, S. Gentles, N. Hamlin, S. Holroyd, T. Hornsby, K. Jagels, A. Krogh, J. McLean, S. Moule, L. Murphy, K. Oliver, J. Osborne, M. A. Quail, M. A. Rajandream, J. Rogers, S. Rutter, K. Seeger, J. Skelton, R. Squares, S. Squares, J. E. Sulston, K. Taylor, S. Whitehead, and B. G. Barrell. 1998. Deciphering the biology of *Mycobacterium tuberculosis* from the complete genome sequence. *Nature* **393:**537–544.

Coombes, B. K., Y. Valdez, and B. B. Finlay. 2004. Evasive maneuvers by secreted bacterial proteins to avoid innate immune responses. *Curr. Biol.* **14:**R856–R867.

Cosma, C. L., D. R. Sherman, and L. Ramakrishnan. 2003. The secret lives of the pathogenic mycobacteria. *Annu. Rev. Microbiol.* **57:**641–676.

Cowley, S. C., R. Babakaif, and Y. Av-Gay. 2002. Expression and localization of the *Mycobacterium tuberculosis* protein tyrosine phosphatase, PtpA. *Res. Microbiol.* **153:**233–241.

Cowley, S., M. Ko, N. Pick, R. Chow, K. J. Downing, B. G. Gordhan, J. C. Betts, V. Mizrahi, D. A. Smith, R. W. Stokes, and Y. Av-Gay. 2004. The *Mycobacterium tuberculosis* protein serine/threonine kinase PknG is linked to cellular glutamate/glutamine levels and is important for growth *in vivo*. *Mol. Microbiol.* **52:**1691–1702.

Cox, J. S., B. Chen, M. McNeil, and W. R. Jacobs, Jr. 1999. Complex lipid determines tissue-specific replication of *Mycobacterium tuberculosis* in mice. *Nature* **402:**79–83.

Cozzone, A. J. 2005. Role of protein phosphorylation on serine/threonine and tyrosine in the virulence of bacterial pathogens. *J. Mol. Microbiol. Biotechnol.* **9:**198–213.

Cozzone, A. J. 1993. ATP-dependent protein kinases in bacteria. *J. Cell. Biochem.* **51:**7–13.

Curry, J. M., R. Whalan, D. M. Hunt, K. Gohil, M. Strom, L. Rickman, M. J. Colston, S. J. Smerdon, and R. S. Buxton. 2005. An ABC transporter containing a forkhead-associated domain interacts with a serine-threonine protein kinase and is required for growth of *Mycobacterium tuberculosis* in mice. *Infect. Immun.* **73:**4471–4477.

Daffe, M., and P. Draper. 1998. The envelope layers of mycobacteria with reference to their pathogenicity. *Adv. Microb. Physiol.* **39:**131–203.

Dannenberg, A. M. Jr. 1991. Delayed type hypersensitivity and cell mediated immunity in the pathogenesis of tuberculosis. *Immunol. Today* **12:**228–233.

Dar, A. C., T. E. Dever, and F. Sicheri. 2005. Higher-order substrate recognition of eIF2a by the RNA-dependent protein kinase PKR. *Cell* **122:**887-900.

Dasgupta, A., P. Datta, M. Kundu, and J. Basu. 2006. The serine/threonine kinase PknB of *Mycobacterium tuberculosis* phosphorylates PBPA, a penicillin-binding protein required for cell division. *Microbiology* **152:**493–504.

Dasgupta, N., V. Kapur, K. K. Singh, T. K. Das, S. Sachdeva, K. Jyothisri, and J. S. Tyagi. 2000. Characterization of a two-component system, devR-devS, of *Mycobacterium tuberculosis*. *Tuber. Lung Dis.* **80:**141–159.

Denu, J. M., and J. E. Dixon. 1995. A catalytic mechanism for the dual-specific phosphatases. *Proc. Natl. Acad. Sci. USA* **92:**5910–5914.

Deol, P., R. Vohra, A. K. Saini, A. Singh, H. Chandra, P. Chopra, T. K. Das, A. K. Tyagi, and Y. Singh. 2005. Role of *Mycobacterium tuberculosis* Ser/Thr kinase PknF: implications in glucose transport and cell division. *J. Bacteriol.* **187:**3415–3420.

Dey, M., C. Cao, A. C. Dar, T. Tamura, K. Ozato, F. Sicheri, and T. E. Dever. 2005. Mechanistic link between PKR dimerization, autophos-phorylation and eIF2a substrate recognition. *Cell* **122:**901–913.

Domenech, P., M. B. Reed, and C. E. Barry, 3rd. 2005. Contribution of the *Mycobacterium tuberculosis* MmpL protein family to virulence and drug resistance. *Infect. Immun.* **73:**3492–3501

Drews, S. J., F. Hung, and Y. Av-Gay. 2001. A protein kinase inhibitor as an antimycobacterial agent. *FEMS Microbiol. Lett.* **205:**369–374.

Dubnau, E., J. Chan, C. Raynaud, V. P. Mohan, M. A. Yu, K. Laneelle, A. Quemard, I. Smith, and M. Daffe. 2000. Oxygenated mycolic acids are necessary for virulence of *Mycobacterium tuberculosis* in mice. *Mol. Microbiol.* **36:**630–637.

Durocher, D., and S. P. Jackson. 2002. The FHA domain. *FEBS Lett.* **13:**58–66.

Echenique, J., A. Kadioglu, S. Romao, P. W. Andrew, and M. C. Trombe. 2004. Protein serine/threonine kinase StkP positively controls virulence and competence in *Streptococcus pneumoniae*. *Infect. Immun.* **72:**2434–2437.

Eiglmeier, K., J. Parkhill, N. Honore, T. Garnier, F. Tekaia, A. Telenti, P. Klatser, K. D. James, N. R. Thomson, P. R.

Wheeler, C. Churcher, D. Harris, K. Mungall, B. G. Barrell, and S. T. Cole. 2001. The decaying genome of *Mycobacterium leprae*. *Lepr. Rev.* **72:**387–398.

Ewann, F., M. Jackson, K. Pethe, A. Cooper, N. Mielcarek, D. Ensergueix, B. Gicquel, C. Locht, and P. Supply. 2002. Transient requirement of the PrrA-PrrB two-component system for early intracellular multiplication of *Mycobacterium tuberculosis*. *Infect. Immun.* **70:**2256–2263.

Fontan P. A., S. Walters, and I. Smith. 2004. Cellular signaling pathways and transcriptional regulation in *Mycobacterium tuberculosis*: Stress control and virulence. *Curr. Sci.* **86:**122–134.

Fratti, R. A., J. M. Backer, J. Gruenberg, S. Corvera, and V. Deretic. 2001. Role of phosphatidylinositol 3-kinase and Rab5 effectors in phagosomal biogenesis and mycobacterial phagosome maturation arrest. *J. Cell. Biol.* **154:**631–644.

Fulop, V., and D. T. Jones. 1999. Beta propellers: structural rigidity and functional diversity. *Curr. Opin. Struct. Biol.* **9:**715–721.

Gaidenko, T. A., T. J. Kim, and C. W. Price. 2002. The PrpC serine-threonine phosphatase and PrkC kinase have opposing physiological roles in stationary-phase *Bacillus subtilis* cells. *J. Bacteriol.* **184:**6109–6114.

Galperin, M. Y., A. N. Nikolskaya, and E. V. Koonin. 2001. Novel domains of the prokaryotic two-component signal transduction systems. *FEMS Microbiol. Lett.* **203:**11–21.

Galyov, E. E., S. Hakansson, A. Forsberg, and H. Wolf-Watz. 1993. A secreted protein kinase of *Yersinia pseudotuberculosis* is an indispensable virulence determinant. *Nature* (London) **361:**730–732.

Gao, L.Y., F. Laval, E. H. Lawson, R. K. Groger, A. Woodruff, J. H. Morisaki, J. S. Cox, M. Daffe, and E. J. Brown. 2003. Requirement for kasB in Mycobacterium mycolic acid biosynthesis, cell wall impermeability and intracellular survival: implications for therapy. *Mol. Microbiol.* **49:**1547–1563.

Garnak, M., and H. C. Reeves. 1979. Phosphorylation of Isocitrate dehydrogenase of *Escherichia coli*. *Science* **203:**1111–1112.

Gautier, J., M. J. Solomon, R. N. Booher, J. F. Bazan, and M. W. Kirschner. 1991. cdc25 is a specific tyrosine phosphatase that directly activates p34cdc2. *Cell* **67:**197–211.

Glickman, M. S., J. S. Cox, and W. R. Jacobs, Jr. 2000. A novel mycolic acid cyclopropane synthetase is required for cording, persistence and virulence of *M. tuberculosis*. *Mol. Cell* **5:**717–727.

Gonzalo Asensio, J., C. Maia, N. L. Ferrer, N. Barilone, F. Laval, C. Y. Soto, N. Winter, M. Daffe, B. Gicquel, C. Martin, and M. Jackson. 2006. The virulence-associated two-component PhoP-PhoR system controls the biosynthesis of polyketide-derived lipids in *Mycobacterium tuberculosis*. *J. Biol. Chem.* **281:**1313–1316.

Gopalaswamy, R., P. R. Narayanan, and S. Narayanan. 2004. Cloning, overexpression, and characterization of a serine/threonine protein kinase pknI from *Mycobacterium tuberculosis* H37Rv. *Protein Expr. Purif.* **36:**82–89.

Graham, J. E., and J. E. Clark-Curtiss. 1999. Identification of *Mycobacterium tuberculosis* RNAs synthesized in response to phagocytosis by human macrophages by selective capture of transcribed sequences (SCOTS). *Proc. Natl. Acad. Sci. USA* **96:**11554–11559.

Greenstein, A. E., C. Grundner, N. Echols, L. M. Gay, T. N. Lombana, C. A. Miecskowski, K. E. Pullen, P. Y. Sung, and T. Alber. 2005. Structure/function studies of Ser/Thr and Tyr protein phosphorylation in *Mycobacterium tuberculosis*. *J. Mol. Microbiol. Biotechnol.* **9:**167–181.

Groisman, E. A. 2001. The pleiotropic two-component regulatory system PhoP–PhoQ. *J. Bacteriol.* **183:**1835–1842.

Grosios, K., and P. Traxler. 2003. Tyrosine kinase targets in drug discovery. *Drug Future* **28:**679–697.

Guan, K., and J. E. Dixon. 1990. Protein tyrosine phosphatase activity of an essential virulence determinant in Yersinia. *Science* **249**:553–559.

Gupta, S., S. Jain, and A. K. Tyagi. 1999. Analysis, expression and prevalence of the *Mycobacterium tuberculosis* homolog of bacterial virulence regulating proteins. *FEMS Microbiol. Lett.* **172**:137–143.

Hakansson, S., E. E. Galyov, R. Rosqvist, and H. Wolf-Watz. 1996. The Yersinia YpkA Ser/Thr kinase is translocated and subsequently targeted to the inner surface of the HeLa cell plasma membrane. *Mol. Microbiol.* **20**:593–603.

Hanks, S. K., A. M. Quinn, and T. Hunter. 1988. The protein kinase family: conserved features and deduced phylogeny of the catalytic domains. *Science* **241**:42–52.

Hanks, S., and A. M. Quinn. 1991. Protein kinase catalytic domain sequence database: identification of conserved features of primary structure and classification of family members. *Methods Enzymol.* **200**:38–62.

Haydel, S. E., and J. E. Clark-Curtiss. 2004. Global expression analysis of two-component system regulator genes during *Mycobacterium tuberculosis* growth in human macrophages. *FEMS Microbiol. Lett.* **236**:341–347.

Haydel, S. E., and J. E. Clark-Curtiss. 2006. The *Mycobacterium tuberculosis* TrcR response regulator represses transcription of the intracellularly expressed Rv1057 gene, encoding a seven-bladed beta-propeller. *J. Bacteriol.* **188**:150–159.

He, H., R. Hovey, J. Kane, V. Singh, and T. C. Zahrt. 2006. MprAB is a stress-responsive two-component system that directly regulates expression of sigma factors SigB and SigE in *Mycobacterium tuberculosis*. *J. Bacteriol.* **188**:2134–2143.

Hilleman, M. R. 2004. Strategies and mechanisms for host and pathogen survival in acute and persistent viral infections. *Proc. Natl. Acad. Sci. USA* **101**(Suppl. 2):14560–14566.

Hornef, M. W., M. J. Wick, M. Rhen, and S. Normark. 2002. Bacterial strategies for overcoming host innate and adaptive immune responses. *Nat. Immunol.* **3**:1033–1040.

Houben, E. N., L. Nguyen, and J. Pieters. 2006. Interaction of pathogenic mycobacteria with the host immune system. *Curr. Opin. Microbiol.* **9**:76–85.

Howell, L. D., C. Griffiths, L. W. Slade, M. Potts, and P. J. Kennelly. 1996. Substrate specificity of IphP, a cyanobacterial dual-specificity protein phosphatase with MAP kinase phosphatase activity. *Biochemistry* **35**:7566–7572.

Hunter, T. 1995. Protein kinases and phosphatases: the yin and yang of protein phosphorylation and signaling. *Cell* **80**:225–236.

Itou, H., and I. Tanaka. 2001. The OmpR-family of proteins: insight into the tertiary structure and functions of two-component regulator proteins. *J. Biochem.* (Tokyo) **129**:343–350.

Johnson, P., H. L. Ostergaard, C. Wasden, and I. S. Trowbridge. 1992. Mutational analysis of CD45. A leukocyte-specific protein tyrosine phosphatase. *J. Biol. Chem.* **267**:8035–8041.

Juris, S. J., A. E. Rudolph, D. Huddler, K. Orth, and J. E. Dixon. 2000. A distinctive role for the Yersinia protein kinase: Actin binding, kinase activation and cytoskeletal disruption. *Proc. Natl. Acad. Sci. USA* **97**:9431–9436.

Kaldor, S. W., V. J. Kalish, J. F. Davies II, B. V. Shetty, J. E. Fritz, K. Appelt, J. A. Burgess, K. M. Campanale, N. Y. Chirgadze, D. K. Clawson, B. A. Dressman, S. D. Hatch, D. A. Khalil, M. B. Kosa, P. P. Lubbehusen, M. A. Muesing, A. K. Patick, S. H. Reich, K. S. Su, and J. H. Tatlock. 1997. Viracept (nelfinavir mesylate, AG1343): a potent, orally bioavailable inhibitor of HIV-1 protease. *J. Med. Chem.* **40**:3979–3985.

Kang, C. M., D. W. Abbott, S. T. Park, C. C. Dascher, L. C. Cantley, and R. N. Husson. 2005. The *Mycobacterium tuberculosis* serine/threonine kinases PknA and PknB: substrate identi-

fication and regulation of cell shape. *Genes Dev.* **19**:1692–1704.

Kaniga, K., J. Uralil, J. B. Bliska, and J. E. Galan. 1996. A secreted protein tyrosine phosphatase with modular effector domains in bacterial pathogen *Salmonella typhimurium*. *Mol. Microbiol.* **21**:633–641.

Kennelly, P. J. 2002. Protein kinases and protein phosphatases in prokaryotes: a genomic perspective. *FEMS Microbiol. Lett.* **206**:1–8.

Kim, C. U., W. Lew, M. A. Williams, H. T. Liu, L. J. Zhang, S. Swaminathan, N. Bischofberger, M. S. Chen, D. B. Mendel, C. Y. Tai, W. G. Laver, and R. C. Stevens. 1997. Influenza neuraminidase inhibitors possessing a novel hydrophobic interaction in the enzyme active site: design, synthesis, and structural analysis of carbocyclic sialic acid analogues with potent anti-influenza activity. *J. Am. Chem. Soc.* **119**:681–690.

Kim, E. E., C. T. Baker, M. D. Dwyer, M. A. Murcko, B. G. Rao, R. D. Tung, and M. A. Navia. 1995. Crystal structure of HIV-1 protease in complex with VX-478, a potent and orally available inhibitor of the enzyme. *J. Am. Chem. Soc.* **117**:1181–1182.

Klein, G., C. Dartigalongue, and S. Raina. 2003. Phosphorylation-mediated regulation of heat shock response in *Escherichia coli*. *Mol. Microbiol.* **48**:269–285.

Koul, A., A. Choidas, M. Treder, A. K. Tyagi, K. Drlica, Y. Singh, and A. Ullrich. 2000. Cloning and characterization of secretory tyrosine phosphatase of *Mycobacterium tuberculosis*. *J. Bacteriol.* **182**:5425–5432.

Koul, A., A. Choidas, A. K. Tyagi, K. Drlica, Y. Singh, and A. Ullrich. 2001. Serine/threonine protein kinases PknF and PknG of *Mycobacterium tuberculosis*: characterization and localization. *Microbiology* **147**:2307–2314.

Kremer, L., A. R. Baulard, and G. S. Besra. 2000. Genetics of mycolic acid biosynthesis, p. 173–190. *In* G. F. Hatfull and W. R. Jacobs, Jr. (ed.), *Molecular Genetics of Mycobacteria*. ASM Press, Washington, DC.

Leonard, C. J., L. Aravind, and E. V. Koonin. 1998. Novel families of putative protein kinases in bacteria and archaea: evolution of the "eukaryotic" protein kinase superfamily. *Genome Res.* **8**:1038–1047.

Liu, K., B. Lemon, and P. Traktman. 1995. The dual-specificity phosphatase encoded by vaccinia virus, VH1, is essential for viral transcription *in vivo* and *in vitro*. *J. Virol.* **69**:7823–7834.

Lois, A. F., M. Weinstein, G. S. Ditta, and D. R. Helinski. 1993. Autophosphorylation and phosphatase activities of the oxygen-sensing protein FixL of *Rhizobium meliloti* are coordinately regulated by oxygen. *J. Biol. Chem.* **268**:4370–4375.

Ludwiczak, P., M. Gilleron, Y. Bordat, C. Martin, B. Gicquel, and G. Puzo. 2002. *Mycobacterium tuberculosis phoP* mutant: lipoarabino-mannan molecular structure. *Microbiology* **148**:3029–3037.

Madec, E., A. Laszkiewicz, A. Iwanicki, M. Obuchowski, and S. Seror. 2002. Characterization of a membrane-linked Ser/Thr protein kinase in *Bacillus subtilis*, implicated in developmental processes, *Mol. Microbiol.* **2**:571–586.

Malhotra, V., D. Sharma, V. D. Ramanathan, H. Shakila, D. K. Saini, S. Chakravorty, T. K. Das, Q. Li, R. F. Silver, P. R. Narayanan, and J. S. Tyagi. 2004. Disruption of response regulator gene, devR, leads to attenuation in virulence of *Mycobacterium tuberculosis*. *FEMS Microbiol. Lett.* **231**:237–245.

Manger, M., M. Scheck, H. Prinz, J. P. von Kries, T. Langer, K. Saxena, H. Schwalbe, A. Furstner, J. Rademann, and H. Waldmann. 2005. Discovery of *Mycobacterium tuberculosis* protein tyrosine phosphatase A (MptpA) inhibitors based on natural products and a fragment-based approach. *Chembiochem* **6**:1749–1753.

Manning, G., D. B. Whyte, R. Martinez, T. Hunter, and S. Sudarsanam. 2002. The protein kinase complement of the human genome. *Science* **298:**1912–1934.

Matsuhashi, M., M. Wachi, and F. Ishino. 1990. Machinery for cell growth and division: penicillin-binding proteins and other proteins. *Res. Microbiol.* **141:**89–103.

Medzhitov, R., and C. Janeway, Jr. 2000. Innate immune recognition: mechanisms and pathways. *Immunol. Rev.* **173:**89–97.

Meggio, F., and L. A. Pinna. 2003. One-thousand-and-one substrates of protein kinase CK2? *FASEB J.* **17:**349–368.

Molle, V., A. K. Brown, G. S. Besra, A. J. Cozzone, and L. Kremer. 2006. The condensing activities of the *Mycobacterium tuberculosis* type II fatty acid synthase are differentially regulated by phosphorylation. *J. Biol. Chem.* **281:**30094–30103.

Molle, V., L. Kremer, C. Girard-Blanc, G. S. Besra, A. J. Cozzone, and J. F. Prost. 2003. An FHA phosphoprotein recognition domain mediates protein EmbR phosphorylation by PknH, a Ser/Thr protein kinase from *Mycobacterium tuberculosis*. *Biochemistry* **42:**15300–15309.

Molle, V., D. Soulat, J. M. Jault, C. Grangeasse, A. J. Cozzone, and J. F. Prost. 2004. Two FHA domains on an ABC transporter, Rv1747, mediate its phosphorylation by PknF, a Ser/Thr protein kinase from *Mycobacterium tuberculosis*. *FEMS Microbiol. Lett.* **234:**215–223.

Morth, J. P., V. Feng, L. J. Perry, D. I. Svergun, and P. A. Tucker. 2004. The crystal and solution structure of a putative transcriptional antiterminator from *Mycobacterium tuberculosis*. *Structure* **12:**1595–1605.

Mossman, K., H. Ostergaard, C. Upton, and G. McFadden. 1995. Myxoma virus and Shope fibroma virus encode dual-specificity tyrosine/serine phosphatases which are essential for virus viability. *Virology* **206:**572–582.

Munoz-Dorado, J., S. Inouye, and M. Inouye. 1991. A gene encoding a protein Serine/threonine kinase is required for the normal development of *Myxococcus xanthus*, a gram negative bacterium. *Cell* **67:**995–1006.

Nariya, H., and S. Inouye. 2002. Activation of 6-phosphofructokinase via phosphorylation by Pkn4, a protein Ser/Thr kinase of Myxococcus xanthus. *Mol. Microbiol.* **46:**1353–1366.

Neu, J. M., S. V. MacMillan, J. R. Nodwell, and G. D. Wright. 2002. StoPK-1, a serine/threonine protein kinase from the glycopeptide antibiotic producer *Streptomyces toyocaensis* NRRL 15009, affects oxidative stress response. *Mol. Microbiol.* **44:**417–430.

Nigou, J., M. Gilleron, and G. Puzo. 2003. Lipoarabinomannans: from structure to biosynthesis. *Biochimie* **85:**153–166.

Noble, M. E., J. A. Endicott, and L. N. Johnson. 2004. Protein kinase inhibitors: insights into drug design from structure. *Science* **303:**1800–1805.

Novakova, L., L. Saskova, P. Pallova, J. Janecek, J. Novotna, A. Ulrych, J. Echenique, M. C. Trombe, and P. Branny. 2005. Characterization of a eukaryotic type serine/threonine protein kinase and protein phosphatase of *Streptococcus pneumoniae* and identification of kinase substrates. *FEBS J.* **272:**1243–1254.

Ortiz-Lombardia, M., F. Pompeo, B. Boitel, and P. M. Alzari. 2003. Crystal structure of the catalytic domain of the PknB serine/threonine kinase from *Mycobacterium tuberculosis*. *J. Biol. Chem.* **278:**13094–13100.

Pallen, M., R. Chaudhuri, and A. Khan. 2002. Bacterial FHA domains: neglected players in the phospho-threonine signalling game? *Trends Micobiol.* **10:**556–563.

Papavinasasundaram, K. G., B. Chan, J. H. Chung, M. J. Colston, E. O. Davis, and Y. Av-Gay. 2005. Deletion of the *Mycobacterium tuberculosis* pknH gene confers a higher bacillary load during the chronic phase of infection in BALB/c mice. *J. Bacteriol.* **187:**5751–5760.

Parish, T., D. A. Smith, S. Kendall, N. Casali, G. J. Bancroft, and N. G. Stoker. 2003a. Deletion of two-component regulatory systems increases the virulence of *Mycobacterium tuberculosis*. *Infect. Immun.* **71:**1134–1140.

Parish, T., D. A. Smith, G. Roberts, J. Betts, and N. G. Stoker. 2003b. The senX3-regX3 two-component regulatory system of *Mycobacterium tuberculosis* is required for virulence. *Microbiology* **149:**1423–1435.

Park, H. D., K. M. Guinn, M. I. Harrell, R. Liao, M. I. Voskuil, M. Tompa, G. K. Schoolnik, and D. R. Sherman. 2003. Rv3133c/DosR is a transcription factor that mediates the hypoxic response of *Mycobacterium tuberculosis*. *Mol. Microbiol.* **48:**833–843.

Peirs, P., L. De Wit, M. Braibant, K. Huygen, and J. Content. 1997. A serine/threonine protein kinase from *Mycobacterium tuberculosis*. *Eur. J. Biochem.* **244:**604–612.

Perez, J., R. Garcia, H. Bach, J. H. de Waard, W. R. Jacobs, Jr, Y. Av-Gay, J. Bubis, and H. E. Takiff. 2006. *Mycobacterium tuberculosis* transporter MmpL7 is a potential substrate for kinase PknD. *Biochem. Biophys. Res. Commun.* **348:**6–12.

Perez, E., S. Samper, Y. Bordas, C. Guilhot, B. Gicquel, and C. Martin. 2001. An essential role for phoP in *Mycobacterium tuberculosis* virulence. *Mol. Microbiol.* **41:**179–187.

Pieters, J., and J. Gatfield. 2002. Hijacking the host: survival of pathogenic mycobacteria inside macrophages. *Trends Microbiol.* **10:**142–146.

Polarek, J. W., G. Williams, and W. Epstein. 1992. The products of the kdpDE operon are required for expression of the Kdp ATPase of *Escherichia coli*. *J. Bacteriol.* **174:**2145–2151.

Poncet, S., I. Mijakovic, S. Nessler, V. Gueguen-Chaignon, V. Chaptal, A. Galinier, G. Boel, A. Maze, and J. Deutscher. 2004. HPr kinase/phosphorylase, a Walker motif A-containing bifunctional sensor enzyme controlling catabolite repression in Gram-positive bacteria. *Biochim. Biophys. Acta* **1697:**123–135.

Potts, M., H. Sun, K. Mockaitis, P. J. Kennelly, D. Reed, and N. K. Tonks. 1993. A protein-tyrosine/serine phosphatase encoded by the genome of the cyanobacterium *Nostoc commune* UTEX 584. *J. Biol. Chem.* **268:**7632–7635.

Prabhakaran, K., E. B. Harris, and B. Randhawa. 2000. Regulation by protein kinase of phagocytosis of *Mycobacterium leprae* by macrophages. *J. Med. Microbiol.* **49:**339–342.

Rajagopal, L., A. Vo, A. Silvestroni, and C. E. Rubens. 2005. Regulation of purine biosynthesis by a eukaryotic-type kinase in *Streptococcus agalactiae*. *Mol. Microbiol.* **56:**1329–1346.

Ray, M. K., G. S. Kumar, and S. Shivaji. 1994. Phosphorylation of membrane proteins in response to temperature in an Antarctic *Pseudomonas syringae*. *Microbiology* **140:**3217–3223.

Raynaud, C., C. Etienne, P. Peyron, M. A. Laneelle, and M. Daffe. 1998. Extracellular enzyme activities potentially involved in the pathogenicity of *Mycobacterium tuberculosis*. *Microbiology* **144:**577–587.

Reed, M. B., P. Domenech, C. Manca, H. Su, A. K. Barczak, B. N. Kreiswirth, G. Kaplan, and C. E. Barry III. 2004. A glycolipid of hypervirulent tuberculosis strains that inhibits the innate immune response. *Nature* **431:**84–87.

Reilly, T. J., G. S. Baron, F. E. Nano, and M. S. Kuhlenschmidt. 1996. Characterization and sequencing of a respiratory burst-inhibiting acid phosphatase from *Francisella tularensis*. *J. Biol. Chem.* **271:**10973–10983.

Remaley, A. T., S. Das, P. I. Campbell, G. M. Larocca, M. T. Pope, and R. H. Glew. 1985. Characterization of *Leishmania donovani* acid phosphatases. *J. Biol. Chem.* **260:**880–886.

Repik, A., A. Rebbapragada, M. S. Johnson, J. O. Haznedar, I. B. Zhulin, and B. L. Taylor. 2000. PAS domain residues involved in signal transduction by the Aer redox sensor of *Escherichia coli*. *Mol. Microbiol.* **36:**806–816.

Rickman, L., J. W. Saldanha, D. M. Hunt, D. N. Hoar, M. J. Colston, J. B. Millar, and R. S. Buxton. 2004. A two-component signal transduction system with a PAS domain-containing sensor is required for virulence of *Mycobacterium tuberculosis* in mice. *Biochem. Biophys. Res. Commun.* 314:259–267.

Roberts, D. M., R. P. Liao, G. Wisedchaisri, W. G. Hol, and D. R. Sherman. 2004. Two sensor kinases contribute to the hypoxic response of *Mycobacterium tuberculosis*. *J. Biol. Chem.* 279: 23082–23087.

Römling, U., M. Gomelsky, and M. Y. Galperin. 2005. C-di-GMP: the dawning of a novel bacterial signalling system. *Mol. Microbiol.* 57:629–639.

Rosenberger, C. M., and B. B. Finlay. 2003. Phagocyte sabotage: disruption of macrophage signalling by bacterial pathogens. *Nat. Rev. Mol. Cell. Biol.* 4:385–396.

Rossolini, G. M., S. Schippa, M. L. Riccio, F. Berlutti, L. E. Macaskie, and M. C. Thaller. 1998. Bacterial nonspecific acid phospho-hydrolases: physiology, evolution and use as tools in microbial biotechnology. *Cell Mol. Life Sci.* 54:833–850.

Saini, D. K., V. Malhotra, and J. S. Tyagi. 2004. Cross talk between DevS sensor kinase homologue, Rv2027c, and DevR response regulator of *Mycobacterium tuberculosis*. *FEBS Lett.* 565:75–80.

Saleh, M. T., and J. T. Belisle. 2000. Secretion of an acid phosphatase (SapM) by *Mycobacterium tuberculosis* that is similar to eukaryotic acid phosphatases. *J. Bacteriol.* 182:6850–6853.

Sardiwal, S., S. L. Kendall, F. Movahedzadeh, S. C. Rison, N. G. Stoker, and S. Djordjevic. 2005. A GAF domain in the hypoxia/NO-inducible *Mycobacterium tuberculosis* DosS protein binds haem. *J. Mol. Biol.* 353:929–936.

Sassetti, C. M., and E. J. Rubin. 2003. Genetic requirements for mycobacterial survival during infection. *Proc. Natl. Acad. Sci. USA* 100:12989–12994.

Sassetti, C. M., D. H. Boyd, and E. J. Rubin. 2001. Comprehensive identification of conditionally essential genes in mycobacteria. *Proc. Natl. Acad. Sci. USA* 98:12712–12717.

Sassetti, C. M., D. H. Boyd, and E. J. Rubin. 2003. Genes required for mycobacterial growth defined by high density mutagenesis. *Mol. Microbiol.* 48:77–84.

Schnappinger, D., S. Ehrt, M. I. Voskuil, Y. Liu, J. A. Mangan, I. M. Monahan, G. Dolganov, B. Efron, P. D. Butcher, C. Nathan, and G. K. Schoolnik. 2003. Transcriptional adaptation of *Mycobacterium tuberculosis* within macrophages: insights into the phagosomal environment. *J. Exp. Med.* 198:693–704.

Sham, H. L., D. J. Kempf, A. Molla, K. C. Marsh, G. N. Kumar, C. M. Chen, W. Kati, K. Stewart, R. Lal, A. Hsu, D. Betebenner, M. Korneyeva, S. Vasavanonda, E. McDonald, A. Saldivar, N. Wideburg, X. Chen, P. Niu, C. Park, V. Jayanti, B. Grabowski, G. R. Granneman, E. Sun, A. J. Japour, J. M. Leonard, J. J. Plattner, and D. W. Norbeck. 1998. ABT-378, a highly potent inhibitor of the human immunodeficiency virus protease. *Antimicrob. Agents Chemother.* 42:3218–3224.

Sharma, K., M. Gupta, A. Krupa, N. Srinivasan, and Y. Singh. 2006b. EmbR, a regulatory protein with ATPase activity, is a substrate of multiple serine/threonine kinases and phosphatase in *Mycobacterium tuberculosis*. *FEBS J.* 273:2711–2721.

Sharma, K., M. Gupta, M. Pathak, N. Gupta, A. Koul, S. Sarangi, R. Baweja, and Y. Singh. 2006a. Transcriptional control of the mycobacterial embCAB operon by PknH through a regulatory protein, EmbR, in vivo. *J. Bacteriol.* 188:2936–2944.

Shawver, L. K., D. Slamon, and A. Ullrich. 2002. Smart drugs: tyrosine kinase inhibitors in cancer therapy. *Cancer Cell* 1:117–123.

Singh, A., R. Gupta, R. A. Vishwakarma, P. R. Narayanan, C. N. Paramasivan, V. D. Ramanathan, and A. K. Tyagi. 2005.

Requirement of the mymA operon for appropriate cell wall ultrastructure and persistence of *Mycobacterium tuberculosis* in the spleens of guinea pigs. *J Bacteriol.* 187:4173–4186.

Singh, A., S. Jain, S. Gupta, T. Das, and A. K. Tyagi. 2003b. mymA operon of *Mycobacterium tuberculosis*: its regulation and importance in the cell envelope. *FEMS Microbiol. Lett.* 227: 53–63.

Singh, A., Y. Singh, R. Pine, L. Shi, R. Chandra, and K. Drlica. 2006. Protein kinase I of *Mycobacterium tuberculosis*: cellular localization and expression during infection of macrophage-like cells. *Tuberculosis* (Edinburgh) 86:28–33.

Singh, R., V. Rao, H. Shakila, R. Gupta, A. Khera, N. Dhar, A. Singh, A. Koul, Y. Singh, M. Naseema, P. R. Narayanan, C. N. Paramasivan, V. D. Ramanathan, and A. K. Tyagi. 2003a. Disruption of mptpB impairs the ability of *Mycobacterium tuberculosis* to survive in guinea pigs. *Mol. Microbiol.* 50:751–762.

South, S. L., R. Nichols, and T. C. Montie. 1994. Tyrosine kinase activity in *Pseudomonas aeruginosa*. *Mol. Microbiol.* 12:903–910.

Sreevatsan, S., K. E. Stockbauer, X. Pan, B. N. Kreiswirth, S. L. Moghazeh, W. R. Jacobs, Jr., A. Telenti, and J. M. Musser. 1997. Ethambutanol resistance in *Mycobacterium tuberculosis*: critical role of embB mutations. *Antimicrob. Agents Chemother.* 41:1677–1681

Stock, A. M., V. L. Robinson, and P. N. Goudreau. 2000. Two-component signal transduction. *Annu. Rev. Biochem.* 69:183–215.

Stock, J. B., A. J. Ninfa, and A. M. Stock. 1989. Protein phosphorylation and regulation of adaptive responses in bacteria. *Microbiol. Rev.* 53:450–490.

Stock, J. B., M. G. Surette, M. Levit, and P. Park. 1995. Two-component signal transduction systems: structure-function relationships and mechanisms of catalysis, p. 25–51. *In* J. A. Hoch and T. J. Silhavy (ed.), *Two-Component Signal Transduction*. ASM Press, Washington, DC.

Stone, R. L., and J. E. Dixon. 1994. Protein tyrosine phosphatases. *J. Biol. Chem.* 269:31323–31326.

Streuli, M., N. X. Krueger, T. Thai, M. Tang, and H. Saito. 1990. Distinct functional roles of the two intracellular phosphatase like domains of the receptor-linked protein tyrosine phosphatases LCA and LAR. *EMBO J.* 9:2399–2407.

Treuner-Lange, A., M. J. Ward, and D. R. Zusman. 2001. Pph1 from *Myxococcus xanthus* is a protein phosphatase involved in vegetative growth and development. *Mol. Microbiol.* 40:126–140.

Tyagi, J. S., and D. Sharma. 2004. Signal transduction systems of mycobacteria with special reference to *M. tuberculosis*. *Curr. Sci.* 86:93–102.

Umeyama, T., P. C. Lee, and S. Horinouchi. 2002. Protein serine/threonine kinases in signal transduction for secondary metabolism and morphogenesis in Streptomyces. *Appl. Microbiol. Biotechnol.* 59:419–425.

Vergne, I., J. Chua, H. H. Lee, M. Lucas, J. Belisle, and V. Deretic. 2005. Mechanism of phagolysosome biogenesis block by viable *Mycobacterium tuberculosis*. *Proc. Natl. Acad. Sci. USA* 102: 4033–4038.

Vescovi, E. G., Y. M. Ayala, E. Di Cera, and E. A. Groisman. 1997. Characterization of the bacterial sensor protein PhoQ. Evidence for distinct binding sites for Mg^{2+} and Ca^{2+}. *J. Biol. Chem.* 272:1440–1443.

Via, L. E., R. Curcic, M. H. Mudd, S. Dhandayuthapani, R. J. Ulmer, and V. Deretic. 1996. Elements of signal transduction in *Mycobacterium tuberculosis*: *in vitro* phosphorylation and *in vivo* expression of the response regulator MtrA. *J. Bacteriol.* 178:3314–3321.

Vieira, O. V., R. J. Botelho, L. Rameh, S. M. Brachmann, T. Matsuo, H. W. Davidson, A. Schreiber, J. M. Backer, L. C. Cantley, and S. Grinstein. 2001. Distinct roles of class I and class III phosphatidylinositol 3-kinases in phagosome formation and maturation. *J. Cell Biol.* **155:**19–25.

Villarino, A., R. Duran, A. Wehenkel, P. Fernandez, P. England, P. Brodin, S. T. Cole, U. Zimny-Arndt, P. R. Jungblut, C. Cervenansky, and P. M. Alzari. 2005. Proteomic identification of *M. tuberculosis* protein kinase substrates: PknB recruits GarA, a FHA domain-containing protein, through activation loop-mediated interactions. *J. Mol. Biol.* **350:**953–63.

von Itzstein, M., W. Y. Wu, G. B. Kok, M. S. Pegg, J. C. Dyason, B. Jin, T. Van Phan, M. L. Smythe, H. F. White, S. W. Oliver, P. M. Colman, J. N. Varghese, D. M. Ryan, J. M. Woods, R. C. Bethell, V. J. Hotham, J. M. Cameron, and C. R. Penn. 1993. Rational design of potent sialidase-based inhibitors of influenza virus replication. *Nature* **363:**418–423.

Walburger, A., A. Koul, G. Ferrari, L. Nguyen, C. Prescianotto-Baschong, K. Huygen, B. Klebl, C. Thompson, G. Bacher, and J. Pieters. 2004. Protein kinase G from pathogenic mycobacteria promotes survival within macrophages. *Science* **304:**1800–1804.

Walters, S. B., E. Dubnau, I. Kolesnikova, F. Laval, M. Daffe, and I. Smith. 2006. The *Mycobacterium tuberculosis* PhoPR two-component system regulates genes essential for virulence and complex lipid biosynthesis. *Mol. Microbiol.* **60:**312–330.

Wehenkel, A., P. Fernandez, M. Bellinzoni, V. Catherinot, N. Barilone, G. Labesse, M. Jackson, and P. M. Alzari. 2006. The structure of PknB in complex with mitoxantrone, an ATP-competitive inhibitor, suggests a mode of protein kinase regulation in mycobacteria. *FEBS Lett.* **580:**3018–3022.

Weide, T., L. Arve, H. Prinz, H. Waldmann, and H. Kessler. 2006. 3-Substituted indolizine-1-carbonitrile derivatives as phosphatase inhibitors. *Bioorg. Med. Chem. Lett.* **16:**59–63.

Wu, J., N. Ohta, J. L. Zhao, and A. Newton. 1999. A novel bacterial tyrosine kinase essential for cell division and differentiation. *Proc. Natl. Acad. Sci. USA* **96:**13068–13073.

Young, T. A., B. Delagoutte, J. A. Endrizzi, A. M. Falick, and T. Alber. 2003. Structure of *Mycobacterium tuberculosis* PknB supports a universal activation mechanism for Ser/Thr protein kinases. *Nat. Struct. Biol.* **10:**168–174.

Yuan, M., F. Deleuil, and M. Fallman. 2005. Interaction between the Yersinia tyrosine phosphatase YopH and its macrophage substrate, Fyn-binding protein, Fyb. *J. Mol. Microbiol. Biotechnol.* **9:**214–223.

Yuvaniyama, J., J. M. Denu, J. E. Dixon, and M. A. Saper. 1996. Crystal structure of the dual specificity protein phosphatase VHR. *Science* **272:**1328–1331.

Zahrt, T. C., and V. Deretic. 2000. An essential two-component signal transduction system in *Mycobacterium tuberculosis*. *J. Bacteriol.* **182:**3832–3838.

Zahrt, T. C., and V. Deretic. 2001. *Mycobacterium tuberculosis* signal transduction system required for persistent infections. *Proc. Natl. Acad. Sci. USA* **98:**12706–12711.

Zhang, W., and L. Shi. 2005. Distribution and evolution of multiple-step phosphorelay in prokaryotes: lateral domain recruitment involved in the formation of hybrid-type histidine kinases. *Microbiology* **151:**2159–2173.

Zhang, Z. Y., B. A. Palfey, L. Wu, and Y. Zhao. 1995. Catalytic function of the conserved hydroxyl group in the protein tyrosine phosphatase signature motif. *Biochemistry* **34:**16389–16396.

Zheng, X., K. G. Papavinasasundaram, and Y. Av-Gay. 2007. Novel substrates of *Mycobacterium tuberculosis* PknH Ser/Thr kinase. *Biochem. Biophys. Res. Commun.* **355:**162–168.

Zhou, M. M., T. M. Logan, Y. Theriault, R. L. Van Etten, and S. W. Fesik. 1994. Backbone 1H, 13C, and 15N assignments and secondary structure of bovine low molecular weight phosphotyrosyl protein phosphatase. *Biochemistry* **33:**5221–5229.

The Mycobacterial Cell Envelope
Edited by M. Daffé and J.-M. Reyrat
© 2008 ASM Press, Washington, DC

Chapter 21

Glycopeptidolipids: a Complex Pathway for Small Pleiotropic Molecules

CAROLINE DESHAYES, DANA KOCÍNCOVÁ, GILLES ETIENNE, AND JEAN-MARC REYRAT

Glycopeptidolipids (GPLs) are major components of the outer layer of many non-tuberculous mycobacterial cell envelopes. GPLs, also called C-mycosides or J substances in the early days (Smith et al., 1960a), are produced by several mycobacterial species, including human opportunistic pathogens (*Mycobacterium avium* subsp. *avium*, *M. avium* subsp. *intracellulare*, *M. scrofulaceum*, *M. peregrinum*, *M. chelonae*, and *M. abscessus*), animal pathogens (*M. lepraemurium*, *M. porcinum*, *M. senegalense,* and *M. xenopi*) and also saprophytic species (*M. smegmatis*). Species that produce GPLs do not produce phenolglycolipids (PGLs) (see chapter 15) and do not produce sulfolipids either (see chapter 16). So far, there is no counterexample, and this mutual exclusion is not yet understood. However, GPLs appear to be functional homologues of PGLs. Indeed, they are both surface exposed, their mechanisms of synthesis are similar, and both molecules interfere with the host immune system. GPLs play a crucial role in bacilli physiology and also during the interaction with the host. Several genes involved in GPL biosynthesis have been recently characterized. This chapter describes the major advances in the understanding of the biology and biosynthesis of GPLs.

THE EARLY PERIOD OF STUDY OF GPLs

Many studies have focused for decades on the mycobacterial cell wall in an attempt to elucidate the structures and functions and to gain knowledge in both the pathogenesis and the intrinsic drug-resistant nature of mycobacteria. It has led to the discovery of an array of metabolites, a number of which are glycolipids. The term mycosides was originally coined to typify species- and type-specific glycolipids, which have been discovered by infrared spectroscopy of chromatographically fractionated ethanol/diethyl ether extracts (Smith et al., 1960a). These types of glycolipids were named J substances and, later, C-mycosides. An additional component termed "Jab" was shown to contain amino acids identical to other J substances but to possess less glycosyl residues (Smith et al., 1960b). In the later part of the 1960s, Schaefer identified antigens useful to serotype clinical isolates of the *M. avium* complex. Biochemical analyses established the GPL nature of these antigens. Later, a series of studies characterized the specificity, diversity, and chemical structures of the various mycosides, unraveling the extraordinary structural variability of these molecules.

STRUCTURE AND LOCALIZATION

Mycobacteria synthesize two classes of GPLs that differ from one another by their sensitivity to alkali. Alkali-stable substances, the so-called C-mycosides, are produced by a number of both rapid- and slow-growing mycobacterial species. The C-mycoside-type GPLs (or C-type GPLs) share the same lipopeptide core and are probably the most studied GPLs. Alkali-labile GPLs have been described to occur only in *M. xenopi* so far. Their structures differ from those of the C-type GPLs and are described below.

Structure of GPLs

The C-type GPLs contain a mixture of 3-hydroxyl and 3-methoxy long chain (C_{26}-C_{34}) fatty

Caroline Deshayes, Dana Kocíncová, and Jean-Marc Reyrat • Faculté de Médecine René Descartes, Université Paris Descartes, and Groupe Avenir, Unité de Pathogénie des Infections Systémiques—U570, INSERM, F-75730 Paris Cedex 15, France. Gilles Etienne • Department of Molecular Mechanisms of Mycobacterial Infections, Institut de Pharmacologie et de Biologie Structurale du Centre National de la Recherche Scientifique (CNRS), and Université Paul Sabatier, 205 route de Narbonne, 31077 Toulouse Cedex 4, France.

acids amidated by a tripeptide composed of amino acids of the unusual D-series (D-phenylalanine, D-*allo*-threonine and D-alanine) and terminated by L-alaninol (from alanine) (Fig. 1) (Brennan and Goren, 1979). These molecules that share the same lipopeptide core differ from one another by the number and the nature of the saccharidyl units linked to the hydroxyl group of *allo*-threonine and/or alaninol (Fig. 1). Indeed, the lipopeptide core is invariably substituted with rare hexoses such as 6-deoxytalose (linked to the *allo*-Thr residue) and L-rhamnose (linked to the terminal alaninol residue) to generate the nonspecific core GPLs (nsGPLs) found in all members of C-mycosides type GPL producing mycobacteria (Aspinall et al., 1995; Belisle and Brennan, 1989; Brennan and Goren, 1979). In addition to sugar moieties, other modifications are decorating the GPL structure. In *M. smegmatis*, the 6-deoxytalose can be O acetylated on positions 3 and 4. The fatty acid and the rhamnosyl residues can be modified with one O-methyl group on position 3 and three O-methyl groups on positions 2, 3, or 4 respectively (Fig. 1). In members of the *M. avium-M. intracellulare* complex (MAC), an oligosaccharidyl unit of unusual composition is linked to the 6-deoxytalosyl unit, leading to the so-called polar GPLs (Brennan, 1988). The unique sugar residues of polar GPLs are type- or species-specific (Chatterjee and Khoo, 2001). The structural variability and the antigenicity of GPLs are the chemical basis of the identification of the different serotypes within the MAC complex, using sero-agglutination assay. The structural biodiversity of GPLs are discussed later.

Structures of GPLs depend on the species but also on the growth conditions. For instance, under carbon starvation, *M. smegmatis* mc²155 produces polar triglycosylated GPLs (Ojha et al., 2002). The additional sugar was later identified as a rhamnose on position 2 of the first rhamnosyl residue (Fig. 1) (Mukherjee et al., 2005; Villeneuve et al., 2003). In addition, polar GPLs containing a succinyl residue acylating the terminal rhamnosyl unit (Fig. 1) were also characterized in *M. smegmatis* (Villeneuve et al., 2003), expending the repertoire of known modification of the sugars.

GPLs: Structural Variation on a Theme

Decoration of the L-rhamnosyl moiety

Strains from the *M. fortuitum* complex contain surface species-specific lipids, allowing their precise identification (Lopez Marin et al., 1991). In *M. fortuitum* bv. Peregrinum, two major GPLs were characterized by a combination of chemical analyses. The disaccharide part linked to alaninol was characterized as either di(3,4)- or mono(3)-OMe-α-L-Rha-3,4-di-OMe-α-L-Rha. Interestingly, another type of GPL was also identified with a single 2-O-sulfated 3,4-di-OMe-Rha*p* attached to the alaninol and a single 3-OMe-α-L-Rha*p* attached to the threonine (Lopez Marin et al., 1992). Sulfatide metabolites are infrequent in mycobacteria (see chapter 16) and the mechanism of sulfatation is currently unknown. In conclusion, the GPLs of *M. fortuitum* bv. Peregrinum are of unusual structures for a C-mycoside because neither 6-deoxytalose nor its derivatives are present.

Five glycoconjugates belonging to the class of C-type GPLs were characterized in two animal pathogens, *M. senegalense* and *M. porcinum* (Lopez Marin et al., 1993). They shared with those described in *M. fortuitum* bv. Peregrinum the same distribution of the disaccharides on the alaninol end of the molecules. Both species also showed the presence of the novel sulfated GPL.

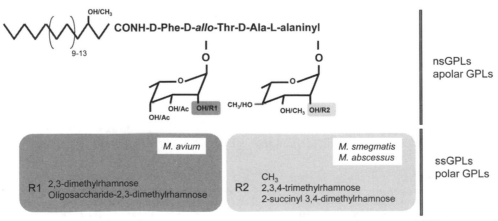

Figure 1. General structures of mycobacterial GPLs. (Adapted from Patterson et al., 2000; and Villeneuve et al., 2003.) Ac, acetyl.

M. abscessus and *M. chelonae* produce the same polar and apolar GPLs as *M. smegmatis*. Both species have emerged as significant pathogens in humans during the last 10 years. They are major causes of skin and soft tissue infections following medical or surgical procedures; *M. abscessus* also causes pulmonary infections and is increasingly recovered from patients with cystic fibrosis (Brown-Elliott and Wallace, 2002; Sermet-Gaudelus et al., 2003). Using an enzyme-linked immunosorbent assay, these GPLs were shown to react with the serum raised against the whole lipid antigens of *M. chelonae*. A comparative serologic study of the native and chemically modified GPL antigens allowed the identification of their epitope as the 3,4-di-OMe-α-L-rhamnosyl residue (Lopez-Marin et al., 1994a). Similar experiments conducted on the GPLs isolated from the serologically cross-reacting species *M. peregrinum* led to the conclusion that the epitope identified in *M. chelonae* and *M. abscessus* was involved in the cross-reactions and demonstrated the existence of a second haptenic moiety in the GPLs of *M. peregrinum*, the 3-OMe-α-L-rhamnosyl unit attached to the *allo*-threonine.

Variation of the 6-deoxytalosyl moiety

The *M. avium*-*M. intracellulare*-*M. scrofulaceum* complex (MAC) is among the most common nontuberculous mycobacteria recovered from clinical specimens and is also a prevalent pathogen in AIDS patients. These mycobacteria commonly cause tuberculosis-like diseases in immunocompromised humans but are not spread from human to human. The lipopeptide cores of MAC's GPLs are diglycosylated at two positions similarly to *M. smegmatis* (a mono- or di-O-methylated rhamnose on L-alaninol and a 6-deoxytalose on the D-*allo*-threonine) (Fig. 1). This is the structure of the so-called nonspecific GPLs (nsGPLs). It has been recognized early that a tremendous diversity exists in the saccharidic extensions of the 6-deoxytalose. The antigenicity allowed researchers to distinguish serotypes that correspond to

structural variants of the nsGPL and that are named serospecific GPLs (ssGPLs). So far, there are 31 recognized serotypes, only 15 of which have been characterized at a structural level. The simplest version of the ssGPLs is serovar 1 (Table 1), which contains a single rhamnosyl residue linked to 6-deoxytalose. This serovar likely represents the starting building block for the biosynthesis of other ssGPLs. Some of the haptenic oligosaccharide extensions of *M. avium* are listed in Table 1, which is adapted from the review of Chatterjee and Khoo (2001).

As stated earlier, in the late 1960s, Schaefer identified GPLs as antigens that are used to classify MAC isolates into various serotypes (Schaefer, 1965). The molecular basis of serological differences between MAC strains observed by Schaefer is due to the ssGPLs and particularly to the unique variable oligosaccharidic sequences (Brennan et al., 1981) present on the 6-deoxytalose. These highly antigenic GPLs have been used to produce monoclonal antibodies to distinguish subspecies of the *M. avium* species, such as *M. avium* subsp. *intracellulare* and *M. scrofulaceum* (Kolk et al., 1989). Surprisingly, very few epidemiological data are available concerning a possible correlation between the serotype, the geographical origin, or the degree of virulence of the strain. However, a study performed in AIDS patients has shown that the prognosis after infection depends on the serotype. Serotype 4 shows an unfavorable prognosis, whereas serotype 16 yields rapid recovery (Takegaki, 2000). Serodiagnostic of MAC strains using GPL is no longer used in identification and classification. This method has been substituted by low-cost and high-resolution techniques such as restriction fragment length polymorphism (Kumar et al., 2006). However, it would be very interesting to examine whether the various serotypes correspond to the clades identified by other methods.

In *M. simiae* and *M. habana*, the 6-deoxytalosyl moiety is extended by complex oligosaccharides characterized by glucuronic acid residue. However, the polar GPLs of *M. simiae* serotype I are clearly

Table 1. Structures of some haptenic oligosaccharide extensions of GPLs from MAC serovars[a]

Serovar	Haptenic oligosaccharide extending from α-L-Rha-(1→2)-L-dTal
1	Non-extended core
2	4-O-Ac-2,3-di-O-Me-α-L-Fuc-(1→3)-core
4	4-O-Me-α-L-Rha-(1→4)-2-O-Me-α-L-Fuc-(1→3)-core
7	4-2′-hydroxypropanoyl-amido-4,6-dideoxy-2-O-Me-β-hexose-(1→3)-α-L-Rha-(1→3)-α-L-Rha-(1→3)-core
8	4,6-O-(1-carboxyethylidene)-3-O-Me-β-D-Glc-(1→3)-core
12	4-(2-OH)propanamido-4,6-dideoxy-3-O-Me-β-D-Glc-(1→3)-4-O-Me-α-L-Rha-(1→3)-α-L-Rha-(1→3)-core
20	2-O-Me-α-D-Rha-(1→3)-2-O-Me-α-L-Fuc-(1→3)-core

[a]Adapted from Chatterjee and Khoo (2001). Rha, rhamnose; Tal, talose; Fuc, fucose; Glc, glucose; Ac, acetyl; Me, methyl.

distinguishable from each of those of *M. habana* by the degree and positions of O methylation on the nonreducing terminal glucuronic acid residue (Khoo et al., 1996). In addition to simple disaccharidic GPLs, the *M. habana* TMC 5135 strain produces three polar GPLs carrying complex oligosaccharides with unique sequences that show migration patterns different from those of *M. simiae* serotype I. Nevertheless, several derivatives of the same sugars (rhamnose, talose, fucose, glucose, and glucuronic acid) are present in the GPLs of both microorganisms, so that, although retaining sufficient specificity, these molecules share common epitopes leading to serological cross-reactivities. Extended analyses of 34 strains by thin-layer chromatography (TLC) demonstrated that most strains presented same polar GPLs. However, a heterogeneous pattern was detected, and some of them contained a possible new compound, designated as GPL-IIα, with apparently the same sugar composition as GPL-II (Mederos et al., 1998; Mederos et al., 2006). Other strains lacked GPLs, and a few of them gave GPL spots closely related to those of *M. simiae*. This study showed that TLC analysis of these compounds is of interest for the rapid identification of these GPL-producing species.

Variation of the lipopeptide

An immunogenic GPL, named GPL X-I, was isolated from *M. xenopi*, a nontuberculous mycobacterium responsible for pulmonary and disseminated infectious diseases mainly occurring in immunocompromised patients (Riviere and Puzo, 1991). It was the first example of a mycobacterial GPL exhibiting a lipopeptidic core differing drastically by (i) the presence of two serines and the absence of D-Ala and L-alaninol; (ii) the COOH-terminal *allo*-threonine in the methyl ester form; and (iii) a shorter fatty acid moiety (C_{12}). The sugar part of GPL X-1 is made up of 3-OMe-α-L-6deoxyTal*p* linked to the first serine, and the tetrasaccharidic appendage attached to the threonine is characterized by the unusual basal oligosaccharide 2,3,4-tri-OMe-L-Rha*p*-α-2-O-lauryl-L-Rha*p*-α-L-Rha*p*-α-2,4-di-O-(acetyl,lauryl)-6-deoxy-α-L-Glc*p*. The GPL X-I structure is so far unique in the mycobacterium genus and can be used as a type-specific probe for *M. xenopi* identification. In later studies, an novel serine-containing GPL termed GPL X-IIb was isolated from an independent *M. xenopi* isolate (Besra et al., 1993; Riviere et al., 1993). Both GPL X-I and GPL X-IIb share a common lipotetrapeptide core (C_{12}-Ser-OMe-Ser-Phe-*allo*-Thr-OMe) but drastically differ in their oligosaccharide appendage. The sugar part of GPL X-IIb differs from GPL X-I by the following disaccharide α-L-Rha*p*-2-O-Lau-α-L-Rha*p* attached to the threonine. Thus, by analogy with MAC, this suggests that *M. xenopi* species can be most probably divided in various serovars characterized by the unique structure of their non C-type GPL oligosaccharide appendage.

The particular case of *M. avium* subsp. *paratuberculosis*

M. avium subsp. *paratuberculosis* is closely related to *M. avium* subsp. *avium* and is responsible for cattle infections (Biet et al., 2005). Although this subspecies has been reported once to produce GPLs similar to *M. avium* subsp. *avium* (Camphausen et al., 1985), a recent study by the same group shows that *M. avium* subsp. *paratuberculosis* produce a lipopeptide of particular structure: C_{20}-acyl–D-Phe–N-Me–L-Val–L-Ile–L-Phe–L-Ala methyl ester instead of GPLs (Eckstein et al., 2006). This recent report is in agreement with older reports showing that several *M. avium* subsp. *paratuberculosis* isolates produce a lipopentapeptide or even a lipooctapeptide (Laneelle and Asselineau, 1962; Laneelle et al., 1965). It is still unclear whether the production of this lipopeptide is a particularity of the strain used in the recent study or a trait that can be extended to the whole subspecies. *M. avium* subsp. *paratuberculosis* is a microorganism that exhibits an important genetic diversity at the strain level (Motiwala et al., 2006). Thus, the production of the lipopentapeptide or the lipooctapeptide should be investigated on a large collection of strains of known origin. It is still unclear whether some particular isolates of *M. avium* subsp. *paratuberculosis* are able to produce GPL. Clarification of these aspects should help in classifying *M. avium* subsp. *paratuberculosis* strains. The structural similarity between GPL and lipopeptide is apparent; it is then very tempting to speculate that one biosynthesis pathway has evolved from the other.

Localization of GPLs on the Cell Surface

Several arguments suggest that the C-type GPLs are surface localized. First, they have been proposed as the receptor of mycobacteriophage D4 (Furuchi and Tokunaga, 1972; Goren et al., 1972). However, this fact has not been demonstrated to date. Second, GPLs have been shown to be antigenic, resulting in the Schaefer typing antigens (Brennan, 1988; Schaefer, 1965). However, the first strong evidence of the surface exposition of GPLs came from Draper's group, which characterized material isolated from *M. lepraemurium* (Draper and Rees, 1973) and *M. avium*

(Draper, 1974) capsules as GPLs (C-mycosides). Fibrillar, capsule-like structures (the electron-transparent zone), often observed in electron micrographs and long considered to be glycolipid, have been shown to be implicated in the intracellular survival and persistence of mycobacteria. Draper (1974) and later Barrow and Brennan (1982) showed a close correlation between rough morphology, the lack of a superficial cell wall sheath, and the absence of C-mycosides. This confirms previous observations made by Fregnan and coworkers, who demonstrated a constant relationship between a particular colony form and the presence of a specific mycoside for *M. kansasii* and *M. fortuitum* (Fregnan et al., 1961; Fregnan and Smith, 1962). More recent studies reported that GPL production is associated with the rough/smooth morphotype variation in several mycobacterial species (Barrow and Brennan, 1982; Billman-Jacobe et al., 1999; Etienne et al., 2002), suggesting the presence of GPLs at the cell surface. In general, species producing large amounts of GPLs tend to produce smooth colonies, presumably because the surface of their cells is covered by hydrophilic carbohydrate moieties of GPLs. This was confirmed by the *M. smegmatis gap* mutant, which produces GPLs at a wild-type level but is unable to transport it at the surface, and which has a rough phenotype (Sonden et al., 2005). Moreover, it has been pointed out above

that some GPLs could exist as micelles or fibrils outside the cell wall bilayer (Brennan and Nikaido, 1995).

To define the chemical nature of the surface-exposed lipids of mycobacteria, a new approach was developed. In this method, the outermost cellular components are released by a gentle mechanical abrasion, followed by an extraction with organic solvents (Ortalo-Magne et al., 1996). This approach coupled with labeling experiments indicated that half of the bacterial GPLs were located on the cell surface and represented 85% of the surface-exposed lipids in *M. smegmatis* (Etienne et al., 2002). However, this percentage is likely to vary from species to species and even from strain to strain. Ultrathin sections of mycobacterial cells stained with ruthenium red, a dye strongly reacting with the surface of mycobacteria (Rastogi et al., 1984), revealed a cell envelope structure of *M. smegmatis* strain mc^{2}155 composed of (i) a plasma membrane, (ii) a thick internal electron-dense layer, (iii) an electron-transparent layer, and (iv) a thick electron-dense outer layer (Fig. 2B). The space observed between the plasma membrane and the thick electron-dense layer corresponds to the hypothetical periplasmic space (see chapter 2). The presence of the thick electron-dense outer layer stained with ruthenium red in all mycobacterial species observed so far, which surrounds an electron-transparent layer believed to correspond to the

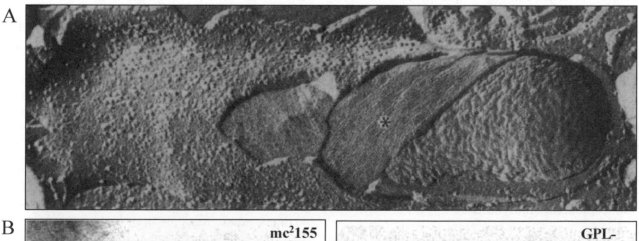

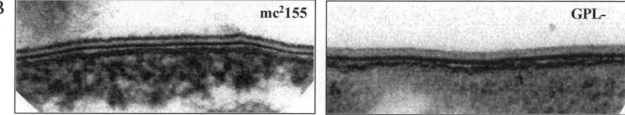

Figure 2. (A) Structural aspect of the multilamellar coat (asterisk) that invests intraphagosomal *M. avium* bacilli in mouse liver cells after 3-month infections as revealed by freeze fracture electron microscopy (magnification, ×62,000) (Rulong et al., 1991). (B) Transmission electron micrograph of *M. smegmatis* strain mc^{2}155 and a non-GPL-producing mutant (Etienne et al., 2002).

mycolic acid part of the cell wall (Brennan and Nikaido, 1995; Daffe and Draper, 1998), corresponds to the surface of the cell envelope, which is composed in majority of PGLs or GPLs. This thick electron-dense outer layer observed on the thin sections of the parent *M. smegmatis* strain mc^{2}155 was unlabeled in mutants that do not produce GPL (Fig. 2B) (Etienne et al., 2002), resulting in a thick electron-transparent outermost layer. Thus, although nonspecific, this method allows the direct visualization of GPLs at the mycobacterial surface.

THE GENETIC BASIS OF THE BIOSYNTHESIS AND EXPORT OF GPLs

Biosynthesis, Export, and Regulation in *M. smegmatis*

Biosynthesis

A decade of microbial genetics, using mostly *M. smegmatis* as a model organism, has unraveled the genetic basis of the nsGPLs biosynthetic pathway. In this species, the GPL locus is compact with approximately 25 genes within a region of 65 kb (Fig. 3). The isolation of GPL-nonproducing mutants after a transposon mutagenesis of *M. smegmatis* was greatly facilitated thanks to the characteristic morphotypes of these mutants. In 1999, Billman-Jacobe and co-workers screened a library of transposon mutants for ones having rough colony morphology (Billman-Jacobe et al., 1999). These mutants were devoid of GPLs and had transposon insertions in a gene encoding a peptide synthetase. Other screens, such as sliding motility, biofilm formation or Congo red fixation, have also been used to identify genes involved in mycobacterial cell wall biosynthesis (Cangelosi et al., 1999; Recht et al., 2000; Sonden et al., 2005). Analysis of the crude cell wall extracts from these mutants revealed that the majority of them were defective in GPL synthesis.

Lipopeptide core biosynthesis

The first gene to be identified is involved in the very early stage of GPL biosynthesis, and encodes a nonribosomal peptide synthetase (Nrp, also called Mps for mycobacterial peptide synthase) catalysing the synthesis of the peptide moiety of the molecule (Billman-Jacobe et al., 1999; Recht et al., 2000). However, the unique *mps* open reading frame (ORF) reported by Billman-Jacobe and colleagues in *M. smegmatis* mc^{2}155 was later shown to correspond to three ORFs, namely *mbtH*, *mps1,* and *mps2* (Sonden et al., 2005). The tetrapeptide moiety is nonribosomally assembled by the product of the *mps1* and the *mps2* genes that belong to the Nrp family. Enzymes belonging to this family are very large (>3,000 amino acids) and are involved in the synthesis of pharmaceutical compounds such as vancomycin and cyclosporine (Walsh, 2004). Nrps are composed of modules that are responsible for selecting (adenylation), modifying (epimerization, cyclization, and *N*-methylation) and elongating (condensation) the peptidic chain of the nonribosomically synthesized oligopeptides (Cane et al., 1998; Lautru and Challis, 2004; Walsh, 2004). The number and order of the modules usually reflect the length and the sequence of the peptide being synthesized (see chapter 6). In *M. smegmatis*, the *mps1* and *mps2* genes collectively encode 4 modules, the first three of which contain an epimerase domain that converts an L-amino acid into the non-natural D-form. Thus, the number and structure of the modules is in agreement with the structure of the pseudotetrapeptide produced by *M. smegmatis*, which is D-Phe-D-*allo*-Thr-D-Ala-L-alaninol. The role of the *mbtH* genes, whose homologues are found associated with *nrp* genes, is unclear to date (Quadri et al., 1998).

The monounsaturated β-hydroxyl fatty acid of the GPL core is probably synthesized by a polyketide synthase (Pks) protein encoded by the *pks* gene present in the locus. An insertional mutant of the *pks* gene does not produce any GPLs (Sonden et al., 2005). This gene is homologue to the *pks10* gene of *M. avium* 104 (Laurent et al., 2003) and to the *pks10* gene of *M. tuberculosis* (Cole et al., 1998), which is so far without any assigned function in this species. These three *pks* genes display a highly similar domain organization. Pks has two domains, one being complete with the four modules required to elongate the acyl chain with a saturated two-carbon unit (Walsh, 2004) and the other predicted to possess only the β-ketoacyl synthase and β-ketoreductase modules; this latter organization is consistent with the GPL characteristic β-hydroxyl fatty acid. The mechanisms of polyketide synthase functions are described in chapter 6. However, the role of this polyketide synthase has only been suggested so far, because no complementation analysis has been carried out. Its size (11 kb) is probably responsible for the difficulty of performing cloning experiments. The functions of other genes involved in lipopeptide formation can be predicted by using a bioinformatic approach. Indeed, FadE5 is homologous to acyl-coA-dehydrogenase protein of various mycobacteria and probably introduces the double-bond in the fatty acid moiety. The lipid attachment to the tetrapeptide moiety probably requires the concerted action of several players like FadD23 and PapA3. The *fadD23*

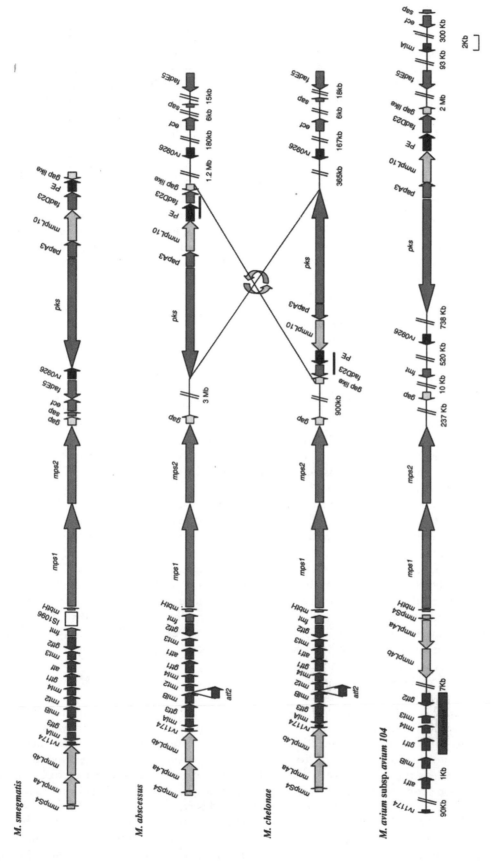

Figure 3. Genetic organization of the GPL locus in various mycobacterial species. The ORFs are depicted as arrows and have been drawn proportionally to their size (Ripoll et al., 2007). Color code: light blue, *mmpL* family; black, unknown; purple, sugar biosynthesis, activation, transfer and modifications; red, lipid biosynthesis, activation, transfer and modifications; green, pseudopeptide biosynthesis; yellow, required for GPL transport to the surface; gray, regulation. (*See the color insert for the color version of this figure.*)

gene is a close homolog to the *fadD28* gene of *M. tuberculosis*, which is involved in acyl transfer of mycocerosic acid (Cox et al., 1999). In general, PapA proteins are conserved polyketide-associated proteins. Two studies have proposed that PapA5 of *M. tuberculosis* catalyzes the diesterification step during phthiocerol dimycocerosate (PDIM) synthesis (Buglino et al., 2004; Onwueme et al., 2004). The authors proposed that PapAs represent a subfamily of acyltransferases primarily involved in polyketide synthesis in mycobacteria.

Substitution of the lipopeptide core

The lipopeptide core can be modified by glycosylation, O methylation, and O acetylation, and each of the genes responsible for these modifications has been characterized. The *rmlA* and *rmlB* genes may be involved in rhamnose biosynthesis, according to their sequence homology with *rmlA* of *M. tuberculosis* and *gepiA* of *M. avium*, respectively (Cole et al., 1998; Jeevarajah et al., 2002; Ma et al., 1997). The *rmlA* gene encodes a putative glucose-1-phosphate thymidylyltransferase and *rmlB* a putative dTDP-glucose-4,6-dehydrogenase. These two genes would be responsible for the production of nucleotide activated rhamnose, a substrate for the glycosyltransferases. In addition to genes involved in sugar biosynthesis, three glycosyltransferase genes (*gtf1*, *gtf2*, and *gtf3*) lie in the GPL locus. Gene disruption has shown that Gtf1 and Gtf2 are responsible for the early steps of GPLs biosynthesis: they respectively catalyze the transfer of the 6-deoxytalosyl and rhamnosyl moieties to the lipopeptide core (Miyamoto et al., 2006). The detection of glycosylated intermediates from both *gtf1* and *gtf2* mutants suggests that (i) the lipopeptide core could be the substrate for both Gtf1 and Gtf2, and (ii) the glycosylated intermediates could also be the substrates for both Gtf1 and Gtf2 (Miyamoto et al., 2006). Interestingly, in both *gtf1* and *gtf2* mutants, the quantity of GPLs is strongly reduced and not even detectable by TLC. It can either be due to GPLs negative regulation by monoglycosylated GPLs or due to the toxicity of monoglycosylated GPLs and thus to the selection of variants having a low production. However, the fact that a rough variant of *M. avium* (104Rg) producing truncated GPLs at wild-type level that lacks the 6-deoxytalose attached to D-*allo*-Thr but retained the rhamnose on the L-alaninol has been isolated (Torrelles et al., 2002) contradicts this hypothesis. It has been shown that Gtf3 is responsible for the production of the triglycosylated forms of GPLs in *M. smegmatis* (Deshayes et al., 2005). Glycosyltransferases have very high donor and acceptor substrate specificities and, in general, are limited to the establishment of

one glycosidic linkage. These enzymes show a limited similarity between them, and it is practically impossible to infer from the amino acid sequence the nature of the sugar that will be transferred, as well as the nature of the molecule that will accept the sugar. The function of the four methyltransferase genes located in the GPL locus has been elucidated by Billman-Jacobe's group. Three of them are responsible for the O methylation of the rhamnose, and one is responsible for O methylation of the fatty acid moiety. Rmt3 (previously named Mtf1) is a rhamnosyl 3-O-methyltransferase (Patterson et al., 2000), whereas Fmt (previously named Mtf2) methylates the hydroxylated fatty acid of the GPL (Jeevarajah et al., 2002). In a third study, they provided genetic and biochemical evidence that the *rmt4* and *rmt2* genes respectively encode rhamnosyl 4-O-methyl- and rhamnosyl 2-O-methyltransferases (Jeevarajah et al., 2004). Deletion of each methyltransferase suggests that O methylation of the rhamnose is initiated by the 3-O-methyltransferase and that the product of this reaction is sequentially methylated by 4-O-methyltransferase and 2-O-methyltransferase. The *S*-adenosylmethionine binding site and three other motifs associated with methyltransferases are present in the *M. smegmatis* methyltransferases. The sequence similarity between the methyltransferases of *M. smegmatis* and other species correlates with their similar enzymatic activities (Jeevarajah et al., 2004). It is interesting to note that, although these proteins are very similar, they display a high acceptor substrate specificity similar to glycosyltransferases. Using a genetic screen based on the inability to form a biofilm, Recht and colleagues (2001) identified the *atf* gene, which is responsible for the O acetylation of GPLs. Atf is predicted to be a membrane protein containing at least 10 transmembrane domains.

The function of a number of other genes of the locus has not yet been proposed, but their presence within the locus suggests that they may be associated with GPL synthesis. No function is predicted for the *Rv1174*, *Rv0926*, and *PE* genes that are located in the GPL locus.

Transport

Although we are beginning to understand the biosynthesis of GPLs, the mechanisms of GPL transport and, more generally, of mycobacterial glycolipids are still unclear. Thus far, only one gene of the GPL locus has been characterized to be required for GPLs transport to the cell surface (Sonden et al., 2005). This study demonstrated that a small integral membrane protein named Gap (GPL addressing protein) is specifically required for the transport of the GPLs to the cell surface (Sonden et al., 2005). Indeed, a *gap* mutant strain produces GPLs at wild-type level but does not export them on the cell surface (Fig. 4A) (Sonden et al., 2005). The Gap protein is predicted to contain six transmembrane segments (Fig. 4B) and possesses homologues across the mycobacterial genus, thus delineating a new protein family (Fig. 4C). The family is specific to mycobacteria because no homologues are found outside of this genus. It is proposed that this protein forms a channel in the cytoplasmic membrane or even in the cell wall pseudomembrane to translocate the GPLs onto the surface. At the present time, it is not known whether this protein acts directly or indirectly in GPL transport or whether this process requires oligomerization of Gap and energy. In the case of *M. tuberculosis*, one *gap* homologue (*Rv3821*) is localized in the immediate vicinity of a locus dedicated to the sulfated glycolipid (SL) (see chapter 16). Therefore, it is very tempting to speculate that the *Rv3821* gene is functionally equivalent to the *gap* gene and, hence, likely to be involved in the transport of SL at the bacillus' surface. Similarly, a *gap* orthologue (*Rv1517*) has been shown to be present in a locus putatively identified to be dedicated to lipooligosaccharide (LOS) biosynthesis (Ren et al., 2007). We postulate that Rv1517 is likely involved in transport of LOS to the surface. The *Map1418c* gene, which is the closest orthologue in *M. avium*, does not complement the *M. smegmatis gap* mutant strain. This is not completely surprising because the GPLs that are produced by *M. avium* are chemically distinct from the GPLs produced by *M. smegmatis*. On the contrary, complementation by the *gap* orthologue of *M. abscessus*, which produces GPLs that are structurally identical to *M. smegmatis*, is effective (Sonden et al., 2005). One may envisage a selectivity between the Gap protein and the substrate that will be transported to the surface. The sequence variation in the central part of the protein that corresponds to the predicted cytoplasmic portion of the protein may be the region of selectivity of the Gap protein (Fig. 4) (Sonden et al., 2005). At present, no function has been assigned to the *gap-like* gene present in the GPL locus.

Four other genes encoding transmembrane proteins present in the GPL locus have been predicted to be involved in the GPL transport, because they share homology with proteins mediating transport. MmpS4 belongs to the MmpS (Mycobacterial membrane protein small) protein family and MmpL4a and MmpL4b to MmpL (Mycobacterial membrane protein large) protein family, both of which are known to be involved in lipid metabolism and transport (see chapter 11). MmpLs are characterized by 12 transmembrane domains, and two extracytoplasmic loops and are part of the resistance nodulation cell division (RND) superfamily. The RND proteins mediate the

transport of substances across the cytoplasmic membrane that is driven by the proton motive force of the transmembrane electrochemical proton gradient. Some members of this protein family have been characterized. Indeed, MmpL7 of *M. tuberculosis* catalyses export of a surface-exposed lipid, the PDIM (Camacho et al., 2001), and MmpL8 is involved in the synthesis of sulfolipid 1 by transporting a precursor of this molecule (Converse et al., 2003; Domenech et al., 2004). MmpL4b of *M. smegmatis*, also called TmtpC, has been shown to be implicated in sliding motility and biofilm formation due to a decrease in GPL production (Recht et al., 2000). The function of these proteins is still not known in detail. However, a recent study performed in *M. tuberculosis* suggests that MmpL proteins (MmpL7) may channel the polyketide product during synthesis by the polyketide synthase, coupling synthesis and export (Jain and Cox, 2005). Nothing is known about the MmpS family of mycobacteria. These MmpS proteins may be the equivalent of the membrane fusion proteins (MFPs), auxiliary proteins that are frequently associated with RND proteins in gram-negative organisms. The proximity of the *mmpS4* gene with the two *mmpL4* genes in the GPLs biosynthesis cluster suggests its role in GPLs biosynthesis and translocation. Because the *gap* gene is not required for GPL biosynthesis per se, it is very likely that the Gap protein is acting downstream of MmpL4 proteins, that is, during export. Interestingly, *mmpL4a* and *mmpL4b* mutants lacked detectable GPLs by TLC analysis (Recht et al., 2000; Sonden et al., 2005), suggesting a role in GPLs synthesis, contrary to MmpL7 of *M. tuberculosis*, which is only involved in PDIM transport (Camacho et al., 2001). Recently, in *Bacillus subtilis*, different Pks proteins have been shown to colocalize to a single spot in the cytoplasm (Straight et al., 2007). This spot is due to Pks proteins that are assembled into a megacomplex of bacillaene synthases, suggesting that the Pks and possibly other proteins form an organelle-like structure. In *B. subtilis*, some proteins (PksJ) have been shown to stabilize and anchor this megacomplex at the cell membrane. A hypothetical function that has been suggested for MmpS/MmpL proteins is that it may stabilize a megacomplex made of Pks, Nrp, Pap, or possibly other enzymes, and couple GPLs synthesis with export. In addition to these three *mmpS/mmpL* genes, a *mmpL10* gene is present at the 3' end of the GPL locus (Fig. 3), but its function is still uncharacterized.

Regulation of GPL synthesis

Like GPL transport, little is known about the regulation of GPL synthesis. Only one gene has been

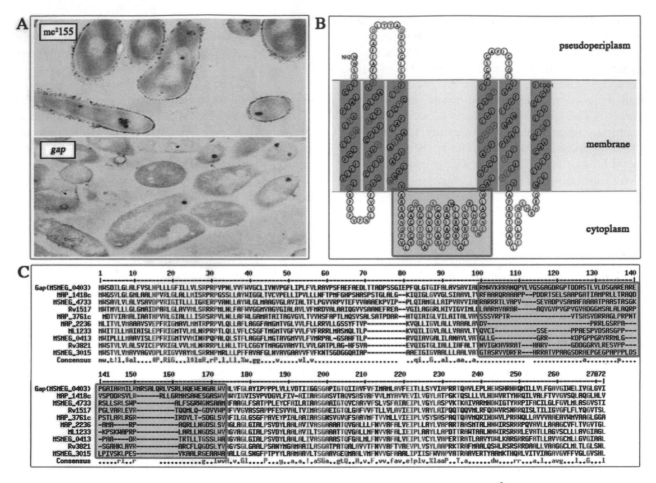

Figure 4. (A) Transmission electron microscopy analysis of the *M. smegmatis* wild-type strain mc²155 and the gap mutant strain after ruthenium red staining (Sonden et al., 2005). (B) Topology of Gap protein predicted by Sosui program. (C) Amino acid alignment of various Gap orthologues found in sequenced mycobacterial genomes: *M. smegmatis* (MSMEG), *M. tuberculosis* (Rv), *M. avium* subsp. *paratuberculosis* (MAP) and *M. leprae* (ML). Color code: red, high concensus (>90%); blue, low consensus (>50%). The central parts bordered in orange correspond to the predicted cytoplasmic portion and may be the region of selectivity of Gap proteins. *(See the color insert for the color version of this figure.)*

proved to be involved in the regulation of GPL synthesis to date. Using an insertional mutagenesis approach in the *M. smegmatis* ATCC607 strain, the *lsr2* gene was shown to negatively regulate GPL synthesis (Kocincova et al., unpublished data). Inactivation of the *lsr2* gene led to the overexpression of the *mps* operon responsible for assembling the tetrapeptide moiety of GPL (see earlier), thus leading to overproduction of GPL. The Lsr2 protein is present in all sequenced mycobacterial species and the amino acid identity between these homologues is high (80% to 94%). The *lsr2* gene is also present in other genera clustered in the *Actinomycetales* order, such as *Streptomyces* or *Nocardia*. In addition, mycobacteriophage Omega possesses genes that are homologue to the *lsr2* gene. Lsr2 has been shown to form dimers and to be localized in the cytoplasm, a property of some DNA-binding proteins (Chen et al., 2006). Although it does not contain any DNA-binding

domain, it is a basic protein (pI = 9.6) and thus is a suitable candidate for binding acidic DNA. However, it is not clear whether Lsr2 influences the *mps* expression directly or indirectly. Lsr2 probably has a pleiotropic effect, because it has also been described to be involved in the positive regulation of biosynthesis of mycolyl diacylglycerols (MDAGs) and in the negative regulation of triacyl glycerols (TAGs) (Chen et al., 2006). It is noteworthy that the negative regulation of GPL production by Lsr2 was not found by Chen and colleagues. This is indeed due to the strain (mc²155) used in this study that contains an insertion sequence (IS1096) in the *mps* promoter region. Indeed, the presence of an IS in the *mps* promoter region abrogates the Lsr2-negative regulation of *mps* expression leading to high constitutive expression (Fig. 5) (Kocincova et al., unpublished data). Similar to mc²155, other natural variants of the ATCC607 strain that contain an insertion sequence in the *mps* promoter region have been isolated. The presence of IS led to the overexpression of the *mps* operon. The activation of the *mps* promoter is regardless of type and direction of insertion sequences and the site of insertion is limited to the region of bp −80 to −210 with respect to *mbtH*, the first gene of the *mps* operon. So far, *lsr2* is the only gene required for GPLs biosynthesis, which is located outside the GPL locus. The mechanism by which this negative regulation operates is not clear and much work will be required to characterize the molecular events leading to *mps* repression.

The regulation of GPL synthesis probably involves both negative and positive regulation. The *mps* operon of *M. smegmatis* contains an *ecf* (extracytoplasmic sigma factor) gene and a *sap* (sigma-associated protein) gene, two candidates possibly involved in the positive regulation of GPL synthesis. ECF is a member of alternative sigma factor family (see chapter 12) that regulates gene expression in response to a variety of conditions, including stress (Helmann, 2002; Waagmeester et al., 2005). Although ECFs are often located close to antisigma factors, Sap does not display a significant homology to antisigma factors. Because the mc²155 strain carries an insertion sequence within the promoter region, it is probably not the best model to study regulation of GPLs synthesis. In the genomes of *M. abscessus* and *M. chelonae* the *ecf* and *sap* genes are separated from the GPL locus, suggesting that the regulatory circuits in these species have probably diverged (Ripoll et al., 2007).

Biosynthesis of the Oligosaccharidic Extending Unit in *M. avium*

As seen earlier, the *M. avium* GPLs have the same lipopeptide core as the *M. smegmatis* GPLs but the alaninol is glycosylated with 3-OMe-Rha or 3,4- di-OMe-Rha (Daffe et al., 1983). The oligosaccharide composition extending the 6-deoxytalose varies between the different serotypes of *M. avium*. A locus named *ser2* has been identified to contain genes encoding the enzymatic machinery necessary for the glycosylation of the lipopeptide core that results in the *M. avium* subsp. *avium* serovar-2-specific GPL (Belisle et al., 1991). It was shown that the simplest nsGPLs naturally found in *M. smegmatis* could serve as starting building block in the biosynthesis of a ssGPLs (serovar 1). Recombinant *M. smegmatis* expressing a glycosyltransferase (RtfA) from *M. avium* (Eckstein et al., 1998) suggested that this gene encodes a glycosyltransferase linking a rhamnosyl unit to the 6-deoxytalose (Eckstein et al., 1998). Later, a deletion mutant of the *rtfA* gene of the serovar 2 TMC724 strain was constructed (Maslow et al., 2003) and proved definitively that RtfA encodes a glycosyltransferase. Complementation of the *rtfA* knockout mutant with wild-type *rtfA* gene restored ssGPL expression, demonstrating that disruption of *rtfA* gene results in the inability to glycosylate the 6-deoxytalose but not the L-alaninol of the lipopeptide core.

Besides the *rtfA* gene, the *ser2* cluster contains genes encoding for instance other glycosyltransferases, methyltransferases, acetyltransferases, D-glucose and D-mannose dehydrogenase, but also several insertion sequences (IS*1245*, IS*1348*, IS*2534*, IS*1601*) (Eckstein et al., 2003). This *ser2* gene cluster of the *M. avium* subsp. *avium* 2151 and TMC724 strains was fully sequenced and compared with the homologous regions of *M. avium* subsp. *avium* 104 (serovar 1), *M. avium* subsp. *paratuberculosis* and *M. avium* subsp. *silvaticum* (Eckstein et al., 2003). This genetic comparison, together with analysis of protein similarities, supports a biosynthetic model in which serovar-2-specific GPLs are synthesized from a serovar-1-specific GPL intermediate that is derived from a

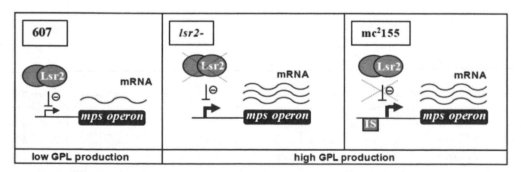

Figure 5. The mechanism arising in the upstream region of the *mps* operon in *M. smegmatis*. In the ATCC607 strain, the *mps* operon is under the direct or indirect negative control of Lsr2, which leads to a low level of mRNA and a low level of GPLs (Kocincova et al., unpublished data). In the *lsr2* mutant, the *mps* operon is highly expressed because of the absence of the negative regulator Lsr2 that leads to high production of GPLs. In the mc²155 strain, the Lsr2-dependent regulation is lost upon insertion of the mobile element, which leads to a major expression of the *mps* operon and consequently to a high level of GPL.

nonspecific GPL precursor. The authors also identified a gene encoding an enzyme that is necessary for the biosynthesis of serovar3- and 9-specific GPLs, but not serovar2-specific GPLs, suggesting that the different serovars may have evolved from the acquisition or loss of genetic information. In addition, a subcluster of genes for the biosynthesis and transfer of fucose, which are needed for production of serovar-specific GPLs such as those of serovar 2, is found in the non-GPL-producing *M. avium* subsp. *paratuberculosis* and *M. avium* subsp. *silvaticum*. To determine phylogenetic relationships among *M. avium* serotypes, selected GPL genes have been used as markers, and the inferred trees have been compared with serotype designations and information obtained from markers used in earlier studies (Krzywinska et al., 2004).

In a recent study, the chemical structure of the serotype 7 GPL from *M. avium* subsp. *intracellulare* was described. This serotype 7 GPL has an unique terminal amido sugar (Table 1) (Fujiwara et al., 2007). The authors also characterized the serotype 7-specific gene cluster involved in glycosylation of the oligosaccharide. A genomic cosmid library of an *M. intracellulare* serotype 7 strain was constructed and the *Escherichia coli* transductants were screened by PCR, amplifying the *rftA* gene coding the glycosyltransferase linking a rhamnosyl unit to the 6-deoxytalose. The insert of cosmid was sequenced and nine ORFs were present in the cluster. Based on the sequence homology, the ORFs are thought to participate in the biosynthesis of the unique amido sugar of serotype 7 GPL. Using in silico metabolic reconstruction, the authors have predicted the biosynthesis pathway of the saccharidic moiety of serotype 7 GPL, in relation with the function of each ORF of the locus.

Biosynthesis in other species: comparative genomics

Most if not all the genes involved in GPL biosynthesis in *M. smegmatis* have close orthologues in other GPL-producing species that have been sequenced (*M. avium* subsp. *avium*, *M. abscessus* and *M. chelonae*). The percentage of identity between orthologues ranges between 30 and 90. The genes having the lowest degree of orthology are the *sap* and *ecf* genes, which are supposed to play a role in regulation of GPL production in *M. smegmatis*. This observation may indicate that the regulation of this pathway is the functional aspect, which is the least conserved between GPL producing species. A genomic study has compared GPL biosynthetic pathways of several GPL producing species (Ripoll et al., 2007).

M. abscessus and *M. chelonae*. As shown in Fig. 3, the global organization of the GPL locus of *M. abscessus* is similar to that of *M. smegmatis*; however, the locus is divided in two regions that are 3 Mb apart. A number of genes (*Rv0926*, *fadE5*, *sap*, and *ecf*) are found elsewhere on the chromosome and are not linked to the other part of the GPL locus. The presence of the *gtf3* gene confirms that *M. abscessus* has the ability to produce triglycosylated polar GPLs via Gtf3 activity, similarly to *M. smegmatis* (Deshayes et al., 2005; Ripoll et al., 2007). In contrast to *M. smegmatis*, two *atf* genes are present in the GPL locus of *M. abscessus*. The acetyltransferases encoded by the *M. abscessus* GPL locus are not redundant but have evolved specificity, being able to transfer one acetyl at once in a sequential manner (Ripoll et al., 2007). In this species, the *PE* and the *fadD23* genes have been subject to an inversion relatively to *M. smegmatis*. In *M. chelonae*, the GPL locus is similarly divided into two regions, separated by only 900 kb, and two *atf* genes are also present. Interestingly, in *M. chelonae*, a block of six genes (*pks* to *gap*-like) is part of a large region that is inverted, relatively to *M. abscessus* and *M. smegmatis*. An attractive hypothesis is that the compact organization of the GPL locus in *M. smegmatis* represents the ancestral form and that evolution has scattered various pieces throughout the genome in *M. abscessus* and *M. chelonae*, although the opposite scenario cannot be excluded.

M. avium subsp. *avium*. In *M. avium* subsp. *avium* 104, which is currently being sequenced by TIGR, the situation is more complex. Indeed, most of the genes that are present in the locus of *M. smegmatis* can be found in *M. avium* subsp. *avium*. However, the compact locus organization is not conserved, and a high number of genes are scattered throughout the chromosome. This is the case for the *rmlA* and *rmlB* genes, for the *fmt* and *rmt* genes as well as for the *gap* orthologue (Fig. 3). However, a number of blocks are reminiscent of the *M. smegmatis* locus: for instance, the *mbtH-mps1-mps2* cluster, the *mmpS4-mmpL4a-mmpL4b* cluster and the group of six genes ranging from *pks* to *gap*-like. Interestingly *M. avium* subsp. *avium* 104 does not have an ortholog of the *rmt2* gene (Fig. 3), which is in agreement with the absence of O methylation of the hydroxyl group located on position 2 of the rhamnosyl unit of the GPLs in this subspecies.

It is somehow surprising that the genetic organization of this metabolite pathway greatly differs from species to species. At the present time, it is not known whether evolution tends to scatter or to group the various actors of a particular pathway. Some species, or isolates, possess additive genes that lead to the modification of the basic backbone of the nsGPLs, and generate chemical diversity and complexity.

GPL: COLONY MORPHOLOGY, SLIDING MOTILITY, AND BIOFILM FORMATION

Impact on Bacterial Phenotype

It was recognized early that *M. avium* grown on solid media forms distinct colony morphologies (Fregnan and Smith, 1962), namely smooth transparent (SmT), smooth opaque (SmO, also referred to as smooth domed) and rough (Rg) (Fregnan and Smith, 1962; Vestal and Kubica, 1966). For any given strain of *M. avium*, the colony morphotypes are typically unstable, with a significant rate of conversion of one morphotype to another (McCarthy, 1970). The frequency of transition from SmT to SmO is rather high (5×10^{-4} cells), whereas the transition of SmO to SmT is low (1×10^{-6} cells) (Woodley and David, 1976). It is very likely that the smooth/rough variation is independent of the opaque/transparent variation. Analyzing the four combinations has probably not helped our understanding of this phenomon. Although the opaque/transparent variation is still not characterized, the smooth variation has been understood in some cases. Detailed chemical analyses of *M. avium* serovar 2 demonstrated that two chemotypes of R variants were formed: one devoid of GPL (Rg-4), and one lacking glycosylation but producing the lipopeptide core (Rg-1) (Belisle et al., 1993). The formation of these R chemotypes is associated with large genomic deletions. Specifically, Rg-1-type organisms arise through a deletion of the *M. avium ser2* gene cluster, the genomic region encoding glycosylation of the serovar 2-specific GPLs, whereas the deletion associated with Rg-4-type organisms encompasses the genes encoding lipopeptide biosynthesis (Belisle et al., 1993). Moreover, a rough variant of *M. avium* 104 producing ssGPL serovar 1 has been described. In this strain, a deletion of the 10-kb region containing genes involved in the GPL synthesis, such as a glycosyltransferase or a methyltransferase, is responsible of this rough aspect (Eckstein et al., 2003). This variant (104Rg) produced truncated GPLs that lacked the 6-deoxytalose attached to D-*allo*-Thr but retained the O-methylated rhamnose on the L-alaninol (Torrelles et al., 2002). The mechanism of the deletions in Rg-1 and Rg-4 was suggested to be mediated by homologous recombination between direct repeats of insertion sequences, but this remains to be demonstrated. Similarly, the deletion event in 104Rg mutant was also suggested to be mediated by IS999 recombination.

Another study reported that a rough *M. abscessus* clinical isolate became attenuated after it spontaneously dissociated into a smooth variant (Byrd and Lyons, 1999). In agreement with this fact, it has been shown that a rough, wild-type human clinical isolate (390R) causes persistent invasive infection in a murine pulmonary infection model, whereas a smooth isogenic mutant (390S) has lost this capability. During serial passage of 390S, a spontaneous rough revertant was obtained and named 390V (Howard et al., 2006). This rough revertant regained the ability to invade human monocytes and to cause persistent lung disease in mice. GPLs were present in abundance in the cell wall of 390S, and were associated with sliding motility and biofilm formation (see later). In contrast, a marked reduction in the amount of GPL in the cell envelope of 390R and 390V was correlated with cord formation, a property traditionally associated with mycobacterial virulence. More recently, a rough variant of *M. abscessus* has been isolated during experimental infection of mice (Catherinot et al., 2007). This variant, which has lost the ability to produce GPLs, is hyperlethal for mice infected intravenously and induces a strong tumor necrosis factor alpha (TNF-α) response by murine monocyte-derived macrophages (Catherinot et al., 2007). These results indicate that the ability to switch between smooth and rough morphologies may allow *M. abscessus* to make the transition between a colonizing phenotype and a more virulent, invasive form.

It has been recently shown using the rough *M. smegmatis* ATCC607 strain as a model that a morphological conversion from rough to smooth morphotype occurs naturally at a low frequency (1×10^{-4} cells). The smooth variant of the ATCC607 strain produced higher level of GPL and acquired several new phenotypes, such as motility, a hypoaggregative growth in liquid culture and the ability to form biofilm. The molecular basis of this conversion was shown to be mediated by insertion of mobile elements (either IS1096 or IS*Msm3*) into the *mps* promoter region or into *lsr2* gene, leading to an increased expression of the *mps* operon and in turn to a increase in GPLs production (Kocincova et al., unpublished data). Modulation of the GPLs production through transposition of insertion elements may provide a selective advantage in particular circumstances at a population scale.

Sliding Motility and Biofilm Formation

Besides contributing to the colony rough/smooth morphotypes, GPLs appear to also be largely involved in other cell envelope properties. Although mycobacteria are nonflagellated, they can spread on the surface of solid media by a sliding mechanism (Martinez et al., 1999). No extracellular structures such as pili or fimbriae appear to be involved in this process. This motility is likely to play a significant role in the ability of mycobacterium to colonize surfaces in the environment or inside the host. All of the nonsliding mutants isolated that have rough morphotype show no detectable levels of GPLs (Etienne et al., 2002; Recht et al., 2000). The group of Kolter also investigated the role of GPLs in biofilm formation (Recht et al., 2000). Although the *M. smegmatis* mc^{2}155 strain forms a clearly detectable biofilm, no biofilm was detected in any of the nonsliding mutants, indicating that GPLs are also required for *M. smegmatis* biofilm formation on a polyvinyl chloride (PVC) plastic surface. In *M. abscessus*, GPLs are also associated with sliding motility and biofilm formation. Indeed, the 390S strain, in which GPLs are present in abundance in the cell wall, is able to slide, contrary to the 390R and 390V strains that have a marked reduction in the amount of GPLs (Howard et al., 2006). In addition, mutations leading to a modification of the GPLs structure (like acetylation or glycosylation) have been shown to affect sliding and/or biofilm formation (Deshayes et al., 2005; Recht and Kolter, 2001). As in *M. smegmatis*, GPLs are also involved in the production of biofilms on PVC plates in *M. avium* (Yamazaki et al., 2006), but the use of diverse in vitro model systems revealed a more complex picture. Indeed, GPL mutants adhered as well as the parent strain to Permanox and silanized glass chamber slides (Freeman et al., 2006). A recirculating system was designed to model drinking water distribution conditions. Non-GPL-producing and wild-type *M. avium* subsp. *hominissuis* bound equally well to the stainless steel coupons, indicating that the GPLs are not required in the primary stages of biofilm formation under these conditions. However, non-GPL-producing mutants were present in relatively small numbers in the recirculating-water (planktonic) phase compared to the wild-type strain. These observations show that GPLs are directly or indirectly required for colonization of some surfaces. Under some conditions, GPLs may play a role by facilitating the survival or dispersal of nonadherent *M. avium* cells in circulating water. Such a function could contribute to waterborne *M. avium* infection (Freeman et al., 2006).

The group of Kolter proposed a model explaining the role of these amphiphilic molecules in both sliding motility and biofilm formation (Recht et al., 2000). This model proposes that the GPLs are anchored through the hydrophilic head to the outermost layer of cell envelope, thereby exposing the hydrophobic tail to the outside. This would result in a hydrophobic cellular surface that cannot interact with the hydrophilic agarose surface. The resulting reduction of friction would lead to sliding on the surface of the agarose. This model seems valid because we observed on growth medium containing hydrophobic components that the wild-type strain is not motile and the GPL mutant is motile (our unpublished data, Fig. 6). Conversely, hydrophobic interactions between the exposed fatty acid tails of the GPLs and the hydrophobic surface of the PVC plastic would mediate attachment to plastic and biofilm formation. This type of interaction has been recognized to play a role in bacterial attachment to surfaces (Rosenberg and Kjelleberg, 1986). We actually favor a model in which the lipid moiety is inserted into the envelope and the sugars exposed on the surface. This model is more compatible with the effect of the glycosylation or O-acetylation observed in biofilm formation or sliding motility. Indeed, as mentioned earlier, it has been shown in several cases that some *M. smegmatis* mutant strains display intermediate phenotypes for sliding motility. This is indeed the case of the strains producing deacetylated (Recht and Kolter, 2001) or triglycosylated GPLs (Deshayes et al., 2005). These kinds of molecules are less hydrophobic than the GPLs produced by the wild-type strain (Villeneuve et al., 2003). Hydrophobic interactions clearly play a role in both biofilm formation and motility. It is more likely that these interactions involved the modified sugars (like O acetylation, O methylation or O succinylation) rather than the fatty acid moiety itself.

ROLE OF GPL IN VIRULENCE AND INTERACTION WITH THE HOST

As many other lipid compounds, GPLs possess a variety of functions in vivo. They induce prostaglandine E$_2$ (PGE$_2$) or TNF-α production but also partic-

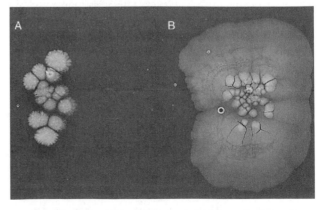

Figure 6. Motility on Tween 80-containing plate of *M. smegmatis* mc^{2}155 (A) and a GPL-nonproducing mutant (B). (Our unpublished data).

ipate in the arrest of phagosome maturation and interact with biological membranes (Pourshafie et al., 1993; Vergne and Desbat, 2000; Villeneuve et al., 2003). However, because GPLs of various structures have been used in conjunction with various in vivo or in vitro models, it is hence very difficult to compare these works and to draw a clear picture of GPL functions in vivo. It is likely that the various molecular forms have distinct activities that may be adapted to a particular ecological niche.

Accumulation of GPL during Macrophage Infection by *M. avium*

GPLs are particularly important because they are not covalently linked to other cell wall constituents and therefore have the ability to accumulate on the bacterial surface during growth in vivo (Barrow et al., 1980; Draper, 1974). Moreover, polar GPLs are poorly degraded after phagocytosis by macrophages (Hooper and Barrow, 1988) and hence may interfere with the bactericidal functions of the macrophages. Freeze-fracture electron microscopy has been used to study the structure of the envelope of *M. avium* cells growing inside mouse liver macrophages (Rulong et al., 1991) and has revealed an onion-like structure (Draper and Rees, 1970; Draper and Rees, 1973; Frehel et al., 1986; Rastogi and David, 1988). These experiments showed a progressive accumulation of GPLs around intramacrophagic bacteria, especially during long-term infections. The lamellar coat is viewed as the result of the accretion of GPL molecules sequestrated inside the phagosomes. This interpretation is in agreement with the evidence of Kim and coworkers (1976), who showed that GPLs present in the outer mycobacterial surface can be abundantly secreted by several mycobacterial species in liquid culture. However, because no specific stain or mutant strain was employed, it is difficult to be absolutely sure that the in vivo coat is indeed the result of GPLs accumulation.

Interaction with Biologic Membranes In Vitro

Because GPLs are produced within infected macrophages (Tereletsky and Barrow, 1983), it has been rapidly postulated that they could help the pathogen to survive by impairing the phagosomal-membrane-linked functions of the host cell. It is well established that some mycobacterial lipids traffic into the host cell during the pathogenesis process (Beatty et al., 2000). However, this has not been investigated in the case of GPLs. Several GPLs differing from one another by the number of saccharidic units have been tested both on mitochondria and on carboxyfluorescein-containing liposomes. Sut and colleagues (1990)

have shown that aqueous dispersions of GPLs added to liposome suspensions increased drastically the membrane permeability, as well as uncoupling oxidative phosphorylation when added to isolated mitochondria. All of the molecules exhibit the same type of activity, but there are differences associated with the various structural forms of GPLs (Lopez-Marin et al., 1994b). This could be due to differences in interactions between the glycopeptidic moiety and membranes, especially with the phospholipidic part of membranes.

To study the interaction of GPLs with membranes, phospholipids monolayers at the air-water interface were used as a model (Vergne et al., 1995). It was proposed that the poor activity of the lipopeptide core of GPL is due to its inability to insert into membranes. When the ability of mono- and triglycosylated GPLs to insert into membranes was compared, the insertion of triglycosylated form was more efficient (Vergne et al., 1995), whereas the permeabilization activity of monoglycosylated GPLs was higher than that of triglycosylated GPLs. To explain these data, the authors proposed a model in which the glycopeptide moiety of triglycosylated GPLs dips into the water phase and, consequently, hinders the molecule to induce alteration of the phospholipids organization; in contrast, the peptide moiety of monoglycosylated GPLs lays at the interface, rendering the molecule more disturbing (Lopez-Marin et al., 1994b; Vergne et al., 1995). The lateral organization of molecules in triglycosylated GPLs differs significantly from that of monoglycosylated GPLs. Although triglycosylated GPLs segregated from the phospholipid phase and self-associate to form intermolecular β-sheets, monoglycosylated GPLs are miscible with phospholipid molecules and increase the disorder of the phospholipid acyl chains. The authors suggested that permeabilization would arise from interactions between GPLs and phospholipid molecules. The difference of permeabilization efficiency of GPLs is likely due to the segregation of highly glycosylated GPLs, which decreases the number of molecules interacting with phospholipids (Vergne and Desbat, 2000). It seems that the GPLs having the highest efficiency to increase membrane passive permeability are the ones that have a sufficient sugar content to present a hydrophobic-hydrophilic balance that favors their insertion into membranes, but not enough sugar residues to form extensive β-structures or with a modified peptidic chain that prevent self-association in β-sheets, structures that exclude GPL molecules from phospholipid regions.

Because of their ability to disturb cell membrane ultrastructure, it is very likely that the GPLs and related lipids interact with immunologically relevant cells located in the vicinity of the macrophage that initially engulfed them. As lipids accumulate, they might be eliminated from the macrophage or released as the result of cell death. Following elimination from host macrophages, the lipids would then most likely interact with other host cells. Although the specific mechanisms that define the pathogenicity of nontuberculous mycobacteria have not been clarified, it is becoming more apparent that surface-associated GPL molecules have a variety of biological activities that could influence host responses during infection (Chatterjee and Khoo, 2001).

Modulation of the Host Cell Response

Modulation of macrophage antimicrobial activity

GPLs from *M. avium*, essentially through its lipopeptide fragment (the β-lipid), have been reported to change the expression of surface receptors of murine macrophages (Pourshafie et al., 1993). Indeed, flow cytometric analysis of peritoneal macrophages revealed that treatment with the β-lipid fragment obtained from GPLs by chemical means decreased the expression of the C3bi complement receptor Mac-1 on macrophages, whereas treatment with GPLs increased the expression of Mac-2 receptor on macrophages (Pourshafie et al., 1993). Treatment of peritoneal macrophages with either GPLs or β-lipid resulted in the release of TNF-α. Perturbation of macrophage membrane ultrastructure by both GPLs and β-lipid was confirmed by electron microscopy, and this may be a possible explanation for the resulting alterations in mononuclear cell function observed in this study. The β-lipid affects the ability of human peripheral blood macrophages (PBMs) to control the growth of *M. avium* serovar 2. The lipid can also induce the release of high levels of PGE$_2$ by the cells and some changes in the membrane ultrastructure (Pourshafie et al., 1993), resulting in the alteration of macrophage functions (Barrow et al., 1993). The polar ssGPLs from *M. avium* serovar 8, but not those from serovars 4 and 20, induced significant levels of TNF-α production, as well as PGE$_2$, in PBMs exposed to a sublethal concentration (Barrow et al., 1995). These results are explained by the difference in the carbohydrate residues of the oligosaccharide moieties. Similarly, lipid fractions extracted from *M. avium* subsp. *paratuberculosis* 18 inhibit the killing of *Candida albicans* by activated bovine PBMs (Hines et al., 1993).

It should be kept in mind that most of the experiments reported in this section were made with gross lipid extract and not with highly purified component. Hence, the possibility of an inhibiting or potentiating effect originating from other components cannot be rigorously excluded.

Inhibition of phagocytosis

More recently, nsGPLs have been reported to play a role in the internalization of non-tuberculosis mycobacteria. Indeed, GPLs of *M. smegmatis* were found to specifically and dose dependently inhibit the phagocytosis of both *M. smegmatis* and the opportunistic pathogen *M. kansasii* by human macrophages derived from monocytes (Villeneuve et al., 2003). Interestingly, mostly the deacetylated GPLs II and succinylated GPLs III exhibit an inhibitory activity and do not affect internalization of control particles. This suggests a specific recognition of structurally defined partners at the surface of macrophages. Consequently, the roles of GPLs as potential ligands of phagocytic receptors was investigated (Villeneuve et al., 2005). It was shown that GPLs III ensure the entry of particles through both complement receptor 3 and mannose receptor, two well-known receptors for the binding and internalization of mycobacteria in macrophages. Hence, the authors propose that GPLs participate in the invasion process of bacilli as ligands of phagocytic receptors previously described to be involved in the internalization of mycobacteria in macrophages.

Inhibition of lymphoproliferation

The immunomodulatory activities of GPLs have been described in the review of Chatterjee and Khoo (2001). An intraperitoneal injection of GPLs from *M. avium* subsp. *intracellulare* serovars 4 and 20 causes the inhibition of the blastogenic response of murine splenic lymphocytes to nonspecific mitogens (concanavalin A, phytohemagglutinin, and lipopolysaccharide) (Brownback and Barrow, 1988; Hooper et al., 1986). In another study, treatment of PBMs with a total lipid fraction derived from *M. avium* serovar 4 resulted in a significant suppression of lymphoproliferative responsiveness to phytohemagglutinin stimulation at concentrations not affecting cell viability (Barrow et al., 1993). Although a similar suppression was not observed when PBMs were treated with purified serovar 4-specific GPLs, treatment with the β-lipid fragment derived from the GPLs did result in a significant suppression of phytohemagglutinin responsiveness. It was also observed that the lymphoproliferative response of murine splenic mononuclear cells to mitogen stimulation was reduced by both the GPLs and their lipopeptide fragment (Pourshafie et al., 1993). Although the responsiveness appeared to be downregulated to a greater extent by the β-lipid, treatment with either GPLs or β-lipid resulted in the release of soluble factors from peritoneal macrophages that caused suppression of the lymphoproliferative responsiveness of splenic mononuclear cells.

Experiments with human PBMs have established that GPLs from *M. avium* can induce production of TNF-α, interleukin-6 (IL-6), and IL-1β, cytokines primarily secreted by monocytes. Other investigations revealed that *M. avium* total lipids have the ability to suppress the secretion of IL-2 and interferon-γ cytokines that are responsible for Th1-type responses (Horgen et al., 2000; Pourshafie et al., 1999). Thus, it has been suggested that because MAC lipids accumulate in chronic infections (Barrow, 1997), they first interact with the host macrophages and then begin to disperse and affect adjacent cells as the infection progresses.

In an effort to test the effect of various GPLs purified from MAC on the phagocytic processes of human PBMs, GPL-coated heat-killed staphylococcal cells were phagocytosed by PBMs and the phagosome-lysosome fusion (P-L fusion) was measured (Kano et al., 2005; Takegaki, 2000). The presence of GPLs was inhibiting the P-L fusion of *Staphylococcus aureus*. Elimination of the oligosaccharide from GPLs abrogated its inhibitory activity. The inhibitory activity of GPLs was competitively reduced by the presence of mannose, suggesting again that inhibition of P-L fusion by GPLs is mediated through mannose receptor (Shimada et al., 2006). Different degrees of phagocytosis and P-L fusion were observed throughout the various serotypes. The treatment with GPLs serovar 4 resulted in modulation of other macrophage functions, suggesting that infection with this serovar could cause severe complications in AIDS patients. This study is in agreement with the fact that in AIDS patients, the prognosis after serotype 4 infection is unfavorable compared with other serotype infections (Takegaki, 2000). Because the serotype-specific oligosaccharide is the only difference between the various serovars, the different activity is likely due to the structural difference of the oligosaccharide part of GPLs.

A group of polar GPLs from *M. chelonae* (which are similar to those from *M. abscessus* and *M. smegmatis*) has also been reported to exert various biological activities. These extracted GPLs have adjuvant activity with regard to protective effects of an inactivated influenza vaccine (Gjata et al., 1994). Moreover, they exhibit properties of hematopoietic growth factor (de Souza Matos et al., 2000; Vincent-Naulleau et al., 1995; Vincent-Naulleau et al., 1997) and stimulate myelopoiesis in normal mice and induce sustained granulocytosis, monocytosis, and thrombocytosis in the peripheral blood. They also increase the resistance of mice against lethal infection with

Candida albicans, apparently mediated by the ability of GPLs to induce hyperleukocytosis (Lagrange et al., 1994, 1995).

In conclusion, GPLs have a variety of immunomodulator properties that likely depend on the chemical structure of the GPLs used, the dose, the degree of purity, the route of administration, and the in vivo model. Further systematic studies are required to understand the molecular events leading to the interference with the host immune system.

Drug Resistance and the GPL Expression Profile

A study investigated the consequence of drug treatment with a regimen of clarithromycin and ethambutol on the chemical alterations of GPLs in *M. avium* (Khoo et al., 1999). Ethambutol inhibits the biosynthesis of the arabinan-containing components of the mycobacterial cell wall and is given in combination with a macrolide antibiotic to treat disseminated infection (Chiu et al., 1990). Two clinical isolates of the same strains originating from an AIDS patient with disseminated MAC infection before and after treatment were investigated. By definition, these are virulent strains, and the post-treatment isolate is naturally and clinically drug resistant. This experiment showed that their GPL expression profiles differed significantly in that the apolar GPLs were over-expressed in the clinically resistant isolate at the expense of the polar serotype 1 GPL. Instead of additional rhamnosylation on the 6-deoxytalose appendage to give the serotype 1 specific disaccharide hapten, there was an accumulation of highly O-methylated 6-deoxytalose and an unusual 4-O-sulfated 6-deoxytalose that was not previously described. The only other sulfated GPLs were found in *M. fortuitum* bv. Peregrinum (Lopez Marin et al., 1992) with sulfated rhamnose on the alaninol. It should, however, be cautioned that sulfatation may be more common than previously thought, albeit at a low level. The authors hypothesized that the altered physicochemical properties of GPLs probably contribute to drug resistance by inhibiting the entrance of some antimicrobial agents, including ethambutol, into the bacillus. It is not clear whether and how the altered expression of GPLs is responsible for the drug resistance of this particular isolate. However, it is well established that GPLs are involved in the wall permeability barrier, because it is known that the absence of C-type GPLs from the cell envelope of *M. smegmatis* has a profound effect on the uptake of chenodeoxycholate, a hydrophobic molecule that diffuses through lipid domains of the mycobacterial cell wall (Etienne et al., 2002). Thus, it is possible that after a drug treatment, the normally regulated GPL biosynthesis pathways may be perturbed as an indirect consequence and selective process.

CONCLUDING REMARKS

It is clear from in vivo data that GPLs exquisitely interact with the immune systems and that particular subspecies display distinct biological activities from another. A systematic approach will help to define the chemical moiety involved in a particular activity. In addition, a better understanding of the biosynthetic pathway should help in altering the GPL backone and hence in creating new derivatives that may have valuable biological activities.

The genetic complexity of the GPL biosynthetic pathway has been unraveled only recently. It is somehow surprising that these mycobacterial species dedicate more than 1% of their genome content to assemble this small metabolite, which plays a key role in bacterial physiology and interaction with the host but is not essential for life. It is also intriguing that GPLs are produced by both rapid and slow growers that are phylogenetically very distinct. Horizontal transfer has been proposed to be involved in the spreading of this locus, but it remains to be demonstrated (Krzywinska et al., 2004). Several structures have been described for the GPLs. The structural variations concern the sugar moieties and the tetrapeptide part of the molecule. Many genes have yet to be discovered in other species as glycosylation and modification (O methylation, O acetylation, and O succinylation) patterns appear to be extensively variable from one isolate to another. It is very likely that investigating other mycobacterial species will by no doubt enlarge the repertoire of structures belonging to this family of glycolipid.

Several key questions remain to be elucidated. A number of genes of the GPL locus of M. smegmatis are still without any ascribed function, and it will be necessary to test their role in the GPL biosynthesis. In addition, there is to date only few data dealing with the regulation of this pathway. There is no reason that GPLs would be produced at the same level whatever the growth conditions. The same remark is also valid for the various molecular forms (O acetylated, O methylated, O succinylated, and triglycosylated) of the GPLs. With regard to virulence, there is increasing evidence that M. abscessus strains producing low levels of GPLs (R strains) are more virulent that strains producing high levels of GPLs (S strains) (Byrd and Lyons, 1999; Catherinot et al., 2007; Howard et al., 2006). The molecular mechanisms of this morphotype variation have yet to be understood, as well as the relation linking GPL production and the increase of virulence. The interactions between the GPLs and the immune system remain to be investigated in vivo.

Small metabolites such as sulfolipids, phenolglycolipids, or glycopeptidolipids use the same building blocks that are Pks, FadD, FadE, MmpL, and Gtf and that have evolved substrate specificity. Understanding the specificity of the various orthologs is probably one of the most challenging projects in this field.

Acknowledgments. C.D. is funded by a doctoral grant of INSERM—Région Ile-de-France and FRM. We gratefully acknowledge Inserm for funding this project under the Avenir program to J.-M.R., Chargé de Recherches at INSERM.

REFERENCES

Aspinall, G. O., D. Chatterjee, and P. J. Brennan. 1995. The variable surface glycolipids of mycobacteria: structures, synthesis of epitopes, and biological properties. *Adv. Carbohydr. Chem. Biochem.* **51:**169–242.

Barrow, W. W. 1997. Processing of mycobacterial lipids and effects on host responsiveness. *Front. Biosci.* **2:**d387–d400.

Barrow, W. W., B. P. Ullom, and P. J. Brennan. 1980. Peptidoglycolipid nature of the superficial cell wall sheath of smooth-colony-forming mycobacteria. *J. Bacteriol.* **144:**814–822.

Barrow, W. W., and P. J. Brennan. 1982. Isolation in high frequency of rough variants of *Mycobacterium intracellulare* lacking C-mycoside glycopeptidolipid antigens. *J. Bacteriol.* **150:**381–384.

Barrow, W. W., J. P. de Sousa, T. L. Davis, E. L. Wright, M. Bachelet, and N. Rastogi. 1993. Immunomodulation of human peripheral blood mononuclear cell functions by defined lipid fractions of *Mycobacterium avium. Infect. Immun.* **61:**5286–5293.

Barrow, W. W., T. L. Davis, E. L. Wright, V. Labrousse, M. Bachelet, and N. Rastogi. 1995. Immunomodulatory spectrum of lipids associated with *Mycobacterium avium* serovar 8. *Infect. Immun.* **63:**126–133.

Beatty, W. L., E. R. Rhoades, H. J. Ullrich, D. Chatterjee, J. E. Heuser, and D. G. Russell. 2000. Trafficking and release of mycobacterial lipids from infected macrophages. *Traffic* **1:**235–247.

Belisle, J. T., and P. J. Brennan. 1989. Chemical basis of rough and smooth variation in mycobacteria. *J. Bacteriol.* **171:**3465–3470.

Belisle, J. T., L. Pascopella, J. M. Inamine, P. J. Brennan, and W. R. Jacobs, Jr. 1991. Isolation and expression of a gene cluster responsible for biosynthesis of the glycopeptidolipid antigens of *Mycobacterium avium. J. Bacteriol.* **173:**6991–6997.

Belisle, J. T., K. Klaczkiewicz, P. J. Brennan, W. R. Jacobs, Jr., and J. M. Inamine. 1993. Rough morphological variants of Mycobacterium avium. Characterization of genomic deletions resulting in the loss of glycopeptidolipid expression. *J. Biol. Chem.* **268:**10517–10523.

Besra, G. S., M. R. McNeil, B. Rivoire, K. H. Khoo, H. R. Morris, A. Dell, and P. J. Brennan. 1993. Further structural definition of a new family of glycopeptidolipids from *Mycobacterium xenopi. Biochemistry* **32:**347–355.

Biet, F., M. L. Boschiroli, M. F. Thorel, and L. A. Guilloteau. 2005. Zoonotic aspects of *Mycobacterium bovis* and *Mycobacterium avium-intracellulare* complex (MAC). *Vet. Res.* **36:**411–436.

Billman-Jacobe, H., M. J. McConville, R. E. Haites, S. Kovacevic, and R. L. Coppel. 1999. Identification of a peptide synthetase involved in the biosynthesis of glycopeptidolipids of *Mycobacterium smegmatis*. *Mol. Microbiol.* **33:**1244–1253.

Brennan, P. J., and M. B. Goren. 1979. Structural studies on the type-specific antigens and lipids of the *Mycobacterium avium·Mycobacterium intracellulare·Mycobacterium scrofulaceum* serocomplex: *Mycobacterium intracellulare* serotype 9. *J. Biol. Chem.* **254:**4205–4211.

Brennan, P. J., H. Mayer, G. O. Aspinall, and J. E. Nam Shin. 1981. Structures of the glycopeptidolipid antigens from serovars in the *Mycobacterium avium/Mycobacterium intracellulare/ Mycobacterium scrofulaceum* serocomplex. *Eur. J. Biochem.* **115:** 7–15.

Brennan, P. J. 1988. Mycobacterium and other actinomycetes, p. 203–298. *In* C. Ratledge and S. G. Wilkinson (ed.), *Microbial Lipids*. Academic Press, London, United Kingdom.

Brennan, P. J., and H. Nikaido. 1995. The envelope of mycobacteria. *Annu. Rev. Biochem.* **64:**29–63.

Brown-Elliott, B. A., and R. J. Wallace, Jr. 2002. Clinical and taxonomic status of pathogenic nonpigmented or late-pigmenting rapidly growing mycobacteria. *Clin. Microbiol. Rev.* **15:**716–746.

Brownback, P. E., and W. W. Barrow. 1988. Modified lymphocyte response to mitogens after intraperitoneal injection of glycopeptidolipid antigens from *Mycobacterium avium* complex. *Infect. Immun.* **56:**1044–1050.

Buglino, J., K. C. Onwueme, J. A. Ferreras, L. E. Quadri, and C. D. Lima. 2004. Crystal structure of PapA5, a phthiocerol dimycocerosyl transferase from *Mycobacterium tuberculosis*. *J. Biol. Chem.* **279:**30634–30642.

Byrd, T. F., and C. R. Lyons. 1999. Preliminary characterization of a *Mycobacterium abscessus* mutant in human and murine models of infection. *Infect. Immun.* **67:**4700–4707.

Camacho, L. R., P. Constant, C. Raynaud, M. A. Laneelle, J. A. Triccas, B. Gicquel, M. Daffe, and C. Guilhot. 2001. Analysis of the phthiocerol dimycocerosate locus of *Mycobacterium tuberculosis*. Evidence that this lipid is involved in the cell wall permeability barrier. *J. Biol. Chem.* **276:**19845–19854.

Camphausen, R. T., R. L. Jones, and P. J. Brennan. 1985. A glycolipid antigen specific to *Mycobacterium paratuberculosis*: structure and antigenicity. *Proc. Natl. Acad. Sci. USA* **82:**3068–3072.

Cane, D. E., C. T. Walsh, and C. Khosla. 1998. Harnessing the biosynthetic code: combinations, permutations, and mutations. *Science* **282:**63–68.

Cangelosi, G. A., C. O. Palermo, J. P. Laurent, A. M. Hamlin, and W. H. Brabant. 1999. Colony morphotypes on Congo red agar segregate along species and drug susceptibility lines in the *Mycobacterium avium-intracellulare* complex. *Microbiology* **145** (Pt. 6):1317–1324.

Catherinot, E., J. Clarissou, G. Etienne, F. Ripoll, J. F. Emile, M. Daffe, C. Perronne, C. Soudais, J. L. Gaillard, and M. Rottman. 2007. Hypervirulence of a rough variant of the *Mycobacterium abscessus* type strain. *Infect. Immun.* **75:**1055–1058.

Chatterjee, D., and K. H. Khoo. 2001. The surface glycopeptidolipids of mycobacteria: structures and biological properties. *Cell Mol. Life Sci.* **58:**2018–2042.

Chen, J. M., G. J. German, D. C. Alexander, H. Ren, T. Tan, and J. Liu. 2006. Roles of Lsr2 in colony morphology and biofilm formation of *Mycobacterium smegmatis*. *J. Bacteriol.* **188:**633–641.

Chiu, J., J. Nussbaum, S. Bozzette, J. G. Tilles, L. S. Young, J. Leedom, P. N. Heseltine, and J. A. McCutchan. 1990. Treatment of disseminated *Mycobacterium avium* complex infection in AIDS with amikacin, ethambutol, rifampin, and ciprofloxacin. *Ann. Intern. Med.* **113:**358–361.

Cole, S. T., R. Brosch, J. Parkhill, T. Garnier, C. Churcher, D. Harris, S. V. Gordon, K. Eiglmeier, S. Gas, C. E. Barry, F. Tekaia, K. Badcock, D. Basham, D. Brown, T. Chillingworth, R. Connor, R. Davies, K. Devlin, T. Feltwell, S. Gentles, N. Hamlin, S. Holroyd, T. Hornsby, K. Jagels, A. Krogh, J. McLean, S. Moule, L. Murphy, K. Oliver, J. Osborne, M. A. Quail, M. A. Rajandream, J. Rogers, S. Rutter, K. Seeger, J. Skelton, R. Squares, S. Squares, J. E. Sulston, K. Taylor, S. Whitehead, and B. G. Barrell. 1998. Deciphering the biology of *Mycobacterium tuberculosis* from the complete genome sequence. *Nature* **393:**537–544.

Converse, S. E., J. D. Mougous, M. D. Leavell, J. A. Leary, C. R. Bertozzi, and J. S. Cox. 2003. MmpL8 is required for sulfolipid-1 biosynthesis and *Mycobacterium tuberculosis* virulence. *Proc. Natl. Acad. Sci. USA* **100:**6121–6126.

Cox, J. S., B. Chen, M. McNeil, and W. R., Jacobs, Jr. 1999. Complex lipid determines tissue-specific replication of Mycobacterium tuberculosis in mice. *Nature* **402:**79–83.

Daffe, M., M. A. Laneelle, and G. Puzo. 1983. Structural elucidation by field desorption and electron-impact mass spectrometry of the C-mycosides isolated from Mycobacterium smegmatis. *Biochim. Biophys. Acta* **751:**439–443.

Daffe, M., and P. Draper. 1998. The envelope layers of mycobacteria with reference to their pathogenicity. *Adv. Microb. Physiol.* **39:**131–203.

Deshayes, C., F. Laval, H. Montrozier, M. Daffe, G. Etienne, and J. M. Reyrat. 2005. A glycosyltransferase involved in biosynthesis of triglycosylated glycopeptidolipids in *Mycobacterium smegmatis*: impact on surface properties. *J. Bacteriol.* **187:** 7283–7291.

de Souza Matos, D. C., R. Marcovistz, T. Neway, A. M. Vieira da Silva, E. N. Alves, and C. Pilet. 2000. Immunostimulatory effects of polar glycopeptidolipids of *Mycobacterium chelonae* for inactivated rabies vaccine. *Vaccine* **18:**2125–2131.

Domenech, P., M. B. Reed, C. S. Dowd, C. Manca, G. Kaplan, and C. E. Barry, III. 2004. The role of MmpL8 in sulfatide biogenesis and virulence of *Mycobacterium tuberculosis*. *J. Biol. Chem.* **279:**21257–21265.

Draper, P. 1974. The mycoside capsule of *Mycobacterium avium* 357. *J. Gen. Microbiol.* **83:**431–433.

Draper, P., and R. J. Rees. 1970. Electron-transparent zone of mycobacteria may be a defence mechanism. *Nature* **228:**860–861.

Draper, P., and R. J. Rees. 1973. The nature of the electron-transparent zone that surrounds *Mycobacterium lepraemurium* inside host cells. *J. Gen. Microbiol.* **77:**79–87.

Eckstein, T. M., F. S. Silbaq, D. Chatterjee, N. J. Kelly, P. J. Brennan, and J. T. Belisle. 1998. Identification and recombinant expression of a *Mycobacterium avium* rhamnosyltransferase gene (rtfA) involved in glycopeptidolipid biosynthesis. *J. Bacteriol.* **180:**5567–5573.

Eckstein, T. M., J. T. Belisle, and J. M. Inamine. 2003. Proposed pathway for the biosynthesis of serovar-specific glycopeptidolipids in *Mycobacterium avium* serovar 2. *Microbiology* **149:** 2797–2807.

Eckstein, T. M., S. Chandrasekaran, S. Mahapatra, M. R. McNeil, D. Chatterjee, C. D. Rithner, P. W. Ryan, J. T. Belisle, and J. M. Inamine. 2006. A major cell wall lipopeptide of *Mycobacterium avium* subspecies *paratuberculosis*. *J. Biol. Chem.* **281:**5209–5215.

Etienne, G., C. Villeneuve, H. Billman-Jacobe, C. Astarie-Dequeker, M. A. Dupont, and M. Daffe. 2002. The impact of the absence of glycopeptidolipids on the ultrastructure, cell surface and cell wall properties, and phagocytosis of *Mycobacterium smegmatis*. *Microbiology* **148:**3089–3100.

Freeman, R., H. Geier, K. M. Weigel, J. Do, T. E. Ford, and G. A. Cangelosi. 2006. Roles for cell wall glycopeptidolipid in surface adherence and planktonic dispersal of *Mycobacterium avium*. *Appl. Environ. Microbiol.* **72:**7554–7558.

Fregnan, G. B., D. W. Smith, and H. M. Randall. 1961. Biological and chemical studies on mycobacteria. Relationship of colony morphology to mycoside content for *Mycobacterium kansasii* and *Mycobacterium fortuitum*. *J. Bacteriol.* **82:**517–527.

Fregnan, G. B., and D. W. Smith. 1962. Description of various colony forms of mycobacteria. *J. Bacteriol.* **83:**819–827.

Frehel, C., A. Ryter, N. Rastogi, and H. David. 1986. The electron-transparent zone in phagocytized *Mycobacterium avium* and other mycobacteria: formation, persistence and role in bacterial survival. *Ann. Inst. Pasteur Microbiol.* **137B:**239–257.

Fujiwara, N., N. Nakata, S. Maeda, T. Naka, M. Doe, I. Yano, and K. Kobayashi. 2007. Structural characterization of a specific glycopeptidolipid containing a novel N-acyl-deoxy sugar from *Mycobacterium intracellulare* serotype 7 and genetic analysis of its glycosylation pathway. *J. Bacteriol.* **189:**1099–1108.

Furuchi, A., and T. Tokunaga. 1972. Nature of the receptor substance of *Mycobacterium smegmatis* for D4 bacteriophage adsorption. *J. Bacteriol.* **111:**404–411.

Gjata, B., C. Hannoun, H. J. Boulouis, T. Neway, and C. Pilet. 1994. Adjuvant activity of polar glycopeptidolipids of *Mycobacterium chelonae* (pGPL-Mc) on the immunogenic and protective effects of an inactivated influenza vaccine. *C. R. Acad. Sci. III* **317:**257–263.

Goren, M. B., J. K. McClatchy, B. Martens, and O. Brokl. 1972. Mycosides C: behavior as receptor site substance for mycobacteriophage D4. *J. Virol.* **9:**999–1003.

Helmann, J. D. 2002. The extracytoplasmic function (ECF) sigma factors. *Adv. Microb. Physiol.* **46:**47–110.

Hines, M. E., II, J. M. Jaynes, S. A. Barker, J. C. Newton, F. M. Enright, and T. G. Snider, III. 1993. Isolation and partial characterization of glycolipid fractions from *Mycobacterium avium* serovar 2 (*Mycobacterium paratuberculosis* 18) that inhibit activated macrophages. *Infect. Immun.* **61:**1–7.

Hooper, L. C., M. M. Johnson, V. R. Khera, and W. W. Barrow. 1986. Macrophage uptake and retention of radiolabeled glycopeptidolipid antigens associated with the superficial L1 layer of *Mycobacterium intracellulare* serovar 20. *Infect. Immun.* **54:**133–141.

Hooper, L. C., and W. W. Barrow. 1988. Decreased mitogenic response of murine spleen cells following intraperitoneal injection of serovar-specific glycopeptidolipid antigens from the *Mycobacterium avium* complex. *Adv. Exp. Med. Biol.* **239:**309–325.

Horgen, L., E. L. Barrow, W. W. Barrow, and N. Rastogi. 2000. Exposure of human peripheral blood mononuclear cells to total lipids and serovar-specific glycopeptidolipids from *Mycobacterium avium* serovars 4 and 8 results in inhibition of TH1-type responses. *Microb. Pathog.* **29:**9–16.

Howard, S. T., E. Rhoades, J. Recht, X. Pang, A. Alsup, R. Kolter, C. R. Lyons, and T. F. Byrd. 2006. Spontaneous reversion of *Mycobacterium abscessus* from a smooth to a rough morphotype is associated with reduced expression of glycopeptidolipid and reacquisition of an invasive phenotype. *Microbiology* **152:**1581–1590.

Jain, M., and J. S. Cox. 2005. Interaction between polyketide synthase and transporter suggests coupled synthesis and export of virulence lipid in *M. tuberculosis*. *PLoS Pathog.* **1:**e2.

Jeevarajah, D., J. H. Patterson, M. J. McConville, and H. Billman-Jacobe. 2002. Modification of glycopeptidolipids by an O-methyltransferase of *Mycobacterium smegmatis*. *Microbiology* **148:**3079–3087.

Jeevarajah, D., J. H. Patterson, E. Taig, T. Sargeant, M. J. McConville, and H. Billman-Jacobe. 2004. Methylation of GPLs in *Mycobacterium smegmatis* and *Mycobacterium avium*. *J. Bacteriol.* **186:**6792–6799.

Kano, H., T. Doi, Y. Fujita, H. Takimoto, I. Yano, and Y. Kumazawa. 2005. Serotype-specific modulation of human monocyte functions by glycopeptidolipid (GPL) isolated from *Mycobacterium avium* complex. *Biol. Pharm. Bull.* **28:**335–339.

Khoo, K. H., D. Chatterjee, A. Dell, H. R. Morris, P. J. Brennan, and P. Draper. 1996. Novel O-methylated terminal glucuronic acid characterizes the polar glycopeptidolipids of *Mycobacterium habana* strain TMC 5135. *J. Biol. Chem.* **271:**12333–12342.

Khoo, K. H., E. Jarboe, A. Barker, J. Torrelles, C. W. Kuo, and D. Chatterjee. 1999. Altered expression profile of the surface glycopeptidolipids in drug-resistant clinical isolates of *Mycobacterium avium* complex. *J. Biol. Chem.* **274:**9778–9785.

Kim, K.-S., M. R. J. Salton, and L. Barksdale. 1976. Ultrastructure of superficial mycosidic integuments of *Mycobacterium* sp. *J. Bacteriol.* **125:**739–743.

Kolk, A. H., R. Evers, D. G. Groothuis, H. Gilis, and S. Kuijper. 1989. Production and characterization of monoclonal antibodies against specific serotypes of Mycobacterium avium and the *Mycobacterium avium-Mycobacterium intracellulare-Mycobacterium scrofulaceum* complex. *Infect. Immun.* **57:**2514–2521.

Krzywinska, E., J. Krzywinski, and J. S. Schorey. 2004. Phylogeny of *Mycobacterium avium* strains inferred from glycopeptidolipid biosynthesis pathway genes. *Microbiology* **150:**1699–1706.

Kumar, S., M. Bose, and M. Isa. 2006. Genotype analysis of human *Mycobacterium avium* isolates from India. *Indian J. Med. Res.* **123:**139–144.

Lagrange, P. H., M. Fourgeaud, T. Neway, and C. Pilet. 1994. Enhanced resistance against lethal disseminated *Candida albicans* infection in mice treated with polar glycopeptidolipids from *Mycobacterium chelonae* (pGPL-Mc). *C. R. Acad. Sci. III* **317:**1107–1113.

Lagrange, P. H., M. Fourgeaud, T. Neway, and C. Pilet. 1995. Mycobacterial polar glycopeptidolipids enhance resistance to experimental murine candidiasis. *C. R. Acad. Sci. III* **318:**359–365.

Laneelle, G., and J. Asselineau. 1962. Isolation of peptide lipids from *Mycobacterium paratuberculosis*. *Biochim. Biophys. Acta* **59:**731–732.

Laneelle, M. A., G. Laneelle, P. Bennet, and J. Asselineau. 1965. On the lipids from a non-photochromogenic strain of mycobacteria. *Bull. Soc. Chim. Biol.* (Paris) **47:**2047–2067.

Laurent, J. P., K. Hauge, K. Burnside, and G. Cangelosi. 2003. Mutational analysis of cell wall biosynthesis in *Mycobacterium avium*. *J. Bacteriol.* **185:**5003–5006.

Lautru, S., and G. L. Challis. 2004. Substrate recognition by non-ribosomal peptide synthetase multi-enzymes. *Microbiology* **150:**1629-1636.

Lopez Marin, L. M., M. A. Laneelle, D. Prome, M. Daffe, G. Laneelle, and J. C. Prome. 1991. Glycopeptidolipids from *Mycobacterium fortuitum*: a variant in the structure of C-mycoside. *Biochemistry* **30:**10536–10542.

Lopez Marin, L. M., M. A. Laneelle, D. Prome, G. Laneelle, J. C. Prome, and M. Daffe. 1992. Structure of a novel sulfate-containing mycobacterial glycolipid. *Biochemistry* **31:**11106–11111.

Lopez Marin, L. M., M. A. Laneelle, D. Prome, and M. Daffe. 1993. Structures of the glycopeptidolipid antigens of two animal pathogens: *Mycobacterium senegalense* and *Mycobacterium porcinum*. *Eur. J. Biochem.* **215:**859–866.

Lopez-Marin, L. M., N. Gautier, M. A. Laneelle, G. Silve, and M. Daffe. 1994a. Structures of the glycopeptidolipid antigens of *Mycobacterium abscessus* and *Mycobacterium chelonae* and possible chemical basis of the serological cross-reactions in the

Mycobacterium fortuitum complex. *Microbiology* **140**(Pt. 5): 1109–1118.

Lopez-Marin, L. M., D. Quesada, F. Lakhdar-Ghazal, J. F. Tocanne, and G. Laneelle. 1994b. Interactions of mycobacterial glycopeptidolipids with membranes: influence of carbohydrate on induced alterations. *Biochemistry* **33**:7056–7061.

Ma, Y., J. A. Mills, J. T. Belisle, V. Vissa, M. Howell, K. Bowlin, M. S. Scherman, and M. McNeil. 1997. Determination of the pathway for rhamnose biosynthesis in mycobacteria: cloning, sequencing and expression of the *Mycobacterium tuberculosis* gene encoding alpha-D-glucose-1-phosphate thymidylyltransferase. *Microbiology* **143**(Pt. 3):937–945.

Martinez, A., S. Torello, and R. Kolter. 1999. Sliding motility in mycobacteria. *J. Bacteriol.* **181**:7331–7338.

Maslow, J. N., V. R. Irani, S. H. Lee, T. M. Eckstein, J. M. Inamine, and J. T. Belisle. 2003. Biosynthetic specificity of the rhamnosyltransferase gene of *Mycobacterium avium* serovar 2 as determined by allelic exchange mutagenesis. *Microbiology* **149**:3193–3202.

McCarthy, C. 1970. Spontaneous and induced mutation in *Mycobacterium avium. Infect. Immun.* **2**:223–228.

Mederos, L. M., J. A. Valdivia, M. A. Sempere, and P. L. Valero-Guillen. 1998. Analysis of lipids reveals differences between 'Mycobacterium habana' and *Mycobacterium simiae. Microbiology* **144**(Pt. 5):1181–1188.

Mederos, L. M., J. A. Valdivia, and P. L. Valero-Guillen. 2006. Lipids of 'Mycobacterium habana', a synonym of *Mycobacterium simiae* with vaccine potential. *Tuberculosis* (Edinburgh) **86**:324–329.

Miyamoto, Y., T. Mukai, N. Nakata, Y. Maeda, M. Kai, T. Naka, I. Yano, and M. Makino. 2006. Identification and characterization of the genes involved in glycosylation pathways of mycobacterial glycopeptidolipid biosynthesis. *J. Bacteriol.* **188**: 86–95.

Motiwala, A. S., L. Li, V. Kapur, and S. Sreevatsan. 2006. Current understanding of the genetic diversity of *Mycobacterium avium* subsp. *paratuberculosis. Microbes Infect.* **8**:1406–1418.

Mukherjee, R., M. Gomez, N. Jayaraman, I. Smith, and D. Chatterji. 2005. Hyperglycosylation of glycopeptidolipid of *Mycobacterium smegmatis* under nutrient starvation: structural studies. *Microbiology* **151**:2385–2392.

Ojha, A. K., S. Varma, and D. Chatterji. 2002. Synthesis of an unusual polar glycopeptidolipid in glucose-limited culture of *Mycobacterium smegmatis. Microbiology* **148**:3039–3048.

Onwueme, K. C., J. A. Ferreras, J. Buglino, C. D. Lima, and L. E. Quadri. 2004. Mycobacterial polyketide-associated proteins are acyltransferases: proof of principle with *Mycobacterium tuberculosis* PapA5. *Proc. Natl. Acad. Sci. USA* **101**:4608–4613.

Ortalo-Magne, A., A. Lemassu, M. A. Laneelle, F. Bardou, G. Silve, P. Gounon, G. Marchal, and M. Daffe. 1996. Identification of the surface-exposed lipids on the cell envelopes of *Mycobacterium tuberculosis* and other mycobacterial species. *J. Bacteriol.* **178**:456–461.

Patterson, J. H., M. J. McConville, R. E. Haites, R. L. Coppel, and H. Billman-Jacobe. 2000. Identification of a methyltransferase from *Mycobacterium smegmatis* involved in glycopeptidolipid synthesis. *J. Biol. Chem.* **275**:4900–4906.

Pourshafie, M., Q. Ayub, and W. W. Barrow. 1993. Comparative effects of *Mycobacterium avium* glycopeptidolipid and lipopeptide fragment on the function and ultrastructure of mononuclear cells. *Clin. Exp. Immunol.* **93**:72–79.

Pourshafie, M. R., G. Sonnenfeld, and W. W. Barrow. 1999. Immunological and ultrastructural disruptions of T lymphocytes following exposure to the glycopeptidolipid isolated from the *Mycobacterium avium* complex. *Scand. J. Immunol.* **49**:405–410.

Quadri, L. E., J. Sello, T. A. Keating, P. H. Weinreb, and C. T. Walsh. 1998. Identification of a *Mycobacterium tuberculosis* gene cluster encoding the biosynthetic enzymes for assembly of the virulence-conferring siderophore mycobactin. *Chem. Biol.* **5**:631–645.

Rastogi, N., C. Frehel, and H. L. David. 1984. Cell envelope architectures of leprosy-derived corynebacteria, *Mycobacterium leprae*, and related organisms: a comparative study. *Curr. Microbiol.* **11**:23–30.

Rastogi, N., and H. L. David. 1988. Mechanisms of pathogenicity in mycobacteria. *Biochimie* **70**:1101–1120.

Recht, J., A. Martinez, S. Torello, and R. Kolter. 2000. Genetic analysis of sliding motility in *Mycobacterium smegmatis. J. Bacteriol.* **182**:4348–4351.

Recht, J., and R. Kolter. 2001. Glycopeptidolipid acetylation affects sliding motility and biofilm formation in *Mycobacterium smegmatis. J. Bacteriol.* **183**:5718–5724.

Ren, H., L. G. Dover, S. T. Islam, D. C. Alexander, J. M. Chen, G. S. Besra, and J. Liu. 2007. Identification of the lipooligosaccharide biosynthetic gene cluster from Mycobacterium marinum. *Mol. Microbiol.* **63**:1345–1359.

Ripoll, F., C. Deshayes, S. Pasek, F. Laval, J. L. Beretti, F. Biet, J. L. Risler, M. Daffé, G. Etienne, J. L. Gaillard, and J. M. Reyrat. 2007. Genomics of glycopeptidolipid biosynthesis in *Mycobacterium abscessus* and *M. chelonae. BMC Genomics* **8**:114.

Riviere, M., and G. Puzo. 1991. A new type of serine-containing glycopeptidolipid from *Mycobacterium xenopi. J. Biol. Chem.* **266**:9057–9063.

Riviere, M., S. Auge, J. Vercauteren, E. Wisingerova, and G. Puzo. 1993. Structure of a novel glycopeptidolipid antigen containing a O-methylated serine isolated from *Mycobacterium xenopi.* Complete 1H-NMR and 13C-NMR assignment. *Eur. J. Biochem.* **214**:395–403.

Rosenberg, M., and S. Kjelleberg. 1986. Hydrophobic interactions in bacterial adhesion. *Adv. Microb. Ecol.* **9**:353–393.

Rulong, S., A. P. Aguas, P. P. da Silva, and M. T. Silva. 1991. Intramacrophagic *Mycobacterium avium* bacilli are coated by a multiple lamellar structure: freeze fracture analysis of infected mouse liver. *Infect. Immun.* **59**:3895–3902.

Schaefer, W. B. 1965. Serologic identification and classification of the atypical mycobacteria by their agglutination. *Am. Rev. Respir. Dis.* **92**:85–93.

Sermet-Gaudelus, I., M. Le Bourgeois, C. Pierre-Audigier, C. Offredo, D. Guillemot, S. Halley, C. Akoua-Koffi, V. Vincent, V. Sivadon-Tardy, A. Ferroni, P. Berche, P. Scheinmann, G. Lenoir, and J. L. Gaillard. 2003. *Mycobacterium abscessus* and children with cystic fibrosis. *Emerg. Infect. Dis.* **9**:1587–1591.

Shimada, K., H. Takimoto, I. Yano, and Y. Kumazawa. 2006. Involvement of mannose receptor in glycopeptidolipid-mediated inhibition of phagosome-lysosome fusion. *Microbiol. Immunol.* **50**:243–251.

Smith, D. W., H. M. Randall, A. P. Maclennan, and E. Lederer. 1960a. Mycosides: a new class of type-specific glycolipids of Mycobacteria. *Nature* **186**:887–888.

Smith, D. W., H. M. Randall, A. P. Maclennan, R. K. Putney, and S. V. Rao. 1960b. Detection of specific lipids in mycobacteria by infrared spectroscopy. *J. Bacteriol.* **79**:217–229.

Sonden, B., D. Kocincova, C. Deshayes, D. Euphrasie, L. Rhayat, F. Laval, C. Frehel, M. Daffe, G. Etienne, and J. M. Reyrat. 2005. Gap, a mycobacterial specific integral membrane protein, is required for glycolipid transport to the cell surface. *Mol. Microbiol.* **58**:426–440.

Straight, P. D., M. A. Fischbach, C. T. Walsh, D. Z. Rudner, and R. Kolter. 2007. A singular enzymatic megacomplex from *Bacillus subtilis. Proc. Natl. Acad. Sci. USA* **104**:305–310.

Sut, A., S. Sirugue, S. Sixou, F. Lakhdar-Ghazal, J. F. Tocanne, and G. Laneelle. 1990. Mycobacteria glycolipids as potential pathogenicity effectors: alteration of model and natural membranes. *Biochemistry* **29**:8498–8502.

Takegaki, Y. 2000. Effect of serotype specific glycopeptidolipid (GPL) isolated from *Mycobacterium avium* complex (MAC) on phagocytosis and phagosome-lysosome fusion of human peripheral blood monocytes. *Kekkaku* **75**:9–18.

Tereletsky, M. J., and W. W. Barrow. 1983. Postphagocytic detection of glycopeptidolipids associated with the superficial L1 layer of *Mycobacterium intracellulare*. *Infect. Immun.* **41**:1312–1321.

Torrelles, J. B., D. Ellis, T. Osborne, A. Hoefer, I. M. Orme, D. Chatterjee, P. J. Brennan, and A. M. Cooper. 2002. Characterization of virulence, colony morphotype and the glycopeptidolipid of *Mycobacterium avium* strain 104. *Tuberculosis* (Edinburgh) **82**:293–300.

Vergne, I., M. Prats, J. F. Tocanne, and G. Laneelle. 1995. Mycobacterial glycopeptidolipid interactions with membranes: a monolayer study. *FEBS Lett.* **375**:254–258.

Vergne, I., and B. Desbat. 2000. Influence of the glycopeptidic moiety of mycobacterial glycopeptidolipids on their lateral organization in phospholipid monolayers. *Biochim. Biophys. Acta* **1467**:113–123.

Vestal, A. L., and G. P. Kubica. 1966. Differential colonial characteristics of mycobacteria on Middlebrook and Cohn 7H10 agarbase medium. *Am. Rev. Respir. Dis.* **94**:247–252.

Villeneuve, C., G. Etienne, V. Abadie, H. Montrozier, C. Bordier, F. Laval, M. Daffe, I. Maridonneau-Parini, and C. Astarie-Dequeker. 2003. Surface-exposed glycopeptidolipids of *Mycobacterium smegmatis* specifically inhibit the phagocytosis of mycobacteria by human macrophages. Identification of a novel family of glycopeptidolipids. *J. Biol. Chem.* **278**:51291–51300.

Villeneuve, C., M. Gilleron, I. Maridonneau-Parini, M. Daffe, C. Astarie-Dequeker, and G. Etienne. 2005. Mycobacteria use their surface-exposed glycolipids to infect human macrophages through a receptor-dependent process. *J. Lipid Res.* **46**:475–483.

Vincent-Naulleau, S., T. Neway, D. Thibault, F. Barrat, H. J. Boulouis, and C. Pilet, 1995. Effects of polar glycopeptidolipids of *Mycobacterium chelonae* (pGPL-Mc) on granulomacrophage progenitors. *Res. Immunol.* **146**:363–371.

Vincent-Naulleau, S., D. Thibault, T. Neway, and C. Pilet. 1997. Stimulatory effects of polar glycopeptidolipids of *Mycobacterium chelonae* on murine haematopoietic stem cells and megakaryocyte progenitors. *Res. Immunol.* **148**:127–136.

Waagmeester, A., J. Thompson, and J. M. Reyrat. 2005. Identifying sigma factors in *Mycobacterium smegmatis* by comparative genomic analysis. *Trends Microbiol.* **13**:505–509.

Walsh, C. T. 2004. Polyketide and nonribosomal peptide antibiotics: modularity and versatility. *Science* **303**:1805–1810.

Woodley, C. L., and H. L. David. 1976. Effect of temperature on the rate of the transparent to opaque colony type transition in *Mycobacterium avium*. *Antimicrob. Agents Chemother.* **9**:113–119.

Yamazaki, Y., L. Danelishvili, M. Wu, M. Macnab, and L. E. Bermudez. 2006. *Mycobacterium avium* genes associated with the ability to form a biofilm. *Appl. Environ. Microbiol.* **72**:819–825.

The Mycobacterial Cell Envelope
Edited by M. Daffé and J.-M. Reyrat
© 2008 ASM Press, Washington, DC

Chapter 22

The Mycolactones: Biologically Active Polyketides Produced by *Mycobacterium ulcerans* and Related Aquatic Mycobacteria

TIMOTHY P. STINEAR AND PAMELA L. C. SMALL

Mycolactones are a family of lipophilic small molecules and virulence factors produced as secondary metabolites by *Mycobacterium ulcerans* and some highly related aquatic mycobacteria. *Mycobacterium ulcerans* is the etiologic agent of a neglected but devastating human disease called Buruli ulcer, a disease that is found in many parts of the world but is most common in rural areas of West and Central Africa, where its incidence now exceeds those of leprosy and in some regions tuberculosis (Debacker et al., 2004). Buruli ulcer often begins as a painless skin nodule or papule. Over a period of weeks to months, bacterial replication results in local extension of the lesion, leading to ulceration that is sometimes preceded by plaque or edema formation (van der Werf et al., 2003). A single ulcer may encompass up to 15% of a person's body surface. In the absence of secondary infection the lesions are usually painless, and although fat necrosis is a prominent feature, pus is generally absent. The pathology of Buruli ulcer is largely attributed to the multiple actions of mycolactone, which possesses cytotoxic, immunosuppressive, apoptotic, and antiphagocytic properties. Victims of Buruli ulcer are frequently left scarred and disfigured. Many questions surround *M. ulcerans* and the role of mycolactones in particular. The aim of this chapter is to describe what is known about mycolactones and their unique role in the pathogenesis of Buruli ulcer, to explain their unusual biosynthetic locus, and to highlight the key questions that remain to be answered.

The histopathology of *M. ulcerans* infection is remarkable. Whereas other pathogenic mycobacteria such as *M. tuberculosis* produce intracellular infections characterized by a robust granulomatous response, Buruli ulcer is characterized by the presence of massive numbers of extracellular bacteria lying within an acellular matrix of coagulation necrosis (Hayman and McQueen, 1985). The presence of extensive tissue damage extending far beyond the site of bacterial colonization led to the hypothesis that pathogenesis of *M. ulcerans* was due to the presence of a secreted toxin (Connor and Lunn, 1965; Connor and Lunn, 1966). More recent work from a careful examination of human tissue adjacent to the area of necrosis showed that *M. ulcerans* has a transient intracellular stage (Torrado et al., 2006). Our current understanding of the infection suggests that bacteria are initially phagocytosed but that toxin production leads to cell death, resulting in an acellular focus of infection. In early studies, cytotoxic activity was demonstrated in *M. ulcerans* culture filtrate using cultured L929 fibroblasts and a guinea pig model of infection (Krieg et al., 1974). The precise chemical identity of the active molecule remained elusive, probably because early attempts at purification were complicated by protein contamination (Hockmeyer et al., 1978). Discovery of the lipid nature of the toxin was followed rapidly by structural characterization and identification of a previously undescribed polyketide that was called mycolactone. It was originally identified as a mixture of *cis-trans* isomers designated mycolactones A and B from a Malaysian isolate of *M. ulcerans* (George et al., 1999; Gunawardana et al., 1999). Since then, it has been shown that mycolactones are a family of molecules produced by *M. ulcerans* and highly related mycobacterial species.

A recent report describes an abundant extracellular matrix produced by M. ulcerans that harbors vesicles that are rich in mycolactones (Marsollier et al., 2007). These vesicles demonstrated enhanced mycolactone-dependent cytotoxicity against cultured

Timothy P. Stinear • Department of Microbiology, Monash University, Wellington Road, Clayton 3800, Australia. Pamela L. C. Small • Department of Microbiology, Walters Life Sciences Building, University of Tennessee, Knoxville, TN 37996-0845.

mammalian cells compared with purified mycolactones.

Both cell culture and animal studies have been used to confirm the role of mycolactones in the virulence of *M. ulcerans* (George et al., 2000; Adusumilli et al., 2005; Coutanceau et al, 2005). Injection of purified mycolactone results in a lesion in guinea pig skin that is histologically identical to that found in human patients (George et al., 1999). Whereas virulent *M. ulcerans* produce an extracellular infection in guinea pigs similar to the human infection, injection of genetically defined mycolactone negative mutants results in a limited intracellular infection characterized by granuloma formation (Adusumilli et al., 2005) (Fig. 1).

MYCOLACTONES: A FAMILY OF RELATED MACROLIDES

Mycolactones are related to antibacterial macrolides such as erythromycin and immunosuppressants such as FK506 because they all share a polyketide-derived macrolactone core. However, unlike antibacterial macrolides, mycolactones are not glycosylated and do not have antibiotic properties (P. Small, unpublished observations). There are five structurally distinct mycolactones and several minor congeners for each of these. These minor cometabolites are thought to arise during polyketide chain extension or through differential oxidative processing of the initial polyketide product (Cadapan et al.,

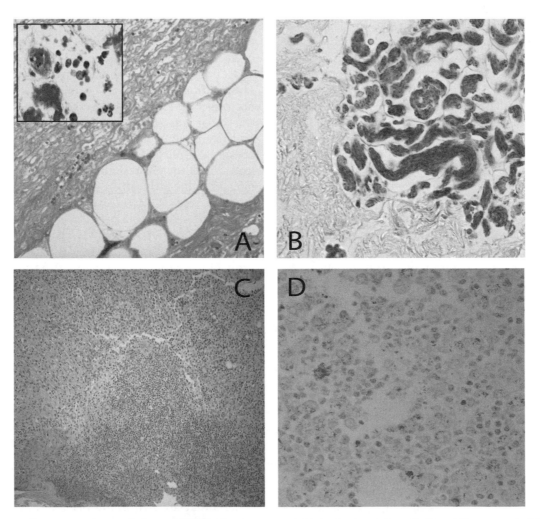

Figure 1. Characteristic histopathology of WT *M. ulcerans* strain 1615 infection in a guinea pig (A and B) compared with *M. ulcerans* mycolactone negative transposon mutant MU 1615::Tn119 (C and D). (A) *M. ulcerans* 1615 infection, showing fatty necrosis, sparse cells, and microhemorrhage characteristic of acute *M. ulcerans* infection in humans (hematoxylin and eosin [H&E] stain); inset, apoptotic cells in an area devoid of bacteria. (B) Extracellular clusters of *M. ulcerans* 1615 (Ziehl-Neelsen stain). (C) *M. ulcerans* mycolactone-negative mutant 1615::Tn119 showing granulomatous response with large influx of inflammatory cells (H&E stain). (D) Ziehl-Neelsen stain showing the intracellular location of the mycolactone-negative mutant.

2001; Hong et al., 2003; Mve-Obiang et al., 2003). Mycolactones A and B were the first discovered and are thus the best studied (George et al., 1999). Their structures are respectively Z and E isomers of a 12-membered macrolactone with an ester-linked polyketide fatty-acyl side-chain. Complete chemical synthesis has established the absolute configuration of mycolactone A/B (Fidanze et al., 2001; Song et al., 2002) and C (Judd et al., 2004) (Fig. 2). Mycolactones A and B are the major metabolites produced by African, Malaysian and Japanese isolates of *M. ulcerans*, mycolactone C is produced by Australian isolates (Mve-Obiang et al., 2003; Judd et al., 2004), and mycolactone D is produced by isolates from China (Hong et al., 2005). The five mycolactones and their congeners are shown in Fig. 2. A striking structural feature is the invariant lactone core, with all the differences occurring within the acyl side chain. Another significant observation is that the major variants correlate with either the geographic origin of a strain or the putative host.

It had been assumed that mycolactone production was strictly associated with *M. ulcerans* and Buruli ulcer. However, recent studies of mycobacteria from around the world have uncovered a broader distribution of mycolactone-producing bacteria than first appreciated that includes mycobacteria cultured from diseased fish in the Red Sea in Israel (Ucko and Colorni, 2005), from fish in the Mediterranean Sea in Greece and in Italy, and diseased fish (Ucko and Colorni, 2005) and frogs (Trott et al., 2004) in the United States. These mycobacteria contain versions of the pMUM001 plasmid (see the following sections) and produce mycolactones E (frog isolates) (Hong et al., 2005; Mve-Obiang et al., 2005) and F (Ranger et al., 2006) (Fig. 2). All mycolactones

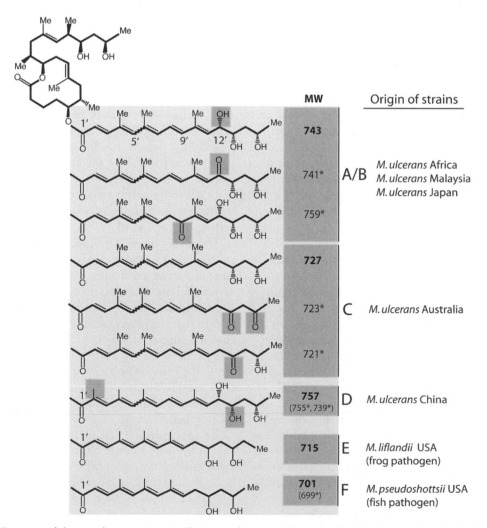

Figure 2. Structure of the mycolactones A to F. All structural variations occur in the side chain and are highlighted by gray shading. Minor cometabolites are denoted by *.

produce cell rounding followed by death through apoptosis in cultured cells. However, mycolactone A/B is the most potent form of the molecule. The apoptotic ability of mycolactone F is significantly less than that of the other mycolactones; a feature that may be associated with its shorter acyl side chain (Ranger et al., 2006). All of these mycolactone-producing mycobacteria, including *M. ulcerans*, form a distinct lineage within a genetically more diverse assemblage of *Mycobacterium marinum* that have clearly evolved from a common *M. marinum*-like progenitor (Yip et al., 2006). This progenitor has since evolved and diversified within different geographic regions or hosts to produce the *M. ulcerans* strains that cause Buruli ulcer and other ecotypes, not linked with human disease, that are pathogens of fish and frogs.

BIOLOGICAL PROPERTIES AND ROLE IN PATHOGENESIS

Mycolactone enters cells by diffusion and does not require the presence of a cellular receptor for entry (Snyder and Small, 2003). Preliminary studies into the effects of mycolactone A/B on cell biology have shown that it causes cell cycle arrest and delayed caspase-dependent apoptotic cell death in murine L929 fibroblasts and J774 macrophage cell lines as well as in infected animals (George et al., 2000). Apoptotic cells are also a characteristic feature of human histopathology (Walsh et al., 2005). Mycolactones induce delayed cell death (2 to 3 days) at concentrations between 12 pg/ml and 1 μg/ml, whereas at concentrations above 5 μg/ml, mycolactone induces cell death through necrosis within 4 hours (Adusumilli et al., 2005).

Mycolactones also have immunomodulatory effects. Experiments with cultured murine macrophages and human neutrophils show that the molecule has antiphagocytic properties because mycolactone negative mutants are internalized by primary human neutrophils and J774 murine macrophages at a significantly higher rate than wild-type (WT) *M. ulcerans* (Adusumilli et al., 2005; Coutanceau et al., 2005). The immunosuppressive properties of mycolactone in cell culture were first demonstrated by using several human and murine cell types; however, only crude acetone soluble lipid extracts were used in this study (Pahlevan et al., 1999). It has more recently been demonstrated that tumor necrosis factor alpha (TNF-α) production from bone marrow-derived murine macrophages is inhibited by WT *M. ulcerans*, whereas infection of macrophages with mycolactone-negative mutants results in stimulation of TNF-α

production (Coutanceau et al., 2005). These results suggest the mycolactone may have a major role in inhibiting the acute inflammatory response of the host to *M. ulcerans*.

Both guinea pig and mouse models have been used to investigate the role of mycolactone in virulence. In guinea pigs, *M. ulcerans* produces an ulcer identical to that found in humans characterized by massive numbers of extracellular mycobacteria and a modest inflammatory infiltrate. Random transposon mutagenesis has been used to generate mycolactone-negative mutants (Stinear et al., 2004). Three different classes of mycolactone-negative mutants were assessed for their ability to produce ulcers in guinea pig skin. These included mutants with transposon insertions in (i) the mycolactone polyketide synthase (PKS) genes, or (ii) putative accessory genes such as MUP045 (see next section), or (iii) mutants that have pleiotropic effects on lipid production including mycolactone. All three classes of mutants are avirulent in guinea pigs and fail to produce an ulcer when injected intradermally (Adusumilli et al., 2005). Furthermore, in contrast to the extracellular infection produced by WT *M. ulcerans*, all mycolactone-negative mutants produce an intracellular infection in which bacteria are found primarily within phagocytic cells. Mycolactone negative mutants also produce a robust granulomatous response (Adusumilli et al., 2005). These results suggest that mycolactone plays a central role in the unique pathology of *M. ulcerans*.

The role of mycolactone has also been studied using a mouse model of virulence (Coutanceau et al., 2005; Oliveira et al., 2005). In the mouse model, *M. ulcerans* produces a primarily intracellular infection in the first few weeks, followed by an increase in extracellular bacteria later in infection. Data comparing different *M. ulcerans* strains, including defined mutants and strains that have undergone partial deletion of the mycolactone plasmid in the mouse model, support the role of mycolactone in virulence (Coutanceau et al., 2005; Oliveira et al., 2005). Most recently, it has been shown that at noncytotoxic concentrations mycolactone can inhibit the maturation of immature murine and human dendritic cells (DC) (Coutanceau et al., 2007). Murine DC pretreated with mycolactone fail to trigger T-cell proliferation and suppressed the production of key proinflammatory chemokines (Coutanceau et al., 2007).

In summary, a growing amount of experimental evidence supports the key role of mycolactone in virulence of *M. ulcerans*. It is unlikely that mycolactone is the only virulence determinant in *M. ulcerans*. However, the inability of mycolactone negative mutants to provoke detectable pathology in mouse or guinea pig experiments or within cultured macrophages emphasizes the central role of this molecule in pathogenesis.

MYCOLACTONE GENETICS AND BIOSYNTHESIS

A major finding of the *M. ulcerans* genome-sequencing project was that it harbored a 174-kb circular plasmid. This plasmid, named pMUM001, contains three very large genes (*mlsA1*, *mlsA2*, and *mlsB*) that encode type I modular polyketide synthases, confirmed by transposon mutagenesis to be the principal enzymatic machinery required for mycolactone synthesis (Stinear et al., 2004; Stinear et al., 2005a; Stinear et al., 2005b).

Type I Modular Polyketide Synthases

Type I polyketide synthases are large, multidomain enzymes, often referred to as "assembly lines" for the production of lipids. For a comprehensive review of these enzymes, see the article by Staunton and Weissman (2001). The enzymology is analogous to fatty acid biosynthesis in which the carbon chain is constructed by the sequential condensation of extender units, usually acetate or propionate, supplied as activated malonyl or methylmalonyl-CoA thioesters. In general, type I polyketide synthases are made up of discrete modules that contain repeating enzymatic domains. Each module is responsible for one round of polyketide chain extension. Within a module, there are three essential domains, (i) the ketosynthase (KS) domain that catalyses the key Claisen condensation reaction between the extender units, (ii) the acyl-transferase (AT) domain that loads a specific extender unit onto (iii) the acyl carrier protein (ACP) of the module. The extender units are linked to the ACP via a flexible phosphopantatheine arm. The reaction proceeds when the extender unit tethered to the ACP domain of an upstream module is passed to the KS domain of the following module. Condensation between this extender and that already loaded by the AT domain onto the ACP of the following module results in chain extension and formation of the characteristic ketone group. Subsequent processing by combinations of reductive domains that may be present within a module between the AT and ACP domains can then modify the newly formed ketone. For example, a ketoreductase (KR) domain reduces the ketone to a hydroxy. A dehydratase (DH) domain reduces the hydroxy to an olefin and an enoylreductase (ER) domain results in full saturation of the growing molecule at that position.

Proposed Biosynthetic Pathway for Mycolactone Production

An outline of the module and domain arrangement of the mycolactone polyketides synthases and their proposed biosynthetic pathways is shown in Fig. 3, and it closely follows the "assembly line" paradigm for type I PKS. MlsA1 (1.8 MDa) and MlsA2 (0.26 MDa) comprise one loading module and nine extension modules that produce the mycolactone core (Fig. 3A). These large proteins are predicted to link together through specific docking domains at the end of MlsA1 and beginning of MlsA2 to form a giant homodimeric multimodular complex. The expected size of this complex is 4.12 MDa. A comparable macromolecule within the cell would be a ribosome (3 MDa). While direct evidence is lacking, it is possible that the terminal integral thioesterase domain of MlsA2 facilitates chain release and cyclization of the fully extended MlsA product (Stinear et al., 2004). MlsB (1.2 MDa) forms a single, homodimeric multienzyme and contains one loading module and seven extension modules that produce the mycolactone acyl side chain (Fig. 3B). The pattern of chain extension and reduction for mycolactone synthesis predicted from analysis of the complete sequences of MlsA1, MlsA2, and MlsB conform almost exactly to the predicted module functions with the exception of module two of MlsA1, where intact DH and ER domains were detected, although the structure of the product does not require these steps. Similar redundancy in reductive domains has been reported for other PKS clusters (Aparicio et al., 1996).

Accessory Enzymes

There are at least three other pMUM001 genes (MUP_038, MUP_045, and MUP_053) that encode enzymes also thought to be involved in mycolactone synthesis (Fig. 3C). The most intriguing of these is MUP_045, whose predicted product is most similar to FabH-like type III beta-ketoacyl synthases. However, MUP_045 has mutations at three positions known to be critical for KS activity. These same mutations in other KS leads to enzymes that promote C-O bond formation rather than C-C (Kwon et al., 2002). Hence, it has been proposed that the MUP_045 product may act as acyltransferase, catalyzing the ester bond that joins the mycolactone core to the side chain (Stinear et al., 2004). Interestingly however, a *M. ulcerans* MUP_045 knockout appears to produce neither the core or side chain, perhaps suggesting a wider role for this gene (Coutanceau et al., 2005). MUP_038 is situated between *mlsA2* and *mlsB*, and encodes a type II thioesterase,

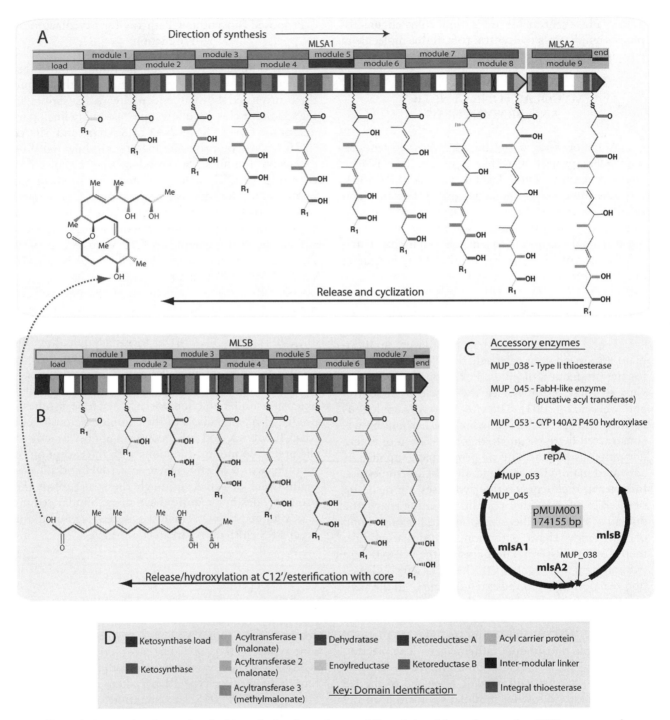

Figure 3. Proposed pathways for the biosynthesis of mycolactone A/B as deduced from the complete DNA sequence of pMUM001 and transposon mutagenesis (Stinear et al., 2004). (A) Module and domain arrangement for synthesis of the core. (B) Module and domain arrangement for synthesis of the side chain. (C) Map of the pMUM001 plasmid showing the relative positions of the *mls* genes and accessory genes. (D) Key for domain function. Identical color indicates both same function and sequence (97 to 100% aa identity). *(See the color insert for color version of this figure.)*

which is predicted to play a role in maintaining the fidelity of the PKS by removing acyl chains from modules where synthesis has stalled (Heathcote et al., 2001). MUP_053, or *cyp140A7* encodes a P450 monooxygenase that is almost certainly responsible

for hydroxylation of the mycolactone acyl side chain at C12′. There is as yet no direct genetic evidence that confirms the role of this enzyme. However, the indirect evidence is strong, as *M. ulcerans* strains naturally lacking MUP_053, such as those from

Australia, predominantly produce a nonoxidized molecule (Mve-Obiang et al., 2003; Stinear et al., 2005) (Fig. 2).

Evolution of the Locus

Although the module and domain arrangement of the mls genes is conventional, there are several extraordinary features within this locus. In addition to the large size of the predicted proteins is the extreme sequence similarity between domains of the same function. Modular PKS domains routinely share 30% to 70% amino acid (aa) identity. For example, intramodule aa identity among the 14 KS domains of the rapamycin cluster is 66% (Aparicio et al., 1996). In comparison, the 16 modules of the Mls locus have an intra-KS domain aa identity of 97% (Fig. 4). Identity scores were even higher among the other Mls domains, ranging from 98.7% to 100%. The repetitive nature of this locus is highlighted in Fig. 3A and 3B, where the identical color for a functional domain and module indicates both identical function and aa sequence (97% to 100%). Most staggering is that the entire 100-kb locus, encompassing *mlsA1, A2 and mlsB*, comprises only 9.5 kb of unique DNA. Thus, one conclusion from these data is that the region has evolved very recently from a series of in-frame recombination and duplication events (Stinear et al., 2004) (Fig. 4). This is illustrated in Fig. 4, where a phylogenetic analysis of the 16 KS domains shows distinct clustering of some of the MlsA (core) KS domains and complete aa identity among some MlsB (side chain) KS domains from side chain modules, indicating perhaps that *mlsB* may have evolved from *mlsA* components.

The high level of sequence identity also suggests that the locus might be prone to recombination-mediated deletion and subsequent loss of mycolactone production. Studies of lab-passaged clinical isolates confirm that this does occur with a frequency that seems to vary between strains (Stinear et al., 2005). Nevertheless, *M. ulcerans* strains separated by time (e.g., isolated from African countries over a 40-year period) and space (e.g., isolated from Africa and Malaysia) all produce the same mycolactone A/B (Mve-Obiang et al., 2003). This paradoxical observation suggests that there is very strong purifying selection acting on *M. ulcerans* populations for preservation of those bacteria producing the molecule. The role of mycolactone for the bacterium is unknown, but possible functions are discussed in the following section.

Domain Swapping

One suggestion from the high level of identity among the Mls domains (and in particular among the critical KS domains that appear to exhibit substrate tolerance) (Fig. 4) was that they might form the basis of a genetic PKS toolbox, where individual domains or modules could be swapped to create hybrid PKS and thus new polyketide metabolites (Stinear et al., 2004). This hope was boosted by the discovery that mycolactone D arises by a naturally occurring domain swap in an *M. ulcerans* isolate from China. In this strain, there has been a swap of the AT domain in module 7 of MlsB from acetate to propionate, leading to the methylene substitution at C2′ of the side chain and thus, mycolactone D (Fig. 2) (Hong et al., 2005). It is likely that some of the other

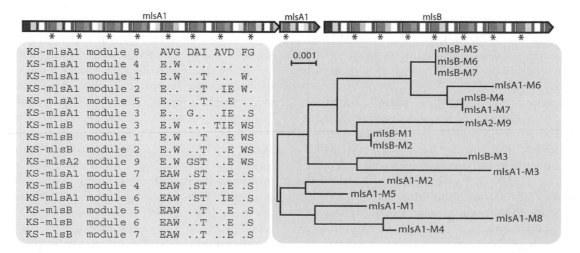

Figure 4. Alignment of the 16 Mls KS domains, showing the 11 variable aa residues and a phylogenetic reconstruction. The scale bar indicates 1 aa substitution per 1,000 sites.

mycolactone variants will also be explained by module and domain rearrangements.

Origin of the *mls* Locus

BLASTP and BLASTN searches give no clues to the immediate origin of the *mls* locus. The highest database matches suggest an actinomycete origin, with best hits to KS domains from Pks8 from *Mycobacterium avium* subsp. *paratuberculosis* K-10 (68% aa identity) and a PKS gene from *Streptomyces avermitilis* MA-4680 (69% aa identity). The apparent uniqueness of the locus has been exploited to develop a new molecular diagnostic tool based on PCR detection KR-B domain (Fig. 3D). This test has recently been used to detect the *M. ulcerans* plasmid in mosquito and soil samples from an area where Buruli ulcer is endemic (Johnson et al., 2007).

Heterologous Mycolactone Expression and Chemical Synthesis

A major impediment to further studies of mycolactones is the slow growth and variable production of the molecules by *M. ulcerans*. It is possible to extract up to 5 mg of mycolactones per g (dry weight) of *M. ulcerans* growing in Middlebrook 7H9 liquid medium supplemented with OADC (Cadapan et al., 2001). Mycolactones are also produced by *M. ulcerans* growing in protein-free culture media (Mve-Obiang et al., 1999).

Little is known about regulation of *mls* gene expression and mycolactone production. In other actinomycetes, such as the *Streptomyces*, the production of polyketides has been shown to be regulated by a number of factors including divalent cations, oleic acid, and phosphate (see Katz, 2003 for a review of this topic). The yield of mycolactone from *M. ulcerans* can be increased by adding egg yolk constituents to Middlebrook 7H9 medium, but these additions also increase the total bacterial mass and thus their effects are probably not restricted to mycolactone (Cadapan et al., 2001).

Transfer of the pMUM001 plasmid to a faster growing bacterium is one strategy currently being attempted to try and obtain a more reliable source of mycolactones. The transfer of a small pMUM001-based replicon to *M. marinum* has been reported (Stinear et al., 2005). However the large size of the *mls* genes and their instability due to their considerable repetitive DNA will present a major obstacle to successful maintenance of the intact 110 kb *mls* locus within a heterologous host. Another option being explored is complete chemical synthesis and this has been reported for mycolactone A/B and C but it is complex and involves more than 70 steps (Song et al., 2002; Judd et al., 2004). New approaches for chemical synthesis are being explored, and several advances show promise (van Summeren et al.; 2005, Alexander et al., 2006; Yin et al., 2006).

THE ROLE OF MYCOLACTONE

Mycolactone has potent biological activities against mammalian cells, but what roles does mycolactone play in the physiology and natural ecology of *M. ulcerans*? The lack of person-to-person transmission of *M. ulcerans* suggests that pathogenesis in humans is unlikely to be its primary function. All mycolactone-producing mycobacteria found so far have been discovered through their ability to cause disease in vertebrates, but they may be acting as "indicator" species and represent only a proportion of the mycolactones that exist in nature. Although mycolactones have attracted interest because of their effects on mammalian cells, it is possible that the molecule has no toxicity in its natural reservoir species. No pathology is exhibited during *M. ulcerans* infection of aquatic insects and snails (Marsollier et al., 2002; Marsollier et al., 2004; Marsollier et al., 2005) and mycolactone has no toxicity for yeasts such as *Candida*, *Saccharomyces* and *Cryptococcus*, for Drosophila S2 cells or for *Dictyostelium discoidium* (P. Small, unpublished observations).

Some insights have been gleaned from comparisons of the complete genome sequences of *M. ulcerans* and *M. marinum*. It is clear that *M. ulcerans* (this species designation includes the novel fish and frog pathogens that also make mycolactones) has recently passed through an evolutionary bottleneck, evolving from the generalist species *M. marinum* to become a specialist, adapting to survive in restricted, more protected, niche environments (Stinear, 2007). A key factor in this evolution was acquisition of the pMUM plasmid and the concomitant ability to produce mycolactone where mycolactone provided the key selective advantage for survival in a new environment. What might mycolactone do? Among many functions, polyketides are known to act as (i) antibiotics, presumably to protect against microbial predation (Katz, 2003); (ii) siderophores, involved in iron accumulation (Quadri et al., 1998); and (iii) signaling molecules (Goh et al., 2002), such as the chlorinated polyketide DIF-1 produced by *Dictyostelium* to trigger cell differentiation (Eichinger et al., 2005). One research group has demonstrated that *M. ulcerans* multiplies actively in the salivary glands of carnivorous water bugs, such as *Naucoris cimicoides* (Marsollier et al., 2002), and that a mycolactone-negative mutant is unable to colonize this niche. *M. ulcerans* has also been shown to form prolific mycolactone-rich biofilms on the surface of aquatic vegetation and algae (Marsollier et al., 2004). One intriguing possibility is that mycolactone may be acting as a colonizing factor, aiding biofilm formation and thus facilitating bacterial persistence in certain environments, an approach used by other bacteria such as *Yersinia pestis*, the plague bacillus, to colonize the proventriculus of the flea (Jarrett et al., 2004).

CONCLUDING REMARKS

Swift advances have been made in our understanding of mycolactones in the 10 years since the discovery of mycolactone A/B. Their potent immunomodulatory and anti-inflammatory activity may lead to the development of new pharmaceutical compounds, and several groups are pursuing research in these areas. Research efforts aimed at establishing the role of mycolactones in the natural ecology of *M. ulcerans* are also a priority.

Acknowledgment. We gratefully acknowledge the financial support of the National Health and Medical Research Council of Australia.

REFERENCES

Adusumilli, S., A. Mve-Obiang, T. Sparer, W. Meyers, J. Hayman, and P. L. Small. 2005. *Mycobacterium ulcerans* toxic macrolide, mycolactone modulates the host immune response and cellular location of *M. ulcerans* in vitro and in vivo. *Cell. Microbiol.* 7:1295–1304.

Alexander, M. D., S. D. Fontaine, J. J. La Clair, A. G. Dipasquale, A. L. Rheingold, and M. D. Burkart. 2006. Synthesis of the mycolactone core by ring-closing metathesis. *Chem. Commun.* 44: 4602–4604.

Aparicio, J. F., I. Molnar, T. Schwecke, A. Konig, S. F. Haydock, L. E. Khaw, J. Staunton, and P. F. Leadlay. 1996. Organization of the biosynthetic gene cluster for rapamycin in *Streptomyces hygroscopicus*: analysis of the enzymatic domains in the modular polyketide synthase. *Gene* 169:9–16.

Cadapan, L. D., R. L. Arslanian, J. R. Carney, S. M. Zavala, P. L. Small, and P. Licari. 2001. Suspension cultivation of *Mycobacterium ulcerans* for the production of mycolactones. *FEMS Microbiol. Lett.* 205:385–389.

Connor, D. H., and H. F. Lunn. 1965. *Mycobacterium ulcerans* infection (with comments on pathogenesis). *Int. J. Lepr.* 33(Suppl.): 698–709.

Connor, D. H., and H. F. Lunn. 1966. Buruli ulceration. A clinicopathologic study of 38 Ugandans with *Mycobacterium ulcerans* infection. *Arch. Pathol.* 81:183–199.

Coutanceau, E., J. Decalf, A. Babon, N. Winter, S. T. Cole, M. L. Albert, and C. Demangel. 2007. Selective suppression of dendritic cell functions by *Mycobacterium ulcerans* toxin mycolactone. *J. Exp. Med.* 204:1395–1403.

Coutanceau, E., L. Marsollier, R. Brosch, E. Perret, P. Goossens, M. Tanguy, S. T. Cole, P. L. Small, and C. Demangel. 2005. Modulation of the host immune response by a transient intracellular stage of *Mycobacterium ulcerans*: the contribution of endogenous mycolactone toxin. *Cell. Microbiol.* 7:1187–1196.

Debacker, M., J. Aguiar, C. Steunou, C. Zinsou, W. M. Meyers, A. Guedenon, J. T. Scott, M. Dramaix, and F. Portaels. 2004. *Mycobacterium ulcerans* disease (Buruli ulcer) in rural hospital, Southern Benin, 1997-2001. *Emerg. Infect. Dis.* 10:1391–1398.

Eichinger, L., J. A. Pachebat, G. Glockner, M. A. Rajandream, R. Sucgang, M. Berriman, J. Song, R. Olsen, K. Szafranski, Q. Xu, B. Tunggal, S. Kummerfeld, M. Madera, B. A. Konfortov, F. Rivero, A. T. Bankier, R. Lehmann, N. Hamlin, R. Davies, P. Gaudet, P. Fey, K. Pilcher, G. Chen, D. Saunders, E. Sodergren, P. Davis, A. Kerhornou, X. Nie, N. Hall, C. Anjard, L. Hemphill, N. Bason, P. Farbrother, B. Desany, E. Just, T. Morio, R. Rost, C. Churcher, J. Cooper, S. Haydock, N. van Driessche, A. Cronin, I. Goodhead, D. Muzny, T. Mourier, A. Pain, M. Lu, D. Harper, R. Lindsay, H. Hauser, K. James, M. Quiles, M. Madan Babu, T. Saito, C. Buchrieser, A. Wardroper, M. Felder, M. Thangavelu, D. Johnson, A. Knights, H. Loulseged, K. Mungall, K. Oliver, C. Price, M. A. Quail, H. Urushihara, J. Hernandez, F. Rabbinowitsch, D. Steffen, M. Sanders, J. Ma, Y. Kohara, S. Sharp, M. Simmonds, S. Spiegler, A. Tivey, S. Sugano, B. White, D. Walker, J. Woodward, T. Winckler, Y. Tanaka, G. Shaulsky, M. Schleicher, G. Weinstock, A. Rosenthal, E. C. Cox, R. L. Chisholm, R. Gibbs, W. F. Loomis, M. Platzer, R. R. Kay, J. Williams, P. H. Dear, A. A. Noegel, B. Barrell, and A. Kuspa. 2005. The genome of the social amoeba *Dictyostelium discoideum*. *Nature* 435:43–57.

Fidanze, S., F. Song, M. Szlosek-Pinaud, P. L. Small, and Y. Kishi. 2001. Complete structure of the mycolactones. *J. Am. Chem. Soc.* 123:10117–10118.

George, K. M., D. Chatterjee, G. Gunawardana, D. Welty, J. Hayman, R. Lee, and P. L. Small. 1999. Mycolactone: a polyketide toxin from *Mycobacterium ulcerans* required for virulence. *Science* 283:854–857.

George, K. M., L. Pascopella, D. M. Welty, and P. L. Small. 2000. A *Mycobacterium ulcerans* toxin, mycolactone, causes apoptosis in guinea pig ulcers and tissue culture cells. *Infect. Immun.* 68:877–883.

Goh, E. B., G. Yim, W. Tsui, J. McClure, M. G. Surette, and J. Davies. 2002. Transcriptional modulation of bacterial gene expression by subinhibitory concentrations of antibiotics. *Proc. Natl. Acad. Sci. USA* 99:17025–17030.

Gunawardana, G., D. Chatterjee, K. M. George, P. J. Brennan, D. Whittern, and P. L. C. Small. 1999. Characterization of novel macrolide toxins, mycolactones A and B, from a human pathogen, Mycobacterium ulcerans. *J. Am. Chem. Soc.* 121:6092–6093.

Hayman, J., and A. McQueen. 1985. The pathology of *Mycobacterium ulcerans* infection. *Pathology* 17:594–600.

Heathcote, M. L., J. Staunton, and P. F. Leadlay. 2001. Role of type II thioesterases: evidence for removal of short acyl chains produced by aberrant decarboxylation of chain extender units. *Chem. Biol.* 8:207–220.

Hockmeyer, W. T., R. E. Krieg, M. Reich, and R. D. Johnson. 1978. Further characterization of *Mycobacterium ulcerans* toxin. *Infect. Immun.* 21:124–128.

Hong, H., P. J. Gates, J. Staunton, T. Stinear, S. T. Cole, P. F. Leadlay, and J. B. Spencer. 2003. Identification using LC-MSn of co-metabolites in the biosynthesis of the polyketide toxin mycolactone by a clinical isolate of *Mycobacterium ulcerans*. *Chem. Commun.* 21:2822–2823.

Hong, H., J. B. Spencer, J. L. Porter, P. F. Leadlay, and T. Stinear. 2005. A novel mycolactone from a clinical isolate of *Mycobacterium ulcerans* provides evidence for additional toxin heterogeneity as a result of specific changes in the modular polyketide synthase. *Chembiochem* 6:643–648.

Hong, H., T. Stinear, P. Skelton, J. B. Spencer, and P. F. Leadlay. 2005. Structure elucidation of a novel family of mycolactone toxins from the frog pathogen *Mycobacterium* sp. MU128FXT by mass spectrometry. *Chem. Commun.* 34:4306–4308.

Jarrett, C. O., E. Deak, K. E. Isherwood, P. C. Oyston, E. R. Fischer, A. R. Whitney, S. D. Kobayashi, F. R. DeLeo, and B. J. Hinnebusch. 2004. Transmission of *Yersinia pestis* from an infectious biofilm in the flea vector. *J. Infect. Dis.* 190:783–792.

Johnson, P. D. R., J. Azuolas, C. J. Lavender, E. Wishart, T. P. Stinear, J. A. Hayman, L. Brown, G. A. Jenkin, and J. A. M. Fyfe. 2007. *Mycobacterium ulcerans* in mosquitoes captured during outbreak of Buruli ulcer, Southeastern Australia. *Emerg. Infect. Dis.* 13:1653–1660.

Judd, T. C., A. Bischoff, Y. Kishi, S. Adusumilli, and P. L. Small. 2004. Structure determination of mycolactone C via total synthesis. *Org. Lett.* 6:4901–4904.

Katz, L. 2003. Regulation of antibiotic biosynthesis in producer organisms, p. 325–335. *In* C. Walsh (ed.), *Antibiotics: Actions, Origins, Resistance.* ASM Press, Washington, DC.

Krieg, R. E., W. T. Hockmeyer, and D. H. Connor. 1974. Toxin of *Mycobacterium ulcerans.* Production and effects in guinea pig skin. *Arch. Dermatol.* 110:783–788.

Kwon, H. J., W. C. Smith, A. J. Scharon, S. H. Hwang, M. J. Kurth, and B. Shen. 2002. C-O bond formation by polyketide synthases. *Science* 297:1327–1330.

Marsollier, L., J. Aubry, E. Coutanceau, J. P. Andre, P. L. Small, G. Milon, P. Legras, S. Guadagnini, B. Carbonnelle, and S. T. Cole. 2005. Colonization of the salivary glands of *Naucoris cimicoides* by *Mycobacterium ulcerans* requires host plasmatocytes and a macrolide toxin, mycolactone. *Cell. Microbiol.* 7:935–943.

Marsollier, L., P. Brodin, M. Jackson, J. Kordulokova, P. Tafalmeyer, E. Carbonelle, J. Aubry, G. Milon, P. Legras, J. P. Saint Andre, C. Leroy, J. Cottin, M. L. J. Guillou, G. Reysset, and S. T. Cole. 2007. Impact of *Mycobacterium ulcerans* biofilm on transmissibility to ecological niches and Buruli ulcer pathogenesis. *PLoS Pathog.* 3:e62.

Marsollier, L., R. Robert, J. Aubry, J. P. Saint Andre, H. Kouakou, P. Legras, A. L. Manceau, C. Mahaza, and B. Carbonnelle. 2002. Aquatic Insects as a vector for *Mycobacterium ulcerans.* *Appl. Environ. Microbiol.* 68:4623–4628.

Marsollier, L., T. Severin, J. Aubry, R. W. Merritt, J. P. Saint Andre, P. Legras, A. L. Manceau, A. Chauty, B. Carbonnelle, and S. T. Cole. 2004. Aquatic snails, passive hosts of *Mycobacterium ulcerans.* *Appl. Environ. Microbiol.* 70:6296–6298.

Marsollier, L., T. Stinear, J. Aubry, J. P. Saint Andre, R. Robert, P. Legras, A. L. Manceau, C. Audrain, S. Bourdon, H. Kouakou, and B. Carbonnelle. 2004. Aquatic plants stimulate the growth of and biofilm formation by *Mycobacterium ulcerans* in axenic culture and harbor these bacteria in the environment. *Appl. Environ. Microbiol.* 70:1097–1103.

Mve-Obiang, A., R. E. Lee, F. Portaels, and P. L. Small. 2003. Heterogeneity of mycolactones produced by clinical isolates of *Mycobacterium ulcerans*: implications for virulence. *Infect. Immun.* 71:774–783.

Mve-Obiang, A., R. E. Lee, E. S. Umstot, K. A. Trott, T. C. Grammer, J. M. Parker, B. S. Ranger, R. Grainger, E. A. Mahrous, and P. L. Small. 2005. A newly discovered mycobacterial pathogen isolated from laboratory colonies of *Xenopus* species with lethal infections produces a novel form of mycolactone, the *Mycobacterium ulcerans* macrolide toxin. *Infect. Immun.* 73:3307–3312.

Mve-Obiang, A., J. Remacle, J. C. Palomino, A. Houbion, and F. Portaels. 1999. Growth and cytotoxic activity by *Mycobacterium ulcerans* in protein-free media. *FEMS Microbiol. Lett.* 181:153–157.

Oliveira, M. S., A. G. Fraga, E. Torrado, A. G. Castro, J. P. Pereira, A. L. Filho, F. Milanezi, F. C. Schmitt, W. M. Meyers,

F. Portaels, M. T. Silva, and J. Pedrosa. 2005. Infection with *Mycobacterium ulcerans* induces persistent inflammatory responses in mice. *Infect. Immun.* 73:6299–6310.

Pahlevan, A. A., D. J. Wright, C. Andrews, K. M. George, P. L. Small, and B. M. Foxwell. 1999. The inhibitory action of *Mycobacterium ulcerans* soluble factor on monocyte/T cell cytokine production and NF-kappa B function. *J. Immunol.* 163:3928–3935.

Quadri, L. E., J. Sello, T. A. Keating, P. H. Weinreb, and C. T. Walsh. 1998. Identification of a *Mycobacterium tuberculosis* gene cluster encoding the biosynthetic enzymes for assembly of the virulence-conferring siderophore mycobactin. *Chem. Biol.* 5:631–645.

Ranger, B. S., E. A. Mahrous, L. Mosi, S. Adusumilli, R. E. Lee, A. Colorni, M. Rhodes, and P. L. Small. 2006. Globally distributed mycobacterial fish pathogens produce a novel plasmid-encoded toxic macrolide, mycolactone F. *Infect. Immun.* 74:6037–6045.

Snyder, D. S., and P. L. Small. 2003. Uptake and cellular actions of mycolactone, a virulence determinant for Mycobacterium ulcerans. *Microb. Pathog.* 34:91–101.

Song, F., S. Fidanze, A. B. Benowitz, and Y. Kishi. 2002. Total synthesis of the mycolactones. *Org. Lett.* 4:647–650.

Staunton, J., and K. J. Weissman. 2001. Polyketide biosynthesis: a millennium review. *Nat. Prod. Rep.* 18:380–416.

Stinear, T. P., H. Hong, W. Frigui, M. J. Pryor, R. Brosch, T. Garnier, P. F. Leadlay, and S. T. Cole. 2005a. Common evolutionary origin for the unstable virulence plasmid pMUM found in geographically diverse strains of *Mycobacterium ulcerans.* *J. Bacteriol.* 187:1668–1676.

Stinear, T. P., A. Mve-Obiang, P. L. Small, W. Frigui, M. J. Pryor, R. Brosch, G. A. Jenkin, P. D. Johnson, J. K. Davies, R. E. Lee, S. Adusumilli, T. Garnier, S. F. Haydock, P. F. Leadlay, and S. T. Cole. 2004. Giant plasmid-encoded polyketide synthases produce the macrolide toxin of *Mycobacterium ulcerans.* *Proc. Natl. Acad. Sci. USA* 101:1345–1349.

Stinear, T. P., M. J. Pryor, J. L. Porter, and S. T. Cole. 2005b. Functional analysis and annotation of the virulence plasmid pMUM001 from *Mycobacterium ulcerans.* *Microbiology* 151:683–692.

Stinear, T. P., T. Seemann, S. Pidot, W. Frigui, G. Reysset, T. Garnier, G. Meurice, D. Simon, C. Bouchier, L. Ma, M. Tichit, J. L. Porter, J. Ryan, P. D. R. Johnson, J. K. Davies, G. A. Jenkin, P. L. C. Small, L. M. Jones, F. Tekaia, F. Laval, M. Daffé, J. Parkhill, and S. T. Cole. 2007. Reductive evolution and niche adaptation inferred from the genome of *Mycobacterium ulcerans,* the causative agent of Buruli ulcer. *Genome Res.* 7:192–200.

Torrado, E., A. G. Fraga, A. G. Castro, P. Stragier, W. M. Meyers, F. Portaels, M. T. Silva, and J. Pedrosa. 2006. Evidence for an intramacrophage growth phase of *Mycobacterium ulcerans.* *Infect. Immun.* 75:977–987

Trott, K. A., B. A. Stacy, B. D. Lifland, H. E. Diggs, R. M. Harland, M. K. Khokha, T. C. Grammer, and J. M. Parker. 2004. Characterization of a *Mycobacterium ulcerans*-like infection in a colony of African tropical clawed frogs (*Xenopus tropicalis*). *Comp. Med.* 54:309–317.

Ucko, M., and A. Colorni. 2005. *Mycobacterium marinum* infections in fish and humans in Israel. *J. Clin. Microbiol.* 43:892–895.

van der Werf, T. S., T. Stinear, Y. Stienstra, W. T. van der Graaf, and P. L. Small. 2003. Mycolactones and *Mycobacterium ulcerans* disease. *Lancet* 362:1062–1064.

van Summeren, R. P., B. L. Feringa, and A. J. Minnaard. 2005. New approaches towards the synthesis of the side-chain of mycolactones A and B. *Org. Biomol. Chem.* 3:2524–2533.

Walsh, D. S., W. M. Meyers, F. Portaels, J. E. Lane, D. Mongkolsirichaikul, K. Hussem, P. Gosi, and K. S. Myint. 2005. High rates of apoptosis in human *Mycobacterium ulcerans* culture-positive buruli ulcer skin lesions. *Am. J. Trop. Med. Hyg.* 73:410–415.

Yin, N., G. Wang, M. Qian, and E. Negishi. 2006. Stereoselective synthesis of the side chains of mycolactones A and B featuring stepwise double substitutions of 1,1-dibromo-1-alkenes. *Angew. Chem. Int. Ed. Engl.* 45:2916–2920.

Yip, M. J., J. L. Porter, J. A. Fyfe, C. J. Lavender, F. Portaels, M. Rhodes, H. Kator, A. Colorni, G. A. Jenkin, and T. Stinear. 2007. Evolution of *Mycobacterium ulcerans* and other mycolactone-producing mycobacteria from a common *Mycobacterium marinum* progenitor. *J. Bacteriol.* 189:2021–2029.

The Mycobacterial Cell Envelope
Edited by M. Daffé and J.-M. Reyrat
© 2008 ASM Press, Washington, DC

Epilogue

The Envelope: beyond the Veil

JEAN-MARC REYRAT AND MAMADOU DAFFÉ

More than 100 years have passed since mycobacteria were first described as an independent genus of the *Actinomycetales* and soon after as a wax factory. This has led to the discovery of many new lipids of esoteric structures such as glycopeptidolipids, phenolic lipids, and sulfolipids. At the same time, biosynthetic studies have revealed the complex machineries needed for producing such compounds. The possible occurrence of an outer asymmetric membrane has been conceptualized very early and later supported by solid biophysical evidence to arrive at the current models for the mycobacterial cell envelope. Although mycobacterial genetics has developed only recently, many players of the various biosynthetic pathways have now been cloned and characterized. Thanks to the sequencing of various mycobacterial species, the genus- or species-specificity of many coding sequences has been established. Most of them have been demonstrated to be essential for envelope homeostasy. In addition, structural projects have been launched, enlarging almost every week the number of structure of regulators, membrane proteins, and enzymes related to the cell surface and, thus, enabling the design of new inhibitors. This is only the beginning of the modern era of the mycobacterial cell wall because many players involved in the tremendous metabolic diversity of mycobacteria remain to be characterized. The combination of biochemistry, genetics, comparative genomics, and metabolic reconstruction will undoubtedly characterize a number of new molecules and their metabolic pathways. Several aspects of the future questions merit, however, a brief discussion.

1. Despite the demonstrated importance of many cell envelope constituents in the pathogenicity of mycobacteria, no obvious difference has been observed in the cell envelope composition of pathogenic and nonpathogenic mycobacterial species. As a consequence, one of the major challenges of the forthcoming years is understanding how pathogenic species use specific substances and a special combination of cell wall compounds recruited in a common repertoire to establish a successful infection. In terms of organization, however, pathogens have been found to cover themselves with a wide outer layer, called a capsule, whereas nonpathogenic mycobacterial species are surrounded by a thin layer. The importance of this difference remains to be experimentally addressed through the construction and analysis of mutant strains devoid of this structure.

2. Cell wall biogenesis is a niche for drug targets development. Some of the new pathways that are regularly characterized represent good candidates for the development of selective inhibitors that may well be useful as antimycobacterial agents. The number of antibiotics effective against mycobacteria is so far very low, and multidrug resistant species are quickly emerging. Consequently, there is an urgent need to design and test such molecules. In this respect, deciphering the impact of environment (dormancy) on cell wall remodeling will help in the selection of putative targets to inhibit mycobacteria in the so-called persistent phase of growth.

3. One of the major problems in the treatment of tuberculosis resides in the early diagnosis in areas of endemicity. Despite the characterization of several molecular markers, they may not be used where needed owing to economical and logistical reasons. Interestingly, many metabolites and proteins of the cell wall are the targets of the host immune system, which enable the development of immunodiagnostic tests, as in the case of phenolic glycolipid I and leprosy. Characterization of new species-specific

Jean-Marc Reyrat • Faculté de Médecine René Descartes, Université Paris Descartes, and Groupe Avenir, Unité de Pathogénie des Infections Systémiques—U570, INSERM, F-75730 Paris Cedex 15, France. **Mamadou Daffé** • Department of Molecular Mechanisms of Mycobacterial Infections, Institut de Pharmacologie et de Biologie Structurale du Centre National de la Recherche Scientifique (CNRS), and Université Paul Sabatier, 205 route de Narbonne, 31077 Toulouse Cedex 4, France.

compounds should help in designing a number of low-cost tests that would be ideal for the diagnosis mycobacterioses in humans and animals.

4. Many proteins and lipid metabolites of the cell surface dramatically stimulate T cells. They consequently represent as such attractive immunomodulator candidates that may be used as adjuvant in vaccine development to improve the protection of humans against various diseases, including tuberculosis and leprosy. In the case of polyketide products, the development of our knowledge in modules and domains organization of nonribosomal peptide synthetases, polyketide synthase, acyl carrier protein synthases, and other complex enzymatic machinery should help the quick development of new derivatives with interesting therapeutic properties.

INDEX